How To Target A Future With Immense Opportunity

Medicines for "My" Body vs. Medicines For "Every" Body

The Molecular Revolution

Gilbert Mertens

Raider Publishing International

New York　　London　　Swansea

First Printing

The information in this book is believed to be correct at the time of publication. However, no responsibility can be accepted by the publisher for its completeness or accuracy. The views, content and descriptions in this book do not represent the views of Raider Publishing International. For clarification of any information in this book, or to point out any inaccuracies or omissions, please contact the author

ISBN: 1-934360-48-1
Published By Raider Publishing International
www.RaiderPublishing.com
New York   London   Swansea

Printed in the United States of America and the United Kingdom
By Lightning Source Ltd.

About the Author

Gilbert Mertens has extensive experience of working within the healthcare industry, including an extended period as a general manager with both a major global pharmaceutical company, and a leading international communications agency. Based on his profound knowledge of the European and US healthcare sector and his personal relationship with many renowned scientists, national healthcare authorities and his excellent access to business leaders, Gilbert Mertens offers strategic healthcare management reflections to key decision makers under constant challenge to redefine a multi-faceted, competitive and constantly changing healthcare environment. His series of Healthcare Management reports, published by the Financial Times, by Reuters Business Insight and also by Nicholas Hall & Co, have been received within the sector as benchmark reports.

For the information provided, for the material received and/or for their encouraging, moral support, the author would like to thank

Dr Hatsumo Aoki -- Robert Béland -- Dr Baerbel Bleicher -- Dr Holger Bengs -- Mariana Brea-Krüger -- Dr Joachim Bug -- Dr Jocelyne Coupat -- Dr Gunter Festel -- Dr Eva Fischer-Mertens -- Dr Spiros Fotinos -- Prof. Dr Serge Gailly -- Ken and Renate Halvorsrude -- Dr Walter Haussmann -- Dr Alain Herrera -- Prof. Dr Trevor M Jones -- Dr Tugrul T Kararli -- Helmut Kerschbaumer -- Dr Johannes Knollmeyer - Klaus Kuhne - Dr Joachim Lach - Leo Mauren - Prof. Dr Reinhard Ricker -- Karl Schlingensief -- Dr Sabine Schuetterle -- Douglas J. Squires -- Hans R. Thoennessen -- Dr Friedrich von Bohlen und Halbach -- Dieter Zander

Because the futuristic broadening of scope in healthcare is made possible by the study of the past and the in-depth analysis of the present, this report, beyond its focus on "what healthcare is likely to be in fifteen to twenty years from now", gives the reader a summarized overview of "where healthcare comes from and where it stands today". The author welcomes suggestions, opinions, alternative views from readers. He can be reached at MertensGilbert@Yahoo.ca

Contents

Chapter Two: Medical Community at a Crossroads

Chapter Three: Medicines for the Masses

Chapter Four: A Healthcare Industry Reinventing Itself

Chapter Five: Finding Gold

Chapter Six: Differentiation Strategies in Drug Delivery Systems

Chapter Seven : An Ongoing Search for Better Medicines

Chapter Ten: Adding a New Dimension in Healthcare

Index of tables

Chapter One

Chapter Two

Chapter Three

15

Chapter Nine

Chapter Ten

Content (Grey Pages)

Overviews:

In detail:

21

Index of tables (blue pages)

Chapter One

The Operating Environment.
Past, Present and Future

I.1. Going for the healthcare economic knockout?

The post-war healthcare systems that have defined the pharmaceutical marketplace for the past 55 years – and are now in their twilight- were rooted in the concept of "health for all"; available to rich and poor according to medical need alone and without charge.

The "Welfare State", a phrase first used in 1941 by William Temple, the Ukase Archbishop of Canterbury, was a historic achievement. Yet the customer, the person who benefited from the products and services, rarely ordered them. And the third party paying the bill neither ordered nor received anything. Thus was created the "passive" patient.

The 1950s and the 1960s saw the glorification of institutionalized healthcare, discouraging individual responsibility for health maintenance and management. In countries with universal access to healthcare, the system produced an "assistance" mentality. The suffering of the people received priority attention. Not the cost of healthcare. To the benefit of the pharmaceutical industry and progress in therapy. The systems guaranteed access to almost unlimited funding to a vast array of hundreds of family-owned pharmaceutical companies.

During the 1970s, the OECD countries accepted the notion of healthcare as a human right, and "Health for All by the year 2000" was adopted as an international slogan. In the US, national, universal health coverage once again looked inevitable. (Health insurance in the US dates back to 1929, when an administrator at Baylor University Hospital in Dallas, Texas received the nation's first health plan. For the sum of 50 cents a week, Dallas schoolteachers received up to 21 days of free hospital care. This eventually led to the creation of Blue Cross/Blue Shield, followed by other familiar names such as Prudential and Travelers).

Twenty years later, as Scrip magazine put it, even the most optimistic defender of the "health for all" concept had to accept that Europe's, and to that matter the world's, healthcare systems were becoming increasingly swamped by consumer demand, as opposed to the patient need that they were originally set up to answer. Concern about the "suffering" of people lost priority to economic aspects of healthcare.

Regarding access to care, the "solidarity principle in health care" means that each citizen contributes according to ability while receiving treatment according to need. Systems that probably best exemplify this principle are financed through progressive income taxes and provide individuals with universal access to a socially defined, usually minimal, package of care.

At the turn of the century and among industrialized democracies, only the United States did not implement the solidarity principle. The country recognizes that the poor and the elderly are entitled to a care package through their Medicaid and Medicare programs. (On July 30, 1965, Lyndon Johnson signed the Medicare and Medicaid acts which put in place the framework for the current US health care financing system. It rests on 3 pillars: an employment-based insurance for workers and their families; Medicare for the elderly; and Medicaid for the poor and unemployed). Nevertheless, based as it is on privately financed care and insurance, the American system denies about 15 percent of its population adequate access to care, despite the rise in total healthcare expenditures, with public financing accounting for 42 percent of these expenditures (With the Democrats gaining control of both the House and the Senate in November 2006, there is again talk of a single public plan covering all Americans for all medically necessary services, including long-term care, mental health and dental services, and prescription drugs and supplies).

In other industrialized democracies universal access to a basic package of state-regulated care is guaranteed. This is the case in Japan, Germany and France.

Australia, Spain, Italy, and, recently, Israel, which previously had systems with traits similar to those in Germany, France, and Japan, changed for universal insurance administered by the state, so that their systems have come more closely to resemble those in the United Kingdom, Canada, New Zealand, and the Scandinavian countries, where the state is assumed to bear ultimate responsibility for health care, even when that care is provided through private institutions.

What the Canadian media report is true for all other countries: "The past decade in health care has been difficult for patients, providers, and governments alike…Recent changes appear unlikely to achieve a sustainable transformation in the organization, delivery, and financing of Canadian Medicare. We foresee continued turbulence…as a growing proportion of Canadians lose patience with health care systems that they perceive as no longer delivering reasonable access to core services."[1]

The past as prologue.
Without remembering where we come from; without reflecting on what shaped today's "modern" healthcare, there may be little or no clear understanding for the (accelerating pace of) progress in biomedicine; for what is expected to become tomorrow's "personalized, predictive" healthcare. A series of "blue pages", gives the reader a selection of highlights from the past and present, together with a glimpse of what researchers, experts, futurologists, expect the future to hold.

---

[1] *Ezekiel J. E., Department of Clinical Bioethics, Warren G. Magnuson Clinical Center, National Institutes of Health. Re-Thinking National Health Care Reform: Universal Healthcare Vouchers. Presentation given at Kellog University Annual Healthcare Conference, 2005.*

The 1950s : Free healthcare for ALL

* **Health for every citizen** becomes the promise of Governments in the industrial world.

*Postwar rapid economic **growth** in which the pharmaceutical industry takes part.

*Freedom of choice for physicians to prescribe. In the USA, the **Durham Humphrey Amendment** (1951) changed medical practice by creating a special class of drugs which requires a prescription (The Act defines the type of drugs that cannot be safely used without medical supervision and restricts their sale to prescription by a licensed practitioner).

***Paternalistic** treatment philosophy restricts patient information.

* Markets are strongly domestic ones and **protected** by national laws for healthcare products. Flourishing family-owned businesses.

***Little government control** on Statutory Healthcare Budgets.

* Geneticists James Watson and Francis Crick decipher (1953) DNA's **double helix structure**.

* **Delaney Committee** (USA 1951) starts congressional investigation of the safety of chemicals in foods and cosmetics, laying the foundation for effective controls over pesticides, food additives and colors.

* 1955, Jonas Salk discovers a **polio vaccine.**

* The **first heart/lung pump** is used in open-heart surgery (1953). The **first resection of aortic aneurysm** takes place in 1951. In 1954,
The technique of organ transplantation was originally tested on a pair of identical twins (kidney graft performed by Joseph Murray).

*Denham Harman advances the theory of **free radical aging**. He describes (1956) research indicating that the various free radicals produced in biological systems could be the chemical substances that cause the damage we see in the aging of cellular proteins, genetic materials and the membranes that hold cells.

* 1951, Gregory Pincus develops the **contraceptive pill**. The birth control pill is approved for general use (USA 1959; **Enovid 10**, a combined oral contraceptive will be launched in 1960).

* Salman A. Waksman receives the Nobel Prize for co-discovery of **streptomycin** (1952).

* **Erythromycin**, a broad-spectrum antibiotic that offers an alternative to people allergic to penicillin (1952).

* The anti-rheumatic **Cortisone** (discovered 1949) starts becoming one of the best-known drugs of all times.

* 1959, **Vancomycin** is introduced; a powerful antibiotic effective against serious hospital-acquired infections resistant to other drugs.

* Lillehei develops the first compact **heart pacemaker** (1958). **Ultrasound** is used by Ian Donald to diagnose fetus disorders (1958).

* Roger Williams, who discovered pantothenic acid —one of the B vitamins- revolutionized medical thinking in the early 1950s with his theory of **"Biochemical Individuality"** and the concept of **"genetotrophic disease".**

* **Psychotropic drugs** start being used to treat schizophrenia and depression.

Postwar healthcare thinking results in generous socialized Statutory Healthcare Systems.

Universal healthcare was a fantastic social ideology based on the willingness of bringing relief to the suffering, whenever illness strikes.

Anglo-American academia started to dominate most domains of medicine. It is to the credit of conventional medicine that, when science revealed progress, it has abandoned products of a former generation of medicines for better ones.

Table 1.1. Leading Healthcare Issues

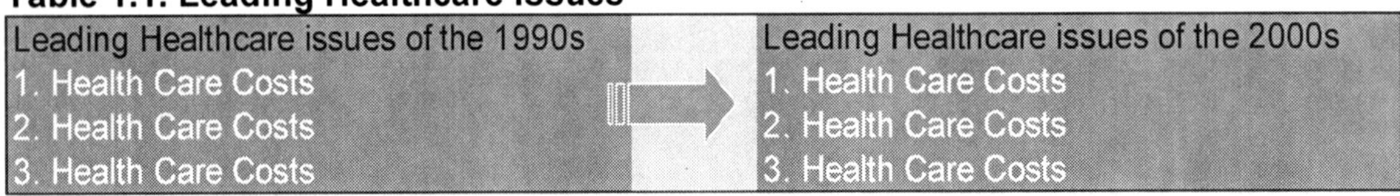

Leading Healthcare issues of the 1990s	Leading Healthcare issues of the 2000s
1. Health Care Costs	1. Health Care Costs
2. Health Care Costs	2. Health Care Costs
3. Health Care Costs	3. Health Care Costs

All countries, whether overseeing a **market-oriented system** (as in the U.S.), a **mixed system** (as in Germany), or a **public-oriented system** (as in the U.K. and Sweden) face rising costs of health care, both in absolute dollar/Euro terms and as a percentage of the GDP. This had a negative effect on labor markets as pressure has mounted on employers to increase their contributions to employee health care. It had an adverse effect on the competitiveness of developed economies in the global marketplace, in particular on countries experiencing already a declining economic growth.

The debate was now about the allocation of healthcare resources and a philosophy of "rational drug therapy" emerged.

Table 1.2. The Healthcare System's Dilemma

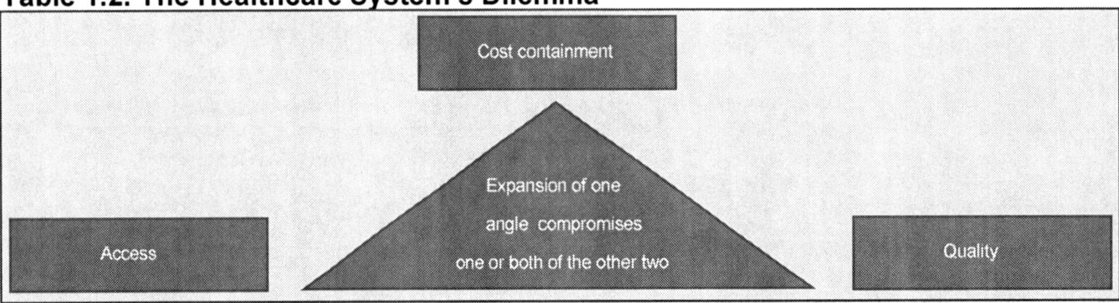

These economic realities not only mean that today's governments are asking their citizens to take on more financial responsibility for their drug treatment. They squeeze the margins of research driven pharmaceutical companies, while at the same time requesting more, better new drugs.

I.2. Challenges to health and equity

The WHO describes a series of challenges to Europe's Social Healthcare Systems.[2] According to the World Health Organization, a few conditions, linked by common risk factors and underlying determinants, are responsible for a large part of the disease burden in Europe (see the table). These leading conditions, their risk factors and determinants are more or less the same for every part and country of Europe and have overtaken communicable diseases in terms of the burden of ill-health, with some countries facing a "double burden" of both NCD and communicable disease.

Cardiovascular disease is the number 1 killer in Europe, causing more than half of all deaths across the Region, with heart disease or stroke the leading cause of death in all 52 Member States. Where gender-divided data are collected, different patterns of NCD become apparent between women and men.

---

[2] *World Health Organization. Gaining Health. The European Strategy for the Prevention and Control of Noncommunicable Diseases Regional Committee for Europe. 56th Session. Copenhagen. 11-14 September 2006.*

From the Postwar Generous Universal Healthcare Philosophy to the Promise of Cost-Effective Personalized Medicine.

The 1960s: Universal Healthcare becomes an uncontested citizen's right.
The pace of scientific advances accelerates. Regulatory focus on efficacy.

* High prices for medicines, **entirely reimbursed** by the Statutory Healthcare System.

* Success for **originator** pharmaceutical companies. Introduction of innovative therapies accelerates. **Generics** in the Rx sector are almost unknown.

* Industry focus moves from anti-infectives to **receptor antagonists**.

* **Thalidomide**, a new sleeping pill, is found to have caused birth defects in thousands of babies born in Western Europe. The scandal is at the basis of a fundamental shift in thinking. The **Kefauver-Harris Drug Amendments** to the US Food, Drug and Cosmetics Act (USA, 1962) are enacted to ensure greater drug safety. They institutionalize the view that randomized, placebo-controlled, double blind trials are the appropriate means to establish the efficacy of a treatment.

* For the first time, manufacturers of drugs (Rx and OTC) must prove to the FDA the **EFFECTIVENESS** of their products. Up to this point, the FDA has only concerned itself with drug SAFETY.

*Wilson & Wilson (1963) advocate "adequate estrogen from puberty to the grave". Introduction of the first human **menopausal gonadotropin** (1963). Robert Wilson's book **Feminine Forever,** describing what women gain from HRT, heralds a dramatic increase in Hormone Replacement Therapy.

* **Diversification** by acquisition outside the core pharmaceutical business (e.g. cosmetics) is the dominating guru-theory for the industry.

* President Kennedy proclaims a **Consumer Bill of Rights.** Included are the right to safety, the right to be informed, the right to choose and the right to be heard.

* Jean-Paul Sartre's theory of **existentialism,** Simone de Beauvoir's ideas on **sexual liberty,** and a radical change in contraception –away from predominantly male-based methods such as the condom or withdrawal- revolutionize Western society.

* **Garamycin** (gentamicin), a broad-spectrum antibiotic effective against both gram-positive and gram-negative infections, is approved by the FDA (launched 1966).

* Christian Barnard performs the world's first **human heart transplant.**

* **UK´s Medicines Act** defines what a medical product is : any substance manufactured, sold, supplied for use wholly or mainly (a) by administration for a "medical purpose", (b) as an ingredient in a substance as in (a).

* The US **FDA** leads the way, on a global scale, towards strong regulatory requirements for medicines . The agency soon becomes the example for the rest of the world's health authorities of how to deal with medicines.

* **Diuretics** offer an effective option too treat hypertension.

* **Hospice** movement for care of terminally ill introduced by Dame Cicely Saunders (1967).

*Charles Brenton Huggins is awarded the Nobel Prize for studies **in hormone treatment** of cancer of the prostate; Francis Peyton Rous for discovery of **tumor-producing viruses**.

* Measles vaccine is introduced. By 1968, there are only 22.000 cases of measles reported in the US- down from 400.000 in 1962.

* African doctors witness cases of **"slim disease"**- later known as AIDS.

* July 1965, Lyndon Johnson signed the Medicare and Medicaid acts which put in place the framework for the US´current health care financing system. It rests on 3 pillars:
~ employment based insurance for workers and their families:
~ Medicare for the elderly;
~ Medicaid for the poor and unemployed.

The biomedical industrial complex thrives on a wave of new drug discoveries and product launches and is supported (protected) by national governments, keen to develop a strong domestic pharmaceutical industry.

2000 years of biblical, Judeo-Christian thinking, blaming women for paradise lost, continues to influence Western society and treat women as a minority group. Little attent is paid to specific women's health concerns, as well as to non-white ethnic groups. **Most of the clinical studies made would be** judged totally unacceptable by today's efficacy and safety standards.

Progress in biomedicine provides doctors with the tools to ease suffering rather than watch patients die. **A tremendous shift in ideas about medicine occurs.** Their impact on regulatory measures sends shockwaves through the industry. The manipulation of power by parties involved in health provision increases.

Table 1.3. Burden of disease and deaths from NCD in the WHO European Region, by cause (2005 estimates)

Group of causes (selected (leading non-communicable diseases/NCDs)	Disease burden (DALYs) (000s)	% all causes	Deaths (000s)	% all causes
Cardiovascular diseases	34421	23	5067	52
Neuropsychiatric conditions	29370	20	264	3
Cancer (malignant neoplasms)	17025	11	1855	19
Digestive diseases	7117	5	391	4
Respiratory diseases	6835	5	420	4
Sense organ disorders	6339	4	0	0
Muskoloskeletal diseases	5745	4	26	0
Diabetes mellitus	2319	2	153	2
Oral conditions	1018	1	0	2
All NCDs	115339	77	8210	86
All causes	150322		9564	

NCDs have a multifactorial etiology and result from complex interactions between individuals and their environment, including their opportunities for promoting health and countering their vulnerability to risks. Individual characteristics (such as sex, ethnicity, genetic predisposition) and health protective factors (such as emotional resilience), together with social, economic and environmental determinants (such as income, education, living and working conditions), determine differences in exposure and vulnerability of individuals to health-compromising conditions. These underlying determinants, or "causes of causes", influence health opportunities, health-seeking and lifestyle behaviors as well as onset, expression and outcome of disease.

Almost 60% of the disease burden in Europe, as measured by DALYs, is accounted for by seven leading risk factors: high blood pressure (12.8%); tobacco (12.3%); alcohol (10.1%); high blood cholesterol (8.7%); overweight (7.8%); low fruit and vegetable intake (4.4%;) and physical inactivity (3.5%). It should also be recognized that diabetes is a major risk factor and trigger for cardiovascular disease (CVD). These are the same leading risk factors in all epidemiological subregions of Europe (Eur- A, -B, -C₁) and in most European countries, although the rank order may differ[3]. In 37 of the 52 European Member States of WHO, the leading risk factor for deaths is high blood pressure; in 31 Member States, tobacco is the leading risk factor for disease burden. Alcohol is the leading risk factor for both disability and death among young people in Europe.

These leading risk factors are common to many of the leading conditions in Europe. Each of these seven leading risk factors, for instance, is associated with at least two of the leading conditions and, in return, each of the leading conditions is associated with two or more risk factors. Furthermore, in many individuals, particularly the socially disadvantaged, risk factors frequently cluster and interact, often multiplicatively[4].

---

[3] *The European health report 2005. Public health action for healthier children and populations. Copenhagen, WHO Regional Office for Europe, 2005(http://www.euro.who.int/InformationSources/Publications/Catalogue/20050909_1).*

[4] *World Health Organization. Gaining health. The European Strategy for the Prevention and Control of Noncommunicable Diseases Regional Committee for Europe.56th Session. Copenhagen. 11-14 September 2006.*

The 70's:
Birth of the Biopharmaceutical industry. Launch of the first "blockbuster" medicine. Emergence of the "natural vs. chemical" concept.

* In Upjohn vs. Finch (1970), the US Court of Appeals upholds enforcement of the **Kefauver-Harris Amendments** by ruling that commercial success alone does not constitute substantial evidence of a drug's safety and efficacy.

* The US Supreme Court upholds the 1962 drug **effectiveness** amendments and endorses FDA action to control entire classes of products through regulations rather than having to rely on time-consuming litigation (1972).

* British engineers invent the **computed tomography scanner** (1971), which assembles thousands of X-ray images into a highly detailed picture of the brain; later models can scan the entire body.

* **Daniel Nathans, Hamilton Smith** and **Werner Arber** are awarded the Nobel Prize (1978) for discovering **restriction enzymes** and their application to problems of **molecular genetics**.

* First signs of **German biotechnophobia**, based on an outdated set of values promoted by the Green party, allow US pharmaceutical companies to surge to the fore.

* The Vitamins and Minerals Amendment prevents the FDA from establishing standards that limit the potency of vitamins and minerals in food supplements or regulating them as drugs based solely on potency. The result will be the creation of what becomes a multibillion dollar "**dietary supplements**" industry, which critics say, is "**unregulated**".

* Success for originator pharmaceutical companies. Introduction of innovative therapies at the rate of **one per year/company.** Generics (of Rx molecules) still unknown or restricted.

* Beecham launches **Cimetidine** (1976), a H2-antagonist that will become **the first billion $ drug** and set off the race for pharmaceutical "blockbusters".

* Two articles in the **New England Journal of Medicine** mentioning the risks of HRT, disturb confidence in **HRT**. Feminist attitudes dominate. In later years Women's criticism will be based mainly on a perceived and/or openly described negative conception of the climacteric as a transition to deterioration and decline. Control is seen as external to women, both through the imposition of new technology and through the outdated physician/woman relationship.

* **Louise Brown**, the world's first **test-tube baby**, is born in England (1978).

* Infant formula Act establishes special US FDA controls to ensure necessary nutritional content and safety.

* The discovery of **recombinant DNA** and **monoclonal antibody technologies** (by **Georges Kohler** and **César Milstein**) in the 1970s marked the birth of the biopharmaceutical industry.

* The US **Saccharin** Study and Labeling Act (1977) stops the FDA from banning the chemical sweetener. But product labels have to warn that the substance has been found to cause cancer in laboratory animals.

* At the Philadelphia Chapter of the American Marketing Association, Juliet Goodfriend launches the idea (1977) of **Direct to Consumer Advertising for Prescription Drugs** (what will become a multi-billion dollar marketing tool in the late 1990s.

* The FDA unveils the **OTC Review** (1971). This program, which focuses on active ingredients rather than products in isolation, is still the basis for the FDA's approval of OTC medicines.

* "**Green, green is the world**". Consumers in western nations become confronted with new food-health concepts: Natural ingredients, pesticide-free, bio-forming. The "Natural vs. chemical" concept will have long-lasting socioeconomic-political relevance.

The **pharmaceutical industry is rewarded** for innovation and progress in healthcare.

Medical progress is celebrated. Focus is put on the **care of the suffering**. First (timid) concerns about the **cost of universal healthcare** arise.

Blocked by **leftist anti-consumerism** and anti-advertising **ideologies**, European governments lack the courage to ease restrictions on medical information, this way encouraging blind dependency of patients on health professionals and discouraging patients of taking self-responsibility for their health.

The WHO report says that diseases also cluster in individuals, so that several co-morbidities can exist at once. At least 35% of men over 60 years of age have been found to have 2 or more chronic conditions and the number of co-morbidities increases progressively with age, with higher levels among women. There are strong interrelationships between physical and mental health, with both related through common determinants such as poor housing, poor nutrition, or poor education, or common risk factors such as alcohol. Depression, for example, is more common in people with physical illness than the healthy, with prevalence of major depression in up to 33% of people with cancer, 29% of those with hypertension and 27% of those with diabetes.

Gender influences the development and course of risk factors and diseases such as obesity, cardiovascular disease and mental health problems. Throughout the life course, women and men are attributed, and take, different roles in society, which are valued differently. This affects risk-taking behavior, exposure to risks and health-seeking behavior; it also determines the degree to which women and men have access to and control over the resources and decision-making needed to protect their health. These result in inequitable patterns of health risk, access to health services, use of health services and health outcomes.

There is an uneven distribution of conditions and their causes throughout the population, with higher concentration among the poor and vulnerable. People in low socioeconomic groups have at least twice the risk of serious illness and premature death as those in high socioeconomic groups[5]. Inequalities in health between people with higher and lower educational level, occupational class and income level have been found in all European countries where measured. The increasing concentration of risk factors in the lower socioeconomic groups is leading to a widening gap in future health outcomes.

When improvements to health do occur, the benefits are unevenly distributed within society, with few exceptions. When all groups in society are exposed to some extent to health interventions, those in higher socioeconomic groups have tended to respond better and benefit more. Mortality rates are declining proportionally faster in the higher than lower socioeconomic groups, particularly for CVD, widening further the differences in life expectancy between the two groups[6].Within Lithuania, increasing inequalities (by education) in mortality due to all major causes of death for men and women were noted between 1989 to 2001, due to a considerable decline in mortality among those with a university education and a sharp increase among those with primary or no education.

These conditions and their causes contribute to differences in healthy life expectancy between and within European countries. While CVD mortality rates have been decreasing in western Europe in recent decades, there is an up to 10-fold difference in premature CVD mortality between western Europe and countries in central and eastern Europe, with the highest rates in the east. Common preventable conditions are

---

[5]Wilkinson R, Marmot M, eds. Social determinants of health: the solid facts. 2nd edition. Copenhagen, WHO Regional Office for Europe, 2003
(http://www.euro.who.int/document/e81384.pdf)
[6] Mackenbach JP. Health inequalities: Europe in profile. Brussels, European Commission, 2006
(http://ec.europa.eu/health/ph_determinants/socio_economics/documents/ev_060302_rd06_en.pdf).

contributing to a 20-year difference in healthy life expectancy across Europe, and to differences within countries[7].

I.3. The 5th hurdle of affordability

Economic arguments dominate the debate about which reforms deemed necessary. One-dimensional cultures, education, the suffering of patients, come to feel the dictate of economics. Those who use the economic-arguments seem to have the better cards. They can come up with figures. It is difficult to avoid falling into the grip of a bank of figures capable of demolishing all opposition.
Economic-financial reasoning developed into a dogmatic model with a build-in refusal of all alternative thinking. What's not economic is labeled as a failure.

Is it? Viewed objectively, what is the "Holy Economic" model based on? It is a reductionist weapon. Complex situations are being reduced.

In reductionism, multiple problems in a system are typically tackled piecemeal. Each problem is partitioned and addressed individually. This approach neglects the complex interplay between all the individual parts of the whole.
And exactly this way of reductionist thinking creates conflicts, and more problems.

Of course: one can ask of a healthcare system to be efficient.
Now, is "efficiency" synonymous with "economy"? Does society no longer have other values? Should social factors of influence be defined as obsolete? Can social, moral, ideological, religious aspects and motivations -to cite just a few- simply be overlooked? What are the priorities society sets in education, culture, and healthcare? What are the values in which society wants to invest? For sure, economy plays a role. It should be restricted to that role and not become a goal in itself[8].

I.4. 27 EU nations…leading to 27 national social healthcare system reforms

In Europe, there is an endless procession of politicians and health experts queuing up to tell citizens that things are getting worse and that reform by unpopular measures is therefore inevitable.

There is no lack of ideas to control budgets. Among them figure:

- The introduction of a reference pricing system for patented medicines, unless they are "really innovative" ("innovative" has still to be defined).
- Medicines not listed under the reference pricing system to be subject to a (short-term, with possibility for extension) price freeze and a certain percentage price reduction.
- A discount on the ex-factory price of medicines reimbursed by the sick Funds.
- The reduction of pharmacy mark-ups for higher-priced medicines.
- The reduction of the trade margin of wholesalers

At the end, they all come down to higher co-payments for the patients, less treatment fees for doctors and hospitals, less margins pharmacists and for the pharmaceutical industry…Who gains? Isn't it a **paradox** that, in a world of continuously growing

---

[7] *World Health Organization. Gaining health.The European Strategy for the Prevention and Control of Noncommunicable Diseases Regional Committee for Europe. 56th Session. Copenhagen. 11-14 September 2006.*

[8] *Mertens G, Mastering the Complexities of Women's Healthcare, Nicholas Hall Strategic Healthcare Management Reports, 2002.*

From the Postwar Generous Universal Healthcare Philosophy to the Promise of Cost-Effective Personalized Medicine.

The 80´s : Emergence of evidence based medicine

* Doctors in San Francisco and NY report the first cases of what later will be termed Acquired Immunodeficiency Syndrome, or **AIDS** (1981)

* Increased competition in the pharmaceutical sector, encouraged by legislation. **Budget restrictions** reduce long-term guaranteed revenues of reimbursed preparations. Instead of courageous action, **Utopian energy** enters the traditional pharmaceutical company boardroom: CEOs compete in "who's best at formulating the company's vision". **Generic** copies of originator products gain in importance.

* Rationalization of the industry starts. Strategic alliances, mergers and acquisitions turn corporations upside down. **Shareholder value philosophy**. European protective governments fruitlessly support proudly uncreative national pharmaceutical companies (for which doomsday neared).

* **Menopause** becomes a crucial social and health problem. Strategies do not yet take into account the fact that self-confident women reject patriarchal dominance in favor of a more individual, woman-focused, considered approach.

* In the face of consumer demand, western medical science is re-awakening, albeit slowly, to the clinical worth of **subjective experience**, as expressed by **Chinese and Ayurvedic** models of health and health care. The scientific biomedical orthodoxy model becomes competition from e.g. acupuncture.

* **Joseph Goldstein**, professor and chairman of the department of molecular genetics at the University of Texas Southwestern Medical Center at Dallas, won the 1985 **Nobel Prize** for his research on **cholesterol metabolism**.

B.1. Emergence of the disease management idea

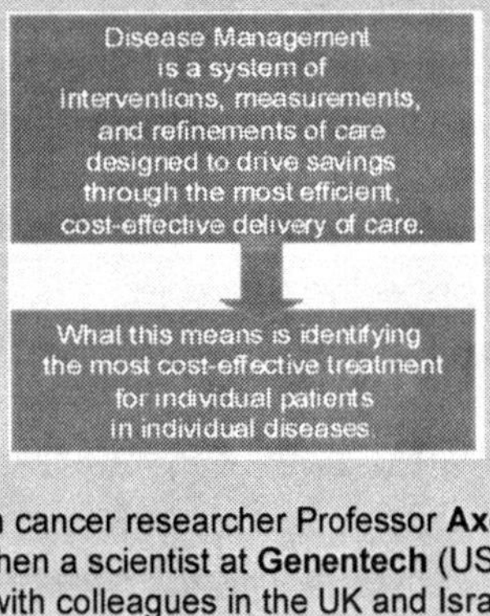

* German cancer researcher Professor **Axel Ullrich**, then a scientist at **Genentech** (USA), working with colleagues in the UK and Israel, succeeded in describing the structure and function of a **receptor** for **epidermal growth factor (EGF)**. Since then, tyrosine kinases and various growth factors have been at the focus of research and development of therapies against tumors.

* Alec Jeffreys devises a **DNA fingerprinting method** (1984).

* The **Orphan Drug Act (**1983) enables the FDA to promote research and marketing of drugs for rare diseases that otherwise would not have been profitable. The 1983 US Orphan Drug Act. The Act provides a seven year period of market exclusivity and financial incentives for the development of drugs against diseases affecting small patient populations (defined as less than 200,000 people). The most important feature of the program and a major factor in attracting venture capital investment is market exclusivity - the FDA will not approve another application for the same drug treating the same indication during the seven year period. (Market exclusivity does not protect orphan drugs from the approval of clinically superior drugs and a company can lose exclusivity if it cannot produce enough products to meet demand). Similar legislation has since been enacted in Japan (1993), Australia (1998), Singapore (1999) and the EU (2000).

* The drug **Price Competition and Patent Restoration Act** (1984) expedites the availability of cheaper generic drugs in the US by allowing the FDA to approve them without the need to duplicate safety and effectiveness studies. Brand-name companies can apply for 5 year **extra patent protection** for newly developed medicines.

* The first **genetically engineered vaccine**, for the prevention of hepatitis B, is approved in the US (1986). Also cleared are the first genetically engineered **interferons** for the treatment of hairy cell leukemia. Later, these powerful drugs will be applied to such diverse conditions as genital warts, AIDS-related Kaposi's sarcoma, hepatitis C, hepatitis B.

* In Germany, the production of human insulin with **genetically modified bacteria** is the subject of fierce discussions about "incalculable risks". The Green party blocks the factory construction plans of Hoechst.

* In the 1970s and 1980s, the **analgesic property of antidepressants was discovered**. Antidepressants were subsequently used for the treatment of a wide variety of **neuropathic pain syndromes**, including diabetic peripheral neuropathy, postherpetic neuralgia, migraine, arthritic pain, cancer-related pain, and chronic lower back pain. **Greenbaum** and colleagues performed a landmark study in 1987. It was the first study to empirically designate the unique effect of **antidepressants** on functional bowel disorders independent of an anticholinergic effect they may have. The lack of correlation between changes in psychiatric status and changes in GI motility also spoke to the unique property of the antidepressants in **improving global well being** in patients with functional bowel disorders. Investigators over the following decade have shown the **benefit** of antidepressants for the **treatment** of **non-cardiac chest pain, functional vomiting**.

* The USA turns to a "Managed Care" system.

economic, technological and geopolitical complexity, the very idea of a nationally governed healthcare system has survived? Global interdependence has not erased the gloriously won or perilously lost "national" borders.

In Europe, epic battles may be locked up in history books. The spirit of politico-socio-cultural competition between regions, however, resurfaces daily. Everyone promotes EU integration...as long as it brings advantages for the home-state. In such a context, and although the European Court of Justice is helpful when it comes to cross-border matters of health –slowly "stealing away" power from national authorities- it should come as no surprise that it will take some more decades before we see the realization of a truly European healthcare system[9].

Every country sticks to its 5th hurdle of affordability.

Table 1.4. Social Heath Security 2000-2040: Financial Resources Needed[10]

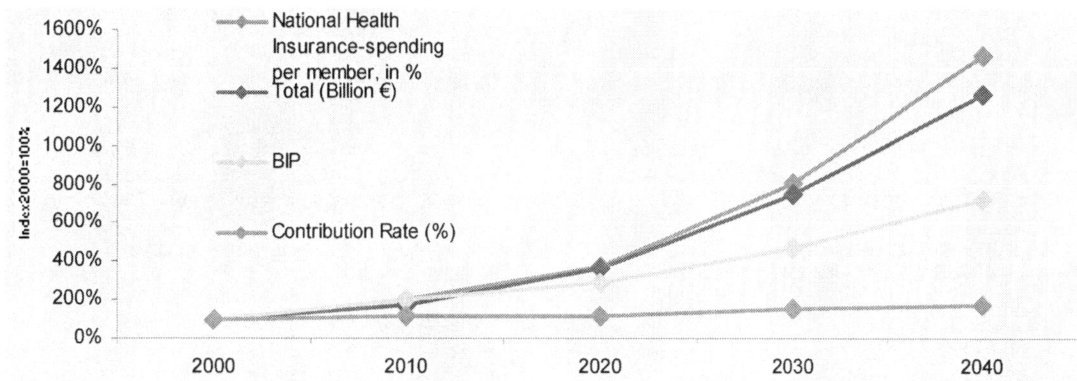

Every country makes it own projections for the financial resources it needs to "rescue" their national health insurance coverage system. Separate national issues, not EU-harmonized structures and solutions continue to dominate the European Union in matters of healthcare.

The projection made by the German Institut fuer Gesundheitspolitik, is given as an example.

The French deficit in the Healthcare branch of its National Social Security System reached 20.2 billion Euro in 2005[11]. The dereimbursement of medicines estimated being of little efficacy, the increase in generics, could not stop the increase in drug spending, according to the Commission des Comptes of the Sécurité Sociale (CCSS).

French doctors prescribe an average of 4.5 drugs per patient compared to 0.8 drugs per patient in Nordic countries. In France, in 90% of physician consultations a drug will be prescribed. In the Netherlands that figure is 43.2%. In drug consumption, France ranks second, after the United States.

---

[10] *IfG Instiut fuer Gesundheitspolitik, Presentation at Handelsblatt/Euroforums Health 2004 Conference, Berlin, Nov 2003*

[11] *Blanchard Sandrine, Le Monde, Dossiers & Documents, No 356, September 2006*

The 1980s *(continued)*

* The US FDA in 1982 approves the first drug developed using **recombinant DNA technology** –a form of human insulin. A vaccine to prevent bacterial meningitis in young children is launched. Over the next 10 years, cases of the disease will drop by nearly 80 percent.

* The first drug produced via **genetic engineering** was human insulin which appeared on the market in 1982.

* Back in 1953, a committee of Britain's top cancer specialists has taken a long, hard look at the connection between **smoking and lung cancer**. They announced an "established relationship and warned of the risk apparently attendant on excessive smoking. It took till the 1980s for authorities to take the matter more seriously and launch (albeit slowly) anti-smoking measures (predominantly based on higher taxes on tobacco products).

* The **physician-centered universe** starts to crumble. Rx industry in the USA timidly begins to exploit communication with the patients (fairly blunt communication). Public relations and what is later called **DTC** –or direct to consumer advertising- are tactically implemented to complement the introductory campaigns of drugs.

* Research links **aspirin** with increased likelihood of **Reye´s Syndrome**, a rare brain and liver disorder.

* Nobel Prizes are awarded to J. Michael Bischop and Harold E Varmus for their discovery of the **cellular origin** of **retroviral oncogenes**; to Susumu Tonegawa for his discovery of the **genetic principle** for **generation** of **antibody diversity**; to Stanley Cohen and Rita Levi-Montalcini for their discoveries of **growth factors**; to Michael S Brown and Joseph L Goldstein for their discoveries concerning the **regulation** of **cholesterol metabolism**; to Niels K Jerne , Georges J F Kohler and Cesar Milstein for theories concerning the specificity in **development** and **control** of the **immune system** and the discovery of the **principle** for **production of monoclonal antibodies**; to Sune K Bergstrom, Bengt I Sameuelsson and Sir John R Vanefor their discoveries concerning **prostaglandins** and **related biologically active substances;** to Baruj Benacerraf , Jean Dausset and George D Snell for their discoveries concerning **genetically determined structures** on the **cell surface** that **regulate immunological reactions;** to Roger W Sperry for his discoveries concerning the **functional specialization** of the **cerebral hemispheres**. And David H Hubel and Torsten N Wiesel for their discoveries concerning **information processing** in the **visual system.**

* Development (1981) of the first oral contraceptive to contain a **selective progestogen.** The Pills of today are vastly different from those available 30 years ago containing just **a fraction of the original dose of estrogen**. In addition, new progestogens which are more selective in their activity and have less metabolic impact have allowed further reduction of the estrogen dose. **Organon's** active **progestogen** developed throughout the late 1970s, desogestrel, was the **first** of the **modern progestogens** to become available. It was combined with a low dose of **ethinylestradiol** (30 mcg) as **Marvelon** which was first launched in 1981 (as **Desogen** in the United States,1992). This **Pill is presently the world's most widely used contraceptive,** in use by more than 70 million women. In 1986, desogestrel was combined with an even lower dose of estrogen (20 mcg) and introduced as **Mercilon**. It takes **12 years of laboratory testing** and clinical trials to develop a new contraceptive product (at a total costs of more than **EUR 300 million**). Safety is of paramount importance given that the products developed will be used by healthy people. This means that clinical trials require very high numbers to study efficacy and safety. Few companies engage in contraceptive research: many are reluctant to commit so much of their research and development resources to an **area of medicine sensitive to litigation.**

* **Gemfibrozil,** the first safe fibric acid derivate, becomes available as a treatment option for lowering LDL and increasing HDL levels. With **Mevacor, the world's first statin, Merck** started the public education about cholesterol.

* **Albuterol,** a short-acting beta-agonist becomes the primary treatment of acute asthma attacks.

* Kary Mullis of **Cetus Corporation** (to become Chiron Corp in 1991) discovers **Polymerase Chain Reaction (PCR)**, a technique of amplifying DNA. Mullis wins the Nobel Prize in 1993 for his method of rapidly multiplying one DNA sequence into many. The process can turn a single sequence into a million identical copies in about an hour. PCR allowed for **DNA fingerprinting to,** for **DNA-based diagnosis of diseases,** for **paternity testing** and **genome mapping** to become reality.

* **Novartis´Sandimmune (cyclosporine),** derived from **soil fungus,** a drug that **selectively suppresses T-cells, revolutionizes surgery** in that it avoids the immune system of patients undergoing organ transplants to attack the new organ.

Economics of care lead to **rational drug therapy philosophy.**

The focus is starting to shift away from universal healthcare for the suffering patient to the care of an ailing healthcare system, which is itself suffering from

In healthcare, the term consumer was first used to describe people who seek mental health services. During the 1980s, US Consumer Activist Ralph

Table 1.5. Total Drug Expenditures USA-France-Germany-UK, per habitant (in $US)[12]

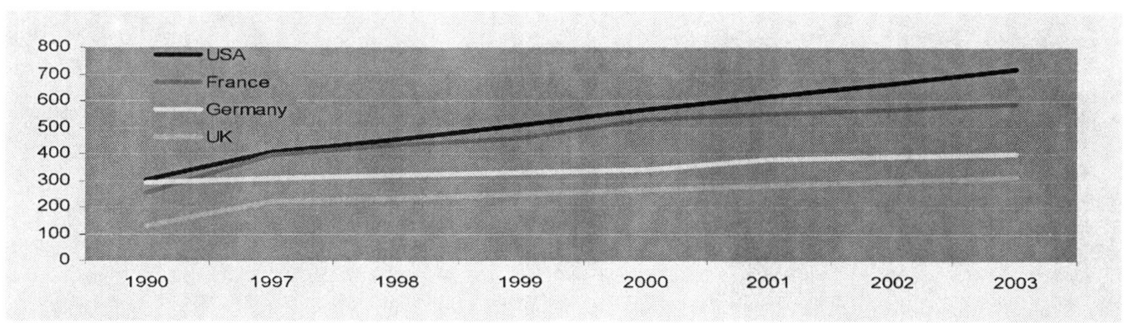

I.5. USA: The "consequences" of individual responsibility

In times where Europe's and Canada's "social" health security systems face inevitable changes because of a chronic lack of financial means to keep the systems going "as always", the USA shows an astonishingly looking developed Western world what it looks like, when a healthcare system focuses on individual responsibility.

Table 1.6. Problems with the US Health Care System	
•15% uninsured	•High administrative costs
•High and escalating costs	•Under investment in information technology
•Inconsistent quality	•Absence of systematic technology
•Frequent medical errors	assessment before dissemination
•Substantial racial, geographic and other disparities	•Personnel shortages

Most people in the US (59.5 percent) were covered by a health insurance plan related to employment for some or all of 2005, a smaller proportion than in the previous year (59.8 percent). As the largest component of private health insurance coverage, this decline in employment-based coverage essentially explains the decrease in total private health insurance coverage, from 68.2 percent in 2004 to 67.7 percent in 2005[13]

US healthcare expenditures amounted to almost $2 trillion, or more than $6000 per person in 2004. Healthcare amounted for 16 percent of GDP[14].

Table 1.7. Development of U.S. National Health Expenditures*	1970	1980	2000	2004
Total (US $ billions)	75,1	254,9	1358,5	1877,6
Hospital care	27,6	101	417	570,8
Professional services	20,6	67,3	426,7	587,4
Home health & Nursing homes	4,3	21,4	125,8	158,4
Retail medical products	10,5	12	120,8	188,5
Prescription drugs	5,5	12	120,8	188,5
*selected calendar years				

---

[12] Mertens G., *Healthcare in Germany: Turning Crisis Into Opportunity. Nicholas Hall Strategic Healthcare Management Reports, June 2003*
[13] *US Census Bureau. Income, Poverty, and Health Insurance Coverage in the United States: 2005. August 2006*

* Growing **dominance of the USA**. In 1999, the US accounted for 40 per cent of the $350bn prescription medicines sold worldwide and –because of higher margins- an estimated 60 per cent of pharmaceutical profits. Europe accounts for about 27% of the global market. Only a decade ago, Europe's share of the global market was over 40%. With Europe's nation-states struggling to limit the cost of medicines, demand has been growing at half the rate of the USA.

B.2. The growing dominance of the USA

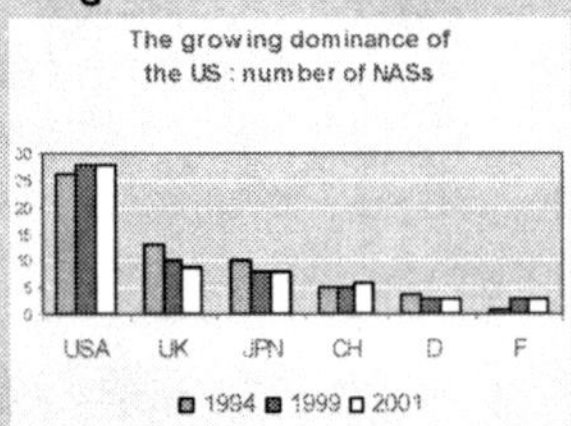

* The world observes **a migration of science to the US**. This is particularly true also for the research based drug industry. Nearly all the (remaining) big European pharma companies increase their R&D presence in the USA. As for the US companies, they decrease their presence outside of their domestic market. The proportion of research investment made outside the US by American companies dropped from 22 percent to 15 percent between 1955 and 1999.

* The principle of **subsidiarity** in healthcare blocks progress in healthcare in Europe. Each member state operates its own pricing system, its own regulatory framework (the **EMEA**, or European Medicines Agency, intended to put a halt on the duplication of corporate effort and expense, is in its infancy) and thus, breaks the medicines market into national fragments.

* The pharmaceutical iIndustry goes from strength to strength. Business is improving by **double digit growth**. Whether this is sustainable became the question.

* Stanley B. Prusiner is awarded the Nobel Prize for discovering a new species of germs, called **prions that** cause degenerative brain disorder.

* **Hoffman-La Ruche's Invirase (saquinavir)** becomes the first drug of the **new class of protease inhibitors** that will go on to **revolutionize AIDS treatment**

* Professor Graeme Clark - Pioneered the **multiple-channel cochlear implant** which has brought hearing and speech understanding to tens of thousands of people with severe-to-profound hearing loss.

* **Biogen Idec´s interferon beta** drug, called **Avonex**, an intramuscular injection becomes the worldwide most prescribed multiple sclerosis product.

* A wave of **Mergers and Acquisition** shakes up the industry. Well known company names disappear, e.g.: **Hoechst, Rhone-Poulenc, Ciba, Marion Merrel Dow, Warner-Lambert, Welcome**. Major mergers include :
1994 **American Home** products acquired all of the outstanding common stock of **American Cyanamid**; **Hoffmann-LaRoche** acquired **Syntex**; **SmithKline** acquired **Sterling Health**; **Rhone-Poulenc** increased ownership to 100% of **Institut Mérieux**.
1995 **Glaxo** Holding acquired **Wellcome**; **Hoechst** acquired **Marion Merrell Dow**; **BASF** acquired **Boots** pharmaceuticals; **Pharmacia** and The **Upjohn** Company merged.
1996 **Ciba** and **Sandoz** merged to create Novartis
1997 **Nycomed** and **Amersham** merged.
1998 **Hoffmann-LaRoche** (the company that brought the world Valium) purchased **Corange** (Boehringer Mannheim)
1999 **Astra** and **Zeneca** Group merged to create AstraZeneca; **Hoechst** and **Rhone-Poulenc** merged to create Aventis; **Sanofi** and **Synthélabo** merged to form Sanofi-Synthélabo.

* Major company acquisitions of the 1990s

Year	Buyer/Target	Price in US$bn
1996	Sandoz / Ciba Geigy	60
1998	Zeneca / Astra	34,6
1999	Pharmacia Upjohn / Monsanto	26,9
1998	Rhone Poulenc Rorer / Hoechst	21,7
1995	Glaxo / Burroughs Welcome	20
1995	Pharmacia / Upjohn	13
1998	Sanofi / Synthélabo	11,1
1997	Hoffmann-LaRoche / Boehringer Mann.	11
1995	Hoechst-Roussel / Marion Merrell Dow	7,1

AstraZeneca claimed that it was the only merged company to have actually increased its global market share (It had a 4,3% share compared with its previous 4,1% share)...Companies cite a variety of reasons for merging or acquiring others, but frequently those boil down to a weakness in R&D relative to their sales growth rates. M&A is expected to boost R&D and obtain industry leadership in several therapeutic areas; it is expected to provide scale to compete globally etc. The debate goes on: Can the **innovation gap** be managed by **M&A** ?

* Professor Peter Doherty, Nobel Laureate 1996 - Discovered that **T-cells**, the foot soldiers in our bloodstream, were expert at **killing cells** that had **viruses locked inside**. This has led to new and better vaccines, healthy organ transplants and better treatment of conditions like Multiple Sclerosis and Diabetes.

* An international team of scientists discovered the **link between folate** intake and **spina bifida**. This led to women being advised to increase folate intake before and during pregnancy and supplementation of some foods with folate.

Total health spending in the U.S. was $1.88 trillion in 2004, or an estimated $6,280 for each person in the U.S. Total health spending increased 7.9 percent in 2004 compared to 2003.

Total health spending totaled 16 percent of GDP in 2004, up from 15.9 percent in 2003. Of the increase in health spending in 2004, 33 percent went to spending on hospitals, 24 percent for physicians, 11 percent for prescription drugs, and 32 percent for all other spending categories. Between 2005 and 2006, premiums for health coverage offered by employers increased 7.7 percent, the third straight year of declines in premium growth. Even so, this was more than twice the growth in the Consumer Price Index. Of every dollar spent on health services in the U.S. in 2004, 45 cents came directly from government sources. Costs for program administration and the net cost of private health insurance were about 7 percent of total health spending in the U.S. in 2004, and grew 9.4 percent, a rate higher than the 7.9 percent increase in total health spending.

Increasing costs in health care are forcing many Americans to go without care. The number of Americans with major medical insurance has decreased, dropping from 81 percent in 2005 to 74 percent in 2006. And the perils of being uninsured are apparent.
A 2001 Harvard study found that illness and medical bills caused half of the 1,458,000 personal bankruptcies in 2001 and estimates that medical bankruptcies affect about 2 million Americans each year. Alarmingly, the American Bankruptcy Institute states that three-fourths of the people who file for bankruptcy because of medical debts have health insurance when the medical problem begins.

A nationwide survey in September 2006 (commissioned by Aflac insurance) found that 89 percent of all Americans say that the rising cost of health care is among the most pressing concerns facing them today.

Table 1.8.   Increase in Workers out-of-pocket healthcare costs*

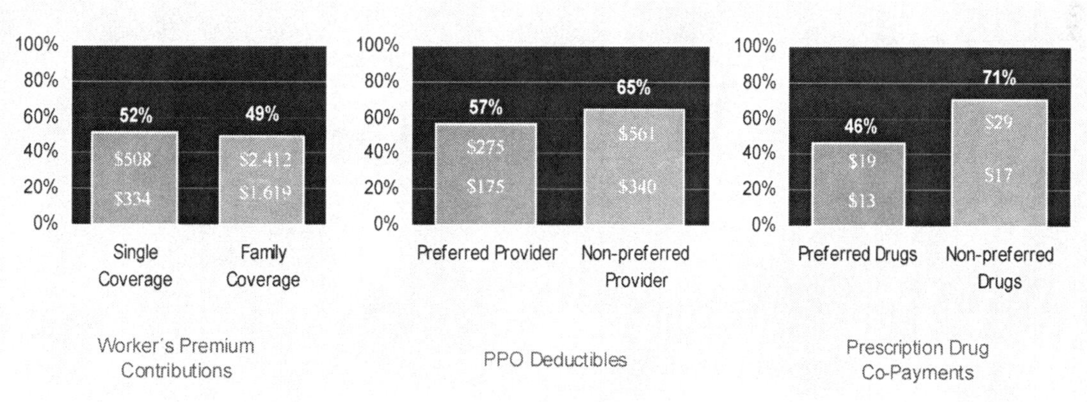

Worker's Premium Contributions
52%
49%
$508
$334
$2.412
$1.619
Single Coverage
Family Coverage
PPO Deductibles
57%
65%
$275
$175
$561
$340
Preferred Provider
Non-preferred Provider
Prescription Drug Co-Payments
46%
71%
$19
$13
$29
$17
Preferred Drugs
Non-preferred Drugs

(*For the period 2000-2003. Family coverage is defined as health coverage for a family of four. Average deductibles include covered workers who do not have a deduction)[15].

Typical family health premiums increased by almost 50 percent during the period mentioned above.

---

[15] Source: Kaiser/HRET Survey of employer-sponsored health benefits:2000-2003; Pharmaceutical Executive, Oct 2003:19

* With the exception of **GlaxoSmithKline**, which created the world's biggest drug group of the late nineties, and of **AstraZeneca**, **Roche** and **Novartis** of Switzerland, to a certain degree also **Aventis** and **Sanofi-Synthelabo** of France (both companies merged in 2004), and, to a certain extend of Germany's **Boehringer** Ingelheim, **Europe** is left with a number of **second-tier players**.

* At the beginning of the decade, about half of all top-selling drugs belonged to European companies. At the end of the nineties, just **4 of the 25 best-selling** drugs had European owners.

* The investor community became obsessed with **"matters of size"**. Achieving critical mass was their all dominating credo. The route they advocated to obtain **critical mass** obviously passed through their vast financial territory and consisted in combining forces between companies (that means: mergers and acquisitions).

* At the end of the decade, the over-inflated valuations for biotech companies, which had been observed in the mid 1990s, spiraled downwards, making deals more attractive to large pharmaceutical companies. **Bayer** acquired **Chiron** Diagnostics (1998), **Abbott** Laboratories acquired **Alza** (1999) for $ 6,6bn, **Warner Lambert** took over **Agouron** Pharmaceuticals for $ 2,2bn, and **Roche** exercised its option to acquire the remaining 33% of the outstanding equity, by investing $ 4,6bn in **Genentech** (while, simultaneously, combining its right with a 15 per cent resale of its investment in the open market).

* In the area of strategic alliances, **platform technologies** are more important to BigPharma than in-licensed products. The most financially lucrative deals have involved **genomics combined with target validation and pre-screened small molecules**. and accounted for 68% of the US$7.3 billion earned by the biotech industry from alliances during the period 1997 through to 2000 (compared to only 32% for products).

* **Regranex** is the first biologic treatment to increase the incidence of complete healing in diabetic foot ulcers and the first medicine to actively stimulate the body to grow new tissue to heal the wounds

* **Purdue's** billion dollar **time-release formulation of oxycodone** becomes controversial, as would-be abusers learn how to defeat the time-release mechanism and obtain pure oxycodone

* **Amgen** succeeds in launching **Neupogen (filgrastim)** in the US and in Europe. The drug is a **recombinant version** of a **human protein** that selectively stimulates the production of infection-fighting white blood cells, called **neutrophils**. From its side, **Genzyme** launches **Cerezyme**, an **orphan drug**, which is a **recombinant version** of **glucocerebrosidase**, the human enzyme **whose lack causes Gaucher's disease**.

* **Safety of drugs** becomes such a critical issue, that authorities may "overreact" : the antibiotic **Trovan** was used by 2.5 million people as of 1999. and the FDA received 14 reports of liver failure. That was enough for the agency to restrict use of the drug.

B.3. "End-of-the-decade" Biotech M&A frenzy
(selected top deals of US $ 150mn minimum)

Bidder company (country)	Target company (country)	Deal value (in US $mn)
Abbott Lab.(US)	Alza (US)	6.559
Hoffmann LaRoche (CH)	Genentech (US)	4.818 for 33% of equity
Warner Lambert (USA)	Agouron (US)	2.196
Bayer (D)	Chiron Diagnostics(US)	1.100
Elan (IRE)	Neurex (US)	824
Pharmacia & Upjohn (US)	Sugen (US)	634
Gilead Sciences(US)	NeXstar (US)	586
Alza (US)	Sequus (US)	577
Watson Pharma (US)	TheraTech (US)	320
Abbott Labs. (US)	International Murex (CAN)	234
Baxter (US)	Somatogen (US)	232
Amersham Pharmacia Biotech (UK)	Molecular Dynamics (US)	222
Mylan (US)	Penederm (US)	205
Merck KGaA (D)	CN Biosciences (USA)	150
Elan (IRE)	Nanosystems E Kodak (US)	150

* The number of **billion-dollar brands** ("blockbusters") raises from 17 in 1995 to 35 in 1999. In 1990, the top-selling drug was **Zantac** (ranitidine), with $2.4bn in sales. The next five biggest drugs , **Capoten** (captopril), **Vasotec** (enalapril), **Tagamet** (cimetidine), **Tenormin** (atenolol) and **Voltaren** (diclofenac) each generated annual sales between $1.5 to$1.1bn.

* Universities, teaching hospitals, and government research institutes have contributed to advances in biotechnology and innovative approaches to disease intervention. Examples include the Cohen-Boyer patents on recombinant DNA technology (**Stanford**), hepatitis B vaccine (**University of California**, licensed to Merck), monoclonal antibody against respiratory syncytial virus (**NIH**, licensed to Medimmune), and a meningitis C vaccine developed at **NRC**.

* The alpha-blockers **sulosin** and **alfusozin** become the medicines of choice for the treatment of benign prostatic hyperplasia (BPH).

* **Clozapine** shows to have superior effects over other antipsychotic medications. The **atypical antipsychotics** (quetiapine, olanzapine, and rosperidone) that were launched after clozapine, show decreased incidence of side effects (such as EPS or extra-pyramidal side effects and TD, or tardive dyskinesia).

* The first clinical trial of gene therapy, for the treatment of **severe combined immunodeficiency** (SCID), was initiated in 1990.

Table 1.9.   Cost of insurance premiums continues rising faster than earnings, inflation, or GDP[16]

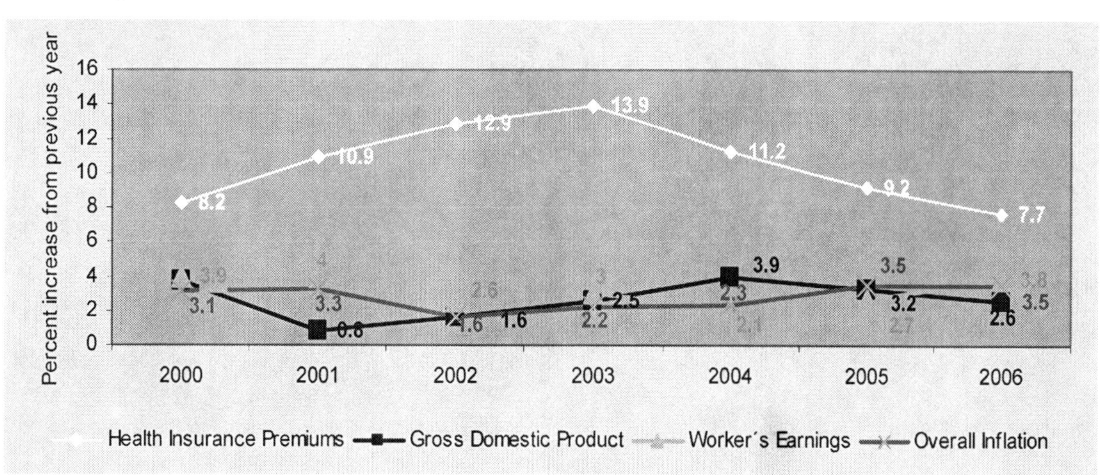

[Note: GDP figure for 2006 is for the second quarter, annually adjusted (2000 dollars)]

Most employers paid three-fourths of the nearly $10.000 annual cost of coverage per family. To avoid reducing eligibility or dropping health coverage, most employers have raised workers contributions to premiums and have offered policies with higher deductibles and co-pays.

A key aim is to reduce outlays for Rx drugs, which 61 percent of employers say are a prime factor in boosting insurance premiums. 86 percent of workers in the USA are in tiered drug plans, and more and more workers are in three-tiered plans, which require individuals to pay more out of pocket for medicines.

I.6. Employment based coverage

Employment based coverage is said to be inequitable, inefficient, increasingly unaffordable.

Inequitable: because the rich receive more substantial tax subsidies than the poor; the workers at larger firms receive better health care coverage than workers at smaller firms; and the workers receive better coverage than non-workers or part time workers or the unemployed.

Inefficient, because it: distorts labor markets/reduces hiring and creates job-lock even wed-lock; is a source of labor-management conflict; has high administrative costs--$120 billion; and leads to discontinuities in coverage as employers change coverage or workers change jobs.

Increasingly unaffordable, because it: costs about $2-5 per hour per worker for coverage; there is a declining[17] proportion of insured. Now just 63% of workforce has employment based coverage with less than half of private, non-

---

[16] Source: Kaiser Family Foundation and Health Research and Educational Trust (2006). "Employer Health Benefits: 2006 Annual Survey". Exhibit1.1.(www.kff.org/insurance/7525/upload/7525.pdf) and US Department of Commerce, Bureau of Economic Analysis (2006). "Gross Domestic Product Percent change from Preceding Period"(www.bea.gov/bea/dn/gdpchg.xls)

* Pharma **R&D costs** soar. Huge increase in spending on R&D, productivity, seen as return on dollars spent, is down. The risk of developing an innovative drug is getting higher.

* R&D costs for the pharmaceutical industry soar : at the end of the decade, it **costs $800 million** to develop a new drug.

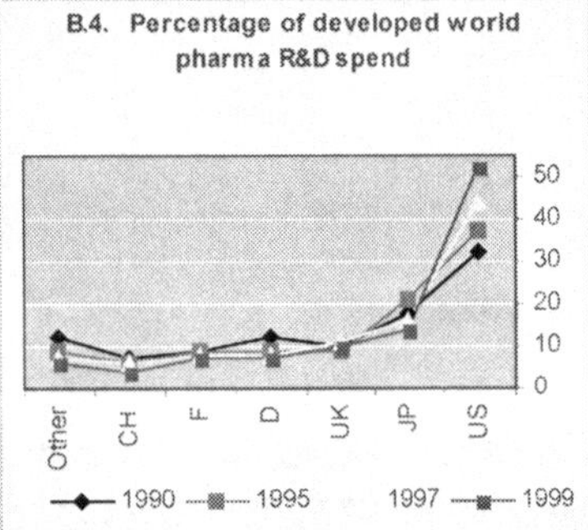

B.4. Percentage of developed world pharma R&D spend

(Source: Prof. Dr. Trevor Jones, BPI, at the Financial Times World Pharmaceuticals Conference, London, 11/2003)

* At the end of the decade, **Pfizer** spends $ 4 billion a year on **R&D**. **GSK, Aventis'** R&D expenditures reach $3,5 and $3 billion respectively.

* **Only 3 of 10 drugs** marketed recover R&D costs.

* **Time to develop** a new drug (15 years) has almost doubled since the 60s.(larger, longer studies, etc).

B.5. Reasons for failure during development	
Undesirable biological effects	60%
Poor results in clinical trials	55%
Poor drug properties (e.g. PK)	33%
Poor results in animal studies	28%
Development too expensive	14%
Better leads found	11%
Too expensive to scale up	5%

(Source: Prof. Dr Trevor Jones, ABPI, Financial Times World Pharmaceuticals Conference, London, 11/2003, based on data from Anderson Consulting)

* In 1999, the **Japanese** government instituted deregulation measures that opened previously unavailable distribution channels to non-prescription products

* With a high number of drug candidates failing in preclinical development and clinical trials due to metabolism, pharmacokinetic problems and/or toxicity, experts turn towards the possible development and implementation of **experimental systems for accurate prediction** of human drug toxicity.

* Average "practical" **patent life** is reduced to 11-12 years.

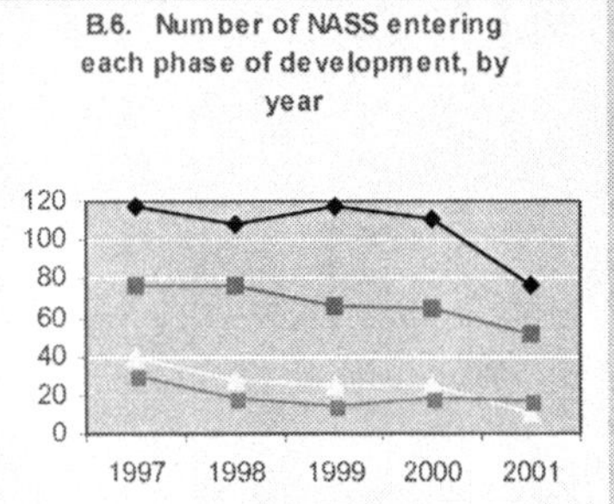

B.6. Number of NASS entering each phase of development, by year

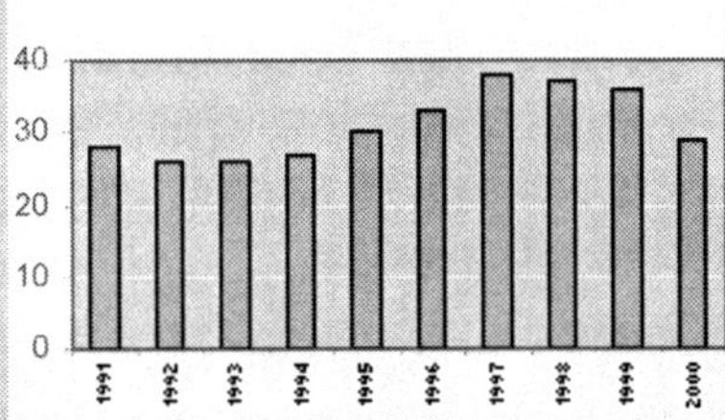

B.7. Percentage of NMEs 1st or 2nd in class; 3 year moving average (from year of first launch)

(Financial Times World Pharmaceuticals Conference, London 11/2003; based on data from national trade assoc.)

*UK, Scandinavia, Germany create regional **Biotech clusters** in order to catch up to US counterparts.

* **Plan B, an emergency contraceptive,** also known as "**the morning after pill**", is approved for prescription sales by the FDA. Plan B functions like a high dose of birth-control pills. When taken within 72 hours following unprotected sex, the **drug reduces the risk of pregnancy by as much as 89%.** The drug is different from **mifestrone (or "RU-486")**, which chemically induces abortion, because the surge of hormones, provoked by Plan B, interferes with ovulation and fertilization or may prevent implantation of a new embryio in the uterus. The medical community generally doesn't consider this to be abortion, but some religious conservatists do. Later, in 2003, when **Barr Laboratories** files for prescription-free status of Plan B, the drug will be at the center a political debate that causes divisions inside and outside the FDA. The agency, under congressional scrutiny for other issues, needed three years to return to its **mission of putting science ahead of ideology**, and finally approved OTC status for Plan B in 2006.

governmental workers; there has been a decline of 5 million jobs with health insurance between 2001-2004.

Table 1.10.   Health Insurance tax breaks[18]

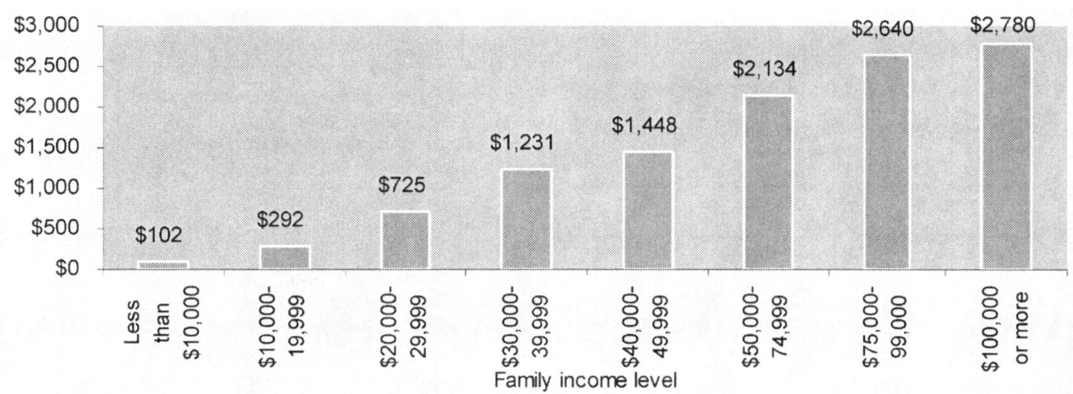

The erosion in employment based health coverage is primarily due to an increasingly competitive economy. Anti-trust actions, deregulation, more foreign competition and the decline of unions; each of these factors had its share of influence. Add to this the effects of globalization, with its outsourcing of jobs to foreign countries and foreign production and the rise in health care costs.

In 2005, the real per capita expenditures for health care were 6 times higher than in 1965. The real average weekly earnings of production workers were lower than in 1965. All increases in productivity have gone to health care. Note also the changes in the health care insurance market, with their ending of cross-subsidization between firms and of cross-subsidization within firms.

Table 1.11. Erosion of employment based coverage	
In 1960s the largest firm AT&T	**in 2000s the largest firm Wal-Mart**
~ 1 million workers all with health insurance	~ 1 million workers, only 38% have health
~ Regulated industry with guaranteed profit	coverage, no workers get coverage in first
~ Health insurance costs passed onto	2 years of employment. Many get Medicaid.
consumers (an implicit tax)	~ Price is king, cheaper even if foreign
	produced is not better.

I.7. Financial Strain from Health Care Costs are a Growing Concern for Americans
The Aflac study revealed that the majority (85 percent) of respondents acknowledged that in addition to caring for their spouses and children, more Americans will have to deal with the financial strain of caring for elderly parents. As the sandwich generation (those sandwiched between caring for aging parents, adult children, and/or grandchildren) grows, apprehension about this strain is understandable. Individuals caring for both aging and young generations face many daunting scenarios, such as caring for a spouse or partner, an aging parent, a sick child, or in the most difficult situations, more than one at once.

---

[18] *Lewin Group estimates using the Health Benefits Simulation Model (HBSM):Average per family is $1,482*

B.8. Consolidation in the healthcare products distribution sector further changes the sector. M&As at the end of the decade include :

Bidder company (country)	Target company (country)	Deal value (in US $mn)
CVS (USA)	Arbor drugs (USA)	967
Bergen Brunswig Co (USA)	Stadtlander Drug (USA)	400
JC Penney (USA)	Genovese drug stores (USA)	398
Thomas H Lee (USA)	Eye Care Centers of America (USA)	300
Henry Schein (USA)	Meer dental supply USA)	144
National Vision Ass. (USA)	New West Eyeworks (USA)	77
Gehe (Germany)	AFM Bologna (Italy)	65
Duane Reade (USA)	Rock Bottom Stores (USA)	61

* Progress in understanding the various aspects of the molecular biology of tumor cells and overall, **tumor pathology,** have led to the discovery of numerous candidate drugs aimed at playing a role in the various existing chemotherapeutic treatments. Regulatory bodies continue to see the risk-benefit evaluation as single therapy or in combination therapy based on the clinical benefit, which may be a variable criterion depending on the stage of the disease and the status of the patient, whereas industry sees the need of using novel surrogate endpoints for clinical efficacy based on **pharmacological or pharmacodynamic** features as well as **population stratification or profiling**.(source : J.M. Alexandre, Hôpital Européen Georges Pompidou, France / S. André, Wyeth Research, at DIA ´s 16th Annual Meeting, Prague, 3/2004)

* **Synercid** (quinopristin and dalfopristin), for life-threatening infections associated with vancomycin-resistant Enterococcus faecium bacteraemia, is the first streptogramin antibiotic on the market.

* Antiviral agents, a class of antimicrobials, treat and prevent influenza. **M2 inhibitors (amantadine and rimantadine)** against influenza A; **Neuraminidase inhibitors (zanamivir and oseltamivir)** are designed to broadly inhibit influenza A and B. **Relenza**, the first inhaled neuramidase inhibitor, delivers antiviral action to the lungs and shortens the course of influenza; it works by interfering with the life cycle of the influenza virus.

*The discovery of the **appetite-controlling hormone leptin** in 1994 led to an explosion of research discoveries elucidating the metabolic and neurobiological mechanisms that regulate appetite, energy expenditure, and energy storage

* UK : Creation of **NICE** (National Institute of Clinical Excellence) to evaluate cost-effectiveness of products. Health Technology Assessment (HTA), and particularly its focus on cost effectiveness, is seen by the industry as the **"fourth hurdle"** (quality, safety, efficacy being the traditional requirements for obtaining marketing authorization of a drug).

* Although the biotechnology discoveries of the 1970s and ´80s benefited the U.S. labs that underwrote them, many **UK and German scientists were instrumental in the developments**. E.g. : The Los Angeles Times wrote (March 13, 2000) that "the best and brightest of German scientists fled to the more supportive environment of U.S. labs, leaving behind a country dominated by its ossified hall of academia, corporatist culture, driven by suspicion of genetic engineering, and contributed to breakthroughs in treating debilitating and fatal diseases (e.g. : when noted biochemistry professor Ernst-Ludwig Winnacker began lobbying the government for a gene research center in the mid 1980s, the very notion of trying to crack the human genetic code was **so taboo** that he became a target of radical environmentalists and the terrorist Red Army faction. So unwelcome was any thought of **"interfering with God"** in rewriting the human body's computer program, that the government had to post armed guards outside his home and sheathe his office in bullet-proof glass.). Some have returned to their homeland, like Axel Ullrich, who **cloned the first gene for human insulin** while at UC San Francisco and helped develop a drug treatment of breast cancer while at Sugen Inc".

* **High profile research projects** for treatments of Alzheimer's –such as inhibitors for **aldose reductase, ACAT, 5 Lipoxygenase, rennin, and etalloproteinase**- failed in clinical trials and **were abandoned**.
* **Beromun** is the first tumor necrosis factor approved for cancer treatment.

* Nobel Prizes are awarded to Alfred G. Gilman and Martin Rodbell for their discovery of **G-proteins** and the role of these proteins in **signal transduction in cells**; to Edward B. Lewis, Christiane Nusslein-Volhard and Eric F Wieschaus for their discoveries concerning the **genetic control** of **early embryonic development**; to Peter C Doherty and Rolf M Zinkernagel for their discoveries concerning the **specificity** of the **cell mediated immune defense**; to Robert F Furchgott, Louis J Ignarro and Ferid Murad for their discoveries concerning **nitric oxide** as a **signaling molecule** in the **cardiovascular system**; Gunter Globel, for the discovery that **proteins** have **intrinsic signals** that **govern their transport** and **localization** in the **cell**; Edmond H Fisher and Edwin G Krebs for their discoveries concerning **reversible protein phosphorylation** as a **biological regulatory mechanism**; to Joseph E Murray and E.Donnall Thomas for their discoveries concerning **organ** and **cell transplantation** in the **treatment of human disease**.

According to the Aflac survey, one in three (33 percent) of Americans have used vacation time to care for a family member and more than a quarter (28 percent) have lost wages because they stayed home to care for a sick child or elderly parent.
- Two-thirds (67 percent) of those with children under 18 at home have missed work to take care of them.
 - One-quarter (26 percent) of Americans have missed work to care for an elderly parent.

I.8. Confronting a growing number of uninsured

What to say about a leading developed country leaving almost one in six people without health insurance coverage?

Table 1.12.   Development in the number of uninsured 1987-2005[19]

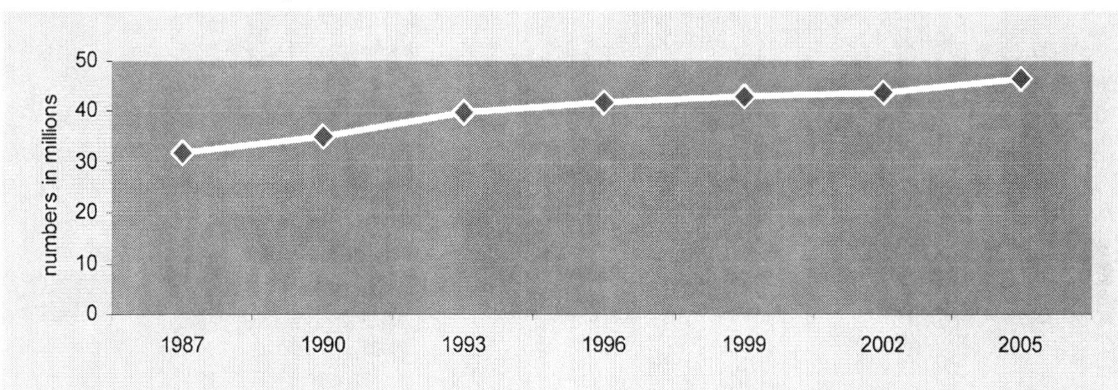

U.S. Census Current Population Survey figures, published August 2006, reveal that not only fewer Americans receive coverage through employers, but that a record 46.6M U.S. residents lacked health insurance in 2005 (in 2000, that number was 40 million individuals), or 15.9% of the U.S. population, compared with 45.3 million individuals, or 15.6% of the population, in 2004.

Table 1.13.   Medicaid: Enrollees and Expenditures[20]

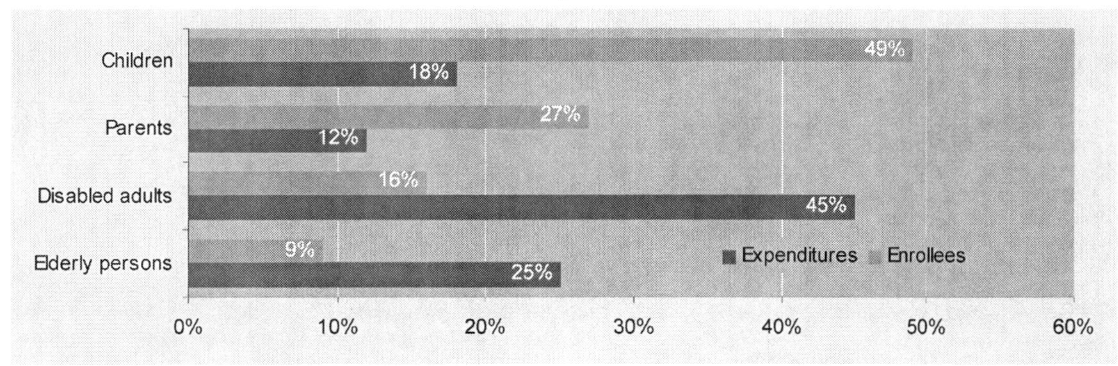

[Note: Totals may not total 100% because of rounding]

[19] Source: US Census, August 2006

[20] Source: Rowland D."Medicaid-Implications for the Health Safety Net", New England Journal of Medicine, Sept 20, 2005 (www.nejm.org)

* Therapies created by their inventor companies **lost out to follow-up** compounds --e.g. **Captopril (Capoten/SmithKline), enalapril (Vasotech/Merck & Co), cimetidine (Tagamet/ SmithKline), ranitidine (Zantac/Glaxo)**—thereby launching a debate about the longtime pharma Rx brand value.

B.9. Exclusivity period for novel drugs: change over time		
Innovative drug & year of launch	Follow drug & year of launch	Years of exclusivity
1968 Inderal	1978 Lopressor	10
1977 Tagamet	1983 Zantac	6
1980 Capoten	1985 Vasotec	5
1985 Seldane	1989 Hismanal	4
1987 AZT	1991 Videx	4
1987 Mevacor	1991 Pravachol	4
1988 Prozac	1992 Zoloft	4
1990 Diflucan	1992 Sporanox	2
1992 Recombinate	1993 Kogenate	1
1995 Invirase	1996 Norvir	1
1999 Celebrex	1999 Vioxx	0

* In the 80s and 90s, lack of vision and courage, hoping that change will arrive slowly enough for being a matter for the successor, **sitzkrieg tactics** of many a manager darkened the perspectives of European pharmaceutical companies, ultimately leading to a complete loss of independence.

* **Bupropion**, a dopamine reuptake-blocking compound becomes the newest antidepressant. The drug will soon be followed by **Venlafaxine**, a selective serotonin/norepinephrine reuptake inhibitor (SSNRI), and **mirtazapine**, a nonadrenergic and specific serotonergic antidepressant (NaSSA). The antidepressant drugs developed from MAO Inhibitors (MAOIs), from the tricyclic antidepressants (TCAs) in the 1950s to more selective treatments which fewer side effects.

* **Valproex** (indicated for epilepsy) is the first medicine to be approved for the prevention of migraines. The drug promptly aborts an attack of an ongoing intractable migraine headache without producing significant side effects.

* In 1990, Professor **Terry Dwyer** confirmed the link between infants' sleeping positions and **sudden infant death syndrome.** This evidence prompted public
health campaigns around the globe which have cut these tragic cases of infant death tenfold.

* US : **failed** Hilary Clinton healthcare reforms

* Globalization, the buzzword of the eighties, has been turned into **"glocalization".** Industry learned that Culture is local. Prescribing habits are local. Distribution patterns are local. The Regulatory & Reimbursement environment is local.

* Western companies continued to eye the **Japanese market**, but, in general, made only modest sales. Barriers to entry remained too strong. It was only in 1999 that **Akzo Nobel** could acquire the Rx division of **Kanebo,** followed by the acquisition by **UCB** of **Fujirebio´s** pharma division and the acquisition of **Mitsui Pharmaceuticals** by **Schering**, that Western companies could break local resistance to M&As.

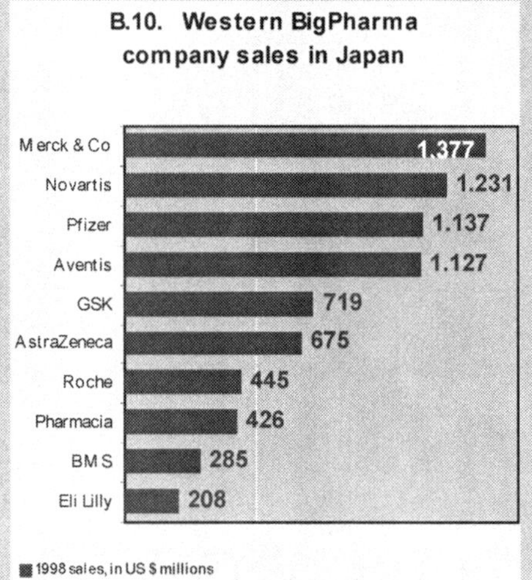

To Western companies, the **Japanese pharma** industry has been relatively protected from international competition by a series of factors including strict clinical development regulation which requires foreign preparations to be re-tested on Japanese patients. What is a time-consuming and costly undertaking for Anglo-American firms, allowing local companies to develop their own versions of new Western products before originator companies reach the market.

* **Generic competition** begins to take a toll on products that have been among the world's most prosperous medicines. Brand companies experience **a decline in sales of up to 80%** after a product comes off patent. In 1999, generics make up about 45% of total prescriptions dispensed.

* The search for a drug that might faithfully mimic natural human sleep has led to **gaboxadol**, which was originally developed in Denmark in the 1970s as a possible anticonvulsant for use in people with epilepsy. At the time, researchers **shelved the drug**, noting that it was **too sedating**.
It was not until the mid-1990s that Dr. **Marike Lancel** at the Max **Planck Institute for Psychiatry** in Munich, Germany, revisited gaboxadol as a possible **hypnotic**, or **sleep aid** (ten years later, 2006, the drug candidate shows positive results in Phase III clinical trials, what makes it a promising future medicine for millions of people).

The number of U.S. residents with health insurance increased by 1.4 million to 247.3 million in 2005.

In addition, the report finds[21] that the percentage of U.S. residents with employer-sponsored. health coverage decreased from 59,8% in 2004 to 59.5% in 2005, the lowest percentage since 1993. By comparison, in 2001, 14.6% of U.S. residents were uninsured, and 62.6% had employer-sponsored coverage. The report also finds that

--The percentage of U.S. residents with any form of private coverage decreased to 67.7% in 2005, compared with 68.2% in 2004[22];
--The percentage of U.S. residents who purchased private health insurance outside of their jobs decreased to 9.1% in 2005, compared with 9.3% in 2004;
-- A total of slightly more than 8.3 million U.S. children did not have health insurance in 2005 an increase from 10.8% in 2004 to 11.2% in 2005 (a most disturbing fact);
-- Minnesota had the lowest percentage of uninsured state residents at 8.7%, and Texas had the highest at 24.6%[23]. (In the otherwise open competition advocating US today, plenty of working families may simply not want coverage for acupuncture, massage therapy and other kinds of care that their states mandate...here, a new law is needed that would allow U.S. residents to purchase health insurance in any state, regardless of their place of residence);
-- 32.7% of Hispanics, 11.3% of non-Hispanic whites and 19.6% of African Americans did not have health insurance in 2005;
--About 961,000 of the 1.3 million increase in the number of people uninsured was among full-time workers (what is especially worrisome because if the USA gets into another economic downturn, there will be even fewer people with access to employer coverage or fewer who can afford it); and
--In households with annual incomes of at least $50,000, 17 million U.S. residents did not have health insurance in 2005, an increase of 1.5 million from 2004[24]

Table 1.14.   Major mandatory/major optional benefits that states must/may cover

Major mandatory benefits that states must cover	Major optional benefits that states may cover
Hospital servies	Prescription drugs
Physician services	Rehabilitation services
Laboratory and X-ray services	Physical therapy
Early and periodic screening, diagnosis and treatment (PSDT) for children under 21	Dental services
	Eyeglasses and hearing aids
Services at Federally Qualified Health Centers	Intermediate care facilities for mentally
Nursing home services for people over 21	retarded persons (ICF-MRs)
Home health care for certain people	Personal care services and home and
Family planning services	community based services (added by the
	Deficit Reduction Act)

According to EBRI (Employee Benefit Research Institute), Health insurance coverage has generally not sustained unbroken trends since 1994. There were crosscurrents: Employment-based coverage expanded significantly in the 1994–2000 period to exceed

---

[21] San Francisco Chronicle, August 30, 2006

[22] Havemann/Alonso-Zaldivar, Los Angeles Times, August 30, 2006

[23] Glauber/Johnson, Milwaukee Journal Sentinel, August 29, 2006

[24] Graham, Chicago Tribune, August 30, 2006

the growth in public programs. Subsequently, the dynamic reversed, as public programs expanded while employment-based coverage declined. 2005 may be the beginning of a new trend that could take the form of erosion in employment-based coverage that is not offset by expansions in public programs, thus leading to long-term growth in the uninsured.

[The percentage of the premium paid by workers has remained stable since 2000 at 16 percent. But in dollar terms, workers' contributions to their health coverage have increased considerably - from $28 a month for single coverage in 2000 to $52 in 2006, and from $135 for family coverage to $248][25].

I.9. American Healthcare Soul searching

According to the Washington Times (Aug 16, 2006), the U.S. has "many wonderful doctors, hospitals and pieces of equipment," but, based on a number of health care indicators, "the US doesn't compare well with any of the countries that have national health care," For example, according to the OECD, French men and women live three years and three and a half years longer than U.S. men and women, respectively, and France has 37.5% more doctors and twice as many hospital beds per capita. In addition, French physicians make house calls on a regular basis, and patients "aren't thrown out of hospitals because the insurance companies decide when it is time to go," and: "The French government pays 75% of all health care costs.

Most of the rest is paid by private insurance, If somebody can't afford private insurance, the government makes up the difference. Americans think that they have 'The Best Health Care System in the World'" and that 'socialized medicine' doesn't work,. The bottom line: US citizens pay 43% more for health care than the French do, and they get a whole lot less for their money".

Paying 43% more for healthcare…Is that an incentive for European countries to follow the US "example"?

I.10. U.S. Health Spending Will Reach 20% Of GDP By 2015

U.S. health care spending will increase by an average of 7.2% annually until 2015, when spending will reach $4 trillion and account for 20% of the gross domestic product, according to a February 2006 report by the National Health Statistics Group at CMS[26]. The report estimates that public and private health care spending will reach about $12,320 per capita in 2015, compared with $6,683 in 2005.

The report, which does not account for recently enacted fiscal year 2006 budget reconciliation bill that will affect Medicare and Medicaid spending, also estimates that prescription drug spending will increase by an average of 8.2% annually until 2015, a lower rate of increase than in recent years.[27] [28]

[25] *Kaiser Family Foundation and Health Research and Educational Trust (2005). "Employer Health Benefits: 2005 Annual Survey."Chart 5. (www.kff.org/insurance/7315/sections/upload/7375.pdf)*

[26] *Girion/Alonso-Zaldivar, Los Angeles Times, February 22, 2006.*

[27] *Lueck, Wall Street Journal, February 22, 2006).*

[28] *Borger, C., et al. (2006). "Health Spending Projections Through 2015: Changes on the Horizon." Health Affairs Web Exclusive, February 22, 2006, p. W61-W73. (www.healthaffairs.org).*

* The **medical technology sector** developed to a US $ 130bn market at the end of the 1990s. **Interventional cardiology and cardiac rhythm management** (pacemakers and implantable cardioverter defibrillators) attracted interest. As did **orthopedics** and **medical imaging** (which moved from analog to digital technology). The 3 sub-sectors reached sales of US $ 12bn,11bn and 10bn respectively.

B.11. Late decade M&A deals in the fast changing medical tech sector include:

Bidder company (country)	Target company (country)	Deal value (in US $mn)
Cardinal Health (US)	Allegiance (US)	5.356
Medtronic (US)	Arterial Vascular Engineering (US)	3.802
Medtronic (US)	Sofamor Danek (US)	3.603
Tyco Internat.(US)	US Surgical (US)	3.394
Boston Scientific(US)	Schneider (CH)	2.100
Stryker(US)	Howmedica (Pfizer)(US)	1.650
Seton Scholl Healthcare(UK)	London intern. group (UK)	997
Synthes (US)	Stratec Holding (CH)	989
GE Medical Systems (US)	Marquette Medical Systems (US)	897
Guidant (US)	Sulzer Medica-Electrophysiology (US)	810
Philipselectronic(NL)	ATL Ultrasound(US)	786

* 1997, the FDA publishes its Guidance for **DTC (Direct-to-Consumer advertising for prescription medicines),** backing the provision of full and well-balanced risk-benefit information to the end-user. It will profoundly impact the way drugs are promoted.

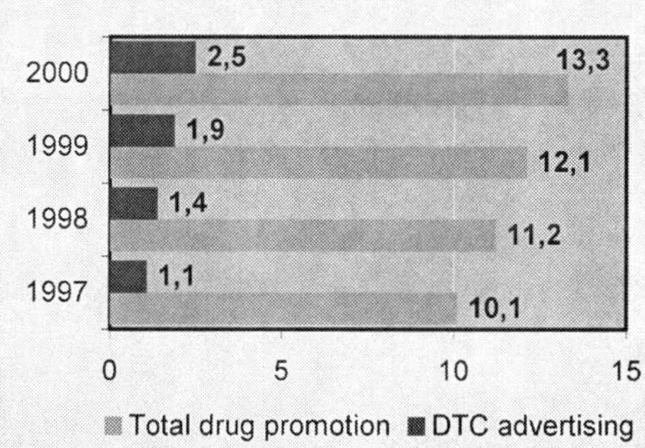

B.12. Development of DTC spending/ total profesional spending

* **Contrast media** and **radio-pharmaceuticals** became a US$ 4bn sub-sector of the medical imaging market.

* Providers of imaging services announced a move away from analog (wet) to **digital** (dry) **technology** to exploit better imaging quality and reduce running and environmental costs associated with chemical film processing. Thereby facilitating information management by speeding up the introduction of **PACS** (picture archiving and communication systems).

*Western society discusses about the influence of obesity and "modern lifestyle" on the strong increase of **diabetes.**

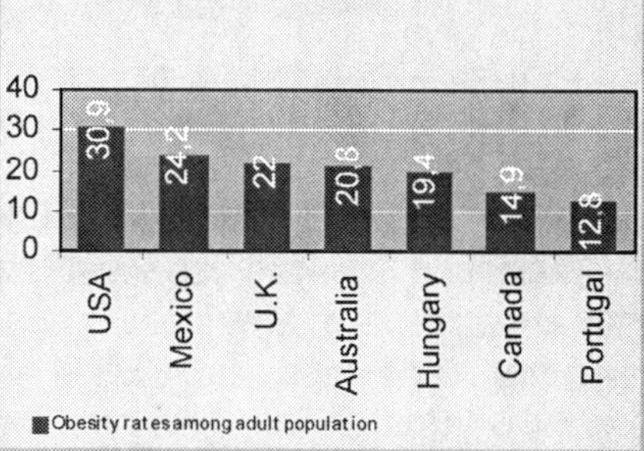

B.13. Growing overweight problem

(obesity rates : body weight of at least 20% over recommended level for person's height and sex. Figures are between 1997 and 2001, and given for the last year available. Source : OECD, Organization for Economic Cooperation and Development.)

*alpha-glucosidase inhibitors** (delaying the absorption of glucose), **meglitinides** (designed to manage meal-related glucose loads; the first drug in this class is **repaglinide**), followed by the **thiazolideniones** (which increase insulin sensitivity and glucose uptake), improved diabetic treatment.

B.14. Chronology of anti-diabetes drugs

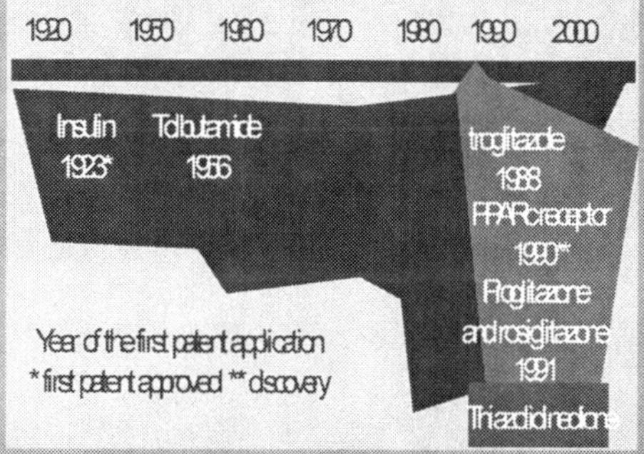

* The **contract biomanufacturing** industry is facing capacity constraints due to the large number of protein-based drugs in development combined with the market success of monoclonal antibodies and their great demands for mammalian cell culture.

Table 1.15. National Health Expenditures (NHE), share of Gross Domestic Product, and Private and Public shares of NHE, selected years 1965-2015

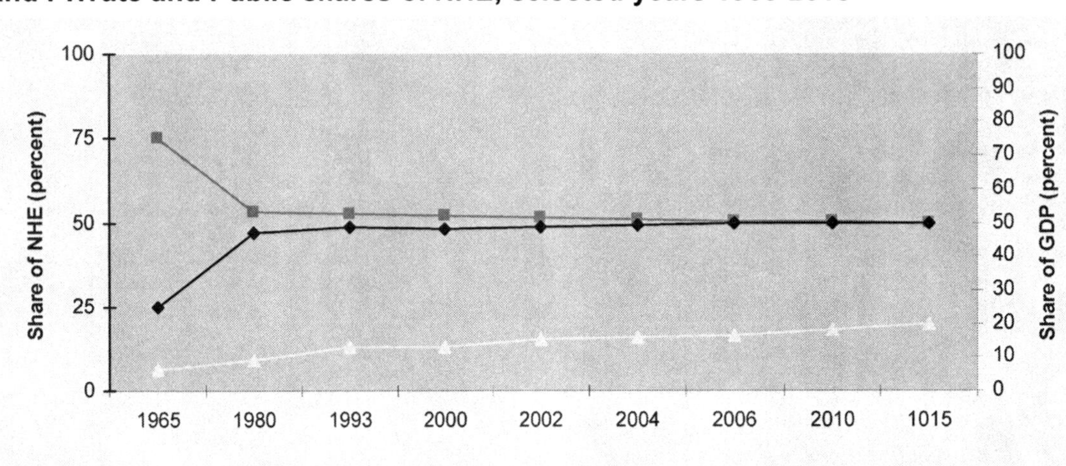

Note: The blue and red lines refer to the left axis (share of NHE). The yellow line refers to the right axis (share of GDP). Data for 2006, 2010 and 2015 are projections[29].

Medicaid, which should not be confused with Medicare, covers three main groups of low-income Americans: parents and children, the elderly and the disabled. In addition to paying for medical care, Medicaid is the primary payer for long-term care in this country. Total spending on Medicaid in 2004 was $293 billion, 7.9 percent higher than 2003, which was the smallest percentage increase in Medicaid spending in six years. The federal government paid $173 billion for Medicaid in 2004 and the states paid most of the rest (in some states, local governments contribute a small amount).[30]

According to the report, prescription drug spending will reach $446 billion in 2015, compared with $188 billion in 2004. The report attributes the lower rate of increase in prescription drug spending in recent years in part to decreased use of medications, increased co-payments for individuals with private health insurance and additional restrictions implemented by employers.

The new Medicare prescription drug benefit likely will not lead to new spending on medications because rebates and discounts offered by pharmaceutical companies are larger than expected, according to the report.

The report estimates that Medicaid spending will increase by an average of 8.6% annually between 2008 and 2015 and that Medicaid spending will reach $670 billion in 2015, compared with $293 billion in 2004.

Hospital spending, which increased by 7.9% in 2005, is expected to double by 2015.

---

[29] *Source: National Health Statistics Group, Office of the Actuary, Centers for Medicare and Medicaid Services & Medicare & Medicaid Cost Estimates Group in Baltimore, 2006*
[30] *Smith, C. et al. (2006). "National Health Spending In 2004: Recent Slowdown Led By Prescription Drug Spending." Health Affairs, January/February, p. 186-196. (www.healthaffairs.org).*

50

B.15. Change in physician's product prescribing habits during the 1990's in 4 major European countries

Germany		Italy	
Ranking 1992	1998	Ranking 1992	1998
1 ginkgo biloba	omeprazole	ranitidine	amlodipine
2 omeprazole	atorvastatin	ranitidine	ceftriaxone
3 ranitidine	amlodipine	enalapril	enalapril
4 ranitidine	simvastatin	ceftriaxone	niemsulide
5 lovastatin	humaninsulin	b-interferon	lorazepam
6 insulin	humaninsulin	ganglioside	ranitidine
7 aciclovir	metoprolol	thymostymulin	ranitidine
8 captopril	hepatitis vac.	thymopentin	nifedipine
9 bezafibrate	budesonide	amlodipine	amoxi-+cl.acid
10 naftidrofuryl	ciclosporin	calcitonin	ciclosporin
11 famotidine	hepatitis vac.	diclofenac	leuroprorelin
12 ciclosporin	enalapril	citicholine	doxazosin
13 metoprolol	tilidin+naloxin	captopril+HCTZ	epopoetin alfa
14 captorpil	lovastatin	captopril+HCTZ	clarithromycin
15 nifedipine	acarbose	indobufen	ciprofloxacin

France		UK	
Ranking 1992	1998	Ranking 1992	1998
1 amoxi.+cl.acid	omeprazole	ranitidine	omeprazole
2 fenofibrate	simvastatin	nifedipine	simvastatin
3 flavonoids	trimetazidine	omeprazole	amlodipine
4 trimetazidine	fluoxetine	beclomethas.	lansoprazole
5 gingkobiloba	fenofibrate	salbutamol	fluoxetine
6 simvastatin	amixi-+cl.acid	diclofenac	paroxetine
7 omeprazole	gingkobiloba	beclomethas.	salmeterol
8 captopril	paracetamol	enalapril	enalapril
9 fluoxetine	paracet.+dext.	captopril	budesonide
10 ranitidine	paroxetine	budesonide	ranitidine
11 enalapril	amlodipine	cimetidine	beclomethas.
12 nicergoline	aciclovir	atenolol	fluticasone
13 naftidrofuryl	gliclazide	spironolactone	lisinopril
14 diltiazem	budesonide	aciclovir	guafen.+sabut.
15 amoxixillin	bioflavonoids	ciclosporin	nifedipine

* Refinements in fiber-optic technology and engineering produced instruments used for **keyhole surgery.** Fine tools are passed into the abdominal and thoracic cavities to allow for operations to be performed through small puncture wounds instead of major incisions for exposure. **Fiber optic** allows the development of instruments that pass along every tube in the body to remove obstructions in major blood vessels, bowel, prostate and other organs. **Promising development** work is done on sophisticated engineering and imaging techniques that enable fine probes to be passed with great accuracy into the brain to deal with tumors, blood vessel abnormalities and other diseases.

* Studies show that the enzyme, **lipoprotein-associated phospholipase A2**, may be as important as cholesterol as a predictor for cardiovascular risk.

* In the nineties, many consultants to industry recommended a model of **outsourcing of R&D** (also of manufacturing, etc.) where at least a quarter of their research and development is contracted out (**Glaxo,** for example, was one of the first multinationals to contract out 50% of its toxicology, 25% of its R&D spend). It was highly praised as a means of forging relationships with dozens of biotech companies and of spreading the risks.

*Sales of **cholesterol-lowering** drugs soar.

B.16. Chronology of cholesterol-lowering agents

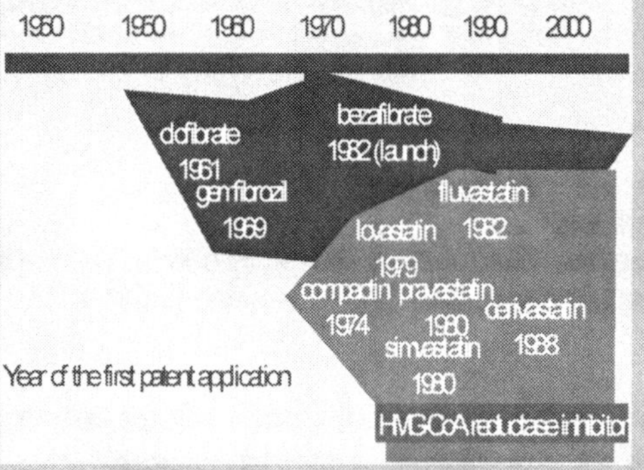

* The applications of **robotic methods in surgery** herald a new era in surgery. A computer laser-scanning system has been designed for the treatment of refractive anomalies of the eye. Computer-controlled tools are being developed to help perform surgery in cases beyond the ability to proceed with hands and eyes. A voice-activated robot surgical assistant is being developed for laparoscopic surgery in the abdominal cavity. Much hope is put in the development of **micro-robots** with applications in surgery, e.g. sensory devices in the circulatory system, drug delivery systems, artificial sphincters. Under development is an intelligent micro-robotic endoscope able to enter organs and videotape them for later study/taking samples of tissue for microscopic examination...

* Ironically, the tremendous progress in medicine leads to disenchantment with chronic disease therapies, their inability to cure everything. Shift to alternative (natural) therapies. Emergence of the **"wellness" factor**.

* **Pharmacoepidemiology**, the study of the use and the effects of drugs, on a large number of people, has become an integral part of drug safety management.

* The economic value of European countries' **pharmacy monopoly** gets questioned.

* Based on advances in technology, **device/drug combination** products become a challenge of the drug & medical devices sector.

* **Concentration** of regional pharma wholesalers in Europe increases.

Table 1.16.   U.S. National Health Expenditures (NHE) by source of funds, amounts, 1993-2015[31]

Source of funds	1993	2004	2006	2010*	2015*
NHE (billions)	916.5	1,877.60	2,163.90	2,879.40	4,031.70
Private funds	514.2	1,030.30	1,148.40	1,544.70	2,116.40
~Consumer payments	442.3	894.2	991.2	1,334.10	1,818.10
~Out-of-pocket payments	145.3	235.7	246.2	316.30	421.00
~Private health insurance	297	658.5	745	1,017.70	1,397.10
Other private funds	71.9	136.1	157.1	210.60	298.30
Public funds	402.3	847.3	1,055.50	1,334.70	1,915.30
~Federal	277.7	600	742	971.40	1,407.80
~~Medicare	184.4	309	420.1	536.00	792.00
~~Medicaid**	76.8	173.1	184	258.90	384.40
~~Other federal	52.5	118	137.8	176.50	231.30
~State and local	124.7	247.3	279.2	371.20	519.40
~~Medicaid**	45.6	119.6	136	191.50	285.30
~~Other state & local***	79.1	127.7	143.2	179.70	234.10

Selected calendar years
*Projected
**Includes Medicaid and State Children's Health Program (SHIP), Expansion (Title XIX)
***Includes Medicaid SHIP Expansion (Title XXI)

Nursing home spending will increase from $121.7 billion in 2005 to $216.8 billion in 2015 and Home Health spending, which increased by 13.2% in 2005, reaching almost $49 billion will increase to $103.7 billion in 2015.

The report also states that increased health care spending "could force payers and providers to re-examine fundamental questions regarding the delivery and financing of health care services".

Critics state that Medicare is inequitable, inefficient, increasingly unaffordable.

Inequitable, because: some are eligible, but many poor Americans—especially working poor—are ineligible; benefits and eligibility criteria differ substantially between states; some states give children at 300% FPL coverage while others—Wyoming—at 133%. Disparity is said to be even worse for adults.

The system is widely viewed as second class medicine compared to employment-based or Medicare coverage: In Massachusetts Medicaid covers just 21 days of hospitalization in a year and In Tennessee Medicaid covers just 5 days of hospitalization in a year.

Inefficient, because: it generates high administrative costs if eligibility is closely monitored. It costs the equivalent of 2 months of insurance to screen and enroll a child in SCHIP. Even more for adults; asks for high tax on increases in income, this way encouraging evasion or avoidance of reported income.

---

[31] *National Health Statistics Group, Office of the Actuary, Centers for Medicare and Medicaid Services & Medicare and Medicaid Cost Estimates Group in Baltimore, Maryland, 2006*

52

* The reputation of the US Food and Drug Administration is pre-eminent. It has long considered itself the **gold standard** for the sector –and convinced many others that they were effectively first among equals. In contrast to Europe, where decisions are taken behind close doors, the FDA observes the **principles of democracy**. The agency is seen as more transparent than its counterparts abroad : not only are its procedures for clinical trials often discussed and agreed in advance, but its system of hearings (open to the public) is a way of involving a wide variety of experts, from patient advocacy groups to scientists.

* Asthma sufferers are brought relief of suffering by new anticholinergics, such as **ipratropium** and the **glycopyrrolate**; drugs that will be used in conjunction with beta-agonists to produce more rapid improvements. Later in the nineties, the **leukotriene** modifiers reduce the need for steroids. They provide oral therapy with fairly rapid onset of action.

* Each year, 100,000 Americans die **of hospital-borne infections, a higher toll than deaths from breast cancer and AIDS combined.** Nearly 2 million patients acquire hospital infections (of a total 35 million stays), costing the healthcare system nearly $30 billion a year according to estimates by the U.S. Centers for Disease Control & Prevention. Since the first reported episode of **MRSA (Methicillin-resistant Staphylococcus aureus)** infection in the U.S. in 1968, the proportion of the bacteria causing infections in hospitalized patients has risen significantly **from 2% in 1974 to about 40% in 1997**.

* Genentech´s biologic **Herceptin** becomes the first therapeutic antibody targeted to a cancer-related molecular marker to receive US FDA approval (in combination with pacitaxel for the first-line treatment of HER2-positive metastatic breast cancer, and as a second- and third-line therapy as a single agent).

* 1997, James Thomson **isolates human embryonic stem cells.**

* **Hoffman-La Roche´s Invirase (saquinavir)** becomes the first drug of the **new class of protease inhibitors** that will go on to revolutionize AIDS **treatment**.

* The first clinical trial of gene therapy, for the treatment of **severe combined immunodeficiency** (SCID), was initiated in 1990.

* 1998, **Craig Venter**, the genetics scientists, founds **Celera Genomics** and declares that he will sequence the human genome within three years (for US$ 300million). In 2001, **Celera** effectively publishes a working draft of the **genome mapping** in the journal **Science** (one day after the **Human Genome Project consortium** publishes its draft in the journal **Nature**).

* **Industrial scale production** of proteins through genetic engineering.

*Although **business deals** outside USA/Europe were less spectacular, from the mid nineties to the end of the decade, interest in healthcare business outside USA & Europe increased. Deals 1998/1999 include :

Bidder company (country)	Target company (country)	Deal value (in US $mill)
Japan Tobacco (Japan)	Torii pharma	339
Fedsure Holdings (RSA)	South African Druggists (RSA)	132
GSK (UK)	Amoun pharma (Egypt)	106
Shangai Industr. Hold.	Chia Tai Health Product	42
Wockhardt (India)	Merind (India)	24

* In **China**, a state-level research program is underway to discover how microbes survive in extreme conditions in the ocean, as part of the **863 High-Tech Program**, initiated in March 1996, to tap into new resources for medicines. The research aims to find natural bio-active substances in marine life that live in extreme environments in deep water. The research on the microbe's life systems and the medical benefits would likely produce large and profound changes in the bio-medicines industry.

* In a meta-analysis examining the incidence of Adverse drug reactions (**ADRs**) in US hospitals, fatal ADRs are **1 of the 4-6 leading causes of death** in the United States *[Bates DW, et al. The costs of adverse drug events in hospitalized patients. Adverse Drug Events Prevention Study Group. JAMA. 1997;277:307-311]*. The scope of the problem of ADRs was brought to public attention in 1999 by a report issued by the Institute of Medicine (IOM), **To Err Is Human: Building a Safer Health System.** *[Kohn L et al., To err is human: building a safer health system. Washington, DC: National Academy Press; 2001;1999-2001]*

* A team led by German cancer researcher Professor **Axel Ullrich** discovered that by **interrupting the oxygen and nutrient supply to tumor cells**, it is possible to **inhibit cancer development**. The Max Planck scientists showed that tumor tissue just a few cubic millimeters in size can create vascular endothelial growth factor (VEGF), which triggers blood vessel development. Ullrich **discovered** that there is **a receptor specific to VEGF**, called **Flk-1/VEGFR2**, found **only in endothelial cells** which are the building stones of new blood vessels. When Flk-1/VEGFR2 is made inoperative, this blocks the development of tumors. If no blood vessels are created in tumor tissue, the tumor fails to grow. This discovery led to the invention of what are called **angiogenesis inhibitors** (angiogenesis is the process of blood vessel formation).

The system leads to discontinuities in coverage. Many who are eligible do not apply because of hassle, embarrassment, or belief that their prospects will improve.

Increasingly unaffordable, because: about 20% of state budgets go to Medicaid other means tested programs and health care for state employees. The problem is the rate of increase. Over the last 4 years health expenses—Medicaid and health coverage for state employees— have gone up 8-10% per year. With flat revenues, this requires cuts in other state services or tax increases. In Georgia, since 2001 state universities lost $211 million while admitting 30,000 more students. This year, $42 million in cuts with 15,000 more students anticipated.

At the 2003 National Governor's Association, it was said that "States are facing a perfect storm: deteriorating tax bases, an explosion in health care costs, and a virtual collapse of capital gains and corporate profit tax revenues."[32]

I.11. Medicare coverage gap

Unlike most forms of insurance, the Medicare Part D prescription drug program has a hole in its middle. This coverage gap, colloquially known as the "doughnut hole," is perhaps the most bizarre and troublesome aspect of the Part D drug program. After beneficiaries reach their initial limit of total drug expenses ($2,250 in 2006), they have no prescription drug coverage until their total drug expenses reach a catastrophic threshold for the year ($5,100 in 2006). While beneficiaries are in the doughnut hole, they must continue to pay their monthly premiums, although they do not receive any drug benefits. Only after they have spent thousands of dollars of their own money to get out of the hole ($2,850 in 2006), in addition to their monthly premiums, does their coverage resume. The doughnut hole makes little sense from a medical perspective, as it financially penalizes sicker individuals who have more substantial drug needs. It has also generated considerable anxiety among seniors and people with disabilities in Medicare, who fear falling into the doughnut hole and being unable to afford their prescriptions[33].

As meaningful coverage through the doughnut hole becomes prohibitively costly—or nonexistent—beneficiaries will have to look for alternative sources of coverage. They are likely to turn to options such as generic-only coverage, coverage for generics and preferred brand-name drugs, or a Medicare Advantage managed care plan. All of these options have serious flaws. Moreover, in the coming years, the doughnut hole will grow, exposing more beneficiaries to greater financial risk.

I.12. Doughnut Hole Problems Will Worsen Looking to the future, it is troubling that stand-alone drug plans with meaningful coverage through the doughnut hole will become increasingly expensive—or disappear entirely. The size of the doughnut holes is not fixed. Rather, it is tied to overall increases in the cost of Part D drugs. As drug prices increase, so does the size of the hole. The table shows how the doughnut hole has grown from 2006 to 2007, as well as its projected size in 2013. In 2007 alone, the doughnut hole will grow from $2,850 to $3,051. It could reach more than $5,000 by 2013.[34]

---

[32] *Ezekiel J. E., Department of Clinical Bioethics, Warren G. Magnuson Clinical Center, National Institutes of Health. Re-Thinking National Health Care Reform: Universal Healthcare Vouchers. Presentation given at Kellog University Annual Healthcare Conference, 2005.*
[33] *Steinberg M, Coverage through the "Doughnut Hole" Grows Scarcer in 2007 Families USA Publication No. 06-107, Families USA Foundation.*
[34] *Congressional Budget Office, A Detailed Description of CBO's Cost Estimate for the Medicare Prescription Drug Benefit (Washington: Congres-sional Budget Office, July 2004), p. 9.*

* Companies start pursuing **licensing-in strategies.** Not to close pipeline gaps but as a strategy of the future. 1999,over 1000 alliances were entered into by the top 25 pharmaceutical companies and leading biotech companies. Most of them focused on **compound discovery**. Others on targeting delivery technologies that allow old and new drugs to be delivered with greater efficacy. Products incorporating **Drug Delivery** already account for some 10% of total pharma sales at the end of the decade. Work started on

- Oral macromol. delivery;
- Oral controlled release beyond once day;
- Stomach retentive technologies
- Effective brain delivery;
- Targeted delivery for cancer and other diseases;
- Effective delivery technologies to get more drug through the skin;
- Lung for macromolecule delivery.

* Also in the late nineties, industry created a huge **laboratory of targets**. But the drug development process remains an imperfectly predictable process.

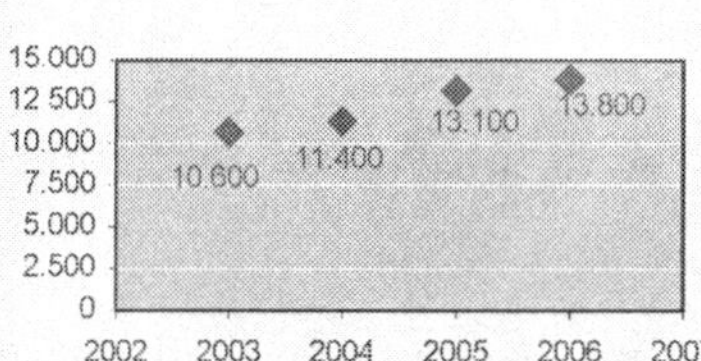

B.17. Number of compounds, from discovery to clinical development (projected)

***Infergen** is a new generation of non-naturally occurring recombinant type-1 interferon bioengineered from a consensus sequence of interferons.

***ReFacto** is the first recombinant factor VIII preparation produced without the addition of human serum albumin to the final formulation.

* **Pfizer's Viagra (sildenafil citrate),** for erectile dysfunction, opened the markets for so-called "life-enhancing drugs". A series of competitive products (**Vasomax, Alibra, Uprima, ...**) is announced.

***Quixil** is the first surgical sealant to contain no animal materials.

***NovoRapid** is available in Innovo, the world's first insulin doser. It allows patients to inject their insulin dose immediately before eating.

* **Synagis** is the first monoclonal antibody for infectious disease. It prevents serious lower respiratory tract disease.

* The **healthcare services sector** does not escape merger mania. This sector   consists of two broad categories of different sub-sectors and company types which either provide care directly to patients or support services that enable patients to receive care. All the players in this category are under .pressure to make economies of scale, while at the same time getting access to new information technology and participate in the emerging e-pharmacy sector. Healthcare service sector.

B.18.   Healthcare sector M&A at the end of the decade

Bidder company (country)	Target company (country)	Deal value (in US $mn)
McKesson (USA)	HBO (USA)	14.320
Healtheon (USA)	WebMD (USA)	7.865
Health Care and Retirement	Manor Care	2.474
Quintiles Transnational (USA)	Envoy (USA)	1.743
Rite Aid (USA)	PCS Health Systems (Eli Lilly) (USA)	1.500
Pax International (Japan)	Questar International (USA)	1.498
Quest Diagnostics (USA)	SmithKline Beecham-Clin Labs (USA)	1.270
Not-for-Profit Consortium (USA)	Columbia/HCA-Hospitals (USA)	1.200
HBO (USA)	Access Health (USA)	1.118
Aetna (USA)	NYLCare Health Plans (USA)	1.050
Welsh Carson Anderson & Stowe (USA)	Concentra Managed Care (USA)	1.105
Aetna (USA)	Prudential Ins-Healthcare Business	1.000

Improved efficiencies at the FDA were due to the1992 US Prescription Drug User Fee Act (PDUFA) which addressed drug approval times and the Food and Drug Modernization Act of 1997 (FDAMA) improved the clinical development process. PDUFA implemented specific review time frames and established user fees. FDAMA helped shorten drug development times through initiatives such as industry-FDA meetings to review clinical research protocols, use of surrogate endpoints such as white blood cell counts in the evaluation of antiviral drugs in AIDS patients as substitutes for survival time, timeline performance goals for each phase, and so called A fast track designation for those products offering a significant improvement over drugs already on the market or for treatments for serious and life-threatening diseases. From 1997 to 2000, about 80 biologicals had requests for fast track designation and were likely to be priority reviewed because of their orphan drug designation. (Source : Health Canada)

A new type of hybrid Part D plan may be emerging in 2007 that claims to offer coverage through the doughnut hole for "generics and preferred brands." In theory, such plans might provide modest help to beneficiaries who take a limited number of brand-name drugs. Upon closer analysis, however, these plans appear to offer little added value.

Table 1.17.   The growing Doughnut Hole[35]

Year initial coverage limit catastrophic threshold overall coverage gap			
	Doughnut hole begins	Doughnut hole ends	Size of doughnut hole
2006	$2,250	$5,100	$2,855
2007	$2,400	$5,400	$3,050
2013	$4,000	$9,600	$5,060

An examination of these plans from across the country could not find a single plan that offered coverage through the doughnut hole for *any* of the 18 top non-generic drugs prescribed to seniors[36].

Datamonitor notes that health care regulations and reforms are working to make generic substitution policies widely implemented. Pharmacists are now incentivized to dispense generics, and ATM (automatic telling machine) type dispensers make distributing copy versions easier, patient co-payments are less for generic products, and some health care providers are giving such products away free. The generic revolution is only limited by physicians that order "dispense as written," which can prevent substitution, notes the Datamonitor report. Carve-out systems (a group of drugs that are not allowed to be substituted for generics due to the generic version being slightly different to the branded product) also prevent substitution, as does the fact that patient awareness of generics has only recently been addressed, it adds[37].

Critics of the current Medicare system find that it is inefficient and increasingly unaffordable.

Inefficient, because: no assessment of quality or cost per benefit is made; it covers "extremely expensive" drugs like Erbitux, which treats metastatic colorectal cancer – without curing patients- and prolongs life on average 1.7 months. Cost: about $60,000 per patient. (This drug typically costs Euro 4,000 a month in Europe[38]). It covers Lung volume reduction surgery at a cost of $300,000 per QALY.

Increasingly unaffordable, because of the continuous increases in premiums. In 2004, there has been an 17% increase in Part B premiums. In the same year, Medicare consumed 2.6% of GNP--$309 billion. In 2006, with the drug benefit Medicare represents 3.4% of GNP—$438 billion—and by 2020 over 5% of GNP. By 2020, Medicare will exhaust the hospital trust fund.

---

[35] *Sources: 2006 and 2007 data are from the Centers for Medicare and Medicaid Services (CMS), Office of the Actuary, Medicare Part D Benefit Parameters for Standard Benefit: Annual Adjustments for 2007 (Washington:CMS, May 22, 2006), available online at http://www.cms.hhs.gov/MedicareAdvtgSpecRateStats/downloads/ 2007_Part_D_Parameter_Update.pdf. 2013 projection by the Congressional Budget Office, A Detailed Description of CBO's Cost Estimate for the Medicare Prescription Drug Benefit (Washington: Congressional Budget Office, July 2004), p. 9. For 2007 & 2013: Projection*

[36] *[Families USA Foundation used the Medicare Prescription Drug Plan Finder to assess the offerings of most plans listed as providing generic and preferred brand-name drug coverage through the doughnut hole. None of these plans provided any doughnut hole savings on the top non-generic drugs].*

[37] *Pharma Marketletter, November 27, 2006*

[38] *Priest L, Dispute blocks cancer drug. The Globe and Mail, 19.06.2006:p1 & A6*

* Bacterial resistance is now found in many different kinds of dangerous bacteria and the emergence of **antibiotic-resistant superbugs** poses a formidable risk. The relentless genetic changes that drive evolution and biodiversity cause the genetic modification underlying the resistance. For every one human generation more than a 100.000 bacterial generations will occur. This allows an endless pattern of bacterial genetic variations. The need to use ever more powerful antibiotics to overcome these new forms of bacterial resistance fuels a vicious cycle of genetic selection in which a descendant of the already resistant ancestor acquires resistance to additional antibiotics. The danger is that some bacteria can transfer their newly found resistance to other completely unrelated bacteria.

* **Generic competition** begins to take a toll on products that have been among the world's most prosperous medicines. Companies experience **a decline in sales of up to 80%** after a product comes off patent.
1999, generics make up about 45% of total prescriptions dispensed.
Late nineties, globally, **8 of the ten fastest growing** pharmaceutical companies are generic manufacturers.

* **Stock market analysts** gain in influence on business fortunes.

* In the field of asthma, 2 new classes of novel therapeutics (anti-immunoglobulin [Ig]E and leukotriene inhibitors) have been successfully developed.

* **Aldosterone** has been identified as a pathogenic stimulus of heart failure. It is produced by cells of the adrenal gland if their intracellular Ca2+ concentration is elevated, such as occurs after stimulation with angiotensin II. Ca2+/calmodulin-dependent kinase II (CaMKII) regulates the Ca2+ channels (such as a1H T-type Ca2+ channels) that must be opened to maintain the elevated concentration of intracellular Ca2+ that is required to sustain the production of **aldosterone**.

*Prescription bound to non-prescription bound switches (**Rx to OTC switches**) of the mid nineties do not live up to expectations.For the most part, the **end-users do not want** the **industry's products**. It is something that is forced on them by disease (or by a **"learned intermediate"** as physicians are called). It is **not a planned expenditure** in most cases.

Companies have difficulties in understanding the distinction between consumer branding for fast moving consumer goods and **branding for pharmaceutical products**; cannot create consumer fidelity.It is said that branding comprises 4 elements : core identity, extended identity, brand personality, and position. With the proliferation of products in any given category, branding (the creation of a unique identity) is even more important today, according to industry experts. However, some argue that you can't brand a pharmaceutical because in 17 years, that brand will no longer exist.

* **Lifestyle enhancers** aimed at the psycho-emotional well-being side of health reach the markets. **Renova** is the first Rx cream with a proven record of diminishing fine lines and wrinkles. Another product used in this field of wrinkles reduction to boost (cosmetic) self-confidence, is **Botox**, a product that dominates the neurotoxin technology market. Technology, the aging population and socio-economic status contribute to the successful development of products that have a profound effect on interpersonal relationships.
* Ironically, the tremendous progress in medicine leads to disenchantment with chronic disease therapies, their inability to cure everything. Shift to alternative (natural) therapies. Emergence of the **"wellness" factor**.
* **Pharmacoepidemiology**, the study of the use and the effects of drugs, on a large number of people, has become an integral part of drug safety management.
* The economic value of European countries' **pharmacy monopoly** gets questioned.
* Besides **Amgen** ($3 billion sales), **Genentech** and **Serono** (over $1 billion sales) become the leaders of a **biotech boom**, driven, in part, by the increased understanding of genomics. Follow a series of companies.
* Whereas the pharmaceutical industry is still very fragmented, the need to improve margins leaded to mega-mergers in the US $21bn in vitro diagnostics business sector (**Dade/Behring, Roche/Boehringer Mannheim, Beckman/Coulter**) The top 10 players reached the 80 percent market share mark.

B.19. Late-decade M&A in the diagnostics sector include :

Bidder company (country)	Target company (country)	Deal value (in US $mn)
Bayer (Germany)	Chiron diagnostics (USA)	1.100
Shield diagnostics (USA)	Axis Biochemicals (N)	438
Sorin Biomedica (I)	Cobe Cardiovascular (USA)	267
Abbott (USA)	International Murex	233
Instrumetation labs (I)	Hemoliance (USA)	130
Thermo BioAnalysis (USA)	Life Sciences-Clinical	104

* **Smoking** has been linked to **approximately 100 million deaths in the 20th century,** and it represents the most significant modifiable risk factor for mortality in the world.

Medicare could be "brought into actuarial balance over the next 75 years by an immediate 108% increase in program income or an immediate 48% reduction in program outlays".[39],[40]

In its May 23, 2004 issue, The New York Times adds to this: "Even in fantasy, no one has yet come up with a way to pay for Medicare."

I.13. Medicare: Pay-for-Performance Approach

Medicare currently provides more than $300 billion in health care benefits annually to around 42 million older and disabled (or who have end-stage renal disease) Americans through a system that reimburses participating providers for the services they deliver.

Table 1.18.   US Population by Primary Source of Insurance (in millions)[41]

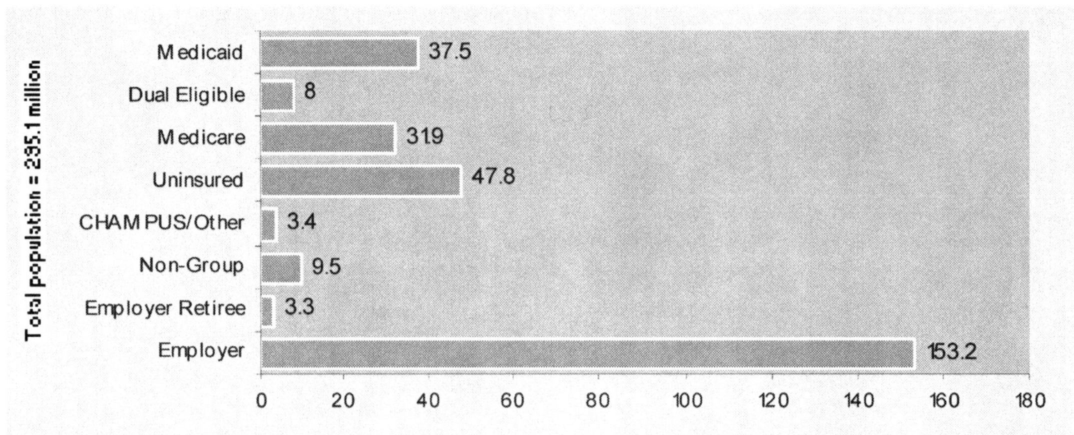

In a report to Congress (December 2006), the Acting Director of the Congressional Budget Office, published the most recent figures on Medicare and designed a premium support system for Medicare. Part A of Medicare (Hospital Insurance) covers inpatient services provided by hospitals as well as skilled nursing and hospice care. Part B of Medicare (Supplementary Medical Insurance) covers services provided by physicians and other practitioners, hospitals' outpatient departments, and suppliers of medical equipment. Home health care may be covered by either Part A or Part B. The Medicare Prescription Drug, Improvement, and Modernization Act of 2003 (MMA) added a voluntary prescription drug benefit beginning in 2006 under Part D. Part A benefits are financed primarily from a payroll tax. Premiums paid by beneficiaries currently cover about 25 percent of the costs of the Part B program, and the rest comes from general revenues. (Premiums will cover a somewhat higher proportion of Part B costs in the future because premiums for high-income beneficiaries will be increased beginning in 2007. Higher premiums will be required of single beneficiaries with annual income over $80,000 and couples with annual income over $160,000 in 2007. Those income thresholds will be indexed to inflation in future years).

---

[39] *Medicare Trustees' Report, 2004*

[40] *Ezekiel J. E., Department of Clinical Bioethics, Warren G. Magnuson Clinical Center, National Institutes of Health. Re-Thinking National Health Care Reform: Universal Healthcare Vouchers. Presentation given at Kellog University Annual Healthcare Conference, 2005.*

[41] *Sheils J, Haught R, Murphy E., Cost and Coverage Estimates for the "Healthy Americans Act". Staff Working Paper, the Lewin Group, December 12, 2006:4*

Those Part B premiums are currently about 11 percent of the total combined cost of the Part A and Part B programs. Enrollees' premiums under Part D are set at a level to cover about one-quarter of the cost of the basic prescription drug benefit, but receipts from premiums will cover less than one-quarter of the total cost of the Part D program because some of the costs of that program (such as subsidies for low income beneficiaries and for employers that maintain drug coverage for their retirees) are not included in the calculation of premiums. The majority of Medicare beneficiaries receive services through the traditional fee-for-service (FFS) part of the program. The rest (about one-sixth) are enrolled in private health plans.

I.14. Fee-For-Service Program
The FFS program is popular with beneficiaries because, unlike many private insurance plans, it does not restrict their choice of providers and does not require prior authorization for any covered service.

The FFS program is said to have several inefficiencies, however. To begin with, it is likely that beneficiaries who are enrolled in the program receive services whose costs exceed their benefits, for two reasons: the program gives providers incentives to increase the volume of services they deliver, and the out-of-pocket costs beneficiaries face are typically much lower than the total costs of providing the services. In addition, the FFS program does not give providers incentives to coordinate the care of beneficiaries who obtain services from multiple providers. Consequently, many beneficiaries—especially those with multiple chronic conditions—may receive care that is fragmented and inefficient.

I.15. Medicare Advantage
In nearly all areas of the country, Medicare beneficiaries have the option of enrolling in Medicare Advantage— the program through which private plans participate in Medicare—rather than receiving their care through the fee-for-service program. (The program through which private plans participate in Medicare is also called Part C). As of October 2006, about 17 percent of beneficiaries were enrolled in private health plans, which accept responsibility and financial risk for providing Medicare benefits[42].

About 75 percent of the Medicare beneficiaries enrolled in private plans are in health maintenance organizations (HMOs). The other main types of available plans are local preferred provider organizations (PPOs), regional PPOs, and private fee-for-service (PFFS) plans. Both HMOs and PPOs have comprehensive networks of providers. Some HMOs offer coverage for services received outside their network (and thus resemble PPOs, which allow beneficiaries to obtain care outside the network if they pay a higher amount), while others require that their enrollees receive all of their nonemergency care within the network. Regional PPOs, an option that became available in 2006, are required to serve broad regions of the country rather than define their service areas on a county-by-county basis, as local PPOs do. PFFS plans allow their enrollees to obtain care from any provider who will furnish it. As of 2006, 80 percent of beneficiaries live in a county served by an HMO or a local PPO, up from 67 percent in 2005[43].

---

[42] Marron D.B., Congressional Budget Office, Designing a premium support system for Medicare, Report to the Congress of the United States, December 2006
[43] Medicare Payment Advisory Commission, Report to the Congress: Increasing the Value of Medicare (June 2006), p. 206.

* The Internet changes relationships : Physicians and patients **"connect"**.

* The accelerating pace of S & T (Science and Technology) confronts society with increasingly complex ethical and legal dilemmas. On an international level, the emergence of NGOs **(Non-Governmental Organizations)** as powerful anti-technology lobbies with well funded and sophisticated media relations (who broadened the coalitions of opposition -together with organized labor and religious lobbies) created distrust o S & T and "bio"-based industries (greed, genetics and globalization). In a climate of concern over cost and access to care, which hightens public scrutiny of industry activities, technical and commercial sectors gave inadequate responses in countering distortion and untruths. **Industry fails to counter** the exploitation o these S&T concerns by dogmatic anti-technology lobbies.

* **Biotech** products top $1 billion/product sales (e.g. **Epogen, Humulin, Neupogen**).

* More Biotech products reach the market. 1999 alone, **15 biotech products** are granted marketing authorization :
Beromun (Tasonermin), Fermathron (Hyaluronan), Forcaltonin (Recombinant salmon calcitonin), Herceptin (Trastuzumab), i.com Comfort Shield (Hylan), Infergen (Interferon alfacon-1), Integrillin (Eptifibatide), Norditropin simplex (somatrophin), NovoRapid (Insulin aspart), Quixil (Human alpha thrombin), ReFacto (Moroctocog alfa), Regranex (Becaplermin), Remicade (Infliximab), Synagis (Palivizumab), Vitravene (Fomivirsen).

* By the end of the decade 84 **biopharmaceuticals** had been approved for marketing with almost half launched sine 1997. Worldwide sales have grown more than seven-fold during the 1990s to reach US$17 billion by 2000. The US represents 48% of the market, compared to 37% for conventional drugs, due to a combination of
  * .earlier regulatory approval,
  * easier market acceptance, and
  * greater pricing flexibility than other countries.
Although biopharmaceuticals comprise only 6% of world prescription drug sales, they account for six of the top 50 selling drugs in 2000, 13% of new medicines approved by the FDA in the 1990s and about 18% of all drugs in development. At the end of 1999 there were 369 biotechnology drugs in US clinical development against 438 disease indications with 25% in Phase III. Europe and Japan are the next most important markets, accounting respectively for 29% and 17%.
* By the end of the 1990s 50% of **biotech drugs** were first introduced on the US market compared to only 20% 10 years earlier.

* The reputation of the US Food and Drug Administration is pre-eminent. It has long considered itself the **gold standard** for the sector –and convinced many others that they were effectively first among equals. In contrast to Europe, where decisions are taken behind close doors, the FDA observes the **principles of democracy**. The agency is seen as more transparent than its counterparts abroad : not only are its procedures for clinical trials often discussed and agreed in advance, but its system of hearings is a way of involving a wide variety of experts, from patient advocacy groups to scientists. And these hearings are open to the public.

* In Europe, even when a product is approved by regulators, it can be launched only once pricing has been agreed with individual countries. It can take months or even years for companies to take that extra hurdle, used by governments as a method of controlling budgets and protecting an **ailing national industry**.

* US :Pharmacy **benefit management**/health maintenance organizations gain prominence.
Large companies have contributed more and more to financing healthcare for their employees as if they were their "citizens", and have done so as a response to the failure of the national system to provide adequate –or affordable- care. These companies are now balking at the increased cost of this healthcare and have provided a push towards managed care.

* The discussion in Europe is about the financing of the system. Not about the suffering of the patients. It is about evidence based medicine. Not about the quality of treatment. A major patient complaint is about a "lack of time", about a disappearing "human factor". Physicians complain about the ever increasing technocratic control of their treatment decisions and the time-consuming administrative work. The **medical profession has lost attractiveness**. E.g. : Since the early nineties, 3500 specialists left Germany. Over 2000 positions at hospitals remain unfilled.

* **Pharmacovigilance** issues lead to a raft of safety reviews for major products.

* The one-size-fits-all medicines inevitably carry in them the risk of rare, unwanted effects on a single individual. However, more widespread use of these medicines increases that risk of a patient complaining about troublesome side effects. With the world becoming a global village (and global branding contributing to the rapidly transnational awareness of particular problems) , the possibility that a few otherwise widely dispersed patients unite, resulting in **class actions**, increases.

European/Japanese governments wake up to the reality of spiraling healthcare costs and **scale back benefits.**

Severe constraints on reimbursement prices – dereimbursements, reference price systems, mandatory generic substitution,…- **erode Europe's industry's ability to compete.** Health economics. **Cost-benefit analysis** becomes a fundamental element

Reimbursement rates do not vary with the quality of the care that patients receive. The current system pays for treating injury and illness -- and encourages use of new, high-tech interventions -- but it does not generally reimburse for preventive services such as patient education. Nor does it pay for coordinating the care of patients whose conditions involve multiple providers, and it offers only a few incentives to improve patients' overall health status.

I.16. Preventive services

The Medicare Modernization Act (MMA) was instrumental in providing drug coverage to seniors and younger people with disabilities and expanding Medicare's preventive services. Medicare now offers a dozen preventive services that empower beneficiaries to engage in healthy behavior so they can avoid, identify and manage chronic conditions that can severely impact their health and quality of life. With national health care costs projected to reach an all-time high of $2.2 billion in 2006, it's time for taking advantage of the ultimate health care cost-saver -- prevention. Preventive services covered by Medicare include a "Welcome to Medicare" physical exam; osteoporosis and glaucoma tests; screenings for breast, cervical, prostate and colon cancers; smoking cessation treatments; cholesterol and diabetes screenings and vaccines such as flu shots. These benefits and others are designed to encourage beneficiaries to be more proactive about their health and become as healthy as they can be.

I.17. Pay-for-Performance: improving the quality of healthcare

Because Medicare's current fee-for-service payment system does little to promote improvements in the quality of health care for the program's nearly 42 million beneficiaries, the U.S. Department of Health and Human Services should gradually replace it with a new pay-for-performance system for reimbursing participating health care providers, says a report from the Institute of Medicine. (Rewarding Provider Performance: Aligning Incentives in Medicare. National Academy Press, September 2006. The study was sponsored by the Centers for Medicare and Medicaid Services of the U.S. Department of Health and Human Services at the request of Congress). Given that pay for performance does not yet have an established track record, the new system should be phased in, so that involved parties can build on successes along the way and avoid unintended negative consequences, said the committee that wrote the report.

For an initial period of three to five years, Congress should reduce base Medicare payments across the board and use the money to fund rewards for strong performance, the committee said. At the same time, efforts should be made to evaluate other ways to fund bonus payments that could be used longer term. Participation by small physician practices should be voluntary for the first three years, at which time the HHS secretary should decide whether to implement broader mandatory participation.

Medicare beneficiaries are not getting the highest possible quality of care because the program's payment system encourages volume rather than efficiency and quality. The urgency of the situation demands that steps be taken now to encourage health care institutions and clinicians to improve their quality.   Pay for performance has demonstrated sufficient promise based on early experience that it should be pursued, albeit cautiously and in a manner that allows for learning and adjustment as needed.

From the Postwar Generous Universal Healthcare Philosophy to the Promise of Cost-Effective Personalized Medicine.

2000-2007

Genomics open the horizons for discovery of treatments for unconquered diseases. Emergence of the targeted medicines specialty blockbuster model.

* **Global growth** driven by the USA. (Stimulus to support large pharma : scientific support, price,…). Europe has become a more difficult place for the industry to operate.(Twenty years ago, leading companies were based in Germany. Today, this country has no leading R&D industry among the world's top companies today).
* **EU Governments** lack the ability (will ?) to create a climate that recognizes that innovation and science puts value into the hands of the patients.

B.20. Development of pharmaceutical sales 2001-2006 in US$bn:

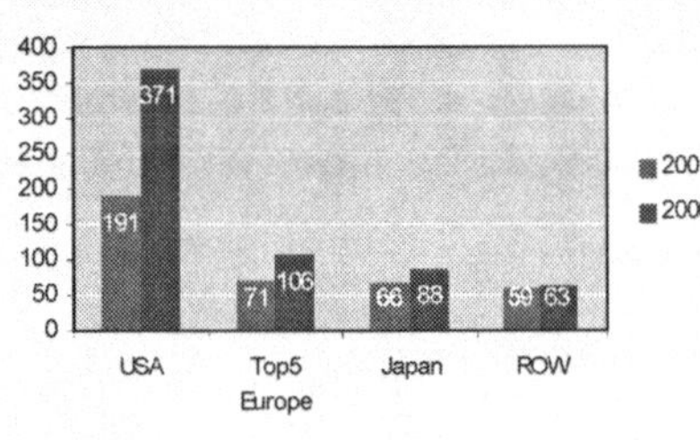

* Within five years from 2001 to 2006, the US increased its world dominance by ten percentage points of global pharmaceutical sales.

B.21. Development of percentage share of global sales, by region:

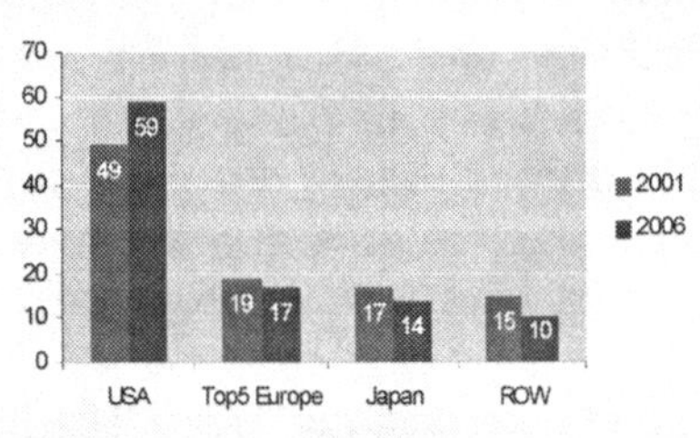

*CAGR for the period 2001 to 2006 is 10%. The USA grows by 13,8%, Europe by 8%, Japan by 1,4% and the rest of the world by 6,5%. *(Source: Professor Trevor Jones, Financial Times World Pharmaceutical Conference, London, November 2003).*
* **Investors** see a lack of R&D productivity. Is 14-20% R&D spend the right amount? In 2002, US companies spent 2 times as much as 1997, 3 times as much as 1992.
* **Innovation hurdle :** numerous branded me-too drugs offering marginal improvements. Strong perception of diminished innovation is related to huge expectations.
*Women and the ethnic minorities **(rainbow society)** claim equal treatment opportunities.

* In 2002 already 20% of drug master files (DMF) come **from India**. India, China, with their tremendous intellectual capital and entrepreneurial skills are expected to build world pharma powerhouses from their home basis, **within the decade.**
* **The cost of a scientist** in India is 1/5 of the US; the cost of clinical trials is 1/10 of he cost in the USA
* The **Internet** undermines government control on public access to medical knowledge and Rx treatment options.
* **Managed care organizations** such as health maintenance organizations, preferred providers and independent physician associations, as well as other payers, further develop alternative payment models to control costs, including procedure-based limitations, contractually approved providers and capitation (a fixed fee for members as payment for all services). The result has been a continuous shift of financial risk from the payers to both the physician/provider and the institutional provider (hospitals, clinics, long-term care providers and re, acute providers and rehabilitative care centers).
* The expansion of the **global generics pharmaceuticals market** increases the pressure on the drugs industry. Business reports indicate that 4 out of 5 "blockbusters", with combined sales of $85bn in 2002, **face patent expiry** by 2007

B.22. Major US originator brands under "generic threat"

Year	Pharmaceuticals	Biopharmaceuticals	Originator Sales (basis 2002)
2004	Diflucan		1.1bn
		Epogen/Procrit	6.5bn
2005	Biaxin/Claricid Zithromax Zocor Zoloft		10.8bn
		Activase Novolin Recormon	2.9bn
2006	Imitrex Paxil Pravachol Premarin		6.5bn

*An example of the growing US interest in generics: the US Center for Drug Evaluation and Research (CDER) approved 362 generic drugs in 2003. They include **Quinapril Hydrochloride** tablets (equivalent of Accupril) for treatment of hypertension and heart failure; **Ganciclovir** capsules and injection (Cytovene) for the treatment and prevention of Cytomegalovirus Retinitis in AIDS and transplant patients; **Paroxetine Hydrochloride** tablets (Paxil-GSK) for the treatment of major depressive disorder; and **Mupirocin** ointment 2% (Bactroban, GSK) for impetigo, an infection of the skin.

Pay for performance is just one part of the solution; other interventions will be needed to achieve the level of quality that Medicare patients deserve.

Little hard data on the effects of pay-for-performance systems are available, the committee acknowledged. Although more than 100 incentive programs have been launched in the private sector in the past few years, fewer than 20 studies have assessed the impact of these programs on quality of care and health outcomes. However, noting that both private- and public-sector groups are eager to move forward with pay for performance, the committee concluded that a gradual implementation would enable officials to assess the program along the way, adapt to knowledge gained, and monitor for unintended negative effects -- such as providers avoiding certain kinds of patients or withdrawing from Medicare.

"Paying physicians and hospitals for delivering quality medical care that is explicitly measured sounds great, but is much harder and less effective than appears at first blush. Assessing quality is very, very difficult especially for multi-dimensional care. One cannot assess just cancer surgeries. Some places do very well on breast surgery but poorly on colorectal surgery and neither correlates with mammography rates. Little data is available on long term outcomes. Skew what physicians and hospitals
focus on to what is measured".[44]

I.18. Reducing base payments to fund bonuses
The committee deferred to Congress to determine by how much to decrease Medicare base payments to create a pool of funds for bonus payments. However, it recommended that the percentage be sufficient to create rewards large enough to motivate health care providers' participation and real improvements.

Physician fees are already scheduled to decline over the next few years under a mechanism that adjusts reimbursements to control Medicare's costs. Congress therefore may need to appropriate some new funds to ensure that the reward pool is sufficient, the committee said.

I.19. Comparative Cost Adjustment Program
Numerous policy analysts have proposed changing the Medicare program by adopting the principles of premium support[45]. The proposals vary in specificity and design, but all envision a system in which private plans would compete on the same terms as the fee for-service (FFS) program and beneficiaries would face incentives to choose plans on the basis of their relative premiums and the quality of care they provide. A demonstration that is scheduled to begin in 2010 and a bill that was introduced in the Congress in 2001 illustrate alternative design options for premium support that have been debated by lawmakers in recent years.

The Medicare Prescription Drug, Improvement, and Modernization Act of 2003 (MMA), Public Law 108-173, requires that the federal government conduct a six-year demonstration of premium support in up to six metropolitan areas beginning in 2010.

---

[44] *Ezekiel J. E., Department of Clinical Bioethics, Warren G. Magnuson Clinical Center, National Institutes of Health. Re-Thinking National Health Care Reform: Universal Healthcare Vouchers. Presentation given at Kellog University Annual Healthcare Conference, 2005.*

[45] *For example, see Henry J. Aaron and Robert D. Reischauer, "The Medicare Reform Debate: What Is the Next Step?" Health Affairs, vol. 14, no. 4 (Winter 1995), pp. 8–30; Bryan E. Dowd, Roger Feldman, and Jon Christianson, Competitive Pricing for Medicare (Washington, D.C.: AEI Press, 1996); and Stuart M. Butler and Robert E. Moffit, "The FEHBP as a Model for a New Medicare Program," Health Affairs, vol. 14, no. 4 (Winter 1995), pp. 47–61.*

* Many companies leave **antibiotics research**. This will become a **serious issue** within the next decade (antibiotics become resistant, and few are on the horizon). Despite increasingly sophisticated technologies to assess drug candidates, research output has become less significant. Translating validated targets into clinical candidates is a major hurdle in discovering new antibacterials.

B.23. Chronology of b-lactam antibiotics

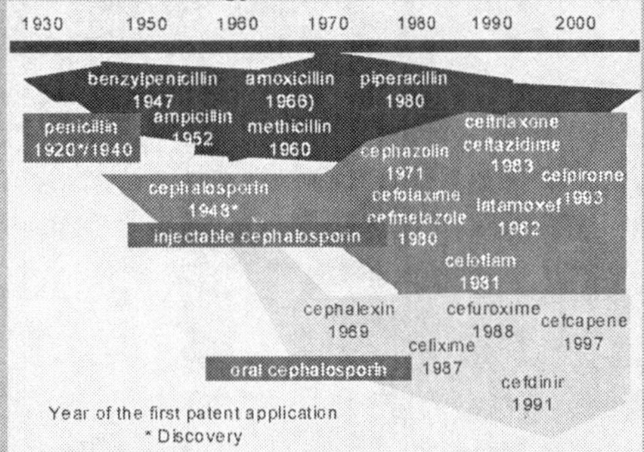

Add to this, that an ever **growing burden of proof and safety** at the **FDA**, and the **short-course of prescriptions** make it economically less interesting to develop anti-infectives.

B.24. Antibacterials : declining number of approvals

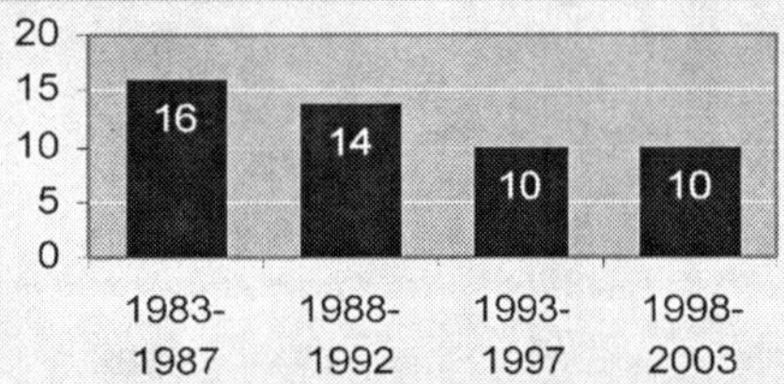

In 2003, only two new antibiotics were approved : **Cubicin** (**daptomycin** from **Cubist**) and **Factive** (**gemifloxacin** from **GeneSoft**). Despite the size of the market ($ 27bn in the US alone) and the huge global need for new antibacterials, in 2004, only five compounds are in development.

* The rate of fetal deaths, also known as **stillbirths**, occurring at 20 weeks of gestation or more declined substantially between 1990 and 2003, according to a 2007 report by the Centers for Disease Control and Prevention (CDC). Although fetal mortality rates declined among all racial and ethnic groups from 1990-2003, the fetal mortality rate for non-Hispanic black women was more than double that of non-Hispanic white women (11.56 per 1,000 vs. 4.94 per 1,000).

* **Richard M. Losick** discovered alternative bacterial sigma factors and fundamentally contributed to understanding the mechanism of **bacterial sporulation.**

* **Ketek** is the world's first in a new class of antibiotics known as ketolides. In a market where improved antibiotics are highly needed, it is expected to largely contribute to combat respiratory tract infections (RTIs). Ketek was designed to deliver an optimal spectrum of activity, with short treatment duration, specifically for upper and lower respiratory tract infections, including those caused by drug-resistant S. pneumoniae pathogens :

B.25. Spectrum of activity of RTIs (1)

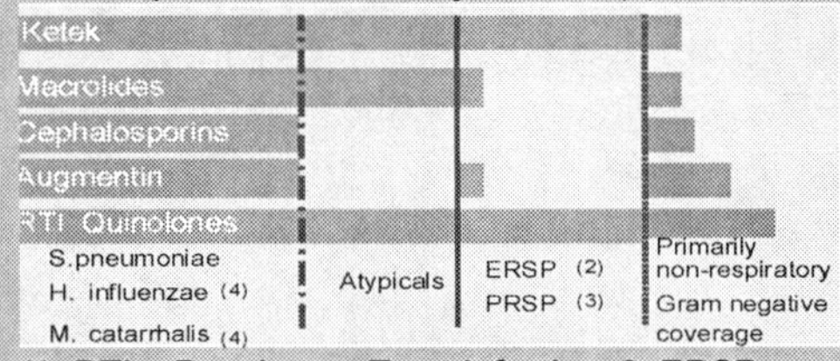

(1: RTI = Respiratory Tract Infection; 2 :ERSP = Erythromycin-Resistant S.pneumoniae; 3 : PRSP = Penicillin-Resistant S.pneumoniae; 4 : Including beta lactamase producing strains)

→→ Streptococcus pneumoniae is the most common respiratory pathogen
_ 45% of community acquired pneumonia
_ 34% of acute sinusitis
→→ PRSP and ERSP resistance levels are high and increasing
_ US: PRSP = 32.6%; ERSP = 31.0%
_ EU: PRSP = 15.6%; ERSP = 25.4%
_ Asia: PRSP = 53.2%; ERSP = 79.6%
→→ Streptococcus pneumoniae is a major cause of morbidity and mortality
_ Fatality rate among patients with pneumoccal pneumonia = 12.3%

* Although Aventis came up an unprecedented set of clinical data; with data from the largest single comparative antibiotic trial ever conducted (more than 24.0000 patients),and with post-marketing experience from more than 1.5 million exposures to Ketek, FDA took a "slow-track" approach for evaluating the drug's marketability.

B.26. Biologics have higher success rates

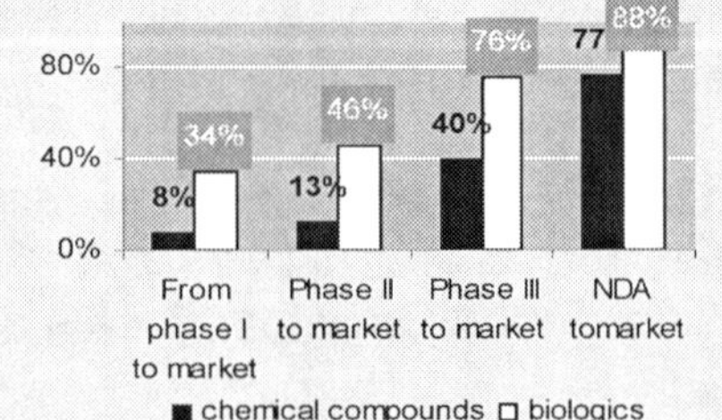

According to CMR international, who compared the success rates of drugs by product type, chemical compounds have less success rates than biologics, especially at earlier stages.

(The legislation uses the term "comparative cost adjustment program" rather than premium support.) The demonstration is a compromise that was achieved as part of the conference agreement. The original bill that passed the House called for the nationwide implementation of premium support beginning in 2010 in all areas that are served by at least two private plans, whereas the Senate bill did not include a provision for premium support. Under the demonstration, the benchmark—the government's maximum payment for an enrollee in a private plan—in each county will be a weighted average of the bids of private plans and local per capita FFS expenditures. (The bid of the FFS program will be weighted by either the percentage of beneficiaries in the county who are in the FFS program or the percentage of beneficiaries nationally who are in the FFS program, whichever is greater. The average bid of private plans in the county will be computed by weighting each plan's bid by its share of local private-plan enrollees, and that average bid of private plans will be weighted by 1 minus the weight of the FFS bid.) The rules for local Medicare Advantage plans will remain in place in the demonstration sites, but they will be subject to the demonstration benchmarks. The fee-for service program in the demonstration sites will be subject to rules that are analogous to those in place for private plans. However, the increase or decrease in the Part B premium for beneficiaries in the FFS program in any year will be constrained to not exceed 5 percent of the national Part B premium. In addition, low income beneficiaries in the FFS program will not be subject to any change in their Part B premium[46].

I.20. Exploring more sustainable long-term strategies

Using a reduction in base payments to fund bonuses should be used initially while other, more sustainable long-term strategies are explored. Sustaining the rewards pool through savings generated by improved efficiency and cost-reducing reforms has great potential, the committee said, and it urged the Centers for Medicare and Medicaid Services to test ways to make this funding source work.

To increase the likelihood of participation by as many health care providers as possible, the program should reward those who improve their performance significantly as well as those who meet or exceed designated thresholds of excellence. As providers increasingly make improvements, the fraction of rewards for excellence will grow; therefore the standards for achieving improvements should be raised appropriately.

Because obtaining the technology and skills needed to collect and submit performance data could impose a burden on providers that discourages their participation, HHS should offer incentives to encourage providers to submit data, the committee recommended. The data should be made publicly available to better inform patients and other stakeholders about the quality of various health care providers.

I.21. "Medicare for all"

US Senator Kennedy (Democrat) outlined a move towards universal health care coverage, by expanding the federal government's State Children's Health Insurance Program, followed by a "Medicare for all" scheme, reducing the qualifying age for Medicare to cover 55-64 year olds, then shifting it upwards to 20 year olds. The changes would extend the Medicare Part D prescription drug plan to the newly-entitled population. House of Representatives member John Dingell (Democrat) is co-sponsoring the proposed bill. They estimate at $600.0 billion *per* year the cost of the Medicare extension,

---

[46] *Marron D.B., Congressional Budget Office, Designing a premium support system for Medicare, Report to the Congress of the United States, December 2006*

* 2003, the **EMEA** gave green light for the approval of **Novartis´ biogeneric growth hormone Omnitrop**; the first copy of a large molecule drug to be made available to patients by a generic company. A **landmark**; a sure sign that the market is going to develop rapidly. The market potential for biogenerics reached **$5bn** in 2006. The real threat to branded players is to come not from copies but from "upgraded" biologics –longer lasting, safer and more efficacious versions.

Safer and **more efficacious** is necessary also because the **many off-label uses** of products such as human growth hormone (hGH) and erythropoietin (EPO).
* hGH rejuvenates the physique of the user and may have real positive effect as it is now thought that aging is –at least partially- caused by the lack of hGH;
* in competitive sports, EPO is well documented; its application in lower doses for the aging, offer positive effects.

* A promising approach to gene therapy involves short DNA fragments (**interfering RNA**) that bind to specific genes and block their "translation" into the corresponding, disease-related protein. Researchers at Stanford University have chosen to use **carbon nanotubes** as their "means of transport". This has allowed them to successfully introduce RNA fragments that "switch off" the genes for special HIV-specific receptors and co-receptors on the cells' surface into human T-cells and primary blood cells. The researchers report that this allows for much better **silencing** effect to the cells than current transport systems based on liposomes.

* **Xiaodong Wang** and **George L. MacGregor** pioneered biochemical studies on **apoptosis**, which have elucidated a molecular pathway leading into and out of the mitochondrion and to the nucleus.

* Researchers have discovered a survival mechanism in a common type of bacteria that can cause illness. The mechanism lets the bacteria protect itself by warding off attacks from **antimicrobial peptides (AMPs)**, which are defense molecules sent by the body to kill bacteria. Little is known about how gram-positive bacteria -- such as those that can lead to food poisoning, skin disorders and toxic shock -- avoid being killed by AMPs. AMPs are made by virtually all groups of organisms, including amphibians, insects, several invertebrates and mammals, including humans. In gram-negative bacteria -- such as those that cause plague and salmonellosis -- a sensory and gene regulation system named **PhoP/PhoQ** protects invading bacteria, and scientists believe if they develop a better understanding of this system they could develop new drugs that are more effective at protecting people from infection. Likewise, scientists are hoping for similar possibilities for gram-positive bacteria with their discovery of **"aps,"** which stands for **antimicrobial peptide sensor**. Aps has three parts: **apsS**, the sensor region; **apsR**, the gene regulation region; and **apsX**, which has an unknown function that they are investigating. Studies show that **all three components of aps must be present** for the system to function and effectively protect bacteria from AMPs.

* Almost eight in ten (78%) family physicians and seven in ten (70%) members of the general public believe men may experience something similar to women's menopause as they get older. Further, a strong majority (71%) of family physicians and (67%) of the general population agree that **male menopause**, also known as **Andropause**, can affect the quality of a man's life as much as menopause can affect a woman's. In the United States, there are 25,172,000 men between the ages of forty and fifty-five who are now going through the **Male Menopause Passage**. Worldwide that number is 408 million. In less than twenty-five years, by 2020, the number of men in the United States going through the Male Menopause Passage will grow to approximately 57.5 million. Research indicates that lowered levels of the following hormones may decrease sex drive, increase depression and weight gain, and contribute to a general decrease in well-being and health.

* In **China**, a state-level research program is underway to discover how microbes survive in extreme conditions in the ocean, as part of the **863 High-Tech Program**, initiated in March 1986, to tap into new resources for medicines. The research aims to find natural bio-active substances in marine life that live in extreme environments in deep water. The research on the microbe's life systems and the medical benefits would likely produce large and profound changes in the bio-medicines industry.

* Future success for the **generic industry** depends on:
* the availability of ingredients from multiple sources;
* reliability of supply to accommodate rapidly-changing demand patterns;
* the flexibility of agreements to cushion against severe price erosion;
* ability to accommodate multiple dose forms and labeling requirements;
* lean production capabilities that result in minimal obsolescence;
* waste, recalls and high capital turnovers;
* reliable lead times;
* flexible stocking programs;
* strong sales and operations planning; and
* a track record with wholesalers and pharmacy chains.

*Clinical trials have suggested that **renin-angiotensin activation** might be a key initiator of **atrial fibrillation**. A post-myocardial infarction trial of angiotensin-converting enzyme (ACE) inhibitor, trandolapril, showed a 50% decrease in AF, in a post-hoc analysis.

* 2004, HHMI researchers identified a gene that appears to have played a role in the expansion of the human brain's cerebral cortex -- a **hallmark of the evolution of humans** from other primates.
By comparing the gene's sequence in a range of primates, including humans, as well as non-primate mammals, the scientists found evidence that the pressure of natural selection accelerated changes in the gene, particularly in the primate lineage leading to humans.

but believe that annual savings would be found of $380.0 billion. Payroll taxes would be increased to help cover these additional costs, they say.[47]

I.22. Europe's Social Security System reform attempts

In contrast to millions of Americans who worry about whether they can afford health care for themselves and their families, the Europeans consider their universal health coverage being as much of a human right as their freedom of speech. "Everyone wants the best available medical treatment, provided someone else pays for it"…that's the mentality. In Germany, sickness funds finance health care mainly from payroll deductions and from payments made by employers (the deduction averages a combined 14%) The funds calculate annually the amount of money they require for self-sustained operation and then set the rate at which employees contribute, ensuring that individuals receive all necessary medical services but contribute only according to their ability to pay.

Germans are struck by the observation that Germany is a stagnant society. For years, they have seen their social-consensus model as the blueprint for the rest of Europe. However, news of such consumers gradually losing any hope of a consistent upturn in the economy in the near future; widespread debate on sites in various sectors being relocated to other countries; the introduction of combined unemployment and social security benefits under the so-called "Hartz IV" law, an unemployment rate of about 10% of the workforce (or much higher, if one adds the 2 million people that do not appear in the statistics), has resulted in citizens seeing themselves as victims of failed policies and subsequent governments.

With the financial shock waves sent through the nation by the consolidation of the east and west German health care systems (the 60 million citizens of the West having to pay for the 20 million citizens of East who never contributed to the former West Germany's health care system), which have not yet been fully absorbed, the changes in the general political and economical conditions are hitting the very foundations of societal solidarity; foundations on which the Statutory Healthcare System relies.

The Government is left with little scope for maneuver. In a climate of: increased international economic competition; slow domestic economic growth with no signs of a rapid turnaround, combined with: the erosion of social and moral foundations of solidarity; the shift in care provided within the boundaries of the family to more costly hospital care; and the occurrence of new diseases and changes in lifestyle, increasing the expectations for health care by putting into the system the added challenge of the altering demographic structure and the huge burden on the nation, having to support millions, and millions of unemployed and forced pre-retirees who, because of the lack of job-related income pay less into the System, the government has no other choice but to put the next reform on the agenda.

Reform attempts have taken place every four years since 1976, all of them two years after each parliamentary election. Because no single reform has yet tackled the roots of the problem, namely the financing of the system based on salary, the reform-debate became a never ending story.

---

[47] *Pharma Marketletter, November 27, 2006*

* In the USA, the **annual cost of diabetes in medical expenditures and lost productivity** climbed from $98 billion in 1997 to $182 billion in 2006. The direct medical costs of diabetes more than tripled in that time, from $44 billion in 1997 to $143,4 billion in 2006.

A US government financed study (2002) found that
- **direct medical expenditures** of $91.8 billion included $23.2 billion for diabetes care, $24.6 billion for chronic diabetes-related complications and $44.1 billion for excess prevalence of general medical conditions.
- **indirect costs** resulting from lost work days, restricted activity days, mortality and permanent disabilities due to diabetes totaled $39.8 billion.
- cardiovascular disease is the **most costly complication** of diabetes, accounting for more than $17.6 billion of the $91.8 billion annual direct medical costs for diabetes in 2002. HHS' Centers for Disease Control and Prevention estimates that 17 million Americans have diabetes, including many who are unaware of their condition. In addition, an estimated 16 million additional Americans have pre-diabetes.

* **58% of diabetes attributable to obesity**
Obesity is typically defined as a chronic disease that develops as the result of an imbalance of energy intake and expenditure, distinguished by a build-up of body fat. The disease poses a **major threat** for serious diet-related chronic illnesses such as type 2 diabetes, cardiovascular disease, stroke and some forms of cancer. In fact, the World Health Report 2002 concludes that around 58% of diabetes, 21% of ischemic heart disease and 8%-42% of certain cancers globally were attributable to obesity that year.
Of particular concern is the increasing incidence of child obesity, already epidemic in some areas of the world and increasing steadily in others. Current estimates put the number of overweight children under five at 17.6 million. In the USA, the number of **overweight children** has **doubled** and that of adolescents has trebled since 1980, according to the WHO.

* The healing potential of every day human brain cells: 2005, researchers published details about how they used **stem-like brain cells** from rodents to duplicate neurogenesis - the process of generating new brain cells - in a dish. In 2006, they showed that **common human brain cells can generate different cell types in cell cultures**. In addition, when researchers transplanted these human cells into mice, the cells effectively incorporated in a variety of brain regions Scientists speculate a small amount of existing progenitors may be emerging from the gray matter of the brain and multiplying in torrents, or perhaps the aging clock of the mature cells actually turns backward when the donor cells are in a new environment, returning them to past lives as progenitors or as stem cells. In addition to using the cells in treatments to repair or replace damaged brain tissue, the ability to massively expand cell populations could prove useful in efforts to test the safety and efficacy of new drugs. It is also possible to genetically modify the cells to produce **neurotrophins** - substances that help brain tissue survive.

* **Physicians** increasingly consider **homocysteine** as a culprit in CVD and combine cholesterol-lowering treatments based on statins with the monitoring and adjusting of homocysteine levels. Epidemiological evidence has accumulated during the past decade that establishes elevated serum levels of the amino acid homocysteine as a **risk factor for CVD**. As many as one-third of people with atherosclerosis have elevated blood levels of homocysteine, and major studies, such as the Framingham Heart Study and the Physicians' Health Study, have linked elevated levels of homocysteine to increased risk for heart attack and stroke. **High levels of homocysteine have been correlated with endothelial damage** (due to its abrasive molecular shape), inhibition of anti-aggregatory activity, and the formation of atherosclerotic lesions.
Hyperhomocysteinemia is caused by genetic and lifestyle influences, including low levels of specific nutrients. Normally, folate and vitamin B12 facilitate remethylation of homocysteine to methionine (the amino acid precursor of homocysteine), while B_6 facilitates transsulfuration, which converts homocysteine into **cysteine**. Maintaining a balance of these critical micronutrients in the diet results in proper homocysteine metabolism and can help prevent the damage associated with an accumulation of excess homocysteine. However, when dietary, genetic, or other factors limit the bioavailability of these important nutrients, an overproduction of homocysteine occurs. Thus, it is important to determine the patient's ability to properly metabolize **methionine** and **homocysteine**. Low levels of folate, vitamin B_6, and vitamin B_{12} have been associated with an increased risk of CVD. A factor that has been implicated in some cases of excessive homocysteine levels is deficiency of enzymes and coenzymes necessary for the conversion of **folate** to its metabolites. By ingesting an intermediate metabolite of folate, such as 5-formyl tetrahydrofolate, **complicated enzyme pathways can be bypassed** resulting in higher levels of folate's end metabolite, **5ME-tetrahydrofolate**, and increased remethylation of homocysteine.
Choline is a direct precursor to trimethylglycine, a substrate of the zinc metalloenzyme: betaine homocysteine methyltransferase (BHMT). **BHMT** catalyzes a methyl transfer from TMG to homocysteine forming dimethylglycine and methionine, respectively. Experimental assay confirms that the human liver contains an abundance of BHMT, emphasizing the importance of liver health in mediating circulating homocysteine.

Bristol-Meyers Squibb and **Gilead Sciences** gain FDA approval of the first once-daily, fixed-dose combination AIDS therapy, **Atripla**.

* Surgeons at Sentara Heart Hospital (Norfolk, Virginia) performed the state's first robotic major thoracic procedures using the **daVinci surgical robot**. The three procedures—two robotic lobectomies removing tumors from lung cancer patients and one surgery to remove a mass from the thymus gland—are innovative new procedures designed to minimize trauma to patients.

The employment-based nature of social insurance has initiated the "standard-debate", in which employers argue that their social insurance contributions are too high, rendering German industry uncompetitive in the global economy and making it a hostile place in which to invest. Employers are, therefore, urging a decrease in the growth of social insurance contributions, or at least the share paid by the firms. Discussions include proposals to roll back the state from social provision, inject market forces into the welfare state, privatize social risks and reduce coverage to serious illness/need.

Nowhere can the crisis in health care (funding) be resolved by politician's short-term strategies or with patchwork efforts of the past. Neither can it be resolved by dealing with only one or several problems currently faced by providers, patients and/or payers. Resolution will require comprehensive health system reform.

I.23. The language of solidarity

Solidarity comes at a price. To some, the continuous increase in financial contributions to provide relief to the suffering –although painful- is the price for the community to pay, if it wants to hold to a system considered to be "sacred", because it is a historical and sacred worker-unions achievement. The "health for all" philosophy cannot be touched. Backed by "social" inspired movements and politicians, pressures to preserve solidarity act as a counterweight to cost-cutting.

Table 1.19.   A German Healthcare Reform (Per Capita Premium Model, 2006)

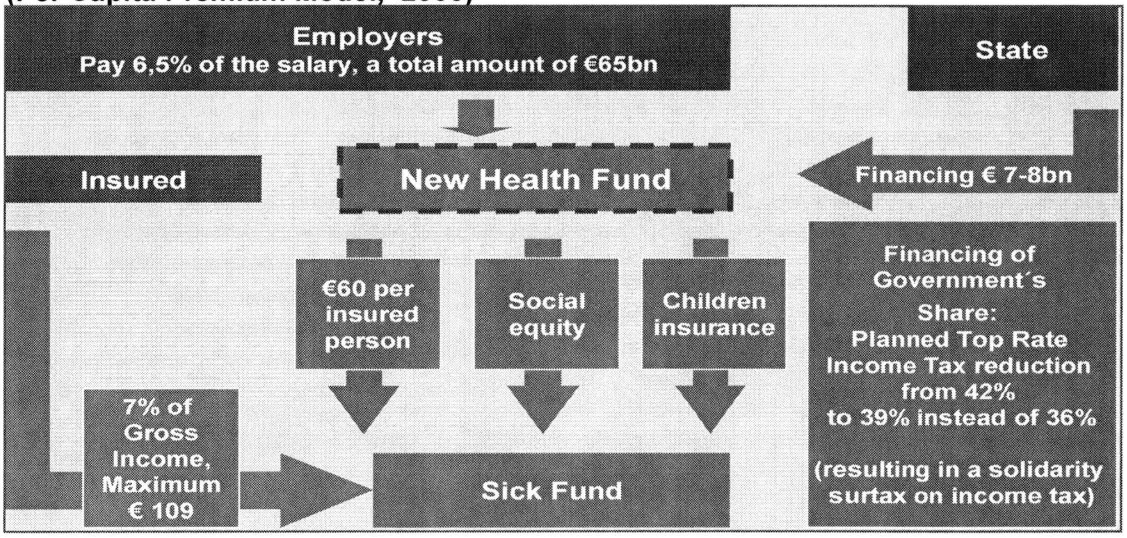

To others, a system based on –and financed only by- an individuals employment-based income, which they consider to be just another tax, is the wrong financial method to safeguard universal access to a comprehensive level of care. They advocate a system that includes more self-responsibility, the payment of extra insurance premiums for high risk sports activities, unhealthy lifestyles, etc.

*** Diagnostic criteria for diabetes have changed** as follows: Normal serum glucose level is 110 mg/dL with HbA1c of less than 7%. Impaired fasting glucose (IFG) is a new term, defined as fasting plasma glucose between 110-125 mg/dL. Diabetes is diagnosed at a serum glucose level of 126 mg/dL, on at least two occasions. For low-risk individuals over age 45 years, the American Diabetic Association, for example, recommends routine screening every three years with a fasting glucose test.

B.27. Activity comparison of insulins

Insulin	Onset	Peak	Duration
Lispro	0.25 hr	0.5-1.5 hrs	6-8 hrs
Insulin aspart	0.5 hr	1-3 hrs	3-5 hrs
Regular	0.5-1 hr	2-4 hrs	13-18 hrs
NPH	1-2 hrs	5-7 hrs	13-18 hrs
Ultralente	2-4 hrs	8-10 hrs	18-30 hrs
Insulin gargline	-	-	24 hrs

New insulins such as **insulin lispro (Humalog)** and **insulin aspart (Novolog)**, when administered intravenously, show pharmacodynamic parameters similar to regular insulin. Only insulin lispro is used for insulin therapy in patients who wear external SC insulin infusion pumps. The newest insulin, **insulin glargine (Lantus),** mimics the patient's basal secretion of insulin and allows for once-daily dosing.

* Multiple pharmaceutical companies are developing new insulin sensitizing drugs for treatment or prevention of type 2 diabetes. One class of drugs acts through a **nuclear receptor to regulate gene expression** in fat and other insulin responsive tissues.
A compound that stimulates insulin-producing cells in the pancreas is being studied. Some of the 30 drugs in development (2005) are aimed at optimizing glycemic control, reducing insulin resistance, reducing obesity, and preventing the complications of diabetes.

* Understanding of **molecular mechanisms** involved in **glucose toxicity** and development of **complications of diabetes** is also yielding new therapeutic strategies. For example, the US NIH (National Institutes of Health) funded research identified a key signaling molecule involved in glucose toxicity. Clinical trials of an inhibitor of this protein for treatment of **diabetic peripheral neuropathy** are underway using this drug to treat neuropathy (nerve damage) and it will also be studied for **retinopathy**.

* Researchers discovered that **insulin blocks signals** that trigger energy release from fat cells. The findings may explain why obesity and type II diabetes often go hand in hand.

* Scientists at Baylor College of Medicine found that here may be a real **code** to **chromosomal organization.** The identification of a cluster of essential genes on mouse chromosome 11 as well as similar clusters on the chromosomes of other organisms - including humans - buttresses the argument that there may be rules as to how genes are structured or laid out on chromosomes.

* Drugs to treat **obesity**, a tremendous and dramatically growing health problem, have historically failed to sell in this market. According to IMS Health, the whole antiobesity market in the USA only managed to generate revenues of **$300 million** in 2003. An attempt by **Roche** to create a wide therapeutic market segment did not result in the expected sales. Partly due to the lack of reimbursement for anti-obesity therapies, and partly due to the large market of all kinds of slimming products –mostly based on testimonials and pseudoscience- also in part by its embarrassing side-effects profile relating to the prevention of dietary fat absorption in the gut, Roche's **Xenical** could not deliver the sales performance its creators had hoped for. Roche's Xenical (**orlistat**) is a reversible inhibitor of pancreatic and gastric lipases and phospholipase A2, all essential to dietary fat hydrolysis in the gut, and prevents absorption of dietary fat by approximately 30%. **Pharma Marketletter** reported that by far the most-prescribed (+50%) antiobesity product in the USA is generic phentermine which mimics the effect of noradrenaline in reducing food intake, and has been on the market since 1959, Unfortunately, in the 1990s the product was subject to bad press; it was shown to cause heart valve failure when taken in combination with **Pondimin (fenfluramine)** and **Redux (dexfenfluramine)** and, although phentermine as a monotherapy had always enjoyed a clean safety profile, the resulting damage was reflected in the low $180 million annual US sales of branded antiobesity drugs achieved in 2003. **Lehman Brothers** analysts expect **Sanofi-Aventis´ Acomplia (rimonabant)** that entered the antiobesity market in 2006, to have potential peak annual sales of $1,2 billion. The company hopes to **disassociate** the product from the **life-style** bracket and "medicalize" obesity by submitting clinical data regarding metabolic biomarkers as well as weight loss findings.

B.28. HHS estimates of prescription drug coverage among Medicare beneficiaries, 2006

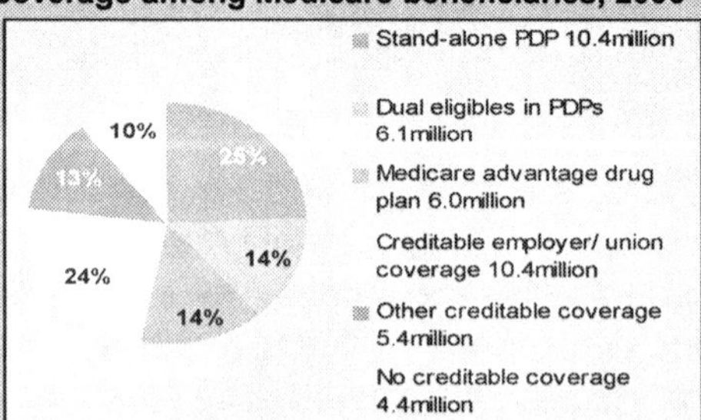

In the US, **43 million people benefit** from **Medicare**. Out of this total 53% or 22.5 million elected to enroll in a prescription drug plan (Part D Plan). For many beneficiaries, Part D coverage replaced benefits from Medicare HMOs, Medigap, and Medicaid. (Source: Kaiser Family Foundation).

Table 1.20.   Health Care cost as a percentage of GNP, selected industrial nations[48]

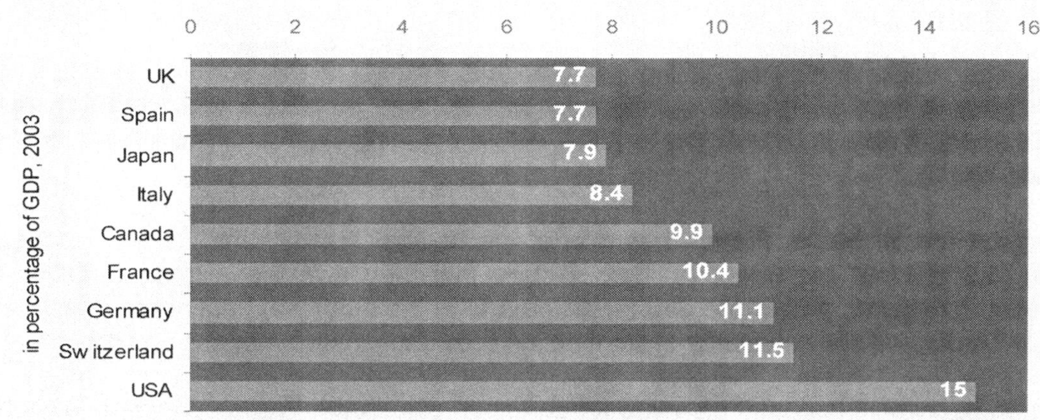

The challenge, as in other western democracies is to find the right balance between the interests of the advocates of state protectionism fearing that purely economic considerations might conflict with the best interest of patients, and an increasingly demanding society turning towards strengthening individuality.

The picture of the German health system portrayed today, is the result of long years of clashes of the interest of different parties where each of these groups has tried and continues to try to defend its stand, by all possible means.

I.24. System reform : The challenge is huge
As in Germany, the French Social Security accounts have been in chronic deficit. Income has fallen, unlike expenditure, in spite of numerous recovery plans and reforms.

Table 1.21.   Social Security Deficit, France 1990-2006[49]

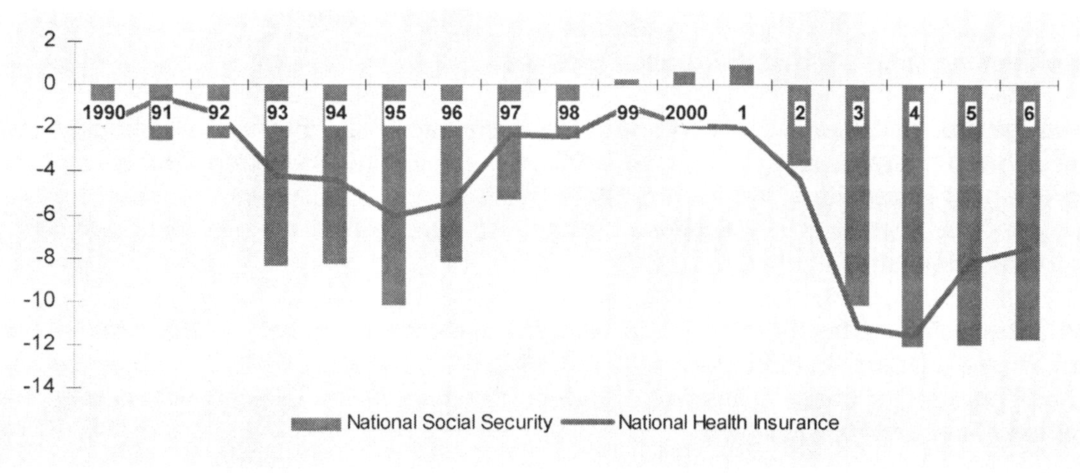

In Euro, billion

---

[48] Source: OECD-Indes; Le Monde Dossiers & Documents, September 2006 ; Figures for UK are for 2002
[49] Source: Sécurité Sociale/FAZ 29.03.06

The challenge is huge. It is a matter of safeguarding a precious institution, to which the French are very attached. France is proud on having a healthcare system that, in 2000/2002 has been ranked by the World Health Organization (WHO) as the best in the world.

Since, the media can't stop telling readers that France is the country in which life expectancy is increasing most quickly - by two to three months a year for more than twenty years. It rose to over eighty in 2004, with a not insignificant number of years lived in good health.

Champion too for births, France has one of the highest fertility rates in the European Union (1.9 children per woman). They declare that they owe these good statistics to progress in hygiene, prevention and medicine and to the quality of healthcare; above all, to their health and social security system which provides cover for illness and maternity

It is also because it is based on some fundamental values the French –be they from the leftist or from the right political ideology- cherish that much: solidarity. Read: solidarity with the redistribution of income by the State who takes the money from the working citizens to redistribute it according to its own principles of equality among citizens...

Table 1.22.   Financial need of the French Social Security System, 2006-2009[50]

	Financing needed, end 2006	Financing needed, 2007-2009
Healthcare (the debt has been taken over by the CADES)	0	4.6
Family	2.9	2.3
Work accidents	0.8	-0.1
Pension Solidarity Fund (FSV)	5.25	3
Farmer Fund (FFIPSA)	3.8	5.5
Total	16.05	21

(In Euro, billion)

France spends more on „welfare" than almost any other EU country: over 30 per cent of GDP. Total social security revenue is around E200 billion per year and the social security budget is higher than the gross national product (GNP), i.e. social security costs more than the value of what the country produces.

Since the first oil crises of the 1970s and the economic crisis, the rise in unemployment and longer life expectancy, leading to a fall in the number of working people, and the improvement in pensions, the French State System, too, has run into financial trouble. Employer's and workers contributions continue to rise. Reforms are a matter of highly controversial debates.

The construction of the French Social Security system was mainly to the work of one man, Pierre Laroque. In the heady days following the Liberation, 1945, his objective was to "achieve greater social justice" and "reduce the inequalities existing between people in terms of security for the future".

---

[50] *Source: Rapport de la Cour des Comptes, Les Echos 11.09.06*

*** Genome sequence** reveals leaner, meaner intestinal parasite :"Cryptosporidium parvum" -- an insidious, one-celled, waterborne parasite that lodges in the intestines of infected people and animals and for which there is currently no effective treatment -- is missing key structures normally found in similar parasites, say researchers supported by the National Institute of Allergy and Infectious Diseases (NIAID), one of the U:S. National Institutes of Health.

"C. parvum" is an extremely hardy parasite found in water supplies throughout the world. For persons with weakened immune systems such as individuals with HIV/AIDS, symptoms may be more severe and can lead to serious or life-threatening illness. Because "C. parvum" could potentially be used as a bioterrorist agent, the NIAID has classified it as a Category B priority pathogen. After reconstructing the predicted genes and resulting proteins of one form of "C. parvum", researchers discovered that "Cryptosporidium" is missing two organelles commonly found in related protozoan parasites. Gone is the apicoplast, a cellular component that provides essential metabolic functions in related parasites, including those that cause malaria and toxoplasmosis, respectively. Also absent is the mitochondrion, the so-called "energy factory" found in the cells of most plants, animals, fungi and one-celled organisms. In addition, the researchers found that **"Cryptosporidium"** has significantly fewer genes than related parasites, and, as a result, can carry out fewer metabolic functions on its own. Because "Cryptosporidium" has been so difficult to study up until now, the decoding of the genome sequence provides valuable opportunities to inform and study the organism's biology. And with an understanding of its biology, researchers are better positioned to find treatments that zero in on unique biological processes essential for the organism's survival.

The results of their genome sequencing project, could help scientists home in on new drug targets that may lead to therapies for the disease.

*** The Chinese government** decided for **China** to **play a global role in pharmaceuticals by 2020**. To help facilitate WTO compliance, the State's drug administration (SDA) brought into effect the new drug registration procedures, Dec.2002. The record number of 6.300 applications for the approval of clinical trials, manufacturing, line extensions and generic drugs in the sole year of 2002, points to the dynamism of the Chinese pharma market.

*** African and African American women** are more likely to die of **breast cancer** than their white counterparts because they tend to get the disease **before the menopause,** suggests research from the University of East Anglia and the Children's Hospital Boston in collaboration with researchers in the U.S. and Italy.

***** In general, the **combination** of a **TNF inhibitor** and **MTX** provides additive or synergistic efficacy and is now considered the gold standard for the treatment of RA. **Anakinra,** a genetically engineered recombinant interleukin (IL)-1ra that competitively binds to the IL-1 receptor, prevents inflammatory substances from activation.

***** Therapeutic modalities for the treatment of established RA are divided in **4** general classes: nonsteroidal anti-inflammatory drugs **(NSAIDs)**/selective cyclooxygenase-2 inhibitors; **glucocorticoids**; disease-modifying antirheumatic drugs **(DMARDs)**; and **biological agents**. There is substantial evidence that **proinflammatory cytokines**, especially **TNF-alpha** and **IL-1**, are **upregulated in Rheumatoid Arthritis**.. Three Tumor **Necrosis Factor (TNF)** inhibitors are approved for the treatment of RA: **etanercept**, a recombinant soluble p75 TNF-receptor-Fc fusion protein; **infliximab**, a chimeric monoclonal anti-TNF-alpha antibody; and **adalimumab**, a human monoclonal anti-TNF-alpha antibody that reduces symptoms of arthritis and inhibits the progression of structural damage in adults. Unlike most **DMARDs** (which work to reduce the body's defense mechanisms and prevent it from attacking its own cells), TNF inhibitors can rapidly improve the patient's clinical status, usually within the first 2 weeks, and potentially halt radiographic progression of joint damage. **Abatacept** is a T-cell regulatory protein that inhibits cells that launch the immune response.

B.29. DMARDs

DMARDs	Biologic Agents
Methotrexate	Etanercept
Sulfasalazine	Infliximab
Leflunomide	Adalimumab
Hydroxychloroquine	Anakinra
Azathioprine	Abatacept
Cyclosporine	Rituximab

***** The group of drugs known as **beta blockers** help slow nerve impulses traveling through the heart in order to reduce the heart's workload. This effect is achieved via their action on beta-adrenergic receptors present in cardiac cells. As such, beta blockers have become a mainstay of the treatment regimen for chronic heart failure. However, doctors have remained puzzled by the variable responses to some beta blockers among heart failure patients. Scientists at the University of Wuerzburg, Germany,found why some heart failure patients may respond better than others to certain beta blockers. The secret lies in a single amino acid change in the **beta1-adrenergic receptor**, that may **differ from person to person,** which alters the receptor's conformation and in doing so may alter the receptor's response to a given beta blocker.

*** Ofatumumab (HuMax-CD20)** is a fully human monoclonal antibody targeted at the CD20 molecule in the membrane of **B-cells** (immunological cells most commonly involved in RA). It is a potential important treatment for patients suffering from **rheumatoid arthritis**. Primary results (evaluated at 24 Weeks) from a, double-blind, randomized, parallel group, placebo-controlled multi-center phase II trial, conducted across several international centers including the UK, France, Denmark, Poland, USA and Hungary, reported positively on the efficacy and safety of Ofatumumab. The study comprised patients with **active RA** who had previously not responded to treatment with disease modifying antirheumatic drugs **(DMARDs)**. *(See also "Towards 2015/continued 34)*

Already, in 1931, laws on social insurance had been passed and covered the risks of illness, old age, permanent disability and death. With the orders of 4 and 19 October 1945, compulsory Social Security was introduced for all employees. It was based on the principle of sharing: on a basis of compulsory and universal solidarity, with social contributions (deductions from income at source, payable by employers and employees alike), Social Security not only provided a substitute income for the sick, but also healthcare and some social services. In 1946, a law established the principle of Social Security cover for all French people, whether salaried or not.

Table 1.23.   Rates and ceilings of Social Security and unemployment contributions [as of 1 January 2006 (Summary)]

Risks	Employee		Employer	
	Monthly ceiling in EUR	Rate	Monthly ceiling in EUR	Rate
Social Security :				
Health Insurance	Total wages	0.75%	Total wages	12.80%
Solidarity independence	-	-	Total wages	0.3 %
Old-age Insurance	2.589	6.65%	2.516	8.3 %
Old-age insurance	Total wages	0.10%	Total wages	1.6 %
Industrial injury	-	-	Total wages	Risk-depending
Family Allowances	-	-	Total wages	5.4 %
Contribution Sociale Généralisée	Total wages less 3%	7.50%	-	-
CRDS	Total wages less 3%	0.50%	-	-
Unemployment:	10.356	2.40%	10.356	4%
Supplementary pensions				
« Non-executive »				
Bracket A	2.589	3%	2.589	4.50%
Bracket B	From "A" to 7.767	8%	From "A" to 7.767	12%
« Executive »				
Bracket A	2.589	3%	2.589	4.50%
Bracket B - C	From "A" to 20.712	7.70%	From "A" to 20.712	12.60%

Services have since been provided to the insured person in the form of reimbursement of charges for medical and hospital treatment and medicines; family allowances (calculated on income and the number of children), daily allowances for illness and maternity; pensions and allowances for disability, industrial accidents, retirement or widowhood.[51]

I.25. A complex system. Even for the French.

The Health system (*Assurance Maladie*) is complex and difficult to understand. Even for the French. It is based on the concept of providing a large amount of help for any medical need, and total help when it is serious : on the basis of a standard cost, medical care (doctors appointments and visits, dental care, etc..) is reimbursed around 80%, medicine from 80% when corresponding to a real medical need to 40% for less needed and of course 0% for others ; standard cost for a general practitioner is 21 Euro (sometimes, they can charge 30 Euro or more: *dépassement d'honoraires*) ; serious

---

[51] *France Diplomatie. Label France. Health for All. N°64, 4th Quarter 2006; M.Gazsi and E. Thevenon*

* The **NIH** (US National Institutes of Health) reports that several compounds that may limit brain damage in stroke victims are now being tested in animal models. Scientists are trying to develop **"neuroprotective drugs"** that prevent strokes from damaging brain cells. Efforts to develop neuroprotective drugs build on very substantial research efforts that are unraveling the complex cascade of harmful events that occur in the brain in the seconds, minutes, and hours following a stroke.

Each step in the cascade presents a potential target for drug intervention. **Excitotoxicity** occurs from excessive release of the normal neurotransmitter glutamate, and, when challenged by stroke, brain cells produce highly reactive and potentially harmful chemicals called "**free radicals.**" Research is developing drugs that intervene at various stages of excitotoxicity, free radical damage, and other aspects of the stroke-induced cascade of events in the brain. For example, anti-oxidants to prevent free-radical damage are being evaluated. As another example, **amapakines** are drugs that act by modulating a subclass of nerve cell receptors for a specific neurotransmitter. Excessive release of this neurotransmitter can cause damage in stroke victims. Testing is being conducted to determine if amapakines can prevent brain damage from stroke or help improve learning and memory following stroke.

Scientists have been encouraged by recent findings that the adult human brain has a surprising capacity to adapt following disease or injury, even to the extent of making new nerve cells. As a result, researchers are trying to develop drug interventions that enhance the brain's capacity **to repair itself.**

* **Tissue engineering :** While the majority of cardiovascular patients benefit greatly from modern drug therapies, there are a growing number who require direct surgical intervention. Heart surgery is one of the major medical advances of the past 50 years. Today **Coronary Bypass Grafting** (CABG) and the repair or replacement of heart valves are among the most frequently performed cardiac operations.

Tissue engineering already produces real human replacement skin for patients. Cardiac products are already being worked on. **Genzyme BioSurgery** and **Myosix** –pioneered the use of a patient's own cells to repair a failing heart. Others are working on growing replacement coronary arteries heart valves and even whole hearts so that cardiac surgeons in the future will have ample human body parts to perform their life-saving surgery.

* The Nobel Prize in medicine was awarded to Andrew Fire and Craig C. Mello for their discovery of "RNA interference – gene silencing by double-stranded RNA."

* A World Health organization report suggests that **Coronary Heart Disease** will be the world's largest single cause of disease by approximately 2020. **Super-statins** (e.g. **Crestor –rosuvastatin**- of **AstraZeneca**), and their possible successors (in form of different approaches ranging from new fibrates, cholesterol-pathway blockers, intestinal cholesterol-uptake blockers, new bile acid binders, and, above all vaccines) are predicted to have **niche blockbuster potential.**

* **Cardiovascular drugs** (includes medicines hat treat conditions such as elevated cholesterol, hypertension, and atherosclerosis) generates sales of well over $ 40 billion in 2006.

B.30. Chronology of vascular agents

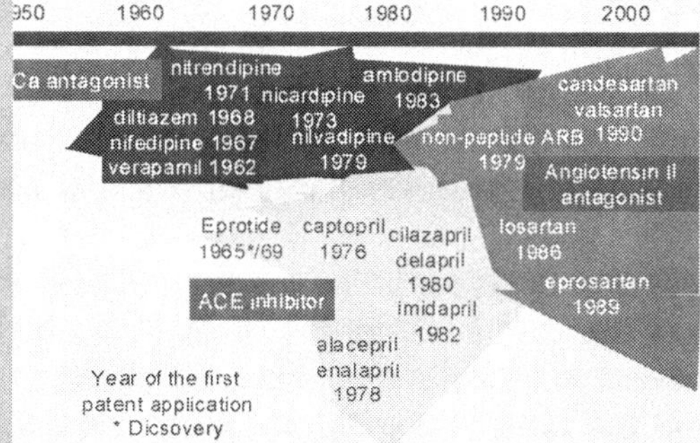

* Le Livre noir de la condition des femmes, [**The Black Book of Women's Condition**], directed by French journalist Christine Ockrent (pub. XO, Paris, 2006) and **including some fifty international contributions from correspondents of all back- grounds,** notably draws up a global inventory of violence against women's bodies, threatening their health, their integrity and even their lives: honor crimes, rape, conjugal violence, excision and prostitution, as well as exclusion from access to education and healthcare.

* 2003/2004, **Pfizer** launched its cardiovascular drug **Inspra (epleronone).** The company, which got approval for this product in 2002, had preferred waiting with this launch (although Pfizer is said to count with just 5 years of patent life, to 2008) until FDA approved a second indication, for congestive heart failure.

* Researchers have discovered how neutrophils, specialized white blood cells, navigate to sites of infection and inflammation. The cells detect and move toward small amounts of chemicals that are released by bacteria and inflamed tissue.

* Research reveals important structural information about the gyrations of DNA during RNA transcription. The discoveries may help in developing new antibiotics.

*Using a systems biology approach, NIGMS-funded researchers have developed a model for the workings of an organelle--the phagosome--responsible for destroying infectious agents.

* In Biotech, by 2003, **platform technology** companies – the darlings of the sector in the 2000 genomics boom – became unfashionable. The trend is towards drug products companies, because of the requirement for immediate investment returns. It's all about delivering value, fast. With pharmaceutical companies looking for suitable, market-ready products, opportunities for start-up technology firms abound. In 2002 alone, the majority of the 474 compounds in clinical trials in Europe were from biotechnology companies.

illnesses, including those due to old age, are covered 100% ; practically speaking each person has a chip card (*Carte Vitale*) which is read by the doctor's or pharmacist's terminal (introduced in 1999, the magnetic card, Vitale, holds medical information for healthcare staff, such as medicines prescribed by other physicians, this way contributing to a better use of medicines, and makes it easier to obtain repayment of treatment costs).

Contributions are collected together by the URSSAF (Union de recouvrement des côtisations de sécurité sociale et d'allocations familiales), which has 105 offices around the country.

The URSSAF then passes the money on to the ACOSS (Agence centrale des organismes de sécurité sociale), which distributes it to the various funds, called caisses, and which are responsible for paying out benefits and making reimbursements.

Unless one is on very low earnings, the French state reimburses only a proportion of the cost of medical treatment. However, if one is in the state system, one is perfectly free, and even well-advised, to obtain complementary health cover - and many French people top up the cover they obtain under the *régime obligatoire* with insurance from a private insurer or *mutuelle*.

I.26. Placing emphasis on prevention

From the 1970s, the financial problems of the health system, the emergence of AIDS, crises threatening health security and the "inability to reduce the rate of premature death" helped to provoke "questions about an almost exclusively treatment-oriented system".

This growing awareness was expressed in France by the creation in the late 1990s of bodies such as the French agency for the health and safety of health products, the French food standards agency, the national health monitoring institute or, in 2002, the national institute of prevention and education for health (INPES).

Prevention is seeing a revival of interest in France. First seen as a way of improving the health of the population through the adoption of behavior designed to prevent the occurrence, indeed the increase, of disease and accidents, it has become one of the pillars of public health policy. National health-education campaigns multiplied, placing great emphasis on prevention: health and nutrition, diabetes, etc. Thus the cancer campaign, announced by the French President as his special project for his five-year term of office, stated the necessity of making up for lost time in the area of prevention and of organizing screening more efficiently. In addition, this priority was enshrined in the Law of 9 August 2004 relating to public health policy. Even though the budget for prevention is currently 3 billion Euro compared to the 150 billion for treatment, the law makes people responsible for themselves and commits the country to running campaigns and establishing priorities.

The setting up, in 2005, of coordinated healthcare routes (the attending physician directs the patient to a specialist) is bringing down the number of worthless examinations and consultations while improving the medical monitoring of patients through better coordination of treatment.

* **Abbott**. D2E7(Adalimubab), from the BASF pipeline (co-developed with **Cambridge Antibody Technology Group** and with **Eisai** for the injectable version).. This antitumor necrosis factor was the **first fully human monoclonal antibody** to enter Phase III trials. It belongs to a new class that neutralizes the activity of tumor necrosis factor alpha, and interrupts the process that can lead to the symptoms of rheumatoid arthritis, including inflammation and swelling.

* **Amgen's** (and Kirin Brewery) Aranesp (a substance named darpoetin alfa), a second generation anemia drug, and expected to become standard therapy for anemia patients, is a once weekly injection and a non-naturally occurring recombinant protein that stimulates the production of red blood cells.

* With a universal **Verigene product platform** for the detection of nucleic acids and proteins, Nanosphere, a nanotechnology-based life sciences company, is rapidly advancing the development of assays for genomic and proteomic research, clinical laboratories and point-of-care markets. In 2005, the company succeeded with a technology that allows for rapid, highly-sensitive and specific **Single Nucleotide Polymorphism (SNP) genotyping**, which is the direct detection of a particular gene and the extent to which it is normal or mutated. The analysis of whole **human genomic DNA** is extraordinarily complex as it requires sifting through the more than one billion base pairs of DNA to find a particular base pair of interest. Once that base pair is located, it is then necessary to determine if either of the bases is mutated (i.e., has SNPs). **Nanosphere's** technology can rapidly, easily, and accurately interrogate both bases in the pair to determine if they are homozygous (i.e., both are mutant or normal) or heterozygous (i.e., one is mutant, one is normal) - the most critical step in correlating the SNP with a disease or drug sensitivity.To do so, Nanosphere scientists employ a two-step process called ClearRead technology. This method sandwiches a **target DNA SNP segment** between **two oligonucleotide sequences** to greatly increase detection specificity and sensitivity. One segment identifies any mutations in the DNA and the probe, a highly sensitive gold nanoparticle, creates a strong signal accurately indicating the presence of a **specific target SNP**.

* Efficient transfection of dendritic cells : A safe and efficient system for transfection of dendritic cells (DCs) is now available to cancer researchers, immunologists, dermatologists and hematologists. **amaxa's Human Dendritic Cell Nucleofector Kit** I, launched in July 2002, offers high-level transient transfection of up to 50% and the safety of a non-viral technique, with transfected cells maintaining their immunological function. DCs play a crucial role in the immune response and are a major target for many clinical applications. In cancer immunotherapy, DCs are a primary target as they can generate tumor-specific immune responses if 'loaded' with antigens ex vivo and then reintroduced into patients. **amaxa's** Nucleofector technology is unique in that it has the ability to transfer DNA straight into the cell nucleus. Thus even non-dividing primary cells are made accessible for efficient gene transfer.

* When setting prices for reimbursable medicines, Governments take into account the prices in other countries. Prices in Germany, UK and to a lesser extend France, have a great impact on drug prices elsewhere.

B.31. setting prices for reimbursable medicines / direct impact

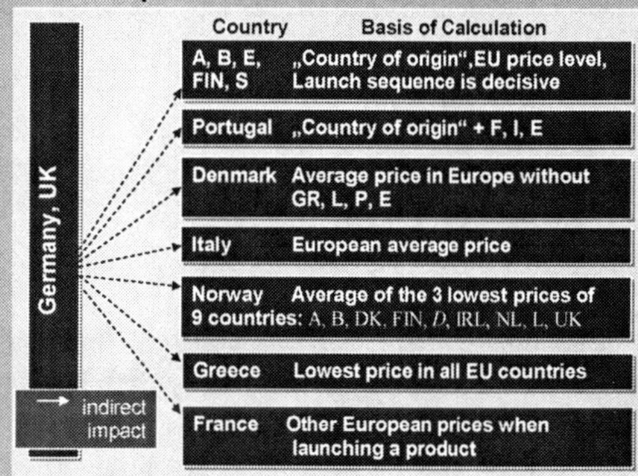

Country	Basis of Calculation
Netherlands	Average prices of B, *D*, F,UK
Switzerland	Average prices of DK, *D*, NL, UK + neighbor countries
Ireland	not more than UK or average prices of DK, *D*, F, NL, UK
Canada	Prices of *D*, F, I, S, CH, UK, USA
Japan	Average prices of *D*, F, UK, USA

Germany, UK → direct impact

B.32. Setting prices for reimbursable medicines / Indirect impact

Country	Basis of Calculation
A, B, E, FIN, S	„Country of origin",EU price level, Launch sequence is decisive
Portugal	„Country of origin" + F, I, E
Denmark	Average price in Europe without GR, L, P, E
Italy	European average price
Norway	Average of the 3 lowest prices of 9 countries: A, B, DK, FIN, *D*, IRL, NL, L, UK
Greece	Lowest price in all EU countries
France	Other European prices when launching a product

Germany, UK → indirect impact

* Researchers at the University of Pennsylvania School of Veterinary Medicine have derived **uniparental embryonic stem cells** - created from a single donor's eggs or two sperm - and, for the first time, successfully used them to **repopulate a damaged organ** with healthy cells in adult mice.

* Scientists at **Roche** have discovered small molecules that activate a key tumor suppression pathway and could eventually provide a novel oral anticancer therapy. The compounds, a series of cis-imidazoline analogues are called **Nutlins**. They are MDM2 antagonists and have inhibited tumor growth in mice by 90% without producing harmful side effects.

* **UVA-1 phototherapy** allows dermatologists to use ultraviolet rays to treat patients with certain skin diseases. It works by resetting the immune system. The new therapy emits a narrow band of light that has the potential to penetrate more deeply into the skin.

Table 1.24.   France: Different reimbursement routes[52]

	Consultation fee	Healthcare Insurance (Assurance-Maladie)	Complementary Fund (the "Mutuelles")	be paid by the patient
Attending physician Sector 1				
coordinated healthcare route	20 Euro	13 Euro	6 Euro	1 Euro
non-coordinated healthcare route	20 Euro	11 Euro	6 Euro	3 Euro
children -2 years	25 Euro	17,50 Euro	7,50 Euro	0 Euro
Specialist physician Sector 1				
coordinated healthcare route	27 Euro	17,90 euro	8,10 Euro	1 Euro
non-coordinated healthcare route	32 Euro maximum amount authorized exceptions	14 Euro	7,50 Euro	5,50 Euro bis 10,50 Euro in function of the amount applicated
Specialist physician direct access				
gynaecologists & ophtalmologists	27 Euro	17,90 Euro	8,10 Euro	1 Euro
Psychiatrists for the 16-25 year old	40 Euro	27 Euro	12 Euro	1 Euro
Specialist physician Sector 2				
coordinated healthcare	free consultancy fee	15,10 Euro	7 Euro maximum	
non-coordinated route	free consultancy fee	12,80 Euro	7 Euro maximum	

This law introduces the method of health targets. Based on the population's state of health, it will determine national plans and programs which, for the sake of effectiveness, will be run locally to take account of specific regional characteristics. In addition, it will reinforce the organization of a partnership between people working in the health system, in order to develop initiatives and promote a culture of consultation and results.

The dynamics of innovation and its diffusion into medical practice is a key driver of quality and cost.

"The time has come to abandon disease as the focus of medical care. The changed spectrum of health, the complex interplay of biological and nonbiological factors, the aging population, and the interindividual variability in health priorities render medical care that is centered on the diagnosis and treatment of individual diseases at best out of date and at worst harmful. A primary focus on disease may inadvertently lead to undertreatment, overtreatment, or mistreatment. The numerous strategies that have evolved to address the limitations of the disease model, although laudable, are offered only to a select subset of persons and often further fragment care. Clinical decision making for all patients should be predicated on the attainment of individual goals and the

---

[52] *Source: Guélaud C., Le généraliste au Cœur de la réforme, Le Monde Dossiers & Documents, N°356, September 2006 :3*

78

* **AstraZeneca**'s **Nexium** (esomeprazol magnesium), the second-generation proton pump inhibitor, developed as an isomer superseded the company's flagship drug Prilosec/Losec (a 5 billion dollar drug). The product is indicated for the eradication of Helicobacter pylori, for the healing of H pylori associated duodenal ulcer, to prevent the relapse of peptic ulcers in patients with H pylori-associated ulcers, for the treatment of erosive reflux oesophagitis, for the long-term management of patients with healed oesophagitis, and for the treatment of gastro-esophageal reflux disease.

* In 2000 the biggest selling drug in the world was **Prilosec/Losec** (omeprazole), which generated –just before patent loss- $6.3bn in sales. **Zocor** (simvastatin) and **Lipitor** (atorvastatin) managed to pass the $5bn mark (resp. $5.2bn and $5.1bn). **By 2000, each of the top 25 drugs had higher sales than the number 2 drug in 1990.**

* The actions of a mutated protein in cells linked to **thyroid cancer** have been uncovered by researchers at Queen's University. The discovery paves the way for the future development of drugs to more effectively target, treat and possibly even prevent both inherited and non-inherited thyroid cancers.

* Gregory J. Hannon, elucidated the **enzymatic engine** for RNA interference.

* While many medical researchers and experts see embryonic **stem cells** as a **potential source of treatments** for **degenerative or otherwise incurable conditions**, religious inspired thinking rejects stem-cell extraction as devaluing nascent human life. Their influence resulted in the US Authorities refusing funding any research involving other than a handful of existing embryonic stem-cell lines (Scientists working without government funding mostly focus on unraveling the mysteries of how the cells produce different tissue types).. Biotechnology researchers from **Advanced Cell Technology**, report that by using a biopsy, they should be able **to derive stem cells from embryos without destroying them**. The team conducted its experiments with single cells extracted from early-stage human embryos, each consisting of eight to 10 cells. By **culturing those extracted cells in just the right way, they were able to nudge some of them into developing as stem cells, the primordial building blocks capable of producing nerves, organs and every other type of tissue.** This technique contrasts with the existing method whereby scientists, looking to derive new lines of embryonic stem cells, to suck out clumps of cells from five-day-old embryos –typically ones frozen during in-vitro fertilization procedures that may otherwise face disposal; a process that destroys embryos. Advanced Cell Technology researchers see a huge **benefit** in the technique, in that the **derived stem-cell line would be genetically match to a child born from the same embryo, making it a theoretically convenient and safe source of cells for later medical treatment.**

* The quality of medical care of chronically ill patients is quite often dissatisfactory. The task of health care research is to find out these lacunae, to explain them and counter-effect them.

➜➜ In Disease Management, the central role of outcome can be seen in the contents and time flow

➜➜ Outcome is more than the sole objective. It consists of several end-points: objective and subjective goals (quality of life) fused into clinical relevance

➜➜ Quality of life can today be measured – it can be portrayed in the case of individual patients as a quality of life profile

➜➜ Patient opinion and patient evaluation are a must

➜➜ Measures for patient awareness, psychosocial, supportive and palliative measures and rehabilitation can be directly substantiated by outcome research.

B.33.The "billion dollar club": Europe's role diminishes :

Global ranking 2001	Brand	Years on the market	Global sales in Euro mill.	European brand owner
3	Prilosec Losec	16	5.684	AstraZeneca
11	Paxil/Serokat	11	2.673	GSK
17	Augmentin	19	2.046	GSK
22	Cipro/Ciprobay	17	1.758	Bayer
24	Allegra/Telfast	7	1.577	Aventis
32	Flovent/Flixotide	8	1.317	GSK
33	Lovenox/Clexane	10	1.301	Aventis
36	Advair/Seretide	5	1.224	GSK
39	Diovan/Co.Diovan	7	1.113	Novartis
40	Zestril	10	1.097	AstraZeneca
41	Imitrex/Imigran	11	1.091	GSK
42	Neoral/Sandimmum	20	1.083	Novartis
48	Rocephin	19	1.005	Roche

* The Nobel Prize in chemistry, 2006, was awarded to Roger D. Kornberg "for his studies of the **molecular basis of eukaryotic transcription**."

* In 2001, the **epilepsy market** exceeded sales of $6bn. This segment is expected to quadruple by 2010, as companies have a series of drug candidates in their pipeline which aims to reduce the known side effects associated with the currently marketed products.

* By the study of rare families with **Mendelian** forms of severely high or severely low blood pressure, scientists at **Yale University School of Medicine** have identified **mutations** in eight genes that **raise blood pressure** and eight genes that **lower blood pressure**. Intriguingly, these genes all converge on the same final **common pathway**, altering net renal salt reabsorption, and provide **definitive evidence that mutations that increase net salt balance raise blood pressure in humans.** These studies have provided new insight into integrated physiology and pathophysiology that have impacted therapeutic approaches to hypertension in the general population; equally importantly, they have identified new targets and pathways for development of novel therapeutic agents.

identification and treatment of all modifiable biological and nonbiological factors, rather than solely on the diagnosis, treatment, or prevention of individual diseases"[53]

I.27. Health Insurance Systems: Escaping the intensified drug cost culture

Statutory Health Insurance system managers in European countries, who continue to see the idea of universal health coverage as the best model for the future, propose five fundamental building blocks for a functioning System

I.27.1. The underlying principle of solidarity is worth maintaining, because it is based on social justice. All persons have access to a wide range of health care services. Young, healthy persons with good income levels are part of the health care system and therefore ensure that the sick and elderly members with low income can avail themselves of healthcare facilities by making a low contribution. To be able to retain and maintain this "natural exchange", it will be necessary to maintain the level of contribution made by the members of the SHI schemes to a bearable limit so that they do not turn away from this solidarity-based insurance systems.

Table 1.25.   The future of Social Healthcare Insurance

I.27.2. Competition at all levels of the health system needs to be intensified. The law in most EU countries prohibits competition in just those areas where competitive elements can contribute towards improving the productivity in the health care system. Either defined competition parameters do not exist or the existing competition incentives are rather wrongly placed. Competition should take place in order to increase efficiency and effectiveness and mainly to accelerate the pace of innovations. In a functional competitive system, only those service providers and health insurance funds that strive to achieve quantifiable success in quality and economic efficiency should be able to gain advantages. Good doctors can be expected to earn well, bad doctors should withdraw from the market. The health insurance funds must also have the option of merging with the aim of expanding competitive parameters.

I.27.3. The economic utilization of reserves is a central and long-term task. Wrong incentives lead to a wasteful use of resources that are already scarce. Double examinations, extensive use of medical equipment, avoidable hospitalization and operations as well as a high number of days of hospitalization are only a few examples to show that there are economic reserves. The concern is not just an improvement in the quality of health care but also the avoidance of wrong health care in order to improve the quality of life for patients.

---

[53] *Mary Tinetti and Terri Fried, The End of the Disease Era. American Journal of Medicine. 116(3):179-185.*

* AstraZeneca's **Viozan,** a dual dopamine receptor and beta-2 agonist for the effective treatment of chronic obstructive pulmonary disease is expected to become a major product for COD disease symptoms as cough and sputum production and breathlessness.
* **Aviron**'s Flumist, the first intranasal (inhaled) flu vaccine for the prevention of influenza and its complications in children and adults, is said to offer the potential for enhanced efficacy through its live-virus characteristics.
* In 2003, FDA launched a broad new agency-wide initiative to speed development of innovative medical technologies. FDA is committed to **reducing total review time** for new drugs and biologics across the board by approximately 10.5%. Such reductions in review time will place much needed treatments in the hands of patients faster.
* Several strategies emerged to increase levels of HDL cholesterol to treat cardiovascular disease, including **nuclear receptor modulation** (Consisting of peroxisome proliferative activated receptorα, or PPARα, and PPARγ, and PPARδ. Each of these nuclear receptors controls a distinct network of target genes. And also consisting of ligand-activated transcription factors; two isoforms, **LXRα and LXRβ** have emerged as central regulators of genes that are involved in lipid metabolism), **inhibition of cholesteryl ester transfer protein** (The discovery of human genetic-deficiency states, characterized by markedly increased HDL levels and moderately **reduced LDL levels**, led to the development of drugs that inhibit CEPT) and **infusion of apolipoprotein/phospholipid complexes (**Genetic deficiencies of ApoA-I, which is the major structural apolipoprotein of HDL lipoproteins, have been observed to result in very low levels of plasma HDL and premature atherosclerosis). In the not-too-distant future, combination therapy (**statins plus HDL elevation**) may replace monotherapy with statins.
* **Intensiv Care Unit (ICU)** patients are very sick and require around-the-clock specialized care, however most ICUs don't have the specially trained physicians (intensivists) available to provide this. With an eICU facility linked via **telemedicine** and computer monitors to their hospital ICU rooms they now can. An eICU facility is staffed with an intensivist-led care team that can monitor and care for hundreds of patients much like air traffic controllers monitor hundreds of planes. In an airplane pilots also use on board sensors to identify problems and intervene to maintain a safe flight. Likewise, the eICU care team uses software alerts to track patient vital trends and intervene earlier—before complications occur. Studies show that this type of care model **can reduce ICU mortality by 25%** and save costs. The keys are constant surveillance, providing the patient with immediate physician access and arming the physician with the patient information needed to make the right decisions, quickly
* 2004, **pagaptamib** sodium, which targets vascular endothelial growth factor, became the first approved therapeutic based on an RNA structure known as an aptamer.

* The cost of bringing a new drug to market, which can take up to 15 years, tops $800mn to $1.7bn (depending on the source), including the drugs that fail.This increase results from a drop in cumulative success rates from 14% to 8% and an increase in R&D and launch costs of about 50% for each of these steps. Of every 5000 compounds, just 250 enter clinical trials. Twenty percent of those are approved.

B.34. R&D becomes more perilous and expensive as ever.

Total	R&D	Spend	in	US $bn
2006				57,0
2000				44,5
1990				20,0

Bristol-Meyers Squibb's Vanlev, said to be the first vasopeptidase inhibitor (omapatrilat) is being developed as an effective and safe new drug to treat hypertension. The company initiated a clinical trial of 25.000 patients to address the angio-oedema side effect question which was raised by the FDA regarding the comparative incidence and severity of this infrequent side effect. The company further researches the product's benefits in congestive heart failure, reducing cardiovascular mortality and morbidity, as well as potential use in the treatment of angina.

B.35. Ten year evolution of a drug's development cost :

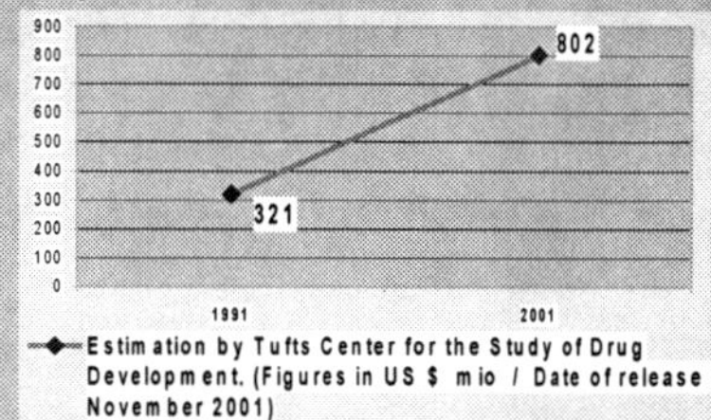

The **average cost to develop a new drug has reached US $ 802 million**, according to **The Tufts Center for the Study of Drug Development**. The Boston Consulting Group estimates this cost at US $ 880 million, bringing it closer to the 1 billion mark (other sources indicate up to US $1,7bn). Included in Tuft's analysis are expenses of project failures and the impact of long development times. In addition, the Tufts estimate accounts for out-of-pocket clinical costs, out-of-pocket discovery and preclinical development costs, clinical success, phase attrition rates, and cost of capital. After adjusting for inflation, the out-of-pocket cost per approved new medicine increased at a rate **of 7,6% per year between 1991 and 2001**. The annual rate of growth in capitalized cost between the two study periods was 7,4% in inflation-adjusted terms.

I.27.4. It is necessary to target for an objective-based services catalogue. The demand for compulsory and optional services may seem innovative at the first sight. It may seem that the fulfillment of this demand would lead to a patent solution for the problem of expenses. However, if one looks at the matter in depth, one easily comes to the conclusion that this, in effect, would mean a full reversal of the basic principles of a Social Health Insurance scheme. This starts with the fact that no-one is really in a position to define a sensible catalogue of additional services. The reason for the same is relatively simple -- the catalogue of services covered by the Statutory Health Insurance scheme does not contain any questionable services that can easily be deleted. The core problem is thus not the definition of a services catalogue as such but whether effective and necessary medical steps are being taken within its' framework. Here, governments must allow for more leeway for health insurance funds.

I.27.5. New technologies are to be used. At present, the healthcare systems suffer from the current restrictive process flows and the existing data protection laws. However, it is more than questionable, whether health insurance funds have the option of bringing together the data of insured persons and their treatment data.

For example, the health insurance card of the future could contain important medical information for the insured person such as emergency information, vaccination data, organ donation card, and other relevant data. In general, it is necessary to support all efforts that are directed towards a breakthrough of telematics in the health care systems. This includes electronic mail orders of medicine through the Internet.

The dangers of focusing a disproportionate amount of resources on high technology, curative care to the near total exclusion of primary health care; the increasing administrative burden on the system; drifting intergenerational relations; the unstoppable "social" healthcare systems financial deficits, amplified by Government's appetite for all sort of taxes on medicines and medical services; the accelerating rhythm in genetic discoveries promising new dimensions in treatment of health disorders...the perception and possibilities of maintaining the welfare state, leading to reform after reform of the Western nation's Health Care systems in an aging world and a globally linked economy is one of today's core social and economic issues, and will remain at the center of the western post-industrialized countries' political agenda for years to come.

The challenges and opportunities for the healthcare systems and the medical community are huge. We are
- facing unprecedented opportunities to enrich the therapeutic means at our disposal;
- operating in a field that is gaining importance by the day;
- being called upon more and more to satisfy the most pressing need of the present-day post-industrial population, i.e. to make people live longer and better.

To be affordable, this care will need to be cheap. Free-of-direct-cost availability will almost certainly create a huge demand for healthcare services, including medicines.

I.28. New medical-technologies
New medical-technological developments will impact on the delivery of healthcare. Genome science, high-throughput screening and combinatorial chemistry are just a few examples of new technologies that are being applied to the intelligent design of new medicines. DNA testing can subdivide patient groups into well-defined cohorts that will

* **Bristol-Meyers Squibb's Orzell** (UFT capsules plus leucovorin calcium tablets) for the first-line treatment of advanced colorectal cancer is announced to be the first oral chemotherapy enabling patients to receive chemotherapy at home..

* **Eli Lilly's Zovant** (recombinant activated protein C) for the treatment of severe sepsis is described as having a unique mechanism of action. It works as an anticoagulant, an anti-inflammatory, and a profibrinolytic.

* **Eli Lilly's Protein Kinase C-Beta Inhibitor**, a leptin analog is said to be the first oral compound for the treatment of diabetic macular edema and diabetic retinopathy. It does not affect blood-sugar levels and improves blood flow in the retina of patients with type I and type 2 diabetes.

* Researchers have decoded the genetic makeup of the parasite that causes **trichomoniasis**, one of the most common **sexually transmitted infections** (STIs), revealing potential clues as to why the parasite has become increasingly drug resistant and suggesting possible pathways for new treatments, diagnostics and a potential vaccine strategy. The genome sequencing project, funded by the National Institute of Allergy and Infectious Diseases (NIAID), part of the National Institutes of Health (NIH), is detailed in the January 12, 2007 issue of Science. Trichomoniasis is a sexually transmitted infection that affects both men and women and results in roughly 7.4 million new cases in the United States each year, according to the U.S. Centers for Disease Control and Prevention. In women, trichomoniasis infection commonly occurs in the vagina, resulting in heavy yellow-green or gray vaginal discharge, vaginal odor, discomfort during sexual intercourse and urination, irritation and itching of the genital area and, in rare cases, lower abdominal pain. In men, trichomoniasis is most common in the urethra; however, infected men often do not have symptoms. Those that do may experience irritation inside the penis, mild discharge, or a slight burning sensation after urination or ejaculation. Symptoms in both men and women generally appear within five to 28 days of exposure to the parasite.

T. vaginalis is pear-shaped with thread-like flagella that propel its movement. Once it attaches to cells lining the host's urinary or genital tract, it flattens out and begins to ingest the cells, as well as white and red blood cells, causing direct damage to the urinary and vaginal tissues and resulting in inflammation. T. vaginalis also consumes bacteria that may be present in the urinary and genital areas, including the bacteria necessary for maintaining a normal healthy environment in the vagina. As a result, women infected with trichomoniasis become more susceptible to becoming infected by HIV and other STIs. Although trichomoniasis is very prevalent as a sexually transmitted infection, it has not received the attention given to other STIs. Women are the population most affected by it, **yet routine gynecological check-ups do not test for this particular infection.**

*(re)**Importation** of drugs from Canada into the USA : the US is confronted with a business model that that flourishes already in certain European countries.

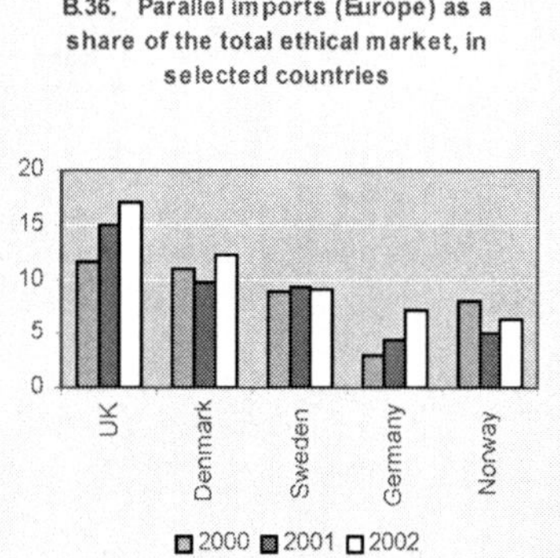

* **Value added taxes** on medicines differ from EU country to EU country. E.g.:
* Austria : 0 percent on reimbursed Rx (all other : 20%);
* Ireland : 0 percent on oral forms (all other : 20%);
* France : 2,1 percent n reimbursed Rx (all other : 5,5 percent);
* Germany : 17 percent;
* UK : 0 percent on NHS prescriptions (17,5% on self-medication);
* Denmark : 25 percent;
* Sweden : 0 percent on Rx products.

* **Parallel trade** is of concern because it has a chilling impact on innovation. Different governmentally-established price levels, when coupled with the cross-border movement of goods (in disregard of intellectual property rights in the countries of importation), create severe market distortions that penalize, rather than encourage, those who invent and develop new medicines. (Dr Henry A McKinnell, Jr, Exec VP, Pfizer Inc, at the Financial Times World Pharmaceuticals Conference, London, March 25/26, 1996)

* The **cost of new medicines** forces third payers and patients alike to adopt a more effective approach of treatments. Already today, monoclonal antibodies are much more expensive than small molecule drugs. They range from US$ 15.000 to $25.000 per year. Ex vivo gene therapy is being described as costing well over US$100.000. Certain orphan drugs e.g. genetically engineered blood clotting factor for hemophiliacs can cost as much as US$150.000 per year.

* **Jean-Pierre Changeux** discovered that fast-acting neurotransmitters mediate their effects through **allosteric regulation of the neurotransmitter protein.**

benefit from a specific drug without suffering the side-effects that are probably caused by applying drugs to too wide a population.

Table 1.26.　The need to think beyond the current systems

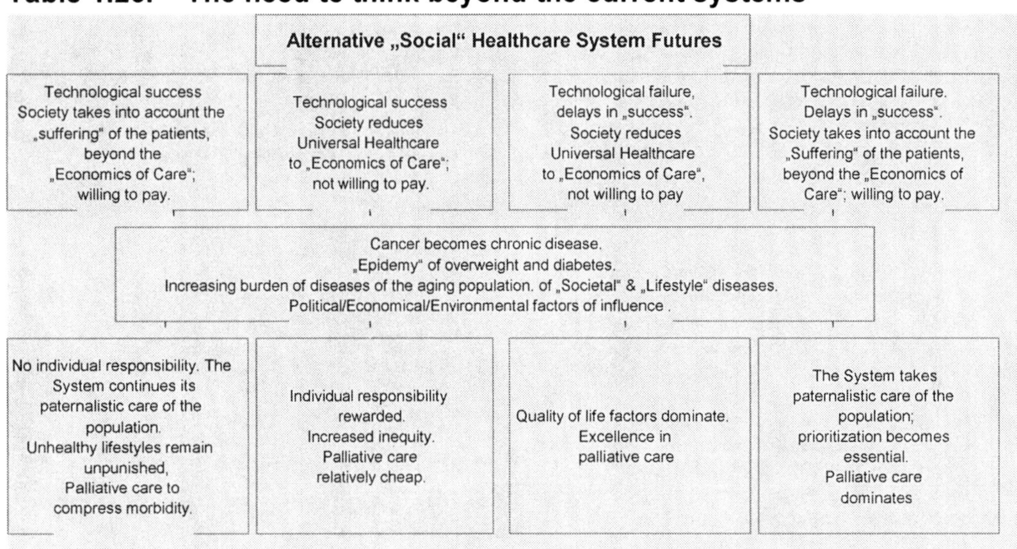

The elucidation of the genetic build-up of human being removes the need to study pathology before deciding on intervention. Genetic information can reveal a predisposition for pathology, which can be treated before the individual suffers the symptoms.

1.29. Long-distance accurate diagnosis

The medical doctor will no longer be the main repository of medical expertise for a sophisticated patient. The accelerating rate of change in medicine makes the "half-life" of knowledge too short. We are not far from the time when the most up-to-date information will be obtained from a database run by a computer expert or dedicated server and where the last missing link to semi-automated medical care is long-distance accurate diagnosis, something that has already been proven to be feasible.

The full application of our data storage and handling capabilities would already make this possible today. When it comes, the change-over will be very fast because the economic and quality considerations are overwhelming and the data transfer facility (the Internet) is already in place. It could also be the only way to provide the universal affordable health provision as mentioned before[54].

I.30.　Data-driven classification scheme for medicine

Computers aren't about to replace doctors, but their immense processing powers will uncover hidden connections between diseases and could lead to better medical care.

Doctors are getting increased access to tests that can distinguish between the most and least aggressive cancers. In turn, more patients will get personalized information about their tumors in the near future, which will help them get treatment tailored to their

---

[54] *Neirinck R. & Mertens G., Mergers and Acquisitions in pharmaceuticals. Why and How? Financial Times Pharmacutical Management reports, London, January 2000:28*

* More than 26 million people worldwide were estimated to be living with Alzheimer's disease in 2006, according to a study led by researchers led by Prof. Brookmeyer at the Johns Hopkins Bloomberg School of Public Health. The researchers also concluded the global prevalence of Alzheimer's disease will grow to more than 106 million by 2050. By that time, 43 percent of those with Alzheimer's disease will need high-level care, equivalent to that of a nursing home..

According to the scientists, interventions that could delay the onset of Alzheimer's disease by as little as one year would reduce prevalence of the disease by 12 million fewer cases in 2050. A similar delay in both the onset and progression of Alzheimer's disease would result in a smaller overall reduction of 9.2 million cases by 2050, because slower disease progression would mean more people surviving with early-stage disease symptoms. However, nearly all of that decline would be attributable to decreases in those needing costly late-stage disease treatment in 2050.

* New potential drugs make it to clinical trials. Much hope is put in an **anti-amyloid** (Alzhemed), **an inhibitor of brain cell death** (Dimebon), an **"Alzheimer's vaccine"** (Immunotherapy Treatment AN1792), and **a drug normally used to treat diabetes (**Avandia)[1].

Alzhemed is an orally-administered amyloid beta antagonist that is currently in Phase III clinical trials to assess its safety, efficacy and disease modifying effects in patients with mild to moderate Alzheimer's. **Alzhemed** binds to amyloid beta protein and interferes with its ability to build plaque and poison brain cells.

Dimebon is an orally-administered drug that has shown ability to stop brain cell death in preclinical testing for Alzheimer's and Huntington's. The compound appears to offer brain cells protection from amyloid build up by blocking something that targets their mitochondria. Mitochondria supply cells with energy and are also involved in programmed cell death (apoptosis) which is associated with neurodegenerative diseases like Alzheimer's and aging.

AN1792 is a synthetic form of the amyloid beta protein which was used in an immunotherapeutic clinical trial that was stopped because 6 per cent of the patients began to suffer from inflammation of the brain (encephalitis). However, the researchers followed the patients after the trial and observed that 4.5 years after immunization with **AN1792**, patients who had developed antibodies to amyloid beta continued to show detectable levels of antibodies and slower decline in daily living compared with patients treated with placebo. They claim that the favorable results on Activities of Daily Living among the antibody responders in their study support the hypothesis that amyloid beta immunotherapy may have long-term benefits for patients with mild to moderate Alzheimer's and their caregivers.

Scientists have speculated that the type 2 diabetes compound **rosiglitazone (Avandia)** may also be able to help Alzheimer's patients because of its effect on brain inflammation and other processes associated with neurodegenerative diseases. *(Presented at the 2nd Alzheimer's Association International Conference on Prevention of Dementia, Washington DC, June 11, 2007)*

* Deficits in **cholinergic neurotransmission**, in particular a reduced number of nicotinic receptors, are an important contributor to the cognitive impairment that is typical with **Alzheimer's** dementia -- a disease that afflicts an estimated 15 million individuals worldwide.

To slow the progression of the symptoms associated with this disabling disease, Janssen recently introduced **galantamine (Reminyl),** a novel therapy that has a unique, dual mode of action. Galantamine inhibits the enzyme acetylcholinesterase, which breaks down acetylcholine, a neurotransmitter that plays a key role in memory and learning. The drug also modulates nicotinic receptors, an action that is believed to result in greater release of this critical brain chemical.

* **Biofrontera Pharmaceuticals** and **Janssen Pharmaceutica** explore modulation of nicotinic receptors in treatment of Alzheimer's disease. Their goal is to further investigate, substantiate and explain how these two mechanisms might contribute to the broad efficacy profile and sustained therapeutic benefit that **Reminyl** has demonstrated in recent clinical trials.

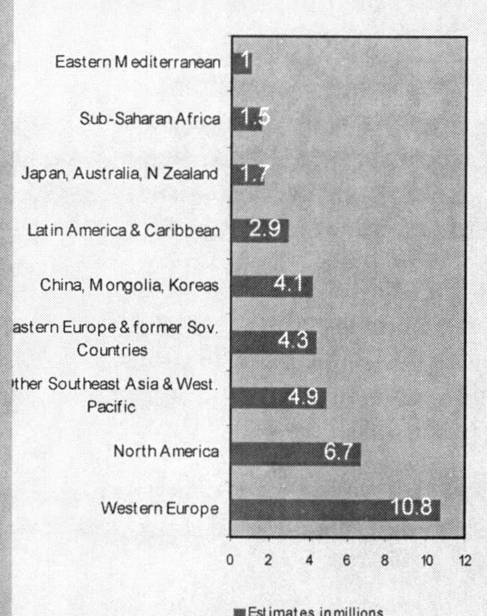

AstraZeneca develops a perisome proliferators-tivated receptor (PPAR); an **oral insulin sensitizer** ving a possible superior potency to the glitazones in proving glycemic control and glucose intolerance.

* Healthy women with no history of heart disease or stroke significantly increase their chances of having a stroke if they have **high cholesterol** *(according to a study of more than 27,000 women published in the February 20, 2007, issue of Neurology, the scientific journal of the American Academy of Neurology.)*

particular cancer. But "doctors currently still classify diseases by a century-old scheme that is more useful for billing than treatment", said Stanford professor Atul Butte. The scheme, for instance, lumps most cancers together, even though they may have little in common besides uncontrolled tumor growth. His lab has used computers to crunch billions of measurements from 75 human diseases and map out molecular similarities. "The idea here is to come up with the first data-driven classification scheme for medicine," explained Butte, who's work is made possible by microarrays, a snapshot of the genome that tells scientists which genes are turned on and off and by how much. With microarrays, scientists can glean details about a disease, such as breast cancer, that could never be noticed examining a biopsy sample under the microscope.

Scientists have used microarrays to study nearly every human disease and made the data available on the Internet, thanks to the open-door policy advanced by biochemistry professor Dr. Pat Brown, and other leaders in the field. Any high school kid today can go to the Web site and download 108,000 microarrays. It's unbelievable how much data that is." So much, says Butte, that the only way to make sense of the numbers is with the muscle of powerful computers. He and other researchers have analyzed that data to connect the molecular dots between diseases nobody suspected were linked. This is how he discovered that muscular dystrophy and heart attack had strikingly similar molecular signatures.

With that knowledge, researchers can test whether any of the myriad drugs available to treat heart attack would work on muscular dystrophy, which has no treatment. "The beauty of this approach is that we can use it for existing drugs," he said. Not to mention that it could sidestep the prohibitive costs often involved in developing new classes of medications[55].

I.31. Regenerative medicine

Regenerative medicine may well be the future of health care: Its goal is to refurbish diseased or damaged tissue using the body's own healthy cells. It's a term that is often used synonymously with tissue engineering, though those involved in regenerative medicine place more emphasis on the use of stem cells to produce tissues.

Scientists are currently working on tissue replacement projects for practically every body part - blood vessels and nerves, muscles, cartilage and bones, esophagus and trachea, pancreas, kidneys, liver, heart and uterus. The long-term hope is that patients someday may not have to wait on the national transplant list for donor organs. Organs could be tailor-made for people.

Imagine that a patient needs a new liver. Rather than waiting for a donated organ, doctors simply take some stem cells from the patient and grow a new one. Sound pretty far-fetched? Maybe not.

Stanford's Professor Gurtner is looking at a five-year time horizon, at least. His laboratory is seeking to create complex tissues for organ replacement, and is now working on creating a synthetic replacement liver for people with liver failure. "The goal is to seed tissue with stem cells harvested from the patient that will differentiate into liver cells"[56].

---

[55] *The year ahead: Stanford experts forecast medical trends to watch in 2007 News release. Stanford University Medical Center, Office of Communication & Public Affairs, December 20, 2006.*

[56] *According to Prof. Geoffrey Gurtner. The year ahead: Stanford experts forecast medical trends to watch in 2007 News release. Stanford University Medical Center, Office of Communication & Public Affairs, December 20, 2006.*

* 2004, researchers from **Howard Hughes Medical Institute** demonstrated that human **astrocytes** –brain cells thought to play a secondary role by providing a nurturing environment for the neuron- can actually function as stem cells.

* The **Japanese Ministry of Health and Welfare** reduced the prices of drugs in April 2000 by an average of 7%. This move, coupled by a background of general financial stringency among health insurers, made competition fiercer in the industry.

* **Genentech**, has already reached such a high level of expertise in the manufacture of protein-based therapeutics that it now controls half the world's bioreactor capacity. This leads to **Amgen**, the world's largest biotechnology firm, to be dependent on Genentech for manufacturing help. This new collaborative model could have important implications for the wider industry, as biopharmaceutical companies specialize in certain core competencies such as batch fermentation, and begin to provide outsourced services to their peers.

* When the **statins** entered the pharmaceutical marketplace, they transformed the cardiovascular product sector from one that **treats symptoms** to one that **slows the progression** of heart diseases and ultimately contributed to reduce mortality. They also became the lifeblood of industry giants like Pfizer (**Lipitor/atorvastatin**), Merck (**Zocor/simvastatin**) and Bristol-Meyers Squibb (**Pravachol/pravastatin**).

B.38. Sales of cardiovascular (hypertension/high cholesterol) medicines in the early 2000s

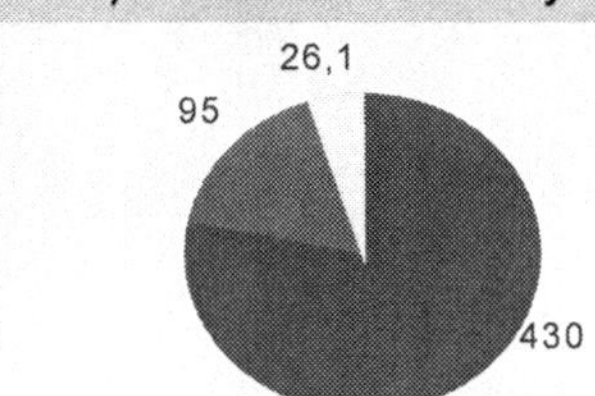

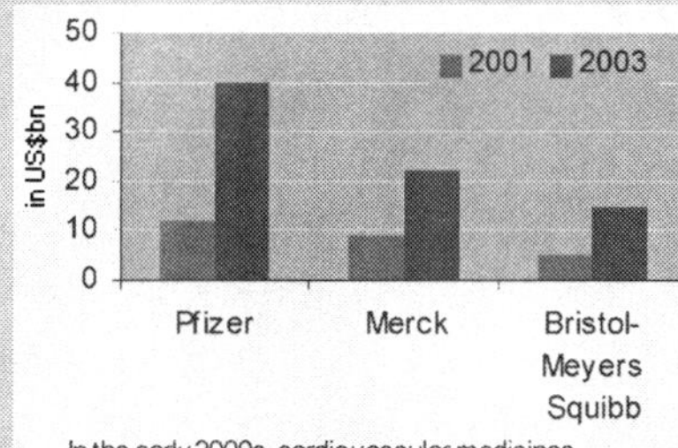

In the early 2000s, cardiovascular medicines became BigPharma's lifeblood

Cordis Cardiology, a unit of **Johnson & Johnson**, offers patients the benefits from a breakthrough technology of the **Cypher stent**; the first and only commercially available drug-device combination to significantly reduce the likelihood of patients requiring treatment for restenosis in the coronary arteries. The company revealed that this significant migration from bare metal stents to drug-eluting stents in the interventional cardiology arena benefit to nearly 500.000 patients that have been treated with the Cypher stent in more than 80 countries around the world in just two years (2002/2003).
In 2003, **Boston Scientific Corp**. arrived on the drug eluting stents market, with extremely good results from its **Taxus** trials. Other players , most notably **Guidant Corp**. and **Medtronic** Inc., announced their intention to heat up the competition. J&J´s **REALITY trial**, comparing its own stents to those of Boston Scientific promises more noise in the medical devices sector.

* In Europe, Sick Funds promote the idea of **prevention** but fear the immediate financial burden. Short-term thinking leads to substantial costs later : e.g. ten percent of the Euros spent by the Sick Funds go into a very expensive acute cardiologic treatment ...because for many years, a large number of people with heart disease remained untreated, resulting in a highly expensive, but inevitable, life-saving intervention later

*A chance for Biotech : BigPharma´s R&D problem. Pharmaceutical companies are seeking to cut development costs and shorten development periods --in part, to reduce the development risk. For that reason, smaller biotechnology companies are more and more often assuming some of the research work for large pharmaceutical groups, pushing ahead with selective developments in their own core areas of competence. Some 8 out of 10 biotechnology companies perform research, development or production for pharmaceutical applications. The drop in R&D output experienced **by BigPharma** is being considered as an opportunity for **Small** and **BigBiotech** alike. The risk for Biotech companies is that they will be bought by the big pharmaceutical companies eager to fill the R&D gap.

Biotechnology is said to be a collection of technologies that capitalize on the attributes of cells and biological molecules, such as DNA, to work for us. The sector generated sales of over US$ 60 billion in 2005, an 18% increase over 2004. Biotechnology's most obvious application is in healthcare. It relies on technological revolutions and breakthroughs and is behind the most promising therapies, such as stemm cell research.

Leading biotechnology platforms that emerged over the last 15 years are the
* recombinant technology,
* the monoclonal antibodies,
* genomics/proteomics.
Smaller biotech companies are based on the three basic technology platforms,

Reports in April 2006 of the first successful transplant of laboratory-grown bladders into children and teenagers at Wake Forest University School of Medicine thrust the field of regenerative medicine into the headlines and stirred up hopes for future successes for other organ transplants. The patients, who suffered bladder disorders, received new bladders grown from their own cells.

For science to advance, researchers need to have a vision of what the future may hold. Many scientists have built their careers on being able to foresee the critical trends in medical science, in fields ranging from stem cells to genetics to health-care policy.

Turning scientific progress in therapies to prevent or cure disease is what the medical community expects from the pharmaceutical industry. What about the vision of those who have the experience of making drugs?

Is the pharmaceutical industry ready to tune in to a changed new healthcare world? What changes will the medical community itself undergo?

Chapter Two

Medical Community
at a
Crossroads

II.1. Healthcare economics

The driving force behind the development of health economics is the increase in healthcare costs worldwide, no matter which health care system is concerned.

Not too long ago, the cost of medicines did not matter that much, and only the efficacy, safety and quality of the pharmaceuticals involved were the parameters any decision was based on. But particularly for pharmaceuticals, costs have become a key issue. And, consequently, pressure on prices for the pharmaceutical industry's products has increased dramatically. Doctors face cuts in their already modest (when compared to other professions and to their responsibility) revenues; the pharmaceutical industry is being blamed for its "excessive " promotion and prices.

Table 2.1. Physician income: Average annual salary, by country[57]

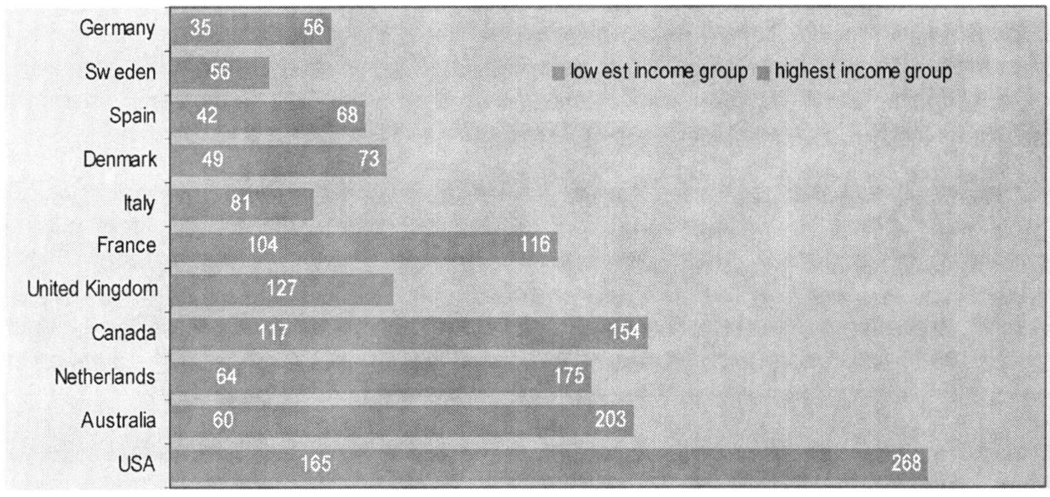

(Note: Figures in 000 US$; estimates for 2002)

---

[57] *Source: National Economic Research Associates (NERA), as published by the Frankfurter Allgemeine Sonntagszeitung, March 19, 2006:42*

Rising healthcare costs have compelled many Health Authorities worldwide to adopt defensive reimbursement policies, combined with more complex (e.g. co-payment, price regulations) market access. A philosophy of health economics reigns. At the end of the day it comes to short term budget relief in order to guarantee long-term access to medical innovation.

II.2. Evidence Based Medicine?

Patients want „care" when they suffer. Not rational budget talk.

The current system makes Evidence Based Medicine not popular in the eyes of the population. There is serious discussion about the pros to avoid immediate costs for the treatment of mild problems versus the cons of later high-cost burden for health problems that have not been treated in time to modest financial intervention. The short-term profit seems to win that philosophical question.

Evidence Based Medicine is not necessarily an obvious concept. Definitions vary. Proponents define it as follows :

'Evidence based medicine is the conscientious, explicit, and judicious use of current best evidence in making decisions about the care of individual patients. The practice of evidence based medicine means integrating individual clinical expertise with the best available external clinical evidence from systematic research".[58]

'Evidence-based medicine (EBM) is an approach to health care that promotes the collection, interpretation, and integration of valid, important and applicable patient-reported, clinician-observed, and research-derived evidence. The best available evidence, moderated by patient circumstances and preferences, is applied to improve the quality of clinical judgments".[59]

It is described as being dependent on the use of randomized controlled trials, as well as systematic reviews (of a series of trials) and meta-analysis, although it is not restricted to these. There is also an emphasis on the dissemination of information, as well as its collection, so that the evidence can reach clinical practice. It therefore has commonality with the idea of research-based practice.

By individual clinical expertise we mean the proficiency and judgement that individual clinicians acquire through clinical experience and clinical practice. Increased expertise is reflected in many ways, but especially in more effective and efficient diagnosis and in the more thoughtful identification and compassionate use of individual patients' predicaments, rights, and preferences in making clinical decisions about their care.

By best available external clinical evidence we mean clinically relevant research, often from the basic sciences of medicine, but especially from patient centered clinical research into the accuracy and precision of diagnostic tests (including the clinical examination), the power of prognostic markers, and the efficacy and safety of therapeutic, rehabilitative, and preventive regimens. External clinical evidence both invalidates previously accepted diagnostic tests and treatments and replaces them with new ones that are more powerful, more accurate, more efficacious, and safer.

Good doctors use both individual clinical expertise and the best available external evidence, and neither alone is enough. Without clinical expertise, practice risks becoming tyrannized by evidence, for even excellent external evidence may be

---

[58] *Sackett, D.L. et al. (1996) Evidence based medicine: what it is and what it isn't. BMJ 312 (7023), 1996, Jan. 13 ;312 :71-72*

[59] *McKibbon, K.A. et al. (1995) The medical literature as a resource for evidence based care. (Working Paper from the Health Information Research Unit, McMaster University, Ontario, Canada*

* **biological systems research** seeks to increase the understanding of the function, control mechanisms and behavior of biological systems. In order to gain a comprehensive understanding of biological systems, complex biological functions will be simulated; experimental, quantitative data generated and models developed that are suitable for the simulation of physiological processes in cells, cell aggregates and whole organisms. The knowledge generated may be used in a wide range of applications, for instance, the development of new pharmaceuticals and therapies.

* **Helsinn Healthcare** SA of Switzerland developed **Palonosetron**, a **selective 5-HT3 receptor antagonist** with strong binding affinity and an extended plasma half-life of approximately 40 hours. Results from Phase III clinical trials demonstrate that a single intravenous dose of palonosetron is effective in preventing both acute and delayed chemotherapy-induced nausea and vomiting in patients receiving moderately emetogenic chemotherapy. **Palonosetron** is the only 5-HT3 blocker to be approved for this indication by the US FDA. It is marketed as **Aloxi** by **MGI Pharma** in the US and as **Onicita** by **Italfarmaco**.

* **Mid-size pharma companies** seldom reach the international news headlines. Schwarz Pharma did. The company could expect and indeed got a boost in US sales with the approval of its omeprazole in the US, late 2002. **Schwarz Pharma** was the only one of four generics manufacturers found not to infringe **AstraZeneca´s omeprazole** (US) patents. It decided to make a deal with **Genpharm (Merck KGaA)** and Andrx that enabled the US Food and Drug Administration to grant immediate approval of its generic omeprazole. (Genpharm and **Andrx** held US marketing co-exclusivity on generic omeprazole, but both were found to infringe AZ´s patents and were thus unable to launch their versions of the product. Under the deal with Germany's Schwarz Pharma – who is represented in the USA by its wholly owned subsidiary **Kudco** – they gave up their 180-day marketing exclusivity).

* HHMI researchers have discovered that a gene commonly mutated in a wide range of cancers can single-handedly trigger pre-cancerous changes in cells. The discovery demonstrates that the oncogene, **K-ras,** can initiate tumor development in ways that were previously unappreciated. The finding that mutant K-ras alone can cause rapid cell proliferation challenges one of the central tenets of cancer biology -- that the cooperative action of two oncogenes is required to initiate the transformation of cells to a cancerous state.

* With the aging baby boomer population expected to place unprecedented demands on the health care system, the US Department of Labor predicts an **additional** 5.3 million health care workers **will be needed** by 2010 (2.2 million replacements, 3.1 million new positions).

* The **pharmaceutical industry** in **Germany** has seen the decline of a giant, **Hoechst,** and the emergence of a new industry leader, **Boehringer Ingelheim;** a company playing by non-stock market/analysts dominated rules. Contrary to stock market listed companies, Boehringer avoids the quarterly "game" of announcing a number of preclinical and phase I drug candidates. Too risky is the business of drug discovery. Too many candidates drop out. Too difficult is the prediction of what the market will be in eleven years.

Boehringer Ingelheim´s R&D focused on NCE and new biological entities, such as **therapeutic proteins or gene therapies**. Projects from the latter category draw particular advantage from the company's in-house biopharmaceutical manufacturing expertise.

The modern era of biotechnology, supported by genetic engineering, began at the end of the 1970s with the manufacture of therapeutically active proteins by genetic engineering. Where Germany discussed issues relating to biotechnology and many of the country's leading figures questioned Germany's role in a technology of which they seemed not exactly to understand the great value for mankind, Boehringer Ingelheim took **a pioneering role**, producing interferon beta, interferon omega and Namalwa interferon from **mammalian cell culture** and interferon alpha, interferon gamma and manganese superoxide dismutase from **microbiological sources**. The company has used recombinant DNA techniques since 1978 and to date has commercially manufactured such clinically valuable proteins **as rt-PA, TNK-tPA** and others :

Front-line research and development projects are the means by which therapeutic goals are achieved. Demonstrating its commitment to innovation, Boehringer Ingelheim's **R & D expenditure** in Prescription Medicines increased from Euro 743 in 1997 **to Euro 1bn+ in 2003** Sales development of Boehringer Ing. **new product introductions** by percentage of prescription medicines gross sales

* Novartis notes that the company´s **Prexige** (lumiracoxib), launched in the UK early 2006, has been extensively studied with a clinical data base of **more than 34,000 patients** (the largest body of evidence supporting the launch of a non-steroidal anti-inflammatory drug). These studies show that the compound decreases pain to an extent similar to some NSAIDs and COX-2 selective inhibitors but has a **gastrointestinal safety profile superior** to that of two much-used NSAIDs, **ibuprofen** and **naproxen**.

* The vast majority of cataracts are the result of the damage that occurs to the lens as part of the aging process. Cataract surgery is the removal of the clouded natural lens and replacement with an artificial lens. Surgeons performed **nearly three million cataract surgeries** in the U.S. in 2005. The U.S. FDA approved **Nevanac** (Nepafenac 0.1% ophthalmic suspension) for the treatment of pain and inflammation associated with cataract surgery.

inapplicable to or inappropriate for an individual patient. Without current best evidence, practice risks becoming rapidly out of date, to the detriment of patients[60].

Criticism has ranged from evidence-based medicine being old-hat to it being a dangerous innovation, perpetrated by the arrogant to serve cost-cutters and suppress clinical freedom. EBM is indeed often directly linked to managed care. Or, better said: Manage Without Care.

When health insurance companies refuse to pay for treatments, for even simple „care", neither doctors nor the patients can be blamed for describing evidence-based medicine as just cost-cutting medicine. The care of the suffering has become the care of the socialized healthcare system.

II.3. The overlooked economic impact of two types of health improvements

A 2006 study calculates the prospective gains that could be obtained from further progress against major diseases. The authors estimate that even modest advancements against major diseases would have a significant impact - a 1 percent reduction in mortality from cancer has a value to Americans of nearly $500 billion. A cure for cancer would be worth about $50 trillion.

The authors distinguish two types of health improvements - those that extend life and those that raise the quality of life. They state that as the population grows, as incomes grow, and as the baby-boom generation approaches the primary ages of disease-related death, the social value of improvements in health will continue to rise."

Many critiques of rising medical expenditures focus on life-extending procedures for persons near death. By breaking down net gains by age and gender, the authors show that the value of increased longevity far exceeds rising medical expenditures overall. Gains in life expectancy over the last century were worth about $1.2 million per person to the current population, with the largest gains at birth and young age.

"An analysis of the value of health improvements is a first step toward evaluating the social returns to medical research and health-augmenting innovations. Improvements in life expectancy raise willingness to pay for further health improvements by increasing the value of remaining life," write the authors, Murphy and Topel. They also chart individual values resulting from the permanent reduction in mortality in several major diseases – including heart disease, cancer, and diabetes. Overall, reductions in mortality from 1970 to 2000 had an economic value to the U.S. population of $3.2 trillion per year[61].

II.4. Efficiency of health

The economics of care have also changed. In the US, insurance firms have made it difficult to get on with the business of actually "treating" the patients. Managed care organizations such as health maintenance organizations, preferred providers and independent physician associations, as well as other payers, have developed alternative payment models to control costs, including procedure-based limitations, contractually-approved providers and capitation (a fixed fee for members as payment for all required services).

---

[60] *David L. Sacket, Professor, NHS Research and Development Centre for Evidence-Based Medicine, Oxford, et al. In an article based on an editorial from the British Medical Journal, BMJ, 1996, Jan.13; 312: 71-72*

[61] *Suzanne Wu, University of Chicago, 5 May 2006. Press release announcing the publication of a new study by two University of Chicago researchers in a May 2006 issue of the Journal of Political Economy.*

* Icos and Eli Lilly' s Cialis, a phosphodiesterase type 5 inhibitor for the treatment of erectile dysfunction has demonstrated that up to 88% of men taking 25 milligrams of therapy reported significantly improved erections in a study of patients with mild-to-severe erectile dysfunction, compared with 28% of this taking placebo. Since its launch in the US, end of 2003, Cialis (tadalafil) achieved $27.9bn in sales. In Europe, it was introduced early 2003, in Mexico in August and in Canada in November 2003. For the Lilly Icos joint venture, and for Eli Lilly in particular, this is a promising start for their erectyle dysfunction treatment, which is competing heavily with other PDE5 inhibitors, sildenafil (Viagra, from Pfizer) and vardenafil (Levitra, from GlaxoSmithKline/ Bayer).

* ``Genomic sequencing and other classical genomics technologies have identified genes and proteins, and have established some aspects of their function. The Tuebingen large scale systematic and comprehensive genetic screens link genes, for the first time, to an actual biological function like organ formation, or patho-physiological process such as cardiac arrhythmia. The next step for Exelixis and Artemis is to see which of these genes are most likely to respond to new drugs to treat diseases related to the heart, bone, blood vessels, immune and nervous systems." *Dr. George A. Scangos., president and CEO,* **Exelixis** (The Tuebingen 2000 Screen is the largest vertebrate genetic screening project ever conducted. More than 10 million zebra fish were examined for mutations affecting several organs and systems. The size of the screen makes it likely that every essential gene in the genome was analyzed several times. More than 6,500 fish were identified with alterations in the heart, blood, blood vessels, bone, cartilage, nervous system and immune system. As a result, for the first time in a vertebrate, it is possible to systematically analyze every gene for its impact on organ structure and function as well as important disease processes)-

* Researchers from Imperial College London, the Royal Brompton Hospital, London, Guys Hospital, London, McGill University Hospital Center, Montreal, and St Barts and the Royal London Hospitals Trust reported that they have discovered an **important link** in the **development** of the **body's response** to **allergic asthma**. They have found that one type of white blood cell, an **eosinophil**, which was known to cause inflammation of lung airways, is also responsible for driving the process which leads to an **excessive 'repair response'** by the body. The response, which is called **airway remodeling**, causes structural changes in the airway walls and can sometimes lead to permanent scarring and narrowing of the airways, resulting in worse and repeated asthma episodes for sufferers.
The team of scientists found that the damaging effects of eosinophils in the remodeling process can be significantly reduced by injection of a single specific antibody. Their research shows that the **monoclonal antibody anti-Interleukin-5 (mepolizumab)** both reduces the number of eosinophils in the bronchi and significantly decreases the deposition of special proteins associated with the remodeling process.

*The era of Innovation Led Therapies** emerges. Drug Discovery : launch of NMEs, NCEs through analog research, NDDS and emerging therapies; R&D services : research services, such as medical chemistry, bioinformatics, protein services / development services, including chemical synthesis, clinical research and data management.

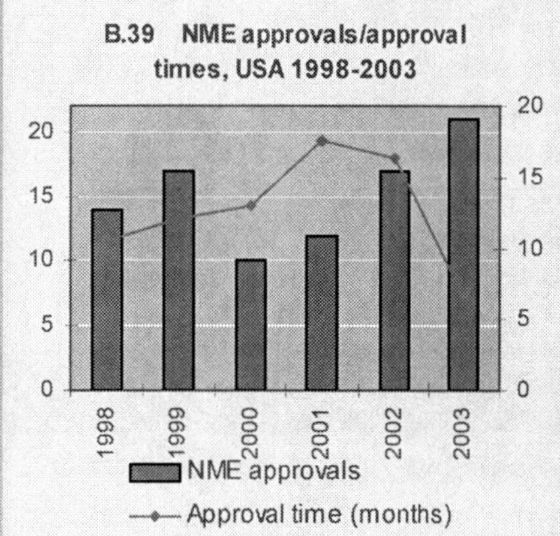

(*sources : Prof. Dr. Trevor Jones, ABPI, based on company data, FDA, Goldman Sachs Research estimates for the period to 2001. FDA data for 2002-2003, released Jan 15,2004*)

* **Biologic response modifiers.** Recent advances in biotechnology have produced novel approaches to treat arthritis. Biologic response modifiers are genetically engineered substances used to reduce the signs and symptoms of rheumatoid arthritis. Drugs in this class include **etanercept**, **anakinra**, and **infliximab. Etanercept**, the first in this new class of drugs, acts by inhibiting tumor necrosis factor (TNF), one of the proteins that play an important role in the cascade of reactions that causes the inflammatory process of rheumatoid arthritis, resulting in significant reduction in inflammatory activity. **Infliximab**, a monoclonal antibody, also inhibits TNF. **Anakinra** is a recombinant IL-1 receptor antagonist that also modifies the inflammatory response.

* **Sir Gustav Nossal** discovered the magic **'one cell-one antibody' rule** which led to the development of effective new therapies for heart disease, breast cancer and severe arthritis.

* **Dr. Gerardo Ferbeyre** and colleagues at the University of Montreal report that the DNA damage response pathway is a **necessary mediator** of oncogene-induced senescence. Using RNAi and a cell culture model of oncogene-induced senescence, the researchers determined that the DNA damage signaling pathways links oncogene expression (either STAT5A, E2F1 or Ras) and cellular senescence. The scientist expects that by identifying DNA damage as an early and persistent response to oncogenes, his work has implications for novel research approaches aimed to prevent cancer formation.

The result has been a continuing shift of financial risk from the payers to both the physician/provider and the institutional provider (hospitals, clinics, long-term care. Acute providers and rehabilitative care centers).

Table 2.2. "Economies of healthcare" vs. "Care for the suffering"[62]

Doctors and patients concentrate on "care for the suffering"		Social Health Security concentrates on "economies of healthcare"	
Most physician prescribed drugs:		Most "expensive" drugs to the System:	
Brandname/Type of drug	units prescribed	Brandname/Type of drug	in Euro, millions
Doliprane (painstiller)	73,3 million	Plavix (cardiovascular diseases)	357,3
Efferalgan (painstiller)	42,5 million	Tahor (anticholesterol)	307,7
Dafalgan (painstiller)	35,5 million	Sérédite (antiasthmatic)	210,9
Levothyrox (hypothyroid)	16,6 million	Elisor (anticholesterol)	162,9
Spasfon (painstiller, mainly stomach)	14,2 million	Inexium (antiulcer)	128,3
Diantalvic (painstiller)	12,8 million	Vasten (anticholesterol)	123,4
Daflon (venotonic)	11,8 million	Triatec (antihypertensive)	122,6
Temesta (anxiolytic)	8 million	Mopral (antiulcer)	115,9
Stilnox (hypnotic)	7,6 million	Zocor (anticholesterol)	106,4
Effexor (antidepressant)	6,4 million	Glivec (anticancer)	101,1
Far over 150 million units were prescribed for painstillers.		With over 1 billion Euro in prescriptions, anticholesterols (statins) top the reimbursement list	

(Note: Figures for France, given as an example 2005)

In response, healthcare providers are aligning themselves with one another in order to form integrated delivery systems in an effort to lower costs and compete more effectively in the changing healthcare environment.

In Europe, universal healthcare systems continue to be subject to various economic, regulatory and operational pressures. The focus shifted away from universal healthcare for the suffering patient to the care of the ailing healthcare system, which is itself suffering from the threat of financial collapse, but is "kept alive" by increased taxes. Meanwhile, improved treatment options continue to become available. They increase the burden already put on Europe's still generous reimbursement schemes (Improvements come at a price!).

In this situation, ethics and human suffering run headlong into direct conflict with a healthcare system in need of general overhaul.

The picture that European health systems portray today is the result of long years of clashes of the interests of different parties where each of these groups has tried and continues to try to defend its stand, by all possible means. Whenever some politician makes a tiny effort to reform, a national outcry and the fear of voter's penalizing vote at the next election, will block the process.

II.5. Protecting citizens from the consequences of their own lifestyle decisions: an infantilizing social democratic grip on the active individual

Europe's social democratic paradise is maternal: protective but also infantilizing. Its high taxes and benefits discourage anybody from doing too well, while ensuring that

---

[62] Source:Delberghe E, Pas d´équilibre de la Sécu à l´horizon 2007.Le Financement. Le Monde Dossiers 6 Documents, No 356, Sept. 206 :8

* **Merck** & Co. **Puts much hope in Invanz.** The effect of this once-daily (long-acting) compound could be superior to third-generation cephalosporins. It is resistant to beta lactamases that break down certain antibiotics before they can exert their microbial effects. Invanz is being developed as IV and IM injections for treating bacterial infections, such as community-acquired pneumonia.

* **Otsuka**'s dopamine receptor agonist/antagonist Aripripazole, a co-development with **Bristol-Meyers Squibb**, is expected to overtake older schizophrenic treatments because of it apparently has a lack of serious side effects (e.g. cardiotoxicity and weight gain) associated with these other drugs.

* The **European Commission** estimates that Europe spent 1.9bn Euro ($2.4bn) in 2004 on **nanotech** research and development, with almost three quarter of it coming from public sources, compared with $3.6bn for the USA (including $1.5bn public funding) and 2$2.8bn for Japan (of which $900m came from public funding). **Nanotechnology** involves the exploitation of materials less than a millionth of a millimeter in scale.

* A list of the 10 most promising treatments for the world's biggest health threats has been put together by the Scientific American journal, based on products that have at least completed Phase I trials. **Alzhemed** (tramiprosate), developed by Canadian biopharmaceutical firm Neurochem, is described as targeting the root of Alzheimer's disease and is possibly "the vanguard of a novel class of treatments." Sanofi-Pasteur, the **vaccines business** of France's Sanofi-Aventis group, is developing a vaccine against dengue fever (live attenuated 17D yellow fever and dengue chimera), a disease which affects 500,000 people worldwide that currently has no cure. Inhalable insulin, such as the **Technosphere Insulin system** developed by the USA's MannKind Corp (Pharma *Marketletter* September 25, 2006), helps diabetics regulate their blood sugar without the use of needles. **Naproxcinod** (HCT 3012) from France's NicOx is considered a potential replacement for Merck & Co's withdrawn COX-2 inhibitor Vioxx (rocecoxib). For lung cancer, **Stimuvax** (BLP25 liposome vaccine; Pharma *Marketletter* July 31, 2006), from Canadian cancer vaccine specialist Biomira, works by "teaching" the body to fight the cancer without the traditional side-effects of older therapies. UK drug major GlaxoSmithKline's **Mosquirix** (RTS,S/AS02A; *Marketletter* October 25, 2004) is a promising malaria vaccine candidate. US firm Nabi Biopharmaceuticals'**NicVAX** (nicotine conjugate vaccine) is intended to be a vaccine against the addictive component of tobacco. The HIV Prevention Trials Network project 046 in four African countries of **Viramune** (nevirapine), from German drugmaker Boehringer Ingelheim, is a Phase III trial to determine the efficacy and safety of an extended regimen of nevirapine in infants born to HIV-infected women to prevent vertical HIV transmission during breastfeeding. The first preventative hepatitis C vaccine, Swiss drug major Novartis' **E1E2/MF59** and US ophthalmic drug company Acuity Pharmaceuticals' visior loss treatment **bevasiranib** complete the list.

*Some of CBER's **priority approvals** included: **Fabrazyme** (agalsidase beta) for use in patients with Fabry disease to reduce GL-3 deposition in kidney cells; **Aldurazyme** (laronidase) for treatment of patients with Hurler, Hurler-Scheie and Scheie forms of Mucopolysaccharidosis I (MPS I); **Bexxar** (tositumomab and iodine I 131 tositumomab) for treatment of patients with CD20 positive, follicular, non-Hodgkin's lymphoma; **Zemaira** (human alpha-1proteinase inhibitor) for use in individuals with Alpha-1-Antitrypsin Deficiency and evidence of emphysema; **FluMist** (an influenza vaccine that is the first nasally administered vaccine to be marketed in the United States); and **BabyBig** (human botulism immune globulin intravenous) for treatment of infant botulism caused by type A or type B Clostridium botulinum. the treatment of complicated skin and skin structure infections.

* Private companies, interested in developing drug therapies, are investing in DNA as well. In Iceland, **deCode Genetics** has pinpointed a gene mutation for type 2 diabetes called **TCF7L2**. One copy of the mutation increases an individual's risk by 40 percent, two copies by 140 percent, says its CEO. A joint effort by the **US National Institutes of Health** and **Pfizer** is searching for genes for a host of diseases, including schizophrenia, bipolar disease and severe depression.

* **Nektar Therapeutics,Sanofi-Aventis, & Pfizer's** pulmonary insulin product Exubera, said to have demonstrated sustained activity and pulmonary safety of inhaled insulin for treating diabetes, is a Type 1 and type 2 diabetes mellitus treatment that improves compliance through an easier mode of administration.

*German drug major **Boehringer Ingelheim** has initiated trials on **flibanserin** in female sexual dysfunction. The agent's ability to increase sexual desire in women was discovered while it was being evaluated as an antidepressant and now Boehringer has begun a 5,000-patient, four-trial program in FSD, hoping to submit a marketing authorization application to US regulators in 2009. While the drug's precise mechanism of action is not known, the agent is believed to act within the brain. The disorder, which is also known as **hypoactive sexual desire disorder**, is characterized by diminished feelings of sexual interest or desire, causing personal distress and/or interpersonal difficulties and currently has no approved treatment options. Up to one in five women are believed to suffer from decreased sexual desire.

* Cancer scientists have shown that a gene that is involved in regulating aging also blocks prostate cancer cell growth. SIRT1 is a member of a family of enzymes called **sirtuins** that have far-reaching influence in all organisms, including roles in metabolism, gene expression and aging. They have shown that by making a prostate cancer with cells overexpressing a mutation for the androgen receptor, which is resistant to current forms of therapy, they can almost completely block the growth of these cells with **SIRT1.**

nobody does badly. Its services are available to all, but are mediocre and inflexible.[63] It is an example of what the British journalist Andrew Neill, following Friedrich Hayek, calls the "fatal conceit": the view that society can be rationally planned and directed.[64]

An important "side effect" of this attitude? Europe sits at the Biotech-sidelines.
Talk to scientists and educators about the future of scientific research, and they will rarely even mention Europe[65].

In March 2000, the EU´s heads of State agreed to make the EU "the most competitive and dynamic knowledge-driven economy by 2010". Words. Just words. By that time, they might realize that they represent structurally economic weak nationstate economies, that they lost precious time and opportunities by discussing subsidies for farmers and how to keep alive a growing number of outdated industries.

In a certain way, the system has an increasing grip on the individual, collecting always more money from the active, entrepreneurial people (who else?), thereby "punishing" them for their efforts, because that system has a growing need of more money to distribute.

Western European countries are the tax and spend champions of the world.

In matters of healthcare, government's, while telling people that they will have to pay more for there healthcare, show no sign of returning responsibilities to their citizens. The rule continues to be: "give us the money; we will decide what you get for it, when you get it, should you ever ask for it."

II.6. Taxing medicines
A trick used by many European Authorities is to impose the VAT (Value Added Tax) on medicines. Without any sign of shame, they divert money from the Social Healthcare Systems, read from the contributors to the system, to the public Treasury instead of spending it for what it has been collected, namely for matters of healthcare.

The Government's trick? First you take away a big chunk of the money and than you claim that there is not enough money to cover all the costs of the system…

A social system that protects people from the consequences of their own decisions is rife with moral hazard: in the long run, it changes not just behavior but even values in a less productive direction[66]

II.7. Healthcare industry image re-building
The pharmaceutical industry has been forced to defend its prices. It missed the opportunities to shift the discussions from costs and side effects to the benefits and value that the products provide. Having become suspicious to the public, regulators and thought leaders, industry motivations have to be re-explained. Rebuild trust, understanding and awareness: we will see the implementation of Reputation Management, an evolving discipline.

---

[63] Wolf, Martin, There is something rotten in the welfare state of Europe, Financial Times, 01.03.2006:13

[64] China and Europe: the Fatal Conceit, Institute of Economic Affairs, London, 28.11.2005, www.iea.org.uk

[65] Zakaria Fareed, The decline and Fall of Europe, Newsweek 20.02.2006:19

[66] Wolf, Martin, There is something rotten in the welfare state of Europe, Financial Times, 01.03.2006:13

* **Pfizer** develops 5-HT1D receptor antagonist (**eletriptan**) for the treatment of migraine including the indication of menstrual migraine). The product's rapid onset of action and high efficacy at early onset of migraine let analysts forecast blockbuster potential sales. The company dominates the premier league in the USA drugs sales rankings.

Pfizer expects from **valdecocib**, a rapidly-acting cyclooxigenase-2 inhibitor, to become the follow-up compound to **Celebrex** (celecoxib) for relieving the signs and symptoms of osteoarthritis and rheumatoid arthritis and reducing the number of adenomatous colorectal polyps.

Its **parecoxib** is an IV and IM injection formulation of Valdecoxib (the second-generation cyclooxygenase-2 inhibitor) for the treatment of acute pain in the hospital setting.

* Using a biochemical version of a computer chip, scientists have shed light on the way cells send and receive signals **through signal transduction**, a process that goes awry in cancer and other illnesses.

* NIGMS-funded researchers have discovered a key enzyme responsible for **triggering** allergic reactions. The research presents an opportunity to prevent, rather than simply control, allergic disease.

* Roche's innovative anti-anemia agent is the first **continuous erythropoietin receptor activator**. Its activity at the receptor sites involved in stimulating red blood cell production is different from that observed with traditional epoetin drugs.

* **Femprox** and **Alista** are in clinical trials for the treatment of **female sexual arousal disorder (FASD)**.Femprox combines the NexACT transdermal penetration-enhancing technology with alprostadil (prostaglandin E1), a synthetic version of a naturally occurring vasodilator. Alista is a proprietary formulation of alprostadil which is applied locally to the female genitalia. It is believed to increased blood flow to the female genitalia, thereby promoting engorgement and other natural processes that occur during stimulation, natural processes that occur during stimulation..

* Pseudomonas, a Gram-negative bacterium, is one of the leading causes of resistant hospital-acquired infections for which treatment options are limited. The Centers for Disease Control and Prevention (CDC) and numerous other agencies and health organizations have identified antibiotic resistance as a serious threat to public health. Of the **2million people** that **acquire bacterial infections** in U.S. **hospitals** each year, 90,000 people die as a result. The CDC estimates that approximately 70 percent of those fatal infections are resistant to at least one drug.

Doripenem, a new carbapenem antibiotic may help for the treatment of complicated intra-abdominal and complicated urinary tract infections. Doripenem has demonstrated activity against a wide range of Gram-positive and Gram-negative bacteria including Pseudomonas.

* The approval (2003) of **Barr Laboratories´** first-ever "extended cycle" oral contraceptive **Seasonale** (levonorgestrel and ethinyl estradiol) marks the beginning of a new era in birth control. It is the idea of continuous dosing that will set the tone for the next wave of pills that will increase convenience. By taking active pills for 12 straight weeks and then dummy pills for a week, women on Seasonale menstruate only 4 times a year.

B.40. Contraception methods used by US women

Age	30-34	35-39	40-44
Fertile women (in millions)	8	8.2	7.3
Method			
Tubal litigation	30%	41%	50%
pill	29%	11%	6%
condom	18%	17%	12%
vasectomy	10%	18%	20%
all others	13%	13%	12%

(source : Conceptus Inc)

B.41. Regional contraceptive preferences:

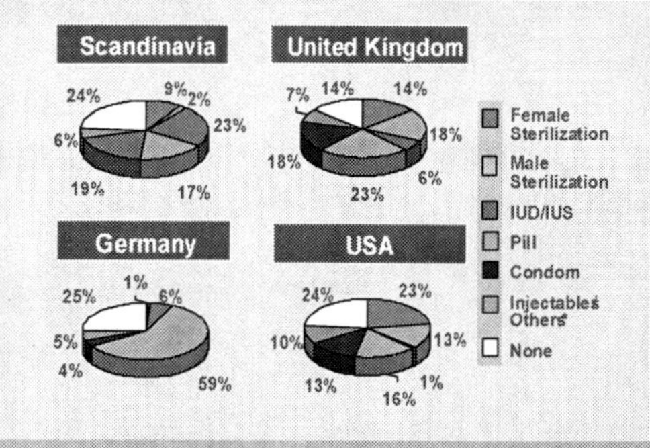

* Scientists have discovered some of the genes and cell-to-cell communication pathways that enable zebrafish to restore their tail fins. The findings may one day help treat human injuries.

* NIGMS-supported researchers have identified a gene – and the molecular function of its protein product – that provides an important clue to further understanding obesity. The gene product, **lipin**, may lead to new drugs to control fat metabolism.

* A class of small RNAs called **microRNAs** influence the evolution of genes far more widely than previous research had indicated.

* Scientists report the discovery of a protein found only in **cerebrospinal fluid** that they say might be useful in identifying a subgroup of patients with multiple sclerosis (MS) or identifying those at risk for the debilitating autoimmune disorder-

Table 2.3. Public spending: General

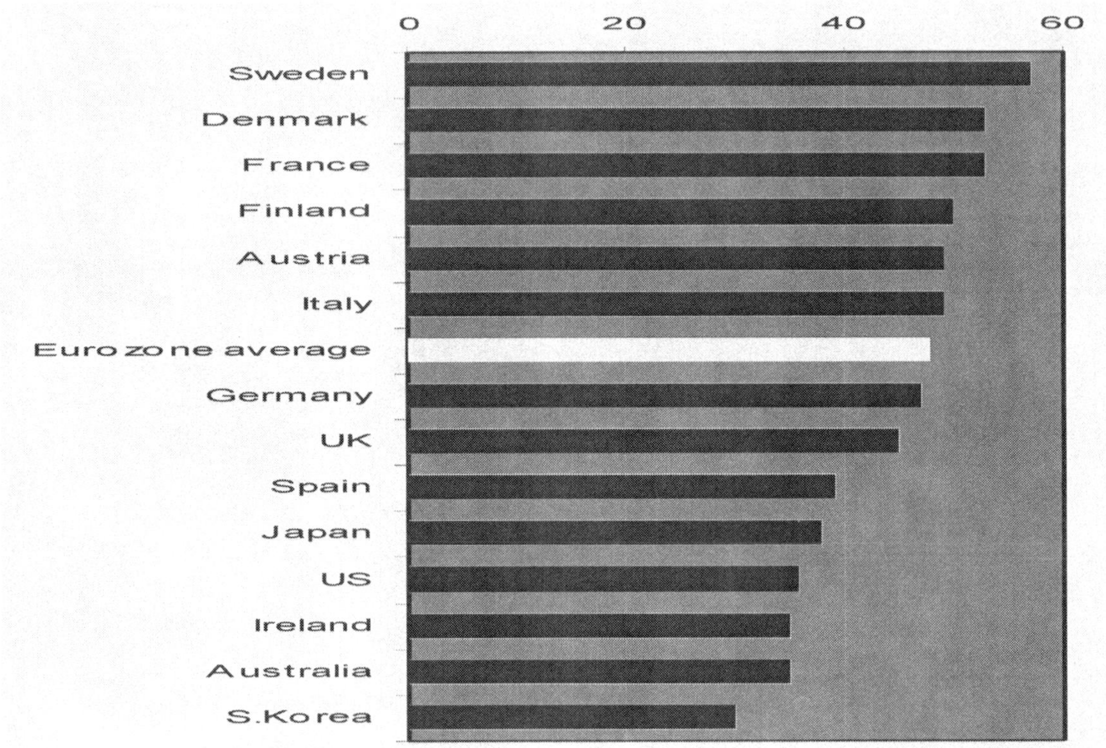

Table 2.4. Productivity growth: government, total outlays as % of GDP (2005) Annual average 1995-2005 (%p.a.)

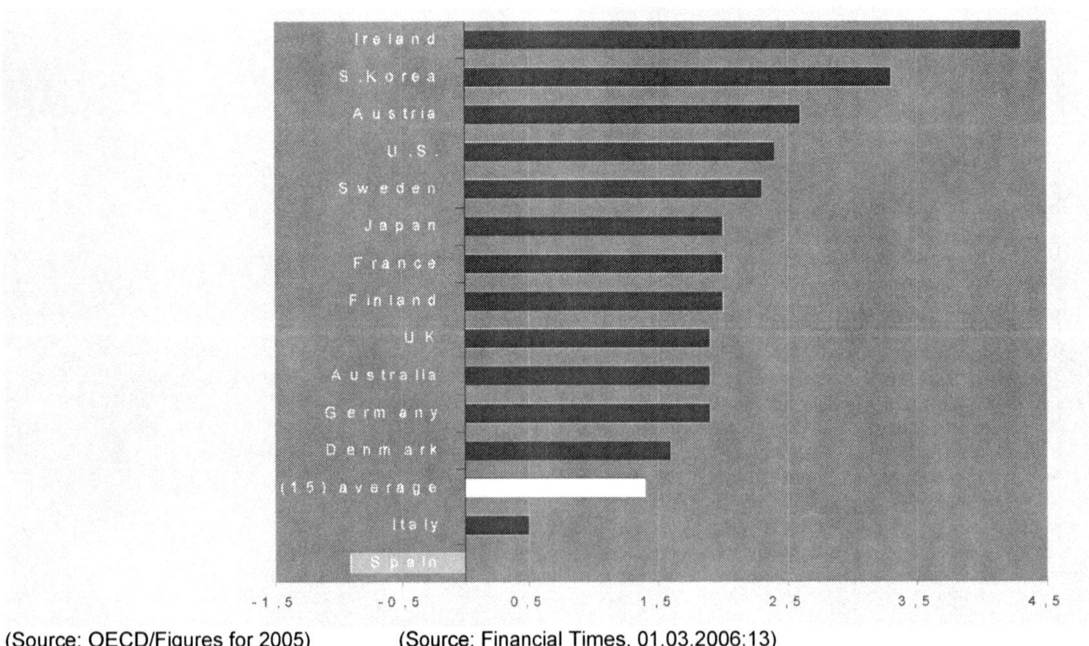

(Source: OECD/Figures for 2005)   (Source: Financial Times, 01.03.2006:13)

Although the proportion of pharmaceuticals is relatively small, authorities in almost all countries have often exclusively focused their actions on cost cutting of prices for drugs. The best current example is the situation in the USA, where Rx drugs take a 10% share of the total healthcare marketing 2003, but a 90% share in media attention (the pharmaceutical industry is almost daily described as being the culprit for increase in healthcare costs)

Table 2.5. VAT rates on medicines, by country

	Value added tax (VAT, in %)	VAT on medicines (in%) prescription	OTC	Country	Value added tax (VAT, in %)	VAT on medicines (in %) prescription	OTC
Belgium	21	6	6	Norway	25	25	25
Denmark	25	25	25	Austria	20	20	20
Germany	19	19	19	Poland	22	7	7
Estland	18	5	5	Portugal	21	5	5
Finland	22	5	5	Slowenia	20	8.5	8.5
France	19.6	2.1-5.5	2.1-5.5	Spain	16	4	4
Greece	19	9	9	Schweden	25	0	25
Ireland	21	21	21	Switzerland	7.6	2,4	2.4
Italy	20	10	10	Tcech rep.	19	5	5
Lettland	18	5	5	Hungary	20	5	5
Lituania	18	5	5	United Kingdom	17.5	0	17.5
Malta	0	0	0	Cyprus	15	0	0
Netherlands	19	6	6				

Table 2.6. Much talk about …10% share of total healthcare costs

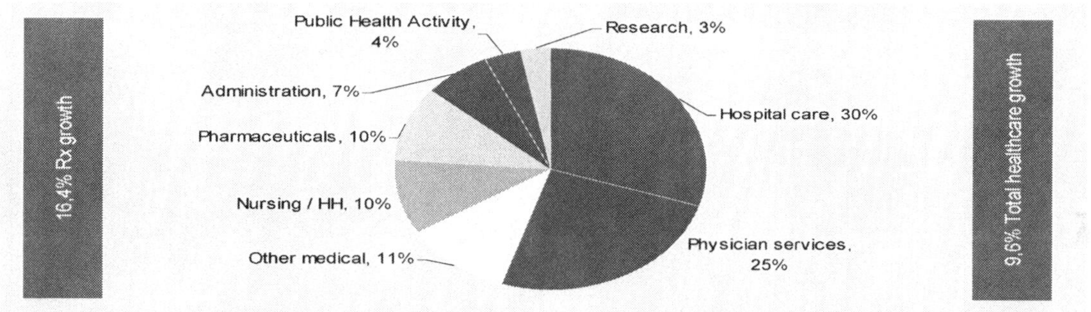

(Total prescription drugs expenditure in the USA increased from 9,1 % to 11% in 1999 and decreased to 10% in 2003.)

Table 2.7. Medicines: ten-year development of the share in total health care expenditures, by country[67]

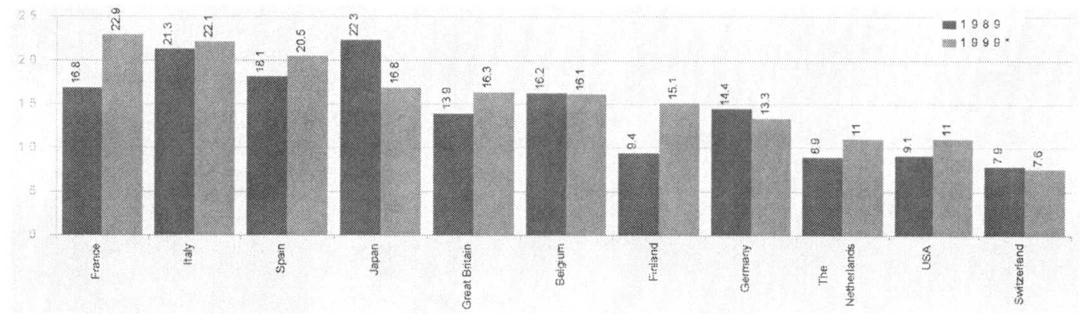

(* Japan, Switzerland figures for 1998; Belgium, Great-Britain and Spain figures for 1997)

---

[67] Source: OECD Health Data; VFA, Verband Forschende Arzneimittelhersteller, Berlin 2002

* **Pegasys** is a pegylated interferon alpha version of **Roferon A** (interferon), marketed already by Roche (to treat hairy cell leukemia, chronic hepatitis C, and Karposi sarcoma). The interferon alpha-2b preparation is intended to treat chronic hepatitis C in non-cirrhotic and cirrhotic adult patients with compensated liver disease.

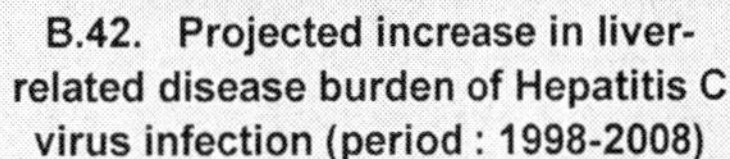
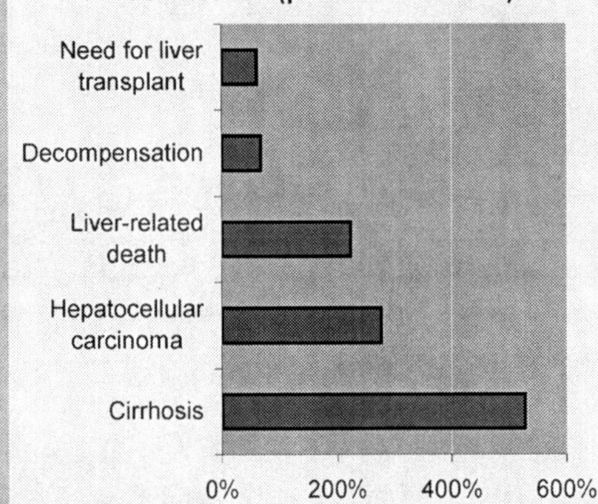

B.42. Projected increase in liver-related disease burden of Hepatitis C virus infection (period : 1998-2008)

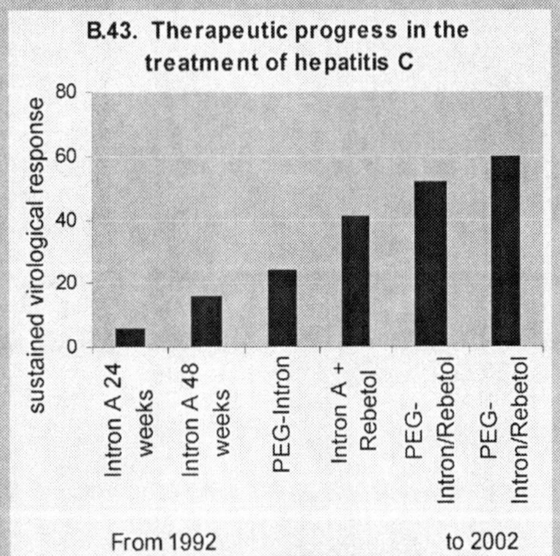

B.43. Therapeutic progress in the treatment of hepatitis C

* Researchers led by Nobel Prize winner Roderick MacKinnon of Rockefeller University have used yeast to express and obtain the structure of a mammalian voltage-dependent potassium ion channel (a type of membrane protein that helps transmit electrical signals into and out of heart and nerve cells).

* Developmental biologists have found that pieces of genetic material called **non-coding RNAs** underlie a previously unknown mechanism for regulating genes.

* The mature **Multiple Sclerose** market can only absorb a new entrant at the expense of existing drugs. The arrival of **Antegren** from **Biogen-Idec** and **Elan** on the Multiple Sclerosis market ahead of schedule is seen by analysts as a new threat for rival interferons, namely **Betaseron** from Schering and **Rebif** from **Serono**. For these two companies, their interferons **represent a high proportion of income**: for 2005, $1.2 billion for Rebif (50% of Serono's sales) and $930m for Betaferon (17% of **Schering's** sales).

* US consumer's out-of-pocket **spending on prescription drugs** rose $6.1 million in 2002, to $48.6 million, out of a total out-of-pocket health spending rise of **$12 billion to $212.5 billion.**

B.44. Medicare enrollment market share by managed care organization, 2002

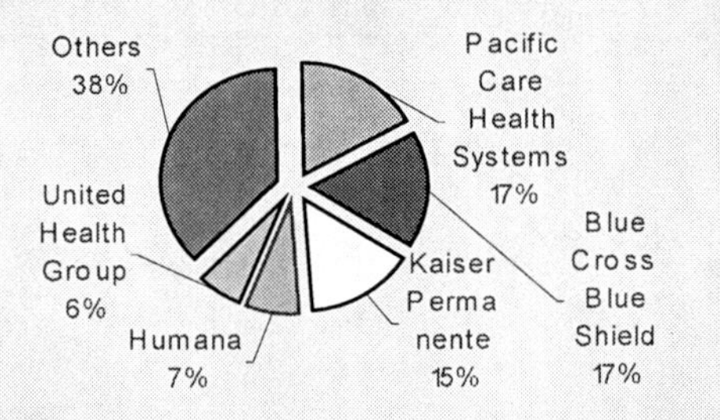

* Given the increasing rates of obesity, **nonalcoholic fatty liver disease (NAFLD)** -- the hepatic consequence of obesity -- has become the most common cause of chronic liver disease, with **a prevalence** in US adults as high as 30%. NAFLD represents a spectrum of hepatic disorders characterized by **macrovesicular steatosis**, with histology ranging from "simple" steatosis to nonalcoholic steatohepatitis (NASH). The latter represents a shift from fatty infiltration to an inflammatory/fibrosing disease that **may progress to cirrhosis.**

*In 2005, Chugai launched **Actemra (tocilizumab)** for he treatment of **Castleman's** disease, a rare condition that causes severe enlargement of the lymph nodes.

*In the same year, **generic drugs** accounted for 30% of prescription volume in Canada, Germany, the UK and the USA, as providers in every country of the world sought to cut costs.

* Scientists at the Kimmel Cancer Center at Thomas Jefferson University in Philadelphia have uncovered a novel pathway by which the **anti-cancer gene p53** springs into action, **protecting a damaged cell from becoming cancer**. The gene can either halt the cell's growth or send it spiraling toward certain death. How this choice is made could have implications for future strategies in chemotherapy drug development.

Big Pharma's quest for ever-larger opportunities in a limited number of high incidence-rate disease/therapeutic areas is going to be replaced by its quest for larger and larger opportunities in multiple specialist disease areas.

The future? Don't ask whether it will really happen; prepare for it. Segmentation into smaller niches is already taking place. The drug industry has not much time left to decide how it will metamorphose itself. Instead of recognizing and responding to the fundamental changes facing the blockbuster industry many companies will stick to their tired ideas. Expect and prepare for exciting times in a health care industry with new dimensions.

Table 2.8. To be successful, pharmaceutical companies must transition

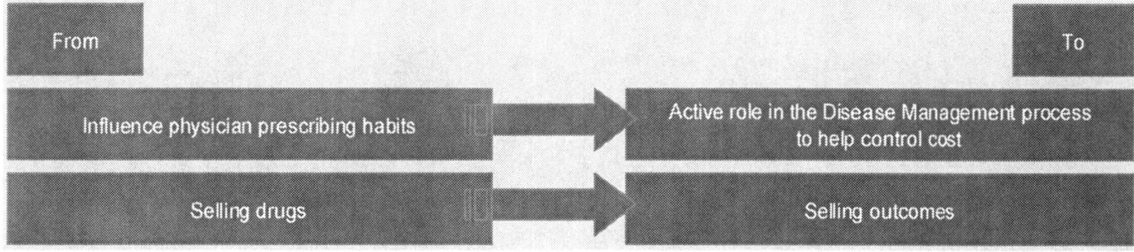

The pharmaceutical industry has not much time left to decide how it will metamorphose to adapt.

*** Pharmacovigilance** is the process and science of monitoring the safety of medicines and taking action to reduce risks and increase benefits from medicines. It is a key public health function.

The legal basis for **pharmacovigilance** in the EU is given in Directive 2001/83/EC (as amended) and Regulation (EC) No 726/20044. In addition, detailed guidance is provided in Volume IX of **Eudralex**. The current EU pharmacovigilance system is organized with functions, responsibilities and accountability shared between the Member State competent authorities, the **European Medicines Evaluation Agency (EMEA) and European Commission.** The EMEA has responsibility for co-ordinating the pharmacovigilance resources and work of the Member States. The exact division of responsibilities changes depending of how a particular medicine is authorized. For medicines authorized through the national authorization mechanisms most (but not all) of the functions, responsibilities and accountability for pharmacovigilance are with the Member States. In contrast, for centrally authorized medicines, that is, those authorized through the central Community authorization mechanism, more of the functions, responsibilities and accountability for pharmacovigilance are with the EMEA and European Commission.

Data sources for the conduct of pharmacovigilance include: spontaneously reported adverse drug reactions (ADRs), periodic safety update reports from pharmaceutical companies, data on the use of medicines, clinical trials and epidemiological studies. Patients and healthcare professionals are central to providing safety data. Industry has legal responsibilities in collecting, assessing and transmitting data. The Member States play a key role in the collection of data, from healthcare professionals, from academic institutions and from pharmaceutical companies. The EMEA also collects data particularly from pharmaceutical companies and the Member States. Although Member States are responsible for many aspects of data management, a Community pharmacovigilance database, Eudravigilance, is operational and being further developed.

*** Oncology drugs** dominate the most-successful commercial launches of 2004. **Avastin, Erbitux, Alimta,** and **Vidaza** all received indications in cancer treatment in 2004 and were among the best-selling products in their first full year on the U.S. market in 2005. **Eli Lilly and Co**. had the most products on the list of the top 10 most-successful product launches. In addition to **Alimta**, the company markets the depression drug **Cymbalta**. Two cholesterol drugs (**Vytorin** and **Caduet**) also made the list. Rounding out the top 10 are **Truvada** for HIV, **Namenda** for Alzheimer's disease, and **Restylane** for facial wrinkles.

*** Japan**, the world's second largest medicines market, which posted slower growth rates since 1991, grew 6.8 per cent to $60.3 billion in 2005.. That performance was fuelled by growth in **angiotensin IIs, antihistamines** and **oncology** therapies, as well as significant uptake in geriatric-related therapies such as **Aricept for** treating **Alzheimer's**, and **Cabaser, Permax** and **Bi Sifrol,** for treating **Parkinson's** Disease.

* Since their introduction into the market in 1993, **proton pump inhibitors (PPIs)** have largely supplanted the other acid-suppressing medications to become the cornerstone of therapy for **GERD** sufferers and other acid-peptic conditions. There is abundant clinical evidence that the acid suppression that is achieved with PPIs is more pronounced and durable than that achieved with **histamine type-2 receptor antagonists (H2RAs).** Defined as reflux of gastric contents into the esophagus, GERD is seen across all age groups and has a prevalence in the adult population of 15% to 20%.

GERD is associated with significant health resource utilization and impairment in the quality of life in affected patients. (Devault K, Castell DO. Updated guidelines for the diagnosis and treatment of gastroesophageal reflux disease. Am J Gastroenterol. 2005;100:190-200). Left untreated, patients with chronic GERD have the potential to develop a number of complications, including but not limited to erosive esophagitis, esophageal strictures, Barrett's esophagus, and esophageal adenocarcinoma.
Expert opinions differ as to the advantages new **acid suppression medications,** such as **the potassium-competitive acid blockers** (These highly lipophilic, weak bases accumulate within the parietal cell at much greater concentrations (> 1000x) than PPIs and act by reversibly blocking potassium entry into the parietal cell, resulting in rapid rises in gastric pH after ingestion) finally will bring to the patients.

*** Hepatitis B** affects more than **350 million individuals** worldwide and is implicated in the development of cirrhosis and 80% of cases of hepatocellular carcinoma. Since the development of an effective vaccine, along with the implementation of perinatal and childhood vaccination programs, the incidence of acute hepatitis B has declined significantly. However, hepatitis B virus (HBV) infection remains a global health concern, as some areas of high **endemicity** still have **seroprevalence** rates of greater than 8%, and data from the Centers for Disease Control have shown an increase in the incidence of hepatitis B among men older than 19 years of age and women 40 years of age and older. [Mast E. et al, Advisory Committee on Immunization Practices (ACIP).

Visit: http://www.cdc.gov/mmwr/PDF/rr/rr5416.pdf]. The 5 treatments for chronic HBV infection are: **standard interferon (alfa-2b); lamivudine; adefovir; and, most recently, pegylated interferon alfa-2a and entecavir.** Existing therapies have reasonable rates of durable **seroconversion** and potent **viral suppression**; however, among medical experts, one can observe growing concerns of viral resistance with long-term HBV-treatment plans.

* US Congress grants six-month patent extensions on drugs whose efficacy is proven (through clinical trials) in **pediatric patients**.

*Scientists have developed a way to predict the ability of small molecules called osmolytes to stabilize cellular proteins. The findings could be useful for studies on a number of diseases, including Alzheimer's and cystic fibrosis.

Chapter Three

Medicines
for the
Masses

III.1. The "Blockbuster Phenomenon"

The "blockbuster phenomenon" is essentially an American situation: the global sales of these drugs are generated mainly in the USA. They range between 67% and 86% of sales of companies that follow this model (with sales levels the industry never experienced in the past).

• In total, some 94 prescription drugs had sales of over $1.0 billion in 2005 (up from 57 in 2000). Of worldwide revenues generated by blockbusters, the USA took a 78% share, the European Union 11% and Japan 6% (others 5%).

• The top seller, Pfizer's Lipitor (atorvastatin), reaching $12.0 billion mega blockbuster global sales in 2006, generates almost 75% of this worldwide sales figure from the product's US market performance. (Within the past two decades, this cholesterol-lowering drug transformed cardiology).

• Merck & Co's Zocor (simvastatin) is in a similar situation: out of a total of $7.0 billion in global sales, the USA took some $5.0 billion.

• The picture was the same for AstraZeneca's Losec/Prilosec (omeprazole) in 2002, the first year after it came off patent: global sales of $5.2 billion of which $3.5 billion were in the USA (a decrease, due to the patent loss, of 21.8% on the previous year);

• For GlaxoSmithKline's Paxil/Deroxat (paroxetine), as another example, global turnover was well over $3.5 billion and US sales reached $2.7 billion.

• New drugs launched in the five years from 2000 to 2005, and which are considered to be megasellers or blockbusters, reached a sales level of $2.0 billion in some 3.5 years.

• At the end of 2005, six out of the list that topped the hit-parade of top 10 global drugs by sales in 2001 lost their patent protection, and five are owned by US companies. In 2006, $17.0 billion in sales were exposed to patent expiration; the industry will bid farewell to mega-sellers including Zocor, Bristol-Myers Squibb's Pravachol, and Zoloft (sertraline),

*The World Health Organization estimates that 35 million people will die for chronic diseases in 2005. According to the WHO, 60% of all deaths are due to chronic diseases.

B.45. Projected main causes of death, worldwide, all ages, 2005 (in Mio)

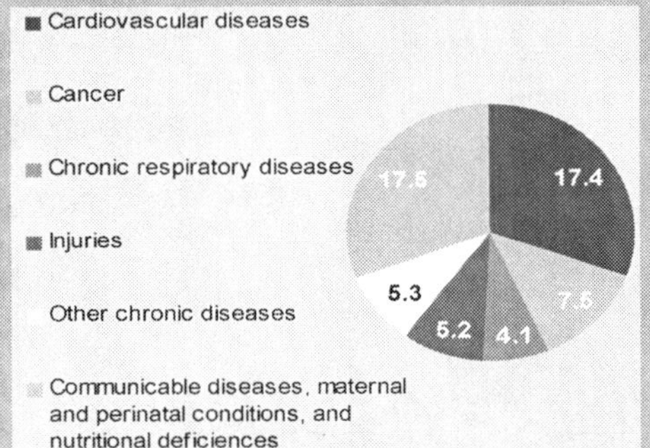

Including 1.2 million deaths due to diabetes, the main causes of death total 58 million. *(Source. WHO 2005. Preventing Chronic Diseases, a vital investment)*

* 2004, **Merck** announces the voluntary withdrawal of **Vioxx (rofecoxib),** its COX-2 painkiller. Less than a year later, a US court rules against the company on all major counts in the first product-related trial, awarding the plaintiff US$253,4million.

* **Genentech's Avastin**, an antiangeniogenesis product is the first to treat tumors by "starving" them of their blood supply.

* In 2005, more than **2,300 products were in clinical development**, A promising range of drugs are in Phase III clinical trials or pre-approval stage, including **96 oncology** products, **51** products for treating **cardiovascular disease**, 37 for **viral infections** and **HIV** and **28** for **arthritis/pain.**

* **FDA** outlines a goal to modernize the scientific process of bringing drugs to market. The new initiative eventually will cover risk communication, clinical trial data collection, and adverse-events reporting.

*Up to **7000 rare diseases** have been described. Developing drugs to treat them is not often an attractive option for pharmaceutical companies. In the EU, **Orphan Drug** designation has been available since 2000. It can be awarded to a product at any stage of development providing it targets diagnosis, prevention of treatment of a serious condition which affects **less than 1 in 2000 people in the EU.**

* The launch of **Roche's AmpliChip CYP450 Test** in Europe and the US, 2005, represents an important step towards more **personalized medicine**. This **DNA chip-based test** marks the beginning of a new generation of diagnostic tools that can identify clinically relevant genetic variations and thus help improve treatment outcomes.

* The **Project BioShield Act** provides US$ 5,6billion to stockpile vaccines and authorizes the US government to expedite R&D on anti-terror medicine.

* The **Medicare Modernization Act** of 2004 provides the first prescription drug benefit to Medicare beneficiaries in the most significant overhaul of the US Medicare system since 1965. 2005 saw the launch of **Byetta**, the first in a new class of medicines for improved blood sugar control in patients with Type 2 diabetes; of **Lunesta** which treats insomnia, decreases sleep latency and improves sleep maintenance; and of **Macugen** which treats neovascular age-related macular degeneration

* Treatment of **Irritable Bowel Disease (IBD)** is advancing at a rapid rate. Due to the success of **infliximab**, most of the attention to improve effectiveness and quality of patient care focuses on biologic agents.

* Of the total new drugs pipeline in 2005, **27** per cent are **biologic** in nature. Biologics also experienced strong growth overall. Led by **Amgen, Roche/Genentech** and **Johnson & Johnson**, this sector grew 17 per cent in 2005, generating sales of **$52.7 billion.**

*Sydney **Brenner**, H. Robert **Horvitz**,John E. **Sulston** receive the **Nobel Prize** for their discoveries concerning **genetic regulation** of **organ development** and **programmed cell death**; Paul C. **Lauterbur**, Sir Peter **Mansfield** for their discoveries concerning **magnetic resonance imaging**; Arvid **Carlsson**, Paul **Greengard**, Eric **Kandel** for their discoveries concerning **signal transduction in the nervous system.**

* **Spending** on **specialty pharmaceuticals** reaches 25% of the US pharmacy bill. The most innovative part of health care has become its single most explosive cost. Examples: **Genzyme's** drug for the treatment of Gaucher Disease is priced at **$200.000** for the average patient in the US; **BioMarin Pharmaceutical/Genzyme** sells a treatment for the rare genetic disease mucopolysaccharidosis-ICQ, or **MPS-1**, at $200.000 for the average patient; **Gleevec** from **Bristol-Meyers Squibb** is priced at **$37.000** annually at the recommended dose *(Source: The Wall Street Journal, Dec 28, 2005)*. Merck KGaA, the company that sells **Erbitux** in Europe, stated that the price publicly listed varies on the specific country, but across Europe the average cost per month for Erbitux is approximately **4.000 Euro.** According to Merck KGaA, the price of Erbitux has been accepted by many pricing authorities in different countries based on rigorous benefit cost analyses, and, although direct comparison cannot be made between other monoclonal antibody treatments across different cancers, Erbitux treatment cost is in the range of other monoclonal antibodies such as Herceptin and MabThera *(Source: correspondence with the author, September 2007).*

* Novel concepts of **collateral efficacy** and **permissive antagonism** in the search for synthetic antagonists are taken into consideration in the search for drugs with unique therapeutic profiles.

Pfizer's Norvasc (amlodipine) and Merck & Co's Fosamax (alendronate). In the absence of an EU patent (same expiration dates of originator products in all member states) and of harmonized price structures, the US-style blockbuster effect (sales level of a given product) could not be duplicated in Europe.

(In 2005, North America, accounted for 47 per cent of global pharmaceutical sales, grew 5.2 per cent, reaching $265.7 billion).

Table 3.1.   Sales exposures to patent expiration, 1990-2006 (USA)[68]

The pharmaceutical industry creates brands, invests in brands, and sees the investment disappear in smoke in a relatively short period of time, due to its own success of constantly creating new, better medicines. 2006 was a year of bidding farewell to a number of „blockbuster" brands, making room for copycat drugs and newer medicines.

Table 3.2.   (Selected) Blockbuster drug sales under threat by patent challenges[69]

	Lipitor	Plavix	Norvasc	Zyprexa	Zithromax
Treats	high cholesterol	blood clots	high blood pressure	schizoprenia	bacterial infections
Manufacturer	Pfizer	Bristol-Meyers Sanofi-Aventis	Pfizer	Eli Lilly	Pfizer
2005 sales (US $)	12.2 billion	6.3 billion	4.7 billion	4.2 billion	2.3 billion
Status	Ranbaxy Labs unsuccessfully challenged the patent (which expires in 2010) in 2006	Apotex sold several billion dollars of a generic before being stopped by an injunction	Mylan labs has FDA approval to sell a generic version in 2007	Barr Pharma is challenging the patent, which expires in 2011	Teva Pharma has FDA approval to sell a generic version. The patent expires in 2007

(Note: Status December 5, 2006)

In the three-year period ending in 2007, the worldwide leader, Pfizer, will see blockbuster drugs with US $14 billion in peak annual sales lose their patent protection.

The different dates at which products lose their patent protection in European countries, together with the varying health systems in Europe, will continue to require:
• the local "tactical" management of patent-protected originator products; and

---

[68] Source: FDA Orange Book, Standard & Poor's
[69] Zehr L, Generic brands spell big change for Big Pharma. The Globe and Mail, December 5, 2006:B3

* Approximately 1.4 million people worldwide developed **GE cancers** in 2002 and 1.1 million died -- figures that make GE tumors as common as lung cancer and almost as lethal.

B.46. Heart disease rates among men, 30 years and over, 1950-2002

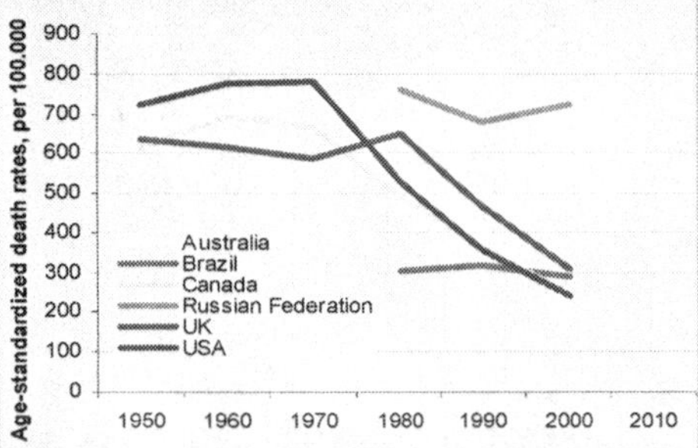

According to the WHO (*WHO 2005, Preventing Chronic diseases : A vital investment*) of all **global coronary heart disease deaths**, 53% affect men.

* Requirements for generic drug approval: The **Abbreviated New Drug Application (ANDA)**, established by the US Food and Drug Administration (FDA), is a tool for the review of generic drugs. The ANDA requires a generic drug and the branded drug to be both pharmaceutically equivalent and bioequivalent. A generic drug is considered pharmaceutically equivalent to a brand drug if it has equal amounts of active ingredients, the same dosage form, identical manufacturing standards of quality and purity, and the same label indications as the branded drug. **Bioequivalence—the central measure of the ANDA in approving generic drugs**—necessitates no significant difference in the rate and extent of availability of active ingredients at the site of drug action between the generic and branded forms of a drug. **Bioequivalence** is determined by the following parameters: the 90% confidence interval of the mean area under the time-absorption curve of the generic agent must be within 80% to 125% of the branded product, and the mean maximum concentration of the generic must be between 80% to 125% of that of the branded agent.

* Cell biologists at Thomas Jefferson University in Philadelphia have found that the protein fragment **endorepellin** blocks both skin and lung cancer tumors from progressing in animal models **by preventing their ability to recruit new blood vessels**, a process called angiogenesis. They showed that endorepellin has surprisingly powerful effects on halting a cancer tumor's ability to move about and spread. The researchers believe that these findings could lead to a new type of tumor inhibitor that might be used to prevent cancer from spreading to other areas in the body.

* Until the 20th century, **men's sexual problems** were dealt with by herbs, ceremonies, incantations, exorcisms, physiotherapy (e.g., ointments, baths, exercises), prayer, pilgrimage, external prostheses, and dramatic surgeries. Interventions combined psychic and somatic elements that involved complex and culturally specific sexual meanings. Group witness and support were often involved. The outcomes, however, are largely unknown. Throughout the first two thirds of the 20th century, a range of **talk-therapies** were developed to deal with human problems, including problems with sexual relations. Group therapies emerged in the 1920s to deal with sexual issues such as **body image and interpersonal trust**. Postwar, group approaches to sexual problems diversified as a result of the human potential movement in the 1960s and involved modalities such as **sensitivity training, encounter groups, gestalt therapy, and bodywork. Couples and family therapies** were developed in the 1950s and 1960s. **Sex therapy clinics** thrived in the 1970s and 1980s but were in decline by 1990 as a result of the growth of **biological psychiatry** and the **medicalization of sexual problems**. In the 1980s, urologists claimed authority over the new area of sexual medicine.A significant contribution to the shift in views toward a biomedical view of erectile dysfunction occurred in 1992 when the National Institute of Diabetes, Digestive and Kidney Disorders organized the NIH consensus conference on sexuality. One of the report's primary conclusions was that "erectile dysfunction increases with age but is not an inevitable consequence of aging," institutionalizing the shift in views about sex and aging and **making the impotence-aging connection sound biomedical** and not related to culture, education, and social values. This event paved the way for the success of the **sildenafil (Viagra)** trials. A series of sexuopharmaceuticals followed at the end of the decade/in the early 2000s.*[The "New View" approach to men's sexual problems, Medscape CME, Leonore Tiefel, Clinical Associate Professor of Psychiatry, NYU School of Medicine, 2005].*

* Since the founding of **Genentech Inc.** in April 1976, the **biotechnology sector** has experienced exceptional growth around the globe, from the maturing U.S. sector to the rapidly emerging Asia-Pacific region. Despite patent challenges, financial insecurities, and changing business models, the biotech industry regularly emerges with record numbers . The revenue of publicly traded biotechnology companies grew 18% in 2005, reaching an all-time high of $63.1 billion. The industry secured **32 new drug approvals** in the United States, including **17 first-time approvals**.

* **Therapeutic vaccines** development is seeing an increasing trend towards the use of multiple antigens (and adjuvants), to attack cancers or HIV from multiple angles. Combinations of therapies, and novel approaches such as prime-boost regimens excite scientists.

• the time-consuming, bureaucratic and outdated establishment of 27 (and more?) networks of good working relations with 27 national health authorities and opinion leaders within an enlarged EU - an expensive undertaking.

The financial community continues to ask the pharmaceutical industry to deliver double-digit growth figures and this has consequences. In 1989, global heavyweights like Pfizer or Merck needed to generate $350.0 million and $200.0 million, respectively, to grow 10%; in 2003, they needed to produce almost $3.0 billion (for Pfizer) and $2.0 billion (for Merck) in order to achieve the same 10% growth rate. To respond to this need, they have to launch two or three new blockbuster-status drugs *per* year and this growth target requires "oversized" promotional budgets.

Take the example of AstraZeneca's Nexium (esomeprazole) and Prilosec: in two years, the promotional budget doubled from $215.0 million to $397.0 million and this produced a growth rate of 4.6% on total sales generated in the second year of Nexium's launch. Also, this was at a time when there were still no generic versions of omeprazole on the market.

The return on promotional investment is divided by two: for each dollar invested, the return drops from $21.56 to $13.35. It is notable that some Big Pharma firms depend on one blockbuster for a third of their profit.

III.2. What to expect in the short-term in USA
"The jackpot" for patients and providers: over the next 10 years, and thanks to blockbusters coming off patent, $60.0 billion in economies will be realized. A generic version of the leading blockbuster drug in its therapeutic class will become the preferred prescription recommendation of health care providers, provided that other products in the same class do not show a notably superior therapeutic benefit.

A report published by the US Federal Trade Commission in July 2002 highlighted the (legal) tactics used by companies to defer the arrival of generic versions. Companies succeed in obtaining delays of four to 40 months after the 30th-month stay. More courtroom battles are to come, as the active defense of intellectual property pays off. For example, Prilosec brought AstraZeneca $6.25 million in sales for each day the generic competitor had to wait to launch its copycat drug.

In the case of a "missing" direct successor to a given blockbuster, manufacturers of originator drugs start using their specific knowledge and category leadership to promote another drug in the same therapeutic field, but with another indication. For example, in the central nervous system field, Eli Lilly focused on its antipsychotic Zyprexa (olanzapine) after its antidepressant Prozac (fluoxetine) came off patent.

As the media increases its interest in an originator product some 18-24 months before its patent protection expires, a "challenger" drug benefits from free press coverage and may be able to accelerate its market penetration.

Prices of blockbuster drugs do not have to decrease during the 180 days' of generic exclusivity; the production capacity of generics can be very limited (retailer Kudco needed more than a year before it could serve 60% of the market).

III.3. Short-term prospects for Europe
Some countries have been strong advocates of generic prescribing, e.g., the UK and Germany. Of the two countries, and even though the UK makes life difficult for pharmaceutical companies with its National Institute for Health and Clinical Excellence,

the fourth hurdle they have before they can market a new product, the UK remains the most attractive to the industry. In 2003, both nations shared more than 50% of the European generic market, estimated at 13.5 billion Euro ($16.40 billion).

Leftist/green political ideologist thinking has negatively influenced German society. Political and societal refusal of "Bio" and "Gene" technology, together with a climate of suspicion surrounding the (chemical) medicines business, plus a series of management errors, contributed to the decline of what was once the world's leading pharmaceutical nation.

Politicians promote generic substitution and re-imports because: "it has to be cheap." Reference prices, substitution rights, fixed pharmacy margins, public campaigns - they all contribute to accelerate the movement to generic prescribing (e.g., France, Spain, Portugal and Italy).

Table 3.3.   Origin of the world's top 100 prescription medicines[70]

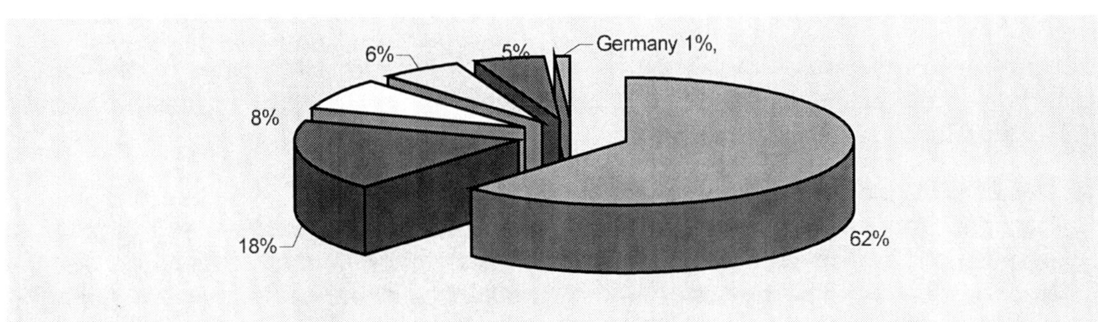

III.4. A hostile environment to a research-driven healthcare industry

The hostile environment for the health care industry in Europe, and also in Japan, casts a shadow on pharmaceutical company growth and largely contributes to a further decline in price-earnings ratios. Ultimately, Europe and Japan's industry transformation will see many groups disappear in continued waves of consolidation. Smaller survivor companies are expected to retrench into the distribution of generics and complementary health care.

Table 3.4. Changing competitive global landscapes (1970-2005)

	1970	1980	1990	1995	2000	2005
1	Hoffmann LaRoche	Hoechst	Merck & Co	Glaxo Wellcome	GlaxoSmithKline	Pfizer
2	Merck & Co	Ciba	Bristol-Myers Squibb	Merck & Co	Pfizer	Glaxo-SmithKline
3	Hoechst	Merck & Co	Glaxo	HoechstMarionRoussel	Merck & Co	Sanofi-Aventis
4	Ciba-Geigy	American Home	SmithKline Beecham	Novartis	Astrazeneca	Novartis
5	American Home	Hoffmann LaRoche	Ciba	Bristol-Meyers Squibb	Aventis	AstraZeneca
6	Eli Lilly	SmithKline	American Home	Hoffmann LaRoche	Bristol-Meyers Squibb	Johnson & Johnson
7	Sterling	Boehringer Ingelheim	Hoechst	Pfizer	Novartis	Merck & Co
8	Pfizer	Sandoz	Johnson & Johnson	American Home	Pharmacia	Wyeth
9	Warner	Bristol-Meyers	Eli Lilly	Johnson & Johnson	Hoffmann LaRoche	Bristol-Meyers Squibb
10	Sandoz	Eli Lilly	Bayer	Pharmacia Upjohn	Johnson & Johnson	Eli Lilly
11	Upjohn	Warner Lambert	Hoffmann LaRoche	Eli Lilly	American Home	Abbott Labs
12	Abbott Labs	Johnson & Johnson	Sandoz	SmithKline Beecham	Eli Lilly	Hoffmann LaRoche
13	Squibb	Bayer	RhonePoluenc Rorer	Bayer	Schering Plough	Amgen
14	Bayer	Schering Plough	Pfizer	Astra	Takeda	Boehringer Ingelheim

---

[70] Note: Data for 2004, as published by the ABPI, Association of the British Pharmaceutical Industry in its 2006 Manifesto

* Since the recognition of non-A, non-B hepatitis as **hepatitis C** in 1989, significant progress has been achieved in complete viral eradication in the infected individual. The current standard of care, **pegylated interferon** given in combination with **ribavirin**, has yielded success rates, measured as sustained viral response, of greater than 50% overall. Research is focusing on new targets in the viral replication cycle that may enhance viral eradication. **Protease inhibitors, polymerase inhibitors**, and other small molecules appear to be on the horizon, although they likely may still be combined with interferon for treatment. Among the novel treatments being studied are:

~**NM283** is a ribonucleoside analog that targets the viral RNA polymerase.

~**Alb-IFN**, a novel recombinant protein consisting of interferon alfa genetically fused to human albumin has exhibited a median half-life of 148 hours, supporting dosing at intervals of 2-4 weeks. This compares with a reported mean elimination half-life of 80 hours (50-140 hours) for pegylated interferon alfa-2a and 40 hours (22-60 hours) for pegylated interferon alfa-2b; thus, theoretically, alb-IFN may improve response rates to treatment.

~**MX-3253** and its active metabolite castanospermine are inhibitors of alpha-glucosidase 1, a host enzyme that alters the processing of glycoproteins. Inhibition of the enzyme prevents the correct folding of the viral envelope glycoproteins, thus inhibiting viral assembly and release

~**CPG 10101** is a toll-like receptor 9 agonist that activates B cells and plasmacytoid dendritic cells, stimulating the innate immune system to produce Th1 cytokines (including interferon alfa).

*Treatments for **age-related macular degeneration** (AMD) are evolving rapidly. Options include laser **photocoagulation of the membrane, photocoagulation of feeder vessels, transpupillary thermotherapy (TTT), photodynamic therapy (PDT)** with or without triamcinolone, and the **vascular endothelial growth factor (VEGF) inhibitors pegaptanib** (under the brandname **Macugen**/a highly selective aptamer –a short strand of RNA that folds into a shape that also binds to VEGF and inactivates it) and **bevacizumab** (brandname **Avastin**/ is an antibody that binds VEGF and prevents it from activating VEGF receptors.). In phase III of clinical trials **ranizumab**, is a fragment of the bevacizuab antibody that nonselectively binds to all forms of VEGF. Even newer agents are under development, such as **squalamine, RNA interference**, and the **VEGF trap.**

* As **general medicine clinicians** retire or leave practice, **35% fewer** medical students are choosing to follow in their footsteps.[1] And this is happening at a time when 77 million baby boomers are entering their peak years of medical care. The number of primary care physicians is simply not keeping up, especially in community health centers and rural areas.[2] [*1/Garibaldi RA, Popkave C, Bylsma W. Career plans for trainees in internal medicine residency programs. Acad Med. 2005;80:507-512. 2/Rosenblatt RA, Andrilla CHA, Curtin T, Hart LG. Shortages of medical personnel at community health centers. JAMA. 2006;295:1042-1049].*

* **Photodynamic therapy** One of the most effective chemotherapy drugs against cancer is **cisplatin** because it attaches to cancer DNA and disrupts repair. However, it also kills healthy tissue. Many scientists are creating alternative drugs or **cisplatin** analogs in attempts to find treatments without side effects. One approach to analog development is light activated drugs, or **photodynamic therapy (PDT).** A **Virginia Tech** chemistry-biology research team that has been working on both **non-cisplatin drugs** and **cisplatin** analogs has combined their findings to create a molecular complex (supramolecule) that exploits **cisplatins** tumor targeting to deliver a light activated drug. The group has developed **supramolecular** complexes that combine light-absorbing PDT agents and **cisplatin** like units. Previous anticancer molecules created by the group have contained platinum-based molecules that bind DNA. They have also developed new light activated systems able to **photocleave** DNA.. **Cisplatin** begins its interaction with cancer DNA by binding to the nitrogen atoms of the DNA bases, typically guanine. The new **supramolecules** use this nitrogen-binding site to hold the light activated drug at the target until signaled to activate. Thus the new **supramolecules** can be delivered to the tumor site but remain inert until activated by a light signal. Light waves in the therapeutic range – that is, those that can penetrate tissue, are used to activate these new drugs. The researchers are also appending other molecules that emit UV light to track the movement of these drugs within cells. **CNS disorders** contribute to as much as 35% of the disease burden in the seven major pharmaceutical markets (US, Japan, France, Germany, Italy, Spain and UK) as measured in terms of daily-adjusted life years. The worldwide patient population with **CNS disorders** is steadily rising, both in terms of prevalence and in terms of treatment, driven by

1. an aging population,
2. improving diagnostic techniques,
3. increasing physician and patient awareness and
4. a gradual shift away from the social stigma traditionally attached to many psychiatric conditions.

The largest CNS indications in terms of prevalence are **depression, migraine and pain**, with 80m, 79m and 158m sufferers respectively across the seven major pharmaceutical markets. The fastest growing indication in terms of prevalence is **Alzheimer's** disease. By 2010 it is forecast that Alzheimer's disease will afflict 15m patients. In 2004 **Parkinson's** disease affected 2m patients across the seven markets, although epidemiological dynamics do not forecast this indication as a major growth area. **Schizophrenia** affected an estimated 8m patients in 2004, with prevalence forecast to rise to 9m patients by 2010. Growth rates associated with the prevalence of this disease are relatively robust compared to many other indications. **Epilepsy** is forecast to have relatively static growth in terms of prevalence of the disease, with 6m sufferers forecast to be affected by the disease by 2010.

Stimulus to support research and development (scientific support, drug prices,), and thus innovation, now comes from the US. For research driven large pharmaceutical companies, Europe has become a difficult place to operate in.

III.5. Political and Bureaucratic Shortsightedness

The view that Europeans lost out to the USA because of politician and bureaucratic short-sightedness, is shared by the American Council on Science and Health: "Those who attack drug companies either don't know or willfully ignore the fact that in addition to saving lives and relieving suffering, effective drugs reduce health-care costs for everyone. If America's drug industry is regulated and negotiated into becoming an arm of government, as some wish, and if importation allows foreign price controls into our country like a Trojan horse, our pharmaceutical innovation will come to resemble that of Europe's during the last 20 years: overly cautious, sluggish, and marked by decelerating progress. When the EU bureaucrats heeded alarmists' cries about "expensive, risky drugs" and socialized their pharmaceutical industry, the result was a reversal of the European dominance of drug discovery that had held sway since the Industrial Revolution. Europe's regulatory overreach allowed America's pharmaceutical industry to rise to dominance"[71].

Table 3.5. Country of origin of major global medicinal products, changes from 1987 to 2005

Rank	1987	1992	1997	2002	2005
1	Zantac (UK)	Zantac (UK)	Losec/Prilosec (SE)	Lipitor (US)	Lipitor (US)
2	Tagamet (UK)	Renitec (US)	Zocor (US)	Zocor (US)	Plavix (US)
3	Adalat (D)	Voltaren (CH)	Prozac (US)	Losec/Prilosec (SE)	Nexium (SE)
4	Capoten (US)	Lopirin (US)	Zantac (UK)	Zyprexa (US)	Seretide/Advair (UK)
5	Tenormin (UK)	Mevacor (US)	Norvasc (US)	Norvasc (US)	Zocor (US)
6	Reitec (US)	Adalat (D)	Renitec (US)	Erypo (US)	Norvasc (US)
7	Naprosyn (US)	Tagamet (UK)	Augmentin (UK)	Ogastro /Prevacid(US/J)	Zyprexa (US)
8	Voltaren (CH)	Zovirax (UK)	Zoloft (US)	Seroxat (UK)	Risperdal (US)
9	Feldene (US)	Cirpoxin (D)	Seroxat (UK)	Celebrex (US)	Ogastro/Prevacid (US/J)
10	Kefral (J)	Cardizem (US)	Ciproxin (US)	Zoloft (US)	Effexor (US)

---

[71] *Ross G.,MD., Heart Care Hurdles. Washington Times, December 23, 2006*

* **Autism** is thought to be the most highly heritable of all neuro-psychiatric disorders. Yet, most cases of this childhood developmental disorder that severely affects social interaction and communication are "**sporadic**" anc come with no family history. New research, led by Cold Spring Harbor Laboratory scientists Sebat, Muthuswamy and Wigler, has found a **distinction between heritable and sporadic forms of the disease**. These findings may influence future autism research and diagnostic testing.

* Virologist Dr Hilary **Koprowski,** reported how he and colleagues used **tobacco plants** to produce **cancer-fighting monoclonal antibodies** that **recognize and hunt down breast and colorectal cancer cells**. While therapeutic uses for such antibodies continue to grow at a rapid rate, production has failed to keep pace. Plants are believed to be safer, less expensive and easier to use than currently used methods in the laboratory and with animals. For mass production purposes, plants would make more sense.
The researchers inserted DNA coding for an antibody against the **Lewis Y antigen**, which is found on breast and colorectal cancer cells, among other cancers, into tobacco plants. The plants, in turn, became factories churning out antibody. The antibodies were able to kill the cancer cells in the laboratory dish and stop tumor growth in mice. Not only that, these plant-made antibodies worked as well as those produced in the standard way, by mammalian cells. Both types of antibodies suppressed tumor cell growth in mice up to 28 days. This particular work had an added twist, which has not been shown before: the researchers also demonstrated that the antibodies attached to the human Fc receptor, a key step in the development of immunotherapy. So-called "**effector cells**" – tumor-cell killing macrophages – recognize monoclonal antibodies through their Fc receptors.
Scientists say that this will be the future, and in the next five to 10 years, it could be the main way that therapeutic antibodies are made.

* Singapore joins the race towards developing **personalized medicines**. The Genome Institute of Singapore (GIS), a member of the Agency for Science, Technology and Research (A*STAR), is a national initiative with a global vision that seeks to use **genomic sciences** to improve public health and public prosperity. As a center for genomic discovery, the **GIS** pursues the integration of technology, genetics and biology towards the goal of **individualized medicine**. In 2003 the GIS, national flagship program for the genomic sciences, opened its new 7,200-square-meter advanced research facility, the Genome, which is set in the Biopolis -- a 180 hectare biomedical city within the Buona Vista Science Hub. GIS plans to ultimately house more than 300 scientists, trainees and staff at this facility. The major technical platforms of high-throughput sequencing, molecular cytogenetics, bioinformatics, proteomics and expression array technologies have been integrated with programs in molecular and cellular biology, computational biology and population genetics.

* Clinical successes such as **Dendreon**'s tumor vaccine **Provenge** are raising hopes that a new class of innovative drugs will establish itself in this area. The Tuebingen-based company **immatics biotechnologies** is following promising approaches in this field. Another opportunity comes from the widespread "off-label use" of cancer drugs. Companies such as **GPC Biotech** use this in order to get quick approval for a drug in a cancer indication where competition is not a problem and approval hurdles are low. In this way, a company can access new market potential, assuming the drug works well, through clinical trials running in tandem on other types of cancer, for example.

* The **metabolic syndrome** consists of a constellation of factors that raise the risk of cardiovascular disease and type 2 diabetes. An estimated 24% of the adult population in the USA is affected. Its clinical identification is based on measures of abdominal obesity, atherogenic dyslipidemia, raised blood pressure, and glucose intolerance. The etiology of this syndrome presumably represents a complex interaction between genetic, metabolic, and environmental factors, including diet. Several new studies also suggest that a proinflammatory state and endothelial dysfunction also associate with the metabolic syndrome. One study revealed that **men with metabolic syndrome had increased prevalence of erectile dysfunction (ED).**

* Attention has been paid to **2 serologic markers**, anti-*Saccharomyces cerevisiae* antibody (**ASCA**) and perinuclear anti-neutrophilic cytoplasmic antibody (**pANCA**), which have been proposed as screening tools for patients who present with signs and symptoms of IBD (**Inflammatory Bowel Disease**).

* By slowing down the snipping process, NIGMS-supported scientists snared a pair **of molecular scissors** and uncovered its identity. The scissors-like protein is key to making the bobbins around which our DNA is wrapped.

Researchers have set a new world's record by performing the first million-atom computer simulation in biology. Using the **"Q Machine"** supercomputer, computer scientists created a molecular simulation of the cell's protein-mak.

*The 2005 Nobel Prize in chemistry Robert H. Grubbs of Caltech and Richard R. Schrock of MIT. The two researchers are honored for developing metal-containing molecules that are now used daily in the chemical and pharmaceutical industries to make important compounds.

* An international consortium has published a **comprehensive catalog of human genetic variation**, a landmark achievement that is already accelerating the search for genes involved in common diseases, such as asthma, diabetes, cancer and heart disease.

Chapter Four

A Healthcare Industry
Reinventing Itself

IV.1. A tremendous responsibility

Thanks to their relentless search for new medicines, both scientists and the pharmaceutical industry have largely contributed to an increased lifetime expectancy.

About 40% of the increase in life expectancy since 1988 can be attributed to improvements in prescription medications

Table 4.1.   New drug discoveries account for longer lives[72]

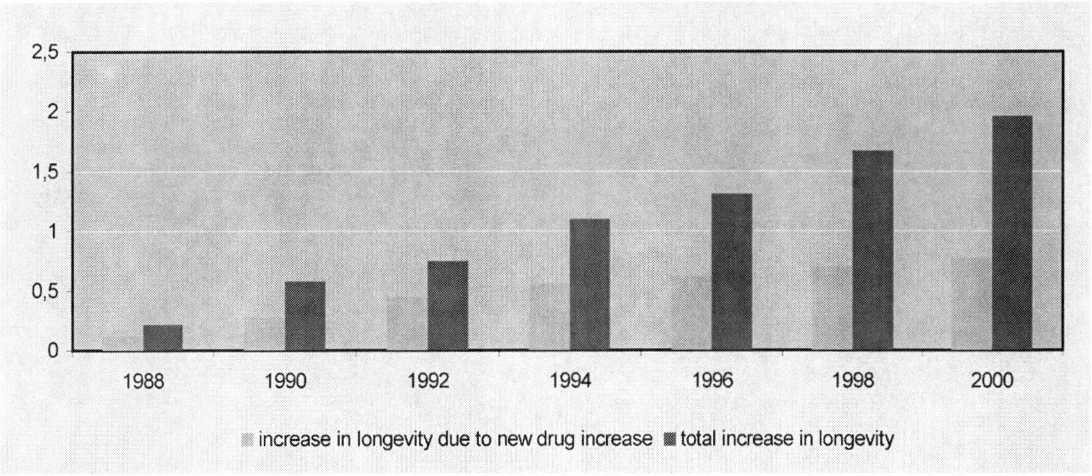

In view of the many unresolved health problems, Society expects the industry to continue playing an important role in the discovery process of new medicines. At an accelerated pace. Is the industry ready?

IV.2. The threat to the health of the research-driven pharmaceutical industry

- Fair market conditions for innovations are being undermined by new restrictive Government policies, in particular in Europe and Japan.

---

[72] *National Bureau of Economic Research Working Paper, as published by Pharmaceutical Executive, Oct 2006*

*** EU Lung Research Project Pulmotension Pulmonary hypertension (PH)** describes a group of **chronic, prolonged crippling and fatal vascular diseases.** It is characterized by high blood pressure in the lung vessels leading to right heart failure. PH often affects young or middle-aged patients, who suffer from progressive loss of exercise capacity and dyspnoea. As a result, this serious lung disease represents a major burden on our healthcare systems. With the formation of **Pulmotension**, European Top centers for scientific and technical competence in PH have taken a decisive step in tackling this large medical problem: Connected into a single project they aim to better understand and find a cure for this serious disease, because the large and complex tasks can only be addressed with the collected multidisciplinary competence and critical mass assembled. This **pan-European "Big Science"-initiative** allows the collaborating researchers to investigate basic science questions in terms of clinical applicability and provides a unique potential for scientific breakthroughs, technological advances and new treatments in the field of pulmonary hypertension. By 2010, scientists aim to uncover **underlying molecular pathways of PH, identify distinct targets for anti-remodeling therapy, foster drug development based on these targets in alliance with industrial partners and exploitation facilities, and carefully test these new treatment options in preclinical and clinical trials.**

The combined expertise in Pulmotension extends from the initial discovery of gene mutations in PH to the establishment of new therapeutic regimen of PH. These include the discovery **of BMPR2 mutations** in PH, or the introduction of **sildenafil (Viagra)** into the treatment of PH.

*** Pfizer's Exubera** becomes the first inhaled form of insulin, and the first insulin option that does not need to be administered by injection.

*At the forefront of **brain-imaging research** are technologies that reveal which brain cells are active during given activities. One technology in particular, functional MRI, has had a starring role in delving into the workings of the brain and mind. Where individuals differ is in which brain cells respond to humor, anger, stress or the sight of a loved one. It's these intimate details that functional MRI, better known as **fMRI**, reveals. Rather than detecting the brain's density, **fMRI** records which part of the brain has greater changes in blood flow during a particular task. This is an indirect measure of which brain cells are active — and therefore require extra blood supply — at that moment. Knowing the brain's functional layout helps determine, for example, if surgery would remove parts of the brain too vital to live without. Likewise, it can help doctors predict the long-term prognosis for a person who has had a stroke. Other studies look at aspects of our brain that are more ethereal, such as the seat of love, cravings or deceitfulness. This field, called cognitive neuroscience, began taking off once the power of fMRI dawned on researchers. *(Source: Amy Adams, Stanford School of Medicine)*

*** Dreams: keeping us in good health?** While the average person might believe that dreams mean something, little is known about **why we dream.** Fifty years ago, scientists discovered that sleep has distinct phases and that dreams usually happen during only one of them — the rapid eye movement or REM phase, during which the sleeper's breathing, heart beat and brain waves speed up and the eyes move quickly behind closed lids. Since then, researchers have made no major breakthroughs nor reached anything close to consensus about the cause or meaning of dreams. Many experts, including Stanford professor **William Dement**, known as the **"father of sleep medicine,"** believe that dreams contain personal messages — and are **linked to our emotions and mood.** Others believe that dreams are simply random byproducts of the brain's chemical changes. There hasn't always been rampant disagreement about dreams. For decades, scientists followed the theories of **Sigmund Freud**, who wrote in 1900 that dreams provide a harmless way to act out our unconscious fears or taboo thoughts. **If people weren't able to dream, Freud said, they would become psychotic.** In the 1950s, scientists tried but were unable to prove a link between a lack of dream sleep and schizophrenia. Studies also showed that people's dreams, **correlated with what was happening in their lives,** which went against Freud's belief that dreams had little connection to daily activities and stresses. **Psychoanalysts** continued to examine dreams as a way to gain understanding of a patient's inner thoughts. The field was rocked by the introduction of the so-called **activation-synthesis model** in the late 1970s. Two Harvard scientists argued that REM sleep is caused by the **ebb and flow of neurotransmitters**, and dreams are merely the brain's attempt to make sense of these **random neurological impulses**. The pendulum swung again when **positron emission tomography scans** and other new **imaging technology** allowed researchers to see which parts of the brain were active during the dream process. A 1998 study published in Science was the first to show that the **portion of the brain that controls emotions, senses and long-term memory is active during dream states.** Experts say the study showed that **dreaming is linked to emotion** and not just a series of random events. It is predicted brain-imaging tools will yield answers. That leaves a lot of work to do for those willing to deal with elusive mysteries. *(Source: Michelle L Brandt, Stanford School of Medicine)*

* Scientists from Imperial College London and Hammersmith Hospitals NHS Trust used injections of **oxyntomodulin**, a naturally occurring digestive hormone found in the small intestine, to reduce body weight and calorific intake in overweight volunteers, providing a crucial breakthrough in developing new drugs to tackle the growing obesity epidemic.

* The **US National Institutes of Health** counts **nanomedicine** as one of its top five priorities. The **National Cancer Institute** commits $144 million to nanotechnology research in 2006 (up more than $2 million from 2005), while the **National Science Foundation** allocates about $50 million of its 2006 nanotechnology budget to biological sciences (also up about $2 million from 2005).

- Strategic and economic pressures of being first to market, faster, are intensifying. Competition has become an increasingly complex game pressed between containing cost at every level and of every kind, and availing the public of new therapies.

- Governments seek to contain healthcare costs by creating industry-hostile savings regulations. The hostile environment to the healthcare industry in Europe and also in Japan casts a shadow on the growth of pharmaceutical companies and largely contributes to a further decline in price-earning ratios. Ultimately, Europe's industry transformation will see many companies disappear in continued waves of transnational consolidation. Smaller survivor companies are expected to retrench into the distribution of generic copies of originator products that have lost their patent, and into complementary healthcare. Into marketing and sales. Not into R&D anymore. European opinion leaders fear that history books will describe how a wealth-creating, medico-scientific based sector of the Continent's national economies, once much envied by the world, how a giant continental pharmaceutical industry has been short-sightedly turned into a global dwarf. *Is history in the making?*

Table 4.2. Development of the overall context

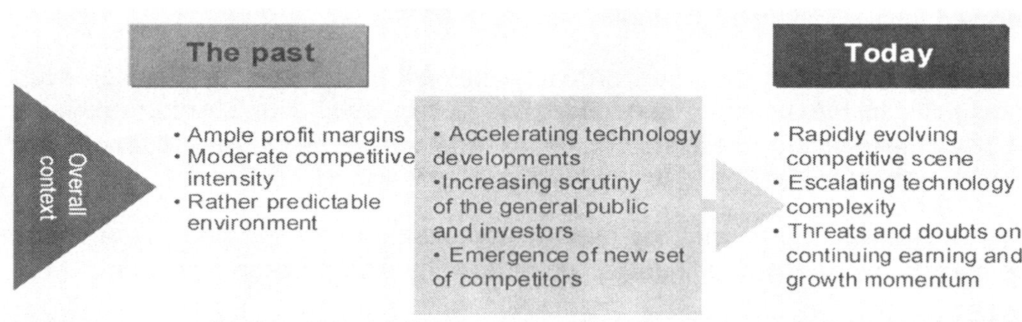

- The traditional (described by some as "paternalistic") role of the medical practitioner is slowly being eroded, forcing manufacturers to reassess their marketing, promotion and selling strategies. Physician's wisdom went unquestioned for decades. Open communication channels and interactive exchange of information require of companies a rethinking of relationship building, integrating the patient. Few companies realize the tremendous, irrevocable impact that a confluence of major societal forces has on their industry.

- Profit erosion caused by the growing strength of generics, & the issue of re-imports of patented drugs (USA/Canada, & Europe).

- Shorter time of effective patent protection. Long handling times for the evaluation of new medicines by governmental regulatory bodies.

- Demands for increasing and more sophisticated production standards. And overcapacities in production.

- The industry faces an avalanche of litigation instigated by users who demand someone be held to account (Vioxx, Bextra, Zyprexa, Celebrex are just a few of the drugs under attack by patients claiming that they are responsible for side effects the industry knowingly did not warn them.

- Increasing impact of media (and public opinion) and their scare-provoking health news.

- Performance gap in Research & Development (R&D). Since the late 1990s, pharmaceutical companies see their investment increase while their output declines.

- Product development costs are rising in the face of more diseases, high technology and more exhaustive regulation, whilst being challenged by Governments as a justification for high prices. [In an era where innovation is dominated by economics rather than by science, companies should become less reluctant of teaming up with third parties to garner economies of scale in selling (diagnostics, prevention programs, new vaccines, there are many opportunities with non-direct competitors). The "not invented here"-syndrome remains uncured].

- Escalating sales & marketing costs caused by more intense competition for share of voice. Companies that waited too long to break out of yesterday's comfortable realities to manage changing market structures and increase the effectiveness of marketing and sales in a diversifying healthcare environment struggle to survive. Forward thinking needs a change in mindset.

- Increasing number of launched products that fail to achieve "break-even", forcing companies to the unusual, uncomfortable re-engineering of key businesses in a constant process, to redefine the focus and scope of internal research and to reassess the value and need for early stage licensing.

- New companies in the emerging economic powerhouses India and China challenge the medico-pharmaceutical industry, dominated by Anglo-Saxon academia.

- Increasing public and investor scrutiny; industry underestimated the need to make its case with the public. Missed chances. Late awakening to realities. We live in the Information-Age. Pharmaceutical have been slow to open the windows, open the doors, reframe their image to salvage their reputation. They have been slow to understand the importance increasingly empowered and sophisticated advocacy groups, state officials, sick funds and the better informed individual. Having become suspect to the public, to regulators, to thought leaders, industry motivations have to be re-explained. Companies have to rebuild trust, understanding and awareness.

- Overall challenge to the profitability of the pharmaceutical industry[73].

IV.3. In need of new business models

On the assumption that new technologies continue to evolve and push up prescription costs, a pharmaceutical industry noted for its conservatism and relentless drive for size (why does size matters that much ?) finds itself at a crossroads.

---

[73] *Neirinckx Rudi & Mertens Gilbert, Mergers & Acquisitions in Pharmaceuticals: Why and How? Strategic Healthcare Management Reports, Financial Times Business Ltd, 2000.*

* **Constipation** is highly prevalent worldwide, with rates ranging from 5% in **Germany**, 8% in **Italy** and the **United Kingdom**, to 14% to 18% in **France**, **Brazil**, **South Korea**, and the **United States**. The pathogenesis of constipation remains to be completely defined, although the condition is commonly categorized as normal-transit constipation, pelvic floor dysfunction, and slow-transit constipation. The role of bacteria in the generation of constipation symptoms has begun to receive considerable attention.Although the precise etiology of **IBS** remains unknown, proposed mechanisms include genetic predisposition, inflammation, infection, stress, and psychological distress. In addition, evidence now suggests that a deficiency in serotonin is associated with IBS with constipation, whereas excess serotonin is associated with IBS with diarrhea. In 2006, some intriguing data suggested that methane may be responsible for some symptoms in IBS with constipation symptoms. New data suggest that **tegaserod** (5-HT receptor antagonist) is safe and effective in treating symptoms of chronic constipation and in treating patients with both IBS with constipation and IBS-M. In studies, **Lubiprostone** (a bicyclic fatty acid metabolite of prostaglandin E1), showed that it can effectively treat symptoms of chronic constipation and that it is safe in the elderly population. Emerging data show that pelvic floor dysfunction needs to be considered in the differential diagnosis of constipation and that the best treatment for this disorder is biofeedback and pelvic floor retraining. **Alosetron** is a viable treatment option for women with severe IBS and diarrhea who have failed other therapies. However, potential risks of severe constipation and ischemic colitis need to be taken into consideration. New medications are currently being evaluated for IBS with constipation (**renzapride**, which has both **5-HT4 agonist and 5-HT3-antagonist** activity) and for chronic constipation (**MD-1100, a guanylate cyclase-C agonist**).

* The significant **unmet clinical need in hematological malignancies** is reflected by an examination of recent FDA approvals of oncology drugs. Of the 31 FDA-approved oncology drugs that received approval up to and including 2005, 14 were approved for the treatment of hematological indications.10 of these drugs received marketing authorization under the US FDA's **accelerated approval program**, a designation assigned to promising therapies for serious or life-threatening diseases.

* **Roger Kornberg**, PhD, professor of structural biology at the Stanford University School of Medicine, was awarded the **2006 Nobel Prize** in Chemistry for his work defining the **universal process of eukaryotic transcription** - copying DNA into RNA in cells with well-defined nuclei. Kornberg was the first to create a picture of how transcription works at the molecular level.

* A report published by **Datamonitor** mentioned the number of approximately 84,000 cases of adult and childhood leukemia in the world's seven major pharmaceutical markets in 2006. **Acute myeloid leukemias (AML)** and **acute lymphoblastic leukemia** (ALL) account for 43% of these cases. and, although this market is relatively small compared to other cancer types, the clear unmet needs and regulatory incentives should help encourage drug makers.

Relapse rates following treatment for AML and ALL are high, and current treatments offer little chance of long-term survival for most patients. The incidence of acute leukemia increases with age, particularly for AML, in which the mean age of diagnosis is 65 years. Thus, **AML is predominantly a disease of the elderly**, with a dramatic rise in incidence beyond the age of 60 years. However, the epidemiology of ALL exhibits a bimodal distribution, with peak incidences in early childhood and then again in old age.

ALL is the most frequently diagnosed childhood cancer and represents 23% of cancer cases in children under 15 years of age. The incidence peaks in early childhood before plateauing throughout adulthood and then gradually increasing after the age of 60.

The significance of this trend is that the biology of both AML and ALL differs between young and old adults and this is reflected in different responses to therapy. As the increased likelihood of co-morbid illness and the decreased capacity of elderly patients to cope with the adverse effects of cytotoxic therapy may restrict available treatment options. *Cytarabine* in combination with an **anthracycline** remains the gold standard for AML remission-induction treatment. While this regime can induce complete remission in an impressive 65-80% of younger patients, this number falls to 40-60% in elderly patients with AML. The duration of this remission, and the ability to maintain long-term control of AML, remains problematic and one of the most significant unmet needs within the **adult acute leukemia** medical market. **Five-year overall survival** is in the range of **5-15% for elderly-AML** and approximately **30% in younger adults**.While **allogeneic stem cell transplantation** offers the only potentially curative treatment strategy in acute leukemia, the high morbidity and mortality associated with the technique - along with the scarcity of eligible donors - means the utility of this approach is limited. Efforts to enhance the rates of disease remission and prolong survival among the broader patient population look likely to rely on the integration of novel developmental agents into existing treatment paradigms. With regard to AML, the most promising of these agents are **Genzyme's Clolar** and **Cephalon's FMS-like tyrosine kinase-3 (flt-3) inhibitor CEP-701**. Therapy improvements could come from the next generation of **Bcr-Abl tyrosine kinase inhibitors,** including **Bristol-Myers Squibb's (BMS) Sprycel (dasatanib)** and **Novartis' AMN107.**

There are currently nine compounds in phase III development for AML and ALL

* Australian professor **Ian Frazer** gained international fame for developing the **world's first vaccine to combat cervical cancer.**

Table 4.3. The very essence of being "research driven", or the cycle of innovation and competition, as has been industry's model for decades

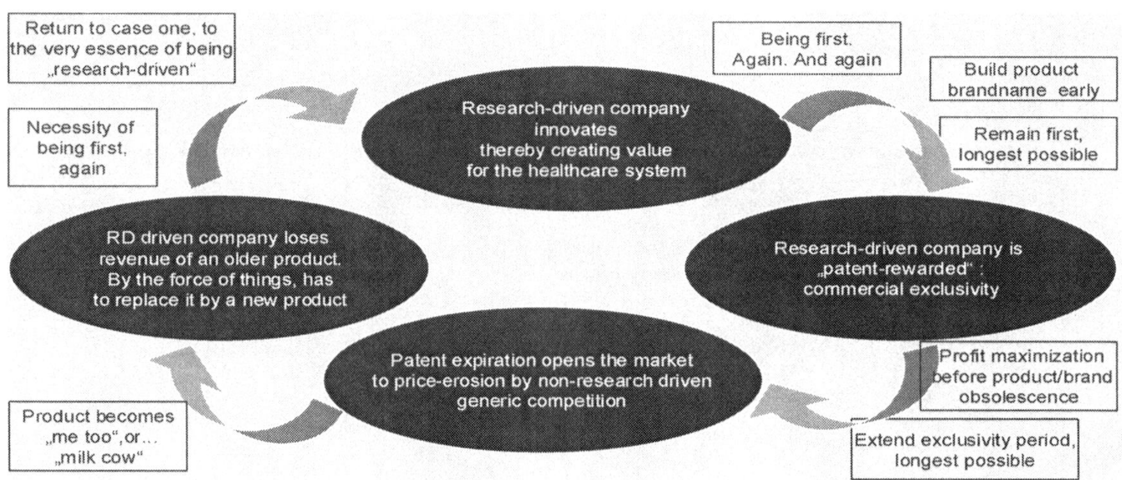

In the early 1990s, one could (surprisingly) observe merger mania as a first reaction to mounting pressure on an industry searching for "size".

In crisis situations, the search for healthy partnerships, the possibility of a merger...any "good" idea to survive inevitably comes up.
Is M&A the right solution, when organic growth cannot be guaranteed anymore?

Table 4.4. Pharmaceutical majors faced strategic options[74].

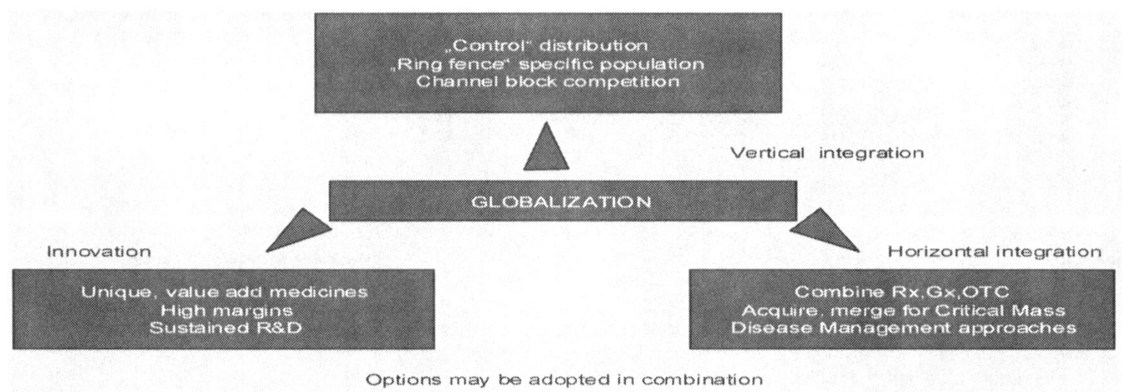

"Statistical proof" of the negative effects of mergers and acquisitions has been proffered several times, e.g. comparisons have been made between companies who have merged and others who have not. In the first category the market share typically remained stable while that of the merged companies has tended to fall. Likewise, calculations have been made of the return to the investor in merged companies as compared to the investors in the non-merged companies. Again, the latter were better off than the former.

---

[74] *Coopers & Lybrandt, Presentation given at the Financial Times Global Pharmaceutical conference, November 1994*

* Researchers from the Massachusetts General Hospital (MGH) Cardiovascular Research Center have discovered what appears to be **a master cardiac stem cell**, capable of differentiating into the three major types of cells that make up the mammalian heart The scientists described identifying these progenitor cells in mice, cloning single cells from embryonic stem cells, and showing that these cloned cells can differentiate into cardiac muscle, smooth muscle or endothelial cells. These cells offer new prospects for drug discovery and genetically based models of human disease. They also give us a new paradigm for cardiac development, in which a single multipotent cell can diversify into both muscle and endothelial lineages.
(Several populations of embryonic cells that develop into the heart and associated structures have previously been indentified. It has been thought that the three types of cells that make up the heart itself - the contracting cardiac muscle cells and the smooth muscle and endothelial cells that make up blood vessels - all develop from different cellular progenitors. Two major groups of cardiac muscle progenitors, called the first and second field, have been identified).

* A research team supported by NIGMS has become the first to clearly **link aggregates** of a protein called **SOD1** with **neuronal cell death** in Lou Gehrig's disease. This evidence could help explain the disease process and eventually lead to new therapeutics.

* Scientists have **synthesized** a **lipid molecule** that holds promise in fighting disease via gene therapy. The new molecule has a **tree-shaped headgroup** and displays unexpectedly **superior DNA-delivery properties.**

* **Riboswitches** are RNA elements that control gene expression in essential metabolic pathways. Researchers in the laboratory of Ronald R. Breaker, the Henry Ford II Professor of Molecular, Cellular and developmental Biology at Yale, showed that a riboswitch controlling vitamin B1 (thiamine) levels is **disrupted** in the **presence of pyrithiamine**, a toxic compound related to the vitamin. Bacteria and fungi fail to grow in pyrithiamine and become resistant by acquiring mutations in their riboswitches.
This work, in combination with the recently solved crystal structures of purine riboswitches, opens a path to the directed design of drugs targeting riboswitches for use as antibiotics. According to the researchers, it is inevitable that bacteria will evolve and that the drugs we use to cure bacterial infections will eventually become ineffective. While it is often possible to alter existing drugs slightly, they suggest that a longer term and more attractive solution is to create entirely new drugs that target bacteria in completely new ways. This approach lengthens the useful clinical lifetime of a drug and makes it possible to tackle **resistant 'superbugs'** with combinations of drugs.

* Figures for 2006 state that less **than half of American adults** are at a **healthy weight**. Approximately one-third of American adults are obese and an additional one-third are overweight and at risk for becoming obese. Alarmingly, approximately **16 percent** of **children** and **teens** ages six to 19 are also overweight. Overweight and obesity increase the risk of developing numerous serious health problems, including heart disease and type 2 diabetes. The prevalence of type 2 diabetes has increased with the national increase in overweight and obesity. Approximately **20.8 million people** -- 7 percent of the United States population -- **have diabetes**. Diseases that once were only seen in adults, like type 2 diabetes, occur in increasing numbers in children, Overweight children tend to become overweight adults, which also puts them at greater risk of high blood pressure, heart disease, and stroke. Children who are obese also are socially ostracized and teased, putting them at risk for depression and other psychiatric conditions. UCSF Researchers found out that Western food environment has become highly **'insulinogenic,'** as demonstrated by its increased energy density, high-fat content, high glycemic index, increased fructose composition, decreased fiber, and decreased dairy content. In particular, fructose (too much) and fiber (not enough) appear to be cornerstones of the obesity epidemic through their effects on insulin. It has long been known that the hormone insulin acts on the brain to encourage eating through two separate mechanisms. First, it blocks the signals that travel from the body's fat stores to the brain by suppressing the effectiveness of the hormone **leptin**, resulting in increased food intake and decreased activity. Second, insulin promotes the signal that seeks the reward of eating carried by the chemical dopamine, which makes a person want to eat to get the pleasurable dopamine "rush."
Calorie intake and expenditure normally are regulated by **leptin**. When **leptin** is functioning properly it "increases physical activity, decreases appetite, and increases feelings of well-being." Conversely, when **leptin** is suppressed, feelings of well-being and activity decrease and appetite increases - a state called "**leptin resistance**." Changes in food processing during the past 30 years, particularly the addition of sugar to a wide variety of foods that once never included sugar and the removal of fiber, both of which promote insulin production, have created an environment in which Western foods are essentially addictive. It will take **acknowledgement** of the **concepts of biological susceptibility and societal accountability** and de-**emphasis of the concept of personal responsibility** to make a difference in the lives of children..

* Gardasil: the first-ever vaccine designed to prevent a cancer, approved by the FDA, works by building immunity against the sexually transmitted human papillomavirus (HPV). Because the vaccine needs to be administered before the virus is present, public health groups push for mandatory vaccinations for young girls; an idea that is fervently refused by conservative organizations.

Whether mergers and acquisitions actually do add to the value of the pharmaceutical company's investor is an open question; it is indeed very difficult to judge. When one compares the share price performance of the acquiring company (expressed in US $) in the 12 months following the announcement of the relevant deal with the Standard's & Poor (S&P) drug index of the same period, a picture emerges of under-performance versus the rest of the S&P index[75] :

The main reasons for a merger or acquisition would be the benefit generated by cutting costs (staff), which makes the bottom line look better immediately. The problem with this, experts repeatedly warn, is that R&D must deliver twice as much. However, this places higher pressure on R&D to deliver, which most companies find themselves unable to achieve. Increasingly, BigPharma has come to accept that mergers do not provide the necessary improvements in company performance and may actually draw resources and attention away from drug discovery and development.

But evidence from analysis of the top 39 pharmaceutical and biotech companies indicates a potential negative relationship between company size and R&D productivity. This data highlights the R&D effectiveness of smaller companies particularly the biotech ones such as Genentech and Gilead, which are expected to demonstrate strong productivity.

Mergers and acquisitions have   changed the global landscape for good. Of the many types of deals, management buy outs have been the exception. Between the two extremes of trying to grow a company purely through organic means (R&D, sales, marketing) and of merging or acquiring another company, lies the possibility of a collaboration with large or small pharmaceutical companies in which the complementary strengths of the partners can be better exploited.

As the final value of a company reflects the value and quality of the products that is sells, most alliances deal with R&D and are made between a large international company and a research-rich start-up company. This allows the large company to spend less R&D money internally and exploit the superior research knowledge of the smaller company.

Table 4.5.   The need for alliances and partnering

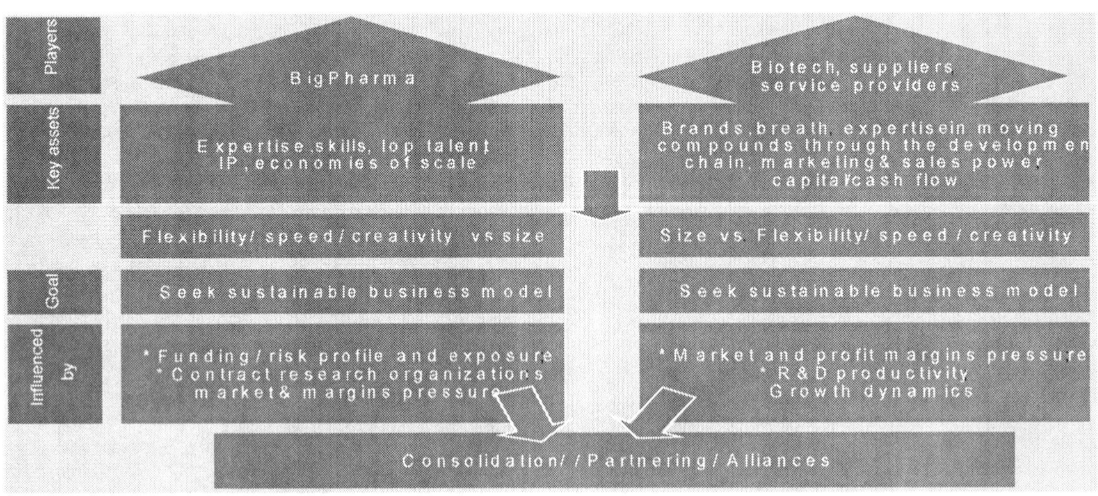

---

[75] *Neirinckx R & Mertens G, Mergers and Acquisitions in Pharmaceuticals. Financial Times Pharmaceuticals Management reports, London;2000:65,151*

* Anti-IgE (**Omalizumab**) is a **humanized monoclonal antibody** and the only therapy which targets **IgE**, the root cause of **allergic asthma**. It is the **first biotechnology derived** treatment for allergic asthma and was approved by the FDA in June 2003 for the treatment of patients diagnosed with moderate-to-severe forms of the disease.

* Xyotax **(paclitaxel poliglumex)** from Cell Therapeutics, Inc., is a **biologically-enhanced chemotherapeutic** that **links paclitaxel**, the active ingredient in Taxol, to a **biodegradable polyglutamate polymer**, which results in a new chemical entity. When bound to the polymer, the chemotherapy is rendered inactive, potentially sparing normal tissue's exposure to high levels of unbound, active chemotherapy and its associated toxicities. Blood vessels in tumor tissue, unlike blood vessels in normal tissue, are porous to molecules like polyglutamate. Based on preclinica studies, it appears that **Xyotax is preferentially distributed to** tumors due to their leaky blood vessels and **trapped in** the tumor bed allowing significantly more of the dose of chemotherapy to localize in the tumor than with standard paclitaxel. Once inside the tumor cell, **enzymes metabolize** the **protein polymer, releasing** the **paclitaxel chemotherapy**. Preclinical and clinical studies support that Xyotax metabolism by lung cancer cells may be influenced by estrogen, which could lead to enhanced release of paclitaxel and efficacy in women with lung cancer compared to standard therapies.

* In the United States over 500,000 people die from **Sudden Cardiac Death** (SCD) with no warning symptoms whatsoever; often times in the prime of life. People at risk for SCD include those who have suffered from a myocardial infarction (heart attack), cardiomy-opthy (weakened heart muscle), a history of ventricular arrhythmias, syncope (fainting) and prolonged QT-interval. A novel technology (the **Micro-T-Wave Alternans technology**) for identifying patients at risk for sudden cardiac death (SCD) has been developed.

* In spite of the issues surrounding gene ownership, and the effect it may have on academic freedom to further investigate genes under patent, well over **2,000 patents** were granted between 1998 and 2003. It is thought that in 2006**, 20% of human genes** were the **subject** of US **patent claims**.

* Bacteria and humans use a number of tools to direct perhaps the most important function in cells—the **accurate copying** of DNA during cell division. Scientists have revealed how cells use these tools to keep the process moving smoothly.

* NIGMS-supported researchers have found that bacteria of different species can talk to each other using a common language—and also that some species can manipulate the conversation to confuse other bacteria.

* Researchers have found that two enzymes that catalyze the **same reaction** have **opposite effects** on cell growth and death. These findings may help researchers develop cancer therapies that target one enzyme, while leaving the other alone.

* The University of Minnesota's "Project EAT" (Eating Among Teens) showed startling results of 2,500 female teenagers studied over a five-year period, from 2001 to 2006 . The study found that high school-aged females' use of diet pills nearly doubled from 7.5 to 14.2 percent. By the ages of 19 and 20, 20 percent of females surveyed used diet pills. Through a greater understanding of the socioeconomic, personal and behavioral factors associated with diet and weight-related behavior during adolescence more effective nutrition interventions can be developed.

* The glo**bal revenues** of publicly-traded **biotechnology** companies exceeded **$60.0 billion** for the first time in 2005 in the **sector's 30-year history**, according to a report from **Ernst & Young,** entitled Beyond Borders; the Global Biotechnology Report 2006. While the US biotech industry remained strong, stable and showed maturation, the Europeans have recovered from several years of consolidation and disappointing financial results and performed significantly better with a record **23** initial public offerings (**raising 52%** or 150 million Euro of **total European IPO capital**) - compared to 13 in the USA - taking the number of public companies to 122. The **UK biotech sector** is in rude health with products, revenues, financing and corporate activities **leading Europe**. Major drugmakers made several large acquisitions as they faced their biggest patent-expiration year ever, with an **estimated $23.0 billion worth** of **pharmaceutical products losing protection**, The future is expected to bring less IPOs and more big pharma M&As, as pricing and the cost of developing new drugs still presents a challenge for the biotech sector. Increasing specialization as well as increased use of alliances, partnerships to leverage off others and specialist skills to bring products to market will dominate the field.

* A U.S. federal study indicated that 37 million older Americans suffer from **frequent sleep problems** that, if ignored, can complicate treatment of other medical conditions, such as arthritis, diabetes, heart and lung disease, and depression. Although **most older adults (67%)** report frequent sleep problems, only about seven million elderly patients have been diagnosed. **Benzodiazepines**—drugs that bind to the benzodia-zepine receptor of the GABA-A receptor complex—have long been considered the gold standard for treatment of insomnia due to their documented efficacy, relative ease of optimization of pharmaco-kinetic properties, and good patient tolerance versus earlier hypnotics. The potential problems with benzodiazepine use have led to development of novel non-benzodiazepine agents that **act** at **similar GABA receptors** and have **comparable mechanisms** of **action**. In the U.S., the nonbenzodiazepines zolpidem (**Ambien**), zaleplon (**Sonata**), and more recently, eszopiclone (**Lunesta**), are marketed as **alternatives** to the **benzodiazepines**. **Ramelteon** (Rozerem) was approved by the FDA on July 22, 2005 for the treatment of insomnia characterized by difficulty with sleep onset (sleep latency); and it is the **first melatonin receptor agonist** to be approved for this indication

Even though the higher research component increases the risk for a failure of the collaboration, the rewards for success are also larger because any breakthrough products are not going to be diluted by the lesser performance of other products[76].

Is it necessary to Have ALL "new technology" knowledge in-house? In the R&D led pharmaceutical industry, establishing more   partnerships is the only way to get better leads.

Table 4.6.   Getting better leads, "New Sciences"

Chemogenics	Bioinformatics	Non-coding RNA	Physiomics
Proteomics	Theragnostics	Addiction biology	Pharmacogenomics
Ageing	Bionics	Zinc biology	"new" viruses
Non-structural chemical patenting	Stem cells and tissue banking	Combinatorial biology	Micro automated physiology
Automated surgery	Brain biology	Nanotechnology	

IV.4. Polarization of the drug developing pharmaceutical industry

The dramatic increase in R&D budgets –see the graph below- has led to the creation of a small elite club of R&D heavyweights. For example, in 2000, only five companies (GlaxoSmithKline, Roche, AstraZeneca, Aventis, and Merck & Co) could come up with over 50% of Pfizer's global R&D budget and only 14 companies shared almost 80% of the global top R&D pharmaceutical research budgets. By 2010, only five companies are expected of being able to invest in expensive new medicine research.

This means that, theoretically, the selection and decision of which disease areas will be "worth" investing in –from an economic point of view- could depend on five to ten company CEO's. A fact, that neither politicians nor society, will be likely to accept.

The current blockbuster model, cheered by the pharmaceutical industry, is unsustainable in the long-term. For example, the average cost to develop a new drug has reached US$ 876 million, according to the Tufts Center of Drug Development. Some sources already put the figure at US$ 1.7 billion.

---

[76] *Neirinckx R & Mertens G, Mergers & Acquisitions in Pharmaceuticals: Why and How?; Strategic Healthcare Management Reports, Financial Times Business Ltd, 2000.*

* **Nystagmus** causes the eyes to move in an uncontrollable manner, so that people with the condition cannot keep their eyes still. Nystagmus can be congenital (occurs at birth or in early childhood) or acquired later in life due to neurological disease. Largely under-researched, Nystagmus is one of a significant number of interests within the **Leicester Ophthalmology Group** concerned with normal and abnormal eye movements. Researchers are looking into many aspects of nystagmus, including possible drug treatments, its epidemiology, impact on visual function, adaptation of the visual system to the constant eye movements, the causes of the condition and its genetic make-up. They are also investigating other eye movement problems, such as reading in schizophrenia and treatment for amblyopia (a lazy eye), including patching therapy with special reward incentives and the education of parents and teachers. December 2006, they have identified for the first time a gene which causes the distressing eye condition.The discovery of this gene will make a **genetic test** for **idiopathic X-linked nystagmus** possible and will lead to greater understanding about the protein which is abnormal in nystagmus.

* Women with **ovarian cancer** are more likely to receive a treatment that will significantly extend their lives. Ir January 2006, results of a major trial were published in the **New England Journal of Medicine** showing that women who received additional chemotherapy pumped directly into their abdominal cavity lived on **average 16 months longer** than women who got the standard care. That is the **biggest improvement in lifespan** seen in the treatment of ovarian cancer in several decades. Due to the impressive results, the National Cancer Institute has urged all doctors to add the **abdominal chemotherapy treatment.**

* Since **Intuitive Surgical's da Vinci system** was approved by the U.S. Food and Drug Administration in July 2000 **robotic surgery** jumped from the pages of science fiction books into the operating room. It has been used in a variety of surgeries from kidney transplants to hysterectomies. The system places the surgeon in front of a console where he or she uses knobs to maneuver tiny surgical instruments with cameras attached to adjustable robotic arms About **36,600 robotic surgeries** were done in 2005, up nearly 50 percent from 2004 In 2005, more than 20 percent of U.S. prostatectomies were done with a robot. Robotic surgery should also gain new ground in advanced **laparoscopic procedures**, along with **esophageal**, **pancreatic**, **colon** and **rectal** surgeries.

* Scientists discovered that a **variant** of a **gene** called **TPMT**, can lead to life-threatening reactions to certain dose₅ of **chemotherapy.**

* The origins of the more virulent strains of **Toxoplasma** were first documented in a 2001 Science paper from Professor **Boothroyd** of Stanford University School of Medicine; the researchers found that the recombination of two relatively benign strains of Toxoplasma can result in a thousand-fold increase in their ability to cause serious disease.

* Disappointing results in the treatment of **glioblastoma multiforme** (the most common form of malignant brain tumor) with surgery, radiation, and chemotherapy have fueled a search for new treatment modalities. Therapies need to act in areas of the brain that are **spatially separated from the site of tumor origin** and **over extended periods of time temporally separated from their introduction**. Over the past decade, laboratory studies and early clinical trials have raised the hope that these therapeutic requirements may be fulfilled by gene therapy. In gene therapy for cancer, viral vectors are administered systemically or intratumorally by surgeons. These viral vectors can be **nonreplicating viruses** that deliver **transgenes** whose expression leads to an anticancer effect, or they can be **replicating oncolytic viruses** that achieve an anticancer effect through **viral replication** leading to cellular lysis. Glioblastoma multiforme represents an excellent target for cancer gene therapy in which viral vectors are used that only propagate in dividing cells, because glioblastoma cells are among the few rapidly proliferating ones in the CNS, where only some microglia and endothelial cells have the ability to divide.

* In 2005, in a comparison of **nanotechnology**-related **patents** based on the type of application, **11 percent** of all nanotechnology-related patents issued by the patent office involved some type of medical application, including **drug delivery, diagnostics, drugs, biologics, nanoemulsions, cosmetics, nanoshells, nanopores, nanocapsules and quantum dots.**

* Scientists discovered that **inherited mutations**, including BRCA1 and 2, contribute to 5 to 10 percent of all breast cancers, and that the main gene variants involved in colon cancer account for 3 to 5 percent of diagnoses.

* In 1999, researchers at UCSF Institute for Regeneration Medicine and UCSF Department of Neurological Surgery, discovered that cells in the Subventricular Zone (SVZ) of the brain region known as **astrocytes** function as **adult neural stem cells** in mice (*Cell*, June 11, 1999) and later discovered similar cells within the **human brain** (*Nature*, Feb. 19, 2004).

* Although South Korean claims to have generated new embryonic stem cell lines using **nuclear transfer** were shown to be false, researchers will continue attempting to create new stem cell lines using that technique. Nuclear transfer, which has been carried out successfully in a variety of other mammals, is necessary for some promising therapies that involve creating stem cells from a person's own tissues.

* Some bacteria and viruses bind to receptors found on the cell surface and use them to gain entry to cells, according to a team of NIGMS-supported researchers. Once inside, they are out of the line of fire of the immune system and can begin to replicate.

* Scientists have uncovered the molecular basis of a defective DNA repair mechanism. This discovery may lead to ways of fixing the process to avoid Huntington's disease and some types of colon cancer.

Table 4.7. Rejuvenating the industry drug pipeline: An elite club. And the "others"

1990 R&D $bn			2000 R&D $bn			2006 R&D $bn		
GlaxoWellcome	0.75	100%	Pfizer	4.21	100%	Pfizer	7.599	100%
Bristol-Meyers Squibb	0.69	93%	GlaxoSmithKline	3.81	91%	GlaxoSmithKline	6.549	82%
Bayer	0.63	84%	Hoffmann LaRoche	3.21	76%	Sanofi-Aventis	5.844	72%
Hoffmann LaRoche	0.62	83%	AstraZeneaca	2.61	62%	Novartis	5.474	72%
SmithKlineBeechem	0.61	81%	Aventis	2.53	60%	Schering AG+Bayer	5.401	71%
Merck & Co	0.56	75%	Merck & Co	2.34	58%	Johnson & Johnson	5.000	66%
Eli Lilly	0.54	73%	Eli Lilly	2.01	48%	Roche, Genentech, Chugai	4.918	65%
Sanofi	0.47	63%	Pharmacia	2.01	47%	Merck & Co	4.783	63%
Abbott	0.45	60%	Novartis	1.91	45%	Amgen	3.366	44%
Pfizer	0.43	57%	Johnson & Johnson	1.91	45%	Eli Lilly	3.129	41%
RhonePoulencRorer	0.38	51%	Bristol-Meyers Squibb	1.87	44%	BMS	3.067	40%
Schering Plough	0.38	51%	American Home	1.62	38%	Abbott	2.255	30%
Warner Lambert	0.37	50%	Schering Plough	1.33	32%	Schering Plough	2.188	29%
Takeda	0.36	49%	Bayer	1.28	31%	Boehringer Ing	2.015	27%
Zeneca	0.36	48%	Boehringer Ingelheim	0.89	21%	Takeda	1.444	19%
Hoechst	0.32	43%	Abbott	0.87	21%	Merck KGaA+Serono	1.352	18%
Akzo	0.26	35%	Sanofi-Synthelabo	0.86	21%	Eisai	1.328	17%
American Home	0.24	32%	Amgen	0.84	20%	Astellas	1.189	16%
Searle	0.22	30%	Sankyo	0.54	13%	Otsuka	833	11%
Eisai	0.21	29%	Dupont	0.53	13%	Altana+Nycomed	860	11%
Astra	0.21	28%	Schering AG	0.52	12%	UCB	811	11%
Fujisawa	0.21	27%	Genentech	0.49	12%	Biogen Idec	718	9%
Sankyo	0.19	26%	Novo	0.42	10%	Novo Nordisk	689	9%
Yamanouchi	0.19	26%	Merck KGaA	0.41	10%	Genzyme	650	9%
Shionogi	0.18	24%	Takeda	0.41	10%	Akzo Nobel	639	8%
Daiichi	0.15	21%	Akzo	0.41	10%	Baxter	614	8%
Tanabe	0.15	20%	Yamanouchi	0.38	9%	Solvay	559	7%
Ares serono	0.11	15%	Eisai	0.38	9%	Teva	495	7%
Synthelabo	0.11	14%	Daiichi	0.32	8%	Allergan	476	6%
Ono	0.07	10%	Biogen	0.32	7%	Mitsubishi	408	5%
Dainippon	0.06	9%	Serono	0.26	6%			

(Sources: Lehmann Brothers at the Financial Times World Pharmaceutical Conference in London 11/2003, Scrip issues from January to March 2006 and Januayr to March 2007; Pharmaceutical Executive monthly issues 2005/2006; Pharmaceutical Executive Europe July/Aug 2007; he Globe and Mail, Dec 6, 2006; Pharma Marketletter weekly issues 2004/2005/2006/2007; Hoffmann LaRoche Press Conferences 2006/2007).

Table 4.8. Making medicines

The cost of development
The development cost for a single drug is subject to a variety of estimates, from US$ 897 million (according to the Tufts Center for the Study of Drug Development) to US$ 1.5 billion (according to industry sources)
The time it takes
A new drug´s gestation period from inception, Phase I, II, III clinicla trials, to marketing authorization by a country´s health authorities takes betw een 10 and 15 years
Success/failure ratio
About one in 5.000 to 10.000 early-stage drug candidates makes it all the w ay through the development process to market...day after day, scientists screen numerous molecules on their potential medicinal properties

* Data from the Integrated Healthcare Information Services, Inc. (IHCIS) National Medicare Benchmark Database, which includes information on more than 30 managed care plans and covers 25 million lives, were analyzed by Naslund and colleagues.[Among 1,134,491 male patients over the age of 50 with a total of 963,425 years of follow-up, they found that **BPH is the fourth most common treated disease** with a **prevalence rate** of **13.5%,** following coronary artery disease and hyperlipidemia, hypertension, and type 2 diabetes mellitus. Incidentally, prostate cancer at a 7.8% prevalence rate ranks number 10 in this population. The authors also analyzed the costs associated with the treatment for these 10 most common conditions and found that **prostate cancer and BPH were among the 10 most costly diseases to treat**, according to this database, with prostate cancer being first and BPH eighth on the list. This finding highlights the enormous socio-economic **importance of prostate diseases, both benign and malignant, to our society and healthcare delivery system.** *[Naslund MJ et al. The prevalence, costs, and burden of enlarged prostate (EP) in men =50 years of age. Program and abstracts of the American Urological Association 2006 Annual Meeting; May 20-25, 2006; Atlanta, Georgia. Abstract 1345].*

* In 2000, researchers from **Chiba University** in Japan and the **University** of **Tokyo** simultaneously discovered a compound in certain **fungi** (clinging to the walls of a New Zealand cave) which might have some tremendous potential as a prototype anticancer agent: it selectively destroyed cells depending upon a gene called *ras*—one of the first known **cancer-causing genes**. They had found **rasfonin**, a compound that seemed tailor-made to knock out *ras*-dependent cancers like **pancreatic cancer**. The **chemical tricks certain cancer cells into suicide while leaving healthy cells untouched**. But no one has been able to study or develop it because it's so hard to get enough of it from natural sources. To bring about a new drug, organic chemists must produce a new chemical in enough quantity to test it under many different circumstances to tease out its modus operandi. No method existed to generate rasfonin, aside from growing more fungus—a time-consuming and terribly inefficient method.

*In 2006 Researchers at **the University of Rochester** revealed a process that produces *67 times* more rasfonin than any previous method. For the first time, scientists can obtain enough rasfonin to conduct proper biological tests on it.

* Researchers have discovered two pathways that **regulate** the organization of structures **inside the nucleus**. Nuclear architecture affects how and when genes are expressed.

* "The field of **developmental biology** is another potential area for big ideas. Many of the genes that serve as the building blocks of development have been identified, so **now we just need to understand how to put them together.** By studying the fundamentals of development, such as how organs are fashioned in the embryo, scientists may begin to determine how to manipulate **stem cells** to adopt specific fates, such as to turn into **neurons** or **heart cells**" *(Eric Olson professor and chairman of the department of molecular biology at UT Southwestern. Science's Next Big Ideas. Symposium speaker at Duke University New York, September 26, 2006)*

* A microscopic green alga helped scientists at the **Salk Institute for Biological Studies** identify a novel function for the **retinoblastoma protein (RB),** which is known for its role as a **tumor suppressor** in mammalian cells. By **coupling cell size** with **cell division,** RB ensures that **cells stay within** an **optimal size range.**
The human body is composed of trillions of cells, each of which must coordinate its growth and division in order to maintain size equilibrium. This process is very tightly regulated and any given cell type will always stay within a very narrow size range, but the means by which cell size is determined remain mysterious. In proliferating cells, control mechanisms termed checkpoints are thought to prevent cells from dividing until they reach a specific size, but the nature of the checkpoints has proved difficult to dissect. Understanding how cells balance the opposing processes of growth and division in order to achieve size control is more than just a fascinating intellectual pursuit for cell biologists: **loss of size control** is a **hallmark** of **cancer cells**, which exhibit severe defects in regulating growth and division.

* In the culmination of four years of work, NIGMS-supported scientists have captured the first 3-dimensional, atomic-resolution images of the motor protein, **myosin V**, as it "walks" along other proteins, revealing new insights that advance the current model of protein motility.

* Driven by advances in biotechnology and computer software, newfound genetic knowledge is swiftly making its way into the clinic. Thousands of cancer patients are already benefiting from a half-dozen **targeted** drugs, such as **Gleevec, Iressa**, and **Tarceva**, that are known to work better in people with certain genetic profiles These therapies represent the very first of a wave of many drugs that are expected to emerge from current research based on **human genetic variations.**

* Research led by Cold Spring Harbor Laboratory may **link viruses** that **have been considered harmless** to **chromosomal instability** (CIN) and cancer. According to lead investigator Dr. Lazebnik, cells can be made cancerous **by viruses** that **fuse cells.** Fusion abruptly unites two or three cells under the same membrane, a change that triggers massive CIN and creates diverse cells whose properties change over time. The researchers found that some of these cells can become **cancerous**. The model that the researchers propose suggests that the best hope for controlling cancer is its prevention. If viruses can cause CIN, **identifying** and **inactivating these viruses** would help to decrease the incidence of the disease. This prognosis is consistent with the successes of **preventative immunization** against hepatitis B and papilloma viruses, which both cause cancer. The scientists estimate that at least 18 of 29 virus families that infect human cells have species that fuse cells, which implies that CIN and cancer might be caused by viruses, even those that are considered harmless. If the model they propose is correct, protecting the body against viruses, or **preventing the cell fusion** that they cause may decrease the frequency of cancers and prevent their progression. *(Media release by Dagnia Zeidlickis. Cold Spring Harbor Laboratory, Feb 22, 2007)*

Included in Tuft's analysis are expenses of project failures and the impact of long development times. In addition, the Tuft's estimate accounts for out-of-pocket clinical costs, out-of-pocket discovery and preclinical development costs, clinical success, phase attrition rates, and cost of capital. After adjusting for inflation, the out-of-pocket cost per approved new medicine increased at a rate of 7.6% per year between 1991 and 2001.

Table 4.9. New medicines development: the time it takes

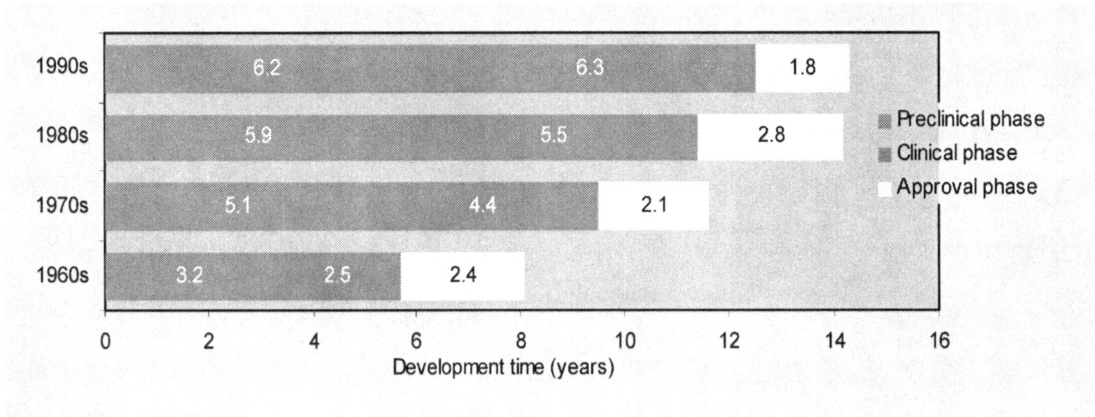

Large pharmaceutical companies have been unable to sustain double digit growth based on a business model of launching ever-larger primary care blockbusters. The blockbuster model focuses on a reductionist strategy of billion dollar brands, promoted by large teams of sales reps -80.000 in the US alone- who are specialized in the promotion of a primary cash cow that is capable of yielding returns on investment (ROI) of 50% to 100% instead of the 5% returns delivered by the "average" drug.

IV.5. Clinical development times are increasing
In the late 1990s, clinical development times were reduced following BPR programs. Since then, they have begun to increase again, mainly because of the greater requirements for safety data, of the hard end-points (e.g. surgical in oncology vs. Tumor shrinkage), and of the treatments targeted at chronic and degenerative diseases.

Table 4.10. Average clinical development times selected by therapy area (1996-2004)

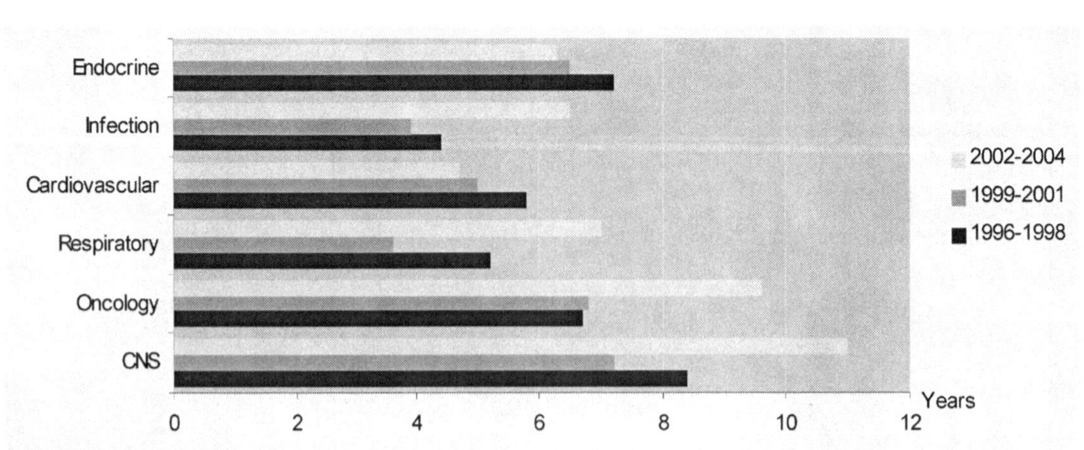

(Based on data from Tufts CSDD)

* **RU486 (mifepristone),** the so-called abortion drug, and one of the most controversial drugs ever, could become the first therapy specifically targeted at PMD, a disease that affects more patients than either schizophrenia or manic depressions.

* A large number of **infants** is **born after 32 to 35 weeks' gestation.** US Health Authorities recommend **RSV prophylaxis** to certain children with a history of preterm birth, chronic lung disease, or congenital heart disease is based. Monthly administration of palivizumab during the RSV season results in a 45% to 55% decrease in the rate of hospitalization attributable to RSV. **Palivizumab** and Respiratory Syncytial Virus Immune Globulin Intravenous (RSV-IGIV) are licensed by the FDA for use in preventing **severe respiratory syncytial virus** (RSV) infections in high-risk infants, children younger than 24 months with chronic lung disease (formerly called **broncho-pulmonary dysplasia),** and certain preterm infants

B.47. Antiviral antibody:

compound	Indication	Target Mechanism of action	Brand name (company)	Sales 2005 %vs. 2004 US$mn
Rec humanized mab	Prevention of RSV in premature infants	Respiratory syncytical virus	Synagis; palivizumab (MedImmune)	1,100 (+13)

* Research has identified **individual biomarkers** related to **cardiovascular risk**. They include CRP, B-type natriuretic peptide, fibrinogen, D-dimer, and homocysteine. Established cardiovascular risk factors include dyslipidemia, smoking, hypertension, and diabetes. In addition to the conventional risk factors, there is substantial interest in the use of biomarkers to identify people who are at risk for the development of cardiovascular disease and who could be **targeted** for **preventive measures**.

* Scientists previously identified two types of T cells, both produced in the thymus: "**effector T cells**", which attack infected cells, and "**regulatory T cells**", which suppress the immune system, protecting the body from inflammatory damage during infection. Regulatory T cells, if given to individuals **receiving transplants,** may help **suppress** the **rejection response**. A team of researchers (led by Prof. Adrian Hayday at the King's College London School of Medicine at Guy's Hospital and Dr Daniel Pennington at Queen Mary, University of London) has discovered a novel mechanism determining whether a maturing T cell is likely to emerge from the thymus as an effector cell or a regulatory cell. The research suggests that new treatments could be developed to **deliberately affect** the type of T cells produced, allowing scientists to tackle a number of diseases which are influenced by these different types of T cells.

* Scientists determined the structure of **ubiquitin bound** to the enzyme, **isopeptidase T.** This could help to develop agents that block ubiquitin binding, and lead to treatments for certain diseases such as cancer.

* Preventive vaccination against **HPV (human papillomaviruses)**: A German-French consortium plans to develop targeted strategies for direct treatment both of precancerous lesions and advanced stages of cervical cancer. A main research focus of the virology alliance is the search for direct and indirect mechanisms by which so-called high-risk HPV types trigger tumor development and influence tumor growth. In addition, the partners work on applications in clinical practice. Here, the researchers endeavor to develop novel molecular HPV tests to supplement or supersede conventional cell-biological tests. This will facilitate a more differentiated diagnosis and early detection of HPV infections than before. In this effort, researchers are interested in new prognostic markers which help medics to assess the individual risk of a patient with HPV infection of developing cancer. **Cervical cancer** is the second most frequent cancer among women worldwide. Each year, about 500,000 women are newly diagnosed with cervical cancer and about 240,000 patients die from it. In Europe, experts estimate the number of new cases at 65,000 and mortality at 28,000 per year.

* Set up in 1964, the French National Institute for Health and Medical Research (**Inserm**) is the only public body in France entirely devoted to the development of **biomedical research**. Specializing primarily in basic research, its researchers are tasked with studying all human diseases, from the most common to the rarest. With an annual budget of 480 million Euro in 2004, the Inserm supports **366 research teams** and employs 13,000 people, 6,040 of them researchers. In addition to the Inserm, the French National Center for Scientific Research (**CNRS**), mainly in its "**life sciences**" department, the French National Institute for Agricultural Research (**Inra**) for the "human nutrition" aspect and the French Atomic Energy Commission (**CEA**), "life sciences" department, are also involved in biomedical research; not forgetting the **Pasteur Institute**, whose researchers **identified** the **AIDS virus** in 1984 and continue to decipher the mechanisms of the disease. They have, for instance, discovered the **mechanism** by which the AIDS virus (HIV) **impairs** the **response** of the **immune system.**

* The **lack** of **inherent rewards** for **implementing socio-behavioral interventions** is in sharp contrast to other parts of the pharmaceutical and health care industries, where new drug discovery, devices, and technologies are rapidly developed and deployed, presumably because there are strong financial incentives to reward the private sector pipeline of discovery, development and delivery. Although the long term benefits of eliminating health disparities are huge and are measured not only in financial terms but also in human terms, there are few inherent incentives and no immediate financial rewards in implementing the **best psychosocial innovations** and **discoveries** that could **eliminate disparities**. *(OBSSR 2006 Prospectus - DRAFT Page 9).*

* A new discovery by NIGMS-supported scientists reveals that the **Pin1 enzyme** plays a **pivotal role** in guarding against the development of amyloid peptide plaques, a brain lesion that characterizes Alzheimer's

It is not exactly the time needed by FDA itself to review the application for the marketing of a new drug. Increasingly stringent regulatory requirements, the increased size of clinical trials, short, it is the need for more data that requires more time for the preparation of the necessary registration files by the applicants.

According to Tuft's CSDD Impact Report[77], which examined new medicinal product approval times by FDA and the European Medicines Agency from 2000 through 2005, EMEA managed to meet its performance goals...but FDA approved 47 products faster than the European Union agency. Fifty-two of the 71 products approved by both agencies were approved by FDA an average of 11.8 months before they received EMEA approval.

Table 4.11. Comparison of average new product marketing approval time needed by authorities to evaluate new medicine marketing applications, worldwide[78]

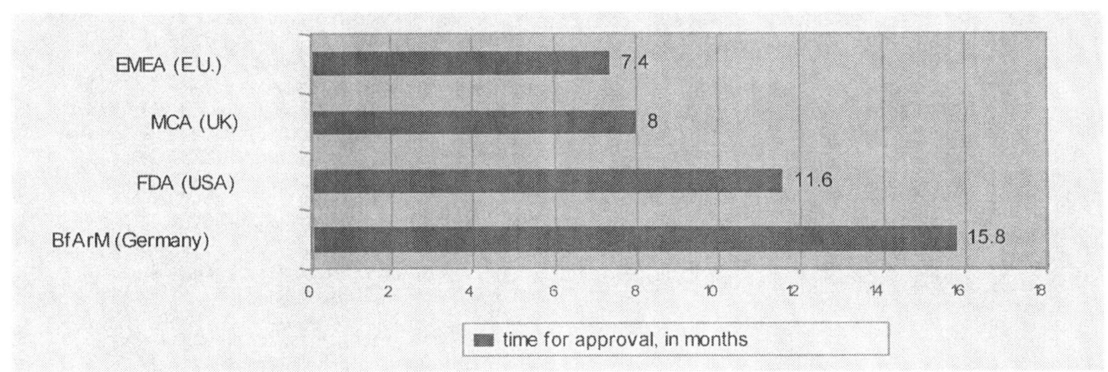

Table 4.12. The New Drug Development Process

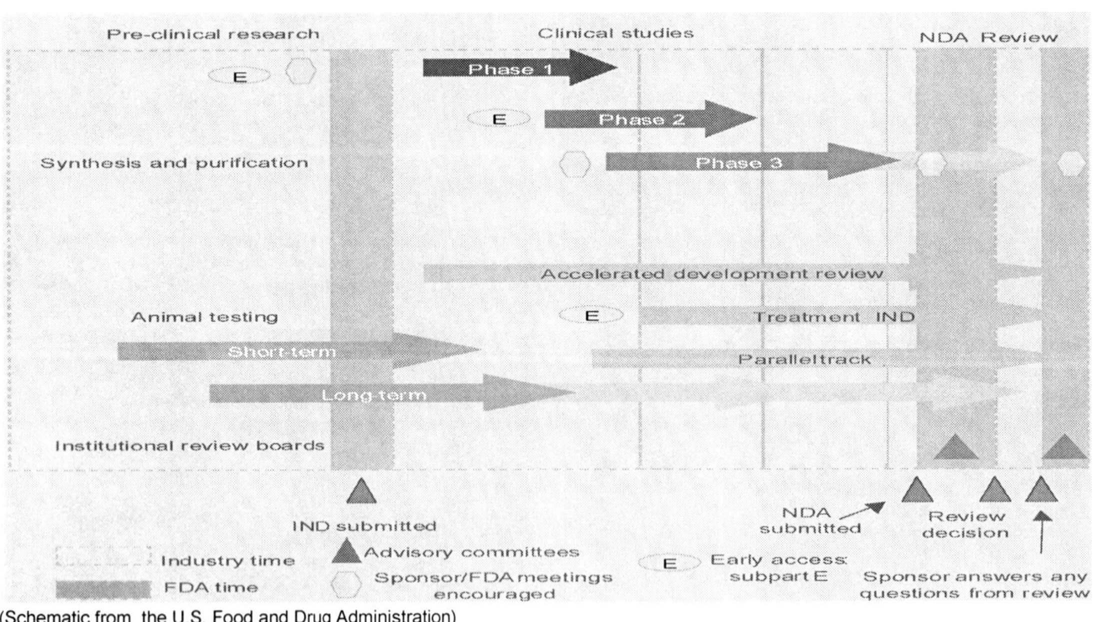

(Schematic from the U.S. Food and Drug Administration)

---

[77] Tufts CSDD Impact Report, January/February 2007.

[78] Source: VfA commissioned report by Boston Consulting Group, 2001

FDA is hardly winning any popularity with pharmaceutical companies. The agency is never fast enough, always too restrictive. Is it? FDA clearly has demonstrated its capability of meeting deadlines when it comes to cutting edge therapies. In 2004, the Agency and the industry agreed on experimenting with a new regulatory process that allows for piecemeal reviews of new drug applications.

Under a conventional review, FDA receives a complete application and all of the agency's review disciplines are working on it at the same time. With the Pilot 1 Rolling Review, only one or two teams are reviewing the application at any one time, significantly limiting interaction between groups critical to the approval. It is part of a "fast track" program designed to expedite review of new drugs intended for serious or life-threatening conditions, and specifically those that demonstrate the potential to address unmet medical needs. This way, FDA shows that it understand industry's concern about the time the agency needs to grant a marketing authorization and shows its willingness to accelerate the approval process, wherever urgency is crucial to safe life[79].

Table 4.13.   Accelerating the approval process

FDA Review Designation	Qualifying criteria
Fast Track	Serves an unmet medical need
Priority Review	Superior to products already on the market and warrants a six-month review
Accelerated Approval	Serves an unmet medical need for serious or life-threatening illnesses. FDA can grant approval without a pivotal Phase III study, using surrogate endpoints that "reasonably" suggest a clinical benefit. Sponsor must commit too a post-approval study to confirm results.
Pilot 1 Rolling Review	Given fast-track status and the sponsor can submit completed pieces of the application as opposed to fiing the submission all at once

A new compound, called ABT-888, has passed the first stage of clinical examination using a new model for drug development that promises to shorten -- by up to six to 12 months -- the timeline for taking anticancer drugs from the laboratory to the clinic[80].

Instead of being tested in the traditional phase I clinical trial, which explores drug safety and tolerance, ABT-888 was tested in a new type of study called an early Phase I trial (also sometimes called a Phase 0 trial). Conducted as part of the pioneering NCI Experimental Therapeutics (NExT) program[81], this early phase I trial shows that using an approach that adds a focus on the mechanism of action (the specific target in a cell that the drug attacks) can reduce the number of patients required for an early clinical study, and the time necessary to gather critical information for development of the drug.

According to its director, With this trial, and according to its Director, the National Cancer Institute (NCI) is doing much more than studying a new intervention; it is blazing a trail for early phase trials, which will hasten our progress across cancer research and drug

---

[79] *Baghdadi R., Flying through FDA:Test runs for "Pilot 1". In Vivo: The business and medicine report, October 2005:50-52*

[80] *According to a team of researchers at the National Cancer Institute (NCI), part of the National Institutes of Health. This result (abstract #3518) was presented June 3, 2007, at the annual meeting of the American Society of Clinical Oncology (ASCO) in Chicago, Ill.*

[81] *The underlying rationale for the NExT program is a guidance issued in 2006 by the U.S. Food and Drug Administration <http://www.cancer.gov/newscenter/pressreleases/FDAGuidance>. The guidance allows human trials to proceed before certain time-consuming and expensive drug development steps, such as extensive preclinical toxicology and bulk production of an agent, occur. Limited toxicity studies are allowed because only restricted doses or shortened courses of drug therapy are permitted in early phase I trials. If an agent succeeds in an early Phase I trial, more comprehensive toxicological studies are performed before the agent undergoes more extensive clinical testing. Early phase I data are anticipated to be useful in improving the efficiency and success of subsequent clinical trials, thereby shortening the overall drug development timeline.*

* According to its **Biotechnology Report 2006**, **Marks & Clerk,** patent and trade mark attorneys, whilst the hostile research environment and controversies that continue to blight stem cell research are not affecting overall investment or the total number of patents filed worldwide, industry has become far more cautious. Biotech companies and the venture capitalist community appear nervous, with uncertainties surrounding return on investment and public reception of such research taking their toll on non-governmental investment. As a result, government and academic bodies make up a higher proportion of patent activity than would be expected of a comparably-aged technology. M&C explores worldwide patent activity related to stem cell technology, RNA interference and genetic diagnostic testing. The report reveals that 2005 has witnessed a marked shift in the pattern of patent filings in the area of **stem cell research**, with **24%** of **patents held** by the **top 20 organizations belonging** to **just three academic and government organizations.** This supports information from another study which found that **government** and **academia** make up **almost half of the top 20 patent holders** and **own seven** of the **16 most "influential" stem cell patents**. The report highlights that over 2,000 patents related to stem cell technology have been granted worldwide since 2000, with a steady growth in the number of patents granted between 2000 and 2003. As a result, patent activity was 41% higher in 2004 than in 2000.

* AstraZeneca gained full approval from FDA for **Arimidex**, the only **aromatase inhibitor** with 5-yr full data set and superior disease free survival over tamoxifen.

* NIGMS-supported researchers report the discovery of an important and completely novel form of regulation of **GLI1**, a gene associated with **severe birth defects** and **several childhood cancers**.

* **Increlex** (Mecasermin injection) from Tercica Inc., received FDA approval for the **treatment of growth failure** in children with **severe primary IGFD** or with **GH gene deletion** who have **developed neutralizing antibodies to GH**. Mecasermin is a recombinant human IGF-1 synthesized in bacteria (*Escherichia coli*) that has been **modified** by the addition of the gene for human IGF-1. Primary insulin growth factor deficiency (IGFD) afflicts an estimated **30,000 children** evaluated for short stature in the U.S. Primary IGFD is a growth hormone (GH)–resistant state characterized by the lack of insulin-like growth factor-1 (IGF-1) production in the presence of normal or elevated levels of endogenous GH.

* Variations in a gene known as **SORL1** may be a factor in the development of late onset Alzheimer's disease, an international team of researchers has discovered. The researchers suggest that faulty versions of the SORL1 gene contribute to formation of amyloid plaques, a hallmark sign of Alzheimer's in the brains of people with the disease. They identified 29 variants that mark relatively short segments of DNA where disease-causing changes could lie. (The genetic clue, which could lead to a better understanding of one cause of Alzheimer's, is reported in "Nature Genetics" online, Jan. 14, 2007, and was supported in part by the US National Institutes of Health)

* Researchers in Italy have demonstrated how **Helicobacter pylori** triggers the **first step** of **cancer** development in cells of the **gastric mucosa**. The bacterium Helicobacter pylori is responsible for the development of stomach ulcers and also of stomach cancer. Dr. **Bagnoli** of Novartis Vaccines, Siena, Italy, is studying the mechanisms how Helicobacter causes the cells of the gastric mucosa to transform. Helicobacter is equipped with special enzymes that enable it to **survive** the **acid attacks** by the stomach. Like a number of other bacterial pathogens, Helicobacter injects a protein into the epithelial cells of the gastric mucosa. While this injection is used by other pathogens to get access into the cell, Helicobacter's protein, called **CagA**, causes a whole range of dramatic changes in the cell biology, as the Siena research group has found out.
Like all epithelial cells, those of the gastric mucosa have two sides with different functions: One faces outward, into the organ lumen, the other is in contact with the blood supply of the tissue. Between the two poles, the cell walls form a tight barrier via close contacts.
Helicobacter dramatically **disrupts this order**. Following CagA injection, epithelial cells lose their polarity, the contact sites break apart. The cells form tiny foot-like extensions that make them mobile and start breaking through the basal membrane that separates them from the blood vessels. CagA causes similar changes in a cell like some cancer genes do. The Italian scientists presume that CagA thus triggers the first step in the development of gastric cancer.

* Scientists of the German Cancer Research Center (Deutsches Krebsforschungszentrum, **DKFZ**) are investigating a key molecule in the development of cancer, infection and inflammation.
The DKFZ believes that a molecule called **DMBT1** plays a key role in cancer development, infection and inflammation. The gene coding for DMBT1 first attracted attention by its absence: Prof. **Mollenhauer** discovered that the genetic information for DMBT 1 is missing in cells of malignant brain tumors. Meanwhile it is known that the DMBT1 gene is completely or partly **lost in 84 percent of tumors** that originate from **epithelial cells**. Numerous results also indicate that DMBT1 plays a role in infection defense. Thus, the protein binds and clots viruses and bacteria, which presumably causes them to lose their infectiousness. In addition, DMBT1 attracts immune cells to the infection site.Recent study results show that DMBT1 is also involved in inflammatory processes. Cells of the intestinal mucosa increase their DMBT1 production as a response to inflammatory stimuli. In cells of the inflamed intestinal mucosa of patients with Crohn's disease, **NOD2**, a protein that is a key sensor of the cell for bacterial infections, gives the signal for a strong increase of DMBT1 production.The scientists conclude from the individual results that complex diseases such as cancer, infection and inflammation, have common underlying molecular mechanisms in which key molecules such as NOD2 and DMBT1 and several others are involved. The scientist speculates that such **"metaproteins"** might serve as central targets for treating a whole range of diseases.

* Relatively **few risk factors**—high cholesterol, high blood pressure, obesity, smoking and alcohol—cause the **majority** of the chronic disease burden. About **75%** of cardiovascular disease can be attributed to these major factors. Up to 80% of cases of coronary heart disease, 90% of type-2 diabetes cases, and a third of cancers can be avoided by changing to a **healthier diet**, increasing **physical activity**, and **stopping smoking**. For over 40 years, unequivocal evidence has accumulated for primary and secondary prevention in coronary heart disease and the array of therapies has grown from aspirin alone to include various antihypertensives, diuretics, statins, thrombolytics, angioplasty, and coronary-bypass graft surgery. The **statin trials** represent some of the **largest clinical trials** in **modern medicine**.*[Banerjee A., Halving the premature death rate. Young voices for Health. Globa Health Forum, 2006]*

* The theoretical approach to cardiovascular-disease prevention has been challenged by the advent of the polypill. The **polypill** consists of a statin (cholesterol-lowering), three antihypertensives (thiazide, beta-blocker, and angiotensin-converting enzyme inhibitor), folic acid, and low-dose aspirin (75 mg). By evaluation of meta-analyses of randomized trials and cohort studies, Wald and Law found that a polypill could **reduce incidence** of myocardial infarction and stroke **by 88%.** *[Wald NJ, Law MR. A strategy to reduce cardiovascular disease by more than 80%. BMJ 2003; 326: 1419].*

* The USA announced the President's Emergency Plan for AIDS Relief (**PEPFAR**) in 2003. It is providing US$ 15 billion in funding over five years to combat HIV/AIDS in more than 120 countries. By the end of March 2006, PEPFAR had supported **antiretroviral therapy** for **561.000** men, women, and children through bilateral programs in 15 of the most afflicted countries in Africa, Asia, and the Caribbean.

* Neurobiologists (Duke University) have discovered why the aging brain produces progressively fewer new nerve cells in its learning and memory center. The scientists said the finding, made in rodents, **refutes current ideas** on how long crucial "**progenitor**" **stem cells persist** in the **aging brain**. The finding also suggests the possibility of treating various neurodegenerative disorders, including Alzheimer's disease, dementia and depression, by stimulating the brain's ability to produce new nerve cells.

* **Neural stem cells** (NSCs) engineered to express therapeutic molecules have shown dramatic efficacy against metastatic cancers in pre-clinical trials. The new technique, described in the launch issue of PLoS ONE (Jan 2007), makes use of the **tendency** of NSCs to **migrate** to diseased and cancerous regions where anti-cancer drugs can then be activated by genes expressed by stem cells at the site of the cancer. The NSCs have been shown to target to even the sub-microscopic cancerous growths believed to cause relapse.

* Researchers have shown that an enzyme that can destroy mRNA to halt protein production can also shelter RNA under stress conditions, allowing quick resumption of protein production when the stress ends.

*The **OECD** estimates indicate that **Iceland** and **Switzerland's** investments in **R&D** for health relative to their **GDPs** are among the **highest** in the **world**.
Globally, most of the **public-sector funding** for health research (more than US$ **56 billion** in 2003) originates in High Income Countries, where it is mainly used to support basic research in universities, medical schools and research institutes. Basic research in the chemical, biological and pharmaceutical sciences gives rise to leads for the creation of new drugs and vaccines. At least in commercially attractive areas, such leads are generally picked up by industry and a process of applied R&D then generates a pipeline of potential candidates for evaluation in clinical trials. In a study, Light argued that **public sources** may be **funding** as much as **84%** of **all basic research** in the health field either directly or indirectly, such as through subsidies and tax breaks to research conducted by the private sector. *[Light,D. Basic research funds to discover important new drugs. Chapter 3 in: Monitoring Financial Flows for Health Research, 2005: Behind the global numbers. Geneva, Global Forum for Health Research, 2005].*

* According to scientists, about a quarter of babies of European ancestry and possibly up to 60 percent of those of Asian ancestry **lack both copies** of the gene called **GSTT1**. Based on their data, published in the January 2007 issue of the "American Journal of Human Genetics", the scientists calculated that if a **pregnant woman smokes** 15 cigarettes or more per day, the chances of her GSTT1-lacking **fetus developing a cleft** increase nearly **20 fold**. Globally, about **12 million** women each year **smoke** through their **pregnancies.**

* Researchers have discovered that an individual microRNA directly regulates a gene implicated in human cancers

* The **accumulation** of **genetic damage** in our cells is a major contributor to **how we age**, according to a study published in the journal *Nature* by an international group of researchers led by Dr Jan **Hoeijmakers**, head of the department of genetics at the **Erasmus Medical Center** in Rotterdam, Netherlands. The study found that mice completely lacking a critical gene for repairing damaged DNA **grow old rapidly** and have physical, genetic and hormonal profiles very similar to mice that grow old naturally. Furthermore, the **premature aging symptoms** of the mice led to the discovery of a new type of **human progeria**, a rare inherited disease in which affected individuals age rapidly and die prematurely.

* Findings described in a new study by Stanford scientists may be the first step toward a major revolution in **human regenerative medicine** - a future where **advanced organ damage can be repaired by the body itself.** In the May 2007 issue of *The FASEB Journal*, researchers show that a human evolutionary ancestor, the sea squirt, can correct abnormalities over a series of generations, suggesting that a similar regenerative process might be possible in people.

* There's a sense of urgency to find new antibiotics. Such urgency is compounded by the speed at which some strains are capable of developing resistance. For example, **bacteria resistant to** one of the newest antibiotics, **linezolid**, appeared **within one year** of the drug's approval for use.

131

discovery. Conducting the ABT-888 trial [82] also required close collaboration and consultation with the FDA, in order to negotiate the prototype for a new regulatory process [83].

ABT-888 inhibits an enzyme critical for repairing damage to DNA. In the absence of an active enzyme, tumor cells are more sensitive to the effects of several cancer drugs.

Abbott Laboratories developed ABT-888 as part of a clinical trial agreement with NCI to study whether performing an early phase I trial would hasten the overall drug development plan for this agent.

A unique aspect of this trial was the development of a rigorous assay to measure a biomarker affected by ABT-888. The study showed that the compound inhibited its targeted enzyme in tumor cells as well as in circulating white blood cells. This latter finding suggests that white blood cells, which are easily accessible, can be used in subsequent trials as an easy way to measure whether the agent is altering its presumed target.

This early phase I trial was a proof-of-principle study aimed to determine whether this new approach to early drug testing could work and that fits in well with the goals of NExT. The goal of the NExT program is to shave up to one year off the typical 10- to 12-year drug-development cycle [84].

IV.6. A new generation clinical trials to save time and money, improve patient care

As we enter the era of personalized medicine, it is time to take a fresh look at how we evaluate new medicines and treatments for cancer. We need to rethink how we design and conduct clinical trials in the United States. Our current system has served us well for the past 50 years, but the demands of 21st century medicine are beginning to put a strain on the current system, and we believe we have something to relieve that strain. [85]

Professor Berry (University of Texas M.D. Anderson Cancer Center) advocates [86] turning the statistical method used to evaluate new drugs on its head. He states that the statistical method used nearly exclusively to design and monitor clinical trials today, a method called frequentist or Neyman-Pearson (for the statisticians who advocated its use), is so narrowly focused and rigorous in its requirements that it limits innovation and learning.

His solution, which he has advocated for more than 30 years, is to adopt a system called the Bayesian method, a statistical approach he says is more in line with how science works. He sites examples of Bayesian approaches being used routinely in physics,

---

[82] *The study was conducted by a team of scientists from NCI-Frederick, Md., CCR, and the Division of Cancer Treatment and Diagnosis (DCTD). The first six patients enrolled in the trial had a variety of cancers and provided the information needed to signal that ABT-888 should proceed to a more comprehensive series of NCI-sponsored Phase I trials, starting 2007*

[83] *Dr. John E. Niederhuber, quoted in a press release by the National Cancer Institute, June 3, 2007*

[84] *According to dr. J.H. Doroshow, DCTD director, the NExT program has grown out of the reality that the number of new anticancer agents reaching human clinical trials has been modest. Even when compounds do proceed to clinical testing, they often fail because of unexpected toxicities or are not effective. NExT not only allows early phase I trials to be conducted in humans, it brings together the teams of scientists necessary to develop and perform assays that can measure the biological effects of potential new anticancer agents. The assays provide a tool for the systematic removal of investigational agents from NCI's drug development pipeline that do not show expected biological effects and inform and expedite decisions about further clinical development. NCI Press release, June 3, 2007.*

[85] *According to Dr. Donald Berry, professor and chair of the Department of Biostatistics and Applied Mathematics at The University of Texas M. D. Anderson Cancer Center, in a statement to the press.*

[86] *Professor Berry outlines his approach to conducting clinical trials in the January 2006 issue of Nature Reviews Drug Discovery.*

geology and other sciences. And he is putting his approach to the test at M. D. Anderson, where more than 100 cancer-related phase I and II clinical trials are being planned or carried out using the Bayesian approach.

The main difference between the Bayesian approach and the frequentist approach to clinical trials has to do with how each method deals with uncertainty, an inescapable component of any clinical trial. Unlike frequentist methods, Bayesian methods assign anything unknown a probability using information from previous experiments. In other words, Bayesian methods make use of the results of previous experiments, whereas frequentist approaches assume we have no prior results.

"Using the Bayesian approach, it is natural to do continuous updating as information accrues," This characteristic makes it possible for us to build adaptive designs in clinical trials" says Berry, arguing that the Bayesian approach is better for doctors, patients who participate in clinical trials and for patients who are waiting for new treatments to become available.

"Doctors want to be able to design trials to look at multiple potential treatment combinations and use biomarkers to determine who is responding to what medication. At the end of the day, when they enroll the last patient in the study they want to be able to treat that patient optimally depending on the patient's disease characteristics. Using a Bayesian approach, the trial design exploits the results as the trial is ongoing and adapts based on these interim results. That kind of thing is an anathema in the standard approach." He says, stating that such flexibility is crucial to clinical trials in the 21st century. "The advances of the 20th century have taught researchers that cancer is a diverse disease, and what works to treat one person's disease may not work for another. In order to have the kind of personalized medicine the 21st century will demand, it will be necessary to be more flexible in how we evaluate potential new treatments."

Of course, the most important factor in whether the Bayesian approach will gain acceptance in clinical trials reporting is whether the U. S. Food and Drug Administration will accept Bayesian approaches in making determination of safety and efficacy of new treatments. The researcher says progress is being made both at pharmaceutical companies and at the FDA in bringing regulators up to speed on the Bayesian approach[87].

In addition, it is possible to reduce the exposure of patients in trials to ineffective therapy using the Bayesian approach. For example, in adaptive clinical trials, if interim results indicate that patients with a certain genetic makeup respond better to a specific treatment, it is possible to recruit more of those patients to that arm of the study without compromising the overall conclusions. Moreover, using the Bayesian approach may make it possible to reduce the number of patients required for a trial by as much as 30 percent, thereby reducing the risk to patients and the cost and time required to develop therapeutic strategies.

---

[87] *For example, the FDA has approved the Bristol-Myers Squibb drug Pravigard Pac for prevention of secondary cardiac events based on data evaluated using the Bayesian approach.*

* 19 of the TOP 20 selling **biologics** achieved sales of over US$ 1bn in the year 2005. The strongest growth rates were found for **therapeutic antibodies** (Avastin: + 141%; Humira: +64 %; Herceptin: +48 %) which made out 6 of the 20 biologics. **Erythropoietins** continue to be the number one selling class of biological products (excluding vaccines), but with overall modest growth and conversion from the native protein to the longer-acting glycoengineered. Sales of the major **cancer antibodies** (US$ 6.77bn) came closer to sales of the TNF-targeting biologics (US$ 7.64bn). **Recombinant insulin-based products** are the number 3 best selling class of biologics. The only antiviral antibody among the TOP 20 biologics achieved for the first time sales of over US$ 1bn.

B.48. 20 Top selling classes of biologics

	Class of preparation	sales 2005	Branded preparations (selected)
		US$ bn	Aranesp, Epogen,
1	Erythropoietins	10.85	Neo-Recormon, Procrit Eprex
2	TNF	7.64	Enbrel, Humira, Remicade
3	Insulin and insulin analogs	7.21	Actrapid, Humolog, Humulin, Lantus, Levemir,
4	Major cancer antibodies	6.77	Avastin, Erbitux, Herceptin, Rituxan/MabThera
5	G-CSF	3.78	Neulesta, Neupogen, Neutrogin
6	Interferon beta	3.77	Avonex, Rebif, Betaferon / Betaseron
7	Human Growth hormone	2.31	Genotropin, Humatrope, Nutropin, Saizen
8	rec. coagution factors	2.19	Benefix, Kogenate, Novoseven, Refacto
9	Interferon alpha	2.11	Intron A, Pagasys, Pegintron
10	Enzyme replacement	1.25	Aldurazyme, Corezyme, Fabrazyme
11	Antiviral antibody	1.1	Synagis
12	follical stimulating hormones	0.97	Gonal-F, Puregon
	Total	49.95	

(Source: LaMerle)

*** RNA splicing antisense technology** allows researchers to influence the ultimate structure and function of proteins. Proteins are synthesized from instructions coded in the DNA through a multi-step process that includes RNA splicing. Information stored in the DNA of genes is transcribed into immature "pre-messenger RNAs" (pre-mRNAs), pre-mRNAs are then spliced into mature "messenger RNAs" (mRNAs), and finally, mRNAs are translated into proteins. In humans and most other organisms, the splicing process thus ensures proper protein production. RNA splicing antisense technology studied at **Cold Spring Harbor Laboratory** (CSHL) effectively **corrected an mRNA splicing defect** found in **spinal muscular atrophy** (SMA) patients, and is now ready to be tested in mouse models. Professor Krainer suggested that SMA patients who suffer from motor-neuron degeneration may benefit from the ability to correct the mRNA splicing defect that makes their *SMN2* **genes** only partially functional.

* Given the emergence of **bio-similars** and increasing numbers of competitor products, the commercial need for **next-generation biologics** has been heightened in a 2006 Datamonitor report, which describes **DME** as a life-cycle management strategy. DME (Directed Molecular Evolution) is a technique that mimics natural evolution to produce optimized protein candidates. In contrast to current techniques for protein optimization, DME is a method that can be applied in the absence of prior structural/functional information. **Datamonitor** says that DME can be thought of as combinatorial chemistry for biologics - instead of (or in addition to) knowledge-based methods of optimizing biologics - DME can be applied to explore the "protein sequence space" of a candidate product.

* In 2003, Dr. **Collins** (Director, Human Genome Research Institute) authored a perspectives paper (Collins FS, et. al., *Nature* 2003;422:835–47) that positioned the HGP(Human Genome Project) as the foundation of a multi-tiered building. The ground floor of such a structure is the application of *genomics to biology*, which encompasses the following goals:
• Define the structure of human variation, the human haplotype map.
• Sequence many additional genomes.
• Develop new technologies for sequencing, genotyping, expression analysis, and proteomics.
• Identify all functional elements of the genome.
• Identify all the proteins of the cell and their interactions.
• Develop a computational model of the cell.

*The **use of stem cells for bone formation** has been studied at **Osiris Therapeutics** since the early 1990s, and numerous preclinical studies have demonstrated the safety and efficacy of adult stem cells for orthopedic applications. The company's product **Trinity** represents the culmination of where bone science has been moving for years. It is unique in the field of **biologics** because it **provides the three functions of autograft necessary for bone formation: osteoconductivity, osteoinductivity and osteogenesis**. Osiris has developed a proprietary process that selectively removes cells that can cause an immune reaction, eliminating the need for donor-to-recipient matching and immune suppression. By providing a safe alternative to autograft, Trinity eliminates the patient discomfort, procedure time, blood loss and risk of complications associated with harvesting autograft.

* **LabCorp** started offering **Comparative Genomic Hybridization** (CGH), a revolutionary new technology that provides physicians with the information necessary to **diagnose over forty genetic syndromes** and **41 subtle chromosomal rearrangements associated with mental retardation and learning disabilities.**
Over six million people in the United States have mental retardation, and in at least thirty percent of those cases, physicians are unable to determine the cause. Many mental retardation syndromes have similar symptoms, and making a diagnosis may often require multiple tests over a long period of time, resulting in significant costs. With a single test, CGH can provide the information that clinicians need to accurately diagnose their patients.

Table 4.14. Expensive high technology drives the R&D process

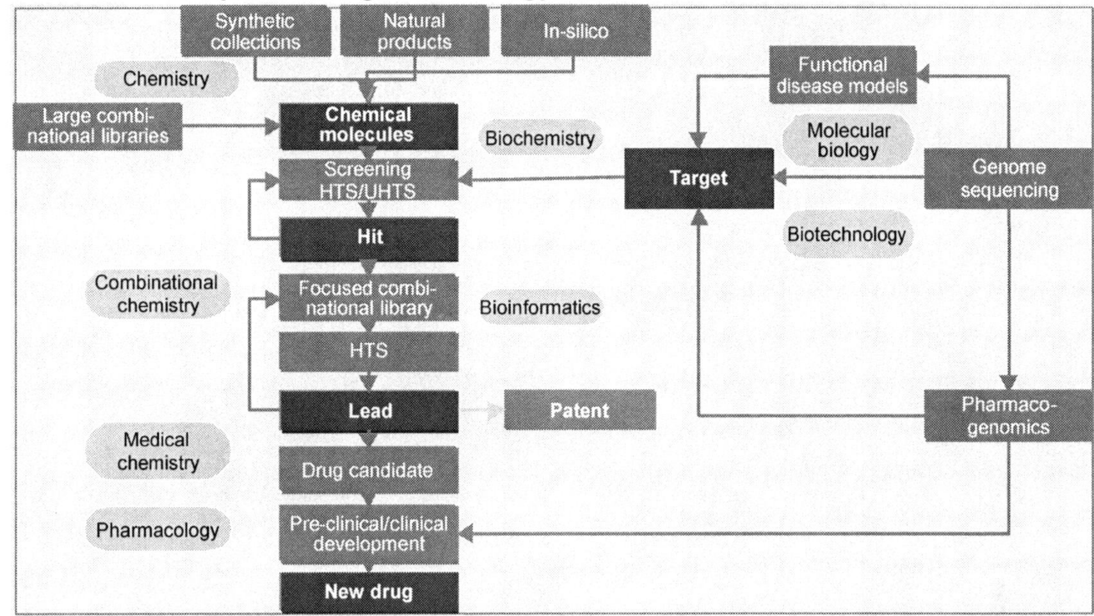

IV.7. Shorter effective patent protection time

A substance discovered 1984 could be launched in 1996, leaving it with an EIGHT year patent protected market exclusivity –to 2004; a short period of time in which full R&D/financial investment has to be earned back before the arrival of generic versions. This situation is far from improving...

A Tufts study included evidence that, after a spike in new drug approvals from 1996 to 1998 — due in part to FDA's clearing of its backlog of applications — total approvals have dropped 47% through the 2002-to-2004 period. Clinical-phase times have increased for most therapeutic areas between 1999-2001 and 2002-2004, and in recent years, drugs for neuropharmacologic diseases have taken the longest to develop and bring to market.

The development process as a whole is lengthening, according to the analysis by the Tufts Center for the Study of Drug Development. This analysis found that new medicines winning approval from FDA in the 2002-2004 period required an average of 8.5 years to move through the clinical and approval phases. This average contrasts with a steady decline in combined clinical and approval-phase times since passage of the Prescription Drug User Fee Act of 1992, from a high of 9.4 years in 1990-1992 to 7.2 years in 1999-2001.

Tufts Director Kenneth Kaitin says lengthening average clinical-phase times had offset the gains made by shorter approval-phase times since passage of PDUFA. He adds that the number of new drug approvals by FDA has fallen and that average clinical times for priority drugs are at their longest since before enactment of PDUFA[88].

---

[88] *PDUFA allows FDA to collect fees from drug companies to be used, in part, to hire additional reviewers and improve the drug-review process. PDUFA was reauthorized in 1997 and again in 2002.*

* **Fighting cancer: They are only a few nanometers in size, but their impact is tremendous**: The tiny particles drive cancer cells to their death in no time at all. **Fraunhofer researchers** demonstrated the great efficiency of **nanoscopic particles** as a vehicle for drug delivery.
They have developed **bio-functional nanoparticles** that **cause necrosis in cancer cells**. The tumor necrosis factor **TNF** for instance, releases a molecule that attaches itself to the receptors of the cancer cell and passes on its deadly message. To introduce the biological messenger TNF into the body, scientists at **Stuttgart University** have developed bio-functional nanoparticles. Known as **nanocytes**, these carry TNF proteins on their surface. In producing these particles, they benefit from the self-organizing capability of the 'building blocks': Once a contact has been established between the particles and the proteins, the proteins grow and envelop the nuclei without any further medical-pharmaceutical intervention. The bio-functional nanoparticles have already proved their mettle in practical applications – as a tool for cell research or as a component in reagents for medical analysis.

* With worries about eroding drug prices due to reforms of health systems in Europe and the U.S., questions about product risk profiles (the Vioxx and Baycol cases) plus the FDA's caution in approving new compounds for the mass market, **generic risk** has become a major issue for the research driven pharmaceutical industry. Mainly in the USA, because of the billion dollar products that come off patent. The "vulnerability" of the European players is considered to be less dramatic.

B.49. Major European Companies´ exposure to generic risk

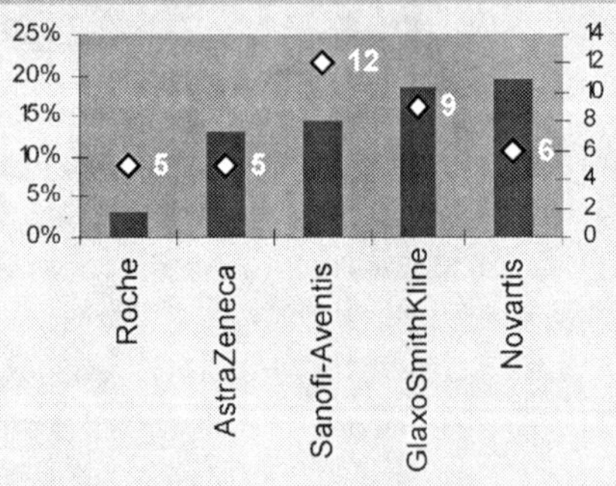

■ Exposure to generic risk (% of 2005 drug sales to go off patent by 2009)

◇ Number of new basic compounds in PhaseIII

* In the USA. the Diabetes Prevention Program demonstrated that **lifestyle interventions** – modest weight loss and regular physical activity – **can reduce the risk of developing type 2 diabetes in high-risk adults by 58%,** compared to 31% reduction with diabetes medication.

* In 2000, *New York Times* headlines stated that **calcium antagonists** were causing something like 85,000 heart attacks per year and cases of congestive heart failure. These headlines were scaring patients and frustrating physicians. At that time, **CCBs** were considered one of the worst options in CV medicine. A new study (published in *Hypertension*. 2006;48:359-361), ended a controversy existing among physicians since 2000 by directly **comparing an ACE inhibitor with a calcium antagonist**. It showed that, if anything, **CCBs (Calcium- Channel Blockers)** are safer and better tolerated. The pendulum has swung from calcium antagonists being the worst possible option to the safest and most efficacious possible option. In their post hoc analysis of cardiovascular and other outcomes in ALLHAT trial with participants randomized to either the dihydropyridine calcium-channel blocker (CCB) amlodipine or the angiotensin-converting-enzyme (ACE) inhibitor lisinopril, the authors found that while rates of fatal coronary heart disease (CHD) or nonfatal myocardial infarction (MI) in older hypertensive patients were similar in CCB- and ACE-inhibitor–treated patients, secondary outcomes differed: CCBs significantly increased the risk of heart failure (HF), while ACE inhibitors increased the risk of stroke, angina, peripheral artery disease (PAD), gastrointestinal (GI) bleeding, and angioedema.

B.50. Comparison between Amlodipine and Lisinopril

Outcome	Amlodipine (%)	Lisinopril (%)	RR (95% CI)	P
Primary end point	11.3	11.4	1.01(0.91-1.11)	0.854
Combined CVD*	32	33.3	1.06(1.00-1.12)	0.047
Stroke	5.4	6.3	1.23(1.08-1.41)	0.003
GI bleed hospitalization	8	9.6	1.20(1.06-1.37)	0.004
HF	10.2	8.7	0.87(0.78-0.96)	0.007
Hospitalized/ Fatal HF	8.4	6.9	0.81(0.72-0.92)	<0.001
Hospitalized or treated angina	12.6	13.6	1.09(1.00-1.19)	0.055
PAD	3.7		1.19(1.01-1.40)	0.036

**Combined CVD indicates CHD death, nonfatal MI, stroke, coronary revascularization procedures, hospitalized or treated angina, treated or hospitalized HF, and peripheral arterial disease. CI = confidence interval.*

The primary end point for the post hoc analysis, as with the main trial, was the **combined incidence of fatal CHD or nonfatal MI by intention to treat:** there were **no differences** in the primary end point between the 2 groups. Secondary outcomes were all-cause mortality, stroke, combined CHD (fatal CHD, nonfatal MI, coronary revascularization, or angina with hospitalization), or combined cardiovascular disease (CVD) (combined CHD plus stroke, treated angina without hospitalization, heart failure, and peripheral arterial disease), end-stage renal disease, cancer, and gastrointestinal bleeding. It is here that key differences emerged. Overall, stroke rates were higher for ACE-inhibitor therapy. Rates of combined CVD were higher in lisinopril-treated patients, driven by higher rates of stroke, PAD, and angina, but slightly offset lower rates of HF. Angioedema also occurred more frequently in the lisinopril-treated patients (38 vs 3; *P* <0.001).

IV.8. Old medicines at low price: healthcare for the poor

Increasing emphasis on generic substitution has led to rapid market share erosion for originator products once their patents expire. Generics will assume a more central role as patients bear a greater percentage of their healthcare costs and payers seek to restrict the growth of healthcare expenditures. Price moderation for branded drugs is likely as a result.

Table 4.15.   The time it takes to bring a substance to market and get a return on investment[89]

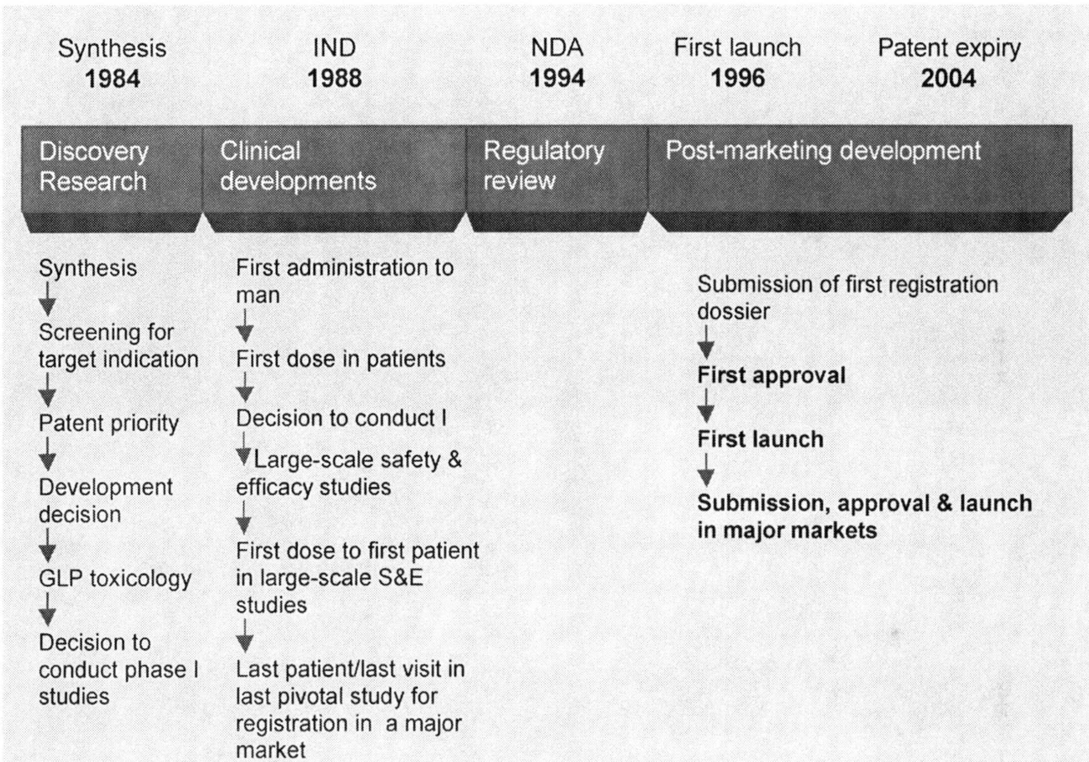

In 2005 sales of generics in the top eight markets (US, Canada, France, Germany, Italy, Spain, UK and Japan) exceeded $55 billion, and are expected to experience double-digit growth over the next five years.

Cost reduction efforts by governments benefit cheaper medicines. Will it lead to "old technology at low price" and raise questions not just about the future of R&D companies, seeing the reward for investment and risk eroded by health authorities with a lack of understanding of the industry's dynamics and value for society, but also about the quality of healthcare for the relatively poor ?

---

[89]   *Professor Stuart R Walker, Director. Center for medicines Research. Financial Times world Pharmaceuticals Conference, London, March 25 & 26, 1996)*

.* The US National Institutes of Health (**NIH**) supports an expansive basic research portfolio on obesity and type 2 diabetes. Among this research are animal and human genetic and developmental studies, and research on nuclear receptors (one of which is already a target for diabetes treatment). Once-a-day **glangine** is the first insulin to offer truly flat insulin levels through the entire day for most users.

B.51. Insulin products

Class of compound	Indication	Target Mechanism of action	Brand name (company)	Sales 2005 % vs 2004 US$mn
Rhu insulin	Diabetes	Insulin receptor	Humalog: insulin lispro (Eli Lilly)	1,198 (+9)
see above	Diabetes mellitus	see above	Lantus: insulin glargine (Sanofi-Aventis)	1,447 (+47,5)
Rhu insulin analogues	Diabetes	see above	Levemir Novorapid, Novomix (Novo Nordisk)	1,165 (+62)
Rhu insulin and related products	see above	see above	Actrapid/Novolin; Insukatard; Mixtard (Novo Nordisk)	2,396 (+4)

* A novel diagnostic tool, called the "GreeneChip", able to simultaneously screen for up to 30,000 different viruses, bacteria, fungi and parasites is being investigated by researchers at New York's Columbia University. Such a tool could help physicians give a more accurate diagnosis and prescribe appropriate treatment. The **GreeneChip** is a glass slide with row after row of DNA or RNA samples from those thousands of viruses. When human fluid and tissue samples are applied to the chip, these probes will stick to any closely related genetic material in the samples. The chip can give an **accurate diagnosis** using urine, tissue, blood and stool samples. The chips were tested on samples from patients with respiratory disease, hemorrhagic fever, tuberculosis and urinary tract infections. The GreeneChip was able to diagnose the underlying pathogen as accurately as more traditional methods, such as culturing or growing the bacteria or virus, or using polymerase chain reaction to look at the genetic material. In addition to suffering with a current lack of suitable, inexpensive diagnostic tools, any of the existing methods such as sinus puncture and aspiration are considered extremely inconvenient. Datamonitor research has found that approximately **68% of patients** with acute bacterial sinusitis continue to be prescribed **empiric treatment** as first-line therapy and **only 35%** of these patients are **tested for the underlying pathogen** prior to initiation of second-line therapy. Advances such as the GreeneChip, which simultaneously screen for multiple pathogens, therefore address a crucial unmet need. An early, accurate diagnosis can help physicians prescribe the appropriate treatment and be vital in containing outbreaks of highly contagious diseases.

* Research has led to dramatic advances in knowledge of the psychosocial determinants of premature aging and effective interventions to **slow degeneration** and **improve cognitive fitness** and **memory** as we age.

* In the USA, **mass media campaigns** draw heavily on research on communication, diffusion, and **behavior change**. For example, the NICHD-sponsored *Back to Sleep Campaign* aims to reduce mortality from sudden infant death syndrome (**SIDS**) by promoting infant back sleeping. Since the campaign was launched in 1994, **back sleeping increased from 26.9% to 72.8% (in 2006) and SIDS has declined by more than 50%.**

* **Clomipramine**, a tricyclic antidepressant (TCA) and an SNRI, was the first drug found to be effective in Obsesssive Compulsory Disorder (**OCD**). Its efficacy has been firmly established in adults since the first finding at the end of the 1960s. In 1989, clomipramine became the first drug to be approved by the FDA for **OCD**. Randomized, controlled trials have shown clomipramine to be superior to placebo. In active comparison studies and meta-analyses of randomized controlled medication trials, no medication has shown to be more effective in treating OCD. **SSRIs** have greater tolerability, but clomipramine may be more effective in OCD. This has been the finding in several meta-analyses, but not in head-to-head comparisons.

Till 2005, five SSRIs have been approved by the FDA for the treatment of adult OCD. They are the **SNRI clomipramine**, and the **SSRIs fluvoxamine, fluoxetine, sertraline, and paroxetine**. Four of these, clomipramine, fluoxetine, fluvoxamine, and sertraline, have also been approved for treatment of pediatric OCD. Citalopram, escitalopram, and the SNRI venlafaxine are also used in OCD.

B.52. SSRIs: dose comparison

Generic name	Brand name	Lowest recommended doses for OCD	Highest FDA approved doses
Citalopram	Celexa	20mg/day	60mg/day
Clomipramine	Anafranil	150mg/day	250mg/day
Escitalopram	Lexapro	10mg/day	20mg/day
Fluoxetine	Prozac	40mg/day	80mg/day
Fluvoxamine	Luvox	150mg/day	300mg/day
Sertraline	Zoloft	50mg/day	200mg/day
Paroxetine	Paxil	40mg/day	60mg/day
Venlafaxine	Effexor	150mg/day	225mg/day

* Scientists at **Yale University School of Medicine** have used **molecular genetic analysis** to dissect physiologic processes that regulate cardiovascular function in humans, with an **emphasis on blood pressure regulation**. By coupling characterization of hundreds of families from around the world with human genetic studies, they have **mapped over 30 human disease genes** and **have identified functional mutations underlying 22 of these**. These have provided **new insight into the mechanisms underlying** hypertension, stroke, osteoporosis, and renal diseases including disorders of electrolyte and pH homeostasis.

Table 4.16.   Worldwide generics market: historical development and growth projection (2010)

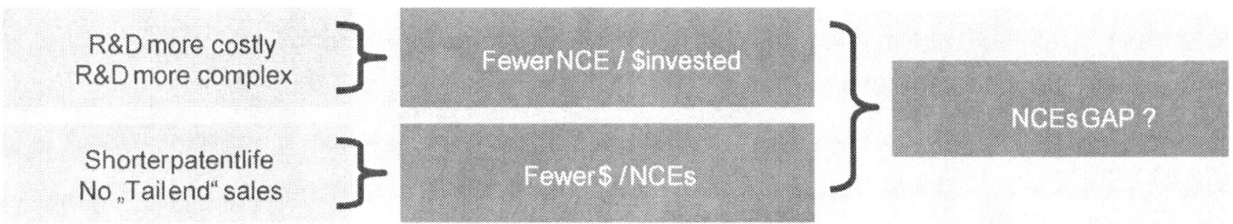

(Industry sources; compounded annual growth rates; biopharmaceuticals excluded; ex-factory prices)

IV.9. The pharmaceutical industry's blockbuster research and commercializing gap

In a remarkable keynote presentation given at the Financial Times World Pharmaceuticals Conference in London, March 25, 1996, Jan Lesley, CEO of SmithKline Beecham plc, explained how the economics of R&D was changing.

Table 4.17.   Changing economics of industry's R&D

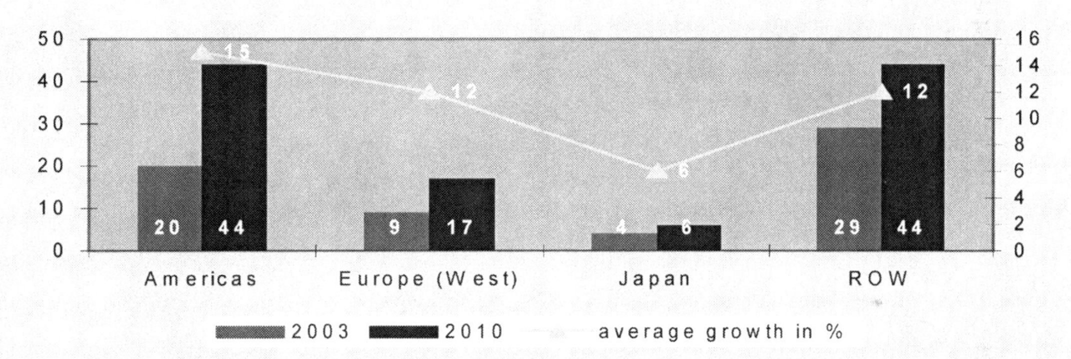

We see fewer NCEs per $ invested AND fewer $´s per NCE, Lesley said.

Table 4.18.   Performance gap in R&D

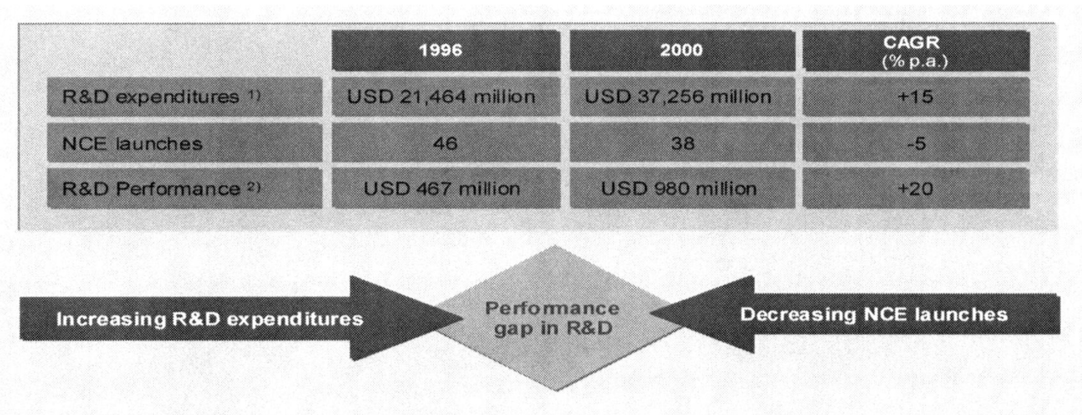

	1996	2000	CAGR (% p.a.)
R&D expenditures [1]	USD 21,464 million	USD 37,256 million	+15
NCE launches	46	38	-5
R&D Performance [2]	USD 467 million	USD 980 million	+20

(*1 R&D expenditures of top 20 pharmaceutical companies worldwide; *2 R&D expenditures per NCE launches. Source: Gunter Festel/Festel Capital)

* **Environmental factors** influencing people's health increasingly make it to news headlines. The identification of important environmental hazards and the improvement of the clinical outcome of environmentally related diseases becomes a broad field of research, having for ultimate goal **to link the effects of these exposures to the cause, moderation or prevention of environmentally-related diseases.**
The 2006 research program funded by the U.S. NIEHS (National Institutes of Environmental Health Sciences) includes e.g.
- research, directed to mechanism by which **cells protect themselves from the toxic effects of arsenic,** a highly poisonous metal that can cause DNA damage and lead to an increased risk for certain cancers.
- exploration of the **effects of environmental agents on telomeres**, small segments of DNA located at the ends of chromosomes, which help control aging and death of cells.
- the examination of the way in which damage to DNA from environmental exposures can trigger **the production of certain proteins that help protect the cell** from toxic agents.
- the study of the way in which certain **airborne pollutants interact with sensory nerve cells** in order to produce eye, nose and throat irritation.
-the examination of the relationship between exposure to airborne chemicals from vehicle exhaust and industrial sources, and **increased susceptibility to respiratory illnesses** such as emphysema and COPD
- the study of the **relationship between outdoor concentrations of ozone,** a form of oxygen that is a primary component of urban smog, and the **incidence of respiratory disease and death** in exposed populations - research on fine particle air pollution -- microscopic particles of dust and soot less than 2.5 microns in diameter -- to determine whether exposure to these tiny particles can **produce changes in immune system function** that could result in **an increased risk for developing asthma.**
-the study of the **effects of fine particle exposure** on blood flow and heart disease risk.

* Researchers in Sydney have discovered that an enzyme known as **PAC-1** (only found in immune cells) plays a key role in promoting rheumatoid arthritis. The work raises the possibility of new and better treatments for the painful and debilitating condition. PAC-1 helps immune cells to respond to the body's signals for help against infection in three critical ways. **It assists the immune cells to survive, to migrate to where the emergency is, and to release potent inflammatory compounds at the site. Rheumatoid arthritis** differs from the more common degenerative arthritis of the elderly, osteoarthritis. It is caused by **over-protective immune cells** which mistakenly attack the body's own cartilage, the material which lubricates the movement of joints. As the cartilage deteriorates, the movement of bone against bone becomes very painful. Inhibition of PAC-1 is therefore a real potential strategy for controlling the overactive immune cells responsible for rheumatoid arthritis without affecting other systems in the body. This is a completely **different approach to existing therapies.**

* As **autoimmune diseases** affect up to 5% of the world's population, it's not surprising that the value of biological drugs indicated for this disease totaled around US$11 billion worldwide in 2004. **Remicade, Enbrel, Humira** and Avonex lead the market in the autoimmune field.

B.53. Leading therapies in the autoimmune field

Class of compound	Indication	Target Mechanism of action	Brand name (company)	Sales 2005 % vs 2004 US$mn
Fusion protein of Ig Ig +sol TNF-R	Rheumatoid arthritis psoriasis	TNF-alpha	Enbrel; etanercept (Amgen w yeth)	2,600 (+59)
Rec chimeric mab	see above	see above	Remicade infliximab Centocor / J&J & S-P)	2,535 (-+18)
Rec fully human mab	see above	see above	Humira; adalimumab (Abbott)	1,400 (-+64)

*According to the U.S. Centers for Disease Control and Prevention, approximately **8% of children** between the ages of 4 and 17 years old, or **4.4 million children**, are diagnosed with **ADHD**, with roughly **2.5 million currently taking medication for this disorder.** Over the past three to four years, drug companies have turned their attention to once-daily extended release oral therapies like **Concerta** and **Adderall XR**, which provide a full day of coverage in treating inattention and hyperactivity.

* **Tuberculosis** kills almost **two million people** worldwide every year and is considered by the World Health Organization to represent a global health emergency. However, the bacillus is much more prevalent in the world's population than the statistics show, because only **5 to 10%** of those **infected** actually **develop tuberculosis.**

* A **CD4+ T-cell type** has been shown to play a critical role in **asthma.** It is the natural killer T (**NKT**) cell. NKT cells constitute a subset of lymphocytes that expresses characteristics of both natural killer cells (part of the innate immune system, the phylogenetically ancient system that has hardwired recognition receptors, such as toll-like receptors and complement receptors) and conventional T cells (part of the adaptive immune system). *[Kronenberg M. Toward an understanding of NKT cell biology: progress and paradoxes. Annu Rev Immunol. 2005;23:877-900].* The idea that NKT cells play an important role in asthma has a number of clinical implications. First, about 10% to 30% of patients with asthma have a steroid-resistant form of asthma, which may be due to the fact that, unlike Th2 cells and eosinophils, NKT cells are **steroid-resistant.** Therefore, therapies that target NKT cells may provide a very effective approach to controlling asthma.

The lack of economies of scale in R&D has led to a performance gap. As a result, stagnant R&D productivity has led to a reorientation towards more in-licensing instead of in-house innovation. This route displays a need for subsidy/survival/support instead of organic growth and includes the use of new technologies (biotech, gene therapy) requiring co-operation, networks and shared decision-making. By taking up new trends late, companies may have missed out on innovation, such as genomics and targeted therapies, in segments that could drive and renew growth.

In-house drug development still dominates all clinical phases, accounting for 68% of all projects. However, its proportional share within each phase differs and shows that BigPharma understood the importance of cooperation in R&D. In-house development dominated the pre-clinical stage, accounting for 91% of all projects. The percentage of in-house sourced drugs falls to 67.8% and continues to fall in Phase I, 63.4% in Phase II and 57.2% in Phase III. A corresponding increase in the proportion of in-licensed drugs can be seen.

The proportion of co-development drugs is broadly similar across the three clinical stages of development accounting for approximately: 19% of Phase III projects, 17% of Phase I and 16% of Phase II projects, while only accounting for 4% of preclinical projects. The lower share of co-developments in the pre-clinical stage indicates that companies mainly focus on early-stage in-house work before seeking R&D collaborations later in the development process.

A study[90], based on an analysis of 83 new drug applications and biologic license applications submitted to the U.S. Food and Drug Administration (FDA) between 2000 and 2005, found that drug sponsors who are more extensive users of CROs tend to complete projects faster, most notably during the study close-out period, while maintaining quality comparable to submissions involving minimal use of CROs. In addition, projects involving high CRO usage typically are submitted more than 30 days closer to the projected FDA submission date than low CRO usage projects. In 2004, the most recent year for which data are available, leading CROs managed nearly 23,000 Phase I-IV studies at 152,000 clinical sites worldwide.

Table 4.19.   Five-year product freshness index

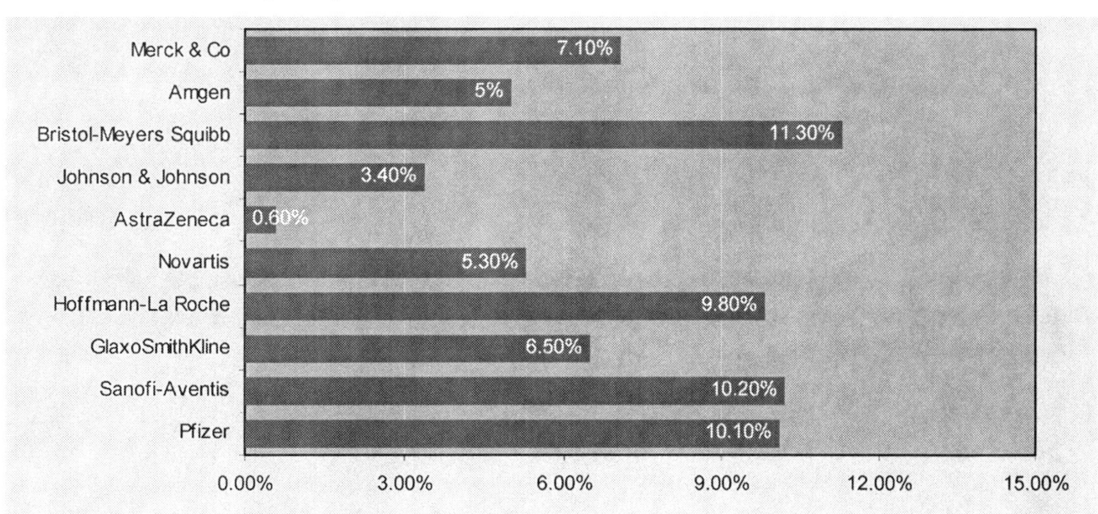

---

[90] *A study by the Tufts Center for the Study of Drug Development, summarized in the January/February Tufts CSDD Impact Report, the first comprehensive quantitative analysis of the overall impact of outsourcing on drug development performance and capacity.*

The study authors state that the results of this study challenge the conventional notion that CROs are simply vendors providing capacity for a specific project and that clinical outsourcing offers a development speed advantage at comparable quality.

As the volume and scope of clinical research activity worldwide continues to grow, CROs increasingly are providing a workforce that is essential to the long-term viability of the enterprise. Tufts CSDD estimates that $5.5 billion, or 15%, of global drug development spending, excluding pass-through fees (e.g., central lab costs and investigator grants), went to contract clinical services in 2004. This compares to 12% in 2001.

Wood Mackenzie calculated a 5 year freshness index, based on projected revenues 2004-2009. The results suggest that despite the industry's heavy investment in R&D, new products will not generate a substantial share of revenues by the end of the decade. This would show either a too small number of new launches or small sales projections for new products, or both. The analysts conclude that the days of the blockbuster-driven pharmaceutical behemoth are over[91].

Among the top 20 companies it is calculated that on average the expected outcome of R&D is only 0.45 NCEs per year and per company –a surprising and very discouraging number, he said. What does all that mean? Lesley used an example:

Table 4.20.   New products gap in the "blockbuster" model

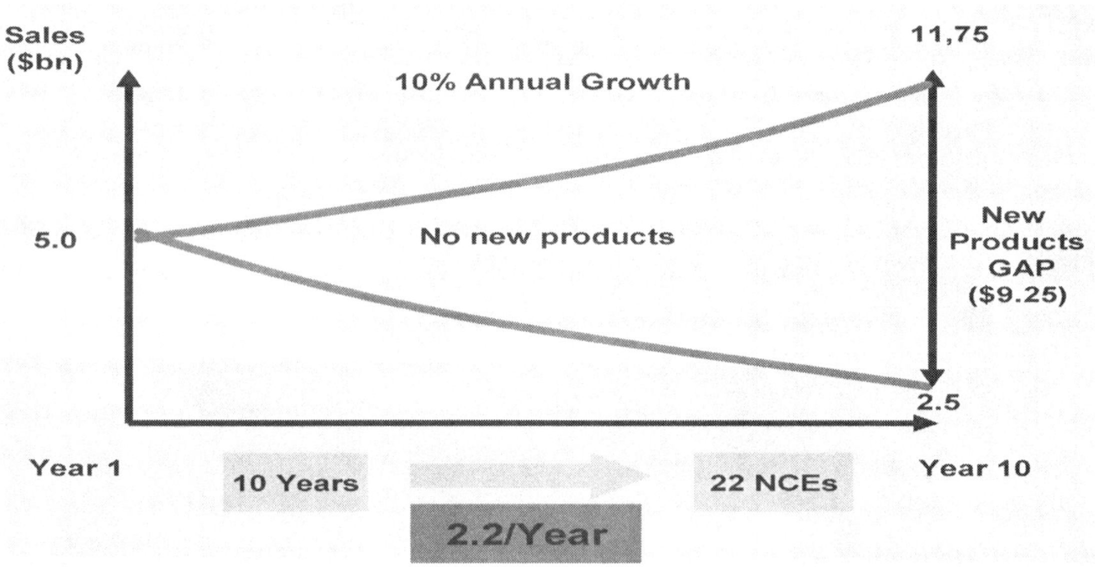

"Imagine a company with annual sales of $5bn. Assuming useful patent life of 10 years and no new product introductions, this company will lose 50% of its sales over a ten year period, from $5bn to $2.5bn.

---

[91] *The Pipeline Payoff. Forecast based on Wood Mackenzie data, Pharmaceutical Executive, June 2005*

*Through basic and clinical research, our understanding of the **bio-behavioral mechanisms** and **treatment** of **mental disorders** has advanced dramatically. **Effective** and cost-effective behavioral and combined behavioral and **pharmacological treatments** have become available for treatment of **depression, anxiety disorders**, and the **abuse** of **nicotine, alcohol and other drugs.**

* The **cardiovascular** market has witnessed new classes of drugs for hypertension and the development of anti-platelet and anti-clotting agents. All these have had a major impact in treating heart attacks. New drug classes such as **endopeptidase/endothelin antagonists, vasopressin-2 inhibitors, brain natriuretic peptide-targeted molecules** and **free radical scavengers** have emerged.

* In **Hypertension**, the development of **combinations of treatments** already on the market is an industry trend. A monotherapy is rarely sufficient to reach targets in arterial pressure. In cases of failure, **doctors prefer to combine approaches that do not exclude each other rather than increase the dose**. This makes fixed combinations a natural and necessary approach in the market for anti-hypertensives: patients take a daily dose (since the disease is asymptomatic, compliance is already low, making ease of use the key factor), and the therapeutic effects are superior (lowering arterial pressure). Seventy percent of patients require a dual therapy to reduce their blood pressure. This observation leaded to **Novartis** filing for **Exforge**, a **fixed combination in hypertension** associating Diovan (valsartan) and amlodipine (the active ingredient in Pfizer's Norvasc, under challenge by generic makers).

B.54. Five classes of therapeutics in hypertension:

Class	Action mode	Main side effect	Major brands or generics
Diuretics	Decrease in blood volume		hydrochloro-thiazide
Beta-blockers	Decrease in heart pace and volume blood flow	Impotence	Coreg (GSK) Toprol XL (AstraZeneca) Concor (Merck KGaA)
Angiotensin conversion-enzyme inhibitors (ACE)	Inhibition of angiotensin2 which fosters hypertension	Cough	captopril enalapril
Angiotensin2 receptor antagonists (ARA2)	Attaches to the angiotensin2 receptor		Diovan (Novartis) Cozaar (Merck) Atacand (AstraZ) Avapro(Sanofi-A)
Calcic canal inhibitors	Vasolidation	Oedema of lower limbs	Norvasc (Pfizer)

(See opposite colon)

* Two groups of classes are often opposed. The older diuretics and beta-blockers and the new ACE and ARA2 share one point: they work on the renine-angiotensin system by inhibiting destruction of bradykinine.

B.55 diuretics, beta-blockers, ACE and ARA2: action on the renine-angiotensin system

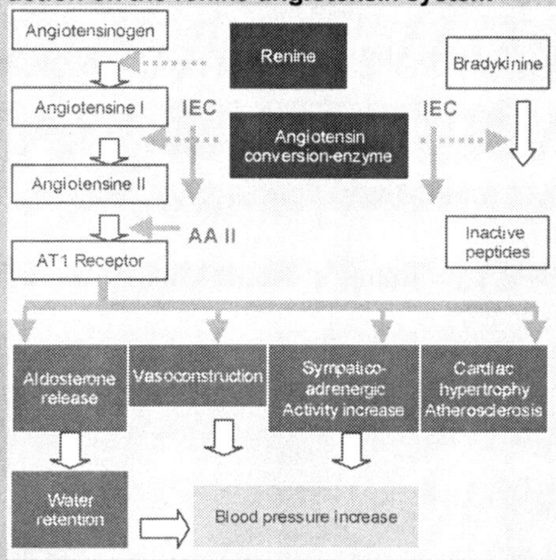

*Despite the dominant market position of **statins** in the treatment of **dyslipidemia**, or cholesterol-lowering, only 81% of patients receive such medicines to treat their condition as a first-line therapy for the disease, according to a report by **Decision Resources**. The study, entitled Treatment Algorithm Insight Series: Dyslipidemia, finds that Pfizer's **Lipitor (atorvastatin)** and AstraZeneca's **Crestor (rosuvastatin)**, both statins, are the leaders of first-line and third-line therapy markets, respectively, in the treatment of dyslipidemia. In fact, Lipitor is the world's largest selling drug, generating turnover of $12.19 billion for the company in 2005. Additionally, US drugmakers Merck & Co/Schering-Plough's **Zetia (ezetimibe)**, a non-statin agent, leads the second-line therapy market for treating high cholesterol. Decision Resources found that, on average, over 97% of first-linetherapy for dyslipidemia is prescribed as monotherapy. Even drugs such as Zetia and **niacin**, which are typically perceived as being adjunct therapies, a represcribed as monotherapy over 90% of the time when they are used as first-line agents.

* In 2004, Serono (later Merck-Serono), the world leader in reproductive health, improved the patient experience by providing a prefilled and ready-to-use pen, **Gonal-f RFF Pen**, indicated for induction of ovulation and pregnancy in anovuluatory infertile women in whom the cause of infertility is functional and not due to primary ovarian failure. It is also indicated for the development of multiple follicles in ovulatory women participating in an Assisted Reproductive Technology (ART) program. Gonal-f is a highly consistent, filled-by-mass recombinant human follicle stimulating hormone **(r-hFSH)** prescribed to supplement or replace naturally occurring FSH, which stimulates the development of follicles in the ovaries.

To be successful, this same company must grow the sales line with 10% per year for the next ten years. This is no different from what you and I would like to see our own companies do, Lesley added. With 10% sales growth per year over the next ten years, sales will increase from $5bn to $11.75bn in year ten. The new product GAP (the difference between $11.75bnand $2.5bn) is $9.25bn. For the company to fill this GAP, it must introduce 22 NCEs each with peak sales of $400mn over that ten year period or 2.2 NCEs per year".

According to Lesley, the challenge for the top tier companies is, therefore, to improve output from 0.45 NCEs to 2.2 NCEs per YEAR. Referring to M&A, Ian Lesley stated that size and consolidation is not the answer. He illustrated the point by comparing the $5bn company A to a $10bn company B –twice the size of A. The $10bn company B needs no less than 44 NCEs over the next ten years or 4.4 NCEs per year to fill the New Products Gap. When a company doubles its sizes through a merger, the hurdle rate becomes twice the size and you have gained little unless you have acquired a healthy new product pipeline and that is very unlikely, he said.

Table 4.21.   Size & consolidation will NOT provide the answer[92].

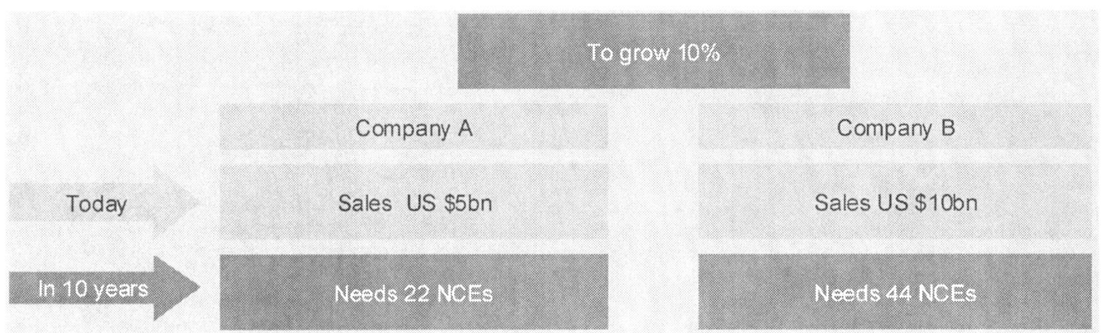

The vulnerability of the blockbuster model means that BigPharma faces radical changes as only one drug candidate currently reaches the market for every thirteen placed in clinical trials will succeed to launch.

Although the pharmaceutical industry is very creative, the number of products required by a company to keep the blockbuster system working, is not matched by the major US companies (the strongest advocates of the global one drug fits all billion dollar mass market segment).

IV.10. A shift in focus

In fact, the unsustainable value of the blockbuster market is already impacting on overall pharmaceutical market growth. In 2000, more than 50% of the revenues of Pfizer, AstraZeneca, Amgen and Eli Lilly came from blockbusters. The compound annual growth rates (CAGT) in revenues from blockbuster products are predicted to fall to 4.3% between 2001 and 2008, from a high of 23.6% between 1994 and 2000[93].

However, the scaled benefits of blockbuster drugs mean that they will continue to be an important source of revenues for BigPharma, striving for scale. The billion dollar drugs effectively serve well markets populated by more than 1m people, and target chronic,

---

[92] *Based on the presentation of Ian Lesley, CEO, SmithKline Beecham, Financial Times World Pharmaceuticals Conference, London, 1996*
[93] *Rao Sanjay K, Rising to the Challenge, Scrip Magazine, 03.2006:18*

* Because of their high cost levels, **biotechnology products** have **become prime targets for cost-cutting measures,** says **Datamonitor**. The report estimates that worldwide sales of biologics by 56 of the leading pharmaceutical and biotechnology companies reached a total of $56.2 billion during 2004, representing a sharp increase of 18.3% over 2003 and, in the USA alone, sales of biologics reached $30.8 billion during 2004. Out of this worldwide market, sales totaling $20.2 billion are derived from just six key product classes that are at immediate risk from **biogenerics**, namely **insulin, human growth factor, epoetin, colony-stimulating factors, interferon alpha and interferon beta.**

B.56. Major Branded G-CSF products

Class of compound	Indication	Target Mechanism of action	Brand name (company)	Sales 2005 % vs. 2004 US$ mn
Pegylated G-CSF	Neutropenia	CSF-receptor	Neulasta (peg-filgrastim) (Amgen)	2,288 (+31,5)
rhu G-CSF	see above	see above	Neupogen (filgrastim) (Amgen)	1,216 (+3,5)
Rhu colony stimulating factor	Neutropenia associated w. chemo therapy	see above	Neutrogin: lenogastim (Roche/ Chugai)	279 (+15)
	Neutropenia	see above	GRAN (Kirin/ Sankyo)	NN

B.57. Branded Interferon Beta Products

Class of compound	Indication	Target Mechanism of action	Brand name (company)	Sales 2005 % vs. 2004 US$mn
Rhu interferon	Multiple sclerosis	IFN-R	Avonex; Interferon beta-1a (Biogen IDEC)	1,506 (+8)
see above	see above	see above	Rebif; interferon beta-1a (Merck-Serono)	1,270 (+15)
Rhu interferon beta-1b	see above	see above	Betaferon; Betaseron Interferon beta-1b (Schering AG)	836 (+9)

* NIGMS-supported scientists have discovered that a protein called **Htz1 is ousted** from a gene when the gene is **activated**, allowing the cellular machinery to read the genetic code and convert it into a protein.

* The **European Medicines Agency** adopted the first positive opinion for a **similar biological medicinal product**. The product, **Omnitrope** (Sandoz GmbH, a subsidiary of Swiss pharmaceutical giant Novartis) contains **somatropin, a recombinant-DNA growth hormone**. It is intended for the treatment of growth disturbance and **growth hormone deficiency in children and adults.** The Agency's scientific committee, the Committee for Medicinal Products for Human Use (CHMP) considered that, in accordance with European Union requirements, Omnitrope has been shown by studies demonstrating comparable quality, safety and efficacy to be **similar to** a reference medicinal product already authorized in the EU, namely **Genotropin**.

The first E.U. guidelines on quality, non-clinical and clinical issues for similar biologic products were adopted by the CHMP in December 2003. A general regulatory guideline on similar biological medicinal products was adopted in September 2005.

Similar biological medicinal products may be **authorized on the basis of appropriate non-clinical and clinical data requirements**. This is based on the experience gained with the reference medicinal product, against which appropriate studies and comparisons are made. However, compared with generics, in the case of similar biological medicinal products, **substantial additional data**, in particular the **toxicological and clinical profile**, have to be provided.

B. 58.   Human growth Hormone products

Class of compound	Indication	Target Mechanism of action	Brand name (company)	Sales 2005 % vs 2004 US$mn
human growth hormone	growth hormone deficiency	regulation of postnatal growth	Genotropin (Pfizer)	808 (+10)
see above	see above	regulation of postnatal growth	Norditropin (Novo Nordisk)	444 (+20)
Rhu growth hormone	see above	see above	Humatrope (Bi Lilly) alfa-2b	414 (-4)
see above	see above	see above	Nutropin, Protropin, Somatropin, Somatrem somatropin, (Genentech/ Roche	365 (+6)
see above	see above	se above	Saizen/ Serostim (Merck-Serono)	206 (+12,5) 70 (-19)

* Every year the consequences of Cardiovascular Diseases (CVDs), such as myocardial infarction and strokes, kill some **19 million people worldwide** (5 million people in Europe), and cost more than **500 billion** dollars in terms of **healthcare expenses** and loss of productivity of those who survive.

low severity disorders. Change will come from a shift in focus: from the me-too compounds that compete in a small number of broad therapeutic classes, from the phenotype model (diagnose and treat) to the genotype approach (predict and prevent) or compounds that will provide highly differentiated therapeutic value.

Table 4.22.   Innovation: Number of new drug approvals, by company, over a 5-year period (USA, 2000 to 2004)

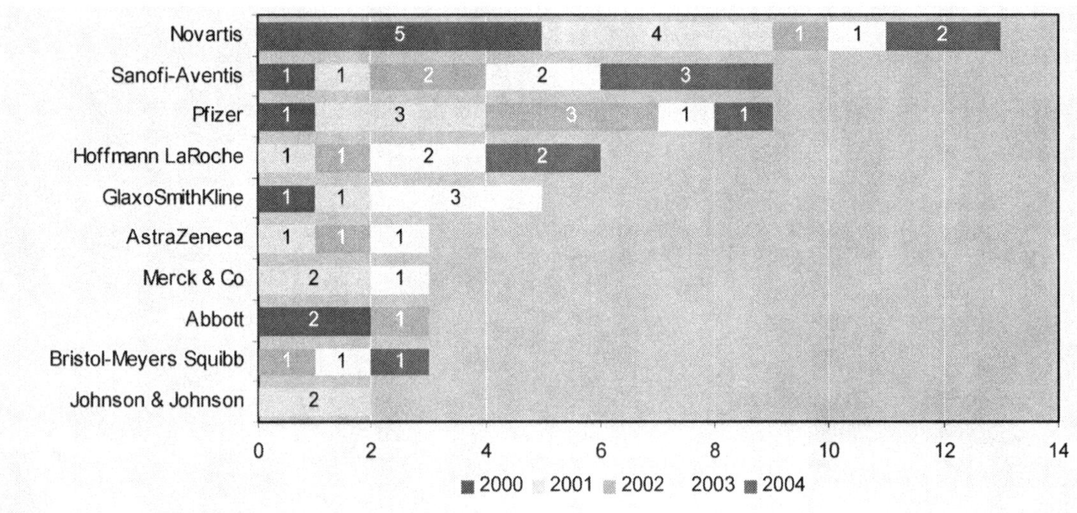

(Includes new molecular entities and biological license applications, excluding vaccines)

IV.11. Reducing late-stage development failures

According to the Tufts Center for the Study of Drug Development's Outlook 2007 report, companies need to make greater use of new technologies to reduce late-stage development failures and contain rising costs, increased reliance on global outsourcing to speed development and reduce costs, and more coordination between and regulators worldwide.

- Companies, acting alone and in consortia, will seek to improve drug discovery by examining pre-competitive data to identify and validate new targets.
- The high cost and low success rates of human studies will lead small/mid-tier pharma companies to adopt outsourcing and other clinical practices that big pharma increasingly uses to control costs and manage risk.
- Biotech firms will escalate biodefense and pandemic disease related R&D, emphasizing translational research and development of effective countermeasures.
- Instead of directly addressing off-label uses of existing prescription drugs, the U.S. Food and Drug Administration (FDA) will evaluate current studies and likely request new ones before deciding whether to further regulate off-label uses.
- Within two to three years, up to 65% of FDA-regulated clinical trials for top pharmaceutical companies will be conducted abroad.
- The European Medicines Agency (EMEA) will focus on implementing new legislation and integrating drug regulatory agencies of new European Union members into a comprehensive pan-European system.
- U.S. third party payers will move away from a binary coverage model in which prescription drugs are either covered or not covered. Instead, more than 90% of prescription drugs will be covered with a variety of limits.

146

* **Infertility** affects about **six million American couples**, approximately 10 percent of the reproductive age population. The Centers for Disease Control reports there were nearly 110,000 cycles of assisted reproductive technology in year 2001. Infertility is defined as the inability to achieve pregnancy after one year of regular, unprotected intercourse (six months if the woman is over 35. About 70% of patients who are treated succeed in having children.

B.59.   Major branded infertility products

Class of compound	Indication	Target Mechanism of action	Brand name (company)	Sales 2005 % vs. 2004 US$mn
follicle stimulating hormone	infertility	FSH-R	Gonal-f; follitropin alfa (Merck-Serono)	547 (-4,5)
see above	see above	see above	Puregon: Follistim follitropin beta Organon	423 (+24)
Rhu grow th hormone	see above	see above	Humatrope (Bi Lilly) alfa-2b	414 (-4)

Follistim AQ Cartridge became the first follicle stimulating hormone (FSH) treatment available in a pre-filled, pre-mixed solution, approved in the U.S., eliminating the need for patients to mix one or more vials of medication. Follistim AQ Cartridge is designed to be used only with the Follistim Pen, an innovative pen device that facilitates accurate delivery of individualized doses of pre-mixed follitropin beta injection, a highly effective and widely used prescription fertility medication made from state of the art recombinant DNA technology. Follistim AQ Cartridge, for use with the Follistim Pen, is prescribed for women undergoing assisted reproductive treatments (ART) such as in vitro fertilization (IVF), and for the induction of ovulation to achieve pregnancy.

* In France's **Alsace** region, the **New Treatments** center is relying on its supremacy in **computer-aided surgery**. In future, on one and the same screen, real-time video images of an operation will be superimposed on synthetic images (magnetic resonance imaging, or MRI scans, etc.) of the patient taken beforehand. This technology will enable the surgeon to see not only the diseased region through the flesh, but also the precise position of vital structures to be preserved - **to within a millimeter.**

* To accelerate our understanding of **mind/body interactions**, such as the relationship of stress to heart disease, decreased immune system functioning and premature aging, five Mind-Body Research Centers have been established by US Health Authorities. Initial findings from these Centers and NCCAM-supported research include **evidence of bi-directional links between stress, social involvement. and cancer.**

* Up until 2004, when **Fabrazyme (agalsidase beta** from Genzyme Corp.) received marketing approval, people with **Fabry disease** have had one certainty...death in their forties or fifties, preceded by years of prolonged suffering. While predominantly affecting males, (inherited through the Fabry gene located on the X chromosome), females can also suffer the serious consequences of Fabry disease, a rare, life-threatening, inherited metabolic disorder caused by inborn errors in metabolism. Due to a deficiency in an important enzyme, **alpha-galactosidase A or alpha-GAL**, Fabry disease causes certain fats to accumulate in the blood vessels over many years. The ensuing buildup of fatty substance globotriaosylceramid (GL-3) compromises the cells of various tissues, including organs such as the kidneys and heart, eventually leading to complete organ collapse. Agalsidase beta is a recombinant form of α-galactosidase A produced by genetically engineered Chinese hamster ovary cells.

B.60.   Major Branded Enzyme Therapy products

Class of compound	Indication	Target Mechanism of action	Brand name (company)	Sales 2005 % vs. 2004 US$mn
Imiglucerase for injection	Gaucher's disease	Enzyme replacement	Cerezyme; ceredase (Genzyme)	933 (+11)
Rhu enzyme	Fabry disease	see above	Fabrazyme; agalsidase beta (Genzyme)	305 (+45)
Rhu enzyme	Mucopoly-sacchari-dosis I (MPS I)	see above	Aldurazyme; laronidase (BioMarin/ Genzyme)	77 (+80)

MPS I is a progressive, debilitating and fatal genetic disease caused by a deficiency of the enzyme **alpha-L-iduronidase**. This deficiency leads to the accumulation of complex carbohydrates in the lysosomes of cells, leading to the progressive dysfunction of cellular, tissue and organ systems. Resulting symptoms can include impaired cardiac and pulmonary function, delayed physical development, skeletal and joint deformities, reduced endurance, and in some cases, delayed mental function. A majority of patients die before adulthood from complications of the disease. The new biotechnology product, **Aldurazyme (laronidase)**, is a version of the human form of the deficient enzyme. It helps prevent the build-up of **glycosaminoglycans** (GAG) in the cells and has been shown to improve lung function and exercise ability; it   is for patients with Hurler and Hurler-Scheie forms of MPS I as well as patients with the Scheie form with moderate to severe symptoms.

* A team from the Institut **Pasteur** has shown that **Mycobacterium tuberculosis**, the bacillus responsible for tuberculosis **can hide**, in a dormant state, **in adipose cells** throughout the body. The bacterium is protected in this cellular environment, to which the natural immune defenses have little access, and is **inaccessible** to **isoniazid**, an antibiotic used to treat tuberculosis.

Table 4.23.   Medicines: From phenotype to genotype

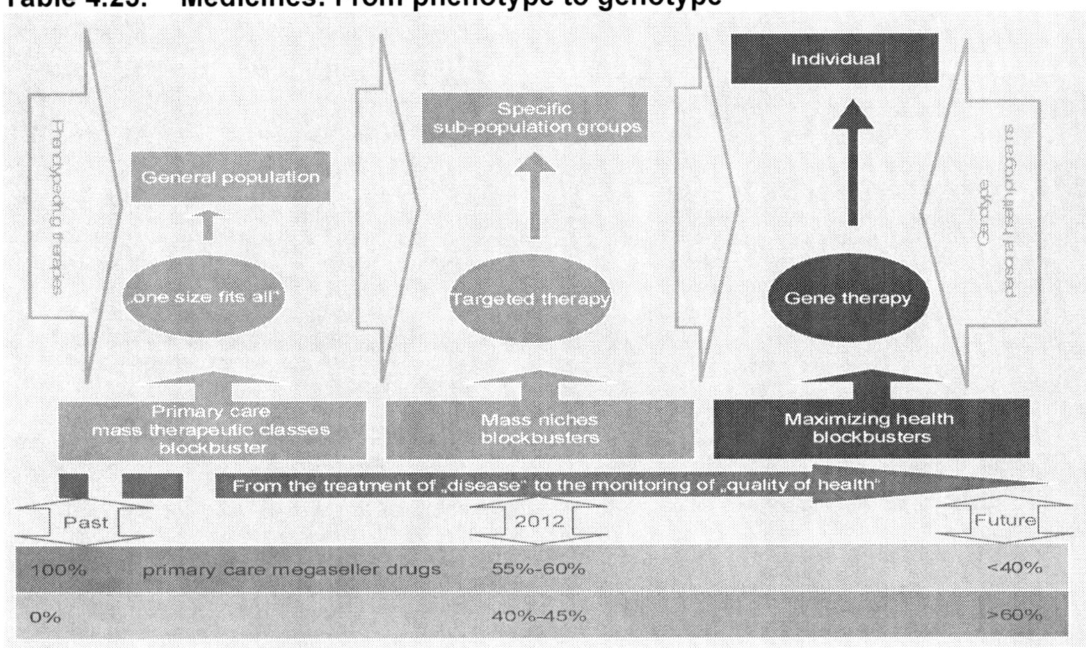

A new generation of mega-seller brands, driven by innovation and addressing individualized patient needs, are needed to segment clearly therapeutic domains into specialized businesses. According to Easton Associates LLC, drugs for higher severity conditions had growth rates between 1990 and 2002 more than twice that of drugs for lower severity conditions, the ones that are served by the traditional "one size fits all" primary care blockbuster drugs.

"If one carries this analysis out to 2012, and keeps compounding the growth of high-severity drugs at 14% and low-severity drugs at 6%, one sees that by 2012, drugs for low-severity conditions will contribute only a 31% share of the industry's dollar growth, while high-severity drugs will contribute 68%"[94]

Executives now concede that scale alone was the dominating concept during the 1980s and 1990s needed to bridge the innovation gap (read: "near-term profit gap") until patents expired and generic versions threatened their existence. It is increasingly acknowledged that a company's size does not correlate with superior performance. Among the top 20 pharmaceutical companies, the largest of them perform no better than the smaller ones.   Analysts observed that active acquirers have posted the same performance as non-acquirers, with each group achieving 12% appreciation in market capitalization since 1992.

Table 4.24.   Profit margins of pharmaceutical companies*

Amgen	29.60%	Wyeth	19.50%	Novartis	16.00%	Eli Lilly	13.50%
Forest	23.80%	Genentech	19.30%	Pfizer	15.70%	Sanofi-Aventis	10.20%
Merck & Co	21.00%	Bristol-Meyers Squibb	16.40%	GlaxoSmithKline	15.40%	Biogen Idec	6.60%
Johnson & Johnson	20.60%	astraZeneca	16.20%	Abbott	15.10%	Schering-Plough	5.50%

Non-exhaustive list of pharmaceutical companies (net profit before taxes after cost of goods sold and operating expenses are deducted)[95]

---

[94] Brian Buxton, quoted in „Blockbusters Then, Now and Later", In Vivo-The business & medicine report, June 2003
[95] Pharmaceutical Executive, Industry Audit, Sept 2006

* **Type 1 diabetes** is a growing health problem among European children. Data indicate that the disease incidence has increased five- to six-fold among children under the age of 15 years after World War II, and there are no signs that the increase in incidence is leveling off. The most conspicuous increase has been seen among children under the age of 5 years. The EU-funded Diabetes Prevention study is generating a wealth of information on breast-feeding practices, infant nutrition and growth in young children in various countries. Newborn infants observed in Northern Europe (NE) had a higher birth weight but a shorter birth length than infants in Central and Southern Europe (CSE). The NE children remained heavier than those from CSE at least up to the age of 18 months. They were also taller than the CSE children starting already from the age of 3 months up to the age of 18 months. Accelerated growth in infancy has been identified as a risk factor for type 1 diabetes later in childhood. Accordingly the observed growth pattern may contribute to the higher incidence of type 1 diabetes in NE compared to CSE. Within the next 10 years the Diabetes Prevention study will generate a definite answer to the question whether early nutritional modification may prevent type 1 diabetes later in childhood. A reduction of 50% in the incidence of type 1 diabetes would have a substantial impact on the quality of life of many European families. [The Diabetes Prevention study is the first study ever aimed at primary prevention of type 1 diabetes. The study is designed to provide an answer to the question whether weaning to a highly hydrolyzed formula in infancy decreases the risk of future diabetes. All subjects are observed for 10 years to gain information on whether the dietary recommendations for infants at increased genetic risk of type 1 diabetes should be revised].

* Over 105,000 people die from **heart disease** in the UK each year. Heart attacks occur in over one million people annually in the US and over thirteen million suffer from coronary artery disease, making it the single largest cause of death in the Western world. A major short-coming of current **angiogenic** therapy in response to myocardial ischemia in humans is that the outcome may be limited to capillary growth without concomitant collateral support of arterioles [terminal branches of arteries]. A study funded by the British Heart Foundation and the Medical Research Council showed that , in mice, **Thymosin beta 4** (or TB4) can promote vessel formation and collateral growth not only during develop-ment but also critically from adult epicardium, thereby suggesting that TB4 has considerable therapeutic potential in humans. Most humans suffering from ischemic cardiac events, either acutely or chronically, do not develop the collateral vessel growth necessary to preserve and restore heart tissue. If, in humans, the same effects occur as seen in mice, **TB4** would be the first drug to prevent loss of [heart] muscle cells and restore blood flow in this manner and provide a new and much needed treatment modality.. [Thymosin beta 4 (TB4) is a synthetic version of a naturally occurring peptide present in virtually all human cells. It is a first-in-class drug candidate that promotes endothelial cell differentiation, angiogenesis in dermal tissues, keratinocyte migration, collagen deposition, and down-regulates inflammation. One of TB4's key mechanisms of action is its ability to regulate the cell-building protein, actin, a vital component of cell structure and movement. Of the thousands of proteins in cells, actin represents up to 10% of the total protein and, thus, plays a major role in the physiology of the cell.]

* Each year, approximately 5 million Americans are rushed to emergency rooms with **severe chest pains**, of which about 1.4 million are identified with **ACS**. ACS can be the precursor to a major heart attack or death. ACS is caused by a blockage in an artery that prevents oxygen-rich blood from reaching the heart.
When people present with ACS symptoms, they are traditionally immediately put on a cocktail of drugs that prevent the formation of blood clots that can cause a heart attack. Two of these drugs are potent anti-clotting agents: **heparin**, which has been in use for decades, and **GPIIb/IIIa inhibitors**. However, heparin can be problematic to use because patients have variable responses to it, and when combined with GPIIb/IIIa inhibitors heparin can thin the blood so much that it can lead to serious bleeding which prolongs the hospital stay, is costly to treat and can be fatal. Excessive bleeding is particularly a risk in patients undergoing balloon angioplasty or surgery procedures to restore blood flow to an injured heart, which are commonly performed in this setting.
A study published in the New England Journal of Medicine (November 2006) led by Columbia University Medical Center and NewYork-Presbyterian Hospital researchers showed that an anti-clotting agent called bivalirudin, when used by itself to treat cardiac emergencies known as acute coronary syndromes (ACS), reduced the risk of major bleeding, a key risk for mortality, by 47 percent compared with the standard combination drugs. The medication was found to be equally as effective as the combination of injectable blood thinners traditionally used to maintain blood flow to the heart during these emergencies. Results of the Acute Catheterization and Urgent Intervention Triage Strategy (ACUITY) trial, one of the largest ACS trials (13.819 participants in 17 countries) to evaluate anti-clotting therapies ever conducted, demonstrated fewer cases of major bleeding in patients who were treated with bivalirudin than there were with patients treated with heparin and GPIIb/IIIa inhibitors. This resulted in a need for fewer blood transfusions. Patients who received bivalirudin alone had similar rates of ischemic complications, such as heart attack, the need for repeat procedures for artery reblockages or death as the other groups, but had significantly lower rates of major bleeding complications compared to the other groups.

* Scientists discovered a protein complex called integrator that helps form a key class of cellular RNA molecules.

* The skin is the largest sensory organ in humans. The sensory innervation of the skin allows us to perceive touch and pain. Christine Wetzel, a researcher with the Max Delbrueck Center for Molecular Medicine (MDC) Berlin-Buch, Germany, and her colleagues have deciphered the function of a molecule necessary for the conversion of mechanical stimuli into neural impulses. They have demonstrated that this molecule, a **protein** called **SLP3**, is essential for the detection and discrimination of fine tactile stimuli. This study provides the first evidence for a **touch receptor gene** in mammals and shows that molecules may in the future prove to be important **therapeutic targets** for the **control** of **chronic pain**.

IV.12. The blockbuster model : delivering just 5% ROI
Analysts at Bain Capital explain to those who want to listen, that based on recent investment levels, success rates and forecasts of commercial performance, they expect the blockbuster model to deliver just 5% return on investment. That's significantly lower than the industry's risk-adjusted cost of capital. Only one out of six new drug prospects will likely deliver returns above their cost of capital. Not exactly the kind of attractive prospect for investors, they say.

Slow or unable to adapt because of the needed massive investments in marketing capabilities providing reach and frequency, without forgetting the huge burden to carry because of the large investments needed in production (over-)capacity, and above all in science, it's BigPharma that faces big problems. Their approach, since the mid-seventies, and increasingly since the late eighties, consisted in developing primary care blockbuster product franchises. At first sight : Nice to have. But: with the severe threat of being forced out of business when the product comes off patent and has no revenue-replacement. A risky business model in times of rising R&D costs coupled with a declining output, increasing cost control measures and shorter exclusivity periods drive returns down to a level as low as 5%.

For several years, industry experts repeat –with an almost religious patience and believe- the factors contributing to the decline of the once saintly followed one-size-fits all blockbuster model. Astronomic R&D costs top the list. Mergers provide only temporary relief and ask for more merger activity to continue building scale. Is bigger really better?

Partnerships that enables economies of scale most often turn out to be a short-sighted tactic that will need to be repeated once the economies-of-scale effect has been fully achieved. It is not the marriage of two past glories that, combined and in a defensive position, creates new wealth.

Companies preoccupied with ensuring they have sufficient scale often take their eyes off new developments, count on existing franchises and/or rapidly available products. The sole focus on economies of scale is to be understood either as the confirmation of highly needed correction of past mismanagement –what is not rare in the continuous battle for power between scientists and business gurus in this industry, unfortunately- or as an indication for expected poor organic growth perspectives inevitably resulting in future under performance.

*Researchers at Mayo Clinic have successfully **isolated nanoparticles** from **human kidney stones** in cell cultures and have **isolated proteins**, **RNA** and **DNA** that appear to be associated with **nanoparticles**. The findings, are significant because it is one step closer in solving the mystery of whether nanoparticles are viable living forms that can lead to disease -- in this case, kidney stones. What causes pathological calcium deposits to build up is not entirely known. Medical scientists at Mayo Clinic are studying **calcification** at the **molecular level** in an effort to determine how this phenomenon occurs. There is a growing body of scientific evidence that links calcification to the presence of nanosized particles, particles so small that some scientists question **whether a nanoparticle can live** and if so, play a viable role in the development of kidney stones.
Approximately 12 percent of men and 5 percent of women will develop kidney stones by the time they reach 70 years old. Some $5 billion is spent in the United States each year to treat patients with kidney stones, but exactly how kidney stones form is not known. Scientists theorize that if nanoparticles become localized in the kidney, they can become the focus of subsequent growth into larger stones over months to years. Other factors, such as **physical chemistry** and **protein inhibitors of crystal growth**, also play a role.

* Study findings show that survival of heart disease patients placed on beta-blocker therapy corresponds to specific variations in their genes.

* Scientists from the Children's Hospital Boston laboratory and the Harvard Stem Cell Institute described the discovery in the first cardiac field of progenitor cells expressing the Nkx2.5 protein that can generate both cardiac and smooth muscle cells.

* NIGMS-supported scientists have determined the structure of a cellular protein whose level rises, for reasons unknown, when people fall ill from several of diseases. The findings may offer insights into these diseases, which include leukemia, diabetes and AIDS.

*Researchers from the Massachusetts General Hospital (MGH) Cardiovascular Research Center have discovered what appears to be a **master cardiac stem cell**, capable of differentiating into the three major types of cells that make up the mammalian heart. They describe identifying these progenitor cells in mice, cloning single cells from embryonic stem cells, and showing that these cloned cells can differentiate into cardiac muscle, smooth muscle or endothelial cells.
These cells offer new prospects for drug discovery and genetically based models of human disease. They also give us a new paradigm for cardiac development, in which a single multipotent cell can diversify into both muscle and endothelial lineages. They additionally suggest a novel strategy for the **regeneration** of cardiac muscle, coronary arterial and pacemaker cells. It now appears that cardiac cells develop in the same way that blood cells do, with a master stem cell giving rise to the entire range of cells. The search is now on for the hormones that trigger expansion of **MIPCs** (multipotent embryonic isl1+ progenitor cells), which would be analogous to the factors that drive blood formation.

* Using massively **parallel sequencing technology**, David **Bartel**, a Howard Hughes Medical Institute Investigator at the Massachusetts Institute of Technology, and colleagues sequenced some 400,000 small RNAs from C. elegans, identifying **18 new microRNA genes** and more than 5,000 other RNAs of a type that had not been previously reported.

* Scientists at The **Forsyth Institute** have discovered that **some cells have to die for regeneration to occur**. This research may provide insight into mechanisms necessary for therapeutic regeneration in humans, potentially addressing tissues that are lost, damaged or non-functional as a result of genetic syndromes, birth defects, cancer, degenerative diseases, accidents, aging and organ failure. Through studies of the frog (Xenopus) tadpole, the Forsyth team examined the cellular underpinnings of regeneration.
The Xenopus tadpole is an ideal model for studying regeneration because it is able to re-grow a fully functioning tail and all of its components.

* Bioengineers have developed a technique using a growth factor and a ceramic and polymer scaffold to encourage the survival and growth of adult stem cells

* Progress in medicines leads to questions about the **safety** and the **correct use** of drugs, but also about doctors ability to make the right diagnosis for ever better known health problems. Although published reports indicate that thousands of deaths occur each year as a result of **medical errors**, the true numbers of deaths attributed to adverse drug reactions (**ADRs**) is unknown. In a meta-analysis examining the incidence of ADRs in US hospitals, fatal ADRs are **one** of the 4-6 **leading causes of death** in the United States. The scope of the problem of ADRs was brought to public attention in 1999 by a report issued by the Institute of Medicine (**IOM**), To Err Is Human: Building a Safer Health System.
Adverse drug reactions (ADRs) or events (**ADEs**) are defined by the World Health Organization as "a response to a drug that is noxious and unintended and occurs at doses normally used for man for prophylaxis, diagnosis, or therapy of disease, or for the modification of physiologic function."
The concept of ADRs is intended to include all responses that place patients at risk or expose them to harm. Serious ADEs are defined by the US Food and Drug Administration as events caused by a drug which result in a patient's death, hospitalization, or disability, or cause a congenital abnormality, a life-threatening event, or require an intervention to prevent permanent damage. ADRs are common, contribute to significant morbidity and healthcare costs, and carry the single greatest risk for harm to patients in **hospitals**. It has been estimated that, each year, over **770,000 people** in the United States who are hospitalized **suffer ADRs**, which costs major hospitals up to $5.6 million per year. This estimate does not include ADRs resulting in admissions, malpractice, and litigation costs, or the costs of injuries to patients. National hospital expenses to treat patients who suffer ADRs during hospitalization are estimated to be between **$1.56 billion** and **$5.6 billion** annually .*[Source: Medscape CME, Aug 30, 2006]*

Chapter Five

*Finding
Gold*

V.1. Back to R&D

Pharmaceutical companies today are at a complex crossroads, with the industry encountering limited expansion and declining revenue growth. The science of drug discovery has moved from chemistry to molecular biology, requiring significant investment in new technologies. These technologies (genomics, proteomics, etc.) have produced an explosion of new poorly validated targets that may actually increase the rate of compound attrition and the costs of R&D. However, this shift in science has precipitated a potentially beneficial focus toward creating targeted medicines, versus the "one drug fits all" approach that has characterized pharmaceutical development for so long. Finally, the blockbusters that have driven industry growth are becoming more difficult to discover and develop at a time when many popular blockbusters are losing patent protection[96]. Faced with increasing pressure to fill drug development pipelines, manage research and development costs, bring more drugs to the market and enhance shareholder value, pharmaceutical companies are investigating ways to control R&D costs while turning their internal focus to late-phase trials, sales and marketing.

Pharma can point to the changes in the industry that began in the early 1990s. The advent of the blockbuster drugs on which big pharma have increasingly relied on also ushered in an era in which the focus shifted to profit-maximization through heavy marketing, possibly at the expense of scientific inquiry. With worldwide sales nearly doubling since 1997 to around $500 billion and a healthy pipeline big pharma had little to complain about.

Fast-forward to the present and it is a different story. The industry finds itself in the dock, faced with an avalanche of litigation instigated by users who demand someone be held to account.

A scientific revolution is taking place. We are reaching new heights in our understanding of life and disease. In the past, the goal was to develop drugs –proteins or small

---

[96] *Jeffrey Jung and Andy E. Wang, "Rx for Pharmaceutical Companies: Internal Collaboration is the Key to Improved Innovation," IBM Institute for Business Value.*

molecules—that helped correct the chemical imbalances within a body suffering from disease. Now, with the availability of genetic information, that paradigm has changed completely and forever. The pharmaceutical industry of the future does not simply produce drugs anymore but advances and applies the use of scientific knowledge itself. This fact transforms not only industry but the practice of medicine as well[97].

V.2. Major innovations come from the biotech industry

Perhaps the greatest potential lies in biotechnology with research of genetically engineered drugs becoming a new dominant factor in the medical industry. New genomic tools allowing researchers to better target medicines, segmenting large and lucrative one-size-fits-all drug markets into smaller ones, seriously ask for a profound rethinking of a mass medicines blockbuster business model which is running out of steam. With a suitable regulatory framework in both Europe and the US, biotech drugs could provide unprecedented treatment breakthroughs with the next decade; a decade that will see the creation of a new era of personalized medicine, in which treatment is expected to be tailored to an individual's unique genetic profile.

The global revenues of publicly-traded biotechnology companies exceeded $60.0 billion for the first time in 2005 in the sector's 30-year history[98].

While the US biotech industry remained strong, stable and showed maturation, the Europeans have recovered from several years of consolidation and disappointing financial results and performed significantly better with a record 23 initial public offerings (raising 52% or 150 million Euro of total European IPO capital) - compared to 13 in the USA - taking the number of public companies to 122. The UK biotech sector is in rude health with products, revenues, financing and corporate activities leading Europe.

Major drugmakers made several large acquisitions as they faced their biggest patent-expiration year ever. The future is expected to bring less IPOs and more big pharma M&As, as pricing and the cost of developing new drugs still presents a challenge for the biotech sector. Increasing specialization as well as increased use of alliances, partnerships to leverage off others and specialist skills to bring products to market will dominate the field.

Table 5.1.   Provenance of all newly approved drugs* (1998-2006)[99]

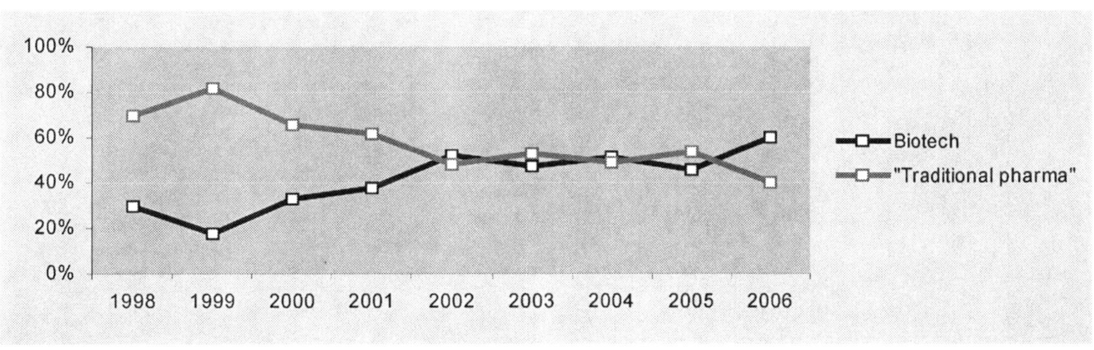

*NCEs and NBEs,

[97] *Kessler Armin M, MD, CEO F.Hoffman-La Roche Ltd, Innovative R&D or vertical integration. Is there a strategic choice? Financial Times World Pharmaceuticals Conference, 21.03.1995*

[98] *According to a report from Ernst & Young, entitled Beyond Borders; the Global Biotechnology Report 2006*

[99] *Source: Center for Medicines Research 2001, Nature Biotechnology to 2003/ DZ BANK analysis, industry data for 2006;*

* More than **71 million** Americans have **1 or more types of CVD(cardiovascular disease)**. CVD affects men and women at similar rates (34% of each group), but the rate of mortality is higher in women than in men (53% vs 47%).[*Thom T. et al., American Heart Association Statistics Committee and Stroke Statistics Subcommittee. Heart disease and stroke statistics -- 2006 update: a report from the American Heart Association Statistics Committee and Stroke Statistics Subcommittee. Circulation. 2006;113:e85-15*]

Similar disparities have been noted among victims of **myocardial infarction** (MI). Each year, more than 1 million people suffer an MI; 50% of cases are fatal, and many of these deaths occur within 1 hour of symptom onset. Data from the **Framingham Heart Study** show that more women than men will die within 1 year of having an MI (38% vs 25%). In addition, in-hospital mortality after elective or emergency percutaneous coronary intervention (**PCI**) is higher in women, while late mortality is similar to that of male patients. *[Lansky AJ,et al. Percutaneous coronary intervention and adjunctive pharmacotherapy in women: a statement for healthcare professionals from the American Heart Association. Circulation. 2005;111:940-953]*

The difference in outcomes between the sexes has been attributed to a complex interplay of many factors, including the underutilization of practice guideline-recommended therapies.[*Wenger NK. Coronary heart disease in women: highlights of the past 2 years -- stepping stones, milestones and obstructing boulders. Nat Clin Pract Cardiovasc Med. 2006;3:194-202*].

For example, more than 1.2 million PCI procedures are performed annually in the United States, but only an estimated **33%** are performed in **women**. Data from the 2002-2003 **Euro Heart Survey** show that, in addition to being less likely to undergo revascularization, women are also less likely to receive aspirin or statins. *[EuroPCR 2006 -- The Paris Course on Revascularization]*

* NIGMS-funded researchers have shown **that editing the sequence of microRNAs**--tiny molecules that are increasingly known for their role in regulating the genome--can **change the genes they target**, with physiologic consequences

* Attention Deficit Hyperactivity Disorder becomes a societal issue.

B.61. Development in the number of prescriptions for ADHD treatment, 2004-2006 (USA)

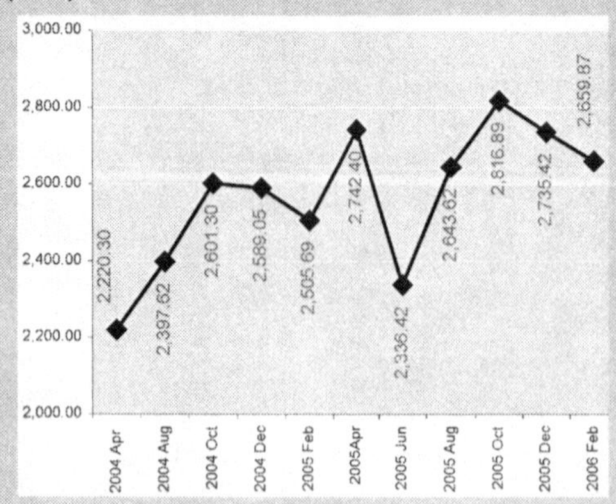

* **Tumorigenesis** is an **evolutionary** process that selects for **genetic** and **epigenetic changes**, allowing **evasion** of **anti-proliferative** and **celldeath-inducing mechanisms** that normally **limit clonal expansion** of **somatic cells**. *[Lowe, S. W.,et al., Intrinsic tumor suppression. Nature 432, 307–315 (2004)].* Most tumors acquire **genetic instability**, but how early this occurs and whether it drives tumor development is unclear. Several mechanisms to **constrain oncogenesis** have been proposed, including **hypoxia** *[Graeber, T. G. et al. Hypoxia induces accumulation of p53 protein, but activation of a G1-phase checkpoint by low-oxygen conditions is independent of p53 status. Mol. Cell. Biol. 14, 6264–6277 (1994)],* **telomere attrition** *[Counter, C. M. et al. Telomere shortening associated with chromosome instability is arrested in immortal cells which express telomerase activity. EMBO J. 11, 1921–1929 (1992).]* and **induced expression** of the **Arf tumor suppressor** (which is caused by the **mitogenic overload** experienced by **incipient cancer cells**)[*Zindy, F. et al. Arf tumor suppressor promoter monitors latent oncogenic signals in vivo. Proc. Natl Acad. Sci. USA 100, 15930–15935 (2003)].* These are all **conditions** that can **activate** the **tumor suppressor p53** *[Vogelstein, B.. et al., Surfing the p53 network. Nature 408, 307–310 (2000).]* However, whether these mechanisms represent the **major force(s)** that **guard against** genetic instability and tumorigenesis is **unknown**. In 2002, another possibility emerged from the observation *[DiTullio, R. A. Jr et al. 53BP1 functions in an ATM-dependent checkpoint pathway that is constitutively activated in human cancer. Nature Cell Biol. 4, 998–1002 (2002)]* that **advanced carcinomas** of the lung and breast show constitutive **activation** of **Chk2**, an **effector kinase** *[Lukas, C.,et al., Distinct spatiotemporal dynamics of mammalian checkpoint regulators induced by DNA damage. Nature Cell Biol. 5, 255–260 (2003)]* within the **DNA damage network** that is activated by the **kinase ATM** (Ataxia Telangiectasia Mutated) in **response** to DNA double-strand Breaks *[Kastan, M. B. & Bartek, J. Cell-cycle checkpoints and cancer. Nature 432, 316–323 (2004).// hiloh, Y. ATM and related protein kinases: safeguarding genome integrity. Nature Rev. Cancer 3, 155–168 (2003)].* Furthermore, oncogenes such **asMyc** cause **DNA damage** in **cultured cells** *[Vafa, O. et al. c-Myc can induce DNA damage, increase reactive oxygen species, and mitigate p53 function: a mechanism for oncogene-induced genetic instability. Mol. Cell 9, 1031–1044 (2002).// Lindstroem, M. S. & Wiman, K. G. Myc and E2F1 induce p53 through p14ARF-independent mechanisms in human fibroblasts. Oncogene 22, 4993–5005 (2003)].* These findings led researchers to hypothesize that **DNA damage checkpoints** might **become activated** in the **early stages** of human **tumorigenesis**, leading to **cell-cycle blockade** or apoptosis and thereby **constraining tumor progression.**[*Bartkova J. et al., DNA damage response as a candidate anti-cancer barrier in early human tumorigenesis. Nature. Vol 434, 14 Apr 2005:1]*

* In preliminary results, researchers have shown that a drug which **mimics** the **effects** of the **nerve-signaling chemical dopamine** causes **new neurons to develop** in the part of the brain where cells are lost in **Parkinson's** disease (PD). The drug also led to long-lasting recovery of function in an animal model of PD. The findings may lead to new ways of treating PD and other neurodegenerative diseases.

* **Antibiotic development** has **slowed** to the point that the US FDA and the EMEA have had few opportunities to approve new agents. In fact, development and approvals of new antibacterial agents have decreased in the US by 56 percent over the past 20 years (1998-2002 vs. 1983-1987).

V.3. Biotechs are Reshaping Drug Development

Biotechnology and smaller, emerging pharmaceutical companies represent one model for reviving innovation and profitability in a challenging R&D environment. These types of companies have been responsible for a disproportionate number of the new drugs introduced to the market in recent years, accounting for nearly 60 percent of NMEs during 2006, Product development gains in this sector are driving growth at twice the rate of growth of traditional pharmaceutical firms (18% versus 9%)[100]. In addition, of the approximately $85 billion USD invested in 2004 by the pharmaceutical, biotech and generic industries in R&D, roughly 14 percent came from the biotech sector

The success of these companies is due to the fact that many have been spun off from university programs where much of the innovation has already occurred. They are usually funded by venture capitalists that are able to spread the investment risk over several projects. Biotechnology and emerging pharmaceutical companies often focus development on a single idea or technology, or only a few compounds. These positive traits, in conjunction with the need to "make it or break it," result in an environment that promotes innovation[101].

Contract research Organizations (CROs) benefit from biotech productivity, as these firms tend to outsource as much as 50% of their R&D[102].

The boom of red and gene technology is becoming evident through its growing share in the research process: of those pharmaceuticals that have become available to patients since 1989 a mere 9% contain substances manufactured through gene technology or biotechnology. However, of the approx. 9.900 substances (as of per January 2007) that are currently still in pre-clinical or early clinical trials one out of four is already based on gene technology.

(2006, the range of drugs that were in Phase III clinical trials or pre-approval stage included 96 oncology products, 51 products for treating cardiovascular disease, 37 for viral infections and HIV, and 28 for arthritis/pain. Of the total pipeline, 27 per cent of the products are biologic in nature. Biologics experienced strong growth overall, adding $7.6 billion in sales to the global pharmaceutical market in 2005. Led by Amgen, Roche/Genentech and Johnson and Johnson, this sector grew 17.1 per cent in 2005, generating sales of $52.7 billion).

V.4.   The most cost-effective form of healthcare

Because 10 to 12 years pass before a new pharmaceutical can be sold in the market, the large investments in R&D of genetically manufactured or altered substances of the past few years will only reach the pharmaceutical market in the form of new pharmaceuticals with a considerable time lag.

---

[100] *Christopher McFadden, Randall Stanicky, Timothy Leahy and James C. Sinclair. Pharmaceutical Services. Goldman Sachs Global Equity Research, September 2003.*
[101] *Nigel Brown, Ph.D., "Capitalizing on a Changing Research and Development Environment," Business Briefing – PharmaTech 2003.*
[102] *Christopher McFadden, Randall Stanicky, Timothy Leahy and James C. Sinclair. Pharmaceutical Services. Goldman Sachs Global Equity Research, September 2003.*

* Rheumatism is not just one disease, but a class of more than 100 different disorders with focus on arthritis, a group of conditions affectig the joints and surrounding tissues. Arthritis is a disease causing pain, swelling and hindered movement of the joints. There are many different forms, each of which has a different cause. The best-known type is osteoarthritis (OA), affecting nearly 80% of people over 50. Osteoarthritis is characterized by degeneration of articular cartilage and reactive changes in surrounding bone and periarticular tissue. The causes of OA are unknown, but its symptoms include joint pain, stiffness or swelling.Other types of arthritis are actually autoimmune diseases, in which the body turns against its own tissues for an unknown reason. Inflammatory arthritic syndromes (i.e. fibromyalgia), systemic rheumatic diseases (i.e. lupus), and metabolic bone and joint diseases (i.e. gout) are all types of arthritis. However, **rheumatoid arthritis** (RA) is the **overall second most common form** of arthritis. The **immune system mistakenly attacks** the joints, **causing inflammation** and **pain**. It most often attacks the joints symmetrically; RA's characteristic chronic inflammation causes deterioration of the joint, pain, limited movement and even deformity.

It is estimated that **the number** of **people suffering from some form of arthritis.** will **increase to 180 million** by **2012.**

The **impact** on **disability attributable** to **knee osteoarthritis is similar to that due to cardiovascular disease** and **greater than that caused by any other medical condition** in the **elderly.**

Conventional treatment for arthritis is fairly standard across types. Nonsteroidal anti-inflammatory drugs **(NSAIDs)** such as ibuprofen or naproxen are used to relieve swelling, while painkillers, including acetaminophen (paracetamol) or aspirin, relive pain. The autoimmune varieties of arthritis are being treated with stronger drugs, such as those used in chemotherapy, tha suppress the immune response. They carry serious side-effects, including blood abnormalities, kidney damage or vision damage. "**Fire in the joints**" has increasingly been treated by an active product ingredient (**glucosamine**) that was virtually unknown in the early nineties, but rapidly became a preferred treatment in the USA, following the publication of **Dr Jason Theodosakis**' bestseller „The Arthritis Cure". In Europe, thanks to **clinical trials** with **Rottapharm/Opfermann's glucosamine sulfate**, scientific evidence accumulated, making of glucosamine a **mainstream product** to **slow the progression of osteoarthritis** –the deterioration of cartilage between joint bones- and **reduce associated pain.** In 2001, the Lancet published a landmark study in OA research. The results of a three-year double blind clinical trial made at the University of Liège (Belgium), proved that patients taking glucosamine sulfate had a 20 to 25% improvement in symptoms of OA. The Liège study found that glucosamine sulfate could be a disease-modifying agent in osteoarthritis, and that the substance had both a bone structure-modifying and a symptom-modifying effect on patients with OA of the knee. Some companies used the popularity of glucosamine to launch products without medical marketing authorization. Severa of these products have been found to offer poor quality and/or low concentrations of glucosamine, causing harm to the reputation of the substance.

Another **remarkable finding** came from Czech scientists in 2002. They confirmed that **long-term treatment** with **glucosamine sulfate** at **once-daily doses** of **1,500mg retarded the progression of knee osteoarthritis**, possibly **determining disease modification**. *[Pavelka K. et al, Glucosamine sulfate use and delay of progression of knee osteoarthritis. Arch Intern Med. 2002;162:2113-2123]*

Clinical trials continue to **prove the efficacy** of **glucosamine sulfate**, notably as an excellent oral symptom-modifying medical therapy with retard effect (a so-called Slow Acting Drug in Osteoarthritis, or SYSADOA). In „**Future Medicine**" Herrero-Baumont & Rovati state that there is a need for medications that offer acceptable short-term symptom control, but in particular have a role in the medium and long-term management of disease (symptom-modifying effect), including the possibility to delay the progression of joint structure changes (structure modifying effect), thereby affecting the evolution of the disease and thus preventing clinically significant disease outcomes (disease-modifying effect). These aims might be achieved by drugs that, unlike unspecific symptomatic agents, might exert specific effects on osteo-arthritis pathogenesis factors. To date, the scientists write, **glucosamine sulfate is probably the drug with the most extensive evidence in this regard**. *[Herrero-Beaumont G. & Rovati LC., Use of crystalline glucosamine sulfate in osteoarthritis, Future Medicine/Future Rheumatol (2006) 1(4),397-414].*

In 2007, Scientists at the **University of Madrid**, reporting on the final results of the GAIT trial, **confirmed** that **glucosamine sulfate** at the **oral once-daily dosage of 1,500mg is more effective** than placebo in **treating knee OA symptoms**. Although acetaminophen (paracetamol) also had a higher responder rate compared with placebo, acetaminophen (paracetamol) failed to show significant effects on the algofunctional indexes.*[Herrero-Beaumont G. et al, Glucosamine sulfate in the treatment of knee osteoarthritis symptoms. Arthr & Rheum. Vol. 56, No. 2, February 2007, pp 555–567]*

* The **exchange** of a **single gene building block** in the **genetic material** of the **tuberculosis bacterium** leads to **resistance** to the antibiotic rifampicin. Scientists of the German Cancer Research Center (Deutsches Krebsforschungszentrum, DKFZ) and the Universities of Heidelberg and Bielefeld, Germany, have developed a highly sensitive **test** for **detecting** this **genetic alteration** at the level of a **single molecule**, thus providing information about the resistance status of an infected person.

* **Drug-eluting stents** (DES) have significantly reduced restenosis after stent implantation in the coronary arteries. Target lesion revascularization (TLR) rates have dropped from an average of 20% to as low as 6%. For this reason, **90%** of all **stents** deployed in the US are **DES**. Similar trends are seen throughout the world.

* Only five medications have been approved specifically for **migraine prophylaxis** by the US Food and Drug Administration (FDA): **divalproex sodium, methyser-gide maleate, propranolol hydrochloride, timolol maleate**, and **topiramate**. Most medications used for migraine prevention have been "discovered" while being used in other therapeutic areas. Very few of the drugs used have been subjected to the rigors of double-blind placebo- controlled clinical trials. Exceptions include **topiramate** and **divalproex** sodium.

Table 5.2. Present-Day environment[103]

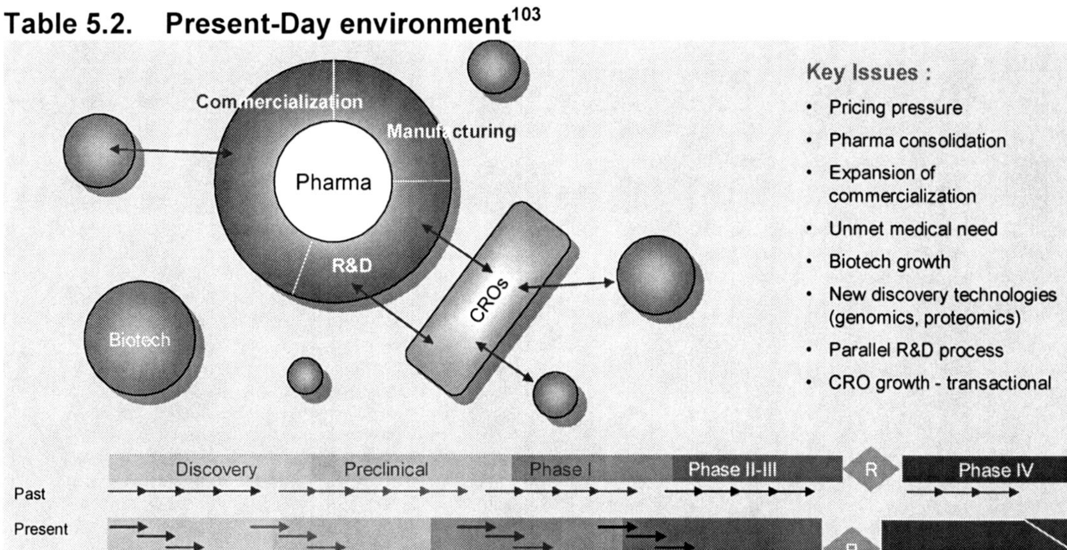

Table 5.3. The future

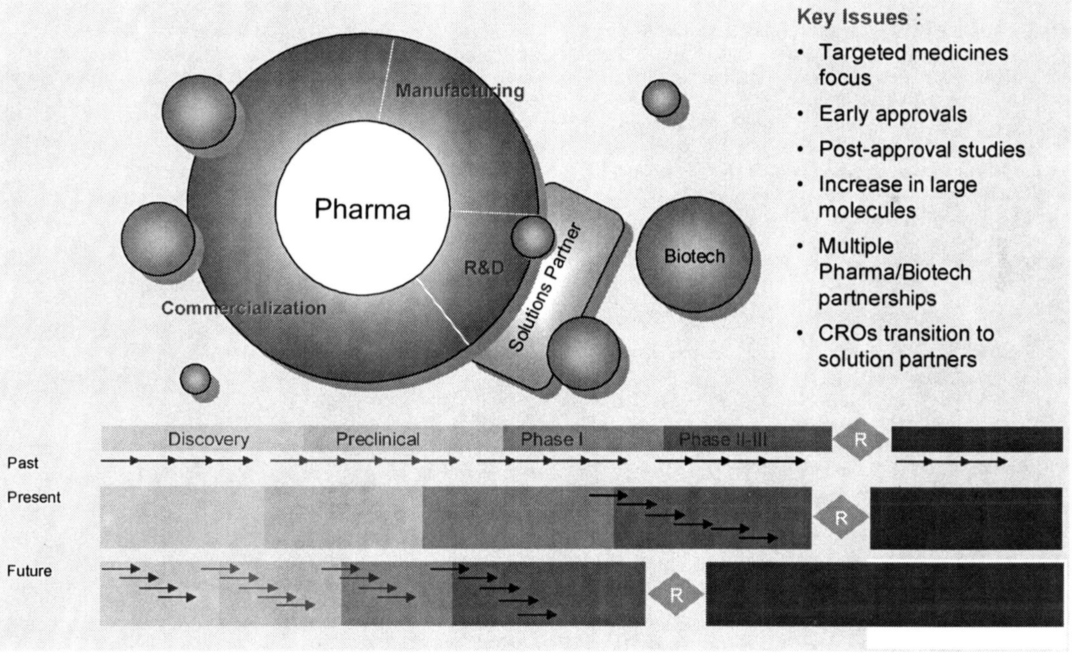

---

[103] Source: 2004 MDS Pharma Services

* The U.S. Centers for Disease Control and Prevention (**CDC**) states that persons infected with drug-resistant organisms are more likely to have longer hospital stays and require treatment with multiple drugs. The **increasing prevalence** of resistant bacteria often necessitates the use of **combinations** of **antibiotics** to fight infections. Antibiotic resistance costs U.S. society between $4 billion and $5 billion annually. According to the CDC, antibiotic resistance has become so widespread that many significant bacterial infections in the world are becoming resistant to commonly used antibiotics.

* A new direction for **psychological science** has emerged since the turn of the century - one that considers confrontation with deep existential issues to be an essential factor in diverse forms of human behavior. **Experimental existential psychology** uses rigorous methods to explore these processes. By illuminating the deeper psychological concerns that underlie human behavior, experimental existential psychology hopes to promote psychological health and more harmonious relationships between individuals and groups. The new insights gained from this research can be used to promote the development of new **scientifically grounded forms** of **psychotherapy**.

* **Statin therapy alone** is said to **fail** bringing patients to the **target levels** of blood cholesterol required by guidelines to minimize their health risk. A ' lower is better' approach increasingly advocated is manifest in the trend towards lower low-density-lipoprotein cholesterol (LDL-C) targets in guidelines. Specialists believe that this requires more intensive therapy earlier. Only **51%** of patients on lipid lowering therapy achieve the goal of total cholesterol <150mg/dL [5mmol/L] in
practice. Standard doses of statins achieve LDL-C <2.6 mmol/L (<100mg/dL) in little more than half of high-risk patients.

* Since the turn of the century, a rapid development in our understanding of the **genetics** behind the **interindividual differences** in drug action has occurred. This encompasses the area of pharmacogenetics (PG) where the interindividual variability in genes related to drug transporters, drug metabolizing enzymes and drug targets is studied in relation to efficacy of drug treatment and adverse drug reactions. A good deal of this variability results from genetic polymorphism, i.e. the occurrence in the same population of multiple
allelic states. With respect to pharmacokinetic (PK) aspects, the highest penetrance of genetic polymorphism is registered at the level of drug metabolism where about 40 % of phase I metabolism of clinically used drugs is affected by polymorphic enzymes. The most important polymorphic cytochrome P450 enzymes are CYP2D6, CYP2C19, and CYP2C9. Regarding phase II enzymes, the genetic variability of UDP-glucuronosyltransferases, N-acetyltransferase-2 and some methyltransferases plays a role in the interindividual variability in PK. The additional contribution of polymorphism in drug transporters has been recognized.

Researchers at Cincinnati Children's Hospital Medical Center **have developed the first gene chip, termed the „jaundice chip",** to use in the early diagnosis of at least five hereditary liver diseases, to detect genetic causes of jaundice in children and adults, and potentially to lead to **personalized treatment options.**
It is the first chip in the world that has been customized to diagnose genetic mutations in patients with **inherited types of liver diseases.**
Previous research on humans identified five genes responsible for inherited forms of jaundice. Until now, the broad array of causes of cholestasis including genetic, metabolic, inflammatory and drug- or toxin- induced disorders, created a challenge for physicians to diagnose a specific disease. Therefore, the treatment of affected children was not disease- specific and aimed at optimizing care to help reduce liver transplantation. With the jaundice chip, however, diagnosis can be simplified by surveying the genetic code for mutations in specific diseases.
It will **improve diagnostic accuracy** for perplexing diseases in infants and children, potentially **decrease** the **need** for **invasive** and costly **studies**, and allow us to **develop specific treatment plans** based on the correct genetic diagnosis.
With further genetic testing of liver disease, there is the potential that medications can be tailored to meet the needs of individual patients taking into account the patient's genetic make-up. The next focus of advances will be the development of medication that may block progression of the disease.

* NIGMS-funded researchers have **deciphered the structure of insulin-degrading enzyme**, which breaks down not only insulin but the protein that builds up in the brains of those with Alzheimer's

* It turns out that sequencing the human genome - determining the order of DNA building blocks -- has not completely cracked the code of how DNA directs various cellular processes. In addition to the sequence of the base pairs, the **instructions** are in the **packaging** - how DNA is folded within a cell.
Virginia Tech researchers used novel methodology and the university's System X supercomputer to carry out what is probably the first simulation that explores full range of motions of a DNA strand of 147 base pairs, the length that is required to form the fundamental unit of DNA packing in the living cells -- the **nucleosome**. Contrary to a long-held belief that DNA is hard to bend, the simulation shows in crisp atomic detail that DNA is considerably more flexible than commonly thought.

* A person's **genetic make-up** is likely to be important in the probability of developing certain diseases, such as diabetes, cardiovascular disease, cancers, schizophrenia and Alzheimer's disease. Although **patterns of inheritance** are not clear-cut, **gene-environment interactions** may play a major role.
Furthermore, patterns of disease differ by **ethnic group**: type 2 diabetes mellitus is up to **six times more common** in people of **South Asian descent** and up to three times more common among those of **African** and **African-Caribbean** origin, for instance. *[World Health Organization,EUR/RC56/8 +EUR/RC56/Conf.Doc./3, 30 June 2006, 60783]*

Table 5.4 Genomic research based products reach the market by 2010[104]

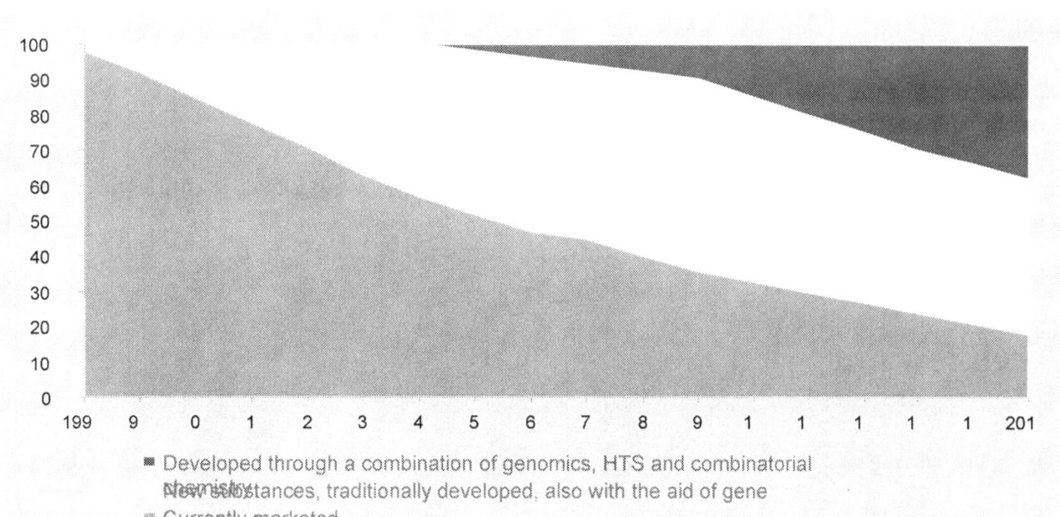

New medical-technological developments will impact on the delivery of healthcare. Genome science, high-throughput screening and combinatorial chemistry are just a few examples of new technologies that are being applied to the intelligent design of new drugs. DNA testing can subdivide patient groups into well-defined cohorts that will benefit from a specific drug without suffering the side-effects that are probably caused by applying drugs to too wide a population.

V.5. Genetic engineering

"New pharmaceutical drug discoveries are the most cost-effective form of healthcare. Every day, prescription drugs create huge savings by reducing the need for surgery, eliminating the need for hospitalization and keeping people healthy and on the job. And in the future, new medicines will save us hundreds of billions of Euro in health care costs"[105].

Table V.5. The promise of genomics

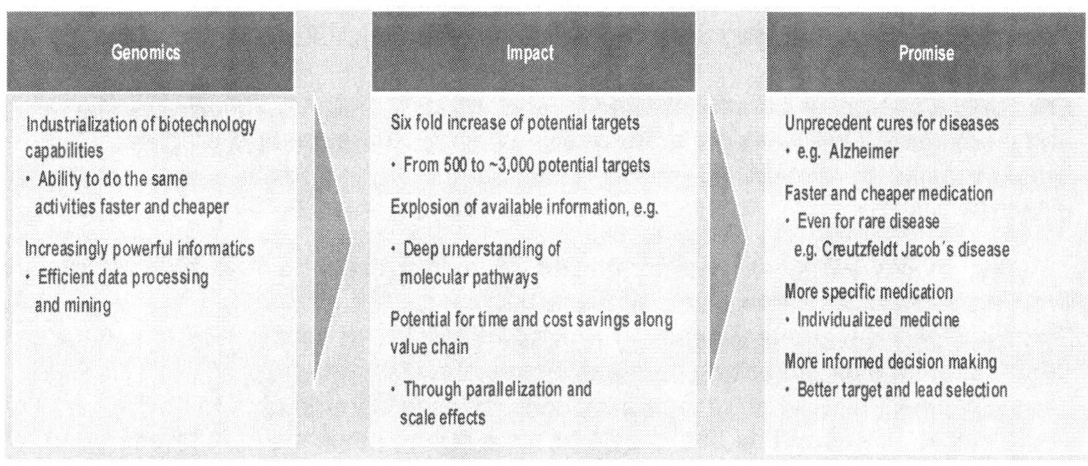

---

[104] Source: Association of Research Based Pharmaceutical Companies of Germany, VfA/BCG commissioned report

[105] Dr Karl H Schlingensief, President of the Board, Hoffmann La Roche Germany in an interview with the author, 2003

Genetic engineering is already revolutionizing medicine. It is enabling the mass production of safe, pure, more effective chemicals that the human body produces naturally.

Pharmaceutical biotechnology's greatest potential for curing chronic and currently "incurable" diseases lies in gene therapy.

Table 5.6. **Genetically manufactured substances in the pipeline** (by percentage share)

Category	Percentage
pre-phase III	26%
in phase III	20%
pre-launch	14%
post-launch	9%

(Source: VfA-Verband Forschender Arzneimittelhersteller, Germany)

The elucidation of the genetic build-up of the human being removes the need to study pathology before deciding on intervention. Genetic information can reveal a predisposition for pathology, which can be treated before the individual suffers the symptoms.

A study made by Batelle Institute summarizes the medical benefits new medicines will provide in the next 25 years :
- Pharmaceutical drug advances will prevent 4.400.000 deaths from heart attacks and strokes;
- Lung cancer deaths will be cut by 409.000 as a result of new drug discoveries;
- New medicines will reduce leukemia deaths by 83.000;
- Prescription drugs will decrease the severity of rheumatoid arthritis cases by as much as 50%;
- For many diseases, such as Alzheimer's, and arthritis, pharmaceutical drug research and development will likely generate nearly all of the future medical progress;
- Breakthroughs in biotechnology will advance the fight against cancer and viral diseases

The market of the future will be composed of ever smaller, but in number rapidly proliferating, niches. The implications of this are legion :
- The traditional government approval processes will prove inadequate to service the demand for an ever increasing number of specialized drugs.
- Monolithic multinational pharmaceutical companies will not be able to dominate all of these niches. They will be better able to have strong positions in the generation of scientific information underlying the development of drugs, as well as on the powerful commercialization of approved drugs, but will not be able to dominate the development of so many smaller drugs.

Table 5.7. The changing medicinal discovery process[106]

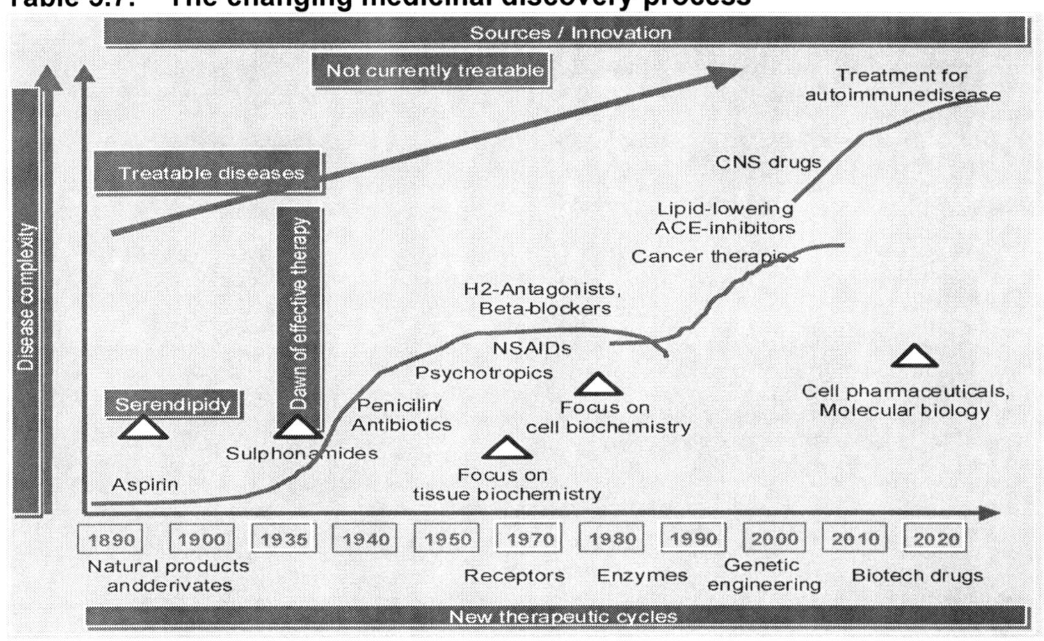

V.6. Frenetically looking for INDs

Their own weakness in New Drug Research & Development is making pharmaceutical companies frenetically looking for company external candidates. Early 2007 deals show that established pharmaceutical companies are prepared to value an Investigational New Drug(IND)-stage drug at US$ 100 million.

Companies	Upfront payment	Milestones	Companies	Upfront payment	Milestones
Roche-Actelion	US$ 75m	Up to US$ 555m	Roche-Intermune	US$ 60m	US$ 150-470m
Roche-Plexxikon	US$ 40m	US$ 150-550m	Biogen-UCB	US$ 30m	Up to US$ 170m
Medimmune-Infinity	US$ 70mn	Up to US$ 430m	Novartis-SGX	US$ 25m	Up to US$ 500m

V.7. The return on R&D investments

BigPharma's search for ever larger opportunities in a limited number of high incidence rate disease/therapeutic areas is going to be replaced by BigPharma's search for ever larger opportunities in multiple specialist disease areas.

Previously, the high revenues of the blockbuster drugs themselves suggested that incremental investment should always focus on maintaining existing brand franchises as long as possible (court actions included) or discovering the next blockbuster ("survival strategy").

---

[106] *Sources : Chiroscience Plc company data based on Boston Consulting Group infomaterial(1996); PhRMA (1999), Lehman Brothers Pharmaceutical Research(1999); Mertens G. "Chronology of drug innovation" in "From quackery to Credibility. Financial Times Pharmaceuticals management report (2000)*

A study conducted by scientists at Children's Hospital Oakland Research Institute (CHORI) identifies a **specific enzyme** that can cause the death of cancer cells. Researchers studied the behavior of an enzyme called **sphingosine phosphate lyase** (SPL), which can regulate cell growth and death by lowering the levels of a natural, growth-promoting lipid called sphingosine-1-phosphate, or S1P.

* A natural chemical that has been ignored by researchers largely because of the runaway success of the blockbuster statin drugs may in fact yield a rare twofer: a prime target for novel cholesterol-lowering drugs and the blueprint for a new generation of antibiotics that can take down Streptococcus pneumonia and Staphylococcus aureus.

As it turns out, whether you are a human who synthesizes cholesterol, a plant making carotenoids or a bacterium just striving to survive, you need the **mevalonic acid pathway**. This biochemical pathway is essential for synthesizing cholesterol, steroid hormones and other essential cellular compounds.

Howard Hughes Medical Institute investigator Dr Noel and his colleagues at the Salk Institute for Biological Studies learned how a chemical compound can shut down this crucial pathway. His group is using an approach that they call "molecular dentistry," to custom design chemicals with unique shapes that can switch off a crucial enzyme in the mevalonic acid pathway. (Proceedings of the National Academy of Sciences, July 17, 2006).

Many organisms -- including people, plants, and some types of bacteria -- use the same set of enzymes to produce compounds known as isoprenoids. These chemicals serve as the starting material for a host of biological compounds, such as vitamin D, various hormones, and cholesterol. Interfering with the mevalonic acid metabolic pathway – the chemical factory that churns out isoprenoids – is the strategy behind the powerful effects of some top-selling, cholesterol-lowering drugs. These drugs, collectively known as statins (Lipitor, Mevacor, and Zocor are examples), prevent cholesterol formation by blocking the activity of the enzyme 3-hydroxy-3-methylglutaryl CoA reductase (HMGR).For the scientists is was important to understand how the enzyme, known as 3-hydroxy-3-methylglutaryl CoA synthase, or HMGS, that works with HMGR in the mevalonic pathway, interacts with F-244, a powerful HMGS inhibitor produced by certain species of fungi. HMGS could be an important alternative target for cholesterol-lowering drugs, but it has been little studied because the statin drugs have been so well developed. Although the statins are effective, they may have side effects sometimes caused by their interactions with other targets in the body besides HMGR. Drugs that target HMGS have the potential to avoid these side effects, which include abnormal liver function and muscle damage. They explored the interaction between HMGS and F-244 using protein x-ray crystallography. The researchers settled on the version of HMGS found in the plant Brassica juncea, a species of mustard plant, because it more closely resembles human HMGS than other forms that had been studied in the past.

(see opposite colon)

The structural analysis of HMGS bound to F-244 revealed important insights into the interaction between HMGS and the natural product that will guide drug development. Trying to improve the selectivity and potency of these compounds using synthetic chemistry, they think of the F-244 molecule as composed of two distinct components.

They refer to the reactive piece that switches HMGS off as the 'warhead,' because it reacts with the HMGS enzyme's catalytic machinery to block its action. This warhead has to remain invariant because the HMGS catalytic machinery is the same from species to species. However, their data revealed that the other component of F-244, a tail-like structure that they call the 'guidance system,' can be readily modified using rational drug design to chemically mold it into specific shapes that can interact with a particular HMGS. The region of HMGS surrounding the guidance system varies across species and kingdoms potentially offering routes to designer inhibitors. Modifying the inhibitor in this way could yield analogs to treat high cholesterol with fewer side effects.

Clinically important pathogenic bacteria such as Streptococcus pneumonia and Staphylococcus aureus also have versions of HMGS that could be targeted with related drugs.

There are no antibiotics that specifically target the mevalonic acid pathway and only a couple in very early development that target the related mevalonate-independent pathway – both of which researchers are exploring in earnest using the tools of structural biology and synthetic chemistry.

* University of Cincinnati (UC) researchers have identified a new way to predict when anti-estrogen drug therapies are inappropriate for patients with hormone-dependent breast cancer.

 The UC researchers found that when a pathway controlling cell growth known as the retinoblastoma (RB) tumor suppressor is disrupted or "shut off," the tumor resists anti-estrogen drugs and the cancer continues to grow in spite of the therapy.

The findings could help physicians more accurately predict which tumors will respond to anti-estrogen therapy and improve long-term survival for breast cancer patients

* Aging, and the processes of deterioration that go with it, are largely attributable to cells that die off in a controlled manner. Therefore, gaining better understanding of this controlled cell death is very important in the fight against deterioration diseases like **dementia**. In this light, researchers from the Flanders Interuniversity Institute for Biotechnology (VIB) connected to the K.U.Leuven, in collaboration with researchers from the Dulbecco-Telethon Institute hosted by the Veneto Institute of Molecular Medicine in Padua

(Italy), have now discovered the function of the **PARL** protein. By studying mice that are unable to produce PARL, the researchers have discovered the significance of this protein in **controlled cell death**. An important step toward a good understanding of the aging processes and of diseases like **Parkinson's** disease.

One of the most interesting factors is that, for European companies, focus on targeted medicines helps redress the balance of spending in comparison to US pharmaceutical companies. Trends indicate that the US companies outspend European ones, giving them some competitive advantage in developing a range of blockbuster drugs. By targeting their research on specific areas, such as oncology, and focusing away from the most intensive areas of competition smaller EU companies could, perhaps, be able to derive higher R&D efficiencies that larger organizations cannot deliver.

In the end, selling drugs is the very reason of existence of a pharmaceutical company and innovative drugs are the most profitable.

Table 5.8. Drug Discovery holds immense possibilities

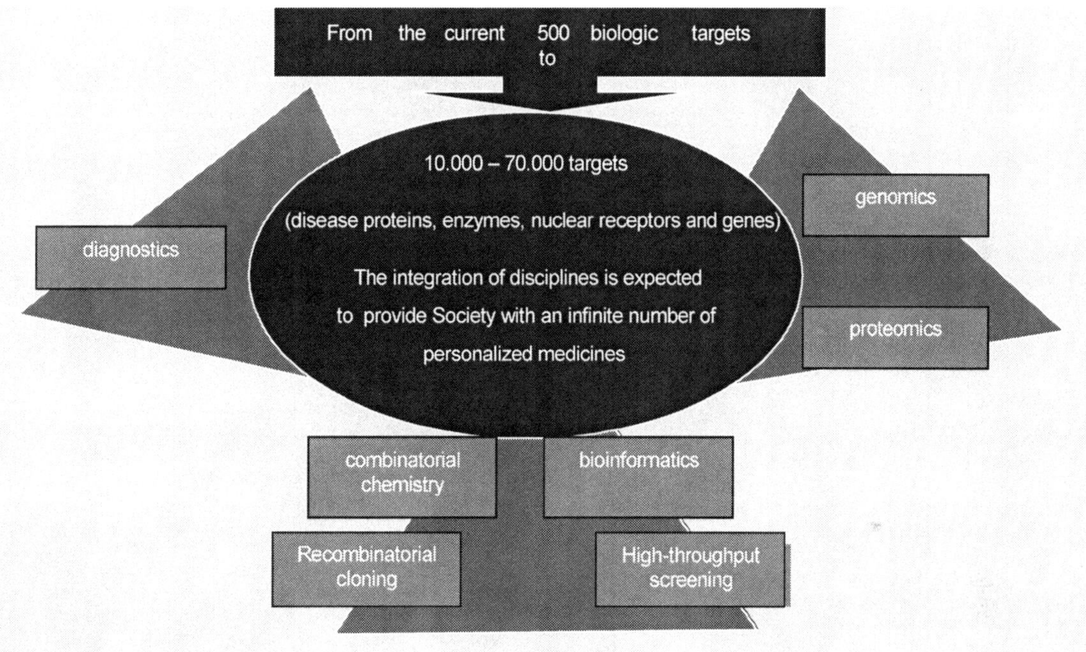

The efficiency of R&D is indicated by its return. The sales increase due to innovation in the industry has averaged 2,3% over the last 20 years. If one integrates this with a 12% discount rate over the anticipated life time of the patent, one arrives at 17% of sales that can be spent on R&D. This is roughly what the big spenders in R&D do actually spend. This is based on the assumptions of a relatively long duration of high sales in the market place for the innovative drug (the period of uniqueness has shortened dramatically over the last ten years and we must assume that –on average- more numerous launches of shorter-lived products should yield the same return as before, that the integral development cost does not increase and that the sales peak out earlier).

In the mid 1990s, Coopers & Lybrandt[107] recommended to accelerate the pace of innovation. The advise given in those days is more actual than ever:

---

[107] *Dr Manfred Karobath, Coopers & Lybrand, Financial Times World Pharmaceutical Conference, March 20-21, 1995*

* Current drug delivery is restricted by the blood barrier and the blood-cerebrospinal fluid barrier. This major obstacle in the treatment of infections and other diseases of the brain, which prevents systemically delivered therapeutic drugs from reaching the brain is a tremendous challenge for the medical research community. Grantees of the National Institute of Allergy and Infectious Diseases, part of the National Institutes of Health, have shown that **a short protein (peptide)** from the rabies virus can carry a **strip of therapeutic material into the brain via intravenous administration.** Once delivered to the nerve cells of the brain, the strip, called a small interfering RNA (**siRNA**), was shown to protect mice from infection caused by the Japanese encephalitis virus (JEV). *("Transvascular delivery of small interfering RNA to the central nervous system," by N Manjunath et al. Nature DOI: 10.1038/nature05901,2007).*
Other novel delivery systems to **transport drugs** from **circulating blood** into the **central nervous system** will emerge. Researchers work on micropsheres, liposomes, polymeric micelles, vector-mediated delivery, niosomes, nanoparticles, solid lipid nanoparticles, noninvasive gene therapy, and nasal administration to try to circumvent the blood - brain barrier.

* Right now **medical treatment exists for about 500 diseases and genetic disorders**, but thanks to the human genome project, before long, **new drugs** will be available for **thousands of other diseases**. A **new way of mass-producing protein-based drugs is needed**; a way which is economical, safe, and fast. Mushrooms might serve as biofactories for the production of various beneficial human drugs, according to plant pathologists who have inserted new genes into mushrooms. The scientists have developed a technique to genetically modify **Agaricus bisporus** -- the button variety of mushroom, which is the predominant edible species worldwide. One application of their technology is the use of **transgenic mushrooms** as **factories for producing therapeutic proteins,** such as vaccines, monoclonal antibodies, and hormones like insulin, or commercial enzymes, such as cellulase for biofuels.

* Scientists provided the first large-scale identification of the **proteins involved in coronary heart disease**. The information will help to better understand the progression of the disease, improve diagnosis, and detect early pathological signs more efficiently.
Coronary heart disease, which is characterized by abnormal thickening and narrowing of the blood vessels, is the first leading cause of death in the United States. But what happens inside the cells of these blood vessels is not completely understood. One way to figure it out is by identifying the proteins present in blood vessels of heart disease patients, and then comparing them with those present in healthy blood vessels.
Scientists developed a technique called **direct tissue proteomics** that identified all the proteins expressed in the coronary arteries of heart disease patients. They found about 800 proteins, some of them not previously known to be involved in heart disease. The investigators also used another highly sensitive proteomics method to detect important cytokines directly from diseased coronary arteries, an approach that could uncover important biomarkers relevant to other diseases.

* Researchers have devised a new way of looking at **proteins that interact with one another inside cells.** Inside a cell, proteins constantly interact and work together: proteins modify other proteins, transport one or more proteins, and associate to form new protein complexes with differing cellular functions. Various methods are available to determine how proteins associate with one another, but these methods can be technically demanding and limited by the strength of the interaction or cannot resolve whether the association indeed occurs inside cells.
The new method takes advantage of a strategy used by a virus called **mammalian orthoreovirus**. When this virus infects a cell, it uses the **cellular machinery** to **synthesize viral proteins** that **replicate the viral RNA** and **assemble new viruses**. This process occurs in large structures called **viral factories made by the virus inside the cell**. Previous work had identified an orthoreoviral protein that can form very similar viral factory-like structures (FLS) when expressed on its own in mammalian cells. The scientists (C.L. Miller, L. Nibert and colleagues) used the viral FLS as a platform to visualize the interactions of two to three other proteins at a time in a light microscope.

* **Eculizumab (Soliris)** was approved in 2007 by the U.S. Food and Drug Administration (FDA) as the first treatment for **PNH**, a rare, debilitating and life-threatening **blood disorder defined by hemolysis**, or the destruction of red blood cells. PNH affects an estimated 8,000 to 10,000 people in North America and Europe. PNH often strikes people in the prime of their lives, with an average age of onset in the early 30's. Ten percent of all patients first develop symptoms at 21 years of age or younger. PNH develops without warning and can occur in men and women of all races, backgrounds and ages. PNH often goes unrecognized, with delays in diagnosis often ranging from one to more than 10 years. The estimated median survival for PNH patients is between 10 and 15 years from the time of diagnosis.

* Dr. Cara T. Pager (Stanford University) is deciphering **the role of** newly discovered **small RNAs in the regulation of hepatitis C virus**, a novel approach with implications for liver cancer antiviral therapies.

* **Neuropathic pain** is associated with peripheral or central nervous system injury. It is often described as „burning" or „shooting" in nature and can be continuous or paroxysmal. It is estimated that up to 5% of the general population of the USA, Europe and Japan are affected by neuropathic conditions including diabetic neuropathic pain and post-herpatic neuralgia.
The global market value for neuropathic pain treatments is estimated at $2.1Bn (2007) and is predicted to rise to $5.5Bn by 2015. Sosei Group Corp. Of Japan is developing SD118 as a potential treatment for ANP.

* Dr. Y. Nam (Harvard Medical School) is determining how **protein molecules are transported into and out of the cell** – a fundamental question with potential impact in the areas of carcinogenesis, drug transport, and drug discovery.

Table 5.9.   Accelerating the pace of innovation

- Abandoning overcrowded targets with low probability of obtaining market exclusivity; fulfilling unmet medical needs at the time of market introduction
- Optimization (Quality, timelines) of preclinical development
- Expand scientific and technological basis through close interaction with external scientists; network of flexible partnerships with risk and reward sharing
- Exploit future technology triggered market chances.
- Be fast, faster, with the transfer of new drug targets to product development

V.8. Licensing grows in importance

With the thin pipeline of in-house developed new medicines candidates at most drug companies, management increasingly turn towards product licensing as a source of new drugs.

Even the most fervent defenders of in-house developments, such as Merck & Co, had to accept that survival means finding new business models.

V.9. Fewer blockbusters to support growth

Blockbuster drugs underpinned the growth of the pharmaceutical market since 1976. The total share of sales generated by these (multi-)billion dollar drugs is expected starting to decline in favor of sales generated by generic copies of these "drugs for the masses" and above all of new, targeted drugs with "nichebuster" status.

Table 5.10.   Share of total pharmaceutical market attributable to sales of blockbusters

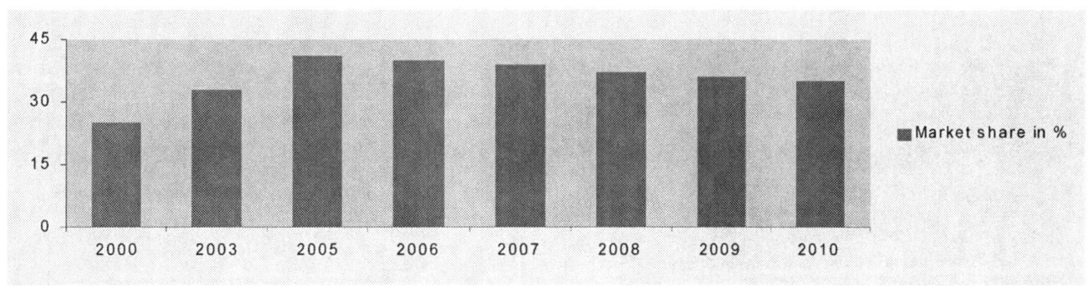

* The discovery of a gene, named **AUF1,** that appears to act as a switch determining when a body will go into septic shock, could offer scientists as a target for treating this dangerous runaway immune response.

* Researchers were able to **crystallize** a bacterial enzyme in the process of building an antibiotic, a finding that is expected to enable the development of new kinds of antibiotics

* **Parkinson's disease** is one of the commonest afflictions caused by damage to the nerve cells of the brain. Patients suffer from muscular tremors and rigidity, and their movements are severely impaired. These symptoms arise as the result of the degeneration of certain nerve cells which normally release the neural transmitter dopamine – causing an insufficiency of dopamine. To directly combat this decline in dopamine levels, researchers at the company **NeuroProgen** have developed a new drug on the basis of stem cells which is now being adapted to the regulatory standards of the pharmaceutical industry in partnership with Germany's Fraunhofer Institute for Cell Therapy and Immunology IZI. If all goes well with the scaling-up of the pharmaceutical process, clinical trials of the cell-therapy drug could commence in 2008 – as the first cell therapy employing neural stem cells.
Although current forms of medication can alleviate the symptoms during the first five to ten years following diagnosis, the results of treatment at a later stage often vary unpredictably. This is because the active drug ingredient is not dopamine itself but various precursor molecules. The symptoms can only be relieved if the patient's dopamine-producing cells are capable of taking up the precursor molecules and converting them into dopamine. A normal healthy individual has 800,000 of these cells, whereas a person who has Parkinson's disease will have lost 80 percent of these **by the time the first symptoms manifest themselves**. The further the disease progresses, the weaker the capacity of the patient's cells to convert the precursor drug. The new cell therapy works in an entirely different way: The patient's degenerated cells are **replaced with fresh human stem cells** that have differentiated into nerve cells. Neurosurgeons can implant these cells in a targeted area of the patient's brain, where they will produce dopamine to compensate for the patient's own insufficiency. The researchers hope that one day a single course of treatment will be all that is needed to cure the disease.

* According to the WHO, **just 10 percent of the global health research budget is spent on diseases that account for 90 percent of the global disease burden**. In an attempt to redress the balance, several organizations have banded together to create the Drugs for Neglected Diseases initiative (DNDi). Partners in this non-profit venture include Médecins sans Frontières, the Indian Council of Medical research, the Institut Patseur, Kenya Medical Research Institute, Brazil's Oswaldo Cruz Foundation, and the Malaysian Ministry of Health, the Who's Tropical Disease research Program and the World Bank. By 2015, DNDi hopes to develop six or seven drugs to treat Ch.agas disease, leishmaniasis, and sleeping sickness. Those three diseases alone affect 350 million people a year.

* NIGMS-funded researchers have reconstructed the signaling pathways that **regulate a molecule**, called **integrin**, that is critical to the control of bleeding and clotting.

* Using a technique called fluorescence resonance energy transfer, researchers have been able to solve the structure of a large complex of proteins and DNA involved in a form of recombination.

* **BigPharma** is on the way to become **BigOncology**. A cure for cancer has become the Holy grail of modern drug research. Contrary to the traditional search for body-external pathogens, cancer research has to focus on how to target and eliminate cancer cells that do not belong to the external invaders but are native parts of the bodies in which they live. limatinib (Gleevec) was the first kinase inhibitor to be used as an anticancer drug (it inhibits the tyrosine kinase KIT. Tyrosine kinase pathways are deregulated in almost in most forms of cancer); erlotinib (Tarceva) was a novel targeted drug with proven survival benefit in advanced non-small lung cancer and pancreatic cancer; Trastuzumab (Herceptin) is designed for a particularly aggressive type of tumor (HER2-positive) that accounts for approximately 20-20% of all breast cancers; Bevacizumab (Avastin) for first line treatment in combination with chemotherapy in metastatic colorectal cancer, and other cancer therapies started to turn pharmaceutical companies away from traditional "blockbuster" products based on a single application in a single clinical area, to "blockbuster" sales by getting cancer therapies registered into multiple indications. **Will society accept the potentially huge financial burden**?

The healthier Western society becomes, the more it regards maximum access to medicine as a right and duty, the more medicine it craves and the more it requires from that super-medicine. **Everyone wants the best medicine...provided someone else pays the bill. The approach that pays dividends** : replacing the primary care mono blockbuster by the specialty blockbuster model/the multidimensional platform. Tremendous opportunities lie ahead because of demographics and medical invention.

Governments see the pharmaceutical industry as "cost". **Industry could be a source of better health** but fails to communicate that it brings "value" to society; underestimates the power of NGOs (criticizing e.g. executive payments, environmental behavior,...).

Society has to decide what it wants to achieve in terms of health care.

From the Primary Care Monoculture -Medicines for Every Body- to a Multidimensional Platform –Personalized Medicine. a Multidimensional Platform

Towards 2015

"Medicine's finest hour is the dawn of its dilemmas. For centuries medicine was impotent and thus unproblematic. From the Greeks to the First World War, its tasks were simple : to grapple with lethal diseases and gross disabilities, to ensure live births and manage pain. It performed these as best it could but with meager success. Today, with "mission accomplished", its triumphs are dissolving in disorientation. Medicine has led to inflated expectations, which the public eagerly swallowed. Yet as those expectations become unlimited, they are unfulfillable : medicine will have to redefine its limits even as it extends its capacities". (Roy Porter, Professor, Social History of Medicine, Wellcome Institute in *"The Greatest Benefit To Mankind"*. *Fontana press, 1997:718)*

* The medicines currently marketed reach only about 10% to 20% of the body's possible drug targets, or 400 to 500 protein receptors and enzymes out of an estimated 5.000. In the next twenty years, all 5.000 targets will be accessible. The companies that will survive and/or emerge as **the winners** will be those that can turn genomics information into useful knowledge for R&D networks (identifying gene and protein function, validated targets, and lead compounds).

* The application of pharmacogenomics and pharmacogenetics in drug development holds great promise to shed scientific light on the often risky and costly process of drug development, and to provide greater confidence about the risks and benefits of drugs in specific populations Pharmacogenomics' promise lies in its potential ability to individualize therapy by predicting which individuals have a greater chance of benefit or risk, thus helping to maximize drugs' effectiveness and safety. Using genomic testing to guide drug therapy will constitute a significant shift from the current practice of population-based treatment towards **'fine-tuning' individual therapy.**Scientific understanding of pharmacogenomics is most advanced in the drug metabolism area, and early results are expected in this field. It is hoped that pharmacogenomic testing will help identify cancers that have a high probability of responding to a particular medication or regimen and that it may also be used to help track down the cause of certain rare, serious drug side effects.

* **Disease**, as diagnosed by symptoms is said to be replaced by genomics, which identifies people who will definitely respond to a specific medicine. The future is said to offer prevention, protection and treatment of disease on a **personalized base.** High-speed computers and DNA filters called microarrays, which made it possible to quickly identify varieties in genes, allow for more forays into personalized medicine. A new **"one-size-fits-one"** approach instead of "one-size-fits-all"-approach will develop with more drugs like Herceptin and Iressa to treat disease.

* **The promise of stem cells and regenerative medicine** discoveries and their potential to transform the treatment of disease, continues to drive biological research and technology development in the life sciences. Analysts predict that a market based on stem cell therapies could grow anywhere from **$10-30 billion** by 2010.

* Scientists put much hope in **active immunotherapy**. This treatment seeks to stimulate the immune system to attack antigens contained within (or expressed by DNA within) a vaccine. The big issue is to find antigens that are expressed solely by tumor cells enabling to direct the immune system to respond to cancers which have so far passed the body's defense.

Antigens chosen to stimulate the immune system can be unique to an individual patient's cancer (autologous therapies). Or they can be from different patient/patients (allogenic therapies). European companies have chosen the allogenic therapy approach.

*"**Contraceptive-liberation**" of women started...just 40 years ago.
Enovid-10, the first PILL, was introduced on May 9, 1960 in the States. The UK followed in 1962.
Regionality :

- 8 per cent worldwide of all married women prefer the Pill for contraceptive use;
- 50 per cent of all married women choose the Pill in „Western" Europe;
- 3 per cent of married women use the Pill in China;
- 45% of unmarried women use the Pill in Europe, within „Eastern" Europe that figure is 36%.

At some time in their reproductive years, almost one in four married women have used the Pill since its introduction

In the USA, this number is four in five for all women born since 1945.

Worldwide, an estimated 13 per cent of women rely on the IUD and 19 per cent prefer sterilization.

Today there are over 400 million contraceptive users worldwide, one in six of whom relies on **the Pill**. Despite the range of effective and safe products currently available, there remains a need for a broader choice of methods The combined oral contraceptive Pill first became available in 1960 but the Pills of today contain a fraction of the dose of estrogen as well as new **selective progestogens**.

The discovery of new and improved methods of hormonal contraception remains among the most important priorities in present and future research programs.

* **US Health spending** is expected to consistently outpace gross domestic product (GDP) over the coming decade, accounting for 20 percent of GDP by 2015. *[Health Affairs 25 (2006): w61–w73 (published online 22 February 2006; 10.1377/hithaff.25.w61)].*

The Increase in number of deals at all stages of development over the last few years is a clear barometer of the failure of in-house new drug productivity. BigPharma's need to find new drugs coming from outside the own R&D structures, has its price: the cost of a late stage deal has increased 7-fold from 2000 to 2005.

Table 5.11.   Number of licensing deals completed, by clinical development stage, 2000-2005

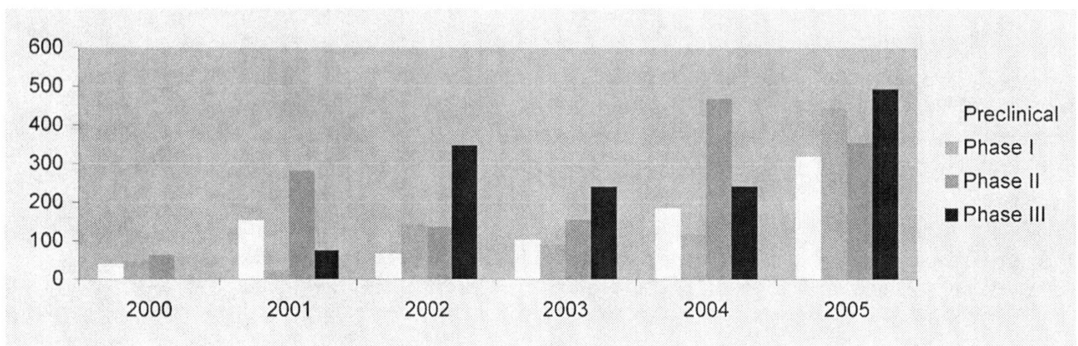

In-licensed products generated $36 billion revenue in 2006 and is forecast to exceed $60 billion by 2010 for the Top 30 pharmaceutical companies.

The in-licensing and acquisitions of the past are already paying off for the pharmaceutical industry, in our view. In the last few years, the proportion of newly approved drugs originating from biotech companies has risen. Although the pharmaceutical industry has gained approval for innovative drugs in the last few years, notably for Gleevec/Glivec (Novartis), Orencia (BMS) and Rozerem (Takeda), the most important innovations such as Macugen (Eyetech), Erbitux (Imclone), Avastin and Tarceva (Genentech) have come from biotech companies.

Table 5.12.   Contribution of licensing to the growth of the pharmaceutical market

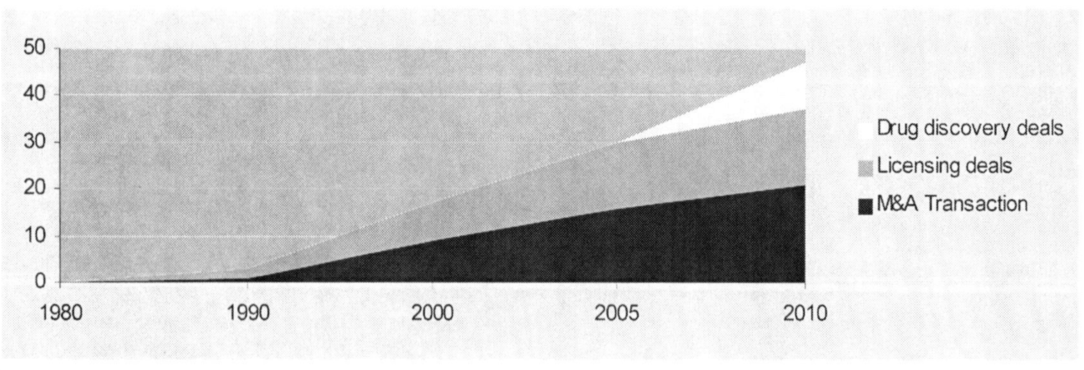

(1980-2010, in US$bn)

In the medium term, we expect the proportion of approved drugs developed by biotech companies to remain at the present high level and that it will continue to represent the bulk of innovations. The following graph shows the growing proportion of newly approved drugs – based on both new chemical entities (NCEs) or new biological entities (NBEs) – originating from biotech companies.

* Significant innovations, as well as commercial activities, are expected from such areas as **nano-enabled drug delivery** for pharmaceutical and medical applications. The use of **nanoparticles, nanoemulsion** and other nanoscale materials for drug delivery is widely expected to become a multibillion-dollar business over the next decade. These delivery systems are rapidly emerging to address multiple challenges, including **disease targeting, solubility, bioavailability, toxicity reduction, cost-reduction and life-cycle extension.**
Another area where there has been a significant amount of activity is **medical diagnostics**. The trend is the advancement and application of nanotechnology toward integrated devices that can incorporate onto a **chip multiple fluidic assay functions**, such as **separation, metering, resuspension, fluid transport and detection**. These chips could soon compete in routine applications with macroscale test devices.
Application of nanotechnology to **biomedical imaging** has also been increasing in recent years. Nanoscale technology has already afforded the ability for
. **intracellular imaging**. The drive is to improve techniques that **can enhance spatial resolution** at the **molecular level** to detect early biological markers can lead to **early diagnosis, more accurate characterization of diseases,** and perhaps opportunities in drug discovery. **Nanosensors** make up another area that can open broad opportunities in many areas of life sciences, including biotechnology and medicine. These sensors either **use nanomaterials as the sensing material or nanoelectronics** to reduce the size and cost of the device to provide a higher level of functional integration. The market for biomedical sensors is expected to reach around $800 millions in 2008 and $1.2 billions by 2012. The impact of **nanotechnology** in the life sciences is likely to be very broad. **Life sciences will be one of the first true nanorelated markets to evolve,** and the **opportunities** that can be generated will be quite significant.

* In the next 10 years, some of the great discoveries could come from the results of the Human Genome Project. In one of **the monumental scientific accomplishments of our time**, researchers sequenced and mapped the approximately 25,000 genes in the genome that comprise the blueprint for human beings. Researchers are now working to translate this achievement into clinical treatments. In the long term, the project will also give researchers the tools to begin understanding basic questions about **human physiology and behavior**, and eventually about the **underlying basis of brain disorders**. *(Bruce Stillman, president of Cold Spring Harbor Laboratory, NY. Presentation at Duke University, September 26, 2006).*

* By 2010, it is likely that **predictive genetic tests** will be available for as many as a dozen common conditions, enabling individuals to take preventive steps to reduce their risks of developing such disorders. Doctors will also begin tailoring prescribing practices to each **patient's unique genetic profile,** choosing medications that are most likely to produce a positive response.

* Psychiatric researchers at The Zucker Hillside Hospital campus of **The Feinstein Institute for Medical Research** have uncovered **evidence of a gene** that appears to **influence intelligence**. Working in conjunction with researchers at **Harvard Partners Center for Genetics and Genomics** in Boston, the Zucker Hillside team examined the genetic blueprints of individuals with schizophrenia, a neuropsychiatric disorder characterized by cognitive impairment, and compared them with healthy volunteers. They discovered that the **dysbindin-1 gene (DTNBP1),** which they previously demonstrated to be associated with schizophrenia, may also **be linked to general cognitive ability.**
 While their data suggests the dysbindin gene influences variation in human cognitive ability and intelligence, it only explains a small proportion of it -- about 3 percent. This supports a model involving **multiple genetic and environmental influences on intelligence.**
The specific role of dysbindin in the central nervous system is unknown, but it is highly present in key brain regions linked to cognition, including learning, problem solving, judgment, memory and comprehension. Scientists speculate that dysbindin plays a role in communication between brain cells in these regions and helps promote their survival. An alteration in the genetic blueprint for dysbindin may ultimately interfere with cell communication and fail to protect brain cells from dying, with a resulting negative impact on cognition and intelligence.
(Burdick KE, et al., Genetic variation in DTNBP1 influences general cognitive ability. Hum Mol Genet. 2006 May 15;15(10):1563-68. Epub 2006 Jan 13).

* **Strokes** are serious pathologies that often lead to death or very severe handicaps. **Cerovive** could provide a considerable gain in patients' chances, allowing them to be given treatment upon arrival at a hospital. Cerovive is a **free radical trapping agent** tested in neurological protection of patients during the acute phase stroke. Three Phase III studies were conducted: SAINT 1 and 2 in ischemic stoke and CHANT in hemorrhagic stroke (15% of all strokes, but with a bleak prognosis). Strokes have very serious consequences in terms of morbi-mortality, and their treatment is limited to **Actiplase (Genentech),** a **thrombolytic** drug for ischemic stoke but no treatment exists for hemorrhagic stroke (Novo Nordisk conducts trials with NovoSeven). These treatments must absolutely be administered within hours of a stroke, but for detection reasons (the time needed for transporting victims to centers and identifying the type of stroke), this time window is frequently exceeded. The aim of **Cerovive** is to increase the length of this timeframe, during which neurons and supporting cells are "put to sleep" before they are killed by lack of irrigation.

* By 2020, the impact of **personalized medicine** is likely to be far more important than any of us can envision today. New **gene-based drugs** will be developed for diabetes, heart disease, Alzheimer's disease, schizophrenia, and many other conditions.

Much more precision in predictive statements about health risks will be possible. Specific diagnostics will accompagny new therapies. Computer based decision support systems will enhance clinical judgement.

Biotechnology is often called the "Third Industrial Revolution. This is probably an understatement. The first two revolutions affected merely products and processes, material aspects of the quality of modern life.

By many measures, medicines have played a major role in the improved health of the world during the 20th century[108]. This becomes obvious when one simply looks at the 1920's and the role medicines played in eradicating diseases during that period. In the 1960s again, medicines played a very substantial role in medical progress. They helped reduce mortality and improve quality of life. The biotechnology revolution addresses the individual himself as well as his individual aspects and perceptions. Biotechnology has already begun to change our society and culture persistently, with so far unseen speed and dynamics.

Table 5.13.   Success rate of in-licensed products varies depending on development phase[109],[110]

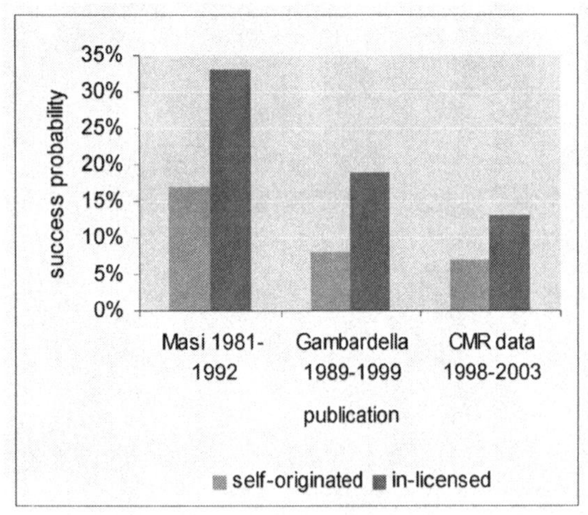

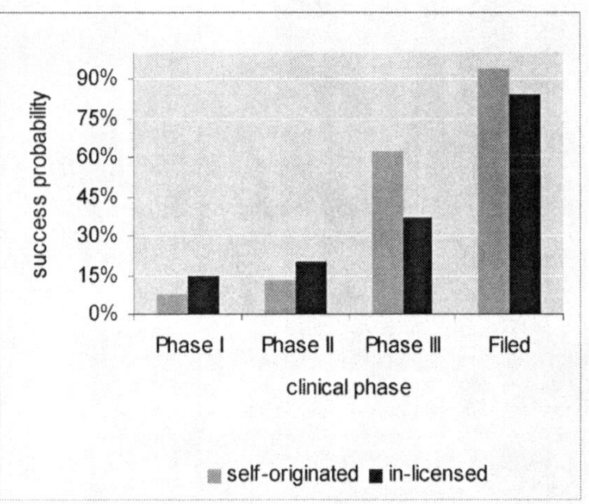

Modern biology right now is in metamorphosis, much like physics at the beginning of the last century. A transition takes place from a rather descriptive subject into a discipline that is coming to understand its molecular foundations. Preconditions to this were technological developments in the 1970s and 1980s that made it possible to register the molecular base for relevant categories of molecules (DNA, RNA, proteins, metabolites) qualitatively and quantitatively. Most fundamental modern technological developments (such as recombinant DNA, sequencing, computer technologies) were developed in the US and are being commercialized over there faster and more consequently than anywhere else.

V.10. Expanding Into Small Molecules For Drug Discovery

Biotech companies traditionally focus on large proteins in their quest to find new drugs. But faced with the complex nature of disease, these companies are increasingly turning to small molecules as well. This new push into small-molecule drug design, which researchers hope will deliver better and more effective disease treatments, is explored by associate editor Lisa Jarvis in Chemical & Engineering News, the American Chemical Society's weekly newsmagazine. Jarvis describes how major biotech companies such as Genentech and Amgen have moved into the small-molecule arena and expanded their

---

[108] *Kessler Armin M, MD, CEO F.Hoffman-La Roche Ltd, Innovative R&D or vertical integration. Is there a strategic choice? Financial Times World Pharmaceuticals Conference, 21.03.1995*
[109] *Source: Nature Reviews Drug Discovery*
[110] *Source: DZ BANK/FDA/Drug Discovery Reviews*

* A report from Decision Resources, Emerging Antibacterial Agents, indicates that premium-priced glycopeptides from Pfizer (**dalbavancin),** Theravance/Astellas (**telavancine**) and Targanta (**oritavancin**), as well as next-generation cephalosporins from Johnson & Johnson/Basilea (**ceftobiprole**) and Cerexa/Takeda (**ceftaroline**), will **partly replace generic vancomycin**, the most widely used antibacterial for resistant gram-positive infections. The predicted uptake of these products will be attributable to their improved side effect profiles and higher potency against drug-resistant staphylococci, namely **methicillin-resistant *Staphylococcus aureus*.** The emergence of drug resistance among bacteria has significantly eroded the effectiveness of current antibiotics. The medical community unanimously agrees that agents with novel mechanisms of action are greatly needed - particularly agents that are effective against multidrug- resistant gram-negative pathogens - to ensure the availability of sufficient options to treat bacterial infections.
Among the most important areas of current and future medical need are antibiotics that are effective against resistant gram-negative pathogens in the **hospital setting** and **novel oral anti-methicillin-resistant *S aureus* agents**. Researchers from Decision resources estimates the market potential of the antibacterial products in the product pipelines of the pharmaceutical companies to be US$ 3bn by 2015.

* Datamonitor estimates that turnover for the **six key biogeneric product types** will reach **$2.20 billion by 2010** within the USA and the five major EU markets.

* Immunologists at Thomas Jefferson University in Philadelphia have used nanotechnology to create a novel "**biosensor**" to solve in part a perplexing problem in immunology: how **immune system cells** called killer T-cells hunt down invading viruses.
They found that surprisingly little virus can turn on the killer T-cells, thanks to some complicated communication among so-called "antigen presenting" proteins that recognize and attach to the virus, in turn, making it visible to the immune system. T-cell receptors then "see" the virus, activating the T-cells.
The researchers showed that different types of presenting proteins cooperate, spreading a signal among only a few T-cell receptors and boosting the T-cell response. This helps explain how only a few virus-infected cells can cause a killer T-cell response. Understanding how the immune system responds to viral threats is critical to finding better ways to manipulate it and could have implications for improved vaccine development.

* All the information point in the same direction: the **end** of the **small-molecule, oral-formulation drug era** and the emergence of therapies based on New Biology, including gene replacement and protein-based therapeutics in combination with various drug delivery systems. Challenging times with big rewards are awaiting the medical world beyond 2010.

* The latest achievements in radiology are revolutionizing gentle, non-invasive heart diagnostics. **Multi-slice computer tomography** (MSCT) can spare patients from invasive cardiac catheter examinations and for example, allow physicians to assess the risk of coronary based on the amount of calcification in coronary vessels. Blood flow rates in the heart muscle can be accurately evaluated today in just 30 seconds using magnetic resonance tomography. Two X-ray tubes rotating parallel to each other (dual source CT) can now also be used to render precise images of fast or irregularly beating hearts. Combining various imaging procedures such as PET and computer tomography or PET and magnetic resonance tomography opens up whole new perspectives. Although clinical use on patients is still in the realm of dreams for the future, initial results already show revolutionary images that contain the widest variety of functional and anatomical information and that enable us to obtain new kinds of insights into cardiac diseases. *(Professor Dr. Claus D. Claussen, University Clinic for Radiology, Tuebingen, at the kick-off press conference of the European Congress of Radiology (ECR), Vienna, March 9 to 13, 2007).*

* Scientists have found a **novel signaling pathway** in cells that is altered by genetic mutations recently identified in **Parkinson disease development**. These new findings show how the mutations affect cellular function and could provide a target for drug therapies to treat the disease. Parkinson disease is a degenerative disorder of the central nervous system resulting from **the loss of neurons in the brain that produce dopamine**. This lowering of dopamine leads to decreased stimulation of the brain's motor cortex. Although scientists have not known the exact cause of the loss of these dopamine-producing neurons, they believe it is related to dysfunctional mitochondria and oxidative stress. Mitochondria are the cell's "**power plants,**" which metabolize oxygen and generate energy. Oxidative stress is the damage caused to cells by reactive oxygen produced during oxygen metabolism.
Although cells have mechanisms in place to protect against oxidative damage, this system can break down in the face of **environmental challenges or genetic mutations**. The investigators found that the **mitochondrial protein PINK1** normally protects cells from oxidative stress and promotes cell survival by **regulating function** of the **protein TRAP1**. When PINK1 is mutated, however, the protective TRAP1 pathway is disrupted, leading to mitochondrial damage.
Other scientists have linked early onset Parkinson disease to mutations in both copies of the PINK1 gene (one from each parent). They also have evidence that **single-copy mutations in PINK1** are a significant risk factor for the development of later-onset Parkinson disease. *[The research by a team of Emory University scientists was published June 18 in the Public Library of Science Biology (PLoS Biology) journal]*

* Dr Maya Capelson, Salk Institute for Biological Studies, is investigating the role of '**gene gates**' or nuclear pores in tumor formation and cancer progression.

Medicines that will make their own way through the body and attack precisely the diseased cells on reaching their destination – such has been the dream of physicians and pharmacists since time immemorial. **Each day, researchers come a little closer to reaching this goal.**

chemistry operations in the process, including new partnerships with drug discovery firms. These efforts have increasingly brought together chemists from a wide range of disciplines, including medicinal, computational and process chemistry.

Biotech's investment in small-molecule research is starting to pay off in the form of new drugs, she notes. A few companies already have commercialized new small-molecule drugs for the treatment of lung cancer and kidney disease, while drugs for multiple sclerosis, Parkinson's and other diseases are making their way through the pipeline. By combining small-molecule approaches with protein research, biotech companies are boosting their odds of producing more effective treatments for some of our most challenging diseases, according to the article[111].

Why are pharmaceutical companies so keen on in-licensing and co-developments with biotech companies rather than just concentrating on their own drug development? One reason for this is an increase in the number of research projects through the use of outside developments. According to Gunter Festel, Festel Capital, pharma companies switch from a sequential to a parallel drug discovery process in the post-genome era.

The result is a complex network of internal and external competencies and expertise. The example of Novartis underscores his statement.

Table 5.14. New post-genome process[112]

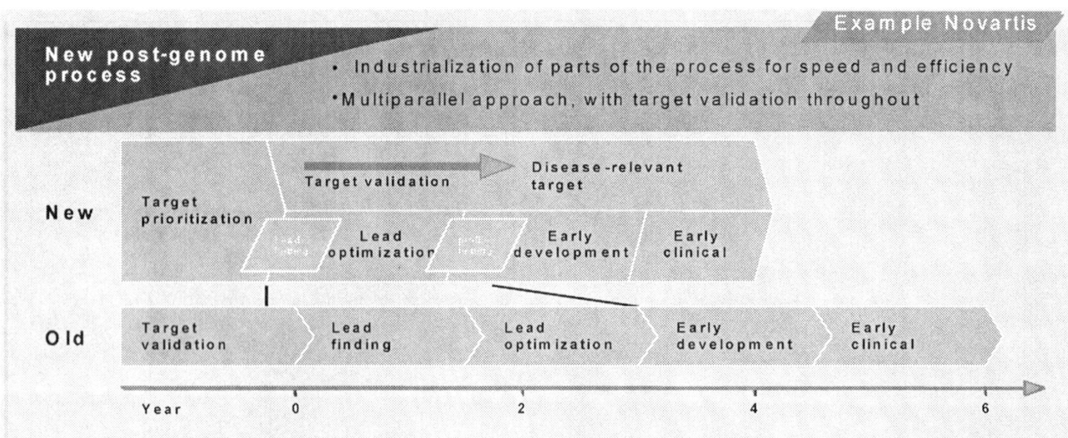

Another reason analysts with Natexis Bleichroeder believe is the high probability of success of in-licensed drugs, which increases the productivity of the research activity. In principle, the drug development of biotech companies carries a similar risk of failure to those of their pharmaceutical counterparts.

The pharmaceutical industry itself reports that 40%-60% of alliances end in disappointment. However, independent studies conclude that in-licensed drug development had a greater probability of making it to the market in the past than in-house development. This could reflect the fact that in-licensed drugs are subjected to a more rigid due diligence process, or that they are back-ups for in-house development

---

[111] "Biotechs take a more agnostic approach to drug discovery by expanding into small molecules" ACS News Service Weekly PressPac - - Oct. 25, 2006

[112] Source: Gunter Festel, Arthur D Little at Euroforum Conference on Management Buy-Outs, Frankfurt 2002 – based on Novartis company data

* Genetic testing is transforming medicine—and the way families think about their health. As science unlocks the intricate secrets of DNA, we face difficult choices and new challenges. Scientific revolutions must be tempered by reality. Genes aren't the only factors involved in complex diseases—lifestyle and environmental influences, such as diet or smoking, are too. And predictions about new tests and treatments may not come to pass as fast as researchers hope— they may not come at all. Still, it's hard not to get excited about the future, especially when you consider the medical competition now underway: NIH has challenged researchers to come up with a **method**, within the next 10 years, to **sequence a single human genome** for **$1,000** (today's cost: $5 million to $10 million). Assuming it works, one day not too far in the future, each of us will go to the doc, hand over our blood and get back our **personalized biological blueprints**. *(Newsweek, Dec. 11, 2006).*

* The **mAb 806** is a monoclonal antibody which acts against a mutant form of the EGF receptor variously called **DeltaEGFR, de2-7 EGFR, or EGFRvIII**. This receptor has been observed to be frequently highly expressed in human malignant gliomas and has been reported for cancers of the lung, breast, and prostate. The antibody does not bind to wild type EGF receptor and possesses significant antitumor activity. To gain a better idea of the mechanism of action of this antibody, scientists from the Ludwig Institute for Cancer Research (Epithelial Biochemistry Laboratory) are attempting to determine the epitope of binding of mAb 806. Preliminary results indicate that the epitope is away from the novel splice site in the protein which is generated as a consequence of the mutation.

* **Decoding data in genomics**: By devising a novel computational approach to examining a large amount of data, scientists have gained insight into the intricate workings of proteins responsible for sensing their surroundings and transmitting that information inward. Despite being studied for decades, little is known about how these methyl-accepting chemotaxis proteins translate their exquisite detection abilities into a signal. The work of Igor Zhulin of ORNL's Computer Science and Mathematics Division and Roger Alexander of Georgia Tech, published in Proceedings of the National Academy of Sciences, represents a giant step in this area and illustrates that genome sequencing can deliver much-needed knowledge when clever **computational approaches** are **applied to decode genomic information**. In addition to identifying a new flexible signaling portion of the protein, Zhulin and Alexander found that changes in the signaling and adaptation domains in any methyl-accepting chemotaxis protein are tightly coupled. This coupling could explain the diversity of chemotaxis mechanisms, including how an attractant of one species could be a repellent of another.

*The US market for **genetic disease therapies** is expected to grow from $4.8 billion in 2005 to $7.3 billion by 2010, according to BCC Research
B. 62.   US Genetic Therapies Market Growth 2000-2010, by segment

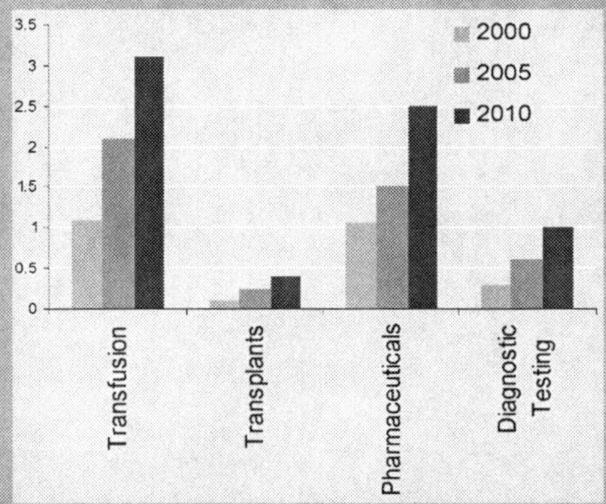

(Source: Pharmaceutical Executive, June 2006:36)

* **Academia and the European industry** follow the example of their American colleagues and develop **excellent cooperations**. An example is given by the collaboration between **Monash University's Center for Vascular Health** and **Bayer HealthCare** in a new exciting field of **sGC activators**. For the first time, selective vasodilation of diseased blood vessels under oxidative stress is possible, according to preclinical study results. The compound BAY 58-2667 developed by Bayer HealthCare is **an activator of the enzyme soluble guanylate cyclase (sGC),** which plays a key role in **signal transduction of the metabolic cascade "nitric oxide (NO) – sGC – cyclic guanosine monophosphate (cGMP).**
Soluble guanylate cyclase (sGC) is activated in the human body by endogenous nitric oxide (NO), which relaxes intact smooth muscle cells in the human organism. sGC plays a key role in signal transduction of the metabolic cascade "nitric oxide (NO) – sGC – cyclic guanosine monophosphate (cGMP)". NO dilates the coronary vessels and improves perfusion of the heart. It also dilates the small veins of the body, with the result that the blood returns to the heart more slowly and in smaller quantities, thereby relieving the burden of the heart. Reduced bioavailability or responsiveness to endogenously produced NO contributes to the development of many cardiovascular and other diseases. **Oxidative stress** is caused by an imbalance between oxidants (reactive oxygen species) and antioxidants in tissues and cells. It is associated with cardiovascular disease conditions such as **heart failure, heart attack, stroke, and hypertension**.

The past century has seen substantial improvement in mortality and morbidity. **Life expectancy increased approximately 30 years with adults now living well into their seventies**. Despite this progress, the behavioral and social sciences have enormous contributions to make in extending longevity and quality of life that have yet to be fully realized *(Centers for Disease Control (2005). Health, United States, 2005, National Center for Health Statistics, URL: http://www.cdc.gov/nchs/hus.htm, accessed November 5, 2006.).*

Table 5.15. External input to early development

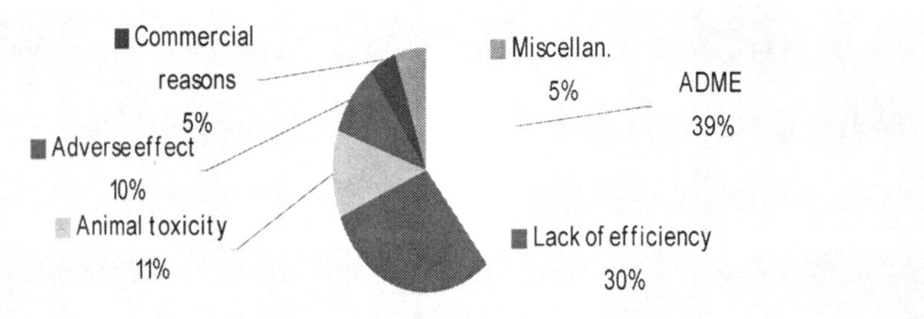

Table 5.16. Reason for failure of early-stage drug candidates

candidates. However, at the same time, the likelihood of success seems to vary depending on the stage of development: in-licensed Phase III drugs and drugs already filed for approval fail more often. One reason could be that Phase II trials of biotech companies might be less stringent or extensive, which increases the risk of failure in Phase III. One prominent example is Serono with the in-licensing of CancerVax's therapeutic vaccine for the treatment of melanomas, Canvaxin: this Phase III drug was in-licensed at the end of 2004 for an upfront payment of $37 million ($253 million in total) and was given up again in 2005 because the independent study committee recommended winding up the study in view of a lack of efficacy.

* **Richard Gibbs** (Baylor College, Houston) who was part of the Human Genome Project team, heads one of the world's largest **gene sequencing** efforts. According to Gibbs "Ten percent of any human disease is caused by inherited **mutations**. Sixty percent of us will be affected by mutations at some time in our lives. Mutations have a substantial social and economic cost. We think there are over **two million mutations directly causing disease in the human genome**. Some 100,000 mutations have been discovered. But there is no systematic global system for collecting and sharing complete and accurate information on these mutations with clinicians around the world. **And 95 percent of human mutations are yet to be discovered.** Fast, reliable access for researchers to information on mutations, and the damage they cause, would transform genetic medicine by: 1) enabling doctors to rapidly diagnose and inform patients with rare diseases; 2) allowing new diagnostics; 3) helping researchers develop new treatments for hundreds of genetic diseases including cystic fibrosis and thalassemia; 4) assisting in uncovering the causes of common diseases such as breast cancer and asthma. It's time for a global human variome (human gene variation) project".

* The **human genome** is comprised of both non-coding DNA and coding regions, or genes. While researchers once believed that only genes were transcribed into messenger RNA (**mRNA**), they discovered that **non-coding DNA is copied into mRNA** as well. Unlike coding mRNAs, which are translated into functional proteins and peptides, the function of most non-coding RNAs is unclear. Although non-coding RNAs fail to produce functional proteins, researchers believe that in some cases **these RNAs may control gene expression**. Synthesized by the pituitary gland, human growth hormone activates growth and cell reproduction. In addition to serving as a major contributor to height growth during childhood, *hGH* plays a role in streng-thening bones and increasing muscle mass throughout life. It was found that the *hGH* gene is controlled by a **non-coding DNA region**, or locus control region. Remarkably, this region is located more than 14,000 base pairs away from the *hGH* gene. At the genomic level, a 14,000 base-pair separation **is equal to the size of 10 growth hormone genes lined end to end**. Researchers found that the locus control region was copied into RNA, and discovered a gene called *CD79b* within this region. Remarkably this *CD79b* gene was also copied into RNA in the pituitary. While the *CD79b* gene normally codes for a protein in blood lymphocytes, researchers discovered that *CD79b* appears to play a very different role in the pituitary gland. Here, *CD79b* was actively transcribed into mRNA, but this mRNA failed to translate into a functional protein. **Instead, the non-coding RNA was suspected to play a role in *hGH* gene regulation.**

* **Cardiovascular disease** is also a major contributor to escalating health care costs. In 2006, the disease will cost Americans an estimated $403 billion in medical expenses and lost productivity. The **aging of the population** is projected to **drive up costs** for cardiovascular disease **54 percent** by 2025.

* **Nobel Laureate Eric Kandel's** work delineates the **molecular changes that underlie learning and memory**. His book, *In Search of Memory*, relates the story of how four distinct disciplines – **behaviorist psychology, cognitive psychology, neuroscience, and molecular biology – converged into a powerful new science of mind.** Through its profound insights into thought, perception, action, recollection, and mental illness, this new science is revolutionizing the understanding of learning and memory while simultaneously showing **great promise for more effective healing**. Dr. Kandel suggests that 21th century **neuroscience** will focus more on the **brain circuits and systems that regulate cognition.** Two major systems problems that needs further study are:
1. central issues in understanding consciousness, and in the recall of memories from different places and times; or the factors that regulate the unconscious processing of sensory information about our environment, and the question of how conscious attention regulates the processes that then stabilize experiential memories.
2. the sociology of cognition, or the relationship between the activity of an individual brain and the corporate activity of a group of brains. This work may shed light on one of the key problems in **understanding addiction**, that is to understand how repeated exposure to drugs 'teaches' the brain to become addicted. *(Kandel E., In Search of Memory: The Emergence of a New Science of Mind. W. W. Norton & Company, 2006).*

* A variety of polymers have been used in gene-delivery studies, but their effectiveness as gene-therapy vectors remains orders of magnitude poorer than the viral vectors. As a result, polymers are generally considered unacceptable for clinical applications. The important extra- and intracellular barriers to efficient gene delivery are known. The lack of efficiency of polymer gene-delivery vectors, nevertheless, results from a lack of functionality for overcoming at least one of these barriers. On the basis of the large number of studies of off-the-shelf gene-delivery polymers, much has been learned about the **structure-function relationships of polymer vectors**. This knowledge has been applied to the design and synthesis of new polymers, tailor-made for gene delivery. With a growing understanding of polymer gene-delivery mechanisms, it is likely that polymer-based delivery systems will become an important tool for human gene therapy *[Pack D.W. et al, Design and development of polymers for gene delivery. Nature Reviews, Vol 4, July 2005:581-591].*

*The **Pathogenomics** network of Germany has the purpose of sequencing the genomes of seven pathogenic bacteria and their apathogenic relatives. Using bioinformatic methods and high-throughput and microarray technologies, the researchers will focus on quick diagnostics methods and the bacteria's ways of evading the immune system. Their findings will be the basis for developing new **cell therapy methods and vaccines.** The network partners have formed three cooperation groups, according to their long-term research interests:
- Gram-positive cocci (Staphylococci, Pneumococci)
- Gram-negative cocci (H.pylori, Enterobacteria)
- Intracellular bacteria (Listeria. Chlamydia)

Personalized medicine will be partly based on genetic background. It will become possible for an individual's genome to be sequenced in one day. There will be an increasing focus on prevention.

Table 5.17. The cost of licensing: a matter of supply and demand

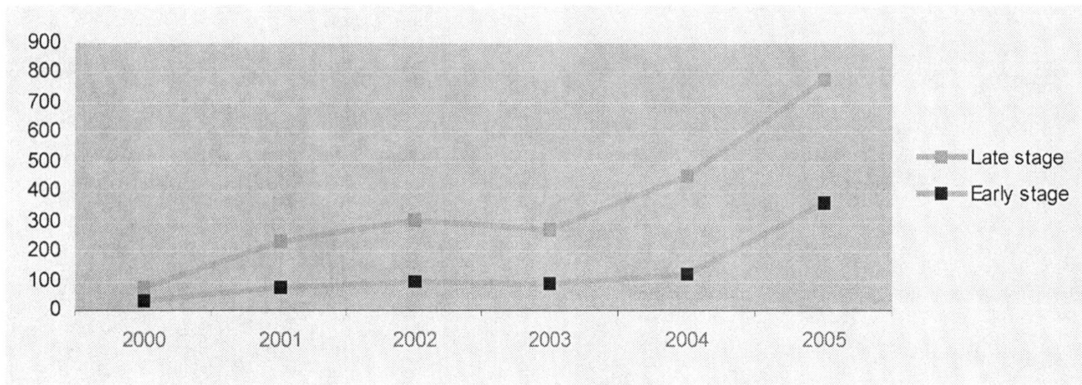

(in US $million / Average price per drug including milestones and royalties)

V.11. Pharma R&D Challenge: too much fragmented data

Highly motivated research scientists committed to accelerate target prioritization have to cope with an enormous amount of data, for example derived from the sequencing of the human genome. In order to manage and analyze the data used in the search for specific molecules for a new medicine, so-called targets, it is necessary to work with many data sources, intelligent software and modern computers within bioinformatics. Exploiting the information to its maximal potential, thereby avoiding repetitive efforts and the time-consuming use of compartmentalized data with multiple tools, is a particular challenge.

Table 5.18. Summary of the most pressing unmet needs of research scientists, specified along the value chain[113]

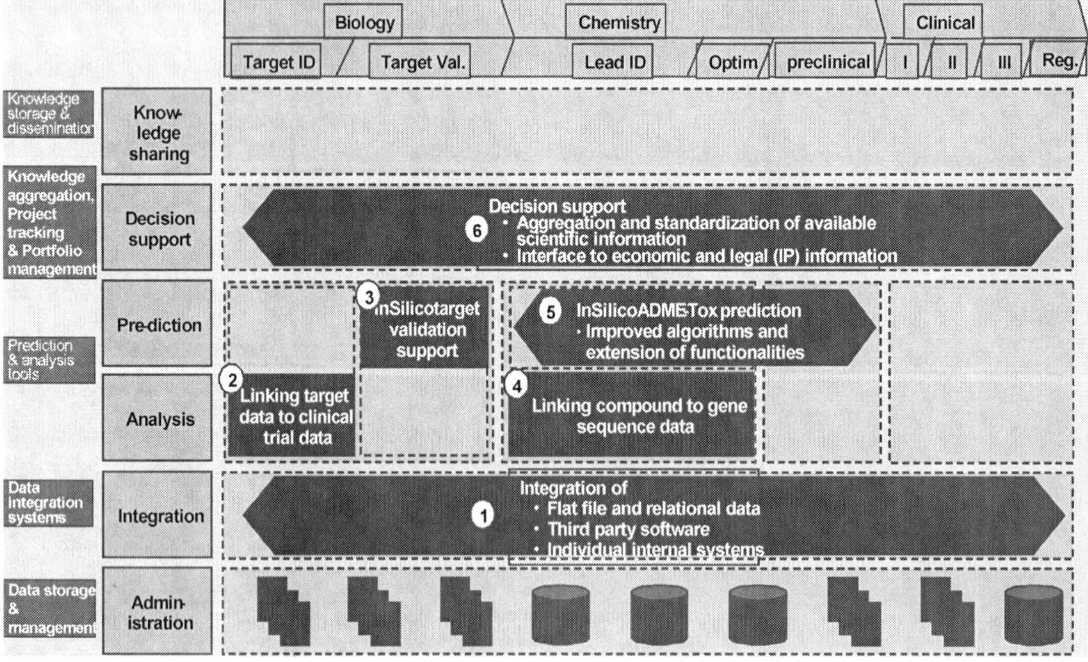

---

[113] Data derived from UBS Global Equity Research report; Deutsche Bank Equity research report

The key needs of researchers have been identified as:

- *Data integration*, including the integration of flat-file and relational-file data as well as the integration of third-party software and individual internal systems.
- *Decision support,* which elevates integration to a higher, more rigorous level by aggregating and standardizing all available scientific information on an enterprise-wide basis and then linking that information with available business and intellectual property information.
- *Interdisciplinary applications,* which link data and information from different disciplines — for example, linking chemical compound information with gene sequence data, and linking *in silico* ADME/Tox predictive tools that can predict the toxicity of a compound in an organism to compound screening data.
- *In silico target validation tools,* which enable scientists to reduce the number of elaborate target validation experiments by identifying and categorizing all the available information associated with specific targets, as well as providing reliable *in silico* simulations that support streamlining of complex validation experiments.

Table 5.19. Genomics generate a flood of data[114]

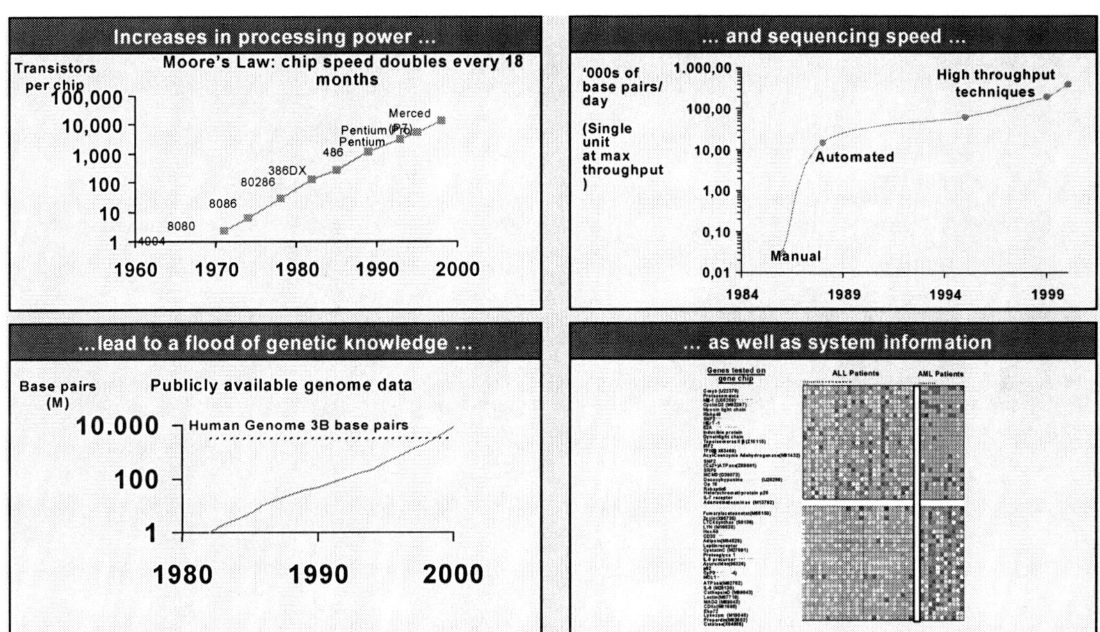

V.12. The interim solution of the industry

Pharmaceutical companies have built an in-house infrastructure on their own. The close relationship of scientists and in-house informaticians has very often led to specific solutions addressing exactly the needs of a department or research group.

Created and implemented easily. But this approach is not without its drawbacks: maintenance of these quickly developed systems can become cumbersome in a very short time. Over time, the need to develop duplicate functionality for multiple user groups can make these "cheap" solutions much more expensive than intended. Additionally, instead of developing specialized tools that enable better understanding of

---

[114] *based on Genbank; Sequenom; 3700.com; Applied Biosystems; Human Genome Project; FAO United Nations; Database on Genome Sizes; Intel Corp.*

* Intelligent administration of active substances and extending the lifecycle of existing pharmaceutical formulations is the goal pursued by Bernina Biosystems. The company uses its internally developed proprietary A-fect technology and other liposome-based technologies to 'administer exact doses of effective substances to the body in a directed, timely and safe manner'. Besides this production-orientated platform technology based on extremophile bacteria's lipids, the company develops chemically modified lipids, nanoparticles and microspheres in co-operation with industry and scientific institutions that will lead to new product formulations. Using these technologies for the intelligent administration of effective substances, Bernina will not only facilitate the administration of pharmaceuticals that are well-tolerated by patients, but will also extend the lifecycle of existing pharmaceutical formulations by enabling companies to apply for new patents.

*The difficulties of European Statutory Health Insurance systems are not only related to costs and quality, but also to the complex and obscure market structures. Compared to the manufacturing sector, the health care sector has invested relatively little money in technology. Industry experts consider this lack of an open platform and digital information to be a shortfall that will jeopardize the quality of medical care and create inefficient information and organizational workflows.
The introduction of an electronic health record (HER) becomes inevitable. It has the potential to improve communication throughout the health care systems. This increased insight into medical information will improve the quality of treatment as well as the patients' quality of life. Furthermore, island solutions and standalone applications will be eliminated and/or avoided. An EHR that involves patients in their own health management also increases the transparency and efficiency of the systems, and have a positive effect on prevention and quality-related costs.

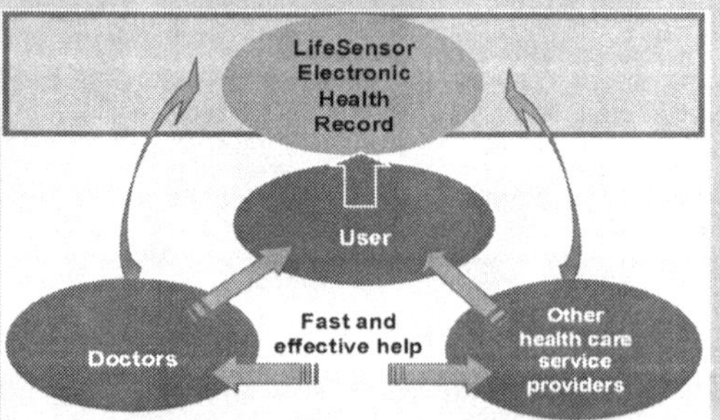

*Joint research by Dr. Leonid Brodsky, of the Institute of Evolution of the University of Haifa, and Dr. Milton Taylor, of Indiana University, led to the discovery of a mathematical method which can **identify which genes in our bodies conduct the battle against the various viruses that attack us**. In their research, they identified 37 genes out of 22,000 possible genes which fight the hepatitis C virus.

* At Purdue University research has opened the door for possible antibiotic treatments for a variety of diseases by determining the structure of **exopolyphosphatase**, a **protein** in **E. coli bacteria** that functions as an enzyme and catalyzes chemical reactions within the bacteria. This enzyme provides the signal for bacteria to enter starvation mode and limit reproduction. *(Sanders D.,August 16, 2006 issue of the journal Structure).* This research is applicable to the treatment of many diseases because that same protein is found in numerous harmful bacteria, including those that cause ulcers, leprosy, food poisoning, whooping cough, meningitis, sexually transmitted diseases, respiratory infections and stomach cancer. In addition, this protein is an excellent **antibiotic target** because it only exists in bacteria and some plants, which means the treatment will only affect the targeted bacterial cells and will be harmless to human cells.
With the ability to control the use of this signal, scientists say they can fool bacteria into thinking they are starving all the time, even when they are not; or that they could never allow them to realize that they're starving, and that would kill them as well.
Researchers could design drugs to bind to the protein and keep it from being used by the bacteria, rendering the bacteria **unable** to **react** to and survive a lack of nutrient supply; the other possibility would be to design a drug to **mimic** the protein, causing the bacteria to react as if it were starving even when in the presence of a plentiful nutrient supply.The protein, which belongs to the **ASKHA** (Acetate and Sugar Kinases, Hsp70, Actin) **superfamily**, also is of particular interest because it is highly **processive**, meaning it is efficient in the chemical reaction it initiates. It is able to latch onto its **substrate**, the substance it uses to fuel its chemical reaction, and to stay tenaciously in place until it has consumed all of the substrate. The next step in this research will be working to develop **inhibitors** for this protein and studying the applications to other bacteria.
* A research team led by scientists at the Broad Institute of **MIT** and **Harvard** announced the development of a new kind of **genetic "roadmap"** that can connect human diseases with potential drugs to treat them, as well as predict how new drugs work in human cells. Called the "**Connectivity Map.**" Their research results show the map's ability to **accurately predict** the **molecular actions** of **novel therapeutic compounds** and to suggest ways that **existing drugs** can be **newly applied** to treat diseases such as cancer. Based on the results, they propose a public project to expand this initial human Connectivity Map - in the spirit of the Human Genome Project - to accelerate the search for new drugs to treat disease. Like other scientific databases, the true value of the Connectivity Map lies in its capacity to be queried by nearly any researcher with a computer. The **genomic signature** of a particular human disease, drug or other biological response of interest serves as the search "word" and **potential functional connections** are revealed through a rank-ordered list of reference compounds in the database (works much like a Google search to discover connections among drugs and diseases) that have matching signatures.
Data from the study are made publicly available at http://www.broad.mit.edu/cmap

living systems, specialists can become mired in activities like scripting and updating databases.

Table 5.20.   ...leading to unprecedented data management challenges[115]

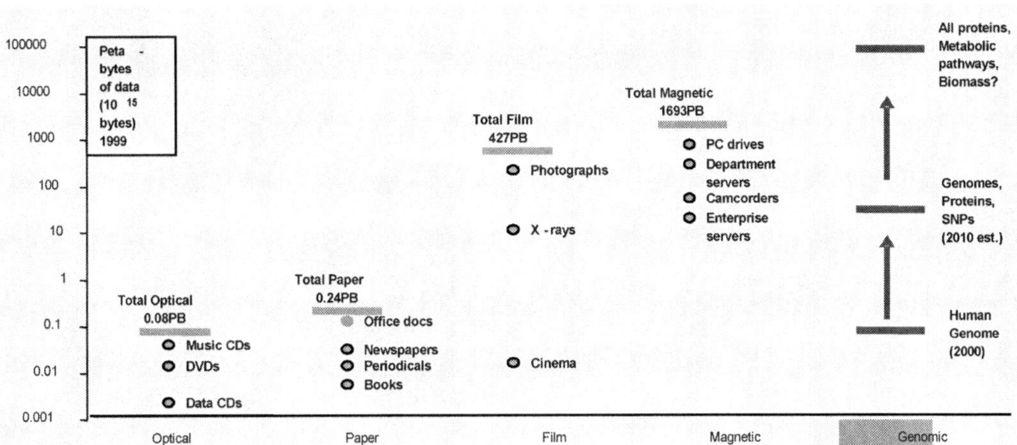

Table 5.21.   To capitalize on these challenges, one has to rethink

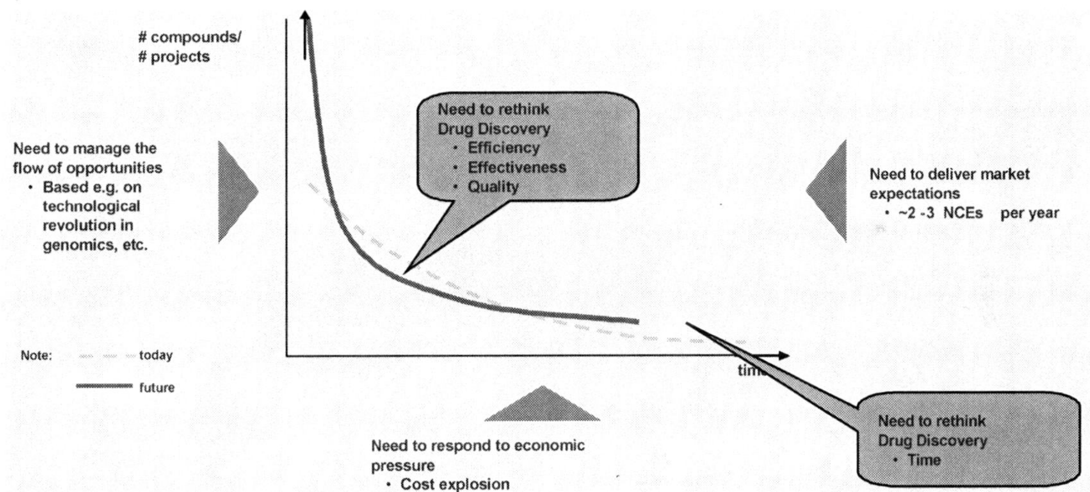

V.13. Using the tools and concepts of the last century to assess this century's drug candidates

The medical product development process is no longer able to keep pace with the tremendous basic scientific innovation. Only a concerted effort to apply the new biomedical science to medical product development will succeed in modernizing the critical path. The new science is not being used to guide the technology discovery process in the same way that it is accelerating the technology discovery process. For medical technology, performance is measured in terms of product safety and effectiveness. Not enough applied scientific work has been done in creating new tools to get fundamentally better answers about how the safety and effectiveness of new products can be demonstrated, in faster time frames, with more certainty, and at lower

---

[115] *based on UC Berkeley Project; Varian, et. al.; Celera Press releases*

*** Socio-economic living conditions in Western society,** where once-dominating family ties and religion have made place for individualism, single-living, divorce, job-losses etc, is expected to give a further boost to the **antidepressants** market. Antidepressants are the third-leading class by prescription sales, totalling $18 in 2003. Seven major brands generate sales in excess of $500mn in 2003. That represents 5% of the 137 drugs, all therapeutic classes combined, that succeed in totalling sales of over $500mn. These 7 major antidepressant drugs account for about 80% of all antidepressant drug sales. Although low generic priced competition caused a slowdown in sales over the last years, estimates for 2010 rank from $ 23bn to $28bn.

Novel compounds, e.g. **neurokinin antagonists, novel noradrenergics, novel serotonergics** have been announced, and should largely contribute to the growth in this segment., where SSRs, TCAs, MAOIs, SNRIs etc have made up much of the RXs during the last decade.

*** Targeting neuropeptide receptors and intracellular messenger systems.** Newer approaches to pharmaceutical treatment of depression target **neuropeptide receptors** and **intracellular messenger systems**. Medications for depression currently under development include: **corticotrophin releasing factor (CRF) receptor antagonists, substance P (neurokinin) receptor antagonists**, and drugs that **modulate glutamatergic transmission.**

Abundant evidence suggests that increased production and/or release of **CRF** within the central nervous system occurs in patients with post-traumatic stress disorder and major depression. Preclinical studies show that CRF antagonists have anti-anxiety and antidepressant properties. **CRF receptor antagonists** appear promising as a **novel class** of antidepressants and anxiolytics

*Rapid advances in **neuroscience** suggest medication development strategies that have yet to be undertaken. These include drugs that interact with **second messenger systems, response elements and transcription factors**, agents that enhance **neuroprotective** and **neurogenic factors**, and compounds that manipulate **cytokine receptor activity.** Researchers are attempting to develop compounds that target novel brain mechanisms suggested by cutting edge research on the causes of depression.

* Medical academia pays increased attention to the concept that obesity, diabetes, hypertension and hyperlipedimia are linked. Back in the 1970s, German researcher H.Haller used the term **Metabolic Syndrome** to describe the cluster of symptoms, which has been classified as ICD9 in the International Classification of Diseases, Ninth Revision. Metabolic syndrome will become of more interest to the general population.

*In laboratory experiments, scientists at Ohio State University found that **nano-sized particles** injected into mice improved the resulting images. This study is one of the first reports (published in the April 2006 issue of the journal Physics in Medicine and Biology) showing that **ultrasound can detect these tiny particles when they are inside the body.** It turns out that **not only can ultrasound waves sense nanoparticles, but the particles can brighten the resulting image.** One day, those bright spots may indicate that a few cells in the area may be on **the verge of mutating and growing out of control.**

Research showed that particles made it into the liver. This suggests that they could be **used to deliver toxic chemotherapeutic drugs that would act locally on a tissue, at the site of a tumor, and not have such a pronounced affect on the rest of the body**. The problem with **chemotherapy** is that the drug affects the whole body, **problems** such as hair loss, diarrhea and anemia. (Nanoparticles are smaller than any cell in the human body, so they may pass through the walls of the leaky blood vessels, or capillaries, of tumor tissue and actually infiltrate the tumor. And despite their miniscule size, **nanoparticles are still big enough to carry a payload of medicine**). The **long-term goal** is to use this technology **to identify disease at its cellular level, at its very earliest stage**. While this research is still in its infancy, the researchers foresee a day when nanotechnology can alert a physician to the beginnings of cancer or heart disease, perhaps in a woman who has a family history of breast cancer: Her doctor could inject the breast with **biodegradable nanoparticles** and the resulting ultrasound image would alert the doctor to any suspicious areas in the tissue, even at the cellular level.

The hope is that **combining ultrasound and nanotechnology may provide a definitive diagnosis in lieu of an invasive procedure like a biopsy.**

(This work was supported by the Susan G. Komen Breast Cancer Foundation, the National Cancer Institute, the National Center for Research Resources and the National Science Foundation).

Creating individual genetic profiles: Progress brings with it new expectations. We have an extended life because we have eradicated diseases like polio and can fight many other infections. But now we are concerned with quality of life as we get older. We also expect to see those who do suffer disabilities given the same chances to follow their dreams as anyone else.

The ability to determine the precise instructions required for each part of the body to grow and work efficiently has given us the potential to study variations in the master plan. While we are all human, we differ in the detail.This is why people react differently to medications, develop allergies or fail to recover as quickly from disease. With further research into the information provided by the human genome project, scientists will be able to compare individual variations against a general blueprint for the human race. **The result will be tailored treatments** and the **ability to pinpoint the mechanisms that can result in problems later in a person's life**. Of equal importance will be the ability to see how individuals interact with environmental challenges, such as alcohol, pollutants, sunlight and foods. Such knowledge will allow the development of customised preventative approaches.

(Source: Imagine the future. Research Australia, 2006)

Table 5.22.   R&D informatics is a prerequisite to materialize this value[116]

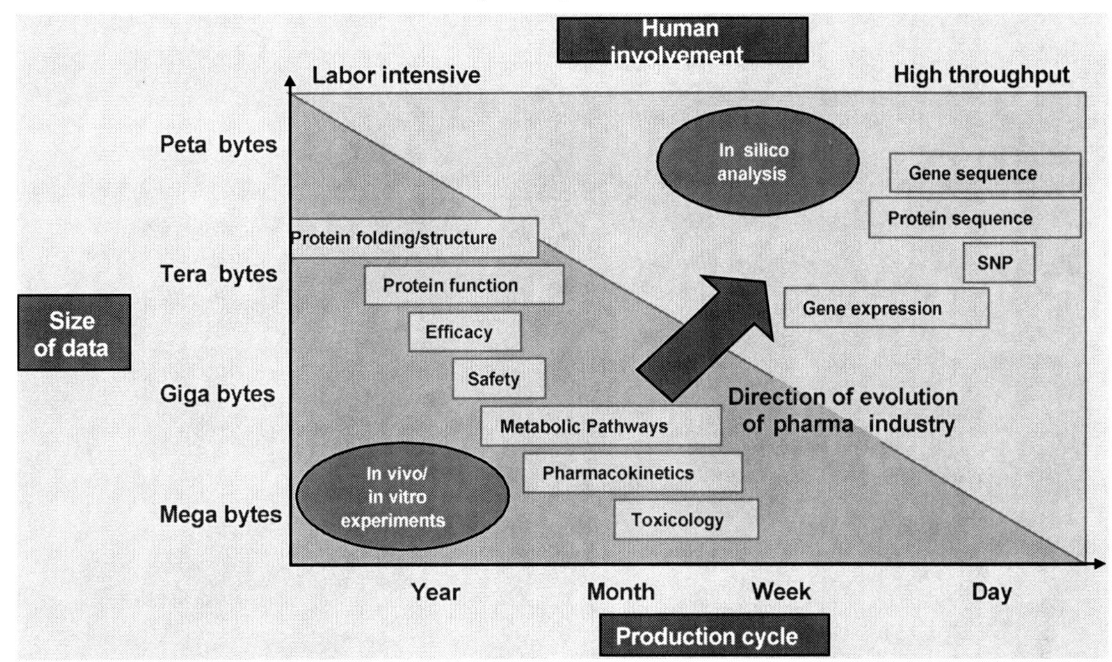

costs. As a result, the vast majority of investigational products that enter the clinical trials fail. Often, product development programs must be abandoned after extensive investment in time and resources. This high failure rate drives up costs, and developers are forced to use the profits from a decreasing number of successful products to subsidize a growing number of expensive failures. In addition, the path to market for successful candidates is long, costly, and inefficient, due in large part to the current reliance on cumbersome assessment methods. In many cases, developers have no choice but to use the tools and concepts of the last century to assess this century's candidates[117].

The decrease of Pharma's R&D productivity implies the need to solve principle issues with a new and more radical approach. Clearly, the R&D organizations have realized that after all, they mostly produce data and only at the end the result is a marketable drug.

So far, neither the R&D organizations, nor the R&D processes, nor the R&D information systems and applications have leveraged these assets. What is needed is an integrated approach, combining biological and chemical data together with preclinical and clinical, and allowing the researcher to use IT applications in order to make better decisions faster.

Such applications can be in-silico simulation and modulation, data and text mining, build on the backbone of an integrated IT architecture. This must be embedded in transformed R&D organizations and process structure.

---

[116] *based on Celera, Gene Bank; Protein Bank; NCBI; Applied Biosystems; Paracel*

[117] *Innovation or stagnation ? Challenge and opportunity on the critical path to new medical products, FDA report, US Department of Health and Human Services/Food and Drug Administration, March 24, 2004, addressing the growing crisis in moving basic discoveries to the market where they can be made available to patients.*

* The issue of Hospital Acquired Infections needs priority attention. Every year, an estimated **6 million people** are affected by Hospital Acquired Infections (HAI) worldwide, of which about **3 million in the European Union**(up to **10% of hospital admissions**), making it a major contributor to morbidity and mortality. HAI affects an estimated **1 in 10 patients in Europe**. It is one of the most prominent reasons of failure of advanced and expensive medical treatment. In the **UK, 5,000 patients die each year** as the result of HAI. In **Europe**, each year an estimated 60,000 persons or **more than 150 persons each day** die due to HAI. Over two million patients a year are infected by antibiotic resistant bacteria in U.S. hospitals each year and **over 90,000 die.** Many HAI are resistant to some kinds of treatment regimens (a phenomenon commonly known as **"antimicrobial resistance"**), thus being difficult to cure. The most common HAI's are:

–**MRSA** or *Methicillin Resistant Staphaureus*

–**VRE** or *Vancomycin Resistant Enterococci*, and

–namely in the USA and also increasingly in Europe, **ESRI** or *enterobacteriaceae* with an *Extended Spectrum Beta Lactamase*. The proportion of these resistant strains is increasing and becoming much more frequent among HAI. Without preventive measures the number of affected patients will also increment gradually. HAI are estimated to **increase at 1.7% per year worldwide.**

Hospital Acquired Infections:

–**Increases mortality**: Patients with a MRSA Blood Stream Infection have a 3 times higher mortality risk than patients with a MSSA (*methicillin sensitive Staphylococcus aureus*) infection.

–**Increases length of hospital stay** on average with 8 days per affected patient: 4 days per "simple" HAI, 10 days per MRSA infection, 16 days per *bacteremia* and 8 days in intensive care (ICU)

–**Are a burden on Health Care Costs**: It is estimated that HAI add more than 10 million unnecessary patient days in Europe. In the UK, only the total cost for treating and controlling HAI is estimated to be £1 billion (€1,476 million) per year and causing 5,000 deaths

The **average cost** for a Hospital Acquired Infection is estimated in **€2,300**, for a **MRSA infection, €8,000**and for a **blood stream infection €40,000**. *(Source: European Diagnostic Manufacturers Association, Screening programs for Hospital Acquired Infections. In Vitro Diagnostics. Making a real difference in health & life quality. September 2006)*

* In general, bacteria become **resistant to antibiotics** by using three major strategies. First is to prevent the drug from reaching its target. Some bacteria accomplish this by having a "**double-membrane**" **layer** to **retard** the **entry** of **the compound into the cell. Other bacteria simply drive out the drugs** from **inside** the cell using **pumps located in the membrane.** The second strategy bacteria employ to become drug resistant is to **produce enzymes** that **inactivate** the antibiotic so it can no longer bind to its cellular target. The third strategy is to **alter the cellular target,** i.e. the **ribosome**, so that an antibiotic can no longer bind and inhibit its function. Bacteria can use one or all of these very effective methods to resist the action of **antibiotics, rendering them therapeutically ineffective.** (see opposite colon)

* **DNA mutations** that allow bacteria to become antibiotic resistant occur as a natural consequence of the bacteria duplicating their DNA. Such mutations typically occur infrequently (10-5 to 10-7), but because bacteria can reproduce every 20 minutes, mutations are much more common than in human cells. In addition to random mutation, bacteria are often able to exchange their DNA with one another thereby transferring resistance determinants, like efflux pumps, to their antibiotic sensitive neighbors. So, by employing these various strategies, bacteria have virtually assured themselves of a way to develop resistance to the antibiotics in use today. *(http://www.rib-x.com/resistance.html / accessed November 5, 2006)*

B.63. Compounds targeting the large subunit of the ribosome

Macrolides	Strepto-gramins	Peptides
Clarithromycin	Streptogramin A	Viomycin
Zithromax	Streptogramin B	Capreomycin
Tylosin	Ostreogrycin G	
Spiramycin	Synercid	
Erythromycin	Virginamycin S1	
Carbomycin	Vernamycin B	
Oleandomycin	Patricin A	
Ketolides	**Oxazolidi-nones**	**Glutarimide**
ABT-773	Linezolid	Cycloheximide
Telithromycin	Dup721	Streptovitacins
CP-642959		Streptimidone
CP-605006		Inactone
		Actiphenol
Thiostrep-tons	**Nucleotide Analogues**	**Trichothe-cenes**
Thiostrepton	Sparsomycin	Trichodermin
Siomycin	Puromycin	Vomitoxin
Sporangiomycin	Anisomycin	Trichothecin
Thiopeptin	Blasticidin S	Nivalenol
Micrococcin	Chloramphenicol	Verrucarin A
	Althiomycins	**Lincosa-mides**
		Clindamycin
		Lincomycin

* **The ribosome as a drug target**: The first atomic level insights into the structure and function of the ribosome have come from crystal structure analysis of **the 50S ribosomal subunit and structures with bound substrates and antibiotics**, have yielded insights into the mechanism of peptide bond formation and its inhibition. The structures of these antibiotic complexes are leading to the design on new antibiotics, now in phase I trials, by **Rib-X Pharmaceuticals, Inc.** *(Steitz TA, "From Understanding the Ribosome Structure and Function to Designing New Antibiotics". Presentation given at "Sciences' Next Great Idea", Duke University New Yrok, Sept 26, 2006)*

Distance surgery will become routine with surgeons operating down the line; monitoring after operations will be carried out at home. Cost control will play an increasingly important role for healthcare payers to avoid inefficient treatments.

V.14. Why outsourcing R&D? The changing CRO Model:
In general, the pharmaceutical industry turns to outsourcing to improve the speed and quality of target validation; the quality of compound leads and candidates; and the efficiency and effectiveness of preclinical and clinical development.

Based on the framework for evaluating strategic outsourcing from McKinsey's 1995 "Market versus Buy: The Wrong Cost Decision" there are several advantages to engaging CROs in R&D outsourcing, including lower labor and overhead costs, improved economies of scale, enhanced R&D output and reductions in capital spending[118].

CROs that offer a global presence and can integrate a wide range of core and ancillary services will seize a significant share of the growing outsourcing market. Total pharmaceutical R&D outsourcing is expected to grow an average of 11 to 12 percent annually from its 2006 level of $10.4 billion, compared to anticipated growth in global R&D spending of only 9.6 percent. With outsourcing growth now outpacing overall growth in R&D, pharmaceutical companies have shown their willingness to contract with CROs for research and development.

In the past 20 years, the CRO industry has matured to the point where full drug development capabilities now exist within a number of CROs. In fact, many small, emerging pharmaceutical companies today rely on CROs for all their drug development needs. Through this practice of 'virtual' drug development, large investments in internal resources can be averted if effective service leasing and product manufacturing arrangements are created and managed effectively[119].

V.15. Moving from Vendor to Strategic Partner
Originally, the pharmaceutical industry turned to contract research companies for added capacity. Today, the growth in outsourcing among CROs is increasingly driven by those companies focusing on innovative solutions. These companies recognize the value of operating as a research partner to pharmaceutical and biotechnology firms and succeed by understanding that, more than data or technology, it is their scientific knowledge, extensive experience and customer service orientation that make them important allies in drug discovery and development[120].

However, relationships between CROs and pharmaceutical and biotechnology companies have yet to meet their shared expectations for strategic value exchange. These relationships have succeeded at achieving tactical and operational value exchange – through improved price, speed, quality and consistency – but have not attained long-term, strategic value.

Alliances hold great promise for both the CRO industry and its pharmaceutical and biotech partners since its current business relationships are, for the most part, fee-for-service, and transactional contracts. On the downside, it means that relationships are often discontinuous, resulting in gaps between projects. This problem is intensified by the fact that customer relationships are likely to occur within individual business units,

---

[118] Christopher McFadden, Randall Stanicky, Timothy Leahy and James C. Sinclair. *Pharmaceutical Services*. Goldman Sachs Global Equity Research, September 2003.

[119] J. Fred Pritchard, Ph.D., and Edward Leung, Ph.D., "Developing Drugs Together – Understanding Virtual Relationships," *Contract Pharma*, October 2003.

[120] Nigel Brown, Ph.D., "Capitalizing on a Changing Research and Development Environment," *Business Briefing – PharmaTech 2003*.

*Researchers reported in the **Journal of the National Cancer Institute** that an **engineered virus** tracks down and infects the most common and deadly form of brain cancer and then **kills tumor cells by forcing them to devour themselves.** The modified adenovirus homed in on **malignant glioma cells** in mice and induced enough self-cannibalization among the cancer cells -- a process called **autophagy** -- to reduce tumor size and extend survival.This virus uses **telomerase**, an enzyme found in 80 percent of brain tumors, **as a target**. Once the virus enters the cell, it needs telomerase to replicate. Normal brain tissue does not have telomerase, **so this virus replicates only in cancer cells.** Other cancers are telomerase-positive, and the researchers showed in lab experiments that the virus kills human prostate and human cervical cancer cells **while sparing normal tissue**. In addition to demonstrating the therapeutic potential of the virus, called **hTERT-Ad**, researchers also clarified the mechanism by which such conditionally replicating adenoviruses (**CRAs**) infect and kill cancer cells. (Autophagy is a protective process that cells employ to consume part of themselves when nutrients are scarce or to destroy some of their organelles to recycle their components. A double membrane forms around the material to be consumed, then everything inside is digested). The researchers showed that **hTERT-Ad** (human telomerase reverse transcriptase promoter regulated adenovirus) infected the glioma cells and induced autophagy by inactivating a molecular pathway -- the **mammalian target** of rapamycin (mTOR) pathway -- that is known to prevent cellular self-cannibalization. The cells showed no sign of having been killed by apoptosis (programmed cell death). A normal biological defense mechanism that systematically kills defective cells, apoptosis is suppressed or dysfunctional in cancer cells. **Many cancer therapies focus on restoring or enhancing apoptosis to combat the disease.** They believe that **autophagy**, but not apoptosis, **mediates the principal anti-tumor effect of conditionally replicating adenoviruses.** Cells killed by apoptosis show specific damage to the cell nucleus and DNA, with other cellular organelles preserved. Cells killed by autophagy have **little damage to the nucleus but heavy degradation of the cells' organelles.**

* In a Nature Reviews Cancer paper September 2005, the same researchers reviewed therapies and molecules that cause or inhibit the **self-cannibalization process** and compared autophagy and apoptosis. **To improve cancer therapeutics**, they concluded that it is vital to identify molecules that regulate autophagy in cancer cells and to understand how autophagy is associated with cell death, a relatively new field in cancer research. *(Source: Marville S., University of Texas M.D. Anderson Cancer Center, May 6, 2006).*

* The market for the treatment **of chronic obstructive pulmonary disease (COPD)** is forecasted to grow from worldwide sales of $5billion in 2006, to $7.5 billion by 2015. Novel dual-action, long-acting bronchodilator combinations will be the most significant development in the inhaled COPD class at the start of the next decade. The term COPD covers a complex group of disorders characterized by a progressive development of airflow limitation. Caused primarily by smoking, it is set to become the third leading cause of death in the developed world by 2020. As no therapy appears to stop the progression of the disease, the strategy is to relieve acute symptoms such as breathlessness and wheezing using bronchodilators. Inhaled short-acting broncho-dilators, such as **Boehringer Ingelheim's Atrovent (ipratropium bromide)** and **GSK's Ventolin (salbutamol / albuterol)**, are effective at relieving bronchoconstriction and are prescribed for patients with few or intermittent symptoms. In patients with moderate to very severe COPD whose symptoms are not adequately controlled with as-needed short-acting bronchodilators, adding regular treatment with a long-acting inhaled bronchodilator, such as **GSK's long-acting b2-agonist Serevent (salmeterol)**, or **Boehringer Ingelheim / Pfizer's long-acting anticholinergic, Spiriva (tiotropium bromide),** is recommended by the medical community. **Spiriva** is the first once-daily inhaled antimuscarinic approved for the maintenance treatment of COPD, which is expected to register global sales of $1.5 billion by 2010. In 2006, there **were 31.5 million COPD sufferers in the US, Europe and Japan** - a prevalence rate of approximately 4.4% of the general population. Thus, the development of novel dual-action, once-daily inhaled therapies should improve adherence to correct treatment regimens.

* Two of the world's leading color vision researchers, Professors Jay and Maureen **Neitz** are also **pioneers** in the field of **functional genomics**. Their studies of **human color vision** have not only identified the genes responsible for colorblindness, but also defined one of the first examples of a nervous system defect for which a person's DNA can predict both the occurrence and the severity of the disorder. Their color vision research has also provided them with unique opportunities to discover the steps in the causal chain from the gene, to protein function, to neural signal. They are applying these lessons to other genetic defects that cause visual impairment. They anticipate that their studies of the basic mechanisms controlling gene expression in the retina, and the structure/functional relationships among proteins involved in signal transduction, may lead to development of new methods for early diagnosis of retinal disorders, and ultimately extend the knowledge of the role genes play in construction of the nervous system.

Behavioral and Social Scientists see enormous potential to improve the health and well-being of all citizens. They envision healthy individuals, living in health-promoting communities, supported by societal policies and economic incentives that maximize the potential to achieve good health – not merely the absence of disease or infirmity, but rather a state of complete physical, mental, and social well-*being (**Preamble to the Constitution of the World Health Organization** as adopted by the International Health Conference, New York, 19-22 June, 1946; signed on 22 July 1946 by the representatives of 61 States (Official Records of the World Health Organization, no. 2, p. 100) and entered into force on 7 April 1948).*

rather than across several business units within the same CRO or pharmaceutical company.

Most attempts to modify these business relationships, such as establishing preferred provider relationships, offering volume discounts, conducting WEB auctions, and creating fee-based models, focus on price savings versus any real value exchange. Value exchange should result in shared services and programs and expanded mutual capabilities through a long-term relationship. Strategic relationships with pharmaceutical and biotechnology companies that provide real impact should: increase revenue, be longer term, connect across business units within both companies, be global in scope, reach higher levels within the client, provide proprietary infrastructure support, offer transformational simplification of the relationship and, most of all, deliver value to both partners[121].

V.16. Realizing Strategic Value

How can CROs and their drug development partners achieve sustainable, strategic value? Effective relationship management is essential, but in order to establish meaningful, sustainable and strategic alliances that build real value, each must find new ways of doing business with existing and potential partners. Partners must move beyond a project-by-project arrangement to focus on relationships built around capacity or programs that share their science, facilities, people and systems to address the sponsor's clinical development needs over a long period of time. Also, focusing on the long-term allows for customization and increased efficiency.

The evolution of the CRO industry due to increased outsourcing is analogous to how environmental challenges have impacted other industries, including banking, manufacturing, insurance and technology.

In particular, parallels may be drawn between the explosive growth of the technology outsourcing market and the potential for the CRO industry to benefit from broader recognition by pharmaceutical and biotechnology companies that certain support functions are not core to their business.

Information technology is the fastest growing area of strategic outsourcing today, with a global market totaling more than $100 billion annually. IBM is the clear industry leader, possessing more than half of the market share[122]. Like pharmaceutical outsourcing, IT outsourcing partnerships exist to: gain access to "next generation" infrastructure and applications; reduce costs; allow companies to focus on core competencies; improve service.

V.17. The Ideal CRO Model for the New Outsourcing Paradigm

Founded in service excellence, leadership in science and technology and an orientation toward creating strategic partnerships, the ideal CRO model for today's rapidly changing pharmaceutical environment is organized around program development throughout the entire drug discovery and development continuum, maximizing its full range of laboratory and clinical service capabilities. This new model has emerged in response to a need for both consulting and implementation expertise to facilitate drug development strategies.

---

[121] Douglas J. Squires, PhD., and James E. McClurg, Ph.D., "A Business Model for Sustainable, Strategic Relationships," Drug Discovery World, Spring 2003.
[122] The Technology Primer 2001. Morgan Stanley.

* In 2006, around 32.000 **renal transplants** were performed in the seven major pharmaceutical markets. Their number is expected to increase to **43.000 by 2015**. In order to prevent the patient's immune system from rejecting the transplanted organ, daily immunosup-pression therapy is a necessity for the lifetime of the graft with a base maintenance drug **(cyclosporine, tacrolimus, sirolimus)** combined with adjunctive therapies **(azathioprine, mycophenolate mofetil, steroids)**. The number of patients with functioning kidney transplants will double to **428,000 by 2015**, creating a large group of patients with unique and complex long-term medical care needs directly attributable to adverse effects of immunosuppressive drugs, including nephrotoxicity, diabetes, hypertension, hyperlipidemia and vulnerability to infection. Although prevention of acute rejection remains a primary treatment goal, agents that do not impair long-term renal function are required. While results from sirolimus based calcineurin-inhibitor withdrawal regimens are inconclu-sive more promising results are emerging from use in the **calcineurin-inhibitor** avoidance or switching setting.

Based on its discovery of a disease-causing molecule found only in the joints of RA patients, **MaimoniDex**, has developed a **murine monoclonal antibody** that has shown bioactivities both in vitro and in vivo. It is the company's goal to produce a commercially viable humanized monoclonal antibody for treating rheumatoid arthritis. Using its antibody engineering technology, **Abmaxis** will humanize and optimize the murine antibody to develop a drug candidate suitable for potential therapeutic use.

* The **antiarrhythmics** market across the seven major markets has experienced a sharp decline in sales, primarily due to the loss of patent protection for Sanofi-Aventis' Cordarone (amiodarone) in the US in 2002, and generic erosion therein. It appears that the antiarrhythmics market is on the brink of rapid expansion due to the expected launch of five novel products between 2007 and 2010, and a billion dollar market looks set to triple in size by 2015. Arrhythmia is defined as an irregular heart rhythm, or an abnormality in the timing or pattern of the heartbeat, causing the heart to beat too rapidly, too slowly, or irregularly. Due to the complex nature of the condition, comprehensive epidemiology data on the prevalence of arrhythmia as a whole is difficult to come by.

Atrial fibrillation (AF) is the most common form of arrhythmia, with an estimated nine million sufferers worldwide. Three new drugs -- **Cardiome's RSD1235** (intravenous), **Sanofi-Aventis'** follow-on to **Cordarone, Multaq (dronedarone)**, and P&G Pharma's **Stedicor (azimilide)**-- could generate sales of greater than $500 million and command over 60% of the entire antiarrhythmics market by 2015.

* In the 1960s, scientists discovered **three different classes of clinical drug**, each of which **recognized DNA in a different way**. Subsequent drugs have used only these three ways to recognize the DNA.
New research, announced early 2006, has isolated a **fourth**, which is completely different and opens up entirely **new possibilities for drug design**.
The scientists, led by a team from **University of Birmingham's School of Chemistry**, have developed a synthetic drug agent that targets and binds to the center of a 3-way junction in the DNA. **These 3-way junction structures are formed where three double-helical regions join together.**
They are particularly exciting as they have been found to be present in diseases, such as some **Huntington's disease and myotonic dystrophy, in viruses and whenever DNA replicates itself, for example, during cancer growth.**
The Birmingham team created a nanosize synthetic drug in the shape of a twisted cylinder. Together with researchers in the UK, Spain and Norway they showed that is had unprecedented effects on DNA.
Molecular level pictures taken by the Barcelona team have shown that it binds itself in a new way to the DNA, by fixing itself to the center of a DNA junction, which had three strands.
It is all held together because the cylinder is positively charged and the DNA is negatively charged. In addition the drug is a perfect fit in the heart of the junction: a round peg in a round hole.
This discovery will **revolutionize the way that we think about how to design molecules to interact with DNA**. It will send chemical drug research off on a new tangent. By targeting specific structures in the DNA scientists **may finally start to achieve control over the way our genetic information is processed and apply that to fight disease.**
(Source: University of Birmingham Press Release 02.03.06)

* **Knowledge** of the **brain's structure** and **function** will translate into treatments for all manner of conditions that affect the brain. For instance, scientists have been able to identify proteins that stop repair following injury. This knowledge will allow the design of drugs to "stop these stoppers" and **promote factors that allow brain cell growth**. Such drugs can be used to treat degenerative conditions, such as **Alzheimer's disease, Parkinson's disease, head trauma and stroke.**
Research into how receptors on the surface of our brain cells transmit messages will lead to the development of **better medicines for schizophrenia, bipolar disorder, major depression and drug addictions.** These medicines will be used in conjunction with behavioral therapies aimed at assisting patients to learn new ways of processing information. *(Health Australia. Imagine The Future. 2006)*

International studies in collaboration with the US Health Institutions FIC, NCMHD and The World Bank, and others have added to our understanding of the role of poverty, social position, culture, and socioeconomic status in the prevention, treatment, and management of diseases. Many preventable diseases that create enormous emotional and financial hardship have their origins in the socioeconomic and built environments. Discoveries in the behavioral and social sciences should contribute to life-saving environmental and policy changes...

Table 5.23.   Drug discovery development[123]

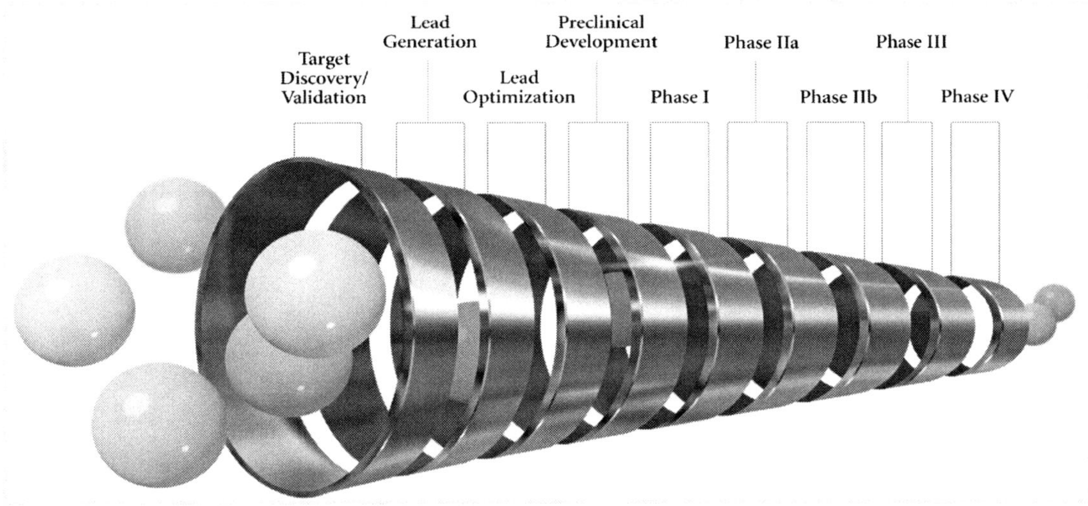

Many CROs offer some degree of integrated drug development services, but the program development mindset brings this concept to a new level, while giving pharmaceutical and biotechnology clients full access to the CRO's scientific and regulatory expertise, experience and resources.

A central element of program development involves assigning a dedicated and experienced program director to manage the planning and execution of each drug development program. In addition, a specialized infrastructure is needed to support the high level of planning and project management capabilities required.

In addition, methods must be in place to capture corporate knowledge about each client and facilitate learning about drug development to maintain delivery of high value work. While certain to appeal to all pharmaceutical organizations, this approach has particular relevance to the special problems encountered by smaller pharmaceutical companies and biotech companies, who often require an integrated drug development program but have limited internal expertise and are left to seek out numerous CROs for their drug development needs[124].

V.18. Models for Outsourcing
In the October 2003 issue of *Contract Pharma* magazine, Dr. J. Fred Pritchard, of MDS Pharma Services, and Dr. Edward Leung, of King Pharmaceuticals R&D, presented a series of outsourcing models for developing a single drug candidate, each one involving a higher level of partnership between CRO and sponsor.

Do It Yourself Model
The sponsor operates like a general contractor, coordinating several vendors and even doing some of the work themselves. This model allows for close control of information generated during the execution of the program, which supports scientific continuity –

---

[123] *Source: Douhglas J. Squires, MDS Pharma Services, 2004*
[124] *Nigel Brown, Ph.D., "Capitalizing on a Changing Research and Development Environment," Business Briefing – PharmaTech 2003.*

* The **U.S.National Human Genome Research Institute** (NHGRI), part of the National Institutes of Health (NIH), granted awards to speed the development of **innovative sequencing technologies** that **reduce the cost of DNA sequencing** and **expand** the **use of genomics in medical research** and **health care**.

There has been significant progress over the last several years to develop faster and more cost-effective sequencing technologies. The NHGRI supports these innovative efforts to benefit scientific labs and medical clinics because **these technologies will eventually revolutionize the way that biomedical research and the practice of medicine are done.**

Since 1990, NHGRI has invested approximately $380 million to develop and improve DNA sequencing technologies. DNA sequencing costs have fallen more than **50-fold** since the mid 1990s, fueled in large part by tools, technologies and process improvements developed as part of the successful project to sequence the human genome. However, it still costs around $10 million to sequence 3 billion base pairs -- the amount of DNA found in the genomes of humans and other mammals. NHGRI's near-term goal is to **lower the cost of sequencing a mammalian-sized genome** to $100,000, allowing researchers to sequence the genomes of hundreds or even thousands of people participating in studies to identify genes that contribute to common, complex diseases.

Ultimately, NHGRI's vision is to **cut the cost of whole genome sequencing to $1,000 or less, which will enable the sequencing of an individual's genome during routine medical care.** The ability to sequence an individual genome cost-effectively could enable health care professionals to **tailor diagnosis, treatment and prevention to each person's unique genetic profile.**

\# One team will use existing enzyme and dye-tagged nucleotide resources, the building block of DNA, in a novel way that will simplify the fundamental, front-end chemistry of massively parallel sequencing-by-synthesis. This method uses the **natural catalytic cycle of DNA polymerase to capture just a single DNA base on an immobilized primer/template.** A fluorescence scanner will be used to scan and identify hund reds of thousand of molecules at once. Then the cycle will be repeated.

\# The goal of the second team is to **fabricate nanoscale channels in which single molecules of DNA will pass between nano-electrodes that are less than 2 nanometers apart, to measure an electric current that will identify individual bases.** A nanometer is one-billionth of a meter, much too small to be seen with a conventional lab microscope. Several groups are developing nanopores (holes about 2 nanometers in diameter) for use as DNA sequence transducers and propose to detect an electrical, or ionic, signal from individual DNA molecules.

\# A third project uses an experimental method for DNA sequencing called **"single molecule sequencing by ligation"**. It aims to develop a method for fabricating high-density arrays of wells with sub-micrometer dimensions for ordering single nanoparticles and DNA molecules. The investigator will attempt to demonstrate that more than 1 billion individual DNA molecules can be sequenced in massive parallel though a process where an **enzyme is used to join pieces of DNA together**.

\# Another group will continue to implement a novel approach in which a **nanopore is used to simultaneously detect electrical and fluorescent signals from many nanopores at one time**. A novel sequencing instrument will be fabricated, along with additional analysis tools, with the aim of producing a viable, low-cost sequencing system.

\# A fifth team of investigators has developed a fully automated instrument capable of sequencing single molecules of DNA on a planar surface. The group will develop a high-throughput version of this technology for the re-sequencing of whole human genomes. The sequencing strategy involves obtaining short reads (about 25 DNA bases) from billions of strands of DNA immobilized on a surface inside a reagent flow cell. The research plan aims to advance this strategy to **achieve high accuracy, re-sequencing of highly variable genomes and assembly of never-before sequenced genomes.**

\# Still another team will apply **force spectroscopy**, a technique used **to understand** the **mechanical properties** of **polymer molecules** or **chemical bonds**, to DNA undergoing arrested polymerization to **initially demonstrate one-molecule-at-a-time analysis** of **changes** in **molecular mechanics** at a **resolution** of a **single base**. Using optical, near-field probes, the methods of force spectroscopy can be advanced into techniques having massively parallel format, where millions of single DNA base additions can be followed at the same time. The identification of bases will be done exclusively on the basis of changes experienced by the molecule as a whole. The team aims to fabricate a low cost table-top setup suitable for use in a majority of biological, chemical and hospital laboratories.

\# Expanding the performance of the **sequencing-by-synthesis technology**, this group will develop a cost-effective method to fabricate **universal DNA nanoarrays using nano-contact printing**. The current photolithography technology can cause damage to DNA probes, which the group will strive to avoid by using nano-contact printing. With the nano-sized features, a **DNA nanoarray** can also improve throughput by offering the ability to accommodate **billions of DNA molecules in a small area**. Hybridization will be detected by atomic force microscopy.

\# Another team will aim to develop **highly integrated arrays of nanopores** that can be fabricated by **lithographic methods**, along with **on-chip silicon-based electronic circuits** and **circuit techniques** that **amplify** and isolate their various electrical signals. This group will also design a **dipole-sensing methodology**, which in principle can distinguish signals from each of the DNA bases. **Arrays of nanopores** will be constructed on silicon substrates using a self-aligned compositional approach. Quadrature dipole moment detectors will be constructed that yield a signal independent of the rotation of the DNA molecule relative to the electrodes.

\# Most current research in **nanopore sequencing** involves the protein pore, a-hemolysin; or artificial pores in inorganic materials. This investigator will explore the use of a different protein pore, **Mycobacterium smegmatis porin A (MspA)**, as a new tool for **nanopore sequencing**.

important to decision making in early drug development. Competitive bidding assures good pricing. However, managing timely information flow from vendor to vendor (often competitors) can be a challenge each contract requires extensive hands-on management. Finally, there is limited opportunity to capture the synergies between experts working at the different companies.

Architect Model
In this model, the sponsor hires an expert consultant to provide consulting and coordination among a number of vendors.   Though potentially costly, the "architect's" expertise can be accessed along with their network of experts. Management of timely information flow among competing vendors remains challenging. Also of concern, the consultant may be seen as an 'outsider' by both vendors and sponsor, making it difficult for the consultant to lead the project.   In truth, the consultant may not be truly independent, maintaining undisclosed relationships with certain vendors.

Integrated Design and Build Model
The sponsor contracts the entire program to one primary vendor who coordinates it internally and with subcontractors. This model provides access to an experienced and integrated team of experts. The vendor designates a 'champion' who will ensure timely information flow between sites and projects, resulting in time and cost efficiencies.   This approach to program management is similar to that used by larger pharmaceutical companies. The smaller client, now with many studies at one vendor, becomes a big client for that vendor.   Serious problems will result, however, if the primary vendor cannot effectively execute the program.   This model requires a high level of trust between sponsor and vendor, with great potential for success.

Risk-Sharing Model
In this model, the sponsor contracts the entire program to one primary vendor in exchange for an equity position with the sponsoring company or other delayed revenue, based on the program's success. Vendor and sponsor share the opportunity to win or lose, creating a strong incentive for the vendor to provide top-notch service and expertise. Since one primary service provider manages the program, much like a large pharmaceutical company would, there is good information flow between studies and sites, resulting in time and cost efficiencies.

The vendor must be able to manage the loss of cash flow. Plus, it is more difficult for the vendor to be totally objective when they have a financial interest in the outcome of the program. Finally, the primary vendor must execute the program well[125].

V.19. Success Factors in Virtual Drug Development

Pritchard and Leung also proposed that there are five key success factors common to successful drug development, and that these factors are central to the emerging pharmaceutical outsourcing environment:

1. A good drug candidate: a product that is proven to be safe, selective and easy to administer. Its mechanism of action should be based on a sound therapeutic concept and ideally meet an unmet medical need.

[125] *J. Fred Pritchard, Ph.D., and Edward Leung, Ph.D., "Developing Drugs Together – Understanding Virtual Relationships," Contract Pharma, October 2003.*

* **Telomeres** protect chromosome ends and distinguish them from DNA breaks in the cell. Telomere DNA consists of tandemly repeated simple repeated sequences. The chromosome termini are maintained by an enzyme called telomerase. Telomerase is specialized reverse transcriptase that contains a catalytic protein subunit and an essential RNA subunit. The telomerase RNA provides the template for the TTAGGG repeats that are synthesized onto chromosome ends. In the absence of telomerase, telomeres shorten progressively and both chromosome instability and cell death can occur. Although many somatic cells in humans do not express telomerase, most human tumors require it for continued growth, suggesting inhibition of telomerase may inhibit the growth of cancer cells. DrGreider and colleagues focused on understanding the telomerase enzyme mechanism and cellular and organismal consequences of telomere dysfunction. We generated telomerase null mice that are viable and show progressive telomere shortening for up to six generations. In the later generations when telomeres are short, cells die via apoptosis. Crosses of these telomerase null mice to other tumor prone mouse models suggest that under some circumstances tumor formation can be greatly reduced by artificial telomere shortening.

The researchers are dissecting the molecular mechanisms of telomere function to understand the subtleties that tip the balance between cell death and genetic instability due to loss of telomere function. In addition, we are using our telomerase null mice to explore the essential role of telomerase in stem cell viability. In the autosomal dominant form of the human genetic disease dyskeratosis congenita, individuals exhibit aplastic anemia and other complications that are likely due to stem cell failure. Mutations in telomerase components cause this stem cell failure. We have developed a mouse model of this disease, which we are using to explore the range of physiological consequences of stem cell loss. *[Greider CW. Daniel Nathans Professor and Director of the Department of Molecular Biology and Genetics, Johns Jopkins University School of Medicine. "Telomerase and the Consequences of Telomere Dysfunction". Science´s Next Big Ideas. Presentation at Duke University, September 26, 2006]*

* An enzyme called **sAC** helps spur the **growth** of **nerve endings** in the **developing embryo**, and might also be used to someday **regrow** these **"axons"** in adults **paralyzed by spinal cord injury.**

Identifying soluble adenylyl cyclase (sAC) as a key player in axonal growth has been like finding a crucial **'missing link'** in the biochemical chain that leads to **nerve cell regeneration**. With this new piece of the puzzle, scientists can begin serious work on introducing sAC directly into damaged spinal cords, where they hope it will encourage axons to seek out vital new connections. The ultimate goal is a treatment that can prevent paralysis or restore movement to paralyzed individuals. The discovery, by a team of researchers at **Weill Cornell Medical College** in New York City, may also have implications for the **treatment** of other **conditions characterized** by **impaired axonal growth**, such as certain develop-mental disorders and diabetes-linked damage to peripheral nerves.

* Many advances in **molecular neuroscience** have stemmed from Dr. Snyder´s (John Hopkins University School of Medicine) identification of receptors for neurotransmitters and drugs and elucidation of the actions of psychotropic agents. He pioneered the **labeling of receptors** by **reversible ligand binding** in the **identification of opiate receptors** and extended this technique to all the major neurotrans-mitter receptors in the brain. In characterizing each new group of receptors, he also **elucidated actions** of **major neuroactive drugs**. The isolation and subsequent cloning of receptor proteins stems from the ability to label, and thus monitor, receptors by using these ligand binding techniques. Classic neuro-transmitters were easy to conceptualize, following the paradigm of acetylcholine, being stored in neuronal vesicles and acting on membrane receptors. Recent **putative transmitters** so change the model that the **term 'neurotransmitter' seems obsolete.**

Gases such as nitric oxide and carbon monoxide cannot be stored in vesicles nor act at membrane receptors. The D-isomer of serine is a transmitter which occurs in glia. Messengers regulating intracellular events of many cells, but especially neurons, are altering signaling conceptualizations. The inositol phosphate IP3 is well known to release intracellular calcium, but other inositol polyphosphates have unprecedented actions. Thus the energetic pyrophosphate IP7 physiologically phosphorylates protein targets and, unlike ATP, it appears to **pyrophosphorylate** them. While nitric oxide signals physiologically, it also initiates a death cascade by nitrosylating the housekeeping glycolytic enzyme glyceraldehyde-3-phosphate dehydrogenase (**GAPDH**) which acquires the **capacity to bind to the ubiquitin-3-ligase Siah, enter the nucleus and kill cells, possibly by enhancing the protein acetylase activity of** p300/CBP. *(Snyder, Distinguished Service Professor of Neuroscience, Pharmacology and Psychiatry. "Novel Neural Messengers". Sciense´s Next Big Ideas. Duke University 75th Anniversary Science Symposium, N.Y. September 26, 2006)*

* Western Society becomes obsessed with the idea of „thin is beautiful". Those women who do not respond to this „ideal body shape" increasingly turn towards beauty clinics and cosmetic surgery, but also to medications. An extra burden to each nation's social healthcare system!. The **role** of the **media** in the development of eating disorders, in setting the standard of what is beautiful and what isn't so nice a physical appearance, is tremendous. Research (published in the November 2006 issue of Psychology of Women Quarterly) explored the relationship between so called **"thin-ideal" images** in the media and body-image issues among young female undergraduates who viewed advertisements displaying ultra-thin women exhibited increases in body dissatisfaction, negative mood, levels of depression and lowered self-esteem. These findings were particularly true for women who have negative views of their current body image and believe themselves to be overweight. The study showed that women who possess these body image concerns are more likely to have those comparisons affect their self-worth, leading to feelings of depression, body dissatisfaction and preoccupation with diet and exercise.

Imaging will dynamically display cellular and molecular abnormalities in every single part of the human body. The healthcare providers, be they public, private or charitable, will be interconnected/integrated to deliver optinal coverage.

2. An effective drug development team, made up of committed people who together cover the required areas of expertise. The team must have a champion, a team leader who is respected for their knowledge, passionate about the program, and who can leverage the right resources to get the job done.

3. A solid plan, based on a decision-gate strategy that manages risk[126]. The plan needs to communicate how the science addresses the medical need and how it fits within the company's business plan. Timeframes for the plan should be challenging but reasonable.

4. Successful execution of the plan.

5. Knowing how to respond to "the unexpected."[127]

V.20. Factors impacting the environment for pharmaceutical development

The environment for pharmaceutical development and increased outsourcing through 2010 will be impacted by the following:

➜ Drug development spending will continue to grow, as will outsourcing, dominated by clinical development.

➜ The industry's approach to clinical drug development will change – increasing its focus on targeted medicines (elements include early, risk-based approvals; more Phase IV studies; use of biomarkers/diagnostics; increased role for pharmacogenomics; improvements in technology and a growing reliance on remote monitoring/data collection capabilities).

➜ Increased demand for integrated development solutions (biotechs need to get compounds to proof-of-concept; pharma increasingly focused on commercialization).

➜ Lead compound selection/earlier qualification will become even more critical for success.

➜ Significant growth in number of large molecule therapeutics developed and approved (though small molecules will continue to dominate).

V.21. How Will CROs Play a Meaningful Role?

CROs forming strategic alliances with pharmaceutical and biotechnology companies represents a natural progression of the CRO business model from order taking through tactical outsourcing to strategic outsourcing and ultimately, in-sourcing, characterized by a "one company" mentality and a high level of integration between CRO and sponsor[128]. Those CROs poised to succeed via strategic alliances have increased their capacity to manage client projects across the drug development continuum, and have invested in the creation of an integrated team of technically and scientifically proficient individuals armed with a strong business sense. A CRO with the capability and expertise to both customize and execute research and development programs offers outstanding value to sponsors seeking accelerated, efficient drug development.

For CROs to add value, they must offer strategies that help their customers adapt to and profit from the rapid changes taking place in drug development. Instead of treating client

---

[126] J. Fred Pritchard, Malle Jurima-Romet, Mark L.J. Reimer, Elisabeth Mortimer, Brenda Rolfe and Mitchell N.Cayen, "Making Better Drugs: Decision Gates in Non-clinical Drug Development," Nature Reviews: Drug Discovery, July 2003, Vol.2.

[127] J. Fred Pritchard, Ph.D., and Edward Leung, Ph.D., "Developing Drugs Together – Understanding Virtual Relationships," Contract Pharma, October 2003.

[128] Douglas J. Squires, PhD., and James E. McClurg, Ph.D., "A Business Model for Sustainable, Strategic Relationships," Drug Discovery World, Spring 2003.

*To avoid wasting resources, the world of **high technology is increasingly organized into specialist centers** in which companies, training centers and research units work together in a spirit of synergy. In the area of health, France is benefiting from the experience of the **Genopôle** at Évry in the Paris suburbs, recognized worldwide in the field of genetic research. On the same principle, seven "**canceropoles**" (cancer research centers) have been established throughout the country. Their aim is to coordinate and network teams belonging to a variety of organizations - hospitals, university laboratories, research institutes - to promote research and speed up patient access to new treatments. They come under the new National Cancer Institute (INCa), founded in 2004, one of whose functions is to facilitate the integration of French research into international programs. The more recent **MediTech Santé** center in Île-de-France intends to occupy a predominant place in the health field by 2010, through the development of a series of innovations. Among the most ground-breaking are a **neuro-protective molecule** to combat Parkinson's Disease and the use of **electronic implants in blind patients** or in those suffering from a **severed spinal cord**. As stated by **Professor Colin Blakemore** (Medical Research Council in the United Kingdom), the biggest breakthroughs will be "…in prevention… and what one might call '**populomics**', the study of the genetic and phenotypic diversity of human populations, and how they interact with their environments and how their behavior influences their health and disease patterns" *(Blakemore, C., An interview with Professor Colin Blakemore. UKWatch, Winter 2005/2006,4).*

*The global **anxiety disorders** market is set to **decline from \$4.5 billion in 2006 to \$2.6 billion by 2015**. This will be primarily due to the launch of numerous generic anxiety drugs from 2006 onwards, which will offset the revenue growth derived from existing drugs and affect the launch of several novel anxiety treatments. In order to maximize drug revenues as a final step before generic incursion, a number of existing market players are seeking approvals in additional anxiety indications. These include Forest/Lundbeck's **Lexapro**, Wyeth's **Effexor XR** and Glaxo's **Paxil CR**. However, manufacturers with existing branded CNS drugs, already approved for indications such as depression, epilepsy and neuropathic pain, are also seeking secondary indications in the anxiety market. It is forecast that these re-branded drugs will reach peak anxiety specific revenues of \$413m in 2013, further **depleting sales of traditional anxiety therapies**. On the plus side however, the greater variety and range of treatments means physicians are likely to use broader combinations of drugs to treat a single patient.

* Scientists observed that In the last two decades, there has been a significant increase in the age-adjusted incidence of **atrial fibrillation** in the Western world from the 1980s to 2000. Atrial fibrillation and its associated complications significantly decrease longevity and quality of life. From a public health perspective, it exacts a major toll on a coutry´s healthcare resources. If atrial fibrillation rates hold steady, then roughly **12.1 million people** will have the arrhythmia by **2050**. If the current trend persists, nearly **16 million cases of the disease will occur.**

* Remarkable **advances in biomedical research** have greatly improved our understanding of genetic, molecular, cellular, and neural underpinnings of physiology and health. For instance, animal studies have demonstrated that maternal care can result in permanent alterations of the expression of behavioral, cognitive, and endocrine stress responses. The availability of whole-genome data makes possible the examination of the links between environmental changes, behavior, physiological responses, and gene expression across the lifespan. **Translating breakthroughs** in genetics, neural circuitry, biomarkers, and neurotransmitters to improve people's health will require increasingly sophisticated and precise behavioral and social methods, measures, and constructs. With such advancements, behavioral and social scientists may be able to address exciting new questions, many of which are described in an ACD Report *(Report of the Working Group of the NIH Advisory Committee to the Director (ACD) on Research Opportunities in the Basic Behavioral and Social Sciences. December 2, 2004)* on research opportunities in the **basic behavioral and social sciences**:

• **Social integration and social capital:** How have advances in technology and mobility affected neighborhood social networks and mechanisms such as resilience and connectedness? What is the impact of these advances on health behaviors and health outcomes?

• **Work-related stresses:** What are the effects of conflicts between work and family associated with women entering the workforce on social stress and health?

• **Gene-Environment interactions:** How are genetic traits and early life experiences linked to physical and emotional health later in life?

• **Biosocial stress markers:** What are the neurological effects of stress, and how do they relate to long-term cognitive and affective reactions? How can these findings be used to understand group behavior in the context of threats such as natural or man-made disasters?

• **Technology, Measurement and Methodology:** How can we apply advancements in biomarker data collection and other technologies to track behavior in real time (e.g., ecological momentary assessment, personal sensors, geo-spatial coding methods) to decipher multi-level pathways linking biology, behavior, environment, and society?

• **Intergenerational transmission of behavior:** How is gene expression related to inter- and trans-generational transmission of behavior and emotion? Conversely, what impact does the transmission of behavior patterns have on DNA?

• **Complex adaptive systems:** How can our growing understanding of complex adaptive systems be used to better understand the process of decision making in health at the personal and systems levels?
(Source: OBSSR Prospectus - DRAFT Page 9, 2006)

interaction as project-by-project, the new breed of CRO should offer a diverse range of services and focus on providing solutions rather than performing a specific type of study[129].

V.22. Emerging nations: "if you can't beat them, join them"

On a global scale, emerging nations catch up, announcing even more challenging times for the established, mostly Anglo-American pharmaceutical companies.

Table 5.24. Catching up: pushing China and India towards global leadership

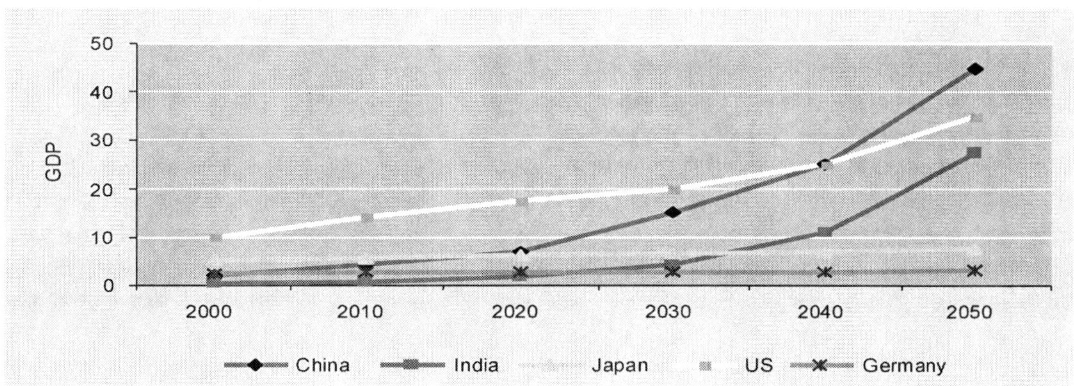

The Chinese government decided for China to play a global role in pharmaceuticals by 2020. To help facilitate WTO compliance, the State's drug administration (SDA) brought into effect the new drug registration procedures, Dec.2002. The record number of 6.300 applications for the approval of clinical trials, manufacturing, line extensions and generic drugs in the sole year of 2002, points to the dynamism of the Chinese pharma market. For the West, China and India will become pharma forces to reckon with.

Whereas India and China literally "push" their industry to worldwide leadership, Europe and Japan seem to take the opposite route. In both these regions, industry   sees itself as the easy primary target for ill-advised Government measures related to the financing of the Statutory Health System.

Table 5. 25. New pharma markets

---

[129] Nigel Brown, Ph.D., "Capitalizing on a Changing Research and Development Environment," Business Briefing – PharmaTech 2003.

China and India's patients will have more income to put toward managing their healthcare. [Goldman Sachs, 2003].

Many multinational pharmaceutical companies are expanding their presence in China because they recognize the significant long-term business opportunities that market presents. Pharmaceutical sales in China grew 20.4 per cent to $11.7 billion in 2005, representing the third consecutive year that market has achieved 20+ per cent growth. Industry experts estimate that China will be the world's fifth largest pharmaceutical market by 2010.

2006, about one hundred contract research organizations are operating in China, but within five years this sector of the pharmaceutical industry will offer fierce competition to its counterparts in other Asian countries like India. BigPharma and biotechnology companies have overcome concerns about data security and intellectual property protection in China, traditionally a barrier to outsourcing this type of work, but now outweighed by the substantially lower costs. Researchers often work with chemical libraries to find potentially useful compounds, but long before a drug reaches the stage where trials test its safety and efficacy in humans, the compounds are usually tested on animals. Meanwhile, the new drugs are constantly being modified, for example, to find the correct dosage, is carried out in China. It would cost a US company $250,000-$300,000 annually to employ a full-time Ph.D. level chemist, whereas for a similar level chemist in work in China averages about half that of the USA.[130]

V.23. Growth in number and scope of clinical drug trials in China

A report[131] by Kline & Co indicates that, while most of the world's leading drug makers are already including China in their research plans to an extent, the Chinese are making systemic improvements that will encourage future growth in the number and scope of clinical drug trials and further enhance the nation's standing as a world hotbed of research activity. As China improves its clinical resources and regulatory processes, it's becoming an increasingly competitive market for major clinical trials.

At the end of 2004, there were already more than 250 trials in progress, all sponsored by multinational companies - an increase of about 25% since 2002. Leading the way are groups such as Swiss drug major Roche and UK-based AstraZeneca, which are now operating their own clinical trial centers, attracted by the expansive 1.3 billion population, as well as the potential for rapid patient recruiting, the report notes.

V.24. India: from the sidelines to the playground

According to the Associated Chambers of Commerce and Industry of India, the country's pharmaceutical industry is expected to grow around 11% to reach a value of 600.00 billion rupees ($13.64 billion) in the year to March 31, 2008This will be due to increased exports of generic drugs to regulated markets in the USA and Europe. Patent expiries on several branded drugs in developing countries in the coming years, coupled with low production costs, will provide adequate opportunities for Indian drug makers to capture a larger share of the US and European markets. Indian companies are expected

---

[130] *China "will compete with other Asian nations in CRO markets within five years". Pharma Marketletter, April 24, 2006*
[131] *Clinical drug research in China "booming," says Kline report. Pharma Marketletter, December 6, 2005*

* The ability to create a complete new organism with a whole set of cells – known as stem cell **totipotency** and best exemplified in humans as a fertilized egg – was recognized in plants as long ago as **1902** when scientists suggested **individual plant cells could give rise to a complete plant.**Key similarities between plant and animal stem cells are only now being identified, with recent discoveries including **specialized organ location**, and **neighboring cells** that **release signals required** for **stem cell regulation. Similarities** in **gene circuitries** regulating stem cells have also been identified.
One of the advantages of working with plant cells, apart from reducing ethical concerns, is that they are largely **"uncommitted"** – meaning they have the potential to change and adapt to areas other than where they originated. In human stem cell University of Melbourne biology experts are trying to find ways to introduce stem cells from somewhere else to repair damaged tissues. Plant tissues and organs show remarkable regenerative ability that is not observed in animal systems and if scientists can better understand the process involved in **plant stem cell regrowth** they can utilize that knowledge in human research.

* Tiny molecular rings with big potential: Ohio State University chemists have devised a new way to create tiny molecular rings that could one day function as drug delivery devices or antibiotics. The rings are made from **polymers** -- large molecules that are made up of many smaller molecules -- and the chemical reaction that creates them is similar to others that create polymer chains. But this new reaction solely makes rings, ones **tailored** to **perform specific functions**. The chemists reported constructing polymer rings of a specific size and binding them to charged sodium atoms -- a first step in a long road that could lead to applications in medicine. These rings could encapsulate certain molecules, transport them somewhere, and release them at a specific time. The technique may eventually be used in drug design. The kind of ring molecules grown in this study, known as **depsipeptides**, are similar to some natural compounds produced by microorganisms that are employed as antibiotics, such as valinomycin. Scientists are also studying depsipeptides as possible anti-cancer agents. There could be potential for future applications in medicine, because these molecules can be varied to perform specific functions.

* Using advanced new microscopy techniques in concert with sophisticated transgenic technologies, scientists at The **Wistar Institute** have for the first time created three-dimensional, time-lapse movies **showing immune cells targeting cancer cells in live tumor tissues**. In recorded experiments, immune cells called T cells can be seen actively migrating though tissues, making direct contact with tumor cells, and killing them. Insights from this new view of the body's on-board defenses against cancer may open the way for **improved immunotherapies** to treat the disease.

* **Dependence on nicotine** drives many of the most preventable causes of death in the Westernized world and is a worldwide health problem.
The unassuming C. elegans nematode worm, a 1-millimeter workhorse of the genetics lab, is quite similar to human beings in its genetic susceptibility to nicotine dependence, according to University of Michigan researchers. This finding should allow researchers to better understand how nicotine dependence works, and perhaps devise new ways to block the craving that keeps humans smoking cigarettes. Nicotine is the addictive substance in tobacco.
The research team found that the genes known to underlie nicotine dependence in mammals are also present in the worms. Having established worms as a model, the Xu team then tried to identify new genes important for nicotine dependence. They found for the first time that TRP channel genes which enable cells to respond to various external stimuli are a part of the nicotine response.
In fact, when they knocked the TRP gene out of worms, the animals no longer responded to nicotine exposures. But when a new generation of worms had that missing gene replaced by a human version of the TRP gene, the worms returned to being nicotine-sensitive. This demonstrates that **human TRP genes have the capacity to mediate nicotine dependence**, suggesting that human TRP genes are important for nicotine dependence in humans. It also makes TRP genes a potential target for the development of drugs to treat tobacco addiction, and the worms can help in that research. C. elegans can also be used to find other unknown genes critical for nicotine dependence.
* Small sections of human liver tissue have been created from umbilical cord stem cells, by using a microgravity bioreactor which creates a weightless environment. According to **Prof. McGuckin** of Newcastle University, UK, this technology could eventually be used to **grow small livers** that could be used for drug tests - doing away with the need for human volunteers to take risks. Early 2006 six volunteers became dangerously ill during a drug trial in the UK. The scientists warned that it will be tens of years before we are anywhere near producing whole new livers for transplants. However, within the next 15 years, tiny livers could be produced and used for treating patients.
* **Road-traffic accidents** are the **leading cause** of **death** by injury and the tenth-leading cause of all deaths globally, thus making up a significant portion of the **global burden** of **ill-health**. An estimated 1·2 million people are killed in road crashes each year, and as many as 50 million people are injured.*[Dinesh Mohan. Road Safety in Less-Motorized Environments: Future Concerns. In J Epidemiol 2002;31: 527–32].*
It is predicted that if the trends continue, road traffic injuries will be the **third-leading contributor** to the worldwide burden of disease by **2020**. *[Murray CJL, Lopez AD, eds. The Global Burden of Disease: A Comprehensive Assessment of Mortality and Disability from Diseases, Injuries, and Risk Factors in 1990 and Projected in 2020. Boston: Harvard School of Public Health, 1996].*

More cancers will be detected prior to metastatic spread. Treatment will start the same week as diagnosis. Surrogate points will be essential to register new medicines. Patients will have. Patients will face high co-payments for new drugs.

to grab around 30% of the increasing generic market in the pharma sector worldover[132].

Table 5.26. Collaboration on a global scale

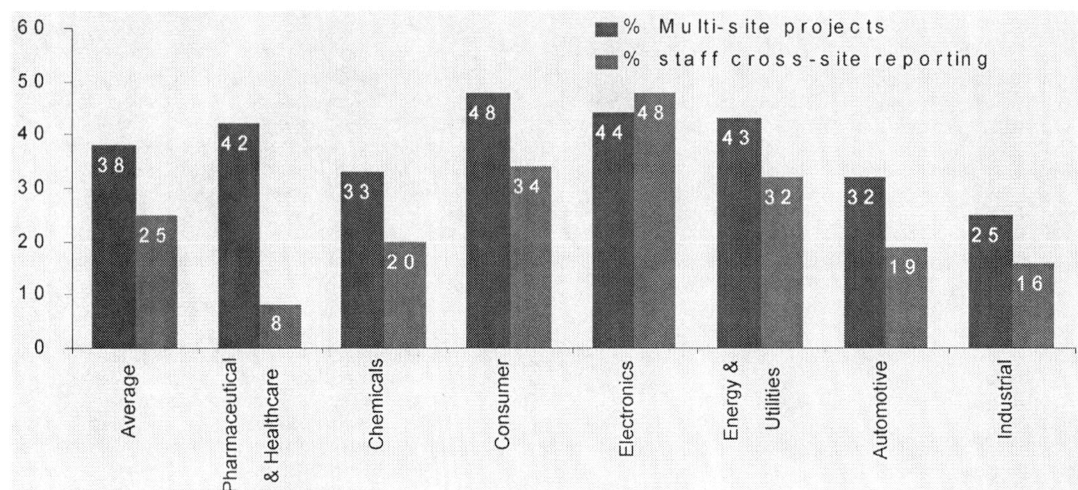

Pharmaceutical companies have a relatively high level of dispersed projects but an extremely low level of cross-reporting.

R&D sites are increasingly moving east, and about 77 percent of planned sites from 2006 to 2009 are concentrated in China and India, according to a report by Booz Hamilton[133]. In 1990, just 3,4 percent of foreign R&D sites were located in China or India, but that number grew to 13,9 percent in 2004, and is projected to grow to 31 percent by the end of 2007. The pharmaceutical industry is likely to increase its R&D in China and India, as the conditions to do so are improving rapidly.

V.25. Research and Development Investment in India

According to management consultants McKinsey, the market for pharmaceutical product clinical trials in India will reach $1.5 billion by 2010. For scientists, the advantage of doing medical research in India for scientists is twofold. There is the low cost of the studies, undeniably attracting researchers . And there is the lack of health care provision, resulting in a pool of patients who have not received any medication before, the clinical value is increased, because there is no chance of a pre-existing treatment reacting to the trial remedy.

The biggest obstruction to more R&D investment in India has been the perception that India's Intellectual Property standards were too lax. Both Japan and Mexico saw USA-based drugmakers expand their R&D in both countries over 200% after those nations brought their patent laws into line with the USA. It is not enough, however, to pass the necessary legislation: the World Trade Organization's TRIPS Agreement, which covers intellectual property issues and provides a global framework for IP law, creating a solid base. What is often lacking is enforcement of the TRIPS rules. India's patent laws were updated in 2002, with the drug industry welcoming the extension of patents from seven to 20 years. However, it also leaves room for "compulsory licensing," which provides a level of uncertainty that is likely to discourage long-term private R&D. The time has

---

[132] *Indian Rx sector "set to grow 11% by 2007-8". Pharma Marketletter, October 3, 2005*

[133] *Pharmaceutical Executive, June 2006:36*

* Advances in the **computer sciences** and **information technology** fields have spawned several **methodological advances** in the **biological** and **molecular sciences** (e.g., **DNA chip technology** and **microarray analysis**), enabled quantum leaps in molecular and submolecular medicine, and catalyzed the emergence of **whole new fields of study** such as **proteomics, phenomics, nutrigenomics, and pharmacogenetics**. Perhaps, in like manner, with the emergence of **eHealth,** the behavioral and population sciences may be on the verge of a similar information technology–based scientific revolution. New eHealth solutions may soon permit the real-time integrative utilization of vast amounts of behavioral-, biological-, and community-level information in ways not previously possible. **Behavioral algorithms** and **decision support tools** for scientists could facilitate the analysis and interpretation of population level data to enable the development of "community (population) arrays" or community-wide risk profiles, which in turn could form the foundation of a new "**populomics.**" This population-level risk characterization could potentially go beyond the limitations of typical geographic analyses and yield insights distinctly different from risk stratification based on current methodologies. *[Gibbons, M.C. (2005). A Historical Overview of Health Disparities and the Potential of eHealth Solutions. J Med Internet Res, 7(5), e50].*

* **Cellectis** SA, founded in 2000 as a spin-off from the **Institut Pasteur**, is the world leader in applying the technology of **Meganuclease Recombination Systems** to **in vivo genome engineering and genome surgery.** The company focuses on the R&D of custom-made Meganucleases for in vivo DNA interventions and also provides new tools for rational reverse genetics and targeted recombination. Cellectis develops Meganucleases that can **induce unique site-directed double-strand breaks in a living cell**, and can be used for biotechnological and therapeutical applications. The company is focusing on bringing "**genome surgery**" to clinic as a genuine new molecular medicine approach.

* A protein with the ironic name "**Srcasm**" can **counteract** the **effects** of **tumor-promoting molecules** in **skin cells**, according to new research by investigators at the **University of Pennsylvania** School of Medicine. Using animal models, the researchers discovered that Srcasm acts like a brake in epithelial cells, preventing uncontrolled cell growth caused by a family of proteins called **Src kinases**. This finding suggests a target for future gene therapy to treat skin, head, neck, colon, and breast cancers.

* Research at the University of Texas Southwestern Medical Center focused on two highly related CNS-enriched transcription factors designated neuronal PAS domain proteins 1 (**NPAS1**) and NPAS3. The genes encoding these proteins are orthologous to the *Trachealess* gene of fruit flies.
Evidence indicates that mice **deficient in the NPAS1 and NPAS3 transcription factors display behavioral deficits reminiscent of psychosis**. These observations are bolstered by recent human genetic studies of a family suffering schizophrenia wherein affected family members have been found to carry **an inactivating mutation in the human *NPAS3* gene.** Studies from the McKnight lab provide evidence that the pathway regulated by the NPAS3 transcription factor is essential for **neurogenesis** emanating from the dentate gyrus of the adult mouse brain. This pathway, as in the case of the pathway regulating branching morphogenesis of the tracheal system in fruit fly larvae, involves **fibroblast growth factor** (FGF), the **FGF receptor**, and **Sprouty** - an intracellular inhibitor of FGF signaling. Additionally, biochemical studies of the mammalian Sprouty proteins have provided evidence that this protein forms a very large, virus-like particle, having the properties of a miniature battery. The biological utility of "**nanobatteries**" formed by the Sprouty and related Spred proteins is under scrutiny. *(McKnight SL, Science's Next Big Idea. ."Curiosities about Growth Factor Signaling in the Adult Brain" Presentation given at Duke University, September 26, 2006)*

* **Cancer Research UK** has an impressive record of developing novel treatments for cancer. Since 1982, their Drug Development Office has taken over **100 potential new anti-cancer agents** into clinical trials in patients, **four of which have made it to market** and many others are still in development. These include **temozolomide**, a drug discovered by Cancer Research UK scientists, that is an effective new treatment for brain cancer. The rate of success is comparable to that of any pharmaceutical company.

* Researchers at the **Stanford** University School of Medicine revealed that tumors that are hypoxic - low in oxygen - make a protein called lysyl oxidase that helps the tumor spread to other organs. **Lysyl oxidase**, or LOX, could be a good target for future cancer therapies. **Hypoxia** is caused when the supply of oxygen from the bloodstream fails to meet demand from body tissues, such tumors. Hypoxic tumors can be found in many parts of the body. A therapy that specifically treats tumors producing LOX would be particularly exciting given that these are often among the deadliest cancers.

New technologies and scientific discoveries are changing our fundamental assumptions. Robust findings are mounting with evidence of how biology, behavior, and the social and physical environment are dynamically intertwined in the ways they produce better health or premature disease, disability, and death. The emerging view is that differences in patterns of health and disease represent the embodiment of a dynamic interaction of genes and environment over time. **Two previously separate, often competing world-views about health and illness may finally be converging: 1) the biomedical view of causation, and 2) the public health or socio-behavioral-ecological view of causation.** The biological "causes" and the socio-behavioral-ecological "causes of the causes" may really be two sides of the same coin. The solutions to some of our biggest health challenges may depend on whether scientists from the biomedical and the behavioral and social science disciplines are able to learn each others' languages, listen across the gulfs that separate their sciences, and forge a new conceptual synthesis across their disciplinary boundaries. *(OBSSR 2006 Prospectus - DRAFT Page 9)*

V.26. Adopting a spirit of change

Major overhauls transforming businesses and culture now happen on a daily basis. The afterglow of today's success fades much faster than it did 20 years ago. In the current decade, companies will be in a race for survival to win patents for gene-derived drugs. And they will be under constant pressure from an unsentimental stock market

The need to initiate change is a priority which many executives find difficult to adopt. Change is disturbing indeed. Plato feared the losses which writing would bring to the memories and speech of those exposed to it. Victorian parents were alarmed by the passivity of children who read.

It is only when the stress level goes way up and when unease builds to the point where it can no longer be ignored that executives feel forced to confront the real issues and surrender to attitudinal changes. Some values were never important; they just assumed that they were.

There is little confusion over the fundamental phenomenon driving the healthcare industry's transformation : far-reaching global changes in the nature of health care delivery. Just a decade ago, success resulted largely from innovative new products characterized by superior efficacy, safety, convenience. Focusing on these essential characteristics is not enough anymore. Evolving market requirements force leading industry players to think in terms of total disease management. Healthcare has become a business where "added value" for the patients now plays the decisive differential role.

In these times of unprecedented change we experience a shift from medicine to *total disease management*.

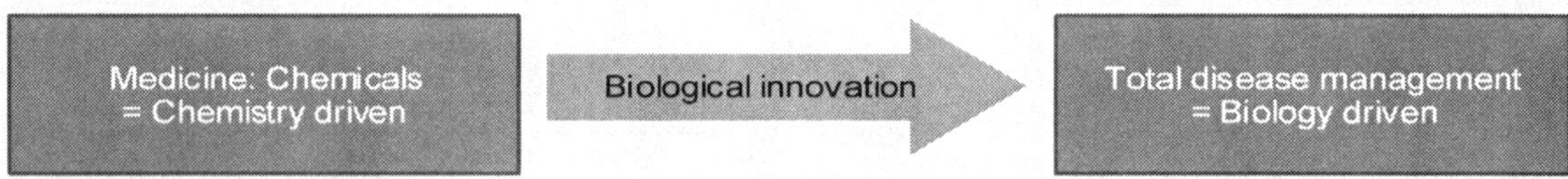

Products need to offer value-adding solutions to broader issues of disease management by improving medical outcomes, reducing costs and improving the quality of patient's lives.

Table 5.27.   Improving the quality of patients' lives

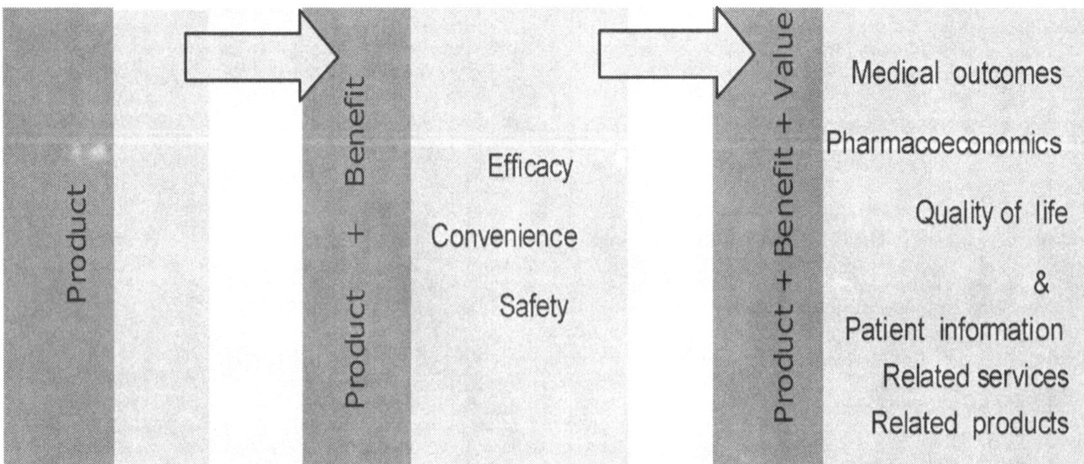

* **Allergy and immunology** is on the threshold of major breakthroughs based on a 40-year history of increasing our fundamental understanding of how the immune system and the allergic reaction part of the immune system works. Scientists at the American College of Allergy, Asthma and Immunology think they are in the early stages of a **revolution in the treatment of allergic disease.** Allergen immunotherapy, a widely used approach, aims to **modulate** an allergic patient's immune response **through administration** of **increasing doses of an extract** which will **eventually reduce or eliminate** the patient's **symptoms**.

Although conventional subcutaneous immunotherapy has been shown to be effective for the treatment of allergic rhinitis, asthma and venom sensitivity, **considerable research** is being **devoted** to the **development** of **therapeutic vaccines** for the **treatment** of **allergic diseases** to **improve efficacy, lessen** the **time** to **achieve effectiveness, reduce** the **inconvenience** and **enhance safety**.

Allergic Asthma: There is a great need for novel therapies, especially in the **10** percent of **severe asthmatics** who account for approximately 50 percent of the health care costs of asthma.. Although novel therapies frequently show promise in pre-clinical studies, only **8 percent of these therapies** which are tested in human subjects prove to be safe and effective in humans with disease. The ability to find novel therapies that are safe and effective is complicated by the fact that **asthma is not a single gene/protein/mediator disorder,** and there are **over 100 genes linked to asthma.** Some novel therapies already in use include **leukotriene inhibitors** and **anti-IgE therapy**. Researchers are also studying the efficacy of promising therapeutics such as antibiotics, probiotics and **TLR** (Toll Like Receptors) **ligands**, **cytokines** (Anti-TNF, Anti-IL-5), and **transcription factors**.

Food Allergy: Researchers are also looking at novel approaches for treating food allergy, which is a major health problem in industrialized nations. It affects **between 6 percent - 8 percent** of **young children** and **4** percent of **adults**. Current management of food allergy includes the avoidance of specific foods and the medical management of acute reactions.

Only a few foods, such as milk, eggs, peanuts, nuts, and fish and shellfish **account for over 90 percent of all food allergic reactions**. Attempts at primary prevention have largely not met with much success.

There are several promising studies being conducted now that likely will result in new treatments for food allergy. One of the new therapies are on the horizon is **Anti-IgE**, which may be able to prevent severe food allergic reactions and can also be used in combination with other new therapies for treatment. Another is the use of **modified allergenic proteins** that may be able to "reverse" food allergy. Routes, other than subcutaneous, for the delivery of allergy immunotherapy for food allergy are being studied extensively. Other Novel Therapies for Allergic Diseases: A number of studies are evaluating the safety and efficacy of non-traditional forms of **immunomodulation**, including sublingual immunotherapy (SLIT) and oral immunotherapy (OIT). *(see next colon)*

But while allergen-specific **SLIT** and **OIT** seem to hold some promise for patients with milder types of allergic disease, there is still a need for blinded, placebo controlled studies of correct dosing, multi-allergen therapy, and in patients with more severe allergic diseases. Yet another novel approach currently under study is the use of an allergen vaccine with **immunostimulatory DNA**. The vaccine, which could be administered in a **few doses**, would have important therapeutic implications for millions of patients with poorly controlled allergic rhinitis and asthma. A novel immunostimulatory DNA vaccine called **AIC**, which is aimed at patients with allergies to ragweed, has already undergone preliminary testing in France, Canada, and the US. The results from clinical studies of patients with ragweed-induced allergic rhinitis demonstrate that AIC is less allergenic than conventional immunotherapeutic products and may offer the potential for an improved safety profile for immunotherapy.

*New findings, providing insight into the **fundamental behavior of the immune system**, also suggest that attention needs to be paid to **chemokines** and **chemokine receptor function** when **designing new vaccine strategies** and evaluating whether **drugs targeting chemokines** might have **unanticipated effects on immune function**. New research shows that when CD8+ T cells enter the lymph node, a combination of specific physical and chemical cues guides them to sites where they receive activation signals. Specifically, two molecules known as chemo-kines help guide them toward the cells that release these activation signals. The body contains hundreds of millions of CD8+ T cells, but only a tiny fraction of them become activated during an infection. These are selected because each CD8+ T cell carries a unique surface protein called a **T-cell receptor**, which **recognizes only specific antigens** (pieces of virus or bacteria that trigger the immune response). During an infection, CD8+ T cells that recognize antigens from the infecting pathogen are activated. These antigen-specific CD8+ T cells expand into a large population of active clones, which then sweep through the body, hunting down and killing infected cells. For CD8+ T-cell activation to occur in the lymph node, the cell must encounter its target antigen -- but that **antigen must be displayed on the surface of another immune system cell, called a dendritic cell.** Usually a **third type of cell**, known as a "**helper" T cell, must be involved** as well. Naïve **CD8+ T cells** do not wander aimlessly through the lymph node but instead **are steered towards areas in which dendritic cells concentrate**. Moreover, CD8+ T cells are **chemically attracted to the cells that might activate them** by the chemokines these other cells produce. When dendritic cells interact in specific fashion with helper T cells, the **activated cell pair releases** the **chemokines CCL3 and CCL4**. It is the **combination of these two chemokines** that the **CD8+ T cells receive best** as a **signal**. By interfering with the action of these chemokines, the scientists demonstrated that CD8+ T cells lost their ability to home in on the dendritic cells interacting with the helper T cells. The result was a marked impairment of memory cell generation.

Healthcare information overflow created by the Internet, and the disappearing „paternalistic healthcare" approach of the late 20th century, will call for new, objective patient information guiding systems.

Improving drug delivery systems can hugely contribute to patient convenience and indirectly to a better quality of life.

Drug delivery is a method that allows a substance to be "delivered" in a form that is convenient to the patient with high efficacy at a specified rate and create relatively few and/or limited side-effects. The history of drug delivery extends as far back as the early 1950's, when "Dexedrine Spansule" (used for weight-loss) was first introduced and again in 1960 when "Contact", a long acting multi symptom slow-release cold medicament premiered[135]. Drug Delivery systems offer the opportunity to extend the viability of many drugs coming off patent [136]. In practice, the majority of the drug products are administered through parenteral, oral, and transdermal routes. The high effectiveness of drug delivery technology can expand the marketing and profitability index through better life-cycle management, attrition rate reduction, increased drug efficacy, reduced side-effects and improved patient compliance. A well known example of life-cycle management using drug delivery technology is Adalat, developed by Bayer. It has been increasing revenue for more than two decades, after being reformulated several times i.e., multi dose, sustained-release, extended-release, etc.[137] Another good example which helped Pfizer in the early 90's propel to its current lead in the pharmaceutical industry is Procardia XL.

Drugs are more likely to be delivered orally as it is one of the most inexpensive and convenient forms of delivery, whereas proteins and peptides have been restricted to delivery by injection. As such, biotechnology drugs will reach the consumer market only if they can be matched with an appropriate means of transportation. A paramount challenge for drug delivery is the barrier that prevents the effective delivery of macromolecules by means other than subcutaneously.

V.27. Drug Delivery Drivers

According to Jim Papanikolaw in "Seeking Growth in Drug Alternative Drug Delivery Systems" an increase in the demand for cost-effective formulations with ease of use will propel the market for alternate drug delivery systems. Patent expiration, cost containment,; patient convenience and compliance with drug regimen, efficacy and safety considerations[138] are the primary drivers for advancement of drug delivery systems in drug products.

• Reformulation / 2nd Indication	• Improved efficacy	• Macromolecule delivery	• Reduced side effects
• Toxicity	• Selective targeting	• Generic drug makers	• Increased patient compliance
• Insolubility			

V.28. Efficacy and Safety

More than fifty percent (50%) of drugs in clinical development typically fail because of toxic effects, insolubility, and miscellaneous safety issues. Drug delivery systems can help reduce this attrition rate because they may provide the mechanisms to manage

[135] Starr, C. "Innovations in Drug Delivery". Patient Care. 15 Jan 2000. Find Articles. 30 Aug 2002 <http://www.findarticles.com/cf_0/m3233/1_34/59247339/p1/article.jhtml?term=Innovations+in+Drug+Delivery>.

[136] Moore, Samuel K. "Drug Delivery Extends Pharmaceutical Reach: Firms Add Technology." Chemical Week. 21 Apr 1999. Find Articles.30Aug2002 <http://www.findarticles.com/cf_0/m3066/16_161/54427290/p1/article.jhtml?term=Drug+Delivery+Extends+Pharmaceutical+Reach>.

[137] Norman, P. "How Delivery Technologies Extend Drug Life Cycles : Case Studies". Spectrum Life Sciences. 8 Feb 2002:1-11.

[138] Moore, Samuel K. "Drug Delivery Extends Pharmaceutical Reach: Firms Add Technology." Chemical Week. 21 Apr 1999.

* Roswell Park Cancer Institute (**RPCI**) researchers played a leading role in clinical studies which led to U.S. Food and Drug Administration (FDA) approval of a new treatment for patients diagnosed with **multiple myeloma** – the second most common blood cancer. Clinical scientists led by **Dr. Asher Chanan-Khan** contributed research which led to the approval of the oral medication, lenalidomide (Revlimid), in combination with dexamethasone, for the treatment for patients with multiple myeloma who have received at least one prior therapy. **Lenalidomide** belongs to a new class of oral cancer drugs that are chemically similar to thalidomide but appear to be more potent and lack some of the more common side effects of thalidomide.

* An international team led by neurologist Ira **Shoulson** of the University of Rochester Medical Center is trying to identify the earliest signs of the onset of **Huntington´s Disease**. The information will help clinicians design better studies of new drugs aimed at alleviating or postponing illness. It also helps researchers understand how patients evaluate potentially life-changing knowledge now available to patients through means such as genetic testing. While the gene that causes the disease is known and can be identified through a blood test, fewer than one in 10 adults at risk for developing the disease have chosen to be tested. People at risk but who have not taken the test have a 50/50 chance of developing the disease. This at-risk group offers physicians a unique opportunity to witness the earliest signs of the disease, before anyone knows whether a person actually has the gene for Huntington's or not. In the **PHAROS** study, one of the largest Huntington's studies ever undertaken, 1,001 healthy people between the ages of 26 to 55 who had at least one parent with the disease have stepped forward to participate. Patients, doctors and nurses from 43 hospitals and medical centers around North America are taking part. **Huntington's** is an **inherited disorder** that affects about **30,000** people in North America; another **150,000** people or so may have the **gene** that causes the disease. The defective gene, a sort of genetic stutter, leads to the destruction of brain cells, causing involuntary movements, cognitive problems, and often psychological problems like depression and paranoia. The disease usually strikes in young- to mid-adulthood, in a patient's 30s or 40s. The study is relevant for many physicians and patients who have **increasing access** to **information** about the **specific genetic causes** of **many diseases**, even for conditions like Huntington's for which there is no cure or approved treatment. The issues these patients face portend concerns that nearly everyone will confront eventually, as researchers ferret out the **genes** that **predispose** some people to **heart trouble, high blood pressure, Alzheimer's** and **Parkinson's** diseases, **stroke**, and **other diseases**. People at risk for Huntington's disease today present a unique opportunity for researchers to examine the **psychosocial, ethical**, and **practical issues** faced by **people** who have chosen **not to** know whether they will develop a serious disease or not.

* **Tissue engineering** is an emerging biotechnology sector at the interface between medicine, cellular/molecular biology, materials science and engineering. It aims at developing cell or tissue-based products to replace, repair or regenerate human tissues. TEPs are manufactured on the basis of cells of human (human TEPs) or animal (xenogeneic TEPs) origin, before being implanted in, applied on or administered to the patient (usually *via* a surgical operation). Current applications of this nascent field of "**regenerative medicine**" include treatment for skin, cartilage and bone diseases or injuries. More complex products – such as heart valves, blood vessels or heart muscle tissue – are currently in pre-clinical and clinical development.

From a regulatory and scientific viewpoint, TEPs lie within a spectrum of **advanced therapies**, which includes other cell-based therapies like gene or somatic cell therapy, with which they share a number of scientific and economic features:

– They are based on complex, innovative manufacturing processes aiming at modifying genetic, physiological or structural properties of cells and tissues. The specificity of the product precisely lies *in* the process.
– Regulatory and scientific expertise for the evaluation of advanced therapies, although increasing, remains limited;
– Traceability from the donor to the patient, long-term patient follow-up and a thorough risk management strategy are crucial aspects to be addressed when evaluating advanced therapies.
– Advanced therapy products are usually developed by innovative small and mediumsized enterprises (SMEs), highly-specialized divisions of larger operators in the Life Science industry (biotechnology, medical devices), hospitals or tissue banks. They are subject to rapid and often radical **innovation**.

One of the longer term perspectives is to be able to regenerate full organs and hence to tackle the issue of donor shortages.

. * The **Microscale Life Sciences Center's** focus is on use of **microscale technology innovation** to solve mysteries about **cell growth** and **death**, answers that will reveal crucial knowledge about fatal diseases. It is developing miniature automated systems designed to rapidly detect and analyze the differences between healthy and diseased body cells to better understand the nature of disease processes. Researchers are aiming to develop **micromodules** with a variety of **optical** and **electronic sensors** to **measure** multiple parameters within single cells. The modules enable live-cell measurements of **physiological** parameters in concert with **genomic** parameters. They also enable measurement of **DNA, RNA** and **proteins inside cells**. The micromodules will be fully automated, allowing precise placement of single cells and controlled experiments on thousands of cells per module that can last hours to days at a time. Data obtained from the experiments will be used to **map out biochemical pathways** in cells and **correlate** it with information on the behavior of mechanisms that determine the progression of disease.

Advanced therapies herald new forms of medical treatment. Tissue engineering, through its focus on the **regeneration** of **human tissues**, is expected to have a major impact on future medical practice.

insolubility and toxicity. Another reason that drug delivery technologies are evolving is the need to be able to sustain "therapeutic" drug levels for extended periods with selective targeting.   This need has arisen due to the limitations and a lack of efficacy with oral drugs.   According to "Innovations in Drug Delivery" the potential advantages of using smaller amounts of active agents by better targeting drugs to diseased organs, fewer doses, less invasive administration modes combined with prolonged circulation of quick acting compounds, particularly "peptides and proteins" will lead to an increase in demand for drug delivery systems[139].

Drug Delivery Technologies

There are at the moment hundreds of delivery technologies in development and/or undergoing clinical trials. A widespread rule to adhere to in developing a drug delivery system is that it be target specific, require less frequent dosing and yield enhanced safety and cost benefits. Not one delivery technology is capable of serving as the only delivery mechanism due to the wide variety and quantity of drugs available today.

V.29. New drug discovery technologies impacting on the industry.
99 % of the genome of each human being seems to be identical. It is only in the remaining 1 % that the enormous diversity of our species is derived. Some of these differences influence physical appearance and are therefore obvious; other differences are much more subtle.

With over 100,000 genes, each being a potential target, the situation in pharmacological research has changed – almost overnight – from one with a dearth of targets to one that is awash with targets. Screening these quickly for relevance has become the main challenge: The DNA probe and biochip companies are counting on a booming trade. The use of biochips by the pharmaceutical companies is expected to increase the sales of the former from $100m in 1998 to $3bn in 200.- The market for DNA probes could increase from $500m to $3bn.

Thus, very rapidly, new technologies such as gene therapy to connect pathological susceptibilities in the genetic build-up, anti-sense technology to block translation of undesirable proteins and DNA-vaccination are – in one form or another – finding their way into healthcare. Their applications are similar to classical vaccinations, an area where the "classical pharmaceutical" giants are not particularly expert.

The technological challenge, i.e. the delivery of the DNA fragment is largely unresolved and again not within the "big pharmaceutical" R&D expertise. They have to deal with start-up high-tech companies in this area.

A number of drug technologies that are expected to change the face of healthcare, and which are not or have not been the natural expertise of the pharmaceutical giants, have been identified as follows :

---

[139] *Starr, C. "Innovations in Drug Delivery". Patient Care. 15 Jan 2000. Find Articles. 30 Aug 2002 <http://www.findarticles.com/cf_0/m3233/1_34/59247339/p1/article.jhtml?term=Innovations+in+Drug+Delivery>.*

B.64. Burden of disease and deaths from Noncommunicable Diseases in the WHO European Region, by cause

Group of causes (selected leading NCD)	Disease burden (DALYs) (000's)	%of all causes	Deaths (000's)	%of all causes
Cardiovascular	34421	23	5067	52
Neuropsychiatric	29370	20	264	3
Cancer (malignant neoplasms)	17025	11	1855	19
Digestive diseases	7117	5	391	4
Respiratory	6835	5	420	4
Sense organ disorders	6339	4	0	0
Musculoskeletal	5745	4	26	0
Diabetes mellitus	2319	2	0	2
Oral conditions	1018	1	0	2
All NCSs	115339	77	8210	86
All causes	150322		9564	

[Preventing chronic diseases: A vital investment. Geneva, World Health Organization, 2005]

Almost 60% of the disease burden in Europe, as measured by DALYs, is accounted for by seven leading risk factors: high blood pressure (12.8%); tobacco (12.3%); alcohol (10.1%); high blood cholesterol (8.7%); overweight (7.8%); low fruit and vegetable intake (4.4%;) and physical inactivity(3.5%).
* Using comparable methods and a common currency (the **DALY**) for health outcomes, the WHO provides a projection of disease attributable to 26 major risk factors. Total **tobacco-attributable deaths** are projected to rise from 5.4 million in 2005 to 6.4 million in 2015 and 8.3 million in 2030. Tobacco is projected to kill 50% more people in 2015 than HIV/AIDS, and to be responsible for **10% of all deaths globally**. Global **cancer deaths** are projected to increase from 7.3 million in 2003 to 11.5 million in 2030, and global **cardiovascular deaths** from 17 million in 2003 to 23.3 million in 2030. The **three leading causes** of burden of **disease in 2030** are projected to include **HIV/AIDS, unipolar depressive disorders** and **ischaemic heart disease**. For some causes, **projected changes in disease burden over the next 30 years are likely to result in dramatic changes in global importance**. Lower respiratory tract infections and diarrhoeal diseases are expected to fall to 9th and 13th place in the global DALYs 'league' table, from 2nd and 5th place respectively in 2003. Malaria is also expected to decline in relative importance, as are congenital anomalies. Conversely, by 2030, HIV/AIDS is expected to be the leading global cause of DALYs, followed by depression, ischaemic heart disease and COPD

* **"Molecular" techniques** have become firmly entrenched in the diagnostic armamentarium of pathologists and clinicians, and are a **bridge to the future of medical diagnostics and therapeutics.** Molecular testing for the **complex genetic diseases can be expected by 2015**. The testing researchers are doing in 2006 is for **single gene defects** that are not very common diseases. For other diseases, like diabetes or atherosclerosis, which also have a genetic compo-nent, they just don't know all the genes yet. ..but are confident that within the near future, they will at least know a majority of them. And probably through some kind of **microarray technology**, they'll be able to do some kind of predictive testing for coronary artery disease, hypertension, diabetes, etc..

B.65. Technologies and their degree of complexity

Technology	Complexity	Technology	Complexity
Antisense	high	Molecular modelling	medium
Chiral chemistry	low		
Combinatorial chemistry	high	Peptidic drugs	low
		r-Abs	medium
Gene therapy	high	Ribozymes	high
Genomics	high	r-Proteins	low
		Transgenics	medium

* Both the **Spanish Nanomedicine Platform** (Nanomed Spain) and the **European Technology Platform on Nanomedicine coincide** in **emphasizing** that **nanotechnology** may have **direct applications** in **medicine** by contributing to improvements in health and quality of life, and moreover at a reduced cost. The Spanish, in addition, estimate that the advances currently being developed in this field will reach the health system in the **next fifteen years.** The two organizations coincide in three application lines for nanomedicine: **diagnosis, drug delivery** and **regenerative medicine**. Devices or sensors will be developed that allow a faster and more **reliable diagnosis** in **early stages** of the disease and at a low cost in most cases, for example, cancer.
Nanotechnological advances will allow the effective introduction of "**personalized therapy**", such as administration of a small dose of a drug to obtain maximum effect; the reduction of secondary effects by means of drugs that act only on the therapeutic target; or the development of new more effective administration routes. For cancer patients, drug delivery systems will be directed exclusively to cancerous cells, thereby **avoiding** the **secondary effects** of present chemotherapy protocols. Another field for the application of nanotechnology is **regenerative medicine**, in which it will allow the substitution of cells or tissues that do not work for equivalent ones that do; as well as tissue regeneration. In the case of cardiovascular diseases, nanotechnology offers the opportunity to develop **intelligent nanobiomaterials** for the regeneration of tissues, for example in the case of myocardial infarction.

Long distance accurate diagnosis: The medical doctor will no longer be the main repository of medical expertise for a sophisticated patient. The accelerating rate of change in medicine makes the "half-life" of knowledge too short. We are not far from the time when the most up-to-date information will be obtained from a database run by a computer expert or dedicated server and where the last missing link to semi-automated medical care is long-distance accurate systems for data storage and handling capabilities would already make this possible today. When it comes, the change-over will be very fast because the economic and quality considerations are overwhelming and the data transfer facility (the Internet) is already in place.

Table 5.28. Drug technologies

Technology	Complexity	Technology	Complexity	Technology	Complexity
Antisense	high	Peptidic drugs	low	Gene therapy	high
Chiral chemistry	low	r-Abs	medium	Genomics	high
Combinatorial chemistry	high	Ribozymes	high	Molecular modeling	medium
Transgenics	medium	r-proteins	low		

A common occurrence in the past has been that the larger pharmaceutical companies have failed to see the importance of the new technologies, or decided to wait until these proved themselves, before buying them[140]. At today's rate of technological development the larger companies are forced to outsource big chunks of their drug discovery process (hoping to increase the number of innovations and spread the risk), as well as to continuously license in new technologies, expertise, and capabilities (for many of them the aggressive implementation of this R&D model has turned out to come at a lower cost and higher quality than those developed in-house)[141].

V. 30. Customized medicine

Although it is difficult to make predictions, the following seem likely to play a part in the future:

- The drug discovery process becomes more automated and, at the same time, based on sounder knowledge.
- As the ease with which genome science advances, we will be able to measure each individual's genetic build-up right before or soon after birth to establish his/her peculiar susceptibilities. This opens the door for proactive intervention or correction of susceptibilities and the accurate prediction of which particular drug will be most suited in the future to treat such pathologies when, and if, they occur. So, from our present world where averaged-out information on a disease state and the action of a drug candidate on a large cohort of patients with individual variations on a disease „theme" are being used to identify a „broad-application" drug, we will find ourselves in a world where particular knowledge of the individual's build-up will allow the custom preparation of drugs. For example, if several analogues of a drug have been found to be active, it is likely that a specific one will be preferable for a defined subpopulation of patients, even though the particular drug may have been less desirable in the bulk of patients and would therefore have been discarded in favor of the more commonly applicable one.

Compounds once discarded because of adverse side effects in a minority of patients may once again be viable if it's possible to identify the small number at risk[142]. For example, Bayer's cholesterol lowering agent, which had to be taken off the market, or an earlier withdrawn product, Thalidomide. This also applies to some older drugs that have been replaced by new generations of compounds, but which will resurface because of their better efficacy in certain patient-types.

---

[140] *Neirinckx R. & Mertens G., M& A in the Pharmaceutical Industry, Strategic Healthcare Management Report, Financial Times Business Ltd, London, January 2000.*
[141] *Mertens G., Healthcare in Germany: Turning Crisis into Opportunity, Strategic Healthcare Management Report, Nicholas Hall & Co, Southend on Sea, June 2003.*
[142] *Fischman J, On Target: A new generation of drugs offers customized cures. US News &World Report, Jan 2003:51-58*

* **Emory University School of Medicine** scientists led by Prof. Scott **Devine** have identified and created a map of more than **400,000** insertions and deletions (**INDELs**) in the human genome that signal a little-explored type of genetic difference among individuals. INDELS are an alternative form of natural genetic variation that differs from the much-studied single nucleotide polymorphisms (**SNPs**). Both types of variation are likely to have a major impact on humans, including their health and susceptibility to disease.

The human genome sequence in our **DNA** contains **three billion base pairs** of **four chemical building blocks** D adenine, thymine, cytosine, and guanine (A, T, C, G), strung together in different combinations in long chains **within 23 pairs of chromosomes**. When the first human genome was being sequenced, it became apparent that **additional human genomes** would have to be sequenced to identify the places in the genetic code that account for human variation. Scientists now know that humans share about 97-99 percent of the genetic code, and the **remaining 1-3 percent** dictates individual differences. These naturally occurring differences, called **polymorphisms**, help explain differences in appearance, susceptibility to diseases, and responses to the environment.

SNPs are differences in single chemical bases in the genome sequence, and INDELs result from the insertion and deletion of small pieces of DNA of varying sizes and types. If the human genome is viewed as a **genetic instruction book,** then **SNPs** are **analogous to single letter changes in the book**, whereas **INDELs** are **equivalent to inserting and deleting words or paragraphs**.

Most polymorphism discovery projects have focused on SNPs, resulting in the **International HapMap Project D** a catalog and map of more than 10 million SNPs derived from diverse individuals throughout the globe. Research instead on INDELs, using a computational approach to examine DNA re-sequences that originally were generated for SNP discovery projects. Thus far they have identified and mapped 415,436 unique INDELs, but they expect to expand the map to **between 1 and 2 million** by continuing their efforts with **additional human sequences**. INDELs can be grouped into **five major categories**, depending on their effect on the genome: (1) insertions or deletions of single base pairs; (2) expansions by only one base pair (monomeric base pair expansions); (3) multi-base pair expansions of 2 to 15 repeats; (4) transposon insertions (insertions of mobile elements); (5) and random DNA sequence insertions or deletions. **INDELs** already are known to cause **human diseases**. For example, **cystic fibrosis** is frequently caused by a three-base-pair deletion in the CFTR gene, and DNA insertions called triplet repeat expansions are implicated in **fragile X syndrome** and **Huntington's disease. Transposon insertions** have been identified in **hemophilia, muscular dystrophy** and **cancer**.

We are entering an exciting new era of **predictive health** where an **individuals personal genetic code** will **provide guidance on healthcare decisions**. Maps of insertions and deletions will be used together with SNP maps to create one big unified map of variation that can identify specific patterns of genetic variation to help us **predict** the **future health of an individual.** (*see next colon*).

(continued from left colon). The next phase of this work is to figure out which changes correspond to changes in human health and develop personalized health treatments. This could include specific drugs tailored to each individual, given their specific genetic code. Ultimately, each person's genome could be re-sequenced in a doctor's office and his or her genetic code analyzed to make **predictions** about their **future health**. Dr. Devine believes the **technology holds** the **promise** of **predicting** whether a person will develop diabetes, mental disorders, cancer, heart disease and a range of other conditions.

* Certain strains of **E. coli** produce a **toxin**, which induces a toxic effect in host cells, characterized by gradual cell enlargement following the arrest of cell proliferation. **INRA** researchers in Toulouse, in collaboration with teams at the German universities of Wuerzburg and Goettingen and the **Institut Pasteur** in Paris, have demonstrated that these bacterial strains possess a **"genomic island"** in their genome, which contains all genes allowing the **biosynthesis** of a new **toxin**, which they have called **"Colibactin"**. The researchers have shown that the bacteria producing this toxin induce serious lesions to the DNA of host cells, causing a **blockade** of the cell cycle of infected cells. Colibactin belongs to a new family of bacterial toxins, which are able to act on the cell cycle of eukaryotic cells. The INRA researchers have proposed to call this family the **"cyclomodulins"**.

Colibactin is a **non-protein toxin**. The genes carried by the genomic island code for several enzymes belonging to the family of **"polyketide synthetases"** (PKS) and **"nonribosomal polypeptide synthetases"** (NRPS). Compounds arising from these biosynthetic pathways constitute a large family of natural products with a very broad range of biological activities and pharmacological properties. This family comprises numerous molecules which are of importance both **agronomically** (anti-parasite substances, such as avermectin) and **medically** (e.g. immunosuppressants, cholesterol-lowering agents, anticancer compounds and antibiotics (cyclosporine, lovastatin, bleomycin, erythromycin, etc.). This is the first time that an enzyme system of this type, producing a molecule active on **eukaryote cells**, has been characterized in E. coli, a bacterial species where genetic engineering is well mastered. This discovery provides a **biotechnological key** to producing new compounds of interest. It opens the way to novel therapeutic approaches as well as preventive opportunities. [*Nougayrède JP et al, Escherichia coli* induces DNA double strand breaks in eukaryotic cells, Science, August 11th 2006, vol.313, n°5788, pp 848-851, DOI: 10.1126/science.1127059]. The work reported in Science also raises an important question for **public health**. DNA double strand breaks are **dangerous lesions** affecting **eukaryotic cells**; if these are not repaired, they give rise to a **high level** of **mutations**, which are the principal triggers of cancer in man. Colibactin is produced by both commensal *E. coli* in the intestinal flora and pathogenic strains which are responsible for **septicaemia, urinary tract infections** and **meningitis**. The presence of these bacteria in the commensal flora may therefore constitute a **predisposing factor** for the development of certain **cancers**. Thus bacterial flora may participate in the development, differentiation and homeostasis of mucosa and hence the development of certain types of cancer, or protection against them.

V.31. A future of unseen opportunity

Healthcare will shift even more to prevention and more of the responsibility onto the patient. It will also splinter the major indications of today into better defined and better treatable subgroups, which may lower the sales for each particular drug but will greatly expand the total drug market, as side-effects will be much less frequent and less severe and as there will be a much larger choice of better targeted drugs. Each genetic variation will slice up the market. It will also lower the cost and time of clinical trials by reducing the number of patients needed to show effectiveness.

This segmentation into smaller niches is already taking place. The suitability of treatment of cancer patients with aggressive, expensive or limited applicability drugs can be assessed on the basis of „cancer markers" which are being used in clinical trials in order to evaluate the effect of the drug. These – arguably sometimes quite limited – niche markets can be addressed in the future if the necessary changes are made within the regulatory agencies in order to facilitate approval for such variants on a known theme. When it comes to targeted drugs, diagnostic tests will play a much more important and profitable role. Tests who identify patients who are most likely to benefit from a drug, will become increasingly useful at predicting who is vulnerable to a disease and what therapy might work best. Without the tests, there will be no clear patients to target.

The "biotechnological sciences" will carry out two different major functions:

- identification of therapeutic proteins and their production (only a handful of obvious ones have been used so far)

- provision of targets for the identification of active small molecules.

Chapter Six

Differentiation Strategies

in

Drug Delivery Systems

VI.1. Attaching therapeutic drugs to molecules

University of Central Florida and University of California Riverside professors Rahman and Stolbov are a step closer to being able to deliver life-saving drugs through tiny molecules that would travel through the bloodstream and destroy only cancer-ridden cells.

The scientists describe[143] how they got an adsorbate molecule (anthraquinone) to pick up two carbon dioxide atoms and carry them in a specific direction on a flat copper surface. Their discovery helps scientists understand how they someday may be able to attach therapeutic drugs to molecules. They consider it as significant because one wouldn't expect atoms to move that way- Atoms tend to move randomly, like dust particles, and getting them to move in a specific direction will help in our understanding and manipulating of the region around atoms.

Right now the only way we can transport atoms is with instruments. Being able to control that at a molecular level would help us create a natural transportation system that could aid us in many ways in the future, they note.
Next, the researchers want to see if they can make the molecule carrier go around corners, rotate its cargo or send out photons to let scientists know where it is located[144].

The research is a step in the direction of what futurologists see as a far away medicine: nanotechnology that sends microscopic creatures to space or the body to make repairs.

VI.2. Differentiation strategies in drug delivery systems
A current differentiation strategy is to find more convenient presentation formats of existing products. For example, the development of next generation Factor VIII products is based on prolonging the half-life of the existing proteins by drug delivery

---

[143] *In a paper published Jan. 18, 2007 in Science Express,*
[144] *University of California Riverside, Jan 17, 2007*

* Medicines are most helpful when they **directly affect the diseased organs or cells** - for example, tumor cells. Scientists at the Max Planck Institute of Colloids and Interfaces in Potsdam, Germany, and Ludwig-Maximilian-University in Munich, have come one step closer to that goal: they have intentionally released a substance in a tumor cell. The scientists placed the substance in a tiny capsule which gets channeled into cancer cells, and is then "unpacked" with **a laser impulse.** The laser light cracks its polymer shell by heating it up and the capsule's contents are released. The vehicle that the researchers used was a polymer capsule only a few micrometers in diameter. The walls of the capsules were built from a number of layers of charged polymers, alternating positive and negative. The capsules were equipped with a kind of **"open sesame"**. But it didn't require any magic - just nanoparticles made of gold or silver atoms. The scientists mixed together charged metal nanoparticles along with the polymers composing the walls of the vesicle. The **tumor cells absorbed the microcapsules** and then the scientists **aimed an infrared laser** at them. Metal nanoparticles -good at absorbing the laser light and transmitting the heat further into their surroundings, heating up the walls -became so hot that the bonds broke between the polymers and the shell and the capsules eventually opened. Besides using a **"thermal opener"**, the scientists have found another way of making the capsules more stable. They simply heat up the newly created microcapsules very slightly, so that the diameter of the hollow capsules becomes smaller. At the same time, the molecules in their shell are located closer to each other, thickening the capsule walls and better protecting their contents. There is still, however, a problem to solve before scientists can use this technology to create medicines which squeeze microcapsules into tumor cells. There is still no way to **"steer"** the microcapsules. Scientists now have to add some kind of feature to the capsules so that they only recognize the target cells. Only these cells would then allow microcapsules through their membrane. [Skirtach, A.G. et al., Laser-Induced Release of Encapsulated Materials inside Living Cells. Angewandte Chemie (July 5, 2006)]

* Scientists from Johns Hopkins and from the University of Milan have effectively proven that they can inhibit lethal human brain cancers in mice using a protein that selectively induces positive changes in the activity of cells that behave like **cancer stem cells.**
The most common type of brain cancer-glioblastoma-is marked by the presence of these stem-cell-like brain cells, which, instead of triggering the replacement of damaged cells, form cancer tissue..
For their treatment experiment, they relied on a class of proteins, **bone morphogenic proteins**, that cause neural stem-cell-like clusters to lose their stem cell properties, which in turn stops their ability to divide. Their idea is to treat patients with **BMP4** or something like it right after surgery to remove glioblastoma in hopes of preventing the regrowth of the cancer and improving survival time. This opens exciting doors to future research into therapies for such a devastating disease.

* **Hormonal contraceptive methods** will continue to be suspected of having harmful effects, e.g. breast cancer. administration. **Transdermal patches** are generally applied once a week for 3 consecutive weeks in the same site, followed by a patch-free week, and have been found to achieve efficacy comparable to combined OCs. Similar efficacy has been demonstrated with **vaginalrings**, which are inserted and worn continuously for 3 weeks, then removed for 1 week. **Subdermal implants**, which are placed under the skin during a minor outpatient surgical procedure, are also thought to have a contraceptive efficacy rate of more than 99% and retain efficacy for up to 5 years. Multiple forms of **intramuscular injections** are also available and must be administered between every 28 and 90 days. *[Women and Epilepsy: Challenges in the Concomitant Use of AEDs and Hormonal Contraceptives CME Sponsored by the University of Cincinnatti School of Medicine. Reviewed by Dr reviewed by Andrew N. Wilner].*
* Scientists have identified genes related to reaching **age 90** with **preserved cognition**, according to a study conducted at the University of Pittsburgh. It is among the first to identify genetic links to cognitive longevity.
They identified **nine genetic regions** that were associated with successful aging, some of which affected men or women, but not both. Historically women have lived longer than men on average, the prevalence of numerous serious diseases differs in men and women, and there are important differences in age-related physiological changes that occur between the sexes over the life span. It would not be surprising if the collection of genes that influences the capacity to reach old age with normal mental capacity differs somewhat for men and women, the researchers explained. The majority of the successful aging or **"SAG"** regions overlapped with gene locations previously reported to show linkage to susceptibility genes for cardiovascular disorders, psychiatric disorders and the accumulation of tissue damage due to oxidative stress. The study also highlighted the detrimental effects of cigarette smoking, excessive drinking and serious mental disorders on successful aging in both sexes. Identifying such genetic and behavioral factors may hold promise for better understanding the aging process and perhaps one day enriching or extending the lives of individuals.*[The study was published online at http://www.ajgponline.org/ in August 2006].*
* While great interest has followed the discovery of neural stem cells and their potential for someday treating diseases and injuries of the brain and spinal cord, research identified "**cancer stem cells**," a small population of cells that appear to be the source of cells comprising a malignant brain tumor. Theoretically, if these mother cells can be destroyed, the tumor will not be able to sustain itself. On the other hand, if these cells are not removed or destroyed, the tumor will continue to return despite the use of current cancer-killing
therapies. Researchers at Cedars-Sinai Medical Center's Maxine Dunitz Neurosurgical Institute, who first **isolated cancer stem cells** in adult brain tumors in 2004, reported (Dec 2006) these cells to be highly resistant to chemotherapy and other treatments. Even if a tumor is almost completely obliterated, it will regenerate from the surviving cancer stem cells and be even **more resistant to treatment than before**.

Global healthcare shopping will become everyday tourism, with informed patients seeking the best possible care. Travel operators and healthcare agencies will cooperate to create new wellness/illness prevention products technologies. This competitive field is pioneered by Bayer Schering Pharma ahead of Baxter and Wyeth with a clinical phase I project and joined by Novo Nordisk as a new entrant into the F. VIII segment. Research stage projects are based on protein engineering and peptide technologies.

The Business Intelligence firm La Merle S.L. reported that the global market of recombinant coagulation factors VIIa, VIII and IX posted strong 2006 sales of more than US$ 4.6bn. The still growing market of recombinant coagulation factors has marginated human plasma-derived products.

Competition between Baxter, Bayer and Wyeth in the recombinant F. VIII segment (sales of US$ 2.8bn) has spurred the launch of safer, animal and human protein free, products formulated with sucrose or trehalose.

These approaches aim at generating long-acting and more potent molecules with Factor VIII activity to allow non-iv, less frequent or even oral administration[145].

Novo Nordisk occupies the recombinant F. VIIa market and has advanced activities in extending the label by new indications, such as bleeding, surgery and brain injury. Novo Nordisk's next generation V. VIIa projects face competition by Roche & collaborator attracted by the monopolistic US$ 1bn market expected to further grow by new indications. Novo Nordisk's recent decision to focus on protein therapeutics also finds its materialization in R&D projects with F. VIII, F. IX and F. XIII. The acquisition of Syntonix, Biogen Idec is entering the field of recombinant next generation coagulation factors by use of Syntonix' Transfuson and Transceptor technologies. Market leader Baxter is opening a new market with its R&D activities of recombinant von Willebrand factor. Purified human and recombinant human versions of thrombin are in the registration process and expected to squeeze out bovine thrombin products.

VI.3. The need for a safe and convenient delivery vehicle for proteins

The need to deliver proteins has already led to the creation of multifarious „drug delivery" companies, all clamoring for attention for their „revolutionary and superior" patented technologies. Nonetheless, there is still no efficient, convenient and safe delivery vehicle for proteins. Finding one remains extremely difficult, as it must circumvent the body's own digestive and defense systems for the drug to be delivered. To do this safely looks like daunting task.

Significant development in proteomics and pharmacogenomics are being fuelled by major advances in high-throughput screening and other enabling technologies. For example, Harvard scientists are developing a full-length expression-ready FLEX-gene repository that would express proteins by an automated technique called recombinatorial cloning. The system will be used to study how proteins, hundreds or thousands at a time, function –the same way that the pharmaceutical industry studies small molecules using combinatorial chemistry, high-throughput screening, and characterization methodologies. These enabling technologies are expected to flood the industry's pipeline with validated drug targets[146]. And these entire future drug (candidates) will not come in the delivery methods we know today, but in a variety of delivery forms.

---

[145] *These results and more were found in a Competitor Analysis conducted by La Merie Business Intelligence. The report can be found at www.pipelinereview.com, La Merle's News Center and Online Store.*
[146] *Kumar S., Post-genomic Drug Development. Is Pharma Ready ?, Pharmaceutical Executive, March 2002:99*

VI.4. An alternative to conventional protein production methods

A large number of today's medicines are made with the aid of biotechnology (and this number should only grow in the future). To do this, scientists use genetically modified bacteria, yeasts, or animal cells that are able to produce human proteins. These proteins are then purified and administered as medicines. Examples of such proteins are antibodies, which can be used, for instance, in the treatment of cancer. The conventional methods for producing antibodies work well, but they are expensive and have a limited production capacity. The high costs are primarily due to the need for well-equipped production labs and to the labor-intensive upkeep of the animal cells, which are needed as production units.

Researchers[147] have made a significant step forward in making protein production in plant seeds a real alternative to current production methods. Using plants to produce useful proteins could be an inexpensive alternative to current medicine production methods. have succeeded in producing in plant seeds proteins that have a very strong resemblance to antibodies. They have also demonstrated that these antibody variants are just as active as the whole antibodies that occur naturally in humans. By virtue of their particular action, antibodies are very useful for therapeutic and diagnostic applications. From this research, it is now also clear that these kinds of antibody variants can be used in medical applications and that it is possible to produce them in the seeds of plants, which can have enormous advantages over conventional production methods. Because production with plants doesn't require expensive high-tech laboratories, scientists anticipate that, by working with plants, production costs will be 10 to 100 times lower. Exhaustive laboratory tests have shown already that the antibody variants produced in plants are just as effective as whole human antibodies in protecting animal cells against infection with the Hepatitis A virus.

VI.5. There is more to innovation than targeting proteins

San Diego, USA-based Lpath, which claims to be the category leader in developing therapeutic agents against bioactive lipids, outlined a novel approach to discovering new drugs for the treatment of cancer, heart disease, inflammation and other human disorders, thereby giving rise to an entirely new class of medicines[148].

This highlights the protein-centric nature of the pharmaceutical development industry: almost all drugs on the market, as well as almost all drug candidates in clinical trials, target proteins, with the next largest category, nucleic acids, comprising only 2% of drug targets.

The paper then discusses the unique value of lipidomics, which involves the study of bioactive lipids and their normal and pathological roles in the various signaling pathways, and how lipidomics provides an exciting new opportunity to identify lipid-signaling molecules as novel drug targets in the treatment of human disorders.

---

[147] *from the Flanders Interuniversity Institute for Biotechnology (VIB) at Ghent University*

[148] *Targeting Sphingosine-1-phosphate for Cancer Therapy. British Journal of Cancer, Nov 6, 2006. See: www.nature.com/bjc/journal/v95/n9/full/6603400a.html. Researchers identify new approach to drug discovery. Pharma Marketletter, November 27, 2006.*

* The U.S. National Human Genome Research Institute NHGRI's **"Near-Term Development for Genome Sequencing"** grants will support research aimed at sequencing a human-sized genome at 100 times lower cost than is possible today. There is strong potential that, in less than five years, several of these technologies will be at or near commercial availability. Grant recipients:
# a team that has developed a novel type of fluorescent nucleotide that is modified for sequencing by synthesis. Their goal is to **improve the chemical subunits**, called **reversible terminators**, for use in a system that will ultimately be used to sequence DNA templates in high-density arrays, using a sensitive fluorescence detection system.
# The main goal of a second project is to develop a high-speed, massively parallel DNA sequencing system using unique base analogues with cleavable dye and reversible terminator groups and the sequencing by synthesis approach. This application is focused on the development of the **subsystems required to construct high-density sample arrays on glass chips** and to run **sequencing by synthesis reactions on them** in an **automated, high-throughput fashion.**

* An estimated **71 million American adults** now suffer from heart disease, stroke and other forms of cardiovascular diseases. Studies suggest that increased rates of diabetes, obesity and other risk factors may reverse four decades of declining mortality. As the baby boom generation ages, **deaths** from **heart disease alone** are projected to **increase** by **130 percent** between 2000 and 2050.

* Researchers, working with colleagues at **St George's, University of London**, are developing drugs designed to **stop allergens from entering the body**, so rendering them harmless. The research takes a completely new approach to the treatment and prevention of allergies. The technology is based on the earlier discovery of how allergens, the substances that cause allergy, enter the body through the surface layer of cells that protect the skin and the tubes of the lungs. Allergens from pollen or house dust mites are inhaled and then dissolve the binding material between the cells that form these protective linings; they can then enter the body by passing between the cells to cause an allergic response. The drugs under development -- called **Allergen Delivery Inhibitors (ADIs)** - are designed to disable these allergens so they can no longer eat through the protective cell layer and block the allergic reaction before it occurs. Current medicines don't act against the allergen at the early stage - they only ease the symptoms - so the development of these ADIs would be a major breakthrough in the fight against allergies. (Datamonitor estimates the market for allergen-prevention drugs to be $26 billion).

* The promise of **regenerative medicine** and the **nanotechnology** catapulting it into the forefront of chemistry were highlighted in two papers being presented during the American Chemical Society's 232nd national meeting (September 2006).
Nanotubes help adult stem cells morph into neurons in brain-damaged rats - Carbon nanotubes - 80,000 times thinner than a human hair - enhance the ability of adult stem cells to differentiate into healthy neurons in stroke-damaged rat brains, according to American and South Korean researchers, who mixed nanotubes with adult rat stem cells and then implanted the mixture into brain-damaged areas of three rats that had suffered strokes. In six other rats that had strokes, they implanted either adult stem cells or nanotubes - but not both - into brain-damaged areas. After following the animals for up to eight weeks, the researchers concluded that neither nanotubes nor adult stem cells alone triggered regeneration or repair in the brain-damaged regions. In fact, when used alone, adult stem cells migrated to healthy areas of the brain. But when combined with nanotubes, adult stem cells not only remained in the brain-damaged regions, **they began to differentiate into functioning neurons**. The finding could have important implications for the treatment of Alzheimer's, Parkinson's disease and other neurological disorders.
Nanostructures promote formation of blood vessels, bolster cardiovascular function after heart attack - Injecting nanoparticles into the hearts of mice that suffered heart attacks helped restore cardiovascular function in these animals. according to the Institute of Bionanotechnology in Medicine at Northwestern University in Evanston, Ill. The finding is an important research advance that one day could help **rapidly restore cardiovascular function in people who have heart disease.** The self-assembling nanoparticles - made from **naturally occurring polysaccharides** and molecules known as **peptide amphiphiles** - boost chemical signals to nearby cells that induce formation of new blood vessels and this may be the mechanism through which they restore cardiovascular function. One month later, the hearts of the treated mice were capable of contracting and pumping blood almost as well as healthy mice. In contrast, the hearts of untreated mice contracted about 50 percent less than normal. In other recent studies using a similar technique, the researchers found **nanoparticles hastened wound healing** in rabbits and, after islet transplantation, cured diabetes in mice. Nanoparticles with other chemical compositions **accelerate bone repair** in rats and **promote the growth of neurons** in mice and rats with spinal cord injuries.

* Dr. David Matus, Duke University, is using a powerful genetic model to identify the **gene networks** that **control cell invasion**- a critical step in understanding how cancer cells spread.

Perhaps one of **the biggest challenges for the future is to broaden the general population's understanding of the fruit of evidence-based research so that they can make informed decisions about their health**. We already have the knowledge to prevent a huge amount of disease. People in westernized countries are responding to knowledge about the impact of factors like smoking and diet on their health but much more needs to be done to reach the entire population and to **carry that knowledge to less privileged countries**. Research already tells us that one of the best predictors of health is the education level of mothers. This tells us that knowledge can be transferred effectively through education. There is important work to be undertaken to translate research findings into affordable treatment and prevention programs, both within the westernized countries and elsewhere. *(Health Australia. Imagine the Future. 2006)*

As the author notes "there may be over 1,000 members of the lipidome, providing many new potential targets for therapeutic interventions." He states that it is possible to generate therapeutic antibodies that bind to and neutralize bioactive lipids, just as Lpath has done with its Sphingomab antibody, which binds to sphingosine-1-phosphate (S1P) with great specificity and extremely high binding affinity. Such an approach can prevent the lipids from binding to receptors on cancer or other cells, thereby stopping them from assisting those cells in the pathogenesis of diseases like cancer, heart disease, inflammation, and various ocular disorders. Sphingomab has previously been tested in several animal models of human cancer. Another advantage of targeting bioactive lipids is, unlike protein targets, lipid structures are generally unchanged from one animal species to another. This suggests an antibody therapeutic like Sphingomab which has shown efficacy in preclinical animal studies also has a much improved probability of showing efficacy in humans.

VI.6. Nanoparticles as a potential drug carrier

A team of scientists in Singapore, led by Dr Yi-Yan Yang, has developed nanoparticles that can carry both small molecular anticancer drugs and nucleic acids simultaneously for improved cancer therapy[149]. The uniqueness of the new technology from the Institute of Bioengineering and Nanotechnology (IBN) lies in the design of a special biodegradable carrier (cationic core-shell nanoparticle), which can enclose drug molecules and allow therapeutic nucleic acids to bind onto it.

It can efficiently introduce DNA into a cell to be incorporated into its genetic make-up, i.e. induce high gene expression level, especially in both human and mouse breast cancer cell lines, and mouse breast cancer model. The co-delivery of small molecular drugs with nucleic acids can improve gene transfection efficiency, reduce side-effects of these drugs, and achieve the synergistic effect of drug and gene therapy for the more effective treatment of cancer.

Results have shown that the co-delivery of an anti-cancer drug (paclitaxel) with a highly potent anti-tumor 'messenger molecule' (IL-12 encoded plasmid. IL-12 is a highly potent anti-tumor cytokine, and may also overcome paclitaxel-mediated T cell suppression.) using the carrier suppressed cancer growth more efficiently than the delivery of either paclitaxel or the plasmid in mice bearing 4T1 breast cancer.

In collaboration with Nanyang Technological University, experiments were also conducted to co-deliver paclitaxel and small interfering RNA (siRNA) targeting a protein that prevents cell death (Bcl-2) to MDA-MB-231 human breast cancer cell line. The cancer cells became more susceptible to the effects of the drug, due to the additional effect of the siRNA targeting Bcl-2 (The suppression of the anti-apoptotic activity of Bcl-2 by the siRNA made the cells more sensitive to paclitaxel, leading to greater cytotoxicity of paclitaxel.)

---

[149] *Y. Wang, S. Gao, W.-H. Ye, H. S. Yoon and Y. Y. Yang, „Co-delivery of drugs and DNA from cationic core-shell nanoparticles self-assembled from a biodegradable copolymer," Nature Materials, published online on September 24, 2006.*

* Although the exact mechanisms are still being worked upon, researchers raise the possibility that **postnatal neurogenesis** can one day be used to **repair brain damage.** Study results show that self-repair and local remodeling can indeed happen along the brain's lateral ventricular wall, and further insights into this process should shed light on whether **SVZ neural stem cells** participate in stroke/trauma-induced brain remodeling and postnatal/adult brain tumor formation. Furthermore, understanding how the SVZ (Subventricular Zone) cells participate in local repair should help bring scientists closer to the goal of using **neural stem cells** as **therapeutic agents** in **neurodegenerative diseases**.

* There's considerable interest in developing new **male contraceptives** . To support this effort, Dr. **Young-Hwan Kim** at Dr. John Herr's laboratory at the University of Virginia Health System has been searching for **proteins** that might serve as **target sites** for **small-molecule drugs**. In a study published online 2006 by **Developmental Biology**, they report the discovery of a new protein within a **sperm's tail** that could prove a key target for male contraceptive drugs. The newly discovered protein is termed **sperm flagellar energy carrier** (SFEC). It is the **fourth** in a family of proteins that **perform transfer processes** to help cells make and use energy. Inside the cell, these proteins operate much like a shuttle bus, binding and exchanging energy-carrying molecules known as **ATP** and **ADP**. The new protein was found in the distal part of the sperm tail, or flagella. This was an unexpected finding because the **three previously** identified ATP/ADP carrier proteins **reside** in cell organelles called mitochondria. In sperm, the mitochondria reside in a region called the midpiece. UVa researchers also found that **SFEC** has extensions at **both ends** of its amino acid sequence. Other ATP/ADP carriers do not have these extensions, which may explain why SFEC is able to perform its **shuttling** function in the distal sperm tail. The distal sperm tail does not have mitochondria, which means that SFEC operates in a different environment than the other ATP/ADP carriers. While mitochondria generate energy through an oxidative process, research indicates that the sperm flagella mainly produces and consumes energy through a process called **glycolysis**.
It is the first time that such an ATP carrier protein has been linked to glycolysis. In experiments where **glycolysis** was **blocked**, the sperm barely quivered, showed no progressive motion and were **infertile**.
By contrast, prior experiments found that when the **oxidative processes** in the sperm's mitochondria are **blocked**, sperm continue to swim and **fertilize eggs**.
The discovery of SFEC has sparked interest among both basic scientists and contraceptive drug developers because of where it is located in the sperm and the kind of energy-making process that occurs there. One approach to male contraception is to **disable sperm** from **swimming**. Here, SFEC may be able to play a role in that process. Because they will aim at specific proteins like SFEC and target a unique sub-domain on the protein's surface, future male contraceptive drugs will be known as **intelligent spermicides**. At UVa, they termed this new drug class: **spermistatics**.

* A new method for **3D imaging** and **quantification** of biological preparations **ten times larger** than the limit for the **traditional confocal microscope** has been presented by researchers from Umea University in Sweden..
Traditional biological imaging techniques are limited by several factors, such as the optical properties of the tissue and access to biological markers.
A major challenge in this connection has been the creation of three-dimensional images of the expressions of specific genes and proteins in large biological preparations. It has been equally complicated to try to measure the mass/volume of cells or structures that express a specific protein in a specific organ, for instance.
The researchers project that it will be possible to use their method to address a great number of medical and biological issues. This may include such diverse fields as the **formation of blood vessels** in tumor models, the **analysis of biopsies** taken from patients (in cirrhosis of the liver, for instance), and **autoimmune infiltration processes**.

* Researchers at the National Institutes of Health have discovered that a previously unexplained fatal form of **Osteogenesis Imperfecta** -- a disorder that weakens bones and which may cause frequent fractures -- results from a genetic defect in a **protein involved** in the **production** of **collagen.**
The affected gene contains the information for cartilage associated protein, or **CRTAP**. The function of CRTAP is not well understood, but it is known to be part of a complex of proteins involved in the chemical transformation of collagen from simple protein "chains" into its final form.
The well-known forms of Osteogenesis Imperfecta (OI) result from a defect in the genes for type I collagen, which serves as a kind of molecular scaffolding that holds together bone, tendons, skin and other tissues. The collagen defects result from dominant mutations, requiring only one copy of a mutant gene to cause bone disease. The NIH researchers discovered that **mutations** in the CRTAP gene accounted for a recessive form of the disorder -- requiring two copies of the affected gene to show a particular trait.
OI is an uncommon disorder which occurs in 1 out of 15,000 to 20,000 births.
The discovery of CRTAP's involvement with OI is exciting to researchers in the bone and genetics fields because it opens a new and unanticipated field of **bone biology** related to **osteoporosis**.

* A new **imaging molecule** that can detect and map **plaques** and **tangles** in the **brains** of people with **Alzheimer's** disease could eventually lead to earlier diagnosis of the devastating disease, researchers at the University of California, Los Angeles report in the Dec. 21, 2006, issue of the **"New England Journal of Medicine"**. The compound developed by UCLA and called **FDDNP**, also holds promise as a research tool to evaluate new treatments for Alzheimer's. FDDNP binds to plaques and tangles, enabling researchers to see these abnormal deposits that form in the brains of people with Alzheimer's disease on PET (positron emission tomography) scans.

This special carrier can also be potentially used to co-deliver therapeutic nucleic acids to prevent cancer cells from developing resistance to multiple drugs. (The therapeutic nucleic acid may be a vector encoding an antisense molecule directed against the P-glycoprotein mRNA in the target cell. Such a system can inhibit P-glycoprotein expression by the target, and hence, incapacitate its ability to establish multi-drug resistance, a common trait among cancer cells. This, together with the cytotoxic effects of the anti-cancer drug, should enhance the therapeutic effect of the system.) This, coupled with the simultaneous delivery of specific anticancer drugs, could enhance the therapeutic effects of such drugs.

Other scientists in this field have tried to use liposomes made from cationic (charged) lipids to transport drugs and DNA. The carrier developed at IBN is self-assembled from a biodegradable cationic copolymer. Hence, it is more easily produced and its size and characteristics are more easily controlled compared to liposomes. More importantly, it can deliver nucleic acids more effectively.

VI.7. Advances in Science

According to Dr Tugrul T Kararli, General Manager, PharmaCircle LLC (a drug delivery knowledge management organization), the quest for novelty and significant advances over the existing products has seemed a natural and desirable progression, a way of making drugs more efficient, better tolerable, with less possible side effects and more convenient to take by patients. The relentless search for improvement focused on

Table 6.1.   Schematic of the IBN nanoparticle, which can carry drugs and nucleic acids simultaneously

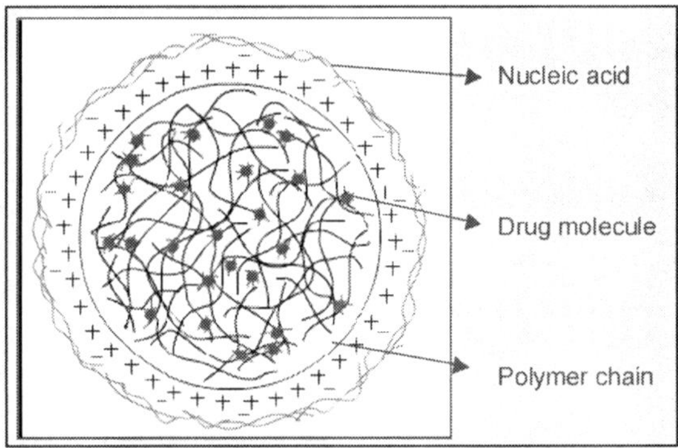

looking for the magic bullet that would work on a precise mechanism. As long as enough new compounds reached the market, few pharmaceutical companies experienced strongly enough constant pressure to invest in finding appropriate new delivery methods. But unrelenting pressure from investors for quarterly results brought about a shift in management objectives and drug delivery systems came into favor in the past decade. Advances in science are helping this sector to expand further.

The total dollar value of the marketed drug products using various drug delivery technologies is estimated to be around $50b, roughly 10-12% total drug product sales today. The use of the patented drug delivery technologies in drug products which

* Researchers led by a **University** of **Utah** medicinal chemist have developed a novel method to make drugs for cancer and other diseases from **bacteria found in sponges** and **other small ocean creatures**. They examined symbiotic bacteria that live only in sea squirts and other marine life. These bacteria are responsible for making a wealth of chemicals, which accumulate in the tissues of sea squirts and may help to defend them against predators. Many of these chemicals have **anticancer properties**, but harvesting them in quantities for large-scale testing and production has been impractical. A new method uses **genetic pathways** in the **bacteria** to produce the small chemicals and to manipulate them to invent new potential drugs. The ability to make these chemicals in the laboratory opens myriad possibilities for **developing drugs** to fight **cancer** and **other diseases**. This represents a new way of attacking the problem. Scientists hope they can use this to find a way to make natural molecules of compounds **through single mutations in DNA**. To synthesize natural compounds, researchers have traditionally made them in the lab using labor-intensive routes. More recently, researchers have begun to use genes to make small molecules within laboratory strains of bacteria. The **promise of genes** is that you can **access** the **tremendous natural diversity** of the **world's organisms** to find **new natural compounds for human health**. One can also use genetic engineering to modify these compounds and **invent new drugs** to target human diseases. By examining the natural chemical and genetic diversity found in sea squirts and their symbionts, researchers identified individual mutations responsible for changing from one compound to another. By **mimicking** this **natural process**, the researchers synthesized a completely new compound. This paves the way to the genetic creation of large **chemical libraries** for **testing against** human diseases.

* **Ageing**, and the processes of deterioration that go with it, are largely attributable to cells that die off in a controlled manner. Therefore, gaining better understand-ding of this controlled cell death is very important in the fight against deterioration diseases like dementia. In this light, researchers from the **Flanders Interuniversity Institute for Biotechnology** (VIB) connected to the K.U.Leuven, in collaboration with researchers from the **Dulbecco-Telethon Institute** hosted by the **Veneto Institute of Molecular Medicine** in Padua (Italy), have now discovered the function of the **PARL protein**. By studying mice that are unable to produce PARL, the researchers have discovered the significance of this protein in **controlled cell death**. An important step toward a good understanding of the aging processes and of diseases like Parkinson's disease.

* **Generics** will assume a more central role as patients bear a greater percentage of their healthcare costs and payers seek to restrict the growth of healthcare expenditures. **Price moderation** for branded drugs is likely as a result. In 2006 sales of generics in the top eight markets (US, Canada, France, Germany, Italy, Spain, UK and Japan) exceeded **$55 billion**, and are expected to experience **double-digit growth till 2010**.

* At **Harvard**, Dr. Rudolph **Tanzi** is using the **Hap Map** to track down gene mutations that cause the common, **late-onset form** of **Alzheimer's**, which could strike as many as 16 million Americans by the year 2050. Tanzi's work is funded by the Cure Alzheimer's Fund, a nonprofit that is investing $3 million to unravel the Alzheimer's genome, which it hopes to complete by the summer of 2008. Tanzi says a prototype genetic chip to test for the disease could be available within five years.

Aging is a global fact of life: average life expectancy in the world today is 66 years compared with 46.5 years only 50 years ago. Not only are people living longer but the way we age is also changing and becoming more **dynamic**. The age barrier between employment and retirement is no longer static with some people leaving work before and some later than pension ages. Older people are **redefining** their **roles** as consumers and citizens.
Increased lifespan is one of the great success stories of our time with people reaching the traditional age of retirement, 65, being expected to live for **another 20 years**. The increase in life expectancy is mainly due to public health measures, such as interventions ranging from routine vaccinations to improved sewage disposal, the control of formerly fatal or debilitating childhood diseases, advances in medical knowledge and medical technology, improved diet and higher standards of living. **Research** has been at the heart of the improvements in life expectancy and is now focused on **improving** the **quality** of people's **lives** as they age.
The New Dynamics of Ageing Program (NDA), the most ambitious research program on aging ever mounted in the UK, was launched in October 2006 in partnership with the UK Funders Forum for Research on Ageing and Older People. The NDA is a collaboration between five of the UK's Research Councils - the Arts and Humanities Research Council, the Biotechnology and Biological Sciences Research Council, the Engineering and Physical Sciences Research Council, the Economic and Social Research Council and the Medical Research Council - which will see the injection of around 20 million pounds into the vital area of **aging research**.
With more people in the UK **aged over 60 than below 16 years of age** for the first time, according to the 2001 census, the NDA program aims to ensure that aging research has the maximum beneficial impact to both the economy and society through enhancing the quality of life, productivity and self-sufficiency of the older generation. .
It is vitally important that we understand the changes taking place in the aging process. This program of research will target resources to look at **all dimensions of aging**, from biology to social and cultural aspects, ensuring that this much needed knowledge is available as quickly as possible for policy makers, practitioners, product designers and anyone in a position to **improve the quality of later lives**.

Table 6.2. Advances in oral drug delivery

- **Absorption models for rational dosage form development – CaCo2, etc**.
  - Membrane or dissolution rate limited absorption, regional drug absorption
  - Structure vs. transport
  - Predictive software/tools: absorption, dissolution, etc.
- **Better understanding of GI tract physiology & biochemistry and function**
  - Peptide transporters,
  - Efflux proteins/Pgp
  - Enzymatic barriers/first pass metabolism
- **Advances in Physical Pharmacy**
  - Solid state/crystallization/amorphous, etc.

started in the 50's with the introduction of meter dose inhalers (MDIs) is rather young. It is no surprise that oral modified release products account roughly half of this total market value. Besides oral, inhalation/nasal, transdermal, injectable depot, liposomal, vaginal, ocular and targeted delivery products make up the rest of the total market.

Table 6.3. Chronology of the DDS market[150]

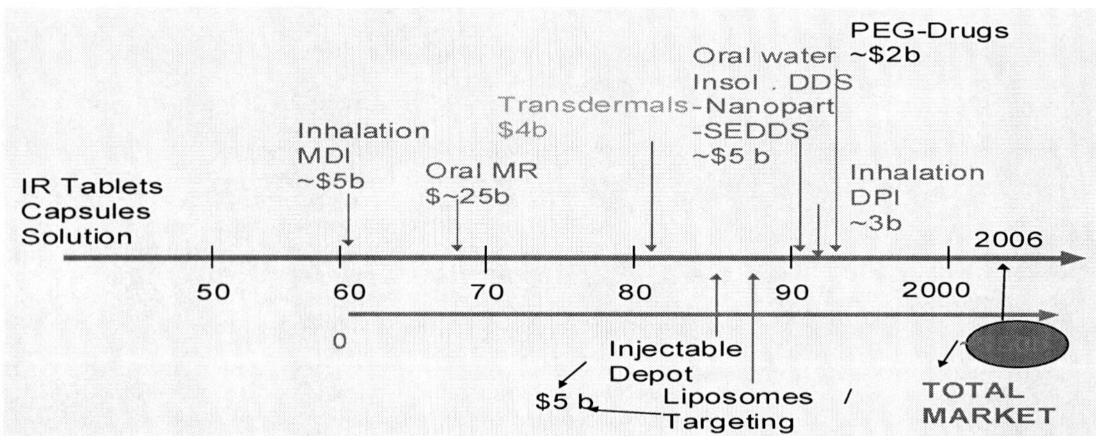

The splintering of the pharmaceutical market will dramatically improve the quality of the drug action, which should then also lead to drug use playing an even bigger role in the management of disease and taking a larger slice of the healthcare dollar/Euro. This will be needed to cover the increased cost of specialization.

Today, there are close to 400 drug delivery companies with well over 600 specific technologies in the oral, parenteral, inhalation/nasal, skin, brain, ocular, vaginal routes.

VI.8. Oral Technology

The delivery of drugs using the oral route is the most preferred delivery system by patients and prescribing physicians and is the largest sector within the drug delivery systems industry.

---

[150] *Source :PharmaCircle LLC*

216

*Researchers report evidence of a promising new example of **personalized medicine**. Using a **lung metagene model**, the group analyzed the molecular characteristics of each patient's tumor and propose that this information may one day be used to tailor treatment decisions. The researchers say they hope genomic tests like theirs will be used not only to predict patient outcomes but also to select individual drugs that will best match the tumor's molecular makeup.(N Engl J Med. 2006; 355:570-580).

* With the aid of complex computer simulations, scientists at the **Max Planck Institute** of Colloids and Interfaces in Potsdam and at the **University of Heidelberg** have **discovered** how the **shape** and **distribution** of certain **sticky areas** on the **cell affect** its **adhesion** in **blood vessels**. According to this research, neither the number nor the size of these adhesive areas are the most important parameters; the most crucial factor is how far they extend from the cell surface. White blood corpuscles and red blood cells infected with malaria are seen to use this spiky hedgehog-like structure for their adhesion strategy. Blood is the universal means with which different types of cells are transported in our bodies. Its movement is determined by hydrodynamic forces. The cells anchor themselves to the walls of the blood vessels in the target tissue with the aid of special adhesive molecules, which are also called receptors. In many cases these receptors are grouped in the cell surface in nanometer-sized patches. The adhesion process is based on the key and lock principle: as a rule, an adhesion molecule only bonds with specific partners. This guarantees that the cells are only brought to a halt where they are to fulfill their biological function. These processes are of **great relevance to medicine**. For example, red blood corpuscles infected with malaria stick to blood vessel walls to escape destruction in the spleen and patrolling white blood corpuscles dock with the blood vessel walls in order to seek out foreign bodies in the adjacent tissue. These "**wandering adhesive cells**" also **include stem cells**, which **move** from the **bone marrow** to **their target tissue,** and **cancer cells** which **metastasize** in the body. Changing the height at which the adhesive patches protrude above the cell membrane has sur-prising results: even small increases give rise to **much faster adhesion**. White blood corpuscles use this effect by covering themselves with hundreds of protrusions called **microvilli**, which stand about 350 nanometers above the cell surface - almost four per cent of the cell diameter. Red blood corpuscles infected with malaria also use this "hedgehog spine" strategy. They have "knobs" that are 20 nanometers high on their surface.

*The scientists suspect that their simulations have helped them to discover a general biological design principle which also occurs in other hydrodynamic contexts - in bacteria, for example, which collect in medical devices through which liquids flow, such as catheters or dialysis equipment. In the future, the software they have developed will allow these situations to be examined more closely than ever before and is another step on the way to "**computational**" **biology.** (Physical Review Letters, 28. September 2006)

* Clinical Development Partnerships (CDP) is a joint initiative between Cancer Research UK and Cancer Research Technology, the charity's development and commercialization arm. It aims to increase the number of successful new treatments for cancer by **taking undeveloped anti-cancer agents from industry** and **putting them into clinical trials**. The initiative is targeted at leading pharmaceutical and biotechnology companies who have a large pool of molecules that may have anti-cancer properties. However, these companies have to **prioritize** which agents they take into clinical development - this leaves potentially effective treatments on pharmaceutical companies' shelves. CDP will take promising but '**deprioritized**' anti-cancer drugs into early stage clinical trials through Cancer Research UK's highly experienced Drug Development Office. Effectively the charity will 'borrow' a drug from a company and conduct early clinical trials at no cost to the company. If the drug looks promising, the company retains the option to develop and market the drug, but with the charity receiving a share of any revenues. Making the leap from something that looks promising in the laboratory to testing it in patients is one of the most challenging steps in drug development. There are many reasons why potential treatments do not make it to market. Science has progressed so rapidly in recent years that there are more compounds available than commercial resources to investigate them. And because drug development is so time-consuming and expensive, anything that doesn't look extremely promising is not developed by a pharmaceutical company. The drug companies have these potential treatments trapped in their pipelines and CDP has the expertise and capacity to release this potential. Their drug candidates coupled with their world-class network of trial centers and scientists, offers a perfect partnership to achieve the greatest impact in the global fight against cancer. .This may include new medicines to tackle the rarer cancers - those that tend to be lower down a business' priority list because they are less profitable.

* The first **vaccine** against **atherosclerosis** is not far away in the future, according to Prof. **Nilsson** at Lunds Universitet in Malmoe (Sweden). Human clinical trials are likely to begin at the end of 2007: they will be aimed at verifying the safety of a preparation, still under investigation in a laboratory model, made of **antibodies** obtained against selected fragments of oxidized Low Density Lipoproteins, or **oxLDLs**. LDLs are the major component of the bad cholesterol: their accumulation in the arterial wall causes inflammation and is a key factor in the onset of atherosclerosis.

* A protein (**caspase-3 protein**) known primarily for its role in killing cells also plays a part in memory formation, researchers at the University of Illinois at Urbana-Champaign reported. When activated, the enzyme caspase-3 triggers a synaptic process essential for **memory storage**.Caspase-3 is best known for its role in a **biochemical cascade** that leads to apoptotic cell death. These new findings demonstrate that the enzyme acts differently under different conditions, and suggest that its regulation in the brain is more complex than previously thought.

Table 6.4.   Technologies used, by delivery area

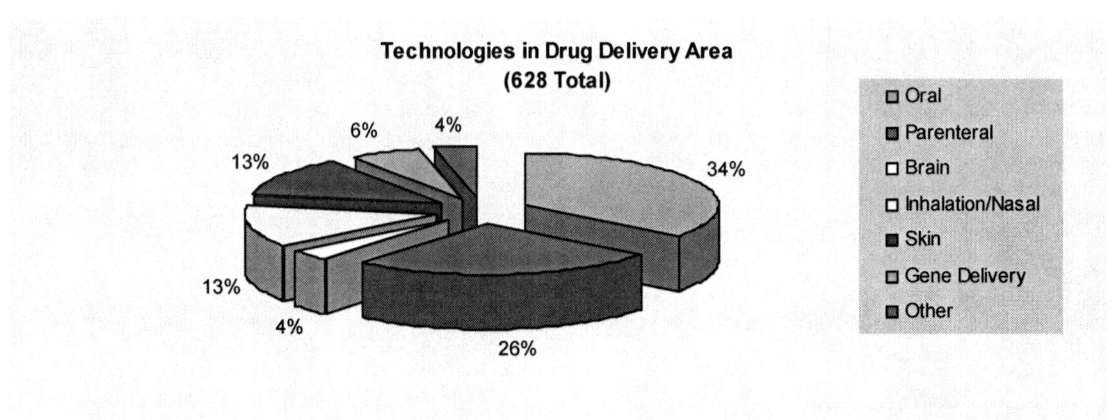

It is the undisputed gold standard in drug delivery because of its low cost per dose, relatively few side-effects, patient convenience and also because it achieves desired outcomes.[151],[152],[153] The one negative aspect of oral delivery is that is lacks efficacy

Table 6.5.   Technologies by type of drug administration

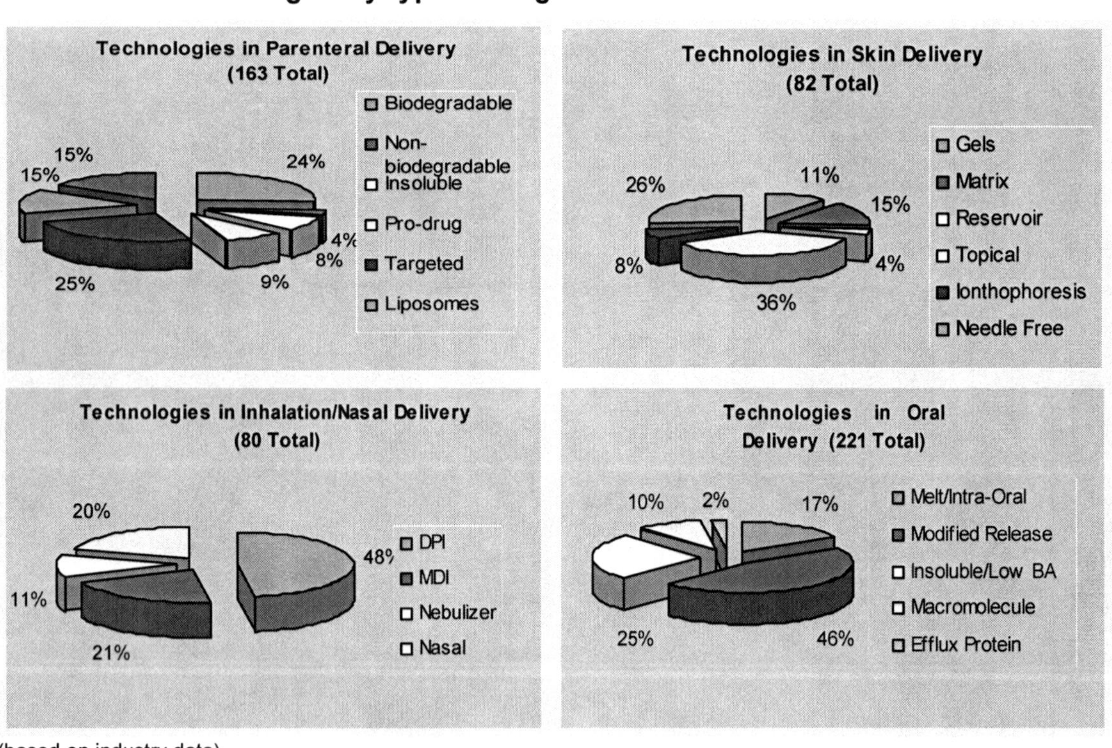

(based on industry data)

control. That is to say, it offers inconsistent absorption rates, undesirable first pass effect, uneven transit times, low bioavailability and poor gastrointestinal tolerance[154].

---

[151] Papanikolaw, J. Seeking Growth in Alternative Drug Delivery Systems. Chemical Market Reporter. 25 Oct 1999.

[152] Grosh, S."Transdermal Drug Delivery – Opening Doors for the Future".European Pharmaceutical Contractor.Nov 2000.3M Healthcare.30 Aug 2002 <http://www.3m.com/us/healthcare/manufacturers/dds/pdf/opening_doors.pdf>.

[153] Erickson, E. "Drug Delivery of the Future Ushers in Individualized Care". Script Magazine May 2000: 6-7.

[154] Evers, P. "Developments in Drug Delivery: Technology & Markets". London: Pearson Professional Ltd, 1997.

Table 6.6.   Oral Drug Delivery systems

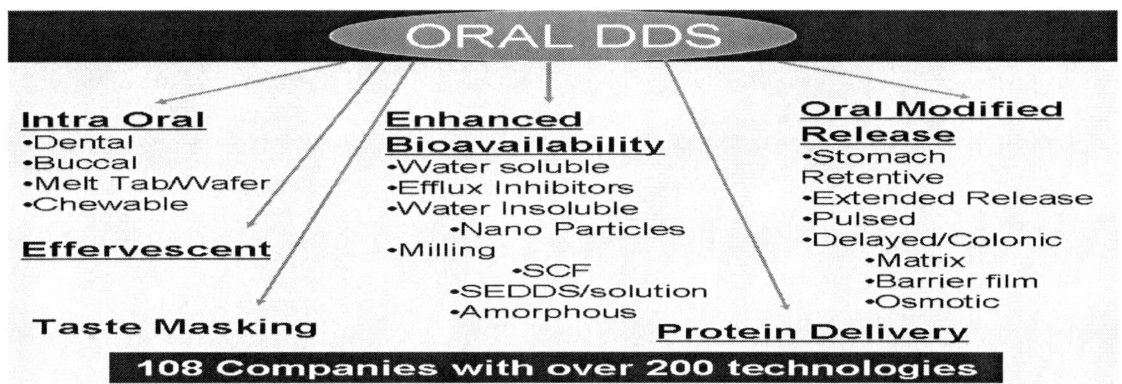

Drug delivery technologies that use modified release formulations, chewable tablets, buccal systems, enhanced bioavailability formulations for water soluble as well as insoluble drugs are presently available on the market e.g., once a day dosage formulas developed from previously three-times-a-day formulas provide a case in point. According to Kermani et al., fifteen percent of R&D budgets in 2001 would be spent on projects incorporating a drug delivery system, and of this number, fifty percent would be for

delivery with oral technologies. Growth in this sector will continue in the short term and encompass the majority of drug delivery projects. The focus in this segment continues to be towards the improvement of drug efficacy through the use of specific drug delivery technologies[155].

VI.9. Modified release
With about 50 companies (and products) active in this oral DDS sector modified release represents the biggest of it. Based on the number of products introduced using its DDS technologies, Alza has dominated this market. Alza's success in this area is driven by Oros osmotic drug delivery technology[156]. But: today's success is based on yesterday's decisions. Other players with hopes of short-term returns on investment are e.g. SkyePharma, Shire, Eurand, Labopharm, Penwest, Flamel, Impax, Elite, Lavipharm, ScolR, Soliqs, West Pharma, Yamanouchi Pharma Technologies.

The sector witnesses the arrival of many drug products from Specialty Pharma: Watson, Andrx, Biovail., KV, Ethypharm, etc.

As BigPharma uses internal modified release technologies routinely, the expert view is that platform technologies that can achieve extended, delayed and pulsed release are most valuable in partnering.

VI.10. Oral delivery of Macromolecule
There has been tremendous interest and efforts for the oral delivery of large molecular weight drugs within both the academia and the industry. Today we know more about the intestinal peptide transporters and enzymes involved in the absorption as well as degradation of peptide and proteins.

---

[155] *Tugrul T. Kararli, Correspondence with the author, May 2, 2004*
[156] *Tugrul T. Kararli, Correspondence with the author, May2, 2004*

* **The Pederson/McIntosh laboratories (University of British Columbia)** have been engaged in the study of mechanisms involved in the control of insulin release; particularly the roles played by the gut   hormones **Glucose-dependent Insulinotropic Polypeptide (GIP) and Glucagon-like Peptide 1 (GLP-1).** These hormones are known as incretins. Using a variety of experimental techniques ranging from the whole animal to studies at the molecular level, the action of incretins on insulin secretion and the regulation of blood sugar levels has been investigated in health and disease states of obesity and diabetes.

The mechanism of action of GIP and GLP-1 on the pancreatic insulin-secreting cell is studied with insulin secretory models ranging from isolated beta cells to the intact animal. These actions are being further defined at the molecular level by studying hormone interaction with, and activation of, cloned incretin receptors. This work has the potential for establishing the way in which GIP and GLP-1 molecules activate the insulin secreting cell in health and disease. Another aspect of the work on **incretin** actions focuses on the metabolism/inactivation of GIP and the related incretin GLP-1 in the circulation. These hormones are rapidly metabolized by the circulating enzyme dipeptidyl peptidase IV (DP IV).
It was established that specific DP IV inhibitors were effective in inhibiting circulating levels of DP IV in experimental animals and, as a consequence, enhanced the insulin-releasing and glucose-lowering actions of incretins released after food stimulation. Subsequently, it was demonstrated that orally administered DP IV-inhibiting drugs had a marked glucose lowering action in an animal model of Type II (maturity onset) diabetes (the most prevalent form of diabetes).

This work has generated a great deal of interest among researchers, highlighting **DP IV inhibitory drugs as potential therapeutic agents in the treatment of diseases characterized by high blood sugar (hyperglycemia)** e.g. obesity and Type II diabetes.

* Several companies are investigating other mechanisms to increase insulin sensitivity (i.e. decrease insulin resistance).Preservation or enhancement of function of the insulin producing beta cells in the pancreas is an important target for therapeutic development for diabetes. Identification of the molecular events involved in beta cell growth and development and in glucose sensing and insulin secretion by this critical cell type has important implications for therapy. Several drugs are under development based on the activity of **glucagon-like peptide-1 (GLP-1)** which appears to enhance growth and function of insulin producing beta cells.
One class of drugs acts through **a nuclear receptor** to regulate gene expression in fat and other insulin responsive tissues. **27 Agents** in this class have been shown to improve glucose control in type 2 diabetes, and also to delay or prevent type 2 diabetes in high-risk women with a history of gestational diabetes. Since earlier drugs in this class had significant side effects, nearly all the major pharmaceutical companies are trying to develop improved drugs that are more potent and less toxic.

* **Diabetes** is a chronic syndrome characterized by hyperglycemia, with disturbances of carbohydrate, fat and protein metabolism resulting from defects in insulin secretion and/or action. Diabetes is one of the world's most common diseases. The long-term damage caused by diabetes in the microcirculation and large blood vessels severely impacts life expectancy and patients' quality of life. Treatment of these secondary illnesses accounts for a large proportion of the costs incurred in the care of diabetics. The objective when treating diabetics is to achieve a normalization of the diabetic metabolism through dietary measures and therapeutic intervention.

B.66. Diabetis: a global health problem

Country	Diabetics (mn)	
	2002	2010 (est)
India	35,5	62,0
China	23,8	47,2
USA	16,0	29,5
Russia	9,7	13,4
Japan	6,7	8,9
Germany	6,3	10,0
Pakistan	6,2	8,7
Brasil	5,7	9,6
Mexico	4,4	7,6
Egypt	3,9	5,5

* Type 2 diabetes generally arises from a combination of **insulin resistance and inadequate production of insulin in the beta cell of the pancreas**; new therapies are targeted at both of these defects. **Molecular mechanisms** involved in glucose toxicity underlying the development of diabetes complications have also been elucidated, yielding new targets for therapy. Moreover, scientists anticipate additional new therapeutic targets will emerge from genetic studies underway to **identify genes predisposing to type 2 diabetes**

B.67. The diabetes market outlook 2015

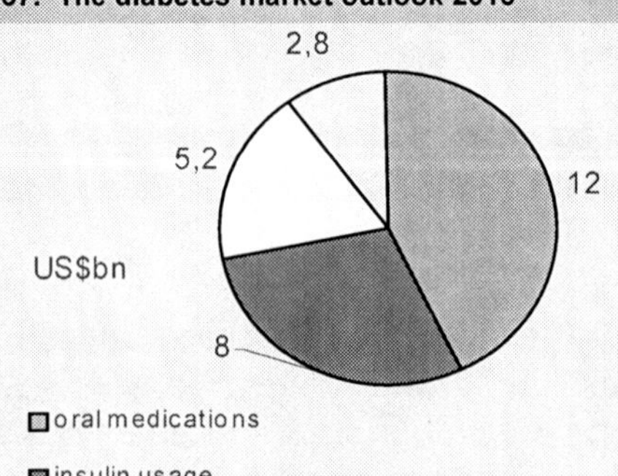

This knowledge is allowing companies to develop mechanism based approaches for the oral delivery such molecules, Within the last 10 years several drug delivery companies such as, Emisphere and Nobex have provided both preclinical and clinical data showing that macromolecular drugs, such as insulin, heparin and others can be absorbed from the oral route in commercially viable amounts.

Almost all of the technologies operating in the field have devised mechanisms in their formulations to prevent enzymatic degradation in the intestinal lumen. The latest technologies available from the start up companies such as, XenoPort, Arizeke, Syntonix and Cellgate take advantage of the natural absorption pathways of the body to get macromolecules absorbed from the oral route. Despite the significant advances in the field commercialization of macromolecule products from the oral delivery is still a big challenge. The total amount of protein delivered in general is limited to a few percent and this provides a big challenge in terms of Cost Of Goods (COGs) . The issues around the frequency of dosing and effect of food on the intestinal absorption must be sorted out before any macromolecular drug can reach market place.

VI.11. Enhanced bioavailability / Efflux inhibitors
It is known in the industry that about 40-50% of NCEs coming from the discovery are water- insoluble drugs. Without the special technologies that can solubilize such insoluble drugs in the intestinal media and help them get absorbed, many efficacious molecules may never get their fair share of testing.

Nanoparticle technologies, lipid based self-emulsifying dosage forms and amorphous technologies which are available from various drug delivery companies (Elan, SkyePharma, Pharmasol, Eurand, Soliqs, Lipocine, etc) are now enabling pharmaceutical companies advance more of their water insoluble molecules into further development. Emend from Merck has been developed using the Nanocrystal technology. Besides Nanoparticle technologies, lipid based systems are also commonly employed to enhance the oral delivery of water insoluble drugs. Neoral as well as many of the HIV protease products on the market utilizes self-emulsifying dosage (SEDDS) forms to enhance to amount of drug absorbed from the oral route.

VI.12. SCF technologies
SCF is a cutting-edge, environmentally friendly technology with multiple applications in the pharmaceutical field ranging from nanoparticle formation for water insoluble drugs, controlled particle generation and formulation to microencapsulating, coating, extraction and fractionation. The particles generated by SCF can have better physical and morphological characteristics to improve drug delivery from various routes, especially from the oral and inhalation routes. Nektar, Thar, Crititech, Lavipharm and other players have signed deals with various pharmaceutical companies in the oral, inhalation and coating applications of their SCF technologies.

VI.13. Inhibiting P-gp Action
It is certain that in the future we will see an increasing number of new NCEs will be coupled with water insoluble drug delivery technologies to facilitate their oral delivery. In the future we should see greater number of drug products reaching to the market both from the NCE as well as the life cycle management side.

* **Biotech** greatly replaces the research function within the pharmaceutical industry. This constantly evolving biotechnology and its existence outside the larger, centralized units of classic R&D companies represent an **applauded source of novel ideas** to the latter.

* **Activating neural growth factors in the brain.** One new approach to treating Alzheimer's disease is to disrupt the formation of plaques, the telltale sign of the disease. In the brains of **Alzheimer's** disease patients, certain proteins cleave the **amyloid precursor protein (APP) into beta-amyloid fragments**, which then aggregate into the characteristic plaques of the disease. Several drugs currently in development are targeted at the steps involved in this process. These include drugs that inhibit proteins that cleave the APP; a variety of agents that are proposed to inhibit the aggregation of beta-amyloid into plaques, including a plant extract from cat's claw; and immuno-therapeutic agents such as beta-amyloid vaccines.
Increasing signalling between nerve cells. A related strategy against Alzheimer's disease is the development of compounds proposed to be neurotrophic (i.e., facilitating the health of nerve cells) or neuroprotective against mechanisms that kill nerve cells. There are several of these types of agents in pre-clinical and clinical trials. (NIH 2002).

* AD is an irreversible disorder of the brain, robbing those who have it of memory, and eventually, overall mental and physical function, leading to death. It is the most common cause of dementia among people over age 65. Recent studies estimate that up to 4.5 million people currently have the disease, and the prevalence (the number of people with the disease at any one time) doubles every 5 years after the age of 65. By 2050, if current population trends continue and no preventive treatments become available, some 13.5 million Americans will have Alzheimer's disease. The annual US national direct and indirect costs of caring for AD patients are estimated to be as much as $100 billion. This suggests that the economic burden will grow as the population ages and the number of AD patients increases.

* **Limiting the neurotoxicity mediated by microglia.** Ideas for drugs that may be useful for the treatment and prevention of the cognitive and behavioral symptoms of **Alzheimer's** disease have come from a variety of sources. Clinical-pathological studies have indicated that there are a variety of brain mechanisms that may lead to or exacerbate the nerve cell dysfunction and death and loss of connections among nerve cells seen in Alzheimer's disease, including abnormal processing of proteins such as the amyloid precursor protein, beta-amyloid; oxidative damage; inflammation; and neurotrophic support of brain cells. Studies in test tubes and in animals have indicated that many of these mechanisms are potential targets for new drug discovery and development. A number of drugs targeting these mechanisms involved in Alzheimer' disease pathogenesis is currently in preclinical development or clinical testing.

* An aging population will result in a **30% increase** in the number of people at risk from osteoporosis by 2010. **Osteoporosis** is largely a preventable and treatable disease. First introduced in the mid-1990s, bisphosphonates, which inhibit bone reabsorption, represent one recent category of pharmaceuticals that effectively treat osteoporosis. **Alendronate** and **risedronate** are in this category of drugs.
B.68. Health "care" in Europe: example of osteoporosis incidence vs. treatment

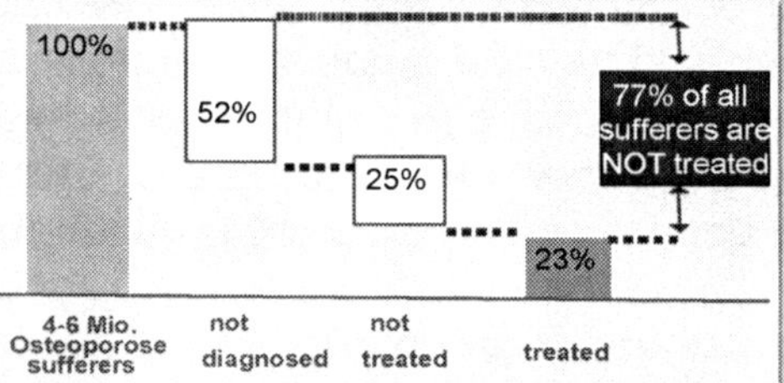

(Example given for Germany. Source: Dr S. Oschmann, MSD Sharp & Dohme)

* Drug candidates for osteoporosis can be divided into two categories: those that prevent bone reabsorption and those that promote new bone formation. Different elements involved in maintaining healthy bone are targeted by these new compounds, including factors involved in bone cell function and regulation, cell membrane receptors and attachment proteins, and cellular enzymes and nuclear transcription factors.

* **Novel approaches** for new drugs to treat osteoporosis target different elements in bone reabsorption and formation. **Selective Estrogen Receptor Modulators (SERMS)** mimic the effects of estrogen and prevent bone loss. Although several SERMs with distinct agonist and antagonist pharmacological profiles have been discovered in the last decade, the search continues for compounds with increasingly selective profiles.
There is a need for SERMs that have agonistic, estrogen-like effects on bone and the cardiovascular system but antiestrogenic, antagonistic effects on the uterus and breast. **Ospemifene** is said to be such a **novel triphenylethylene SERM** that is a major metabolite of toremifene. It is a tissue-specific estrogen and antiestrogen that binds to estrogen receptors alfa and beta with affinity comparable to tamoxifen. The compound has been shown to prevent estrogen depletion-induced bone loss in animal models, in addition to inhibiting the growth of human growth of human breast cancer MCF7 and DMBA-induced mammary tumors. (Drugs of the Future 2004, 29(1):38-44)

* Researchers at MIT develop a **biodegradable drug delivery system** with the potential to **release pulses of different drugs at various intervals after implantation** by using materials of different molecular masses for the membranes covering the drug-containing reservoirs. The microchip device is designed to achieve **multi-pulse drug release over periods of several months**, at required times, without requiring a stimulus to trigger that drug release.

Table 6.7. SCF principle[157]

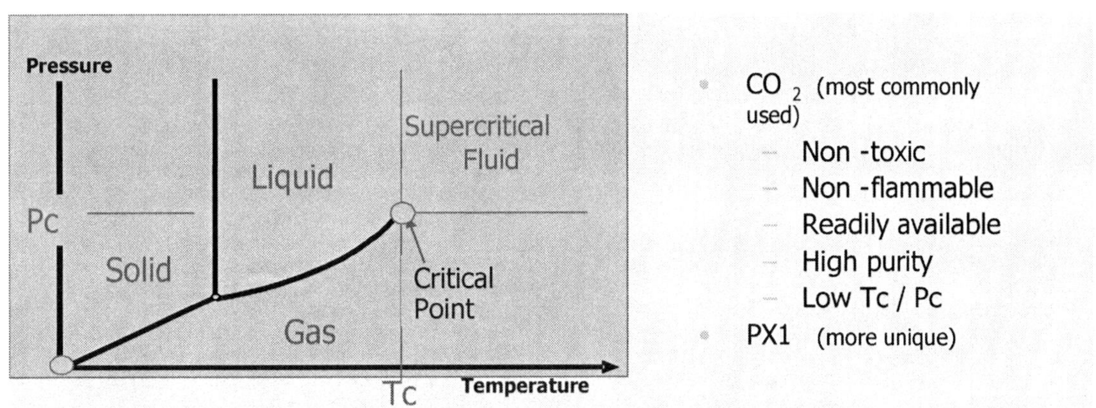

The discovery that p-Glycoprotein limits intestinal transport of many classes of compounds by secreting absorbed drugs back into the intestinal lumen can be considered as another significant advancement in the drug delivery field. This discovery is allowing companies to design improved dosage forms that shows less variable and higher absorption from the oral route. Companies such as Eurand, Shire, Supratek, Bioavailability Systems have been searching for safe excipients to inhibit P-gp action.

Because of extensive drug-drug interactions that P-gp substrates show, the use of such inhibitors may best be suited at the early development stage for NCEs

Table 6.8. Particles generated by SCF technologies[158]

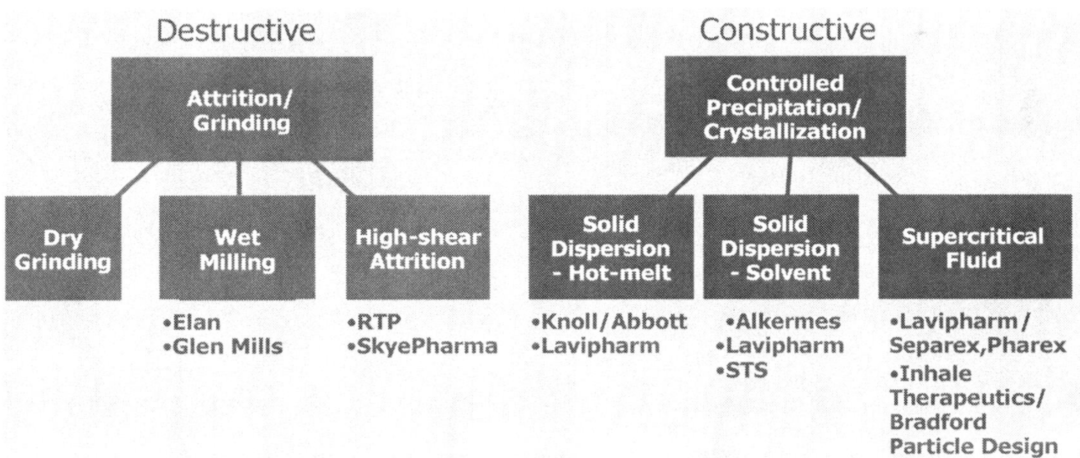

(Destructive methodology has limited applications, in that it is only to be used for particle sizing. Constructive methodologies have a broad application, in that they can be used for particle sizing, formulation, and small molecules and proteins/peptides.)

VI.14. Transdermal DD Technology

A high level of interest in transdermal drug development technology exists because of its high patient acceptance and improved compliance. Although transdermal products are

---

[157] Source :Spiros Fotinos, Lavipharm, Greece/USA, 2004
[158] Source: Spiros Fotinos, Lavipharm Greece/USA, 2004

* Impressive advances in the treatment of **arthritis** are expected to come from emerging oral therapies based on inhibitors of alpha4 integrin and P38 MAP kinase. Work on **BIP** (a new immunomodulator with anti-arthritic properties) shows that BIP (an endoplasmic reticulum chaperone), as an intra-cellular chaperone, protects cells from stress; as an extra-cellular protein, stimulates IL-10 production from monocytes. It both prevents and treats collagen induced arthritis (in the mouse). Work on **IL-6 blockade** (IL-& is a pleiotropic cytokine regulating immune responses and acute phase reactions) shows that **deregulation of IL-6** has been implicated in the pathogenesis of a variety of diseases; the functions of IL-6 are mediated through a receptor system comprising a signal transducer and a binding molecule (the IL-6 receptor); and that **blockade of IL-6 binding** to its receptor using a novel monoclonal antibody has been efficient in rheumatoid arthritis.

A potential new approach to treat inflammatory diseases is seen in a fully **human monoclonal** antibody designed to **block ligand-induced activation of VEGFR1** (VEGFR1 mab inhibits disease progression in mouse models of arthritis and atherosclerosis. VEGFR1 mab inhibits influx of inflammatory cells to sites of inflammation. The mechanism of this action is thought to be the inhibition of recruitment from bone marrow of hematopoietic precursor and inflammatory cells).

* In developed countries, **allergies** are increasing because of **defective immunoregulatory function. Potential immunotherapeutic approaches** for the treatment of allergic diseases include :
- Allergen-specific immunotherapy;
- Recombinant allergens;
- T-cell peptides;
- DNA vaccines;
- CpG ODN;
- Bacterial vaccines,
- Toll-like receptor-based treatment strategies in asthma

* Using genomic testing to guide drug therapy will constitute a significant shift from the current practice of population-based treatment towards '**fine-tuning**' individual therapy.

* Newer approaches to pharmaceutical treatment of depression target **neuropeptide receptors** and **intracellular messenger systems**. Medications for depression currently under development include: **corticotrophin releasing factor (CRF) receptor antagonists, substance P (neurokinin) receptor antagonists**, and drugs that **modulate glutamatergic transmission.** Abundant evidence suggests that increased production and/or release of **CRF** within the central nervous system occurs in patients with post-traumatic stress disorder and major depression. Preclinical studies show that CRF antagonists have anti-anxiety and antidepressant properties. **CRF receptor antagonists** appear promising as a **novel class** of antidepressants and anxiolytics

* **Pharmacogenomic** research is expected to benefit of rapid, sensitive and inexpensive methods for detecting sequence polymorphisms. One such method already in use involves the use of DNA **microarray** chips. Because it permits the simultaneous monitoring of the expression of tens of thousands of polymorphisms that may be present in the genome of the individual patient, microarray technology   is revolutionizing drug design, genetic diagnosis, and genomics research.

Microarray technology will allow to establish a "**signature**" of DNA sequence variants that are characteristic of an individual. It is hoped that pharmacogenomic testing will help identify cancers that have a high probability of responding to a particular medication or regimen and that it may also be used to help track down the cause of certain rare, serious drug side effects.

* The German Federal Government's **functional proteomics initiative** is aimed at the development of rapid, standardized and automated processes and methods for functional proteome analysis. It focuses on the analysis of quantitative and spatial protein patterns under defined conditions in cells or tissues; the functional analysis of proteins and their interaction with ligands and low molecular weight agents; and the interaction and control mechanisms of complex protein networks. Projects are required to cover the whole analytical process - from the provision of samples to the interpretation and visualization of the analytical results. The implementation, use and further development of bioinformatics applications also form an important part in the projects.

The term "proteome" is being used to define a complete set of proteins encoded by the genome of a given organism. "Proteomics", as a research area, addresses the analysis of quantitative changes of the protein content of a cell, tissue or body fluid (it studies how the proteins respond, interact and change) and can provide valuable insights and approaches for diagnostics and therapy of diseases.

It will take researchers time to understand the proteome. But they will be rewarded with one of the most important **medical breakthroughs.**

* A newly developed transfection system, based on a protein of hyperthermophilic bacteria called **TmHU**, has the potential to substitute common transfection systems. Transfection systems like the TmHU are important tools in molecular medicine, functional genomic studies and High Throughput Screening (HTS).

*Rapid advances in **neuroscience** suggest medication development strategies that have yet to be undertaken. These include drugs that interact with **second messenger systems, response elements and transcription factors**, agents that enhance **neuroprotective** and **neurogenic factors**, and compounds that manipulate **cytokine receptor activity.** Researchers are attempting to develop compounds that target novel brain mechanisms suggested by cutting edge research on the causes of depression

more expensive to make (compared to oral delivery systems), they offer ease of use, reliability, non invasiveness, reduced side-effects and are needle free i.e., pain free[159],[160]. The use of patches is ideal for pediatric and older patient populations according to Erickson's "Drug delivery of the Future ushers in Individualized Care". Erickson also believes that genomics, gene therapy, circadian rhythm, and the prevention of antibiotic resistance will increase the speed of which drug delivery technologies are developed. A key benefit is its reliability in delivering a drug at constant levels over time.

According to Dr. Tugrul Kararli, PharmaCircle LLC, transdermal drug delivery is ideal for low dose drugs (less than 10 mg dose) and drugs that have extensive first pass intestinal or liver metabolism. Current transdermal technologies available from Alza, Noven, Watson LTS Lohmann, Nitto Denko, etc. provide elegant, thin and small patches that can be worn for up to 7 days.

Currently, majority of the hormone replacement therapy and smoking secession products and nitroglycerin for the prevention angina are marketed using transdermal patches, World-wide, well over 25 products are currently marketed by some 75 companies. Today, new technologies available from some new comers, such as Dermatrend and TransPharma are allowing the delivery significantly higher amounts of drugs from this route (40-50 mg range). In addition to above passive transdermal delivery technologies, skin poration technologies available from companies such as, Alza, Becton Dickinson, Norwood Abbey, Altea, Sontra and TransPharma is creating a realistic option for the delivery for small as well as large molecules from the skin. These technologies rely on the removal of the top impervious layer of the skin, stratum corneum using multiple but short needle arrays and pulses of laser, electrical current, ultrasound and pressure pulses. In the future we may see some macromolecular drug products delivered from the skin using some of these poration technologies.

Table 6.9.   Typical transdermal drug delivery system designs

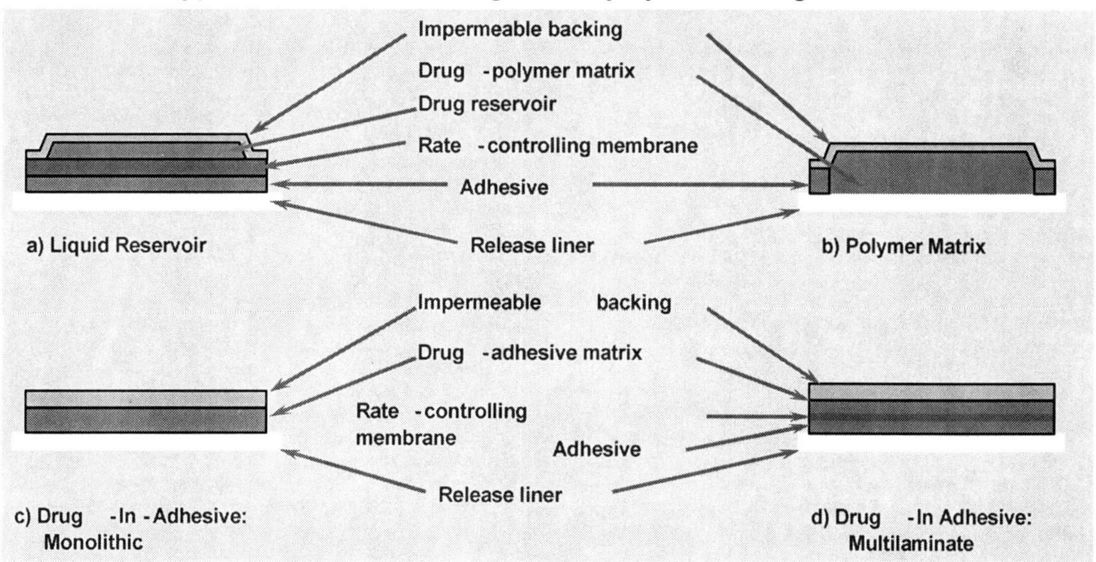

---

[159] "More than Skin Deep – Absorbing Drugs Gently through the Skin Beats Injecting, Swallowing, or Sniffing them". The Economist Newspaper.20Sep2001.TheEconomist.30Aug2002 <http://www.economist.com/displayStory.cfm?Story_ID=S%26%28%280%25QQ%5B%26%0A>.
[160] Papanikolaw, J. Seeking Growth in Alternative Drug Delivery Systems. Chemical Market Reporter. 25 Oct 1999.

*** Socio-economic living conditions in Western society,** where once-dominating family ties and religion have made place for individualism, single-living, divorce, job-losses etc, is expected to give a further boost to the **antidepressants** market. Novel compounds, e.g. **neurokinin antagonists, novel noradrenergics, novel serotonergics** have been announced.

* A great number of drug candidates do not reach the next stage of clinical trials, and only very few ever make it to market, what makes it a difficult undertaking to publish updated lists. Out of the following antidepressant drug candidates, **many will disappear.** The selection made, is intended to show the activity of the research-driven pharmaceutical industry:

~**Blonanserin** (Dainippon pharmaceutical) is a compound with potent D2 and 5-HT2A receptor-antagonist properties.

~**Deramciclane** fumarate (Egis) is an anxyolitic agent with antagonist activity at 5-HT2A and 5-HT2C receptors, which also acts as a GABA reuptake inhibitor.

~**Iloperidone** (Titan/Novartis) is a serotonin/dopamine receptor antagonist (SDA), a broad-spectrum antipsychotic agent.

~**Lesopitron** dihydrochloride (Esteve) is a new nonbenzodiazepine anxiolytic agent with potent 5-HT1A receptor-agonist activity and lacking sedative effects.

~**Nemifitide** ditriflutate (Innapharma) is a low-molecular-weight pentapeptide compound for the treatment of depression.

~**Org-522** (Organon) is a highly potent dopamine D1/D2 and 5-HT2C receptor antagonist in evaluation for the treatment of psychosis.

~**Agomelatine** (Servier) combines melatonin receptor antagonist and selective 5-HT2C receptor-antagonist activity.

~**CP 122721** (Pfizer) is said to be a potent and selective tachykinin NK, receptor antagonist for the treatment of depression.

~**OPC-14523** (Otsuka), is a potent sigma/5-HT1A receptor agonist and 5-HT reuptake inhibitor for major depressive disorder.

~**(R)-Sibutramine** metabolite (Sepracor) is an isomer of an active metabolite of sibutramine, a potent monoamine (5-HT, noredraline and dopamine) reuptake inhibitor.

~**Talnetan** (GlaxosmithKline), a selective, nonpetide tachykinin NK3 receptor antagonist for the treatment of e.g. schizophrenia.

~**SR-58611** (Sanofi-Synthélabo) is a selective beta3-adrenoreceptor agonist for the treatment of major depressive disorder.

~**SM-13496** (Sumitomo) is a drug candidate with 5-HT2A receptor and dopamine D2 receptor-antagonist effects.

~**DU-125530** (Solvay) is a selective 5-HT1A receptor antagonist for the treatment of major depressive disorder

~**Bifeprunox** (Solvay/Lundbeck) is an atypical antipsychotic agent that combines partial agonist/antagonist effects at the dopamine receptor with 5-HT1A receptor-agonist effects.

* The medium-term market volume for **innovative nucleic acid diagnosis (NAD)** is about three to five billion EURO with a forecasted growth rate of 25 percent p.a. for the worldwide NAD-market.

*The invention of compounds that **inhibit protein-protein interactions** is critical to the development of new drugs for the treatment of cancer that modulate intracellular signaling pathways such as those involved in apoptosis. One such a compound could be **MI-147**, designed and developed in the laboratory of Dr. **Shaomeng Wang** at the University of Michigan Comprehensive Cancer Center..A potent, highly **selective small molecule inhibitor** of the **MDM2-p53 interaction**, MI-147 is representative of a new class of molecules that is being developed for the treatment of the approximately **50%** of human cancers that express the **wild-type p53 protein**.

* Human cells function through the **concerted action** of thousands of proteins that control their **growth** and **differentiation**. Yet, the specific function of most human proteins remains either unknown or poorly characterized. Diseases being often due to aberrations in the **function** of **key cellular proteins**, numerous large-scale research initiatives have been launched internationally to crack the function of all human proteins. **Defining the maps of protein interactions that regulate cell growth, differentiation and disease progression** is the overall goal of the **Human Proteotheque Initiative (HuPI),** a forward-looking project conducted in the Coulombe laboratory at the Institut de recherches cliniques de Montréal. Central to the HuPI project is its experimental platform, termed the "HuPI discovery engine", which ultimately generates maps of protein interaction networks. New research reports on the **first generation** of this technology platform that is currently being improved by a multi-disciplinary team of scientists. It describes a powerful **proteomics** approach that promises to have a profound impact on our current understanding of the human proteome and the function of its **individual proteins**. The unique property of proteins, exploited by the ICRM research team, is the fact that proteins **rarely work alone**, but rather assemble with other proteins into complexes to concertedly exert their function. The initial guess of the scientists was that proteins interacting together are likely to be partners in the **same biological pathway** and, consequently, to serve the same (or related) function(s). By systematically identifying the interaction partners of **32 human proteins** known to exert specific functions in gene transcription and RNA processing, they defined an intricate network of **805 high-confidence interactions** that connect together **436 different proteins**. Among them, many proteins of previously unknown function can now be inferred putative functions based on their association. *(Jeronimo C. et al., Systematic analysis of the protein interaction network for the human transcription machinery reveals the identity of the 7SK capping enzyme. Molecular Cell 27, July 20th, 2007. Research led by Dr Benoit Coulombe)*

Biological systems research seeks to increase the understanding of the function, control mechanisms and behavior of biological systems. In order to gain a comprehensive understanding of biological systems, complex biological functions will be simulated; experimental, quantitative data generated and models developed that are suitable for the simulation of physiological processes in cells, cell aggregates and whole organisms. The knowledge generated may be used in a wide range of applications, for instance, the development of new pharmaceuticals and therapies.

VI.15. Ocular drug delivery technologies

The local delivery of drugs to the posterior segment of the eye with ample efficacy and nominal side-effects is at the moment a major concern. The ocular delivery sector is radically changing with new techniques such as iontophoresis, which uses an electrical field to pass medicaments through normally impermeable tissues. According to a UBS-Warburg study, a very small percentage of pharmaceutical sales are derived from drugs to treat disorders associated with the posterior segment of the eye. With iontophoresis technology, this delivery sector could boom to be a $10 billion a year market, however. Acient, Controlled Delivery Systems, Insite Vision, Oculex provide technologies for the sustained delivery of drugs into the eye in the form of inserts or implants or iontophoretic systems.

VI.16. Nasal and inhalation Technologies

The nasal delivery route is a particularly patient friendly system that provides quick drug delivery and is inexpensive compared to parenteral technology[161]. Intranasal delivery could also provide a mechanism of bypassing the blood brain barrier due to the connection between the olfactory and trigeminal nerves. This therapeutic area is still early in development and requires further advancement. Some examples of future products now in development are intranasal morphine and apomorphine (Nastech Pharmaceuticals) for the relief of pain and sexual dysfunction, respectively. West Pharma Services also provide delivery technologies that allow both small as well large molecular drugs to be delivered from the nasal cavity.

Aerosol delivery systems (pulmonary) for the treatment of Asthma, although not new, are available in various forms to include but not limited to metered dose inhalers, dry powder inhalers (DPIs), and nebulizers. Inhalation technologies offer considerable benefits over conventional products as they provide effective therapy for local diseases of the lung, such as asthma and COPD as well as systemic delivery of small and large molecules. For the systemic applications drugs delivered from the lung bypasses first pass elimination associated with the oral delivery route, says Mariana Brea-Krüger of IMI Consulting.

Two main benefits of this technology are that it requires less dosage and offers fewer side-effects. Nevertheless, development work in this area continues with mechanisms for large molecule delivery not far away (e.g. inhaled insulin, such as Exubera). The largest problem associated with the nasal and pulmonary delivery is due to the aggressive nature of the immune system in preventing foreign objects from entering the bloodstream e.g., an immune reaction. Nonetheless, both the nasal mucosa as well as the alveoli of the lungs can be targeted for drug delivery use. A large focus in this sector involves work with reducing particle sizes for improved penetration and also for large molecule delivery[162].

VI.17. More innovation to come in inhalation technologies

The ban on the use of CFCs in new MDIs opened the door to innovation; DPIs and nebulizers. There are currently 65 companies active with MDI, DPI and nebulizer technologies (e.g. Glaxo, Boehringer, AstraZeneca, Aventis, Chiesi, Orion, 3M, Nektar, SkyePharma, Ventaira, Alkermes, Aradigm, Focus, Directhaler, Innovata Biomed, SkyePharma, Microdose, Respirics, Aerogen Vectura, Sofotec, Chrysalis, Meridica, etc).

---

[161] *"Scripts Drug Delivery Companies Fact File 2000". PJB Publications Ltd. 2000:5-15.*

[162] *Rhodes III, W. "Pulmonary & Nasal Delivery of Protein Drugs". Drug Delivery Tech. Jun 2002. Drug Delivery Technology. 30 Aug 2002 <http://www.drugdeliverytech.com/article_arch.shtml>.*

* In addition to advances in **human genomics**, great strides have been made in the discovery of specific genetic markers linked with tumor development, **infectious diseases** such as mycobacterium tuberculosis, markers of well defined fetal anomalies like Tay-Sachs Disease and Fragile X Syndrome. In general, it is now possible to define, with great precision, specific genetic markers that then can be used to **find foreign biological material that may be inside the body**. An important recent discovery found that nucleic acid markers present in blood circulation pass through the kidneys and are **readily detectable in urine**. This discovery shows that kidneys perform the most important and costly steps in isolating genetic markers from the blood stream, and that the markers are efficiently collected in a simple milieu of urine.

Sientists at **Xenomics** were the first to ask whether DNA from cells throughout the body can appear in urine. It was generally taught by scientists and clinicians that DNA was simply too large and had the wrong electronic charge to get through the kidney's filtration apparatus. Xenomics' scientists were well aware of the fact that over one hundred billion cells normally die each day by this process. Along with many cellular components such proteins and lipids, genomic DNA is disposed of by breaking the very large molecules into small, uniform segments of about 180 base pairs during apoptosis. These DNA segments are known as **nucleosomes**, and can be detected in plasma. Until the team of Xenomics scientists made this discovery, scientists were unaware that these DNA segments passed through the kidney barrier and could be found in urine. This newly found DNA was called "**transrenal DNA**," or **tr-DNA** for short. It is especially interesting to note that the genetic signatures of foreign material in the body is not to be found in the sediment of urine, the source of all current urine DNA diagnostic material, but rather in the soluble portion of urine, the portion of each specimen that is currently discarded by laboratories when they attempt to detect DNA markers. The **discovery of tr-DNA** provides a new outlook for the future of molecular diagnostics, one that is expected to have great influence on the way medicine is practiced in the future. Instead of traditional reliance upon blood specimens, tissue biopsies, and other tests, a simple, safe urine specimen can be collected and analyzed using the most sensitive, advanced technologies. Currently the Polymerase Chain Reaction, known as PCR, is the most common technique used today in genomic studies. PCR is also used in tr-DNA based tests to amplify and detect specific DNA markers for molecular diagnostics. Clinics will be able to **reduce exposure to hazardous biological specimens, making the clinic environment safer for both patients and healthcare workers**, since urine is safe whereas blood and tissue are not. Testing laboratories will now be able to run a broad array of molecular diagnostic tests on small urine specimens, using a single platform. (see opposite colon)

Tr-DNA tests can be significantly less expensive to run than either blood or tissue, which require costly manipulation to extract and identify DNA markers. New genetic forms of infectious disease agents are emerging in the USA and worldwide, that are more virulent and often resistant to conventional therapies. These may be detectable in one basic tr-DNA diagnostic test such as for *Mycobacterium* tuberculosis or HIV Proviral DNA. Similar advances are expected in the field of cancer detection and treatment.

The detection of tr-DNA is clearly the only completely safe, non-invasive means to obtain access to the genetic information from foreign material in the body. **Xenomics** intends to make an array of tr-DNA tests available commercially in the coming years. (Source: *L. David Tomei, Ph.D. CEO and founder, Xenomics Inc. New York, N.).*

* Many common diseases are sufficiently **multifactorial** that their causes are poorly understood, and in some cases even the organs in which primary abnormalities lie have been unknown. In this setting, identification of extreme genetic outliers in the population has the potential to lead to identification of **rate-determining pathways** whose variation impacts these traits. These key pathways may identify the key sites at which therapeutic intervention will have the greatest impact on disease treatment or prevention. By collection of such extreme outliers from around the globe, scientists have identified more than **20 disease genes**, including genes that have large effects on blood pressure, bone density, pH and electrolyte homeostasis. These have established the key role of variation in **renal salt handling** in blood pressure homeostasis, the role of **Wnt signaling** in bone density, and the role of **claudin proteins** in paracellular electrolyte flux. These findings have implications for the treatment of both these rare disorders and their common forms. *[Richard Lifton. Chairman of the Department of Genetics at Yale University School of Medicine. "Genetics, Genomics, and the Future of Medicine". Science´s Next Big Ideas. Duke University 75th Anniversary Science Symposium. New York, September 26, 2006]*

* The human body is home to many different kinds of fungi. While the majority normally do not harm us, some fungi can cause unpleasant infections of skin, nails or lungs. A new mechanism to attack hard-to-treat fungal infections has been revealed by scientists from the biotech company **Anacor** Pharmaceuticals Inc. and the **European Molecular Biology Laboratory** (EMBL) outstation in Grenoble, France. They described how a new compound called **AN2690** kills fungal pathogens by blocking their ability to make an enzyme crucial for their protein synthesis. AN2690 interferes with an enzyme called **leucyl-tRNA** synthetase. The process begins when the cell makes an RNA version of the gene's code, called messenger RNA. Ribosomes then translate the messenger RNA into protein by stitching together the amino acids in the order specified by the message. This requires the help of molecules called tRNAs, which link the code of the messenger RNA to the correct amino acid.

Cancer transforms into chronic disease. Metastatic cancer may not yet be cured, but medical oncologists will be able to keep it under control for long periods of time. New, specific genotypes of tumors or viruses and bacteria will be discovered, giving an ever-sharpening focus to attempts to not only detect diseases, but also more accurately define the patient's prognosis and response to drugs.

With already many approved DPI products & many in pipeline, this delivery form is predicted to expand. In the future we should expect to see more products marketed with better devices which are more patient friendly, provide more efficient drug delivery to the lung, are moisture proof, provide multiple doses of drugs, monitor patient breathing, more efficient in terms of delivering drug dose out of the device into deep lung.

VI.18. Parenteral drug delivery technologies

One of the first reported uses of parenteral technology used to relieve neuralgia dates back to the 19th century. While it may be a very old technique, it remains one of the most widely used and least preferred delivery methods by patients because of its painful administration, be it real and/or perceived. Frequently, opponents of the parental delivery system quote the need for a medical professional to administer the drug, the accompanying costs and the inconvenience of doing so. These arguments, fortunately, are not strong enough to render this technology obsolete.   It continues to enjoy moderate growth a relatively large share of the delivery market[163].[164]

Advanced technologies that will allow specific targeting and sustained/controlled release of parenteral medications are in ongoing development. They include but are not limited to liposome, monoclonal antibodies, niosomes, nanoparticles, cyclodextrins, emulsions, polymer drug conjugates, implant systems and needle free injection systems[165]. In a report by The Freedonia Group, the parenteral sector is expected to experience moderate growth with its market share increasing to $23 billion in 2006.   It is the principal mechanism for the delivery of proteins and peptides today and it will continue to be so in the short term until competition emerges from other delivery technologies such as transdermal and/or pulmonary. According to PJB Publications, the peptide and protein pharmaceuticals markets was valued greater than $10 billion (in 2000) with significant expectations for market share enlargement. This should not be a concern as hundreds of biotechnology companies are working to produce large molecule drugs which are ideal for parenteral delivery.

The current focus for parenteral technology is improvement in targeting and in the bioavailability of drugs[166].

VI.19. Injectable Biodegradable Depot Drug Delivery technologies

This sector should be of great interest to pharmaceutical and biotechnology companies since it may be one of the surest way of providing added value to products and convenience to patients. In early 2004, there were 35 companies (45 technologies) active in this field. Among the important Rx products currently marketed are Lupron Depot, Zoladex, Nutropin Depot, Sandostatin, Risperdal Consta, Plenaxis and DepoCyt. It is fair to say that, In the future, we should see more injectable depot technologies as well as products on the market.

Injectable and implantable biodegradable depot technologies in the form of microcapsules, rods and lipid dispersions provide patient convenience, and enhanced efficacy and safety. . Many chronic medications that are injected on a daily basis could be delivered with depot formulations much less frequently, in some cases once every six months.

---

[163] Thomas J, Brea-Krüger M, IMI Consulting, Burgwedel, Germany. Company data on file

[164] Thomas J, Brea-Krüger M, IMI Consulting, Burgwedel, Germany. Company data on file

[165] Uchegbu, I. "Parenteral drug delivery".   PJ Online.   28 Aug 1999.   Pharmaceutical Journal.   30 Aug 2002 <http://www.pharmj.com/Editorial/19990828/education/parenteral.html>.

[166] Evers, P. "Developments in Drug Delivery: Technology & Markets". London: Pearson Professional Ltd, 1997.

VI.20. Sustained release implants, circumventing the digestive tract

An unquestioned advance in the daily routine of patients, a new technique might become popular with the major pharmaceutical companies, as it would constitute an added-value for traditional drugs which could be presented in a sustained-release formulation[167].

A drug is always the result of a subtle marriage between a compound (the active substance which treats a patient) and an appropriate excipient (a neutral substance in which the active substance is incorporated so that it can be absorbed by the body). But what would happen if biodegradable materials were used instead of these neutral excipients? This was already possible with some active substances, and can now be applied to many others. Researchers[168] have indeed managed to develop a novel synthetic process for these materials which significantly increases their diversity.

Some biodegradable polymers such as polyesters have already been employed as excipients in pharmacology. But this was only possible when they were mixed with certain active substances such as anticancer drugs or growth hormones. And in the field of surgery, the secret of absorbable sutures does indeed reside in the use of these same polyesters.

The biodegradable excipient containing the active substance can take the form of an implant - a 1 cm-long rod about one millimeter in diameter - which is inserted just under the skin. This procedure is performed by a doctor and only takes a few minutes. The specificity of these polyesters is that they can be hydrolyzed; in other words, broken down by water, which is unlucky for them, as our bodies are full of this substance. Thus the excipient gradually breaks down, depending on its type, releasing the active substance it contains. Hence the major advantage of the technique: biodegradable excipients enable the controlled administration of sustained-release drugs. This is of considerable benefit in the setting of chronic diseases, as it avoids frequent, repeated intakes of medicines. This method also reduces side effects; by circumventing the digestive tract, the active substance passes directly into the bloodstream. Thus the quantity of drugs administered can be reduced, as there is no longer any need to allow for its partial destruction as it passes through the digestive tract.

Until now, the knowledge to make these biodegradable polymers was limited to the use of two monomers (the basic components of polymers), lactide and glycolide. The polyesters obtained cannot be combined with just any active substance. Without accounting for the fact that when using these little reactive lactide and glycolide monomers, industrial preparation of the polymers requires lengthy reaction times at high temperatures (e.g. several hours at 140°C-160°C).

In collaboration with Isochem, Didier Bourissou's team has developed a new synthetic process for these polymers. This involves new elementary building blocks, the O-carboxy anhydrides, which are much more reactive, so that the polyesters can be prepared under much less harsh laboratory conditions (e.g. a few minutes at 25°C). And above all, a much wider variety of polymers is available, thus multiplying the chances that an active substance will find its appropriate biodegradable excipient.

---

[167] *Thillaye de Boullay O. et al, An Activated Equivalent of Lactide toward Organocatalytic Ring-Opening Polymerization O. Thillaye du Boullay, J. Am. Chem. Soc. 27 Dec. 2006, 128, 16442.*

[168] *A research team led by Didier Bourissou in the Laboratory for Fundamental and Applied Heterochemistry (CNRS/University Toulouse 3, France),*

* Blood cells have limited lifespans, which means that they must be continually replaced by calling up reserves and turning these into the blood cell types needed by the body. Claus **Nerlov** and his colleagues at the European Molecular Biology Laboratory (**EMBL**) unit in Monterotondo, Italy, in collaboration with researchers from **Sten Eirik Jacobsen's laboratory** at the University of Lund in Sweden, have now uncovered how an intracellular communication pathway contributes to this process. Their study proves that **beta-catenin** plays a central role in determining whether blood cells form or not. On the other hand, an overactive **Wingless** pathway (a sequence of molecules which manage the flow of information within the cell) doesn't seem to damage cells that already exist. Thus beta-catenin seems to be a decision-maker, a selector of how information gets routed within the cell, rather than something which maintains the vitality of existing cells.

* Typically, **small molecule drugs** work by binding to a target protein and either preventing it from functioning or ensuring it functions better or at different times. Antisense drugs hoped to do the job of these inhibitors but by **preventing** the protein targets from **ever being produced**. These types of drugs are highly specific. They discriminate between protein targets based on genetic information, as opposed to small molecule drugs that often bind to several proteins with similar structures. The more specific a drug, the fewer side effects can be expected. Also, as they are all made of nucleotides, **antisense drugs** should have **predictable metabolism mechanisms**, which should lead to shortened development times and a reduced number of failures in the early stage of development. However, till early 2007, this has not proved to be the case. It is said that there are at least 12 different ways that mRNA is destroyed and these mechanisms are not fully understood. Only one antisense drug, Vitravene (**fomivirsen**), a treatment for cytomegalovirus retinitis (CMV) in immunocompromised patients, has ever been approved by the US FDA: Other antisense-like technologies use double stranded oligonucleotides to bind to mRNA, an approach known as **RNA interference**. Drugs based on this RNAi technology are still in their infancy but are believed to have more promising results. *[Nagle M.,Antisense drugs stop making sense? BiopharmaReporter.com.Accessed Jan 4,20079].*

* Danish and Belgian researchers have made a crucial breakthrough finding in the understanding of heritable disorders, and a breakthrough for systems biology as a research strategy in the field of genetics and disease. They found a **computer key that maps genes underlying heritable disorders**, such as breast cancer, multiple sclerosis, and Alzheimer's disease. These results will possibly ease the discovery of new medicines and improve treatment in various disorders. Heritable disorders will be easier to interpret for clinicians using the new results. Furthermore, the identification of new genes likely to be involved in disorders will help patients with defects in these genes. For example, if one is at a high risk carrier of a gene that underlies a disease such as Type 2 diabetes, physicians could prevent or delay the manifestations of the disease by dietary guidance early in life.

* Determination of the **human genome sequence** and the development of **microarray technologies** allowing the rapid measurement of all genes in the genome have generated new perspectives for **biomedical research**. **Gene expression analysis** will make a major contribution to the insight into the **underlying biology** of disease and will lead to improved methods for diagnostics, prognosis and treatment. Microarray studies create the possibility to subclassify patients with diseases such as rheumatoid arthritis, diffuse large B-cell lymphoma and breast cancer, with both prognostic and therapeutic consequences. The simultaneous quantification of the activity of all genes in tissues or cells from patients by microarray technology, linked to the clinical parameters, creates a large number of data points, which cannot be analyzed without the aid of the advanced application of **bioinformatics**. As a result, genomic research has become, in part, a **bioinformatics discipline** that will be integrated with **clinical medicine**. The microarray technology makes it possible to develop **personalized medicine**, with a more accurate diagnosis and prognosis for every patient and subsequently a **tailored treatment strategy.** *[Van der Pouw Kraan CT. Molecular unraveling of disease by means of DANN-mocroarrays. Ned Tijdschr Geneeskd. 2005 Mar 19;149(12):618-22.]*

* Rutgers researcher Richard H. **Ebright** and his collaborators have resolved key questions regarding **transcription**, the **fundamental life process** that was the subject of the **2006 Nobel Prize in Chemistry**. Transcription is the first step in the process cells employ to read and carry out the out instructions contained in genes. Transcription is carried out by a molecular machine known as **RNA polymerase**, which synthesizes an RNA copy of the information in DNA. Ebright and collaborators defined for the first time the mechanisms by which the machine begins synthesis of RNA and then breaks free from its initial binding site to move along DNA to continue synthesizing RNA. The results establish that during transcription initiation the machine remains stationary at its initial binding site and **"reels in"** adjacent DNA segments, unwinding these segments and pulling the unwound DNA strands into itself. This remarkable mechanism, termed **"DNA scrunching,"** enables the machine to acquire and accumulate the energy it needs to break its binding interactions with the initial binding site, and to begin to move down the gene.
The findings were made possible by newly developed, single-molecule methods. They enabled the researchers to analyze and manipulate individual molecules of the machine, one-by-one, as they carried out reactions. The discoveries significantly advance the understanding of the structure and function of the molecular machine that carries out transcription, setting the stage for new opportunities in combating the bacterial diseases that kill 13 million persons each year worldwide.

* **Unmet patient need** is the major driver of innovation in both cancer therapeutics and diagnostics. There is significant need for high-sensitivity diagnostic methods to detect the presence of early-stage disease.
Small interfering RNAs are emerging as potential innovative cancer therapies. Whilst the delivery of these molecules remains limited, the number of candidates in development is increasing, with over 25 cancer vaccines in development..

231

VI.21. PEGylation technologies to deliver protein and cancer drugs

Chemical conjugation of polyethylene glycol (PEG) polymers to large and small molecules has several benefits from the formulation as well as the therapeutic outcome sides. In general, PEGylation of large molecules reduces immunigenicity, prolongs plasma half life and improves targetability leading to safer and more efficacious drugs. For small molecules PEGylation has been used to improve water solubility as well as targeting aspects of drugs. In general, the proven benefits of PEGylation technologies in drug safety and efficacy exemplify the true benefits of drug delivery technologies[169].

The PEGylated protein and cancer drugs segment is dominated by Enzon and Nektar. Valentis and Prolong are the new comers. A growing number of important prescription drugs can be noted (e.g. Neulast, PEG-Intron) as well as deals between the technology companies offering these technologies and pharmaceutical and biotechnology companies.. A next generation of PEG substitutes (e.g. NanoPharm) is actively being developed.

VI.22. Targeted drug delivery

"This is an area where drug delivery technologies will revolutionize drug therapy. Targeted products for cancer, immune and other diseases add real value to patients by making drug products less toxic and more efficacious and convenient . This segment should be expected to generate new markets for the existing highly efficacious but toxic cancer drugs as well as future drugs for cancer and immune diseases", says Dr Kararli There are well over 42 companies with targeted drug delivery technologies. We have already seen significant progress in this field from companies with liposome and nanoparticulate technologies. Liposomal formulations available from Alza , Neopharm, Inex, Gilead and other companies, for example, limit the free drug in the blood by keeping them encapsulated in the liposome particles. Furthermore, due to leaky nature of blood vessels in cancer tissues, nanosize lipid vesicles can be taken up preferentially by cancer tissues. Doxil, for example, which is a liposomal doxorubicin product from Alza has been enjoying market growth in Ovarian cancer and KS applications in AIDS patients .

VI.23. Challenges / Future opportunities

It should be expected that oral drug delivery would still be the favorite route unless drugs can not be delivered from this route.   Within oral drug delivery we may one day see technologies that will allow once or twice a weekly delivery of drugs.

Stomach retentive technologies of the future will allow us to develop best modified release drug products in terms of the timing and the speed of drug release. In the oral delivery field we should expect to see more products using melt tablets as well as water insoluble drug technologies. Because of its importance, investment and efforts in the development of effective delivery technologies for the oral delivery of macromolecules will continue.   In the near future, if FDA approves Nektar's inhaled insulin we may see insurgence of other inhalable macromolecular drugs entering into market place and this will have a positive effect as a whole for the Drug Delivery industry. Effective drug delivery to brain although is a challenge currently will also see its share of success in the not too distant future. Targeted drug delivery has the chance to revolutionize drug treatment by providing safer and more efficacious delivery for cancer and immune diseases . Injectable depot formulations will keep adding real value to injected

---

[169] Tugrul T. Kararli, Correspondence with the author, May 2, 2004

* Using a microscope and some extreme "snapshot" photography with shutter speeds only a few nanoseconds long, researchers from the National Institute of Standards and Technology (NIST) and Cornell University have uncovered the traces of ephemeral **"nanobubbles"** formed in boiling water on a microheater. Their observations* suggest an added complexity to the everyday phenomenon of boiling, and may affect some proposed cancer therapies. The findings may impact proposed **thermal cancer therapies** in which **nanoscale objects** are designed to **accumulate** in tumors and are **subsequently heated remotely** by **infrared radiation** or **alternating magnetic fields. Each particle acts as a nanoscale heater,** with nanobubbles being created if the applied radiation is sufficient. The **bubbles** may have a **therapeutic effect** through additional heat delivered and mechanical stresses they may impart to the surrounding tissue. *[R. E. Cavicchi and C.T. Avedisian. Bubble nucleation and growth anomaly for a hydrophilic microheater attributed to metastable nanobubbles. Phys. Rev. Lett. 98, 124501 (2007)].*

* **Erythropoietin (EPO),** a naturally occurring hormone that stimulates the production of red blood cells, is used to treat some forms of anemia. In the fetus EPO is produced by the liver. By contrast, in the adult most EPO is produced by the kidneys but some is produced by liver cells known as hepatocytes, particularly when levels of oxygen in the body are low. The role of two related proteins, **HIF-1** and **HIF-2** in EPO production by hepa-tocytes has been controversial. But now, researchers from the University of Pennsylvania, Philadelphia, have shown that in mice HIF-2 regulates hepatocyte produc-tion of EPO. This finding will impact the development of new approaches to treat individuals with **anemia.**

* Based on clues provided by a study with transgenic mice, a research group at the National Human Genome Research Institute (NHGRI), part of the National Institutes of Health (NIH), has developed a strategy that will be tested as the first treatment for people with **hereditary inclusion body myopathy (HIBM).**

HIBM is a genetic muscle disease with non life-threatening symptoms that emerge in adulthood and lead to slowly progressive muscle weakness. Most patients develop symptoms while in their early 20s and become wheelchair-bound by the time they reach 40, as their arm, hand, leg and core muscles progressively weaken. It is caused by a mutation in the "GNE" gene, which codes for two enzymes that produce sialic acid, a sugar important to muscle development and kidney function. The new research focused on a form of HIBM **in Iranian-Jewish families,** which is caused by a specific mutation in the **"GNE" gene** known as **M712T.** Numerous other "GNE" mutations can cause HIBM and occur in populations worldwide. At this time, there is no treatment available for any type of HIBM. In their search for potential HIBM treatments, the researchers drew upon previous evidence that **impairment of the enzymes** that **promote sialic acid** production causes low sialic acid levels in muscle proteins. They hypothesized that a compound called N-acetylmannosamine, or **ManNAc,** a sugar that is naturally converted to sialic acid, might have an impact on the muscle weakness caused by HIBM.

B.69. newer anticoagulant agents

class of compounds		company	launch
parenteral pentasaccharides			(est.)
SR-34006	idraparinux	Organon/Sanofi	2008
SSR-126517		Sanofi-Aventis	2011
oral direct thrombin inhibitors			
BIBR-1048	dabigatran	Boehringer	2008
tissue-Factor VIIa complex inhibitors			
NAPc2			
oral direct Factor Xa inhibitors			
BAY-597939	rivaroxaban	JNJ/Bayer	2008
BMS-562247	apixaban	BMS	2011
YM-150		Astellas	2013
DU-176b		Daiichi	2013
PRT-054021		Portola	2013
LY-517717		Eli Lilly	2013
813893		GSK	2013

New **anticoagulants** are under development to overcome the limitations of **heparin** and low-molecular-weight heparin, whereas novel orally active anticoagulants have been designed to provide more streamlined therapy than **vitamin K antagonists.** Newer anticoagulant drugs such as parenteral pentasaccharides (idraparinux and SSR-126517-E), oral direct thrombin inhibitors (dabigatran), oral direct Factor Xa inhibitors (rivaroxaban, apixaban, YM-150 and DU-176b) and tissue factor-Factor VIIa complex inhibitors (NAPc2) are tailor-made to target specific procoagulant complexes and have the potential to greatly expand our antithrombotic armamentarium for both acute and long-term treatment of VTE, especially as non-monitored parenteral and oral anticoagulants with a wide therapeutic window and a predictable anticoagulant response.*(Spyropoulos AC, Investigational treatments of venous thromboembolism.Expert Opin Investig Drugs, 2007 Apr;16(4):431-40).*

B.70. Factor Xa

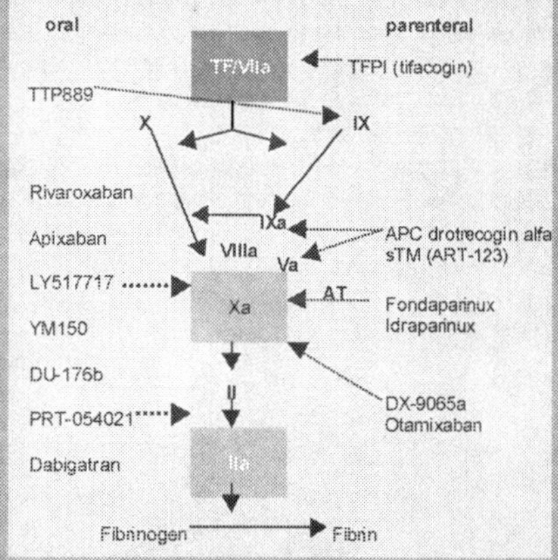

Adapted from Weitz&Bates, J Thromb Haemost, 2005

(Factor Xa is a validated target (one for which there are approved drugs on the market), and inhibiting its activity is believed to have superior anticoagulant properties compared to other targets such as thrombin)

macromolecular drugs and provide both convenience and improved efficacy and safety to patience. Finally delivery of large molecular drugs from the skin by poration technologies may be a reality within the next 5-10 years[170].

With the speed of innovation, DDS is one of those sectors with considerable investment opportunity (as does biotechnology), but determining value in a product pipeline that is often far from generating immediate revenue remains a challenge for investors.[171] Big Pharma's lack of pursuit for products with <$100-200 Mil. Is becoming a blessing for Specialty Pharma which will introduce an increasing number of enhanced drug products formulated with advanced drug delivery technologies with new indications/routes. Scientists worldwide contribute to the creation of a whole new world of medicines.

What follows are some examples of research findings (RNA, Nanotech, Microarrays, Stem cell, Genetics, and many more researched areas of high interest) from a sampling of scientific developments and fascinating trends to watch in the coming years.
1.   Representing the ingredients of life
2.   Understanding genome stability in humans
3.   Engineering RNA for medical purposes
4.   The sense/antisense madness: a role for dueling RNAs
5.   RNAi Shows Promise In Gene Therapy
6.   siRNA delivery into human T cells and primary cells with carbon-nanotube transporters
7.   Watery Nanoparticles Deliver Anticancer Therapy
8.   Tumor-targeting capability of nanoparticles
9.   Increasing the in-vivo lifetime of polymeric nanoparticles
10.   Magnetic nanocrystals carry and release anticancer agents
11.   Nanocrystaline coating
12.   Pairing nanoparticles protein
13.   New drug-delivery system using nanomaterials
14.   A new approach to treating intractable tumors
15.   Nanobodies: the race to develop a new class of therapeutics
16.   Quantum dots and quantum rods
17.   Molecular 'cages' to deliver drugs
18.   One more step closer to tailor-made molecular medicine for patients.
19.   Microarrays: Making a family tree for disease
20.   Microarrays: cancer to get personal
21.   Driving the discovery of genes associated with a host of common diseases.
22.   Natural electric fields in Regenerative Medicine
23.   Refurbishing diseased or damaged tissue using the body's own healthy cells
24.   'Sticky' proteins, adult stem cells, and repairing hearts
25.   Stem cell future shaped by epigenetics
26.   Jumonji enzymes and cell regeneration
27.   Tackling diseases influenced by different types of T cells
28.   Autoimmune disease and the early development of immune system B cells
29.   Vaccines for people with compromised immune systems
30.   Genomics for treating people with depression
31.   Taking heart failure to the MAT1
32.   Changing the way heart disease is detected and treated in women
33.   Detecting heart disease long before the disease presents itself
34.   Unlocking genetic risk factors for tobacco addiction disorder
35.   Different types of proteins in sperm and the mysteries of infertility
36.   Using bi-functionalized dendrimers to discover disease-causing proteins
37.   International stem cell research

---

[170] *Tugrul T. Kararli, Correspondence with the author, April 27, 2004*
[171] *Mariana Brea-Krüger, IMI Consulting GmbH, Burgwedel/Germany, in a Meeting with the author, March 2004*

Scientists at Penn State University revealed the first **complete high-resolution map** of important structures that control how genes are packaged and regulated throughout an entire genome. The map pinpoints the locations of certain **key gene-controlling nucleosomes** - - spool-like structures that wrap short regions of DNA around a protein core. The research suggests how these nucleosomes, positioned at important transcription-promoter sites throughout the cell's DNA, control whether or not a gene's function can be turned on in a particular cell.

The study's many surprising findings together reveal an intimate relationship between the architecture of nucleosome structures and the **underlying DNA sequences** they regulate. The scientists now know exactly where these nucleosomes are positioned on the DNA molecule and which DNA building blocks they have wrapped up under their tight control. Among those building blocks, the research revealed the architecture of a critical gateway, controlled by the nucleosome, which must be unlocked before a gene can be transcribed. The study revealed that almost all genes have the same kind of structure where transcription begins, that this beginning contains a critical gateway for transcription, and that the transcription gateway of each gene almost always is **located at the same place** on a nucleosome.The study also revealed that the nucleosomes at the transcription-promoter control centers occupy several overlapping positions on the DNA molecule, typically 10 base pairs apart, which exactly matches the periodic rotation of the DNA double helix. This result powerfully simplifies previous theories about the possible architecture of gene packaging. There is a certain DNA sequence that shapes the gene's architecture in the same way, producing the same structure in every gene.The overall sequence of DNA building blocks is different in each gene, but the underlying architecture is the same.

Another discovery is that transcription-control centers tend to be located on the **outside edge** of the **nucleosome** and tend to **face outward** on the **DNA helix**, allowing the cell's transcription proteins to find them more easily. According to the scientists, this arrangement makes sense, because when signaling proteins arrive at a control center they are well situated to help push the nucleosome out of the way so the reading of the gene can begin.

The researchers think that the function of the nucleosome is to **control the gateway to transcription**. The paper by professor Pugh's team marks the leading edge of a new wave of anticipated discoveries about **gene regulation**, made possible by recently developed laboratory equipment for high-volume, or massively parallel, DNA sequencing. Traditional DNA sequencing methods processed one DNA strand at a time, but now researchers can sequence hundreds of thousands of DNA strands at once, rapidly learning incredible amounts of new information. *Barbara K. Kennedy, Penn State University press release. "Scientists Reveal Structure Of Gateways To Gene Control". The study is published in the March 29, 2007 issue of Nature).*

* A **genetic mutation** that **does not** cause a **change** in the **amino acid sequence** of the **resulting protein** can still **alter** the protein's **expected function**, according to a study conducted at the National Cancer Institute (NCI). The study shows that mutations involving only single chemical bases in a gene known as the multidrug resistance gene (**MDR1**) that do not affect the protein sequence of the MDR1 gene product can still alter the protein's ability to bind certain drugs. Changes in drug binding may ultimately affect the response to treatment with many types of drugs, including those used in chemotherapy.

The genetic mutations examined in this research are known as **single nucleotide polymorphisms** (SNPs) and are very common. Some SNPs do not change the DNA's coding sequence, so these types of so-called 'silent' mutations were not thought to change the function of the resulting proteins.

This study provides an **exception** to the **silent SNP paradigm** by showing that some minor mutations can profoundly affect normal cell activity. According to the scientists, these results may not only change the thinking about mechanisms of drug resistance, but may also cause us to reassess our whole understanding of SNPs in general, and what role they play in disease. *[Kimchi-Sarfaty C. et al, A "silent" polymorphism in the MDR1 gene changes substrate specificity. "Science Express", December 21, 2006].*

* Researchers at **Emory University's Winship Cancer Institute** have identified a **novel biomarker** for **brain tumors** and have uncovered a **potential role** the **marker may play** when the **tumor spreads** or **comes back after treatment**. The biomarker, a protein known as "**soluble attractin**," is normally **absent** in the central nervous system (**CNS**) and is **undetectable** in cerebral spinal fluid (**CSF**), **unless malignant astrocytomas**--the most common form of intracranial tumors--**are present** in the **CNS**. The CSF is a liquid that bathes the brain and acts as a reservoir, which can be sampled for analysis of proteins secreted by CNS tumors.

This newfound ability to reliably identify biomarkers for malignant astrocytomas means that physicians will have a **new minimally invasive method** to track the **success** of **treatments**. These biomarkers, singly or in combination, will provide a **fingerprint** of the disease and be able in the future to better define the disease, predict what kind of treatment to use and allow doctors to monitor how well the tumor responds to treatment.

Equally important, the researchers also found that attractin plays a key role in the **motility** of the cancer cells, **which influences their ability** to **spread** in the brain, a major cause of recurrence of malignant astrocytoma. Secreted by the tumors themselves, attractin **induces** cancer cells to **migrate**, although that mechanism is not well understood. Knowing that attractin modulates the migration and possibly the recurrence of these tumors makes it a promising target for **future therapeutic intervention**.

Complex genetic-environment interactions will be identified. As long as people continue to smoke, health inequity will continue to exist in smoking related diseases. Will society turn to lifestyle changes in order to limit the burden of disease for the individual and the cost of treatment to the healthcare providers?

Chapter Seven

An Ongoing Search
for
Better Medicines

VII.1. Representing the ingredients of life

Researchers led by professor Wishart at the University of Alberta, Canada, and Principal Investigator at NRC, National Institute for Nanotechnology, have announced the completion of the first draft of the human metabolome, the chemical equivalent of the human genome[172].

The metabolome is the complete complement of all small molecule chemicals (metabolites) found in or produced by an organism. By analogy, if the genome represents the blueprint of life, the metabolome represents the ingredients of life.

The scientists have catalogued and characterized 2,500 metabolites, 1,200 drugs and 3,500 food components that can be found in the human body.

The researchers believe that the results of their work represent the starting point for a new era in diagnosing and detecting diseases.

They believe that the Human Metabolome Project (HMP), which began in Canada in 2004, will have a more immediate impact on medicine and medical practices than the Human Genome Project, because the metabolome is far more sensitive to the body's health and physiology. Metabolites are the canaries of the genome. A single base change in our DNA can lead to a 100,000X change in metabolite levels.

This $7.5 Million project funded by Genome Canada through Genome Alberta, the Canada Foundation for Innovation (CFI), Alberta Ingenuity Center for Machine Learning, and the University of Alberta will have far reaching benefits to patient care. "The results of this research will have a significant impact on the diagnosis, prediction, prevention and monitoring of many genetic, infectious and environmental diseases," stated Dr. David Bailey, President and CEO of Genome Alberta.

---

[172] „Canadian Researchers First To Complete The Human Metabalome". University of Alberta press release, Jan 29, 2007. The research is published in the journal Nucleic Acids Research

* In almost all forms of **heart failure**, the heart begins to express genes that are normally only expressed in the fetal heart. Researchers have known for years that this fetal-gene reactivation happens, yet not what regulates it. Research at the University of Pennsylvania resulted in the discovery that an enzyme important in fetal heart-cell development regulates the enlargement of heart cells, known as cardiac hypertrophy, which is a precursor to many forms of congestive heart failure (CHF).

It was found that by inhibiting the enzyme **HDAC** in adult mice the fetal-gene program can be prevented from restarting. In various mouse models of cardiac hypertrophy and heart failure, treatment with chemical HDAC inhibitors or genetic deletion of HDAC2 **prevented the beginning of the downward slide to progressive heart failure**.

HDAC is an enzyme switch that regulates how DNA is packaged inside the cell, and therefore how large groups of related genes are turned on and off. During development HDAC normally regulates proliferation of heart cells in the embryo. According to the scientists, this makes sense if a molecular pathway in which HDAC has a major role is **re-expressed**--the adult heart instead makes the cells it already has bigger since it is unable to make more cells very easily. The researchers also found that HDAC works in the heart in part by regulating expression of another enzyme called **Inpp5f**, which is **involved in a pathway that controls the growth and multiplication of cells**. Inpp5f is also related to tumor-suppressor genes involved in cancer. HDAC and Inpp5f gives medical research **new targets** for **regulating cardiac hypertrophy**. HDAC inhibitors are already **in trials for cancer** and one, **valproic acid,** has been used for years to treat seizures. Most CHF medications are aimed at regulating blood pressure, but very few are targeted at the heart-muscle cells themselves. The study paves the way for new targets for treating cardiac hypertrophy and heart failure.*[Dr. Jonathan A. Epstein, the W.W. Smith Endowed Chair for Cardiovascular Research at Penn (University of Pennsylvania School of Medicine) in a statement to the press, „Fetal Heart-Cell Enzyme And Onset Of Heart Failure", The study appeared in an advanced online publication of Nature medicine, February 2007. february 8, 2007. PENN Medicine consists of the University of Pennsylvania School of Medicine (founded in 1765 as the USA´s first medical school) and the University of Pennsylvania Health System]. It is a $2.9 billion enterprise dedicated to the related missions of medical education, biomedical research, and high-quality patient care.*

* Growth in molecular diagnostic technologies is a driving factor leading major pharmaceutical companies to incorporate a new category of medicine -- **Theranostics** -- into their drug development strategies, according to a prediction in a new industry Trend Report from FIND/SVP. The result of these efforts is tests that reveal how well patients respond to particular therapies.

Theranostics is key to improving the efficiency of drug treatment by helping doctors **identify patients who are the best candidates for the treatment in question**. In addition, the adoption of theranostics could very well eliminate the unnecessary treatment of patients for whom therapy is not appropriate, resulting in significant drug cost savings for these patients.

* Although **EGFr testing is** employed in many clinical trials to select patients for EGFr-targeted therapies (e.g., MAb, TKI), and it is required by the FDA for selection of patients for treatment with Erbitux, its utility as a predictor of medication response is not yet clear. Testing for gene amplification or overexpression of HEr2, an EGFr family receptor, is well established, required by FDA, and offers **theragnostic value** for treatment of women with breast cancer with Herceptin (trastuzumab; Genentech). Most EGFr studies to date, however, do not supply data that indicates which subsets of patients may derive the most benefit from EGFr-targeted agents. EGFr is overexpressed in varying degrees in a wide variety of malignancies. Many **new tests of EGFr expression and mutations are in development** and a large number of clinical trials is underway to elucidate their roles in **determining patient selection for EGFr-targeted therapies**.*(New Medicine's Oncology KnowledgeBASE, April 4, 2007)*

* New research published by scientists from The Institute of Cancer Research, funded by Cancer Research UK and the Wellcome Trust, reveals for the first time the biological properties of a promising new class of synthetic potential anti-cancer drugs.

The compounds are from a new group of inhibitors known as **HSP90 (Heat Shock Protein 90)** inhibitors. HSP90 is a key cellular protein that has been shown to be a particularly appealing target for anti-cancer drugs as its inhibition causes the breakdown of multiple cancer causing proteins. Previous HSP90 research has demonstrated that these inhibitors could be developed into effective treatments for a range of cancers including: prostate, breast, colon, ovarian and skin. The research provides the first detailed description of the lead compound - CCT018159 - **a 3,4 diaryl pyrazole resorcinol HSP90 inhibitor** discovered by the team. These molecules work by binding into a key region of the HSP90 protein in place of an ATP (Adenosine Triphosphate) molecule. CCT018159 is particularly exciting as it delivers a **multi-pronged attack** on cancer cells, causing them to 'commit suicide' and also preventing the spread of the cancer cells and the formation of new tumor blood vessels (angiogenesis). *(Cancer Research, Volume 67, issue 5).*

* Researchers at the **National Cancer Institute** (NCI),have shown in mice that just a few **modifications** to a **shortened version** of a **bacterial toxin**, called **PE38,** may be able to improve the efficacy of an immunotoxin that contains PE38 as a **new therapy for cancer.** As part of this new study, scientists have modified the PE38 toxin in an effort to broaden the range of the types of tumors that can be treated successfully. Immunotoxins, which are **antibodies linked to toxic proteins**, can selectively bind to cancer cells and kill them and immunotoxins containing PE38 have already been used as effective treatments for certain types of leukemias and lymphomas. **Immunotoxin therapy** is one of a growing number of treatments in which scientists engineer molecules to target cancer cells and leave healthy cells unharmed.

Significant barriers remain before theranostics can be broadly applied to medical treatment. Patient privacy concerns and the threat of discrimination by insurers are issues that need to be addressed. The associated costs for maintaining all of the information from personalized medicine will also require a significant investment.

According to the researchers, the metabolome is exquisitely sensitive to what a person eats, where they live, the time of day, the time of year, their general health and even their mood. The HMP is aimed at allowing doctors to better diagnose and treat diseases. Most medical tests today are based on measuring metabolites in blood or urine. Unfortunately, less than 1% of known metabolites are being used in routine clinical testing. If you can only see 1% of what's going on in the body, obviously a lot is going to missed.

Table 7.1.   Drug targets[173]

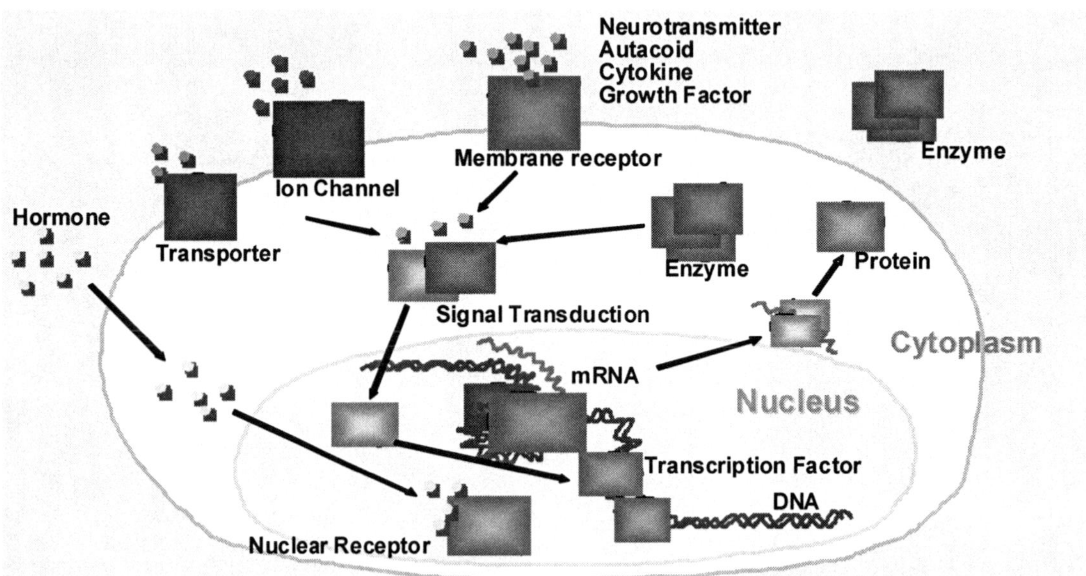

By measuring or acquiring chemical, biological and disease association data on all known human metabolites, the HMP Consortium, which consists of some 50 scientists based at the University of Alberta and the University of Calgary, compiled the remaining 95% of all known metabolites in the human metabolome. Detailed information about each of the 2500 metabolites identified so far can be found on the Human Metabolome Database (HMDB) at http://www.hmdb.ca. With the data in the HMDB, anyone can find out what metabolites are associated with which diseases, what the normal and abnormal concentrations are, where the metabolites are found or what genes are associated with which metabolites.

By decoding the human metabolome, the researchers state that they can identify and diagnose hundreds of diseases in a matter of seconds at a cost of pennies.

VII.2. Understanding genome stability in humans

An organelle called the nucleolus resides deep within the cell nucleus and performs one of the cell's most critical functions: it manufactures ribosomes, the molecular machines that convert the genetic information carried by messenger RNA into proteins that do the work of life.

---

[173] source: Dr. Hatsumi Aoki, Annual Financial Times Global Pharmaceutical Conference, 2003

An interdisciplinary collaboration between **molecular biologists** and **physicists** at the University of Copenhagen resolved a virus puzzle. The researchers investigated how a virus **exploits the machinery of human cells to produce the proteins the virus needs in order to replicate** to billions of new vira. The virus penetrates into the host cell where it liberates its RNA which is a copy of the heritage material, DNA. RNA is like a 'cook book' which contains the recipes of which proteins the virus needs for replication.

The work process of a virus: The cell has ribosomes, a kind of 'molecular motors', which move along the RNA and read the code for the proteins to be produced to fulfill the needs of the living cell. The task of the ribosomes is to read the code of the host cell, but the virus has the special trick that its RNA resembles that of the host cell, and hence, the ribosomes of the host cell will start reading the viral RNA and produce the proteins requested by the virus. In order words, the virus can be viewed as a parasite, exploiting the human cell to live and replicate in.

Viral RNA resembles human RNA, but it has a tendency to curl up into 'pseudoknots', a three dimensional structure. When the ribosome walking along an RNA encounters a pseudoknot it needs to unravel the pseudoknot before it can proceed. The question is: how does it do that? Lene Oddershede at the Niels Bohr Institute, University of Copenhagen has developed optical tweezers which can investigate and manipulate molecules at the nano-meter scale. Using a tightly focussed laserbeam this instrument can grab the ends of the RNA tether and follow the process of how the pseudoknot is mechanically unfolded.

A crucial slip of the cellular motor: In their investigations the researchers use a pseudoknot which is related to bird flu. When the ribosome encounters a pseudoknot it has to unravel the knot before the reading can proceed. During this process the ribosome sometimes slips backwards and, like the letters making up a word, it now reads a new RNA sequence and hence uses another recipe to construct the protein. The researchers have found that the stronger the pseudoknot the more often this backwards slipping happens. The different protein formed is the protein needed by the virus, with possible serious consequences for the hosting organism. This is the manner in which many vira, e.g. HIV, trick the cell into producing something that it never would have done otherwise. Understanding the role of the pseudoknots can be an important step in developing a viral vaccine. *(The research finding has been published in the April 3, 2007 issue of PNAS, Proceedings of the National Academy of Sciences. For more information, contact: Lene Oddershede, Associate Professor, The Niels Bohr Institute, Copenhagen University, http://www.nbi.ku.dk)*

* The **monoclonal antibody market** is one of the fastest growing and most lucrative sectors of the pharmaceutical industry, According to Datamonitor, It has the potential to triple in value over the next six years and reach $30.3 billion in 2010, driven by technological evolution from chimeric and humanized to fully human antibodies. The development focus of the industry is moving away from murine and chimeric antibodies, to humanized and, in particular, fully human technologies.

* When **India's biotechnology** sector matures, it will make a significant impact on the world stage.

Actemra (tocilizumab) is the first humanized monoclonal antibody to the interleukin-6 (IL-6) receptor, blocking the activity of IL-6 , a protein that plays a major role in the RA inflammation process, a disease with a high unmet medical need. Roche and Chugai have initiated a collaborative phase III clinical development program in RA running outside Japan, with more than 4000 patients enrolled in 41 countries including several European countries and the USA. In Japan, Actemra was launched in June 2005 as a therapy **for Cattleman's disease** and in April 2006 filed for the additional indications of **rheumatoid arthritis** and **systemic-onset juvenile idiopathic arthritis**. January 2007, Roche announced that **'OPTION'1**, the first multinational phase III study of Actemra outside of Japan, successfully met its primary endpoint in the group of patients with moderate to severe rheumatoid arthritis (RA) who had an inadequate response to methotrexate. The OPTION **(Tocilizumab Pivotal Trial in Methotrexate Inadequate respONders)** study was an international study which took place in 17 countries with 73 centers entering 623 patients with moderate to severe RA.

One of the most important drivers for growth at Roche over the next few years is expected to be the company's emerging franchise in **autoimmune diseases** with rheumatoid arthritis as the first indication. Following the launch of MabThera (rituximab) there are a number of projects in development. **MabThera** is the first and only selective **B-cell therapy** for RA, providing a fundamentally different treatment approach by targeting B cells, one of the key players in the pathogenesis of RA. Additional projects creating a **rich new medicines pipeline** include compounds in Phase I, II and III clinical trials., include **ocrelizumab, a fully humanized anti-CD20 antibody**. [The ACR response is a standard assessment used to measure patients' responses to anti-rheumatic therapies, devised by the American College of Rheumatology (ACR). It requires a patient to have a defined percentage reduction in a number of symptoms and measures of their disease. For example, a 20 or 50% level of reduction (the percentage of reduction of RA symptoms) is represented as ACR20, ACR50 or ACR70. An ACR70 response is exceptional for existing treatments and represents a significant improvement in a patient's condition].

* Dr. Sarah E. Siegrist at the University of California, Berkeley, is uncovering **novel genes** that **control cell growth and neurogenesis** in the adult brain –key processes for understanding the origins of brain cancers.

* Dr. Lea A. Goentoro at Harvard Medical school is looking for clues as to how a key regulator of cell-to-cell communication commonly mutated in colon, liver, skin, and ovarian cancers called the **Wnt pathway**, transmit signals.

* By using a **cylindrical-shaped carrier** of synthetic polymers researchers were able to sustain delivery of the anticancer drug paclitaxel to an animal model of lung cancer ten times longer than that delivered on spherical-shaped carriers. These findings have implications for drug delivery as well as for better understanding cylinder-shaped viruses like **Ebola** and **H5N1 influenza.**

Dr. Ryan C. Heller at the Massachusetts Institute of Technology is identifying the **determinants of proper DNA replication** –a key issue for understanding how mistakes in DNA duplication contribute to mutations common in cancer cells.

defects and cancer. The organization of structures in the cell's nucleus has profound effects on such essential functions as how and when genes are expressed. When regulation fails, genome aberrations accumulate, including repeated sequences of DNA or even entire extra chromosomes.

The project continues to point to an understanding of genome stability in humans[176]. Controlling functions of the cell and organism through nuclear architecture and spatial rearrangements is known as epigenetics from the Greek for "on, over, or at" the genes, instead of by the DNA sequence. The chromosomal material known as heterochromatin mediates gene silencing, chromosome inheritance, and other processes. Karpen and Peng have identified the molecular pathways that regulate two of heterochromatin's important functions.

One of these is control of repeated DNA sequences in and outside the heterochromatin. The other is the organization and structure of the nucleolus, the ribosome factory, which is situated at the specific site in the heterochromatin where ribosomal DNA -- consisting of large genes, repeated 300 to 400 times -- codes for the production of the RNA from which ribosomes are built. Even though the gross organization of chromosomes and other nuclear elements is well known in cell biology, learning about the regulation of nuclear architecture is in its early stages. The most striking feature of any nucleus is its chromosomes. These are made of chromatin, which combines DNA with a set of proteins known as histones; four similar histones join together to form a cylindrical spool around which the DNA wraps. Each of these bundles is called a nucleosome, and many nucleosomes are bound together by the continuing strand of DNA, which forms a string of beads that further coils to form one of two kinds of chromatin, either euchromatin or heterochromatin.

Most genes reside in euchromatin, which is of relatively low density and where the DNA is more accessible to the machinery of gene transcription. By contrast, heterochromatin is dense and contains relatively few genes; most of the DNA in heterochromatin, including numerous short repeated sequences, does not code for proteins.

Heterochromatin is typically found at the ends of a chromosome, where it abuts the telomeres -- chromatin structures best known for limiting, by their diminishing length, how many times a cell can replicate. Heterochromatin also flanks the centromere in the central region of the chromosome, the chromatin structure that plays a crucial role in chromosome segregation during cell division. What other functions abundant heterochromatin may perform are still an open question.

Even epigenetics ultimately has its roots in the genes; the researchers' first step was to identify which genes affect organization of the heterochromatin, then to identify the proteins expressed by these genes, and finally to learn how they act on the chromosomal material.

Several genes previously identified in Drosophila as having a role in the suppression of gene expression, known as suppressors of variegation, were named Su(vars). The gene

---

[176] *Statement to the press,   Dec 14, 2006, by Professor Gary Karpen , who heads Berkeley Lab's Department of Genome and Computational Biology. Karpen is also codirector of the Drosophila Genome Center (sponsored by the National Human Genome Research Institute, the National Cancer Institute, and the Department of Energy) and an adjunct professor of molecular and cell biology at the University of California at Berkeley.*

*** Metagenomics to transform modern microbiology:** The **emerging field** of metagenomics, where the DNA of entire communities of microbes is studied simultaneously, presents the **greatest opportunity** -- perhaps since the invention of the microscope -- to **revolutionize understanding of the microbial world.**

Microorganisms are essential to life on Earth, transforming key elements into energy, maintaining the chemical balance in the atmosphere, providing plants and animals with nutrients, and performing other functions necessary for survival. There are billions of benign microbes in the human body, for example, that help to digest food, break down toxins, and **fight off disease-causing microbes**. Microbes are used commercially for many purposes, **including making antibiotics**, remediating oil spills, enhancing crop production, and producing biofuels.

Historically, microbiology focused on the study of individual species of organisms that could be grown in a laboratory and examined under a microscope, but most of the life-supporting activities of microbes are carried out by **complex communities of microorganisms**, and many cannot be grown in laboratory culture. **Metagenomics will transform modern microbiology** by giving scientists the tools to study **entire communities of microbes** -- the vast majority of which are likely to be **previously unknown species** that **cannot be cultured** -- and how they **interact** to perform such functions as balancing the atmosphere's composition, **fighting disease,** and supporting plant growth.

Metagenomics lets us see into the previously invisible microbial world, **opening a frontier of science that was unimaginable until recently**. Shedding light on thousands of new microorganisms will lead to new biological concepts as well as practical applications for **human health**, agriculture, and environmental stewardship.

Metagenomics studies begin by extracting DNA from all the microbes living in a particular environmental sample; there could be thousands or even millions of organisms in one sample. The extracted genetic material consists of millions of random fragments of DNA that can be cloned into a form capable of being maintained in laboratory bacteria. These bacteria are used to create a "library" that includes the genomes of all the microbes found in a habitat, the natural environment of the organisms. Although the genomes are fragmented, new DNA sequencing technology and more powerful computers are allowing scientists to begin making sense of these metagenomic jigsaw puzzles. They can examine gene sequences from thousands of previously unknown microorganisms, or induce the bacteria to **express proteins** that are screened for capabilities such as **vitamin production** or **antibiotic resistance.**

A report from the **National Research Council** calls for a new **Global Metagenomics Initiative** to drive advances in the field in the same way that the Human Genome Project advanced the mapping of our genetic code. (see opposite colon)

The goal of the large projects should be to **characterize in great detail** carefully chosen microbial communities in habitats worldwide, the report says. These studies could unite scientists from multiple disciplines around the study of a particular sample, habitat, function, or analytical challenge, the report adds. For example, one large project could focus on the microbial community associated with the human body, while others could focus on soil and seawater, or managed environments such as sludge processing sites.

The large projects would be **"virtual"** centers collecting data from scientists working at many locations around the world, and would probably need to be **sustained for 10 years**, the report notes. They would also serve as incubators for the **development of novel techniques** and community databases that would inform investigators running smaller experiments. In addition, the large studies would provide the **"big science"** appeal that is often useful in igniting public interest.

The metagenomics research community will include scientists funded by many government agencies working on many different habitats. These scientists should be encouraged to work together to disseminate advances, agree on common standards, and develop guidelines for best practices, the committee said. Such collaboration between government agencies should be facilitated by a group like the **Microbe Project**, an interagency committee that was formed in 2000 to advance **"genome-enabled microbial science."**

Noting that the data generated by the Human Genome Project were quickly made available in a public database, making them more valuable to researchers, the committee urged that metagenomic data likewise be made publicly available in international archives as rapidly as possible. The databases should include not only gene sequences but also information about sampling and DNA extraction techniques, as well as the computational and algorithmic methods used to analyze the data.

*. Future growth in sales of pharmaceutical treatments for cancer will primarily come from the emergence of **innovative biologics** and the **monoclonal antibody** class of drugs. With pharmaceutical companies Amgen, Genentech and Roche predicted to be marketing the five leading cancer drugs in 2010, as well as comprising the lead positions amongst cancer drug manufacturers, the cancer treatment market will enter a new age over the next five years, dominated by biotechnology- derived products marketed by biotechnology- focused companies. (The global cancer market grew to over $60billion in 2006. Depending on the sources, this therapeutic class is said to reach or top the $100bn mark in 2010).

* Researchers are optimistic that there may be long-term savings to patients, and to the health care system, due to **MRI's ability to** detect cancer in **both breasts** prior to therapy -- which may result in **fewer rounds** of **chemotherapy** and **breast surgeries**.

Biomarkers of risk will enhance the validation of preventive therapeutics. Cancer diagnostics will assume far greater importance in healthcare delivery. Precise prediction of toxicity of specific cancer drugs in an individual will become possible. With improved early kowledge of disease, the demand for psycho-social care will increase.

its ninth amino acid residue, which is a lysine. In the histone code the methylation target is known as H3K9, where K stands for lysine. Methylation of chromatin causes it to condense.

Another Su(var) gene, Su(var)2-5, makes a protein called HP1 that binds methylated H3K9 and Su(var)3-9 proteins; the effect is to further close up the chromatin, silencing genes by making access to DNA more difficult.

Other genes called E(var)s act to enhance variegation in Drosophila. One way they do this is by acetylation of the H3 histone's fourth amino acid, also a lysine. The effect is to open the chromatin at that site, making the DNA accessible.

"Acetylating H3K4 and demethylating H3K9 can activate a gene; deacetylating H3K4 and methylating H3K9 can silence it," Karpen explains. "Multiply modified at various locations, these histones are a major factor in changing the functions of chromatin, independent of DNA sequence."

The scientists studied fruit flies with mutant Su(var)3-9 genes, flies that could not produce HP1 proteins or the proteins necessary to methylate H3K9. The results were dramatic. Because the ribosomal DNA genes could not be silenced, rDNA proliferated. These and other repeated DNAs became dispersed and disorganized throughout the cell nucleus. Instead of a single well-formed nucleolus where genes for ribosomal DNA were normally located in the heterochromatin, multiple nucleoli appeared in the nucleus.

Mutations in another pathway, that of RNA interference (RNAi), produced related results. One way RNAi works is by unleashing the Dicer-2 enzyme to chop strands of RNA into short lengths, which can degrade messenger RNA and prevent gene expression.

The researchers explains that "the RNAi pathway is a sort of immune system against viral DNA and transposons." Transposons are pieces of DNA that can "jump" to different locations in the genome, either directly or by copying themselves, inserting themselves in new sequences where they can cause mutations.

Short repetitive pieces of DNA that keep copying themselves could conceivably expand and take over the whole genome, so some balance must be achieved, between the need for replication processes on the one hand and the need to repress or control or regulate those processes on the other.

Since DNA pieces that copy themselves must use RNA intermediates, these can be disabled by RNA interference. Transposons are also silenced by accumulating in the transcriptionally-silent heterochromatin, which the scientists call "a graveyard of transposons."

While the gene for the Dicer-2 enzyme, dcr-2, is a key player in the RNAi pathway, the RNAi process also targets H3K9 methylation. Sure enough, Karpen and Peng found that Drosophila with mutant dcr-2 genes showed multiple nucleoli and other symptoms of heterochromatin deregulation, just as Su(var)3-9 mutants did.

One effect of both Su(var)3-9 and RNAi mutations was to increase the amount of extrachromosomal DNA, typically small loops of repeating sequences, in cell nuclei. This observation led Karpen and Peng to propose a specific mechanism by which H3K9 methylation and RNAi pathways regulate nuclear organization.

When the H3K9 methylation pathway is functioning properly, heterochromatin remains condensed and a single nucleolus forms around the ribosomal DNA. But lack of H3K9 methylation -- whether from a mutation in Su(var)3-9; or in the gene that codes for the HP1 protein; or a mutation that disables RNA interference -- allows the heterochromatin to open up.

The decondensed, repeated DNAs are then free to interact; DNA recombination and repair processes excise the DNA from the chromosome to form extrachromosomal DNA. If these extrachromosomal loops consist of ribosomal DNA, they further accelerate the production of misplaced nucleoli.

Curiously, say the scientists, this chaotic state of affairs is not, by itself, fatal to the organism. He wondered why the flies with mutations in these important pathways didn't die. But replication and cell division checkpoints slow down the cell cycle when damaged DNA is detected, and DNA repair mechanisms move in to fix the damage. However, if these replication checkpoints are crippled in flies that also have mutations in the H3K9 methylation and RNAi pathways, genome aberrations accumulate "and they do die."

These findings have broad implications for genome stability. Karpen plans to investigate whether the H3K9 and RNAi pathways he and his team have identified in Drosophila play the same role in humans; if so, misregulation of heterochromatin could be one cause of the rampant genome instability observed in cancer. Beyond these specific pathways, the scientist is interested in the larger question of how such pathways affect the organization of the chromosomes in the cell nucleus.

VII.3. Engineering RNA for medical purposes

RNA enzymes, also known as ribozymes, accelerate chemical reactions inside cells, just as their better-known protein counterparts do. And just as a protein enzyme is not a static structure, a ribozyme also changes shape, cycling back and forth between active and inactive forms (called conformations).

Research[177] found that modifications made anywhere on the ribozyme molecule---even far from the site where the chemical reaction occurs---affect the rates at which the enzyme changes conformation and catalyzes the reaction. Something similar had been seen in protein enzymes, but never before in RNA enzymes.

The finding suggests that information about changes in distant parts of the ribozyme travels through some sort of network to the core of the molecule, where chemical reactions take place. They also discovered that water molecules trapped inside the ribozyme's core are essential components of that network.

The network acts like a jostling crowd at a cocktail party, where hydrogen bonds---weak, electrostatic attractions between molecules or parts of molecules---take the place of handshakes. Water molecules trapped in ribozymes can form hydrogen bonds with other water molecules or with parts of the ribozyme molecule. Walter and coworkers also

---

[177] *A University of Michigan team, led by Nils Walter,*

Scientists at Weill Cornell Medical College have identified a protein called **Rab10** as an important partner in the **insulin-mediated uptake** of glucose by cells, opening the way to potential new drug targets for the prevention and treatment of type 2 diabetes.

The glucose found in food is a main source of energy. The body uses complex mechanisms to maintain an ideal level of glucose in the blood. However, conditions such as **obesity** put such a strain on this system and cells no longer react as they should in the presence of insulin, the hormone that directs glucose uptake.

Because insulin sensitivity is so important, scientists have studied it for decades. "When it's working right, insulin, which is released from the pancreas after a meal, stimulates glucose uptake into fat and muscle cells, thereby lowering the blood glucose level". That sounds like it should be simple, but researchers quickly learned that this messaging process involves many different molecules inside the cell. The main **"switch"** allowing for the influx of glucose into cells is the glucose transporter protein called **GLUT4**. When it's not active, GLUT4 is sequestered in vesicles within the cell. In healthy non-diabetics, insulin sets off a chain of biochemical signals that releases GLUT4 to the cell surface. Once there, it forms a kind of **'gate'** through which glucose can enter the cell. Scientists at Dartmouth Medical School have already discovered that another protein, called **AS160,** has a role in insulin signaling to the GLUT4 glucose transporter. AS160 was found to work on a family of enzymes called **rab proteins**. In their study, when Rab10 was eliminated or its activity switched off, the scientists found that insulin was no longer able to properly trigger the recruitment of the GLUT4 glucose transporter to the surface of cells. They concluded that the recruitment of GLUT4 to the cell surface increases glucose movement from the blood into cells, where the glucose is stored for future use and that Rab10 is involved in the insulin regulation of blood glucose levels, whereby a disruption in the regulation of blood glucose levels, a condition called **'insulin insensitivity'**, is a hallmark of type 2 diabetes.

The finding moves scientists one step closer to understanding how **insulin signaling** works and how insulin insensitivity someday might be prevented.

In their experiments using **cell cultures**, Dr. McGraw's team at Weill Cornell Medical College, keen to know which rab was involved, discovered that cells engineered to lack the Rab10 protein **suffered an immediate decline in insulin sensitivity.**

Rab10 now appears to be another **key middleman** in the insulin-GLUT4. This discovery gives one more link in the cellular chain of events linking insulin signaling and the regulation of glucose transport. These results might eventually lead to a better understanding of insulin-resistance and type 2 diabetes.

The identification of AS160 and now Rab10 means that the potential number of therapeutic drug targets to prevent or treat insulin resistance will continue to expand. *(The study is published in the April 2007 issue of Cell Metabolism).*

For the first time, scientists from the University of Washington School of Medicine, Indiana University Bloomington and the University of Cambridge have determined how a plant hormone -- **auxin -- interacts** with its hormone receptor, **called TIR1,** also may have important implications for the treatment of human disease, because TIR1 is **similar to human enzymes** that are known to be involved in cancer.

Some of TIR1's human relatives play a role in different human cancers, and it is possible that this work on plants will eventually lead to new cancer drugs. Until now it was believed enzymes like TIR1, called **ubiquitin ligases**, could only be controlled through protein-protein interactions. Ubiquitin ligases influence growth and light response in plants, poison mitigation in yeasts and also cancerous cell division in humans. Although ubiquitin ligases have long been recognized as potential drug targets for treating cancers and other human diseases, it's been a bumpy road for scientists to come up with a feasible approach. The mechanism by which auxin works points out a new direction for researchers to develop therapeutic compounds **targeting ubiquitin ligases.** The scientists extracted and purified TIR1 from the common plant model **Arabidopsis.** By x-raying crystals of the protein, the scientists determined the enzyme's three-dimensional structure -- a first for plant hormone receptors. They then soaked the crystal in a solution containing auxin and repeated the x-ray treatment to determine where the auxin had bound. Finally the scientists added a peptide that TIR1 is known to bind and modify. They learned that auxin is a sort of **"molecular glue"** that improves the ability of TIR1 to bind its peptide target. In the absence of auxin, TIR1 does not bind its target as tightly. A number of human disorders including Parkinson's disease, and colon and breast cancers, are **caused by defective interactions between ubiquitin ligases and their substrate polypeptides**. What the plant hormone tells scientists is that it might be possible to rescue these interactions using small molecules.Because the architecture of **TIR1** is **highly conserved** among other ubiquitin ligases, including those in human cells, the scientists expect other ubiquitin ligases may be affected by small molecules like auxin. Organic chemists could, hypothetically, **synthesize such molecules as a new type of cancer drug.** *("Mechanism of Auxin Perception by the TIR1 Ubiquitin Ligase," Nature, Vol. 446, no. 7136, April 2007)*

* Some day, **heart attack survivors** might have a patch of laboratory-grown muscle placed in their heart, to replace areas that died during their attack. Children born with defective heart valves might get new ones that can grow in place, rather than being replaced every few years. And people with clogged or weak blood vessels might get a **new "natural" replacement**, instead of a factory-made one. Technology has advanced so much in recent years, they write, that scientists are closer than ever to **"bioengineering"** entire areas of the heart, as well as heart valves and major blood vessels.

Patient advocacy groups will require the development of drugs for short-course prescriptions. Social change will create new challenges of unprecedented nature: access to care will depend on cost (there will be conflicting demand on resources...medicine for the rich/medicine for the poor?) and political will.

found evidence that water is directly involved in catalyzing reactions in the ribozyme's core.

The situation in ribozymes contrasts with what happens in protein enzymes, which repel water from their cores and rely on direct contact, rather than a network of hydrogen bonds, to communicate structural changes from one part of the molecule to another.

Once thought to be only a passive carrier of encoded genetic information, RNA is now known to regulate gene expression and other important cellular processes and to act as a sort of sensor---detecting cellular signals and carrying out appropriate reactions in response. In fact, there are many more so-called non-protein coding RNAs in the cell (around 100,000 in humans), which are not translated into protein, than there are protein coding messenger RNAs (about 25,000), making these vast numbers of RNA molecules central players in our bodies.

(With protein-coding genes, a copy of the gene's DNA is transcribed into RNA, and this messenger RNA, after processing, arrives at the cell's protein-making units where its information is used to manufacture a specific kind of protein molecule.

In the case of RNA, the RNA transcript of a gene is processed by a baroquely named troika of enzymes known as Dicer, Slicer and Argonaute, the end result being a slew of short fragments of RNA some 20 or so units in length. Despite their brevity, these snippets of RNA are long enough to match specific sequences of RNA units within the cell's many different kinds of messenger RNAs. When the snippets find a match to messenger RNA, the micro-RNA's curb the activity of the messenger RNA's and the silencing RNA's destroy their messenger RNA targets. This means that much less or none of a protein gets produced in the cell. Biologists discover more about the versatility of RNA and treat RNA with more respect than they did before, when they considered DNA as the big star of molecular biology, viewing RNA as a mere copyist of the genes encoded in the famous double helix.
"Anything DNA can do, RNA can do better" was the slogan on a slide shown by one biologist at the Cold Spring Harbor Laboratory)[178].

Work is also underway in academic and industrial labs around the world to engineer RNA for medical purposes. The engineered molecules, called RNA aptamers, are selected for their ability to bind to particular proteins involved in certain diseases, blocking key steps in the disease process.

Table 7.2.   Biochemical classes of drug targets of current therapies (1965-2000)[179]

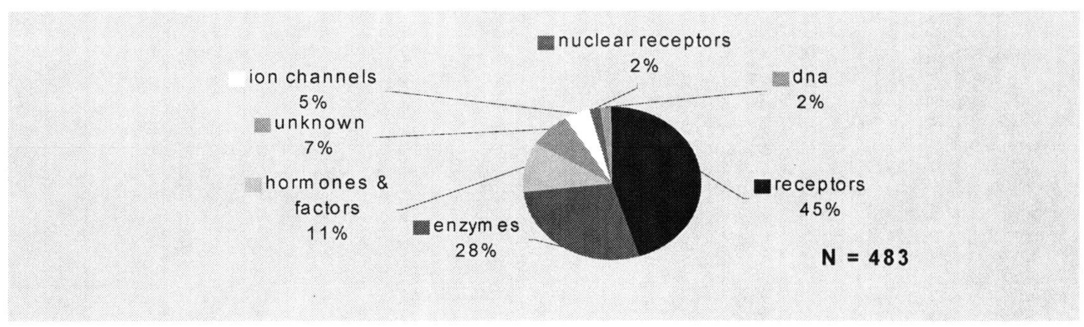

---

[178] Wade N., Dna or RNA? Versatile player takes a leading role in molecular research. The New York Times, June 20, 2006:D3

* **BiDil** is an example of a medication whose efficacy has been examined with respect to **race and ethnicity**. The compound failed to find relevant efficacy in the general population, but later trials showed BiDil therapy to be significantly associated with **lower heart attack risk** in **African-American** patients. The drug was approved in June 2005 on the basis of outstanding clinical results from NitroMed Inc.'s **African American Heart Failure Trial** (A-HeFT). Ongoing analyses of patient data from A-HeFT continue to provide valuable insight into how and in whom BiDil may provide treatment benefit.

One notable BiDil finding **demonstrates longer-term compliance** with the fixed-dose combination drug, as well as continued efficacy, safety and tolerability. A second finding highlights a **particular genotype** of **the G protein beta 3-subunit** (GNB3 TT) that is closely associated with positive results in A-HeFT, providing a basis for further study for the role of the genotype in targeting therapy. The **addition** of genotype to the diagnostic criteria might show that the drug is a candidate for treatment in other population groups. Scientists try to find out if there are genetic modifiers of BiDil response that are found at higher frequencies in African-American patients; to what extend they are found and if non-African-Americans carry the same genetic biomarker. Although the mechanism of action of BiDil has not been completely established, the drug is believed to **stimulate** and **maintain nitric oxide signaling**. BiDil is found to significantly decrease cardiac death by decreasing pump failure deaths in the moderately severe A-HeFT heart failure population. This is consistent with previously reported significant reductions in all-cause mortality and heart failure hospitalizations for BiDil treated patients in A-HeFT. Another study finding suggests that while BiDil was superior to placebo with or without use of beta blockers or angiotensin converting enzyme (ACE) inhibitors, there appears to be further benefit in the group treated with BiDil and beta blockers than without beta blockers.

* According to the Los Angeles Times, the high prices of the new generation of cancer drugs, which can cost as much as $100,000 for a year's supply, and their relatively limited health benefits "have sparked arguments among policymakers and medical professionals about what to do with the growing number of people who are depleting their life savings on the drugs, or worse, who can't get them at all." Drug companies and some patients **"insist even incremental gains"** in their health **"are worthwhile,"** even as the cost burden of such drugs grows for patients, The prices of these drugs will become a "burning" issue to society... Limited budgets for healthcare will put pressure on Society to decide what - and to what cost- it wants to achieve in healthcare. *(March 19 issue of LA Times)*

* **Aurora A and B** are a type of enzyme known as protein kinases; they modify other proteins by chemically adding phosphate groups to them. In cancer, both these protein kinases are **'overexpressed'**. A lot of current cancer drugs, while effective, are also toxic; by contrast, the **toxic effects** of Aurora inhibitors **has been relatively mild** and so could provide a revolutionary new way to treat cancer in the future

* **Protein kinases** comprise a large and medically important family of enzymes that are key regulators of cellular signaling pathways in all higher organisms from yeast to man. **Hundreds of kinase inhibitors are in development** by pharmaceutical companies for diseases ranging from Alzheimer's to diabetes to arthritis to cancer. Designing selective and safe kinase inhibitors is a daunting task because there are **over 500 human protein kinases** and the structural similarities among them are strikingly high. Profiling a drug broadly and accurately against as many kinases as possible is the Holy Grail for kinase profiling technologies. The best profiling platforms claim assays for 220 or so kinases well short of the 500 known human kinases. **KiNativ's** profiling capability currently stands at ~ 430 kinases and is said to represent a paradigm shift in how kinase activity and inhibitor selectivity will be determined in the future.

* All diseases are complex, the result of different genes and environmental risk factors acting together in concert. By searching the genome, researchers at St George's, University of London, the University of Colorado at Denver and Health Sciences Center (UCDHSC) and the Barbara Davis Center for Childhood Diabetes discovered that **NALP1** (a gene that controls part of the immune system that serves to alert the body to viral and bacterial attacks was a key gene involved in predisposing to vitiligo and all the other autoimmune diseases that ran in these families) turns out to be **one of the major genes involved in numerous autoimmune diseases**, and if they can interrupt its negative effects, doctors may have the chance to treat many different chronic autoimmune disorders like vitiligo, lupus and psoriasis and perhaps eventually eliminate them altogether.

* While **MRI** is known to be more sensitive than mammography and can find smaller tumors, its specificity causes concern for some. **More false positive results** are likely. Additional tests, including biopsy, may be necessary to determine the significance of MRI-derived results to rule out cancer. This adds expense and stress for the patient.The **cost is significantly more for MRI** screening than for a mammogram. Many insurers have been reluctant to cover the cost of even a diagnostic breast MRI, which could portend a lack of coverage for screening MRI.

* Scientists are working now to create **novel nanostructures** that serve as new kinds of drugs for treating cancer, Parkinson's and cardiovascular disease; to engineer nanomaterials for use as artificial tissues that would replace diseased kidneys and livers, and even repair nerve damage; and to integrate nanodevices with the nervous system to create implants that restore vision and hearing, and build new prosthetic limbs.

* **Public perceptions** of **safety and quality,** which find their origin in the basic concept of „one size fits all" medicines, can slowly be overturned by the arrival of more **targeted medicines**, highly efficient therapeutics for the treatment of severe diseases in **specific population groups.**

The fact that many well-know medicines are effective in only 40-70% of the population, underlying the complexity of human populations. Medicine's goal is to determine how an individual's genetic makeup controls the ways a person responds to a particular drug and to use that knowledge to determine a drug's potential for each patient.

VII. 4.   The sense/antisense madness: a role for dueling RNAs

Researchers[180] have found that a class of RNA molecules, previously thought to have no function, may in fact protect sex cells from self-destructing. Central to this discovery is the fundamental process of gene expression. When a gene is ready to produce a protein, the two strands of DNA that comprise the gene unravel. The first strand produces a molecule called messenger RNA, which acts as the protein's template. Biologists call this first strand of DNA the "sense" or "coding" transcript. Even though the other strand doesn't contain a protein recipe, it may also, on occasion, produce an "antisense" RNA molecule, one whose sequence is complementary to that of the messenger, or sense, RNA. Antisense RNA has been detected for a number of genes, but is largely considered a genetic oddity.

Using common baker's yeast, the researchers discovered that in the case of a gene called IME4, the antisense RNA blocks the sense RNA[181]. In other words, the gene disables its own ability to make protein. According to the researchers, this is the first case where a specific function in a higher cell for antisense RNA has been found. It points to an entirely new process of gene regulation that we've never seen before in eukaryotic cells.

There is a method to this sense/antisense madness, one that has a kind of yin and yang quality. When conditions around yeast cells are good and rich in nutrients, the cells divide by mitosis--that is, the DNA duplicates so each daughter cell receives exactly the same number of chromosomes as the original cell. However, when the yeast cells are starving, IME4 switches on and activates a process called meiosis. Here, the cells divide into germ-cell spores that, like mammalian egg and sperm cells, have half the number of chromosomes. Yeast spores withstand this harsh environment far more ably than the larger cells from which they originate. But in some cases, flipping the meiotic switch can be catastrophic. If a cell with only one copy of each chromosome (a haploid cell) is forced into meiosis, the progeny won't survive. Fortunately, such destructive meiotic division is avoided in haploid cells because they continually produce IME4 antisense RNA, blocking the production of sense RNA. Antisense IME4, then, safeguards against meiosis in cells that can't handle it.

For years scientists have evaluated genomes by measuring the sense RNA, with antisense transcripts thought to have no meaning at all. Now, they have found a process in which antisense RNA regulates sense RNA. This same process may occur in the sex cells of mammals. In fact, considering how widespread these antisense transcripts are, I wouldn't be surprised if these findings eventually lead us to discover an entirely new level of gene regulation.

VII.5.   RNAi shows promise in gene therapy

In 2004, after ten years of work on a gene therapeutic approach against viral hepatitis, Mark Kay, professor of genetics and of pediatrics at the Stanford University School of Medicine published the first results showing that a biological phenomenon called RNA interference could be an effective gene therapy technique. Since then he has used RNAi gene therapy to effectively shut down the viruses that cause hepatitis and HIV in mice.

---

[180] *Cintia Hongay, a postdoctoral researcher in the lab of Whitehead Member and MIT Professor Gerald Fink*

[181] *David Cameron, Whitehead Institute for Biomedical Research, announcing the publication of the research findings of Dr Hongay and Dr Fink in the November 16, 2006 issue of the journal Cell*

* Because of the frequent use of **antibiotics in hospitals**, infections acquired there are particularly problematic; more than 70% of the bacteria causing infections in people while they are patients in hospitals are resistant to at least one of the drugs commonly used to fight them. The **problem is largely underestimated.**

* Stevens Proof Of Concept Inc., a technogenesis company founded at the Stevens Institute of Technology, has developed a revolutionary proprietary point-of-care medical diagnostic system, consisting of a medical device and methodology that pinpoint the specific myofascial (muscle) trigger points causing pain. The device is the first use of electroneural stimulation for diagnostic purposes. SPOC's diagnostic system will benefit patients by helping to eliminate treatments that prove to be ineffective, such as surgical procedures, and by allowing physicians to locate and treat more effectively muscles that generate pain. The **diagnostic tool may become as common as the stethoscope.**

* A new member of the TNF super-family, named BlyS, also known as **BAFF**, and its homologue **APRIL** have emerged as a candidate target for therapies of various autoimmune diseases and hematological cancers.

* The **REPEAT** trial will assess the efficacy, safety, and tolerability of three repeated dosing cycles of Promacta (a non-peptide thrombopoietin receptor agonist that has been shown in pre-clinical research and clinical trials to stimulate the proliferation and differentiation of megakaryocytes, the bone marrow cells that give rise to blood platelets, and thus maybe considered a platelet growth factor) in patients previously treated for chronic ITP, examples of which include splenectomy, corticosteroids, immunoglobulins, cyclophosphamide or rituximab. The primary objective is to evaluate the effect of Promacta on platelet counts during three consecutive treatment cycles. Secondary objectives include evaluating incidence and severity of bleeding, as well as the frequency of adverse events.

* For several types of cancer, **persistently high levels** of the **soluble factor TGF-beta** in the blood after surgery, chemotherapy, or radiation therapy correlate with increased risk of early metastasis and a poor prognosis. Using a mouse model of breast cancer, researchers from Vanderbilt University have now generated evidence to suggest that treatment with TGF-beta inhibitors might help such patients. This study has important clinical implications as it suggests that monitoring TGF-beta levels after primary therapy might indicate the patients most at risk of developing tumor metastases and that treatment with **TGF-beta inhibitors** might be of clinical benefit to these individuals.

* Spotting epigenetic markers like **lost IGF2** in humans could be used in future cancer-prevention strategies. It could be possible to screen for colon cancer risk by looking at the **epigenetic changes** in colon cells of healthy people.

* Very early and reliable diagnosis, intelligent drug delivery systems and smart regenerative medicine are key issues in the approach **towards effective treatment** of plagues of mankind such as cancer, cardiovascular diseases, diabetes, Alzheimer's.

* **Systems biology** concepts and tools can be applied to translational medicine in getting a better understanding of disease processes, drug reactions and the predictivity of in vitro and in vivo models.

* New research strongly support the notion that, under physiological conditions, a decrease in eIF2a (described as being a bidirectional point of memory control) phosphorylation constitutes a critical step for the activation of gene expression that leads to the long-term synaptic changes required for memory formation. These findings also raise the interesting possibility that regulators of translation could serve as therapeutic targets for the improvement of memory, for instance in human disorders associated with memory loss. [*Costa-Mattioli et al.: eIF2a Phosphorylation Bidirectionally Regulates the Switch from Short- to Long-Term Synaptic Plasticity and Memory. Cell 129, April 6, 2007. DOI 10.1016/j.cell.2007.01.050*]

* Researchers have identified a new compound called **CDDO-Im** that protects against the development of liver cancer in laboratory animals. The compound appears to stimulate the enzymes that remove toxic substances from the cells, thereby increasing the cells' resistance to cancer-causing toxins. Its effectiveness at very low doses suggests it may have similar cancer-fighting properties in humans. It is believed to be particularly effective in preventing cancers with a strong inflamma-tory component, such as liver, colon, prostate and gastric cancers. The compound could eventually play a preventive role in a wide range of other illnesses such as neurodegenerative disease, asthma and emphysema.

* The latest achievements in radiology are revolutionizing gentle, non-invasive heart diagnostics. **Multi-slice computer tomography (MSCT)** can spare patients from invasive cardiac catheter examinations and for example, allow physicians to assess the risk of coronary based on the amount of calcification in coronary vessels. Blood flow rates in the heart muscle can be accurately evaluated today in just 30 seconds using magnetic resonance tomography. Two X-ray tubes rotating parallel to each other (dual source CT) can now also be used to render precise images of fast or irregularly beating hearts. **Combining the various imaging methods** is a **promising approach** for the future, as initial results have already shown.

* 85 percent of cancers involve solid tumors, but **only 25** percent of approved antibody therapies are directed at **solid tumor targets**. Scientists suggest manipulation of the physical properties of a monoclonal antibody as a potentially effective way to **improve antibody** treatments for **solid tumors.**

* Knowledge of predisposing factors in the development of BRAF mutations, such as **MC1R,** might aid prevention and therapeutic strategies in the future.

Medical schools need to overcome educational barriers to the adoption of future therapies. Only 40% of medical schools today offer medical genetics courses. How will the practicing physicians ever be competent to discuss genetic information when they are not trained to do so by medical schools that do not keep up with the pace of innovation?

With three human RNAi gene therapy trials under way in 2007- two in macular degeneration and one in RSV pneumonia - the technique Kay pioneered may be among the first to find widespread use for treating human diseases[182].

RNAi is a biological phenomenon in which a strand of RNA in the cell can cause the destruction of another strand of RNA that is relaying a protein-coding message from a gene. With that protein-coding message removed, the gene's message is effectively destroyed.
When used as gene therapy, RNAi turns off genes that are overactive in such diseases as cancer or macular degeneration, or disables genes needed by an invading virus. With key genes shut off, viruses such as hepatitis or HIV are unable to multiply and cause disease.

In early RNAi experiments, researchers saw some hints that the technique could induce an immune reaction or switch off the wrong gene or genes. In work 2006, the scientist confirmed those findings but also showed a possible way around those toxic effects by selecting particular RNA sequences.

One benefit of RNAi gene therapy is that it uses the body's own machinery, making it an effective approach. However, the detriment of RNAi gene therapy turns out to be that it uses the body's own machinery." The researcher said he expects the current trials will help him and others figure out the best way to bring RNAi gene therapy safely to humans.

VII.6. siRNA delivery into human T cells and primary cells with carbon-nanotube transporters

Short pieces of RNA, known as small interfering RNA (siRNA), have the potential to become a new class of anticancer drugs if researchers can solve the problem of how to deliver these fragile molecules to cancer cells.

One possible solution [183] is to use carbon nanotubes to transport siRNA agents through the bloodstream and into cells.

Carbon nanotubes are adept at passing through the cell membrane, and Hongjie Dai, Ph.D., and his colleagues used that property to deliver siRNA molecules into human cells. Starting with commercially available single-walled carbon nanotubes, the investigators made the nanotubes water soluble by coating them with the biocompatible polymer poly(ethylene glycol), or PEG. They then attached siRNA molecules to the PEG coating using a mild chemical reaction to link the therapeutic agent to the PEG molecules.

After purifying the nanotube-siRNA combination, the investigators added it to cultured human cells. Within 24 hours, production of the protein targeted by the siRNA agent dropped by approximately 90 percent, indicating that the nanotubes were ferrying their therapeutic agent into the cultured cells. In contrast, the researchers observed no change in protein production when they delivered the same siRNA molecule to the cells using liposomes. The investigators also noted that they conducted toxicity tests and

---

[182] *Presentation given by professor Kay about future uses for RNAi gene therapy at the American Association for the Advancement of Sciences annual meeting in San Francisco, February 2007, during a session titled "RNAi for emerging pandemics and biosecurity. Press release from Amy Adams,Stanford University Medical Center, February 27, 2007*
[183] *identified by investigators at Stanford University's Center for Cancer Nanotechnology Excellence Focused on Therapy Response,*

* In an important advance in the battle against **Alzheimer's disease**, physician-scientists at NewYork-Presbyterian Hospital/Weill Cornell Medical Center have **identified naturally occurring antibodies** in human blood that may help to defend against this form of dementia as well as other neurodegenerative diseases.

* Scientists currently are working to **create novel nanostructures** that can serve as new kinds of drugs for treating cancer, Parkinson's, and cardiovascular disease. They also are seeking ways to **engineer nanomaterials** for use **as artificial tissues** that could replace diseased kidneys and livers, and even repair nerve damage. In addition, although the research is still exploratory, scientists are beginning to build nanostructures that **mimic complex biomolecules**. Some of these engineered structures appear to have regenerative powers that could potentially lead to therapies for conditions such as Alzheimer's, nerve injury and brain damage from stroke.

* **AD 923** is an optimized, sublingual formulation of the strong opioid analgesic fentanyl for the treatment of cancer breakthrough pain.

It has been specifically designed to provide rapid onset of analgesia in a device that is easy to use by either the patient or their care giver. An additional benefit is the lockout system that prevents inadvertent overdosing. Japanese Sosei group has concluded a range of studies that confirm the potential of this novel product.

* A synthetic version of a molecule found in **the eggs** of the **Northern Leopard frog** (Rana pipiens) could provide the world with the **first drug treatment for brain tumors** Known as **Amphinase**, the molecule recognizes the sugary coating found on a tumor cell and binds to its surface before invading the cell and **inactivating the RNA it contains**, causing the tumor to die. Although it could be used as a treatment for many forms of cancer, Amphinase offers greatest hope in the treatment of brain tumors, for which complex surgery and chemotherapy are the only current treatments.

Ribonucleases are a common type of enzyme found in all organisms. They are responsible for tidying up free-loaded strands of RNA cells by latching on to the molecule and cutting it into smaller sections. In areas of the cell where the RNA is needed for essential functions, ribonucleases are prevented from working by inhibitor molecules. But **because Amphinase is an amphibian ribonuclease**, it can **evade** the **mammalian inhibitor molecules to attack the cancer cells**. Amphinase is the second anti-tumor ribonuclease to be isolated by Alfacell Corporation from Rina pipiens oocytes. The other, **ranpimase**, is in late-stage clinical trials as a treatment for **unresectable malignant mesothelioma**, a rare and **fatal form of lung cancer**, and in early trials in non-small cell lung cancer and other solid tumors.

* Scientists at Fox Chase Cancer Center are investigating possible roles for **HEF1 and Aurora A** in **PKD**. They are intrigued by the fact that a study showed that an important gene, PKHD1, commonly mutated in PKD has also been found as a target of mutation in colorectal cancer.

Polycystic kidney disease, or PKD, arises from genetic mutations that cause flawed kidney-cell ciliary signaling. PKD is the **most common serious hereditary disease,** affecting more **than 12.5 million people worldwide.** In this incurable syndrome, patients develop numerous, fluid-filled cysts on the kidneys. The two proteins with important roles in cancer progression and metastasis, HEF1 and Aurora A, have an unexpected **role in controlling the temporary disappearance of cilia during normal cell division**, by turning on a third protein, HDAC6Commonly, cancer cells have entirely lost their cilia, and this absence may help explain why tumors fail to respond properly to environmental cues that cause normal cells to stop growing. Hence, the discovery that too much HEF1 and Aurora A cause cilia to disassemble provides important hints into what may be happening in cancers. Defects in cilia have already been identified in one disease that represents a significant public health burden. PKD leads to kidney failure in about half of cases, requiring kidney dialysis or a kidney transplant.The proteins involved in dismantling the cilia are no strangers to Dr. Golemis and her team. She has been studying HEF1 for over a decade, since **she first identified the gene.** She first discovered that HEF1 has a role in controlling normal cell movement and tumor cell invasion. her laboratory has also shown **that Aurora A and HEF1 interact to initiate mitosis (chromosome separation) during cell division.** Suggestively, many cancers produce too much of the Aurora A protein, including breast and colorectal cancers and leukemia. **Excessive production of HEF1** (also known as **NEDD9**) was found to drive metastasis in **over a third of human melanomas,** while HEF1 signaling also contributes to the aggressiveness of some brain cancers (glioblastomas). This complex HEF1 and Aurora A function may mean the increased levels of these proteins in cancer affect cellular response to multiple signaling pathways, rather like a chain reaction highway accident.

* Scientists are working to create **nanostructures** that can serve as new kinds of drugs for **treating cancer, Parkinson's, and cardiovascular disease**. They also are seeking ways to engineer nanomaterials for use as **artificial tissues** that could **replace diseased kidneys and livers, and even repair nerve damage**. In addition, although the research is still exploratory, scientists are beginning to build nanostructures that **mimic complex biomolecules.** Some of these engineered structures appear to have regenerative powers that could potentially lead to **therapies** for conditions such as Alzheimer's, nerve injury and brain damage from stroke.

New advances continue at an astonishing rate. In just the last few years we have seen: the sequencing of the human genome, providing a catalogue of all human genes; the development of RNAi technology, permitting the sophisticated loss of function analysis of human cells, and the identification of cancer stem cells, defining a potential new paradigm for cancer etiology. The next decade will turn much of the fascinating new knowledge into improved therapeutic care.

found no evidence that the water-soluble carbon nanotubes had any adverse effect on the cultured cells.

Researchers worldwide see huge potential for the use of nanotechnology in medicine. New research findings are being published almost on a daily base.

VII.7. Watery nanoparticles deliver anticancer therapy

Ultrafine nanoparticles made of a lacy web of polymer and tiny pockets of water may prove to be an ideal vehicle for delivering light-activated drugs to tumors. Preliminary experiments, show that cancer cells die quickly when treated with these nanoparticles and exposed to light.

Researchers[184] developed a versatile chemical technique for creating ultrafine nanosized hydrogels, essentially a network of polymer chains that absorb as much as 99 percent of their weight in water. The researchers used the well-studied polymer known as polyacrylamide as the foundation for creating 2-nanometer-diameter nanoparticles that have no charge on their surfaces. This lack of charge prevents blood proteins from sticking to the surface of the nanoparticles. Combined with the fact that these nanoparticles are too small to be recognized by the immune system, the result is a nanoscale drug delivery vehicle with the ability to remain in circulation long enough to reach and permeate tumors before being excreted through the kidneys.

The investigators' first test of these new nanoscale hydrogels was to use them as a drug delivery vehicle for a water-insoluble light-activated drug known as a photosensitizer. In particular, the researchers chose a compound known as meta-tetra(hydroxyphenyl) chlorin, or mTHPC, which was recently approved by European regulators for use in treating head and neck cancer. mTHPC produces cell-killing reactive oxygen when irradiated with red light, but not without serious side effects resulting from the method now used to deliver this drug to tumors.

When added to the chemical mixture used to create the nanoparticles, mTHPC becomes trapped within the polymer framework. Characterization experiments showed that this photosensitizer does not escape from the nanoparticles, yet is still capable of producing the same amount of reactive oxygen as if it were free in solution. When added to human brain cancer cells growing in culture and irradiated with red light, this formulation kills the cells rapidly. Empty nanoparticles had no effect on the cells. Neither did drug-loaded nanoparticles added to the cells that were kept in the dark[185]

VII.8. Tumor-targeting capability of nanoparticles

A well-established fact in cancer therapy is that early tumor detection improves the odds that a patient will survive the disease. Now, using nanoparticles targeted to the tiny blood vessels that surround even the smallest tumors, researchers[186] have developed a radioactive imaging agent that was able to identify human tumors in rabbits[187].

---

[184] *Dr. Raoul Kopelman and colleagues from the University of Michigan*

[185] *Gao D. et al, Ultrafine Hydrogel Nanoparticles: Synthetic Approach and Therapeutic Application in Living Cells. Angew Chem Int Ed Engl. 2007, Feb 20.*

[186] *Researchers at the Siteman Center for Cancer Nanotechnology Excellence (CCNE). Researchers Gregory Lanza and Samuel Wickline, both at Washington University in St. Louis, led the group of investigators that created perfluorocarbon nanoparticles, each containing an average of 10 atoms of the radioactive element indium-111, an agent used in a variety of biomedical imaging applications.*

[187] *Hu G. et al, "Imaging of Vx-2 rabbit tumors with alpha(nu)beta(3)-integrin-targeted (111)In nanoparticles." Int J Cancer. 2007 Feb 2;120(9):1951-1957. This work was supported by the National Cancer Institute's Alliance for Nanotechnology in Cancer. Investigators from Philips Medical Systems, Dow Chemical Co., and the University of Texas at Dallas also participated in this study.*

*, Specialized pulsed lasers have been used to inject individual cells with a variety of materials, but little has been known about how this type of injection might **affect living cells**. For the first time, researchers have analyzed this nanoscale injection process on living cells and discovered that minor changes in the **intensity of the laser** could mark the difference between a healthy cell and a dead one. This makes discoveries at the cellular level extremely important.

The new findings could serve as a **set of guidelines** for **future research** that requires **precise microinjection of live single cells**. Such research ranges from testing drugs for toxicity to targeting tumor cells with chemotherapy. The technique will allow researchers to use unprecedented precision to microinject cells or even perform nanosurgery on cells.

* Dr. Melissa R. Junttila at the University of California, San Francisco is designing new mouse models to **challenge** long standing **dogma about how p53**, the most commonly mutated gene in human cancer, contributes to tumor formation and aging processes.

* Dr. L'szl' K'rti at Harvard University is devising a chemical synthesis strategy for the natural compound **Cortistatin A,** a **novel anti-cancer** drug that targets metastatic disease.

* Dr. Katharina Schlacher at the Memorial Sloan-Kettering Cancer Center in New York is defining **DNA repair mechanisms** that contribute to breast and ovarian cancers.

* Dr. Karsten H. Siller at the University of California, Berkeley, California, is using simple genetic systems to investigate how molecules that are mis-regulated in cancer **act to control neuron to neuron signaling**.

* Biodesign Institute researchers have received nearly US$9 million in grants to develop a **preventive vaccine against cancer**. The focus will be on breast cancer. The goal, based on some promising preliminary results, is to see a vaccine can be made that would be given to all adult women to **prevent the occurrence** of breast cancer. Scientists note that the **most successful medical intervention in history** has been the **development of vaccines against infectious disease.** Therefore, developing a **cancer vaccine** would be the **perfect solution**, not only in eliminating cancer mortality, but also potentially some of the costs of diagnosis and treatment.

It's been well-established that cancers create foreign proteins that the immune system can recognize. If it would become possible to **pre-immunize an individual with a collection of proteins** that effectively **represent any foreign protein that a breast tumor would produce**, the **immune system would arm itself against** breast cancer. And, if the platform technology proves successful, it could be applied to other cancers (*Arizona State University press release, July 14, 2007*).

* Scientists have developed a new platform for regulating the expression of therapeutic genes. This is particularly relevant for the emerging field **of stem cell gene therapy,** in which genes are delivered into a cell that can give rise to many distinct cell types.

* Researchers in Slovakia have been able to derive **mesenchymal stem cells** from **human adipose**, or fat, tissue and engineer them into **"suicide genes"** that seek out and destroy tumors like tiny homing missiles. This gene therapy approach is a **novel way** to attack small tumor metastases that **evade current detection techniques and treatments.** Mesenchymal stem cells help repair damaged tissue and organs by renewing injured cells. They are also found in the mass of normal cells that mix with cancer cells to make up a solid tumor. Researchers believe mesenchymal stem cells **"see"** a tumor as a damaged organ and migrate to it, and so might be utilized as a **"vehicle"** for treatment that can find both primary tumors and small metastases. These stem cells also have some plasticity, which means they can be converted by the micro environment of a given tissue into specialized cells. These fat-derived stem cells could be exploited for **personalized cell-based therapeutics.**

* A **molecularly engineered therapy selectively embeds** a gene in pancreatic cancer that **shrinks or eradicates tumors, inhibits metastasis, and prolongs survival with virtually no toxicity**, researchers from The University of Texas M. D. Anderson Cancer Center reported.

This vehicle, or vector, is so targeted and robust in its **cancer-specific expression** that it can be used for therapy and perhaps for imaging. The researchers call the system a **versatile expression vector** -- nicknamed VISA. It **includes a targeting agent**, also called a promoter, two **components that boost gene expression** in the target tissue, and a **payload** -- in this case a gene known to kill cancer cells. It's all packaged in a fatty ball called a liposome and delivered intravenously.

Researchers are working to move the system, developed and tested in mouse models of pancreatic cancer, to a Phase I clinical trial.

Early diagnosis of pancreatic cancer is extremely difficult, so the disease is often discovered at a late stage after it already has spread, or metastasized. **Fewer than 4 percent** of pancreatic cancer patients **survive five years after diagnosis**, one of the lowest cancer survival rates.

* Little has been known how genes are copied and repaired so efficiently and precisely. These processes always involve a so-called DNA polymerase, an enzyme that performs the actual new growth of genes. The genes consist of two DNA strands, but scientists have not known what polymerase copies the two DNA strands.

Researchers **mutated DNA polymerase epsilon**, creating an enzyme that makes a particular error when it copies genes. This means that the enzyme leaves a signature at all sites where it copies the genes in the cell. By reading where this signature is left, the scientists have then been able to determine that DNA polymerase epsilon copies one of the strands, the so-called **"leading"** strand. For decades researchers have been wondering what enzyme synthesizes this particular part, and now proof has been found. These research findings, which describe fundamental biological functions, pave the way for an **enhanced understanding of how mutations occur in genes**, mutations that can lead to cancer, for instance.

To target tumors, the researchers added a molecule that recognizes and binds to æ$_v$ß$_3$-integrin, a complex molecule found on the surface of new blood vessels.

When injected into rabbits bearing human tumors, these nanoparticles quickly accumulated in and around the tumors. Using a standard gamma camera, the investigators were able to detect the nanoparticles at the site of tumors within 15 minutes after injection. The gamma signal persisted over the course of two hours. In contrast, the researchers detected no nanoparticle accumulation in the muscle, a tissue rich with blood vessels, demonstrating the tumor-targeting capability of these nanoparticles.

The researchers note that while the sensitivity of this method is high, its spatial resolution is low. In practice, then, this method would best be used in conjunction with imaging techniques such as fluorine-19 magnetic resonance imaging (^{19}F-MRI). Gamma imaging would then provide sensitive detection of small tumors, while ^{19}F-MRI would pinpoint the location of the tumors for subsequent surgical removal.

VII.9. Increasing the in-vivo lifetime of polymeric nanoparticles

Researchers[188] have discovered that attaching polymeric nanoparticles to the surface of red blood cells dramatically increases the in vivo lifetime of the nanoparticles. The research[189] could offer applications for the delivery of drugs for a variety of conditions such as cancer, blood clots and heart diseases. It may also have important implications for future treatment of hematologic disorders.

Polymeric nanoparticles are excellent carriers for delivering drugs. They protect drugs from degradation until they reach their target and provide sustained release of drugs. Polymeric nanoparticles, however, suffer from one major limitation: they are quickly removed from the blood, sometimes in minutes, rendering them ineffective in delivering drugs.

The research team found that nanoparticles can be forced to remain in the circulation when attached to red blood cells. The particles eventually detach from the blood cells due to shear forces and cell-to-cell interactions, and are cleared from the system by the liver and spleen. Red blood cell circulation is not affected by attaching the nanoparticles. According to the investigators, attachment of polymeric nanoparticles to red blood cells combines the advantages of the long circulating lifetime of the red blood cell, and their abundance, with the robustness of polymeric nanoparticles. Using red blood cells to extend the circulation time of the particles avoids the need to modify the surface chemistry of the entire particle, which offers the potential to attach chemicals to the exposed surface for targeting applications.

The researchers have learned that particles adhered to red blood cells can escape phagocytosis because red blood cells have a knack for evading macrophages. Nanoparticles aren't the first to be piggybacking on red blood cells; the strategy has already been adopted by certain bacteria, such as hemobartonella, that adhere to RBCs

---

[188] *at the University of California, Santa Barbara, led by Samir Mitragotri, a professor of chemical engineering, and Elizabeth Chambers, a doctoral graduate*

[189] *published in the July 07 issue of Experimental Biology and Medicine*

* Nanotech fights cancer: The future of drug design lies in finding ways to **target a drug specifically to a diseased cell,** or even a **molecule within that cell,** while **leaving healthy cells** and **molecules unharmed**.
Researchers face multiple challenges in developing new treatments for cancer, including a lack of selectivity in the cells the drugs kill and difficulty getting through the cell membrane. They can add a target to the drug and tell it to go to the liver, for example, but that doesn't solve the challenge of getting into the diseased cells and leaving healthy cells alone.
Nanomedicine is already delivering significant benefits through new diagnostics, imaging agents and even nanomedicines themselves, with examples being **biosensors** from Oxford Biosensors, **imaging systems** from Philips and Schering AG, and **polymer based cancer therapeutics** from Celltech. One of the first **nanomedicines** to achieve a launch is a suspension of the anticancer drug **doxorubicin** packed into **liposomes,** which extended the drug's lifetime in the body, lowering its cost and toxicity while increasing its effectiveness. The potential for new drug-delivery systems made from **nanoparticles,** pointing out that they have many advantages over classical methods of drug delivery has been underlined by Prof. Gabizon. For example, they can dramatically alter the absorption, distribution and persistence of drugs in the body, while the targeted delivery of drugs to diseased sites spares sensitive tissues, reducing side effects.**Treatments** for atherosclerosis, HIV/AIDS, diabetes and Alzheimer's disease could be **revolutionized by nanomedicine,** too. Research is currently advancing in many different areas, with new developments including: the **imaging** of **single cells** and **proteins**; the **construction** of **miniaturized analytical devices**; and the **development** of **nanomaterials** for **tissue regeneration** and **drug delivery**.*(Source: European Science Foundation, 2006)*

* **Nanotechnology** promises to have a **profound impact** on society. Defined as science and engineering done at the **scale of a billionth of a meter,** nanotechnology has been heralded by many scientists, futurists and investors as the **next industrial revolution**. Optimistic forecast see **nanobots** to perform **microsurgery** or be used as **in-body sensors** to **monitor human health,** According to the **Biodesign Institute,** US federally supported nanotechnology R&D in 2005 was $1 billion, and the future global marketplace for goods and services using nanotechnologies will grow **to $1 trillion** by **2015**. However, nano- scientists and engineers are still working out the rules and techniques for imaging, manipulating and manufacturing matter at this minute atomic scale, **10,000 times** smaller than the width of a human hair. And social scientists and humanists are just starting to understand how such inquiries and technologies interact with the broader society.

* In 2006, cancer was responsible for **14%** of the 60m deaths world wide. The World Health Organization predicts **a rise in cancer incidence by 50%** by the year 2020, largely due to lifestyle factors.

* Dr. Danielle S.W. Benoit at the University of Washington, Seattle, is developing novel drug delivery systems for a new class of biomolecular anti-cancer agents called **siRNAs**

* A tiny **implant** which is being developed at **MIT** could one day help doctors rapidly monitor the growth of tumors and the progress of chemotherapy in cancer patients. The implant contains **nanoparticles** that can be designed to test for different substances, including **metabolites** such as glucose and oxygen that are associated with **tumor growth**. It can also track the **effects** of **cancer drugs**: Once inside a patient, the implant could reveal how much of a certain cancer drug has reached the tumor, helping doctors determine whether a treatment is working in a particular patient.
The device can be **implanted directly** into a **tumor,** allowing researchers to get a more direct look at what is happening in the tumor over time. With blood testing, one can't be sure that drugs present in the blood have also reached the tumor, because the system of blood vessels surrounding tumors is complicated. The new technique, known as **implanted magnetic sensing,** makes use of detection nanoparticles composed of iron oxide and coated with a sugar called dextran. Antibodies specific to the target molecules are attached to the surface of the particles. When the target molecules are present, they bind to the particles and cause them to clump together. That **clumping** can be detected by **MRI** (magnetic resonance imaging). The nanoparticles are trapped inside the silicone device, which is sealed off by a porous membrane. The membrane allows molecules smaller than 30 nm to get in, but the detection particles are **too big to get out.** In addition to monitoring the presence of chemotherapy drugs, the device could also be used to check whether a tumor is growing or shrinking, or whether it has spread to other locations, by sensing the amount and location of tumor markers. Next, researchers will be looking for a hormone, **human chorionic gonadotropin** (HCG), that can be considered a marker for cancer because it is **produced by tumors** but not normally found in healthy individuals (unless they are pregnant).

* **Clinical trials** are underway with many additional novel therapeutic agents directed against targets identified through their **genomic lesions** in cancer. Examples include **inhibitors** of the **serine-threonine kinase** encoded by the **BRAF gene,** found to be mutated in 80% of melanomas; **inhibitors** of the **receptor tyrosine kinase** encoded by the **FLT3 gene,** found to be mutated in 25% of patients with acute myelocytic leukemia; and inhibitors of **downstream components** suppressed by the product of the **VHL gene,** which was found to be mutated in renal cell carcinoma.

* Research has shown there is a **connection** between **reproductive factors**-such as age at first birth, number of births, and breastfeeding-and **a woman's risk of breast cancer.** Yet to be established is how these factors interact, and whether they have differing effects on risk for breast cancers that are estrogen and progesterone receptor positive (ERPR-positive) versus those that are not (ERPR-negative).

* Dr. Michael D. Gordon at the University of California, Berkeley, California, is defining the **neural circuits that control feeding** - providing fundamental knowledge to frame the **growing health crisis of obesity.**

and can remain in circulation for several weeks. The researchers say that it may be possible to keep the

nanoparticles in circulation for a relatively long time, theoretically up to the circulation lifetime of a red blood cell - which is 120 days - if the binding between particles and the red blood cells is strengthened. The methodology is applicable to drugs that are effective while still attached to a red blood cell, although the researchers say that slow release from the red blood cell surface is also feasible[190].

VII.10. Magnetic nanocrystals carry and release anticancer agents

Imagine using a focused magnetic field to concentrate anticancer drugs in and around tumors, and then turning off the magnetic field so that the drugs then leave the body. That possibility may become a reality as a result of work showing that magnetic nanocrystalline iron-nickel alloys can effectively carry and release anticancer agents.

The researchers[191] first prepared magnetic nickel ferrite ($NiFe_2O_4$) nanocrystals coated with the biocompatible polymer polymethacrylic acid (PMAA) and developed methods of hooking the anticancer agent doxorubicin to the ends of the PMAA chains. Characterization experiments showed that the modified nanocrystals retained their strong magnetic properties. The polymer coating prevents the nanocrystals from clumping together, which would degrade their magnetic properties. Moreover, the polymer coating prevents blood proteins from accumulating on the nanocrystals, thereby reducing clearance of the nanocrystals by the immune system.

Drug delivery experiments demonstrated that the nanocrystals slowly and steadily released doxorubicin over 200 hours. However, when the investigators applied a magnetic field to the nanocrystals, drug release rates increased dramatically, with the particles releasing 2.5 times more drug in the first two hours and over the next 24 hours. After 60 hours in a magnetic field, the nanocrystals released approximately 75 percent of their drug payload[192].

VII.11. Nanocrystalline coating

Silver, platinum and gold, which are elements of the noble metals group, have long been known to have medicinal properties. For example, platinum is the primary active ingredient in cisplatin, a prominent cancer drug. Similarly, gold is the active agent in some treatments for rheumatoid arthritis. There are numerous silver-containing advanced wound care dressings and silver-coated medical devices available from a variety of health care companies.

Nucryst Pharmaceuticals Corp. developed a technology to convert silver's microcrystalline structure into an atomically disordered nanocrystalline coating that they believe enhance silver's natural antimicrobial properties by providing for the sustained release of an increased quantity of positively-charged particles called ions.

They selected silver as the first noble metal for the application of their proprietary nanotechnology based on silver's antimicrobial properties. Although silver's medicinal

---

[190] *Experimental Biology and Medicine is a journal dedicated to the publication of multidisciplinary and interdisciplinary research in the biomedical sciences.*

[191] *Dr. Devesh Misra, Ph.D., and colleagues at the University of Louisiana at Lafayette*

[192] *Rana S, et al, "On the suitability of nanocrystalline ferrites as a magnetic carrier for drug delivery: Functionalization, conjugation and drug release kinetics". Acta Biomater.2007 Mar;3(2):233-42. Epub 2007 Jan 16*

properties have been well known for centuries, the company says   its use in its microcrystalline form has been limited due to its slow release of relatively small quantities of silver ions and the widespread use of antibiotics. Silver is composed of large microcrystals, usually of one or two microns in diameter or greater. These microcrystals dissolve slowly, thereby limiting the rate and amount of silver released over time. By converting silver's microcrystalline structure into an atomically disordered nanocrystalline structure, they believe that their nanocrystalline silver combat infection longer than other silver-based wound care products and offer a broader spectrum of antimicrobial activity than many topically applied antibiotics. In addition, their nanocrystalline silver structures have exhibited potent anti-inflammatory properties in preclinical studies.

Antibacterial agents inhibit or kill bacterial cells by attacking one of the bacterium's structures or processes. Common targets are the bacterium's outer shell (called the *"cell wall"*) and the bacterium's intracellular processes that normally help the bacterium grow and reproduce. However, since a particular antibiotic typically attacks one or a limited number of cellular targets, any bacteria with a resistance to that antibiotic's killing mechanism could potentially survive and repopulate the bacterial colony. Over time, these bacteria could make resistance or immunity to this antibiotic widespread. Unlike antibiotics, silver has been shown to simultaneously attack several targets in the bacterial cell and therefore it is thought to be less likely that bacteria would become resistant to all of these killing mechanisms and thereby create a new silver-resistant strain of bacteria. This may be the reason that bacterial resistance to silver has not yet been widely observed despite its centuries-long use. This can be particularly important in hospitals, nursing homes and other healthcare institutions where patients are at risk of developing infections.

The medical device and pharmaceutical industries are intensely competitive. Major companies in the advanced wound dressing market in which this type of products is sold, are Smith & Nephew's (Acticoat of Nucryst Pharmaceuticals) Convatec, a Bristol Myers Squibb company, Johnson & Johnson Wound Management, a division of Ethicon, Inc., Argentum Medical, LLC, Coloplast Corp., AcryMed, Inc., 3M Company, Kinetic Concepts Inc., Mölnlycke Health Care Group AB and Paul Hartmann AG. To the extent that they develop pharmaceutical products to treat dermatological and gastrointestinal conditions, they will face competition from pharmaceutical companies developing alternative drugs to treat these diseases. A number of parties are beginning to compete in the medical device coating area, including AcryMed, Inc., Covalon Technologies Ltd., C.R. Bard, Inc. and St. Jude[193].

VII.12. Pairing nanoparticles with protein

In groundbreaking research, Brookhaven[194] scientists have demonstrated the ability to strategically attach gold nanoparticles -- particles on the order of billionths of a meter -- to proteins so as to form sheets of protein-gold arrays. The nanoparticles and methods to create nanoparticle-protein complexes can be used to help decipher protein structures, to identify functional parts of proteins, and to "glue" together new protein complexes.

---

[193] *Adapted from NUCRYST Pharmaceuticals Corp · 10-K · For 12/31/05 SEC Filings*

[194] *One of ten national laboratories overseen and primarily funded by the Office of Science of the U.S. Department of Energy (DOE), Brookhaven National Laboratory conducts research in the physical, biomedical, and environmental sciences, as well as in energy technologies and national security. Brookhaven Lab also builds and operates major scientific facilities available to university, industry and government researchers. Brookhaven is operated and managed for DOE's Office of Science by Brookhaven Science Associates, a limited-liability company founded by the Research Foundation of State University of New York on behalf of Stony Brook University, the largest academic user of Laboratory facilities, and Battelle, a nonprofit, applied science and technology organization.*

* After **binding DNA segments** to tiny **iron-containing spheres called nanoparticles**, researchers at the Philadelphia Children's Hospital have **used magnetic fields to direct** the nanoparticles **into arterial muscle cells**, where the DNA could have a **therapeutic effect**. The nanoparticles might find broader application, such as delivering gene therapy to tumors, or carrying drugs instead of or in addition to genes. Another possibility is that after preloading genetically engineered cells with nanoparticles, researchers could use **magnetic forces to direct the cells to a target organ**.

* Researchers at the University of York are using an understanding of the special cells that line the bladder to develop ways of **restoring continence** to patients with **serious bladder conditions**, including cancer. The longer term aim for this research is to help patients who have lost bladder function or have had all or part of their bladder removed because of cancer.

* **Cell division** is among the most **fundamental processes of biology,** and if something goes wrong at any point during the meticulously orchestrated process, the **mistake could result** in the **misdistribution of chromosomes and lead to cancer or other diseases**. So researchers are concentrating on every step of the process, trying to learn as much as they can about how cells divide and which molecules are involved. Of perticular interest to researchers is how chromosomes are responsible for directing cell division and how microtubules form the bipolar spindle; the critical step that allows the chromosomes to align, and involves at least three cellular pathways, one of which, the **chromosomal passenger complex,** or CPC, has been discovered by Dr Funabiki of Rockefeller University. The CPC is a group of proteins that bind to chromosomes as they're lined up along the cell's center in preparation for division. New research shows that the ability of the bipolar spindle to assemble only in the presence of chromosomal DNA can be pinned on a specific enzyme in the CPC, the **Aurora B kinase.**

* Communication is critical. Garbled in, garbled out, so to (mis-)speak. Workers who get incomplete instructions produce an incomplete product, and that's exactly what happens with the stem cells in our aging muscles, according to researchers from the Stanford University School of Medicine. Their study found that, **as we age,** the **lines of communication to the stem cells of our muscles deteriorate** and, **without the full instructions**, **it takes longer for injured muscles to heal**. Even then, the repairs aren't as good. The key to the whole process is **Wnt,** a protein traditionally thought to help promote maintenance and proliferation of stem cells in many tissues. In this instance, **Wnt** appears to **block proper communication.** The scientists theorize that, theoretically, given the number of ways to block Wnt and Wnt signaling, one could envision this becoming a **therapeutic**, one could potentially enhance the healing of aged tissues by reducing this effect of Wnt signaling on the resident stem cells. In addition to helping the elderly heal faster and better from muscle injuries, the potential benefits could include tissues such as skin, gut and bone marrow, or for that matter, potentially any tissue, such as liver and brain, in which stem cells contribute to normal cellular turnover.

* Dr. Yarden is investigating the. Dr. Yosef Yarden of the Weizmann Institute of Science in Rehovot, Israel has been involved in many crucial developments in the field of **epidermal growth factor receptor** (EGFR/ErbB) family of proteins with the goal of designing therapies that inhibit their oncogenic properties, including isolating EGFR/ErbB-1, identifying several proteins that bind to EGFR/ErbB-1, and understanding the role of another member of the EGFR family (HER2/ErbB-2) **in cell signaling and tumor development**. His future research will focus on **developing therapies that target ErbB**. This work will be carried out through development of: 1) **novel molecules** that **bind EGFR/ErbB receptors**; 2) **engineered soluble portions** of these receptors; and 3) **inhibitors of EGFR/ErbB receptor activation** that **promote** receptor degradation. Immunotherapy strategies will also be explored.

* Research by Drs Oviedo and Lewin of the Forsyth Institute, with the flatworm, planaria, highlighs the importance of **direct cell-cell transfer of small molecules between stem cells and their neighbors**, and provides an important **roadmap** for learning about **regeneration**. These findings suggest that similar mechanisms may be extraordinarily relevant for **controlling the behavior of migratory, plastic cells.** Further analysis in both planarians and in vertebrates will provide crucial opportunities for **understanding what drives stem cell behavior** and may help medical science identify novel therapeutic targets.

* Researchers at Purdue University led by Dr Irudayaraj, hope that by 2012, their newly created **gold nanorods for cancer detection** will **replace** the about three times more expensive current analogous technology, called **flow cytometry**. .This method works by attaching fluorescent probes to cancer cells, whereas the nanorod technology has its basis in sensing plasmons, or sub-atomic particles present in the gold nanoparticles. The gold nanoparticles, or **nanorods**, are tiny rod-shaped gold particles, even smaller than viruses, which are equipped with antibodies designed to bind to a specific marker on cell surfaces, making the tiny particles a potential tool to better diagnose and treat cancer.Researchers analyze these surface markers, **proteins on a cell's exterior**, because they can contain valuable information about what type of cell they belong to or what state that cell may be in. The nanorods also require **only a few cells**, whereas flow cytometry requires hundreds to thousands of cells. This could be **advantageous** when dealing with **scarce sample sizes**.

* A new **minimally invasive** and **gentle treatment method**, developed by an Innsbruck-based research team of the University clinic for Urology, promises considerably improved success rates in the treatment **of stress incontinence**: the Urocell **cell treatment. This cell therapy** directly works on the cause of the weak bladder, on **the bladder sphincter**: the **patient's own muscle stem cells** are **cultured** and **then re-injected**, so they will form new muscle fibres and strengthen the urethral closure mechanism. In this way, the patient's own muscle cells work as a doctor in the patient's body: with their help the sphincter muscle is able to restore its function.

Applications envisioned by the researchers include catalysts for converting biomass to energy and precision "vehicles" for targeted drug delivery.

This new research[195] shows that precisely engineered gold nanoparticles can be used to "glue" enzymes together to form oriented and ordered single layers, and that these monolayers are mechanically stable enough to be transferred onto a solid surface such as an electrode.

The scientists attached gold nanoparticles to an enzyme complex that helps drug-resistant tuberculosis bacteria survive. They suggest that gold nanoparticles might also be tailored to inactivate this enzyme complex, thereby thwarting drug-resistant TB -- a research avenue they may explore in future studies.

In another part of the study, the researchers used proteins found on the surface of adenovirus, a virus that causes the common cold. Previous studies have characterized how this virus binds to the human cells it infects, and have suggested that modified forms of adenovirus could be used as vehicles to deliver drugs to specific target cells, such as those that make up tumors.

One key to this approach would be to enhance strong binding to the target cells. Toward that end, the research group[196] attached multiple viral proteins to the gold nanoparticles. Such constructs should have increased binding affinity for target cells and their larger size should extend blood residence time for improved drug delivery.

In another application, this new research showed that gold nanoparticles can enhance scientists' ability to decipher the structures and functionally important regions of protein molecules - the workhorses that carry out every function of living cells and whose dysfunction often leads to disease. With added nanoparticles, the "signal-to-noise ratio" and resolution of an imaging technique known as cryo-electron microscopy were significantly increased. This method might enable analysis of small biological macromolecules and complexes that are currently intractable to analyze by cryo-electron microscopy or x-ray crystallography.

VII.13. New drug-delivery system using nanomaterials
A critical obstacle and challenge for cancer therapy is the limited availability of effective biocompatible delivery systems.

Since the poor solubility of many effective anticancer drugs is one of the major problems in cancer therapy because the drugs require the addition of solvents in order to be easily absorbed into cancer cells (unfortunately, these solvents not only dilute the potency of the drugs but create toxicity as well), the development of novel delivery systems for these molecules without the use of organic solvents has received significant attention. In order to be used on humans, current cancer therapies such as CPT or Taxol, which are poorly water soluble, must be mixed with organic solvents in order to be delivered into the body. These elements produce toxic side effects and in fact decrease the potency of

---

[195] conducted at the U.S. Department of Energy's Brookhaven National Laboratory, and published in the July 2, 2007 issue of the journal Angewandte Chemie.

[196] lead author Minghui Hu, James Hainfeld, and colleagues in the Biology Department at Brookhaven Lab. It was funded by Brookhaven's Laboratory Directed Research and Development program, the Office of Environmental and Biological Research within the U.S. Department of Energy's Office of Science, and by the Institute of General Medical Sciences within the National Institutes of Health.

From the Primary Care Monoculture -Medicines for Every Body- to a Multidimensional Platform –Personalized Medicine.

"It is not the strongest of the species that survives, not the most intelligent, but the one most responsive to change"
Charles Darwin

* While the Japanese & European (political) objective is to retain the social, universal health care system and to more efficiently follow the cost/benefit balance approach to healthcare the US is practicing, the aging population is expected to **bear more of their own healthcare costs**.

* Regulators become more concerned about the **safety of new drugs with novel targets**. The genomics revolution in the field of life sciences will raise hopes and expectations, but also fears, anxieties and doubts.

* To overcome the high failure rate of new drug candidates, the search is on for a suitable preclinical model of new drug development to **mimic the human response.**

* Health Insurance Funds are to make lifestyle changes a condition for health risk coverage.

* **A new wave of vaccine products** (e.g. DNA vaccines, cell-culture vaccines, protein vaccines, viral-vector vaccines) represents the opportunity to benefit millions -of people.

* **Embryonic stem cells** wish usher in a second biotech revolution, resulting in an unprecedented series of remarkable medicines to help patients suffering from life-threatening or debilitating illness. Research that validates new drugs/new technologies is an expensive undertaking. **New models of collaboration** across the pharmaceutical industry, between industry and government-funded researchers will emerge.
Long Term results of today's research:
* Genetic and molecular testing will be the **standard of care;**
* Gene-based therapies will **prevent** and **treat** illness.

* Over half of healthcare expenses will be spent on people aged 75+. The need for more **accurate outcome measures** will dominate the healthcare debate.
* The recognition of electrical activity would provide scientists the possibility to see the **'heartbeat'** of an individual cell, allowing the daily functioning of the cell to be monitored in a similar way to a cardiograph showing the workings of a human heart, and a new way to test drugs.
* **Individualized drug therapy** is to revolutionize the healthcare world as we know it today.
 Medicine may soon be able to look at individual patients instead of diseases. Rather than handing out diagnoses, doctors may soon be able to dissect how a given disease progresses in a particular individual. The philosophy of predicting care, of preventive care (which is more efficient than treating disease after the fact) starts changing society's view on healthcare.
* The growing danger of bioterrorism or a destabilizing infectious disease pandemic creates the unprecedented need for the prevention and treatment of severe infectious diseases that present **global security challenges**. This asks for rapid development, production, and stockpiling medical countermeasures (medicines and vaccines) for the global community.
* More **multicomponent drugs** will be developed from separate single-compound medicines that already exist to treat the target disease on the basis of a clinical rationale.
* **Genomic technologies** make it increasingly possible to identify patients most likely to benefit from a molecularly target drug. Drugs will be developed for **disease subtypes** with the aid of **molecular diagnostics.**
* pharmacogenomics will essentially define a **multitude of diseases within diseases**, creating many more **niche patient populations.**

For decades, scientists have looked for new therapies that target cancer. Researchers now know that cancer cells have several unique characteristics -- they require new blood vessels to deliver oxygen and nutrients (angiogenesis), they grow uncontrollably (proliferation), they travel throughout the body (metastasis), and they escape programmed cell death, a natural process by which the body rids itself of damaged or unwanted cells (apoptosis). The medical research community is dedicated to finding the next generation of small molecule and biologic therapies that will have an impact on this horrible disease.

Targeted medicines will dramatically improve the quality of the drug action, which should then also lead to drug use playing a role of "precision medicine" in the management of disease.

Science is helping towards using genetics to move from treating the disease after it happens, to preventing the worst symptoms of the disease before it happens. The road to predictive healthcare will be difficult, and still far away, but there is no final alternative to reach optimal health.

Society is to witness breakthroughs in understanding causes of major diseases within the next decade.
Advances in technology create new measurement possibilities of predisposition and progression. The treatment of diseases with underlying biological causes that remain largely mysterious, such as cancer, diabetes, depression and arthritis, will be facilitated by the identification of the genetic material that predisposes to these difficult-to-treat illnesses. This knowledge will allow physicians to tailor medical treatment to patients more exactly than it has ever been possible. Although the large majority of drugs that are brought into clinical trials never reach the market, the medical potential of new decade gene-based drugs is enormous.

the cancer therapy. To overcome these problems, drug delivery systems using pegylated polymers, liposomal particles or albumin-based nanoparticles have been developed.

Researchers at UCLA[197] have successfully manipulated nanomaterials to create a new drug-delivery system that promises to solve the challenge of the poor water solubility of today's most promising anticancer drugs and thereby increase their effectiveness. They report a novel approach using silica-based nanoparticles to deliver the anticancer drug camptothecin and other water-insoluble drugs into human cancer cells.

Camptothecin (CPT) and its derivatives are considered to be among the most effective anticancer drugs of the 21st century. Although studies have demonstrated their effectiveness against carcinomas of the stomach, colon, neck and bladder, as well as against breast cancer, small-cell lung cancer and leukemia in vitro, clinical application of CPT in humans has only been carried out with CPT derivatives that have improved water solubility.

The new research findings show that mesoporous silica nanoparticles offer great potential and a promising approach to the delivery of therapeutic agents into targeted organs or cells. The pores in the nanoparticles could be closed by constructing an appropriate cap structure. This provides the ability to control the release of anticancer drugs by external stimuli.

VII.14. A new approach to treating intractable tumors
Administering a small amount of a potent but potentially toxic anticancer agent along with nanoparticles loaded with a second anticancer agent produced a dramatic inhibition of tumor growth in normally intractable cancers. These findings suggest a new approach to treating malignancies such as pancreatic cancer and diffuse gastric cancer.[198]

The researchers[199] described their experiments using an inhibitor of transforming growth factor beta (TGF-ß) to boost the antitumor activity of doxorubicin-containing nanoparticles.

TGF-ß inhibitors have shown promise in stopping tumor growth and metastasis, but some experiments suggest that these inhibitors may trigger serious side effects, including the development of new tumors unrelated to the original cancer.

The investigators reasoned that sub-therapeutic doses of a short-acting TGF-ß inhibitor might interact with the blood vessels surrounding tumors in a way that might sensitize the tumors to a second therapeutic agent, in this case nanoparticles loaded with the anticancer drug doxorubicin. Indeed, experiments showed that very low doses of a short-acting TGF-ß inhibitor had no measurable effect on tumor biochemistry, but did lower the number of cells known as pericytes that normally coat the inside of newly growing blood vessels. With fewer pericytes present, the blood vessels surrounding tumors become

---

[197] *Results of a study by UCLA scientists appeared in the nanoscience journal Small in June 2007. The study was led by Fuyu Tamanoi, UCLA professor of microbiology, immunology and molecular genetics and director of the Jonsson Cancer Center's Signal Transduction and Therapeutics Program Area, and Jeffrey Zink, UCLA professor of chemistry and biochemistry.*

[198] *Kano et al., "Improvement of cancer-targeting therapy, using nanocarriers for intractable solid tumors by inhibition of TGF-ß signaling." PNAS Proceedings of the National Academy of Science of the United States of America 104 (9): 3460. Investigators from the Osaka City University Graduate School of Medicine also participated in this study.*

[199] *A team of investigators led by Dr. Kohei Miyazono, and Dr. Kazunori Kataoka, both at the University of Tokyo, reported their work in the Proceedings of the National Academy of Science.*

even leakier than normal, allowing doxorubicin-loaded nanoparticles to escape from these vessels and accumulate at high levels within the tumors.

Experiments in animals with normally intractable cancers showed that this combination therapy was effective at inhibiting tumor growth. In somewhat of a surprise, the researchers also found that co-administering the TGF-ß inhibitor with a small molecule anticancer drug had no effect on tumor growth. Only the nanoparticle-inhibitor combination produced the desired boost in antitumor activity.

VII.15. Nanobodies: the race to develop a new class of drugs

Nanobodies are a novel class of therapeutic proteins based on single-domain antibody fragments, for a range of serious and life-threatening human diseases, including inflammation, thrombosis, oncology and Alzheimer's disease.

Nanobody-based therapeutics represent a major commercial opportunity as they combine the beneficial features of conventional antibodies, with desirable properties of small-molecule drugs. Because they are derived from naturally-occurring heavy-chain antibodies, Nanobodies have unparalleled stability and can be administered in a variety of ways (injected, orally, in sprays or creams), thus overcoming the delivery issues associated with full-sized antibodies, that can only be delivered by injection. In addition, because of their unique structure they can also address therapeutic opportunities that are beyond the reach of conventional antibodies or their fragments, for example targeting epitopes such as receptor clefts, enzyme active sites and viral canyon sites. Nanobodies manufactured in micro-organisms also presents a significant cost advantage in comparison to production methods for conventional antibodies.

Biopharmaceutical company Ablynx (Belgium) has generated Nanobodies against more than hundred different disease targets. The company has shown the absence of any detectable immunogenicity for its Nanobody development candidates in advanced primate studies.

Table 7.3. Acquisitions prove the importance of antibodies to the major pharmaceutical research driven companies:

Buyer/Target	Year	Buyer/Target	Year	Buyer/Target	Year	Buyer/Target	Year
Roche/Glycart	2005	Pfizer/Rinat	2006	GSK/Domantis	2006	Eisai/Morphotek	2007
Pfizer/Bioren	2005	AstraZeneca/CAT	2006	Novartis/NeuTec	2006	Roche/THP	2007
Amgen/Abgenix	2005	Merck/GlycoFI, Abmaxis	2006	Genentech/Tanox	2006	AstraZeneca/Medimmune	2007

The Nanobody technology was originally developed following the discovery that camelidae (camels and llamas) possess fully functional antibodies that lack light chains. These heavy-chain antibodies contain a single variable domain (VHH) and two constant domains (C_H2 and C_H3).

Importantly, the cloned and isolated VHH domain is a perfectly stable polypeptide harboring the full antigen-binding capacity of the original heavy-chain antibody. These newly discovered VHH domains with their unique structural and functional properties form the basis of a new generation of therapeutic antibodies which Ablynx has named Nanobodies.

Table 7.4. The speed of discovery

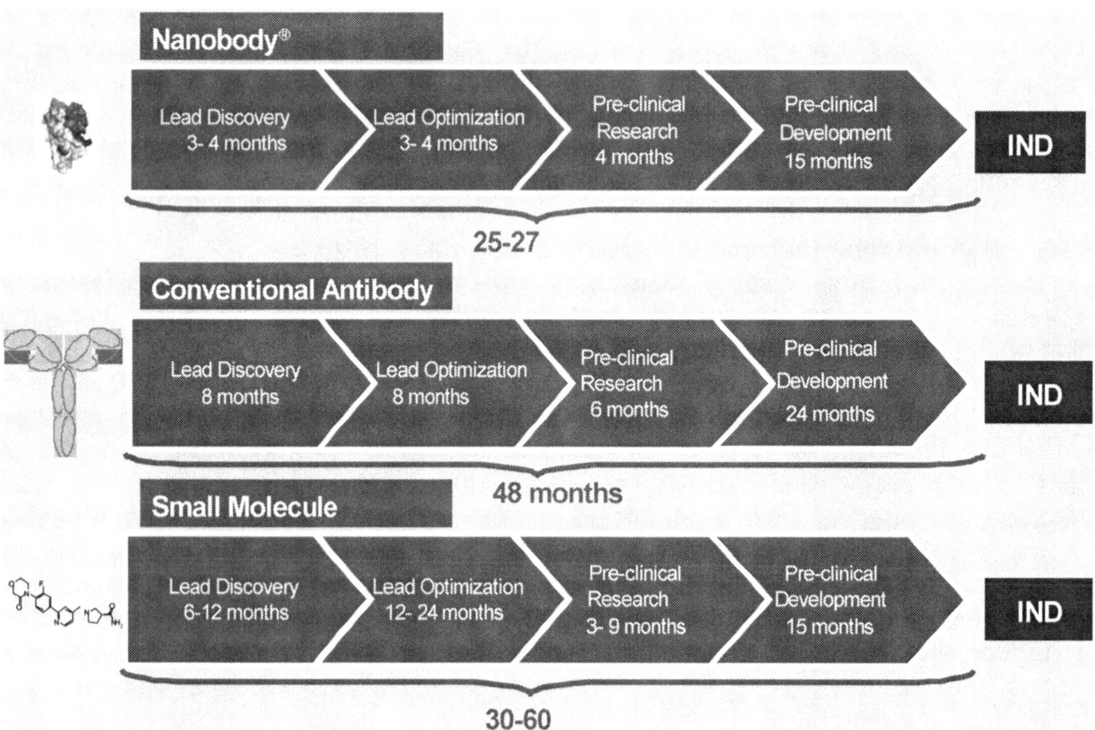

Ablynx has ongoing research collaborations and significant, multi-target partnerships with several major pharmaceutical companies, including Boehringer Ingelheim, Wyeth Pharmaceuticals, Novartis, Centocor (J&J), Kirin and P&G Pharma. Ablynx is building a diverse and broad portfolio of therapeutic Nanobodies through these collaborations as well as through its own internal discovery programs.

The company announced positive interim results from the ongoing Phase I study of its lead development program, ALX-0081, an anti-thrombotic treatment. ALX-0081 is a novel 'first-in-class' therapeutic Nanobody targeting von Willebrand Factor (vWF), which can reduce the risk of thrombosis in patients with acute coronary syndrome. Through its novel, highly selective mode of action, ALX-0081 is intended to prevent arterial thrombosis, without interfering with the desired hemostatis (wound healing) in the patient which results in less bleeding complications. Pre-clinical in vivo studies confirmed Ablynx's belief that ALX-0081 has unique potential to set a new standard in anti-thrombotic therapy based on its immediate onset of action, its high potency and significantly improved safety compared to the currently marketed therapies in the form of significantly reduced bleeding complications. Ablynx has demonstrated a large therapeutic window for ALX-0081 based on the high efficacy and low bleeding demonstrated in primate studies, indicating a highly attractive drug profile.

GlaxoSmithKline has committed to becoming a leader in biopharmaceuticals by dedicating one of its Centers of Excellence for Drug Discovery to this endeavor. Large pharmaceutical companies traditionally specialize in medicines that are small molecules, administered orally. Biopharmaceuticals, by contrast, are large molecules typically administered by injection or infusion, though research continues in other delivery technologies. Examples of biopharmaceuticals are monoclonal antibodies, therapeutic vaccines, and recombinant therapeutic proteins.

Through its acquisition of Domantis (2006), GSK is in the race to develop antibody therapies, based on the smallest functional binding units of human antibodies. These units, termed domain antibodies (dAbs), may be administered in inhaled, topical, and, potentially, oral formulations as well as by injection and infusion. The Domantis technology also enables dAbs to serve as building blocks for therapeutics simultaneously directed at more than one disease target. Research programs at Domantis address diseases including rheumatoid arthritis, asthma, chronic obstructive pulmonary disease, and multiple myeloma.

Table 7.5. Antibody engineering technologies that will be used in the coming years include e.g.:

Humanized Technologies	Ribosome Display Technologies
* Protein Design Labs' SMART Humanization Technology	* Dyax's Phage Display Discovery Tool
* XOMA's Human Engineering Technology	**Antibody Fragment Technologies**
* Aeres Biomedical's Humanization Technology	* Enzon's Single Chain Antibody Technology
Fully Human Technologies	* Domantis' Domain Antibodies
* Medarex's UltiMAb Technology	**Conjugated Antibody Technologies**
* Abgenix's XenoMouse and XenoMax	* Immunogen's Tumor-Activated Prodrug (TAP) technology
* Cambridge Antibody Technology's Phage Display	* Seattle Genetics' Antibody-drug Conjugate (ADC)
	& Antibody-Directed Enzyme Prodrug Therapy (ADEPT) Technology

The Australian biopharmaceutical company, Peptech Limited announced the commencement of a Phase I trial of its lead compound PN0621, which marks the first administration to humans of a domain antibody derived product.

PN0621 has been tailored for greater potential in reaching parts of the human body that current drugs are not able to reach, potentially resulting in quicker and better treatment of inflammatory diseases.

PN0621 is an anti-TNF, a name given to a class of drugs being developed for the treatment of auto-immune diseases such as rheumatoid arthritis, psoriasis and Crohn's disease. It works by blocking the action of TNF (tumor necrosis factor) which is involved in this attack.
Peptech has successfully completed pre-clinical trials of PN0621.

There are three anti-TNF drugs on the market, Remicade, Humira and Enbrel. In 2006 these drugs had combined global sales of US$10.8 billion, with a projected market value of greater than US$20 billion in 2012.

It is well documented that patients can develop a resistance to one antibody treatment, but respond to another. The list of indications that can be treated with anti-TNF drugs is increasing to include, asthma, atherosclerosis and atopic dermatitis and there is also a trend developing in diagnosing and treating the inflammatory conditions at a much earlier stage.

VII.16. Quantum dots and quantum rods

Brightly fluorescent quantum dots and quantum rods are quickly becoming important tools for identifying specific molecules and cells in living systems. Two reports demonstrate some of the ways in which cancer researchers are using these nanoscale imaging agents.

The investigators [200] used antibody-labeled quantum dots and a high-sensitivity fluorescence microscope fitted with a video camera to make 30-frame-per-second movies of these nanoparticles as they traveled through the bloodstream to tumors in mice. The investigators identified six distinct steps in the process by which quantum dots labeled with the HER2 monoclonal antibody travel from the site of injection to the space surrounding the cell nucleus[201]. The HER2 monoclonal antibody binds to a protein found on the surface of certain breast and other tumors.

Using the labeled quantum dots, the researchers obtained quantitative measurements of these six steps. From these experiments, the investigators were able to determine that each stage of the delivery process proceeds in a stop-and-go manner. The researchers note that a better understanding of each of these steps could improve the ability of nanoparticles to deliver drugs specifically to tumors.

In another study researchers[202] showed that they could create water-soluble quantum rods that can be used as targeted probes for imaging cancer cells using a technique known as two-photon fluorescence imaging. Quantum rods, like spherical quantum dots, can be made to fluoresce with a wide range of colors, but the larger dimensions of quantum rods make them easier to excite with incoming light than their spherical cousins. This work was conducted as part of the Multifunctional Nanoparticles in Diagnosis and Therapy of Pancreatic Cancer Platform Partnership[203].

The investigators first developed a new method for making quantum rods that would remain well-dispersed in water, and then refined their synthetic technique to also include the ability to attach targeting molecules to the surface of the quantum rods.

In the reported experiments, the investigators attached a protein known as transferrin to the quantum rods. Transferrin binds to a receptor that is overexpressed on many types of cancer cells.

Experiments with these labeled quantum rods showed that they were only taken up by targeted cells, and that they accumulated within the targeted cells. The quantum rods were readily visible within the cells using low-intensity near-infrared light. The ability to use low-intensity light to detect the quantum rods helps protect the integrity of the targeted cells.

---

[200] *Dr. Hideo Higuchi and colleagues at Tohoku University in Japan,*

[201] *Tada H., In vivo real-time tracking of single quantum dots conjugated with monoclonal anti-HER2 antibody in tumors of mice, Cancer Res. 2007 Feb 1;67(3):1138-44*

[202] *, Dr. Paras Prasad and his colleagues at the State University of New York in Buffalo*

[203] *Yong KT et al, Quantum Rod Bioconjugates as Targeted Probes for Confocal and Two-Photon Fluorescence Imaging of Cancer Cells. Nano Let. 2007 Feb 9*

264

VII.17. Molecular 'cages' to deliver drugs

By combining DNA macromolecules with polymers containing iron, molecular 'cages' can be made: porous structures capable of carrying and delivering drugs or DNA-fragments. By using small molecules as keys, the cage can be opened or part of the DNA can be freed[204].

DNA, being the carrier of genetic information in living creatures, can also be used in man-made technology, for instance in bioinformatics and DNA-computing. Scientists Yujie Ma and Mark Hempenius of the University of Twente managed to combine DNA macromolecules with synthetic polymers containing iron. The result is a novel way of creating porous structures, spherical 'cages' for example.

The walls of these cages are built step by step. The scientist therefore ingeniously use the different properties the two types of molecules have. DNA has a negative electrical charge while the polymer containing iron is positively charged. Another essential features of DNA is that the molecule is much more rigid than the polymer. The polymer wraps around the DNA and forms a very stable couple with it. What binds them together are electrostatic forces.

The spherical cage can transport medicine and deliver it locally. The cage can be opened by letting small molecules function as 'keys': they oxidize the iron and break the bond between the DNA and the polymer locally. In the same way, it is possible to free DNA-fragments from the cage, and apply them in gene therapy. Genes are then inserted into cells and tissue to treat inherited disease.

Macroporous materials like the new cages, with pore sizes larger than 50 nanometers, have a wide range of possible applications, but they are not easily fabricated until now. The DNA-polymer combination is an example of 'self-assembly' in which molecules organize themselves. It is a powerful new method to create the materials and an important step towards innovative applications.

Rather than forming a hypothesis about a specific gene or protein and designing experiments to test it, molecular profiling takes a more global approach to cancer research. This technique surveys the expression of thousands of genes in a single experiment to map the changes in the human genetic blueprint associated with cancer. The molecular profiling approach will accelerate our understanding of the molecular basis of cancer and will lead to new insights for the treatment, detection, and prevention of these diseases, say scientists at the National Cancer Institute.

Cancer is a term that encompasses at least 200 different diseases characterized by genetic changes that alter the normal, controlled growth and division of cells. If cancer research in the pre-genomic era -- before the sequencing of the human genome -- was a cottage industry dedicated to the study of individual molecules and processes, then the post-genomic era is an industrial revolution. This new age is defined by technological advances, such as microarray platforms, that allow for the global, simultaneous study of the 20,000 to 25,000 genes that make up the human genome.

---

[204] *Scientists led by Prof. Julius Vancso of the MESA+ Institute for Nanotechnology of the University of Twente in The Netherlands and prof. Helmuth Moehwald of the Max Plnack Institute in Golm, Germany, report about this in Angewandte Chemie International Edition, in their cover article on February 26, 2007.*

The real value of molecular profiling will be realized when biomedical scientists with a particular expertise are able to integrate and use the data fluently for hypothesis generation, hypothesis-testing, and what would be termed 'hypothesis-enrichment', according to the Genomics and Bioinformatics Group at NCI.

Scientists described that they used a panel of 60 human cancer cell lines, known as the NCI-60 panel, to analyze the actions of L-asparaginase (L-ASP), a bacterial enzyme that has been used since the 1970s to treat acute lymphoblastic leukemia. L-ASP scavenges the blood, chewing up molecules of free asparagine, one of 20 amino acids needed to build proteins in a cell. Normal cells can use the enzyme asparagine synthetase (ASNS) to make their own asparagine, but L-ASP selectively starves cancer cells that cannot produce enough of the amino acid for their own needs.

Since recent studies have suggested a link between L-ASP activity and ASNS, the NCI research team analyzed activation of the ASNS gene in the NCI-60 cancer cell lines, the most extensively-profiled set of cells in existence. Each cell line originated from a single cell type taken from a cancer patient and was then transformed in the lab to grow indefinitely outside the body. The NCI-60 panel of cells has been used by NCI's Developmental Therapeutics Program to screen more than 100,000 compounds for anti-cancer activity since 1990.

To examine this relationship, the researchers used microarray analysis. In this study five different microarray platforms used in the molecular profiling[205] of the NCI-60 revealed a strong correlation between the anticancer activity of L-ASP and reduced activation of the ASNS gene in ovarian cell lines. Subsequently, the researchers and their collaborators used RNA interference, a recently developed genetic technique, to reduce the activation level of ASNS five-fold in one of those cell lines. As a result, L-ASP became over 500 times more effective at killing the cancer cells, suggesting that ASNS levels are the principal determinant of L-ASP activity. Furthermore, this increased activity was maintained in ovarian cancer cells that had developed classical multi-drug resistance to other forms of treatment[206].

They are hopeful that the level of ASNS expression may one day be useful as a tool for selecting ovarian cancer patients[207] who will most benefit from the use of L-ASP.

The Cancer genome Atlas (TCGA)[208] aims to use tissue samples derived from cancer patients to systematically explore the universe of genomic changes involved in several types of human cancers. Cell lines, including the NCI-60, are different from the clinical tumors that will be the focus of TCGA. Cell lines are unlimited in number, easy to manipulate, and valuable for repeating experiments under the same conditions, but they do not necessarily reflect all of the properties of tumors found in patients.

This emphasis on molecular profiling reflects a shift in research from small-scale to large-scale efforts, which are necessary because the genetic changes that lead to

---

[205] *Weinstein JN, et al. Spotlight on molecular profiling: "Integromic" analysis of the NCI-60 cancer cell lines. "Molecular Cancer Therapeutics" 2006; Online November 7, 2006.*

[206] *Ikediobi ON, et al., DNA sequence analysis of 32 known cancer genes in the NCI-60 cell lines "Molecular Cancer Therapeutics" 2006; Online November 7, 2006.*

[207] *Lorenzi PL, et al.,Asparagine synthetase as a causal, predictive biomarker for L-asparaginase activity in ovarian cancer cells. "Molecular Cancer Therapeutics" 2006; Online November 7, 2006.*

[208] *The Cancer Genome Atlas (TCGA) is a three-year, 100 million dollar collaborative pilot project launched in December 2005 by NCI and the National Human Genome Research Institute, also part of the NIH.*

cancer occur in the context of whole genomes. Not all genetic changes are the same, not all cancers are the same, and they should not be treated as such.

VII.18.   One more step closer to tailor-made molecular medicine for patients.

Invented in 1991, DNA microarrays have become one of the most powerful research tools. Scientists are able to perform thousands of experiments with incredible accuracy and speed. According to MarketResearch.com, by 2009, sales of DNA microarrays are projected to be more than $5.3 billion a year.

A novel invention developed by a scientist from New York Institute of Technology (NYIT) School of Health Professions, Behavioral and Life Sciences could revolutionize biological and clinical research and may lead to treatments for cancer, AIDS, Alzheimer's, diabetes, and genetic and infectious diseases.

Since the discovery of DNA, biologists have worked to unlock the secrets of the human cell. Professor Claude E. Gagna, a molecular biologist-pathologist who performs research in the structure-function of DNA in normal and diseased cells, discovered how to immobilize intact double-stranded (ds-), multistranded or alternative DNA and RNA on one microarray. This immobilization allows researchers to duplicate the environment of a cell, and study, examine and experiment with human, bacterial and viral genes. According to Gagna, this invention provides the methodology to analyze more than 150,000 non-denatured genes. A leap forward from conventional DNA microarrays that use hybridization, this will help pharmaceutical companies produce new classes of drugs that target genes, with fewer side effects. It will lower the cost and increase the speed of drug discovery, saving millions of dollars.

The patented "Gagna/NYIT Multi-Stranded and Alternative DNA, RNA and Plasmid Microarray," discovery will help scientists understand how transitions in DNA structure regulate gene expression (B-DNA to Z-DNA), and how DNA-protein, and DNA-drug interactions regulate genes. The breakthrough can aid in genetic screening, clinical diagnosis, forensics, DNA synthesis-sequencing and biodefense.
In the near future, practical applications of the patent[209] will include enabling researchers to directly target and inhibit mutated genes and/or proteins that are responsible for pathologies, making it easier to treat or even cure disease.
Additionally, Gagna has developed a novel surface that increases the adherence of the DNA to the microarray so that any type of nucleic acid can be anchored. Unlike conventional microarrays, which immobilize single-stranded DNA, scientists will now be able to secure intact, non-denatured, unaltered ds-DNA, triplex-, quadruplex-, or pentaplex DNA onto the microarray.

VII.19.   Microarrays: Making a family tree for disease

Computers aren't about to replace doctors, but their immense processing powers will uncover hidden connections between diseases and could lead to better medical care.

Doctors currently classify diseases by a century-old scheme that is more useful for billing than treatment[210]. The scheme, for instance, lumps most cancers together, even though they may have little in common besides uncontrolled tumor growth.

---

[209] *A discussion of the patent and two new applications -- known as transitional structural chemogenomics and transitional structural chemoproteomics -- was published in the May 2006 issue of Medical Hypotheses (67:1099-1114).*
[210] *Dr. Atul Butte, assistant professor of medicine and of pediatrics at Stanford University Medical School, in a Stanford University press release/statement to the press*

The investigators used computers to crunch billions of measurements from 75 human diseases and map out molecular similarities. The idea is to come up with the first data-driven classification scheme for medicine. The work is made possible by microarrays, a snapshot of the genome that tells scientists which genes are turned on and off and by how much. With microarrays, scientists can glean details about a disease, such as breast cancer, that could never be noticed examining a biopsy sample under the microscope.

Scientists have used microarrays to study nearly every human disease and made the data available on the Internet, thanks to the open-door policy advanced by biochemistry professor Pat Brown and other leaders in the field. Any high school kid today can go to the Web site and download 108,000 microarrays. It's unbelievable how much data that is. So much, the scientist said, that the only way to make sense of the numbers is with the muscle of powerful computers.

Researchers have analyzed that data to connect the molecular dots between diseases nobody suspected were linked. This is how he discovered that muscular dystrophy and heart attack had strikingly similar molecular signatures. With that knowledge, researchers can test whether any of the myriad drugs available to treat heart attack would work on muscular dystrophy, which has no treatment. The beauty of this approach is that it can be used for existing drugs. Not to mention that it could sidestep the prohibitive costs often involved in developing new classes of medications.

VII.20. Microarrays: cancer to get personal

A rose is a rose is a rose, but the same cannot be said for tumors, as will become ever more apparent in the coming years. Doctors are getting increased access to tests that can distinguish between the most and least aggressive cancers. In turn, more patients will get personalized information about their tumors in the near future, which will help them get treatment tailored to their particular cancer.

The impact of the DNA tests has been across cancer types and is now growing nationally. These tests, based on work pioneered at Stanford, can help guide decisions about treatment specific for subtypes of surgery. People with the most aggressive cancers, as indicated by the genes that are active in the tumor samples, can also get the most aggressive treatment[211].

Such therapeutic applications are what biochemistry professor Pat Brown had in mind when he helped develop the microarrays – also known as gene chips - that underlie the genetic tests of cancers. He first showed the power of the microarray in 2000, when they used it to find genes that were either more or less active in breast tumors than in normal breast tissue. They also found groups of genes that characterized the tumors that were most likely to return. Since that landmark paper, researchers across the world have carried out similar experiments with almost every tumor type. The challenge has been figuring out which subsets of genes are best at predicting a tumor's aggressiveness, and refining the algorithms to provide the best information on patient outcomes.

---

[211] Dr. *Beverly Mitchell, deputy director of the Stanford Comprehensive Cancer Center, in a statement to the press.*

VII.21. Driving the discovery of genes associated with a host of common diseases.

Leading causes of disability and death worldwide are today a step closer to being identified thanks to a new map of a vital region of the human genome, the Major Histocompatibility Complex (MHC). An international consortium[212] has developed a detailed map of common variation in the MHC, a region that contains genes linked to common diseases such as diabetes and multiple sclerosis, as well as bestowing protection from pathogens and being vital in transplant medicine.

The new map of variation is the finest that can be produced with current technology and in a pilot study the team have shown it can be used to identify variants that predispose to two diseases. The MHC map will help researchers to pinpoint genes involved in many more conditions.

The extended MHC is a region of around 8,000,000 bases on chromosome 6 that contains over 250 genes that code for proteins. As well as being one of the most gene-dense regions of our genome, it is one of the most variable - our immune system and its ability to fight a host of invaders rely on variation in products of the MHC.

This was one of the reasons to focus on the region but also the most challenging part of the project. No existing project had the power or depth to uncover and map the variation in the MHC. This is the definitive map that will drive the discovery of genes associated with a host of common diseases[213].

Studies over many years have shown that combinations of variants, called haplotypes, in the MHC can predispose to diseases as diverse as multiple sclerosis, rheumatoid arthritis, coeliac disease and diabetes, while other variants can help to protect against infections such as malaria, TB and AIDS . Associations between genes and disease have been extremely difficult to pinpoint because the high level of individual variation means that it is very difficult to find sequence variants common to many or all of the affected people.

In the new study[214], the large amount of common variation is simplified by selecting a reduced set of 'tag' variants that can act as surrogates for most of the variants in the region. This reduced set provides a cost-effective resource to investigate disease association in greater detail and with more efficiency.

The project is complementary to HapMap in which we mapped common human variation across the whole genome, but focused to provide a refined and definitive resource for this critical region. We are already using the MHC map in our to find genes involved in

---

[212] *Participating centers were: Program in Medical and Population Genetics, Broad Institute of Harvard and MIT; Center for Human Genetic Research, Massachusetts General Hospital, Boston; Department of Statistics, University of Oxford; Wellcome Trust Sanger Institute;Wellcome Trust Centre for Human Genetics, University of Oxford; Complex Genetics Section, Department of Medical Genetics, University Medical Center, Utrecht;Laboratory of Genomic Diversity, SAIC-Frederick, Inc. and National Cancer Institute; Illumina, Inc., San Diego; Center for Human Genetics, Duke University Medical Center, Durham;Brigham and Women's Hospital, Department of Neurology, Boston; Juvenile Diabetes Research Foundation/Wellcome Trust Diabetes and Inflammation Laboratory, Cambridge Institute for Medical Research, University of Cambridge; Cambridge Institute for Medical Research, Addenbrooke's Hospital;Imperial College of London; Université de Montréal, Department of Medicine;Montréal Heart Institute, Montréal.*

[213] *Explanation given in a statement to the press by Dr Stephan Beck, Project Leader at the Wellcome Trust Sanger Institute, where one-half of the mapping was carried out*

[214] *De Bakker PI et al., A high-resolution HLA and SNP haplotype map for disease association studies in the extended human MHC. Nat Genet. 2006;38;1166-1172. PMID: 16998491 Published online in Nature Genetics on September 24, 2006.*

eleven common diseases, but also in fine mapping studies such as the search for genes implicated in Type 1 diabetes and the cancer of the nasopharynx[215].

The MHC map places a marker approximately every 1500 bases. Mining this comprehensive collection of common variation in detailed studies has allowed the team to select the tag SNPs, which are up to fivefold more efficient. To test the tag approach, the team examined whether previously discovered risk haplotypes in two diseases - coeliac diseases and systemic lupus erythematosis - could be identified using the new MHC map. The results were remarkable, with sensitivity and specificity both 97% or greater for common HLA alleles (e.g. HLA-A, B and C), confirming the map's value in screening for new genes implicated in disease. Conventionally, finding risk-associated haplotypes within this region must be carried out using a process known as HLA typing: this new study found that SNPs can be used instead, in a much simpler assay, saving time and expense.

The new map of the MHC is an example of how cutting-edge research could help patients by enriching our understanding of this vital region influencing transplantation treatments, and the outcome of a number of diseases. This remarkable, detailed study will tell us much about common disease and, it is hoped, will speed progress to new benefits in healthcare[216].

Table 7.6. Examples of diseases associated with genes in the extended MHC

Autoimmunity	Other immune disorders	Infections	Other diseases	Hypersensitivity
Ankylosing spondylitis	Behcet's disease	AIDS progression	Cervical cancer	Abacavir (HIV)
Graves disease	Coeliac disease	Kaposi's sarcoma	Gastric cancer	Allopurinol
Multiple sclerosis	Crohn's disease	Leprosy/Tuberculosis	Hepatitis B virus non-response	(gouty arthritis)
Myasthenia gravis	Psoriasis	Lyme disease	to vaccine	Carbamazepine
Pemphigus vulgaris		Malaria	Hepatitis C virus sustained	(anticonvulsant, epilepsy)
Rheumatoid arthritis		SARS	response to therapy	Pollen-induced
Type 1 diabetes			Hypertrophic cardiomyopathy	allergic rhinitis
			Nasopharyngeal carcinoma	

VII.22. Natural electric fields in Regenerative Medicine

Scientists at Forsyth may have moved one step closer to regenerating human spinal cord tissue by artificially inducing a frog tadpole to re-grow its tail at a stage in its development when it is normally impossible. Using a variety of methods including a kind of gene therapy, the scientists altered the electrical properties of cells thus inducing regeneration. This discovery may provide clues about how bioelectricity can be used to help humans regenerate.

This study, for the first time, gave scientists a direct glimpse of the source of natural electric fields that are crucial for regeneration, as well as revealing how these are produced. In addition, the findings provide the first detailed mechanistic synthesis of bioelectrical, molecular-genetic, and cell-biological events underlying the regeneration of a complex vertebrate structure that includes skin, muscle, vasculature and critically spinal cord. Although the Xenopus (frog) tadpole sometimes has the ability to re-grow its tail, there are specific times during its development that regeneration does not take place

---

[215] *Dr Panos Deloukas, Project Leader at the Wellcome Trust Sanger Institute, in a statement to the press.*

[216] *Comment given in a press release by Dr. Marcela Contreras, Professor of Transfusion Medicine, RFUCHMS, National Director of Diagnostics, Development & Research at the National Blood Service.*

(much as human children lose the ability to regenerate finger-tips after 7 years of age). During the Forsyth study, the activity of a yeast proton pump (which produces H^+ ion flow and thus sets up regions of higher and lower pH) triggered the regeneration of the frog's tail during the normally quiescent time[217].

According to the researchers, applied electric fields have long been known to enhance regeneration in amphibia, and in fact have led to clinical trials in human patients. However, the molecular sources of relevant currents and the mechanisms underlying their control have remained poorly understood. To truly make strides in regenerative medicine, we need to understand the innate components that underlie bioelectrical events during normal development and regeneration. The ability to stop regeneration by blocking a particular H^+ pump and to induce regeneration when it is normally absent, means the scientists have found at least one critical component.

The research team has been using the Xenopus tadpole to study regeneration because it provides an opportunity to see how much can be done with non-embryonic (somatic) cells during regeneration, and it is a perfect model system in which to understand how movement of electric charges leads to the ability to re-grow a fully functioning tail. Furthermore, tail regeneration in Xenopus is more likely to be similar to tissue renewal in human beings than some other regenerative model systems. The Forsyth scientists previously studied the role that apoptosis, a process of programmed cell death in multi-cellular organisms, plays in regeneration.

Through experimental approaches and mathematical modeling, they examined the processes governing large-scale pattern formation and biological information storage during animal embryogenesis. The lab investigates mechanisms of signaling between cells and tissues that allows a living system to reliably generate and maintain a complex morphology. The team studies these processes in the context of embryonic development and regeneration, with a particular focus on the biophysics of cell behavior. The Forsyth Institute is the world's leading independent organization dedicated to scientific research and education in oral, craniofacial and related biomedical sciences.

VII.23.   Refurbishing diseased or damaged tissue using the body's own healthy cells

Imagine that a patient needs a new liver. Rather than waiting for a donated organ, doctors simply take some stem cells from the patient and grow a new one. Sound pretty far-fetched? Maybe not.

We're looking at a five-year time horizon, at least.[218]. His laboratory is seeking to create complex tissues for organ replacement, and is now working on creating a synthetic replacement liver for people with liver failure. The goal is to seed tissue with stem cells harvested from the patient that will differentiate into liver cells. The key stumbling block is how to use these cells to create a three-dimensional organ. But the researcher is optimistic it will work.

Reports in April 2006 of the first successful transplant of laboratory-grown bladders into children and teenagers at Wake Forest University School of Medicine thrust the field of

---

[217] *This research, by the team of Michael Levin, Ph.D., Director of the Forsyth Center for Regenerative and Developmental Biology, was published in the April 2007 issue of Development and will appeared online on February 28, 2007.*
[218] *Geoffrey Gurtner, associate professor of surgery. Stanford University School of Medicine. The year ahead: Stanford experts forecast medical trends to watch in 2007. Press release, December 21, 2006*

regenerative medicine into the headlines and stirred up hopes for future successes for other organ transplants. The patients, who suffered bladder disorders, received new bladders grown from their own cells.

Regenerative medicine may well be the future of health care: Its goal is to refurbish diseased or damaged tissue using the body's own healthy cells. It's a term that is often used synonymously with tissue engineering, though those involved in regenerative medicine place more emphasis on the use of stem cells to produce tissues.

Scientists are currently working on tissue replacement projects for practically every body part - blood vessels and nerves, muscles, cartilage and bones, esophagus and trachea, pancreas, kidneys, liver, heart and uterus. The long-term hope is that patients someday may not have to wait on the national transplant list for donor organs. Organs could be tailor-made for people.

There will never be enough transplants. One just has to look at the numbers. It increases everyday. If society wants a solution that will solve the problem once and for all, regenerative medicine is the best solution, the scientists say.

VII.24. 'Sticky' proteins fuse adult stem cells to cardiac muscle, repairing hearts

Cardiologists are increasingly using adult stem cells in clinical trials to repair hearts following heart attacks, but no one has understood how the therapy actually works. Now, in animal experiments, researchers at The University of Texas M. D. Anderson Cancer Center have deconstructed the process, describing how the stem cells fuse with heart muscle cells to create new cells that repopulate the ailing organ[219].

Investigators found that this fusion is only possible if two cell adhesion proteins that stick to each other like Velcro are available to attach a stem cell to a heart muscle cell. They show in cell and mice studies that if either protein is blocked, the two cells don't blend. They also discovered that these new cells, once fused, divide again in an attempt to produce enough cells to help the heart contract.

The accepted dogma is that heart cells cannot divide, but we show that fusing stem cells onto muscle cells bestows these cells with a new and wonderful ability to divide again to repair the heart. It is marvelous that adult stem cells can help heal a heart, and by understanding the mechanisms involved, we may be able to refine and optimize the process[220].

But there are not enough natural stem cells available in a body to mount an effective repair response to a heart attack, which is why researchers and clinicians are focused on boosting that response.

And in the future, given what the researchers also have discovered about how stem cells can build new cells to line blood vessels, it may be possible to "choose to either augment rebuilding of heart muscle or restoration of blood vessels, depending on what is therapeutically best for the patient."

---

[219] *Scott Merville. University of Texas M.D. Anderson Cancer Center. Press release. February 21, 2007. The study was funded by a Multidisciplinary Research Program by M. D. Anderson. Co-authors include Sui Zhang, M.D., Ph.D., M. D. Anderson Department of Cardiology, and Elizabeth Shpall, M.D., of the M. D. Anderson Department of Stem Cell Transplantation; and James T. Willerson, M.D. of the Texas Heart Institute and The Brown Foundation Institute of Molecular Medicine at The University of Texas Health Science Center at Houston. Yeh has appointments at both the Texas Heart Institute and the Brown Foundation Institute of Molecular Medicine.*

[220] *Dr. Edward T. H. Yeh, professor and chair of the Department of Cardiology at M. D. Anderson Cancer Center, in a statement to the press.*

In 2003, the researchers demonstrated that adult stem cells circulating in blood can be used to repair hearts, and that it is not necessary to take the stem cells from bone marrow. In 2004, they found stem cells use different methods to morph into the two kinds of cells needed to restore heart function. In animal studies, they showed that to make new heart muscle cells, the human stem cells fuse onto cardiac cells to produce new muscle (myocyte) cells. But to form new blood vessel cells, the stem cells "differentiate" or mature by themselves to provide new endothelial cells that patch vessel damage.

In this study[221], they looked into the mechanism by which stem cells fuse to cardiac myocytes. In laboratory experiments, they added adult human stem cells (those that express the CD34+ protein known to be associated with stem cells) to cardiac muscle cells from mice. After 24 hours, some fusion occurred spontaneously - cells were created that had both human and murine protein signatures - but this occurred at a very low rate. They then created conditions that reflect an ongoing heart attack, such as exposing the cells to low oxygen, and saw production of two cytokine molecules, IL-6 and TNF-a, that are part of an inflammatory reaction and which are known to be released when a heart attack occurs. Next, the researchers exposed the cells to all three conditions simultaneously (hypoxia, and extra IL-6 and TNF-a) and found that cell fusion increased 10-fold. It went from .2 percent of cells becoming fused to 2 percent.

The researchers noted that as a consequence of the experimental "heart attack," expression of two cell surface adhesion molecules was increased. Expression of the protein a4ÃŸ1 was induced in the mouse cardiac cells and VCAM-1 was expressed by human stem cells. The two molecules act like a pair, and stick to each other. This is the first step to the fusion process.

To double-check that production of a4ÃŸ1 and VCAM-1 were critical to cell fusion, the research team added antibodies to each protein into the cell culture, and found fusion couldn't occur. They then tested the fusion process in mice that don't have an immune system, so they cannot reject the human stem cells. They induced a heart attack in the animals, injected stem cells and saw production of fused cells with signatures donated by each species. However, fusion was markedly reduced in mice given antibodies to a4ÃŸ1 and VCAM-1.

The researchers tested whether the adhesion molecules interrupted production of new blood vessel endothelial cells, and found that because fusion is not involved, formation of these cells was not affected. Then they blocked vascular endothelial growth factor (VEGF), a protein known to spur the formation of new blood vessels, and found that if VEGF isn't available, stem cells will not differentiate into new endothelial cells. Finally, they tested the newly fused heart muscle cells to see what they did after they formed. Heart muscle cells do not divide, so the researchers did not know whether the new fused cells were an endpoint in themselves, designed to replace dying cardiac muscle, or whether they could give rise to other new cells. They discovered that fused cells took on some "stemness" - they divided, and continued to do so as long as new tissue is needed, but not long enough to produce a tumor. They showed in animal experiments that

---

[221] *This study is the latest undertaken in a focused research program conducted by Yeh and a team of researchers at M. D. Anderson, the Texas Heart Institute at St. Luke's Episcopal Hospital and The University of Texas Health Science Center at Houston to investigate stem cell repair of heart and vascular tissue. In an effort to understand and treat cardiotoxicity related to cancer treatment, M. D. Anderson has one of the largest cardiology programs at any cancer center.*

human adult stem cells can form new blood vessels and heart muscle cells, and knowing how these two different processes can be blocked could be very useful in determining the relative contribution of each toward heart repair.

VII.25. Stem cell future shaped by epigenetics

Everyone hopes that one day stem cell-based regenerative medicine will help repair diseased tissue. Before then, it may be necessary to decipher the epigenetic signals that give stem cells their unique ability to self-renew and transform them into different cell types.

The hype over epigenetic research is because it opens up the possibility of reprogramming cells. By manipulating epigenetic marks, cells can be transformed into other cell types without changing their DNA. It is simply a question of adding or removing the chemical tags involved.

Stem cells rely heavily on epigenetic signals. As a stem cell develops, chemical tags on the DNA or its surrounding histone proteins switch genes on or off, controlling a cell's fate.

Epigenetic research has benefited tremendously from genome technology, and work in the field is advancing at break-neck speed. "If you think that the first enzymes controlling histone methylation were found in 2001, the acceleration is tremendous. We are making good use of past investments in genome sequencing. In the next five years the technology will be ten times faster than it has been so far[222].

New high-throughput approaches and refined analytical techniques promise to fill in some big gaps in understanding how epigenetic tags define a stem cell and how they can be manipulated. With this knowledge on board, researchers will be boosting the odds that one day stem cell therapies will reach the clinic.

VII.26. Jumonji enzymes and cell regulation

Researchers from Biotech Research & Innovation Center (BRIC) at University of Copenhagen have identified a new group of proteins that regulate the function of stem cells[223].

All living organisms, including human beings, consist of a number of specialized cell types that all originate from the same type of primal cell; the embryonic stem cell. Stem cells can develop into any type of cell through a carefully regulated process referred to as cellular differentiation. During differentiation, specific genes are switched on while other genes are switched off. The genes that are activated during differentiation determine which type of cell the stem cell will become. The result is that cells in a particular organ, e.g. a liver, only express genes specific to that organ.

In 2006, BRIC researchers described how Jumonji proteins regulate the growth of cancer cells and their involvement in the development of specific cancer types.

---

[222] *Robert Feil, a EuroSTELLS researcher based at the CNRS Institute of Molecular Genetics in Montpellier. Sofia Valleley, European Science Foundation, February 27, 2007. To build on existing expertise and stimulate the exchange on novel technologies, the European Science Foundation organized the EuroSTELLS workshop 'Exploring chromatin in stem cells.' Established in 1974 as an independent non-governmental organization, the ESF currently serves 75 Member Organizations across 30 countries. EuroSTELLS is the European Collaborative Research (EUROCORES) program on "Development of a Stem Cell Tool Box" developed by the European Science Foundation.*

[223] *Kristian Helin, "How stem cells are regulated". University of Copenhagen. Press release, 24 Feb 2007. Director of BRIC, Professor Kristian Helin led the research team consisting of Jesper Christensen, Karl Agger and Paul Cloos.*

The study results show that a different subgroup of Jumonji proteins is essential for cellular differentiation. The Jumonji enzymes can turn off, or inactivate, particular genes that play an important part in embryogenesis. The conclusions are based on studies of the nematode (roundworm) C. elegans and studies of mouse embryonic stem cells. The BRIC researchers are currently developing inhibitors to the Jumonji proteins. Their aim is to use these inhibitors to treat cancer patients with increased levels of the Jumonji proteins.

VII.27. Tackling diseases influenced by different types of T cells

A study funded by the Wellcome Trust has suggested a novel way of combating diseases related to the immune system, including cancer and autoimmune diseases such as type I diabetes and arthritis[224].

T cells are produced by the body to fight infection. Scientists previously identified two types of T cell, both produced in the thymus: "effector T cells", which attack infected cells, and "regulatory T cells", which suppress the immune system, protecting the body from inflammatory damage during infection. Regulatory T cells, if given to individuals receiving transplants, may help suppress the rejection response.
A team of researchers has discovered a novel mechanism determining whether a maturing T cell is likely to emerge from the thymus as an effector cell or a regulatory cell. The research suggests that new treatments could be developed to deliberately affect the type of T cells produced, allowing scientists to tackle a number of diseases which are influenced by these different types of T cells.

The team has shown that a process known as 'trans-conditioning', which is known to be involved in T cell development, actually has a profound influence on whether a T cell becomes an effector or a regulatory cell. This may be clinically significant; if we can find a way to influence this process, it may be possible to make the body produce effector T cells in a cancer patient or regulatory T cells in someone suffering from autoimmune disease, both of which are caused by the immune system malfunctioning[225].

The scientists believe that the findings may also answer one of medical research's mysteries: why autoimmune diseases in women commonly go into remission in pregnancy.

They believe that trans-conditioning is less active during pregnancy. This means that most T cells emerging at that time will be regulatory. Regulatory T cells prevent an over-active immune system from causing inflammatory damage to the body. This may be one of the key steps in preventing the mother from rejecting the fetus growing inside her.

VII.28. Autoimmune disease and the early development of immune system B cells

Doctors have long wondered why, in some people, the immune system turns against parts of the body it is designed to protect, leading to autoimmune disease. Researchers at the U.S.National Institutes of Health's (NIH) National Institute of Arthritis and

---

[224] Source: the Wellcome Trust.November 18, 2006. The research was carried out at the King's College London School of Medicine at Guy's Hospital and was co-lead by Dr Daniel Pennington at Queen Mary, University of London. Collaborating researchers were based at Faculdade de Medicina de Lisboa, Lisbon; University College, London; Yale University School of Medicine; Institute for Animal Health; and Imperial College London.
[225] Explanation given by Professor Adrian Hayday of King's College London,in a press release of November 18, 2006

Musculoskeletal and Skin Diseases (NIAMS) have provided some new clues into one likely factor: the early development of immune system cells called B cells.

B cells are formed in the bone marrow and produce antibodies. Antibodies are generated from the cutting and splicing of immunoglobulin genes early in B-cell development, and have a strong and highly specific affinity for different pathogens. When an infectious pathogen (a disease-causing agent) enters the body, B cells are activated and release antibodies into the bloodstream to combat the pathogen. When antibodies encounter the pathogen, they bind to it, rendering it incapable of causing further harm. Antibody molecules also serve as receptors on the surface of B cells.

The problem occurs when the random cutting and splicing of immunoglobulin genes results in an antibody that recognizes a component of one's own body. While the body has a built-in mechanism to correct these errant cells, the NIAMS researchers discovered this doesn't always work the way it was intended.

What happens is that, if the body ever produces a cell with a self-reactive antibody molecule, that cell will get arrested in development at the point where it is actually combining and creating an antibody receptor. Often, rather than killing off the cell, the body edits -- or corrects -- the receptor, like one might edit a paper. In normal circumstances, this new, good receptor replaces the bad, but what investigators[226] found was that about 10 percent of the body's B cells retain both receptors: a good, useful one! and a faulty self-reactive one that the good receptor was designed to replace. This means that the aberrant B cells have escaped the body's mechanism to correct them. This research goes against the theory that B cells should only express a single receptor.

Their new findings raise the question of how this knowledge might eventually help people with autoimmune disease. A question that will take time to answer. For now, the step forward to understand where these self-directed cells are coming from is a big one.

VII.29.   Vaccines for people with compromised immune systems

The tiny fruit fly has a lot to teach humans. Researchers at the Stanford University School of Medicine have found for the first time that flies' primitive immune systems may develop long-term protection from infection, an ability previously thought impossible for insects[227].

The findings could have implications for new ways of developing human vaccines, especially for people with compromised immune systems.

The evidence that a fruit fly's immune response can adapt to - or retain memory of - an earlier infection contradicts the long-held dogma that immune memory cannot exist in invertebrates such as insects. Such memory of a specific pathogen, known as adaptation, is supposed to be a hallmark of the higher-level immune system response of humans and other vertebrates. The Stanford work raises the possibility that humans could make use of this rudimentary immune response if their higher-level system is

---

[226] *Dr. Rafael Casellas, an investigator in NIAMS's Genomic Integrity and Immunity Group and his colleagues at the Oklahoma University of Health Sciences*

[227] *Stanford University school of Medicine media release. "Insight on fruit fly immune system could lead to new types of vaccines, Stanford researchers say". March 8, 2007. The study findings are published in the March 9 issue of Public Library of Science-Pathogens.*

crippled. The scientists call it "a springboard to looking at the immune system in a whole new way" .

One of the two arms of the immune response in higher organisms is similar to that of flies. This arm is known as innate immunity, and it is thought to be a primitive first-line, nonspecific response to a pathogen "invader." The other arm found in higher organisms is adaptive immunity, which has a memory that retains an internal record of contact with an invader and - employing T and B cells - springs into action once it encounters the same invader again. This adaptive immunity explains why a vaccine provides protection.

Harnessing the potential power of adaptation in the innate immune system might also be a boon in the body's defense against bioterrorism or disease pandemics. The B and T cells of the adaptive immune system take a long time to react. But you might be able to speed things up if you could snort something up your nose that would make your innate immune system ready to fight[228].

While inviting novel ways of thinking about future vaccines and treating AIDS patients (who are like fruit flies in the sense that they don't have properly functioning T cells) , the new finding immediately stirs up the field of insect immunology. Existing publications about the fruit fly's immune system explicitly state that it has no memory, and no ability to make specific long-term changes prompted by its exposure to pathogens. Because immune memory was defined as nonexistent, no one ever did the experiment to question whether the fly's immune system could adapt.

In the past decade or so, work done on fruit fly immunology has always been done on a fly infected only once - and that's not how things happen outside a lab, where a fly would be continually exposed to microbes.

The researchers found a bacterium - Streptococcus pneumoniae - that infected the flies but didn't kill all of them. They essentially vaccinated over a million flies, typically doing 7,000 in a day, in numerous experiments. In a key experiment, they injected some flies with killed bacteria and others with just saline solution, waited a week, then reinjected both groups with what should have been a lethal dose of live bacteria. Then, it was time to calculate the percentage of how many survived, compared with the flies that been injected only with saline. The results were clean-cut. Within two days, the second dose killed almost all of the flies that had initially received just saline solution. Those that had been vaccinated lived just as long - about one month - as a separate group that had not been infected.

To ensure it wasn't a fluke of the bacterium they chose, other organisms were tested; they identified a fungus that infects fruit flies in the wild, Beauveria bassiana, that elicited a similar protective effect. In the study, the researchers conclude that a much-studied receptor called Toll is involved, as are other processes. They hope their work encourages the search for a similar adaptive response in the innate immune systems of humans or other vertebrates. And that it might be a way to develop a vaccine that modulates the innate immune system.

---

[228] . Dr Schneider, assistant professor of microbiology and immunology at Stanford University in a statement to the press, March 8, 2007.

VII.30. Genomics for treating people with depression.

Researchers are now better able to predict which patients will respond to treatment for depression through the presence of genetic markers, according to results from a major NIH study on treatment resistant depression released at the annual meeting of the American College of Neuropsychopharmacology.[229]

Medications to treat depression are widely available, but no one treatment works for everyone. Additionally, it can be difficult to predict which patients will experience harmful or unpleasant side effects. We are seeking to better understand why this is the case, and, using genetic markers, develop personalized treatments that give patients the best chance at remission[230].

The research is part of a landmark clinical trial known as Sequenced Treatment Alternatives to Relieve Depression, or STAR*D. The STAR*D is the only large sample of patients who suffer from major depression, and who are treatment resistant with the same drug for a significant period of time.

The investigator examined the effects of polymorphisms (common differences in DNA sequences) of 68 genes on treatment effectiveness and incidence of side effects. Analysis of the data showed that polymorphisms in a gene that regulates serotonin was positively associated with treatment outcome. McMahon concluded that individuals who carried two copies of the polymorphism associated with response were 18% more likely to respond to treatment than those who did not.

Polymorphisms in 2 other genes - a receptor for the brain chemical glutamate and a protein involved in neurogenesis - were also associated with treatment effectiveness. Neurogenesis is a dynamic process in the brain through which neural connections are formed and lost. Patients who carried all 3 response-associated polymorphisms were 40% more likely to respond to treatment than those who carried none of them.

Other investigators have knocked down genes involved with neurogenesis or blocked neurogenesis directly in rodents, which eliminated the animals' ability to respond to anti-depressants. This supports the hypothesis that neurogenesis is involved in the response to antidepressant treatment in humans.

In addition to providing valuable information that may be ultimately useful in a clinical treatment setting, the study is part of a larger movement in depression research.

This is the beginning of a new generation of studies to help clinicians personalize treatment. I predict that genomics will be an important tool for future psychiatrists treating people with depression just as it is being used today by oncologists selecting treatments for breast cancer or lymphoma[231].

---

[229] *Sharon Reis, Ross Tomlin. Media release. ACNP, American College of Neuropsychopharmacology. Annual Meeting December 3 - 7, 2006, in Hollywood, FL. ACNP, founded in 1961, is a non-profit professional organization of more than 700 leading scientists, including three Nobel Laureates. The mission of ACNP is to further research and education in neuropsychopharmacology and related fields in the following ways: promoting the interaction of a broad range of scientific disciplines of brain and behavior in order to advance the understanding of prevention and treatment of disease of the nervous system including psychiatric, neurological, behavioral and addictive disorders; encouraging scientists to enter research careers in fields related to these disorders and their treatment; and ensuring the dissemination of relevant scientific advances.*

[230] *Dr. Francis McMahon, Chief of Genetic Basis of Mood & Anxiety Disorders, National Institute of Mental Health*

[231] Comment given by *Dr. Thomas Insel, Director of the National Institute of Mental Health (NIMH) in a statement to the press.*

For the scientists, this success is just the beginning. Ultimately, their goal is to put together a panel of genetic markers that can guide treatment decisions and help doctors choose an antidepressant that will work best for an individual patient.

VII.31. Taking heart failure to the MAT1

A gene called menage-a-trois 1, or MAT1, plays a crucial role in the function of a master switch for production of energy in the heart cell -- a finding that has important implications for understanding and maybe even treating heart failure[232].

When researchers[233] studied infant mice that lacked this gene in their heart muscle cells, they found that the hearts grew normally. This was surprising, in view of some postulated functions of MAT1. But when the animals reached five weeks of age, they began to succumb to catastrophic heart failure, and all of them were dead by two months.

Using "gene chip" technology, the researchers looked for abnormal patterns of gene expression in hearts from which the MAT1 gene was deleted. They found that genes controlling energy production in cells were particularly affected and that the cells had correspondingly low levels of the proteins required for energy production. The mitochondria -- the cell's energy factories -- were defective.

Further research showed that a particular protein called peroxisome proliferator-activated receptor-1 coactivator, or PGC-1, which is a known master regulator of energy production by cells, did not function in cells that lacked MAT1. Even when the scientists artificially increased the amount of PGC-1 in the cells, its function was decreased if there was no MAT1. Ultimately, the investigators proved that MAT1 binds to PGC-1 and forms a physical complex with it, providing a direct biochemical explanation for the ability of MAT1 to serve as an essential partner to PGC-1, facilitating its role in regulating cell metabolism.

In fact, two forms of PGC-1 exist - alpha and beta - both of which have been reported by other groups to be vital to the heart. Both forms of PGC-1 were shown to depend highly on MAT1 and to turn on the ordinarily responsive genes for energy production in heart tissue.

One of the problems in failing hearts is that energy production is deficient. Drugs that act on the PPARs (peroxisome proliferator activated receptors) and other nuclear receptors to promote better metabolism are a very active area of study. Finding an essential partner of PGC-1 alpha and beta that enables them to switch genes on via these receptors should be helpful in that kind of work.

VII.32. Changing the way heart disease is detected and treated in women.

Results of a landmark study suggest that many women with heart disease don't get a proper diagnosis because they have a form of the disease that doesn't show up on the usual diagnostic tests. The new research shows that heart disease is not one but several

---

[232] Baylor College of Medicine, February 12, 2007. Taking part in this investigation were scientists from Baylor College of Medicine, from the Molecular Cancer Biolgy Program & Institute of Biomedicine, Biomedicum, Helsinki, Finland; from the University of Texas Health Science Center at Houston, the University of Utah School of Medicine, the Scripps Research Institute in La Jolla California, and from the Institute of Medical Science at the University of Tokyo.

[233] Dr. Michael Schneider, professor of medicine, molecular & cellular biology, and molecular physiology & biophysics at Baylor College of Medicine, and his colleagues

disorders and may shed light on why heart disease often behaves differently in men and women[234].

In the Women's Ischemia Syndrome Evaluation (WISE) study, women with chronic chest pain underwent standard diagnostic procedures, including stress tests and coronary angiograms. Earlier studies had found that among people who show signs of trouble on stress tests, women are far more likely than men to appear free of blockages on follow-up angiograms. This was also true in the WISE study. But newer tests, including ultrasound of the blood vessels, revealed heart problems the angiograms didn't pick up. Many of these women had a condition called vascular dysfunction, in which the blood vessels supplying the heart don't expand properly to accommodate increased blood flow. Vascular dysfunction may affect not only the large coronary arteries, but also the smaller vessels that serve the heart -- a problem dubbed microvessel disease.

The WISE results may help explain why women with heart disease are often underdiagnosed and undertreated. In men, the main problem may be a blockage in a large coronary artery, which shows up on an angiogram, while women are more likely to have microvessel disease that can't be seen. With this in mind, the WISE investigators are working to develop a new system for screening women for heart disease

VII.33.   Detecting heart disease long before the disease presents itself.

Scientists at Children's Hospital Oakland Research Institute (CHORI)[235], the University of Iowa and Roche Molecular Systems are the first to identify a new gene variant that makes women more susceptible to developing heart disease. The affected gene is called Leukotriene C4 Synthase (LTC4S) and its variant could be identified through a genetic test at birth. The use of such a test would allow physicians to initiate preventative treatments to reduce or even eliminate the risk of heart disease in those women possessing the variant gene.

The study[236] began in 1971 with 11,377 children in Muscatine, Iowa. During the study, researchers[237] periodically evaluated the participants' risks of developing heart disease starting in their teens and into their 40's. Their weight, height, blood pressure, cholesterol and other health factors and risks were recorded between 1971 and 1996. The women and men in the study were selected because they live in the City of Muscatine, Iowa where residents rarely move, which is an ideal component to conduct a multi-year study[238].

The researchers hypothesized that inflammation was an important predictor for the development of heart disease. Inflammation is necessary to repair and heal nicks to the lining of blood vessels, which occur daily. The variant form of the LTC4S gene however, leads to an excessive inflammatory response at the site of blood vessel injury. As a

---

[234] *February 2007 issue of "Harvard Women's Health Watch.", Harvard Health Publications.*

[235] *CHORI, February 5, 2007. Children's Hospital Oakland Research Institute (CHORI) has made significant progress in areas including pediatric obesity, cancers, sickle cell disease, AIDS/HIV, hemophilia and cystic fibrosis. Children's Hospital & Research Center Oakland is a designated Level I pediatric trauma center and the largest pediatric critical care facility in the region. The hospital has 181 licensed beds and 166 hospital-based physicians in 31 specialties, more than two thousand five hundred employees, and an operating budget of $287 million. The hospital's research institute has an annual budget of $41 million with more than 300 basic and clinic investigators. Children's Hospital & Research Center Oakland, 747 Fifty Second St., Oakland, CA 94609, United States*

[236] *The study was conducted by CHORI Scientists David Iovannisci and Edward Lammer*

[237] *Researchers from the University of Iowa include: Larry T. Mahoney, Patricia H. Davis, Ronald M. Lauer and Trudy L. Burns. Researchers from Roche Molecular Systems include: Lori Steiner and Suzanne Cheng. The study was published in the February 2007 issue of the American Heart Association journal Arteriosclerosis, Thrombosis, and Vascular Biology.*

[238] *The identification and monitoring of study participants was led by Ronald Lauer, M.D., from the University of Iowa. Scientists at CHORI were responsible for genotyping DNA samples and drawing the study's conclusions.*

result, people who inherit this gene variant don't repair damage to their blood vessels as well as others. Until now, the LTC4S gene variant was only known to cause asthma.

This is the first direct evidence that a gene known to be linked to asthma is also tied to adult heart disease. Our work is significant because we made allowances for other risk factors such as smoking, cholesterol and blood pressure levels. Consequently, we believe that the risk is genetic and not significantly influenced by a person's environment[239].

Based upon their theory, researchers focused on two key measurements in the study. The first measurement was to identify study participants who had the LTC4S gene variant. Both women and men participated in the study, but the LTC4S gene variant was only linked to early signs of coronary artery disease in women. The second test was to assess two early measurements of coronary artery disease and stroke: 1) calcium build-up within coronary arteries commonly seen in heart attack victims and 2) the thickening of the carotid artery associated with stroke victims. The test results confirmed that those women with the LTC4S gene variant were also at the beginning stages of heart disease. It is important to note that none of the participants in the study have yet been diagnosed with heart disease.

This is one of the first studies to track the development of heart disease in young people and not the aftermath of the disease in older populations. Our research provides critical clues to help us identify people who are susceptible to heart disease long before the disease presents itself and when treatments may be most effective[240].

These research results add to growing evidence suggesting that inflammatory responses are very important for development of coronary artery disease and stroke. Consequently, researchers believe their findings could assist in the development of personalized medicine for people with this altered inflammatory response. New drugs that target inflammatory responses may prove to be effective for preventing adult heart disease and stroke.

VII.34.   Unlocking genetic risk factors for tobacco addiction disorder.

Results of a new genetic study bring scientists one step closer to understanding why some smokers become addicted to nicotine, the primary reinforcing component of tobacco. The research represents the most powerful and extensive evidence to date of genetic risk factors for tobacco addiction. The study not only completed the first scan of the human genome to identify genes not previously associated with nicotine dependence (or addiction), it also focused on genetic variants in previously suspected gene families[241].

This genome wide association scan is an important step in a large-scale genetic examination of nicotine addiction. As more genomic variations are discovered that are associated with substance abuse, including smoking, we will be better able to understand how to prevent and treat human addictive disorders.[242].

---

[239] *Dr. Edward Lammer, Geneticist at Children's Hospital Oakland Research Institute in a statement to the press, February 5, 2007.*

[240] *Dr. Iovannisci, Assistant Staff Scientist at Children's Hospital Oakland Research Institute in a statement to the press, February 5, 2007.*

[241] *The research results appeared December 1 in the online issue of the "Journal of Human Molecular Genetics".*

[242] *Dr. Elias A. Zerhouni, Director of the NIH U .S. Department of Health and Human Services. U.S. National Institutes of health. National Institute on Drug Abuse (NIDA), NIH news release, December 4, 2006*

Smoking behaviors, including the onset of smoking, smoking persistence (current smoking versus past smoking), and nicotine addiction, cluster in families. Studies of twins indicate that this clustering partly reflects genetic factors. To identify those genes that could potentially contribute to nicotine dependence scientists combined a comprehensive genome-wide scan with a more traditional approach that focuses on a limited number of candidate genes, using unrelated nicotine-dependent smokers as cases and unrelated non-dependent smokers as controls. A candidate gene has one or more variant forms, which, according to current scientific evidence, appear to be linked to a genetic disease.

When two teenage friends experiment with smoking at the same age, one can become addicted and the other might not. We want to know why. This systematic survey of the genome coupled with the ongoing identification of variants in candidate genes brings us closer to understanding what factors increase a person's risk of transitioning from experimentation to nicotine addiction[243].

Tobacco use, primarily in the form of cigarette smoking, is a leading contributor to death and disability worldwide. Each year, approximately 440,000 Americans die of smoking-related illnesses and about 5 million deaths are attributed to tobacco worldwide. Although the prevalence of cigarette smoking in the United States has decreased over the last 30 years, adolescents continue to initiate cigarette use, with more than 20 percent of high school seniors reporting cigarette smoking in the last month.

Efforts to understand nicotine addiction are important so that new approaches can be developed to reduce tobacco use. Tailoring of smoking cessation medications to an individual's genetic background may significantly increase the efficacy of treatment[244].

The hope is that continued identification of these genes that are associated with risk of addiction will not only help us predict who is more likely to become addicted but will also help identify who will respond best to specific cessation therapies

New technologies related to the study of the human genome have helped us collect new information related to nicotine addiction. We must now determine how to translate these findings into approaches that will reduce smoking related disease and death[245].

The term "genome" refers to the total genetic information of a particular organism. The normal human genome consists of about 3 billion base pairs of DNA in each set of chromosomes from one parent. The term "genetic variation" is used to describe differences in the sequence of DNA among individuals. Genetic variation plays a role in whether a person has a higher or lower risk for getting particular diseases.

VII.35. Different types of proteins in sperm and the mysteries of infertility
The first ever catalogue of the different types of proteins found in sperm could help reveal the origins of sex and explain some of the mysteries of infertility, say scientists. Research[246] describes 381 proteins present in sperm of the fruit fly, Drosophila melanogaster. Whilst more proteins may be identified as research progresses, this study

---

[243] *NIDA Director Dr. Nora D. Volkow in a statement to the press, February 5, 2007.*

[244] *Study leader Dr. Laura Jean Bierut, of Washington University School of Medicine in Saint Louis, Missouri in a statement to the press, February 5, 2007*

[245] *NIDA Director Dr. Nora D. Volkow in a statement to the press, February 5, 2007.*

marks the first substantial 'whole-cell' characterization of the protein components of a higher eukaryotic cell (a cell in which all the genetic components are contained within a nucleus).

This so-called 'proteome' contains everything the sperm needs to survive and function correctly, and scientists can use it to investigate the factors that make some sperm more successful than others.

Around half of the genes of the fruit fly sperm proteome have comparable versions in humans and mice, making it a useful model for studying male infertility in mammals.

By comparing the sperm proteome of the fruit fly with other species, scientists will also be able to rewind evolution and work out the core sperm proteome - the most basic constituents a sperm needs for sexual reproduction. This will shed light on how sex itself evolved.

This is the first catalogue of sperm proteins for any organism, and it offers a tantalizing glimpse into how we might begin to answer some of biology's most fundamental questions[247].

Amazingly we know very little about what is in a sperm, which probably explains why we don't really understand sex, let alone how it evolved. Before we catalogued the sperm proteome, we only knew a few specific proteins in the Drosophila sperm.

Being able to compare the structure and content of the proteomes of sperm from different species should help us understand the evolution and origin of sperm. We now know of at least 381, which is a greater than 50-fold increase in our knowledge base. Now that we have identified them, we should be able to study the function of all of these[248].

Proteins carry out an immense range of functions, from forming structural materials to catalyzing chemical reactions, so knowing exactly what proteins are in sperm is a great step forward in understanding.

The research involved purifying fruit fly sperm and developing methods to study their protein content. Previous estimates for the protein content of sperm were based on counts of proteins separated into 'spots' on a special gel matrix. However, these only identify the total number of proteins in sperm - rather than identifying the specific identity of each protein constituent.

The sperm proteome provides a basis for studying the critical functional components of sperm required for motility, fertilization and possibly early embryo development. It should be a valuable tool in the study of infertility as more targeted studies can now be established in model organisms[249].

---

[246] *Andrew Mc Laughlin, University of Bath. Sperm Proteome Gives 'Tantalizing Glimpse' Towards The Origin Of Sex, Could Help Explain Some Of The Mysteries Of Infertility. Research published in Nature Genetics on 12-Nov-2006. The research is funded by the Royal Society and the National Science Foundation with additional support from the Biotechnology & Biological Sciences Research Council and the U.S.National Institutes of Health. The University of Bath is one of the UK's leading universities, with an international reputation for quality research and teaching. In 20 subject areas the University of Bath is rated in the top ten in the country.*
[247] *Comment given by Dr Tim Karr from the University of Bath who led the study.*
[248] *Comment given by Dr Tim Karr from the University of Bath who led the study*
[249] *Dr Steve Dorus, from the University of Bath, who collaborated with Dr Karr on the project.*

The researchers indicate that they can start to look for the 'core' sperm proteome - that is, the most basic required constituents of sperm. This will not only shed light on the evolutionary origins of sperm, but may advance our understanding of the evolution of sex itself.

The research will also help further our understanding of sperm competition - the attributes within a sperm that make one sperm more successful at reaching and fertilizing the egg than its peers. If we can work out what makes one sperm more successful than another, we might be able to apply this knowledge to clinical therapies for the treatment of sperm that are not functioning properly.

The findings are particularly timely as a variety of research is beginning to highlight the increasingly important role of sperm. Scientists are discovering that as well as carrying the DNA that spells out the male's contribution to a new life, sperm carries RNA and proteins which have a direct influence on fertilization and embryo development.

VII.36.   Using bi-functionalized dendrimers to discover disease-causing proteins
A complex molecule and snake venom may provide researchers with a more reliable method of diagnosing human diseases and developing new drugs[250].

Purdue University researchers bound a complex nanomolecule, called a dendrimer, with a glowing identification tag that was delivered to specific proteins in living venom cells from a rattlesnake. The scientists want to find a better way to ascertain the presence, concentration and function of proteins involved in disease processes. They also hope the new method will facilitate better, more efficient diagnosis in living cells and patients.

Most diagnostic methods must be done on minute dead blood or tissue cell samples in a laboratory dish.[251]. Because molecular interactions and protein functions are disturbed when samples are collected, researchers can't obtain an accurate picture of biochemical mechanisms related to illnesses such as cancer and heart disease.

The research team used dendrimers because they can pass through cell walls efficiently with little disturbance to the cells and then label specific proteins with isotopic tags while cells are still alive. This allows the scientists to determine the activities of proteins that play roles in specific diseases. Proteins carry genetic messages throughout the cell causing biochemical changes that can determine whether a cell behaves normally or abnormally. Proteins also are important in directing immune responses.

The problem with the current method of using proteomics - protein profiling - is that we use very small sample amounts so sensitive that we can't effectively use existing technologies to study them. In addition, to study a specific protein and its function, we want to preserve its natural environment and see where two molecules meet and what the interaction is when they bind. Taking small samples of blood, cells or tissue to study extracted proteins in laboratory dishes damages the sample and the natural environment is destroyed. The dendrimers would carry one of the stable isotopic or fluorescent labels to identify the presence or absence of a protein that can be further developed for use as a disease indicator, or biomarker.

---

[250] *Susan A. Steeves, Purdue University news release. The Purdue scientists report on their new strategy to discover proteins and protein levels, called soluble polymer-based isotopic labeling (SoPIL), in the February 2007 issue of the journal Chemical Communications. The study also is featured in the journal's news publication Chemical Biology*
[251] Comment given by *Andy Tao, a Purdue biochemist and senior author of the study..*

Snake venom cells were used because they have a very high concentration of proteins similar to some found in human blood. The proteins apparently are part of the biochemical process that affects blood clotting or hemorrhage. Understanding how the proteins behave could help determine predisposition to heart disease and cancer and also be useful in diagnosis and drug development. In future research, Tao plans to investigate how dendrimers are able to enter the cell so easily, what happens to them once they are in the cell and whether there are any long-term effects.

VII.37. International stem cell research

International stem cell research has yielded important new findings in recent years, especially in research on human embryonic stem cells. It has extended and enhanced our knowledge of the properties of stem cells, for example in connection with regenerative cell treatment or the investigation of genetic diseases. However, ethical reservations and legal considerations ended in restrictions on stem cell research in a few countries. One country in particular, Germany, took a conservative stand. As a result of the legal framework conditions, science in Germany can only make a limited contribution to this field. Due to the key date regulation and the penalties established in the Stem Cell Act of 2002, German researchers are denied access to new cell lines and, to a large extent, prevented from working in international projects. These new cell lines, which are available as standard in the stem cell banks in other countries, enable research on an international level and, in the long term, circumvent the production of further stem cell lines. For German researchers, however, access to these stem cell banks is currently prohibited. The Deutsche Forschungsgemeinschaft (DFG, German Research Foundation)[252] therefore argues that the 2002 Stem Cell Act stands in urgent need of revision and makes the following recommendations:

1. The key date regulation should be revoked. German research should be given access to new stem cell lines that are produced and used abroad, so long as these originate from "surplus" embryos.
2. The introduction of cell lines should also be permitted, if these are to be used for diagnostic, preventative or therapeutic purposes.
3. The threat of penalty for German scientists should be removed and the scope of the law should be limited unambiguously to the national territory.

In a Commentary on the recommendations, the DFG writes:
On 1): The cells available in Germany are contaminated with animal cell products or viruses and have not been extracted or cultivated under standardized conditions. In recent years, new stem cell lines have been established that are free from contamination and that can be licensed in the EU. These cell lines have been partly acquired by the "International Stem Cell Forum," which releases them for research. In the view of the DFG, it is urgent that scientists in Germany be granted access to cell lines that have been established since 1 January 2002 and thereby meet the current standards in science and technology. At the same time, it should be ensured that the cell lines are acquired exclusively from surplus embryos. The revocation of the key date regulation would significantly improve the competitiveness of German scientists in the field of stem cell research.

On 2): The Stem Cell Act states that cell lines may only be imported into Germany from

---

[252] Dr. Eva-Maria Streier, Deutsche Forschungsgemeinschaft, *New Recommendations For Stem Cell Research Put Forward By German Research Foundation*, 14 Nov 2006

abroad for purposes of research. As the development of new application-oriented procedures is slowly becoming a reality, importation for diagnostic, preventative and therapeutic purposes should also be allowed.

On 3): The current stem cell law, and the penalty threat implied by it, poses a significant legal risk for German researchers, for example when they are involved in international cooperative projects (including EU-financed projects); when they conduct research in foreign laboratories on cells that would be forbidden in Germany; and when they publish papers based on such research. This situation has led to the increasing isolation of German researchers, especially considering the fact that more and more European countries are removing legal restrictions in the field of stem cell research. In order to avoid the criminalization of German researchers and to achieve legal security, the DFG proposes the removal of the threat of penalty and the explicit restriction of the scope of application of the Stem Cell Act to the home country.

Chapter Eight

Coming Up with Medicines

not

Imagined Even a Decade Ago

VIII.1.   Uncovering a new reason why patients respond differently to the same drug dose

Why does the standard dose of certain medications prove dangerously high for some patients and too low to produce beneficial effects in others? Scientists have added a previously unrecognized factor to the list of explanations (such as age, gender, diet and genetics) for this common problem of individual variability in response to drugs.

Researchers[253] reported that variations in the body's production of hydrogen peroxide -- believed to serve as a signaling molecule at low levels -- can affect accumulation of drugs inside cells. The study[254] involved cultured human cells and a common anti-cancer drug.

The demonstrated correlation between hydrogen peroxide exposure and the concomitant increase in drug accumulation represents a substantial finding," the study reports. "The results suggest that patients experiencing oxidative stress (or an increase in hydrogen peroxide levels) may have an increased response to a given dosage of a drug relative to a patient with decreased oxidative stress, or those patients chronically taking antioxidants [such as vitamins C or E] with their medications. (Hydrogen peroxide's effects could be especially important in about two dozen so-called narrow therapeutic index drugs (such as aminophylline, carbamazepine, lithium carbonate, phenytoin, theophylline and warfarin) for which very small changes in dosage level could cause either subtherapeutic or toxic results, they note). This could be a very important factor in our continued efforts to provide more individualized dosing of drugs.

Personalized medicine will be the right answer.

Pharmaceutical biotechnology's greatest potential for curing chronic and currently "incurable" diseases lies in gene therapy. Currently disease is diagnosed by symptoms.

---

[253] *Jeffrey P. Krise and Ryan S. Funk, at the University of Kansas,February 5, 2007*

[254] *American Chemical Society.January 18, 2007. The study "Exposure of Cells to Hydrogen Peroxide Can Increase the Intracellular Accumulation of Drugs" appeared in the Feb. 5, 2007, issue of the ACS's Molecular Pharmaceutics, a bi-monthly journal*

Genomics could identify people who will definitely respond to a specific medicine, limiting side-effects, ineffective treatments and costs.

The gene therapy of the 21st century will enable medicine to come up with treatments not imagined even a decade ago.

VIII.2. Understanding links between genes and disease for individuals, and across populations

New findings from a global survey[255] of the consequences of small and large DNA variants in our genome will accelerate the search for genes involved in human disease and provide a first genome-wide view of how the unique composition of genetic variation within each of us leads to unique patterns of gene activity.

By defining those genetic variants with a biological effect, the results will help to prioritize regions of the genome that are investigated for association with disease. This is an important step to understanding links between genes and disease for individuals, and across populations.

The Human Genome Project gave us the instruction manual for building a human. The HapMap and Copy Number Variation (CNV) Projects developed indices of where to find differences in the manuals of different people. One of the challenges for research into variation and disease is that most variants have no consequence for our wellbeing.

The new study[256] gives a global view of the *consequences* of those differences for gene activity. The work shows that activity of more than 1000 genes is affected by sequence variation and is the first map of human populations that identifies the most important fraction of DNA variation, that which directly affects gene activity.[257]

Using the HapMap series of cell samples from four populations, they measured the activity of more than 14,000 genes in cells grown in culture. The cell samples provide a snapshot of genetic activity in one cell type. The activity of each gene was then correlated with genetic variation nearby, as defined by the HapMap, an index of single-base changes (single nucleotide polymorphisms, or SNPs) and the new index of copy number variants (CNVs).

We've been able to look back into our history and find changes that are older and likely to be shared among populations. But we also find many that are newer and less widespread. These are part of our recent evolution and a step along the way to understanding the origin and personal consequences of genetic change, not least for our wellbeing. This is a first generation map of biologically important DNA sequence variation[258].

---

[255] *Participating centers were: Wellcome Trust Sanger Institute, Wellcome Trust Genome Campus, Hinxton; Department of Oncology, University of Cambridge, Cancer Research UK Cambridge Research Institute; Istituto di Tecnologie Biomediche-Sezione di Bari, Consiglio Nazionale della Ricerche (CNR), Bari; Department of Pathology, Brigham and Women's Hospital and Harvard Medical School, Boston; Broad Institute of Harvard and Massachusetts Institute of Technology, Cambridge; Centre for Applied Genomics and Program in Genetics and Genomic Biology, The Hospital for Sick Children, MaRS Centre, Toronto; Department of Molecular and Medical Genetics, University of Toronto, Toronto, Ontario; Program in Molecular and Computational Biology, University of Southern California, Los Angeles.*

[256] *Stranger BE. Et al, Relative impact of nucleotide and copy number variation on gene expression phenotypes. Science. 2007;315;848-53. PMID: 17289997 DOI: 10.1126/science.1136678*

[257] *The research was led by scientists from the Wellcome Trust Sanger Institute, together with colleagues from the University of Cambridge, Hospital for Sick Children/University of Toronto and Harvard Medical School/Brigham and Women's Hospital. Funding was provided by The Wellcome Trust, National Institutes of Health,Cancer Research UK, Leukemia and Lymphoma Society, Brigham and Women's Hospital Department of Pathology, UK Medical Research Council, Royal Society, Genome Canada/Ontario Genomics Institute.*

[258] *Explanation given by Dr Manolis Dermitzakis, senior author and Project Leader at the Wellcome Trust Sanger Institute*

The understanding of the genetic basis of gene activity will help medical research to provide individuals with information about their personal predisposition to disease.

The study was a massive undertaking: it included HapMap genotype data on 700,000 SNPs located close to genes, as well as 25,000 sites interrogated for potential structural variation to examine copy-number differences, looking at the activity of 14,000 genes in 210 unrelated individuals.

SNP and CNV variation correlated with altered activity in almost 900 and 240 genes, respectively. The HapMap has been invaluable in detecting variants involved in many diseases and these results suggest that the CNV index will prove similarly useful.

Table 8.1.   Genome variants matter. Identifying genetic variants that influence gene activity in different populations

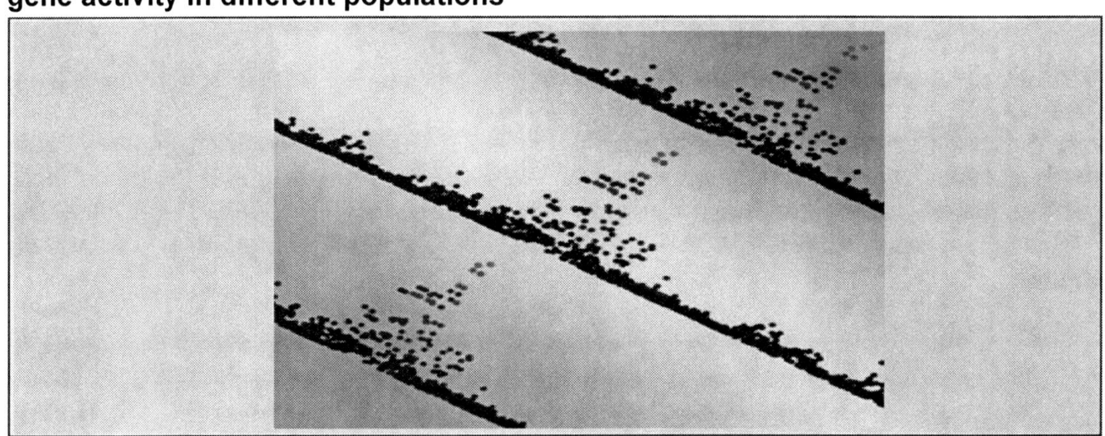

Identifying genetic variants that influence gene activity in different populations: a region where the same group of SNPs are associated with altered levels of gene activity in four populations, each shown by differently colored points that lie off the line[259].

The remarkable finding was that there is such little overlap in the genes found by using the two indices. Only about 10% of the activity variants associated with a CNV were also associated with a SNP. This suggests that we must include CNV studies in our searches for genetic variation associated with disease or we will be missing a lot of the important genetic effects[260].

The results show that at least 10-20% of heritable variation in gene activity is due to CNVs. The team found associations that included previously known examples, such as *UGT2B17*, which has been associated with prostate cancer, proving that the new approach works well.

They also showed for the first time that activity of other genes, located close to *UGT2B17*, was affected. Finding other effects in this way will enhance the search for critical genes within a region of genetic possibilities.

Some associations were not found in all four populations, two-thirds (CNVs or SNPs) being found in only one population. A gene implicated in Spinal Muscular Atrophy

---

[259] *Matt Hurles, Wellcome Trust Sanger Institute, February 8, 2007*
[260] *Comment given by Dr Matthew Hurles, co-leader of the project at the Wellcome Trust Sanger Institute.*

showed an association in three populations, but not in Yoruba from Ibadan, Nigeria. Understanding population differences can help us understand our history.

Variation in copy number can affect gene activity by altering the 'dose' of a gene, by disrupting the active parts of a gene that contain the code for protein, or by disrupting the regulatory regions of the genome that control gene activity - the on/off and dimmer switches in our genome.

Although the simplest model for a CNV affecting gene activity is where the variant is a deletion of a gene or part of a gene, we found examples where activity is affected from a distance. This may occur when the CNV reduces the effectiveness of a region that works to switch the genes on or off[261].

The survey gives the first global view of the effects of SNPs and CNVs on gene activity. The methods and resources developed will help researchers better understand the link between differences - large and small - in our genome and our health.

VIII.3. The organization and function of the human blueprint: ENCODE´s surprising revelations.

Since the completion of the Human Genome Project in 2003, research efforts have been aimed at analyzing the functions of various sequences in the genome, using both experimental and computational strategies. The goals of the ENCODE (ENCyclopedia Of DNA Elements) Project[262] is to characterize all functional elements in the human genome.

An international consortium of researchers sifted through 1 percent of the genome looking for pieces of DNA that are copied by the cell or help control gene activity.

A genome consists of only four different nucleotide bases, or DNA subunits, arranged in a particular sequence. The publication of the human genome revealed its sequence-the significance of which remains a mystery. In particular, genes account for only 1.2 percent of the genome's three billion bases.

The four-year effort to build a "parts list" of all biologically functional elements in 1 percent of the human genome was organized by the National Human Genome Research Institute (NHGRI), part of the U.S. National Institutes of Health. The analysis was led by the European Molecular Biology Laboratory's European Bioinformatics Institute (EMBL-EBI), drawing on expertise from 35 groups from 80 organizations around the world. The project served as a pilot to test the feasibility of a full-scale initiative to produce a comprehensive catalogue of all components of the human genome crucial for biological function.

The findings promise to reshape our understanding of how the human genome functions. They challenge the traditional view of our genetic blueprint as a tidy collection of independent genes, pointing instead to a network in which genes, regulatory elements and other types of DNA sequences interact in complex, overlapping ways.

By integrating 200 datasets generated by various high-throughput methods scientists

---

[261] *Barbara Stranger, first author and post-doctoral fellow at the Wellcome Trust Sanger Institute.*

[262] *The June issue of Genome Research (http://www. Genome.org) is devoted to the ENCODE Project. Genome Research is an international, monthly, peer-reviewed journal published by Cold Spring Harbor Laboratory Press. Launched in 1995, it is one of the five most highly cited primary research journals in genetics and genomics.*

now have a very good idea what 1 percent of our DNA might be doing. The results reveal important principles about the organization of functional elements in the human genome, providing new perspectives on everything from DNA transcription to mammalian evolution. In particular, we gained significant insight into DNA sequences that do not encode proteins, which we knew very little about before[263].

The ENCODE consortium's major findings include the discovery that the majority of human DNA is transcribed into RNA and that these transcripts extensively overlap one another. This broad pattern of transcription challenges the long-standing view that the human genome consists of a small set of discrete genes, along with a vast amount of "junk" DNA that is not biologically active. The new data indicate that the genome contains very little unused sequences; genes are just one of many types of DNA sequences that have a functional impact. The consortium identified many previously unrecognized start sites for transcription and new regulatory sequences that contrary to traditional views are located not only upstream but also downstream of transcription start sites.

Other surprises in the ENCODE data have major implications for our understanding of the evolution of genomes. Until recently, researchers had thought that most DNA sequences with important biological function would be constrained by evolution making them likely to be conserved as species evolve. But about half of the functional elements in the human genome do not appear to have been constrained during evolution, suggesting that many species' genomes contain a pool of functional elements that provide no specific benefits in terms of survival or reproduction.

The major findings of the ENCODE investigations span the areas of chromatin and replication, gene transcription and regulation, and evolutionary constraint.

Although genes make up only 3 percent of the ENCODE sequence, the consortium found that 93 percent of the sequence is transcribed. Scientists at Lausanne University[264] conducted a series of experiments to annotate all 399 protein-coding genes in the ENCODE regions. In doing so, they found that 65 percent of the genes produced transcripts that contained sequences mapping outside of the known boundaries of these genes. Interestingly, these transcribed sequences often overlapped with other genes, were located a significant distance from the main portion of the coding sequence, and spanned large genomic segments.

According to the investigators, their results modify our current understanding of the architecture and regulation of protein-coding genes. Furthermore, some sequence polymorphisms hitherto considered to be located in 'non-coding' regions may ultimately be related to disease[265].

Using an integrated computational and experimental approach, a team of researchers[266] estimated that at least 35% more promoters exist in the human genome than are currently annotated. Based on the transcript sequences, the researchers identified 1,437

---

[263] Dr. Ewan Birney, head of genome annotation at the European Molecular Biology Laboratory's European Bioinformatics Institute (EMBL-EBI), who led ENCODE's massive data integration and analysis effort

[264] Denoeud F. et al. 2007. Prominent use of 5' transcription start sites and discovery of a large number of additional exons in ENCODE regions. Genome Res. 17: 746-759. (doi:10.1101/gr.5660607)

[265] Dr. Alexandre Reymond, University of Lausanne, Switzerland. Cold spring harbor Laboratory press release, June 18, 2007

[266] Led by Dr. Zhiping Weng, Boston University and Dr. Richard M. Myers, Stanford University School of Medicine.

new promoters-short DNA sequences where transcription begins-in or between genes, on top of the 1,730 promoters they knew of. That is nearly ten promoters per gene,

Interestingly, about one-fourth of the newly identified and validated promoters were located on the antisense strand of annotated transcripts, mostly in terminal exons. The authors speculate that these promoters may regulate transcription that occurs in the reverse direction (antisense) to protein-coding genes.

Future research will be necessary to determine whether these newly identified promoters are alternate transcription start sites of known genes, or whether they represent the first evidence of heretofore unidentified genes[267].

Other researchers sequenced the ENCODE regions in 23 mammalian species, aligned the sequences, and identified regions of evolutionary constraint (in other words, sequences that have changed little during evolutionary time). They used four different methods to align the sequences, and three different algorithms to assess constraint. The comparison among the different approaches, as well as the newly generated genomic data from the 23 mammalian species, will be valuable resources for the genomics community.

They also determined which evolutionarily constrained regions overlapped with ENCODE experimental annotations. The project leader explained that The signature of conservation was most apparent in protein-coding regions- In other regions, the situation was more complex, with different annotations showing different patterns of constraint. The investigators[268] were quite surprised by the fact that many experimental annotations had no evidence of mammalian sequence constraint. Their manuscript describes several possibilities for this low correlation, the most intriguing of which suggests that smaller portions of annotated regions than expected are evolutionarily constrained.

By integrating ENCODE experimental data, investigators developed and employed a computational approach to classify genomic regions as "active" or "repressed." Remarkably, they found that the pattern of active versus repressed domains was strikingly conserved between different cell types, and thus may be a universal feature of human genome architecture.

They explained that for over four decades, they have speculated that human chromosomes are partitioned into discrete functional territories that either facilitate or inhibit gene activity. But for the first time, the availability of the ENCODE data has allowed us to systematically evaluate this concept at high resolution and to produce the first human 'domain map[269].

The methodology developed in this study simplifies the interpretation of large amounts of

[267] Trinklein, N.D. et al. 2007. Integrated analysis of experimental datasets reveals many novel promoters in 1% of the human genome. Genome Res. 17: 720-731. (doi:10.1101/gr.5716607)

[268] A team of scientists led by Dr. Elliott H. Margulies, National Human Genome Research Institute (NHGRI). Margulies, E.H. et al. 2007. Analyses of deep mammalian sequence alignments and constraint predictions for 1% of the human genome. Genome Res. 17: 760-774. (doi:10.1101/gr.6034307)

[269] Dr.John A. Stamatoyannopoulos, Depts. of Genome Sciences and Medicine University of Washington, Seattle. Cold Spring Harbor Laboratory press release, June 18, 2007.

genomic data and will be employed in future genome-wide analyses aimed at understanding the large-scale functional organization of complex genomes[270].

The ENCODE Project produced an enormous amount of data on transcriptionally active regions (TARs). Because TARs are difficult to wrangle with, scientists[271] constructed the Database of Active Regions and Tools (DART) which is a Web resource[272] for classifying, storing, manipulating, and visualizing TARs.

Using the DART classification system, the scientists categorized 6,988 unannotated TARs based on expression profiles, sequence composition, relatedness to similar sequences from other organisms, and genomic location. Of the new TARs identified, approximately 20% were produced from previously unidentified potential genes. In addition, many of the TARs associated with known genes were found to have the potential to form functional secondary structures[273].

A major part of the project was identifying sequences that cells copy, or transcribe, into RNA molecules. Cells make proteins from RNA they copy from genes, but some RNAs play roles by themselves. In addition, some studies have found evidence that species from flies and worms to humans copy large amounts of RNA from noncoding DNA, with no apparent purpose[274].

A team of researchers[275] has uncovered a major secret in the mystery of how the DNA helix replicates itself time after time. It turns out that it is not just the sequence of the bases (building blocks) in the DNA, but also how loosely or tightly the chromatin (the material that makes up chromosomes) is packed at different points of the chromosome that is critical.

The ENCODE project studied a cervical cancer cell line, the HeLa cell line, which is well studied by biochemists. It has disturbances in the genome structure and some genes are scrambled or exist in extra copy numbers. As a control, the researchers examined the chromatin packaging in a normal cell line, a lymphocytic cell line from a patient.

Where chromatin is packed more loosely, the genes are replicated earlier than other genes and are expressed at high levels. Where chromatin is dense, these genes are replicated later and are not expressed.

This finding held true for both cell lines studied, the HeLa cervical cancer cell line and a normal cell line (lymphocytic cells). What the HeLa cells were predicting in terms of chromatin packaging held true in the lymphocytes, even though the HeLa cells are cancer cells with scrambled genes. The chromatin packaging predictions were approximately comparable[276].

---

[270] *Thurman, R.E. et al. 2007. Identification of higher-order functional domains in the human ENCODE regions. Genome Res. 17: 917-927. (doi:10.1101/gr.6081407)*

[271] *Dr. Mark Gerstein and Dr. Joel Rozowsky, both of Yale University.*

[272] *dart.gersteinlab.org*

[273] *Rozowsky, J. et al. 2007. The DART classification of unannotated transcription within the ENCODE regions: associating transcription with known and novel loci. Genome Res. 17: 732-745. (doi:10.1101/gr.5696007)*

[274] *Dr. Mark Gerstein of Yale University, quoted by JR Minkel in a scientific American online article "Tiny slice of genome reveals bustling activity in the gaps between genes" June 15, 2007*

[275] *led by University of Virginia Health System geneticist Dr. Anindya Dutta, M.D., UVA professor of biochemistry and molecular genetics, who headed the replication portion of the major ENCODE project*

[276] *Overall, the consortium looked at just 1 percent of the human genome, using HeLa and lymphocytic cells for the study. Groups that were performing other experiments found that their data about genetic expression perfectly correlated with what the Dutta lab would have*

Chromosomes, then, are not just a framework of DNA but also are influenced by the proteins that pack the DNA, particularly histones. This packaging determines how the cell's enzymes get access to the DNA to read off the DNA and replicate, ultimately to create all of the proteins needed by a given cell.

Another interesting finding shows that a different longtime idea about genetics is apparently false. Humans get copies of a gene from both mother and father. The expectation was that each copy would behave the same way. The researchers thought that both copies would either be expressed and replicated early or repressed and replicated late. Instead, they discovered that 20 percent of the pairs of genes have different (asynchronous) replication due to the two copies in the pair being in different chromatin environments.

What this means, unfortunately, is that one of the copies of a gene X might be inactive because of chromatin packaging, losing the redundancy that comes from both copies being active. If tumor suppressor genes, the ones that turn off cancer cells, reside in areas of the genome with asynchronous replication, the cancer cell can turn such genes off merely by mutating the one active copy instead of going through the trouble of separately mutating both copies.

The lead report defined in detail which regions of the genome are actively copied in the cell, revealed the location and studied evolution of elements that control gene activity, and defined the relationship between DNA-associated proteins and gene activity and DNA replication.

Our understanding of genome biology from the Human Genome Project gave us an overview of a 3-billion-base genome, peppered with some 22,000 discrete genes and the sequences that regulate their activity. These were estimated to occupy perhaps 3-5% of the genome, though this number is expected to be an underestimate. This understanding may have to evolve.

The new view transforms our view of the genomic fabric". The majority of the genome is copied, or transcribed, into RNA, which is the active molecule in our cells, relaying information from the archival DNA copy to the cellular machinery. This is a remarkable finding, since most prior research suggested only a fraction of the genome was transcribed. But it is our new understanding of regulation of genes that stands out. The integrated approach has helped us to identify new regions of gene regulation and altered our view of how gene regulation occurs.[277]

From the earliest studies of gene activity in bacteria, a picture emerged that suggested control regions were most often located at or near sites from which gene transcription started. The new work identifies many previously unknown control regions and shows that control regions are as likely to be beyond the end of the gene.

Alterations in control regions are increasingly thought to be of significance for human disease.

---

predicted, in terms of gene expression timing. In the ENCODE study, the next project is to examine the remaining 99 percent of the genome, using at least 10 more cell lines.

[277] . Dr Tim Hubbard, from the Wellcome Trust Sanger Institute. Sanger Centre, Cambridge, press release June 16, 2007

294

The research team showed that transcription of DNA is pervasive across the genome, and that RNA transcripts overlap known genes and are found in what were previously thought to be gene 'deserts'.

Although much that is new has been discovered, much yet remains to be understood. Similarity of DNA sequence between species is often a sign of the value of that sequence, yet a function has not been found for many DNA sequences that are conserved. The role of the massive new numbers of RNA transcripts is unknown. And the function of the large number of control elements is yet to be elucidated[278].

In addition to coordinating the analysis and integration of the ENCODE data, EMBL-EBI researchers in collaboration with the BioSapiens Network of Excellence have investigated as part of the ENCODE effort how RNA transcripts are processed in human cells. They reported that alternative splicing, the phenomenon that the same RNA transcript can be cut at two or more different positions to make different products, is very common in humans. It is unlikely, however, that alternative splicing adds substantially to the variety of functions and structures among proteins.

VIII.4. Towards a proliferating number of therapeutic niche areas

The new pharmaceutical environment will have place for the large (virtual) multinationals, coordinating a network of satellite enterprises, as well as for an increasing number of expertise providers. Because monolithic multinational groups will not be able to dominate all the niches (in the future, their strongest role will be in the generation of the scientific information that underlies the development of drugs, as well as the powerful communication of approved drugs. They will not be able to dominate the development of so many "niche" drugs), a continued –worldwide- proliferation of small drug companies, each trying to serve a particular niche, can be expected. For the others, the lights are being switched off.

These changes are also reflected in an analysis of pipeline activity and forecasts of the growth of future oncology blockbusters. Cardiovascular and central nervous system (CNS) therapy areas are expected to continue to dominate the market for medicines "for the masses" over the next few years (cardiovascular drugs, hyperlipidemia and dyslipidemia, the two key indications behind its success are highly crowded and competitive pharmaceutical markets).

Pharmaceutical companies will find more opportunities in other disease indications where there are fewer therapies and a high degree of unmet need enabling highly innovative drugs to compete strongly. The thrombosis, arthritis and respiratory disease markets will see strong sales growth. Indications within these markets are associated with a high degree of unmet need, offering pharmaceutical companies opportunities for new drugs to drive growth over the next five to ten years; the period needed for innovative companies to develop more targeted cancer therapies, the future drivers of growth (one of the most important factors has been the trend of declining R&D success rates, which have served to highlight the exposure of BigPharma, in particular of US

---

[278] *Over the next couple of years the ENCODE project will be scaled up to the entire genome. The Ensembl project, a joint EMBL-EBI and Sanger Institute project, jointly headed by Ewan Birney, head of genome annotation at EMBL-EBI, has already generated some initial genome wide datasets with early full scale datasets. This integration has lead to the identification of just over 110,000 regulatory elements across the human genome.*

BigPharma, caused by dependence on blockbuster drugs, and the shift towards new research focuses and business models).

Table 8.2.   Molecular approach

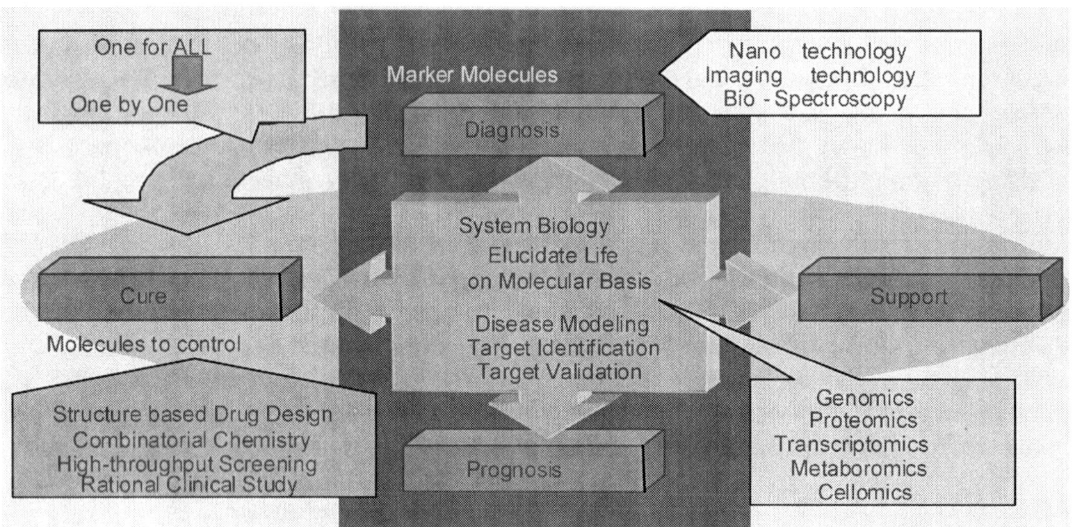

In the pharmaceutical industry today, research scientists have only one goal: to speed up the drug development process by finding more disease-related and drug able targets and by producing more high-quality leads in less time. Part of reaching that goal requires establishing a "decision support" infrastructure: getting all of the right information at the right time to make the right decisions. In addition, in today's "holistic" drug development environment, scientists need to collaborate and share knowledge with interdisciplinary team members to become more creative, more efficient and more productive in terms of being able to select the best drug candidate as early as possible.

Four platform technologies have turned the drug discovery process upside down:
- genomics
- combinatorial chemistry
- molecular screening technology
- bioinformatics

In the past, companies went from function to structure. Today, they go from structure to function.

Table 8.3.   Revolutionary progress in biomedical science changes paradigm of discovery and development

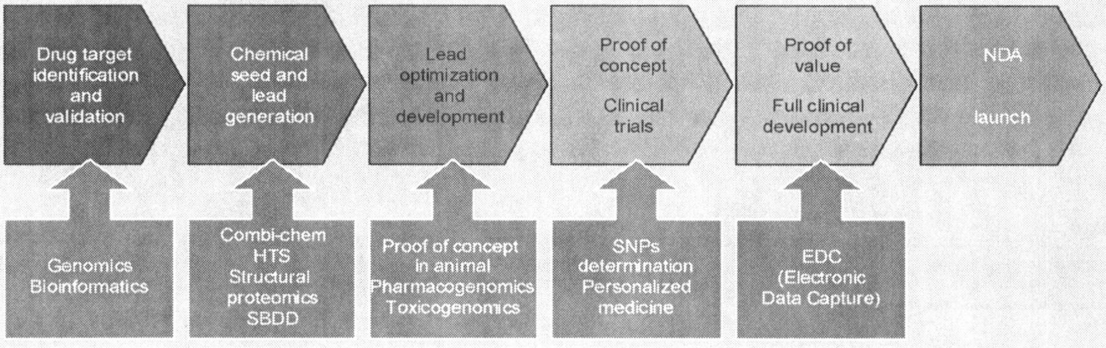

VIII.5.   The golden age of medicine discoveries lies ahead of the transformed industry

Despite predictions on the fate of the blockbuster drugs, *the golden age of medicine discoveries is...still to come!* The healthier Western society becomes, the more it regards maximum access to medicines as a right and duty, the more medicine it craves and the more it requires from that medicine. For example, 50 years ago, the face of healthcare was very different to that of today, with no third payers in the US, and universal health coverage had yet to become an uncontested "human right", tetracycline was the new miracle drug, cancer was a death sentence, and life expectancy was much shorter than it is today. Scientific medicine of today is no longer comparable with that of only half a century ago, and with it, treatment patterns have also evolved. However, some of the key changes are due to occur in the future as targeted medicines and personalized health maintenance programs will be used to treat a wider mass of less prevalent diseases.

Table 8.4.   Drugs: from the past to the future[279]

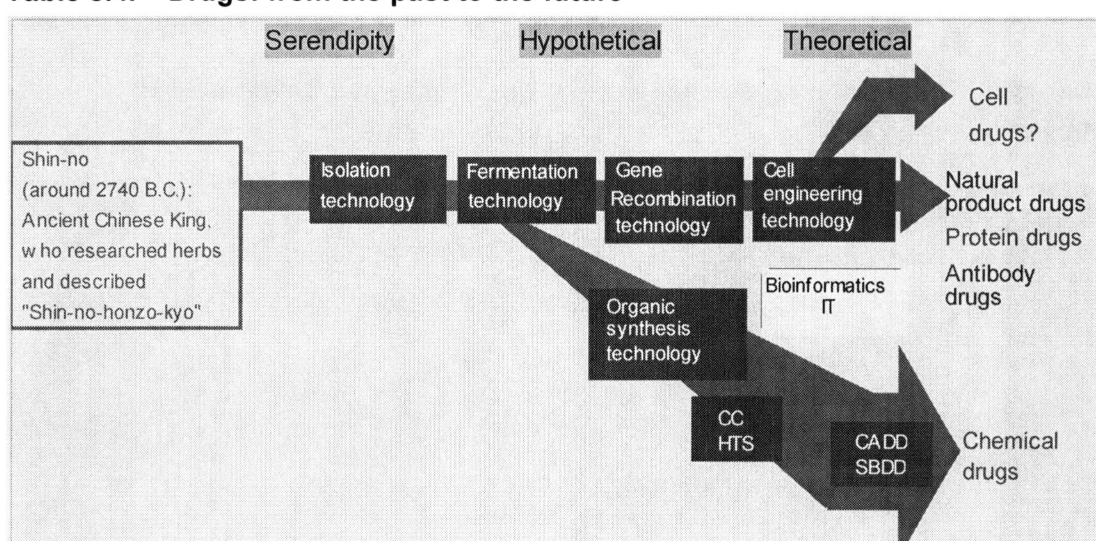

VIII.6   Medicines for "every body"

Since 1976, and based on the US $ 1bn success of SmithKline's Tagamet (cimetidine), BigPharma has increasingly used a business model based on primary care "blockbusters" supported by ever bigger sales forces and, in the US, also by sophisticated mass marketing. *That era is inevitably coming to an end.*

Non-responders, adverse reactants will be eliminated from target populations. However, there will be much less "waste", as targeted therapies will work for some 85 to 95% of the target −or "niche"- segment of the market/population group.

We are witnessing the emergence of the "nichebusters", of multiple busters in a more selective, better defined segment of the mass market, where only some 50% of medicines work well for "Every" body. As a result, profit margins can only grow. *It makes the future extremely exciting.*

---

[279] *Based on the presentation of Dr Hatsumo Aoki, Financial Times Global Pharmaceutical Conference, London 2003*

Table 8.5.   The current system is based on :
"primary care blockbuster medicines for the masses; on "one size-fits-all drugs"

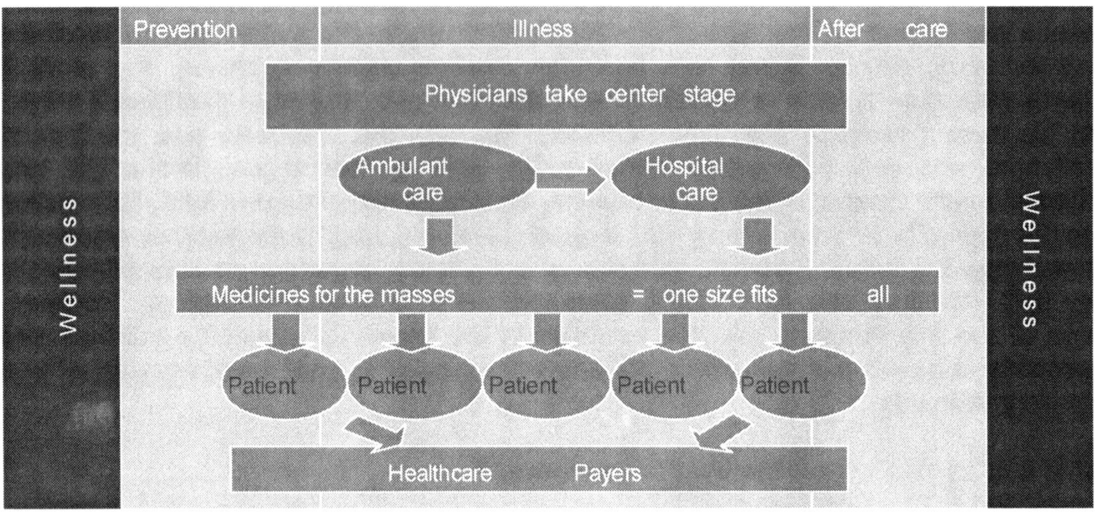

Table 8.6.   Tomorrows system will be based on : targeted, individualized medicine

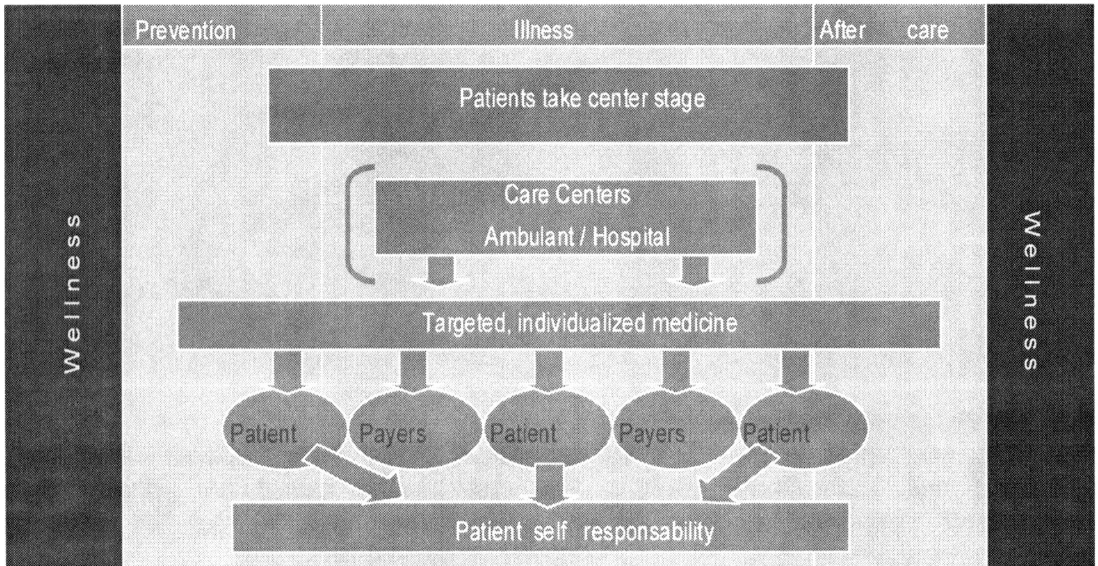

VIII.7.   Coming up with treatments not imagined even a decade ago

The figure shows how these changes will affect the provision of healthcare: the focus of drug discovery will move away from a small number of mostly chronic diseases that have been traditionally targeted by BigPharma and towards other, numerous areas of less prevalent diseases. This will, in turn, create a targeted medicines sector and, in time, lead to the development of personalized health maintenance programs. The golden age of medicine discoveries is not behind, but ahead of us. The promise of yet another medical revolution makes it difficult to predict what the pharmacogenomic market will be like in ten year's time. However, individualized medicine will make it possible to pay more attention to specific ethnic groups and to particular health needs of men and women in different sub-phases of their live.

Table 8.7. The shift towards personalized medicine

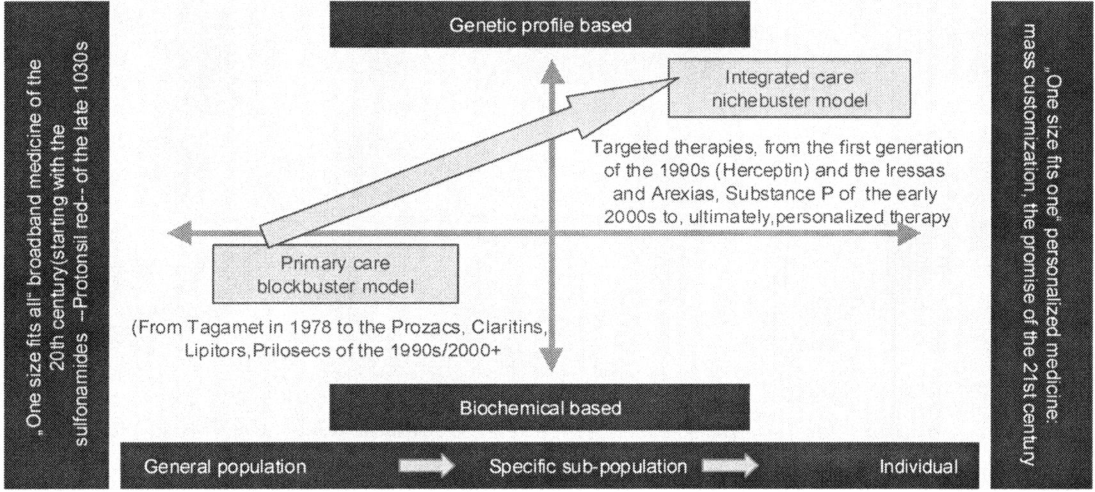

At first, we will see a split up of the existing major therapeutic areas in which BigPharma invested huge sums of money, which is based on the idea to find "volume diseases". And not on so-called "rare diseases" such as Chagas disease, leishmaniasis, sleeping disease (despite the fact that these three combined affect 350 million of people a year). But, by capitalizing on their knowledge and expertise in personalized medicines, and encouraged by patient advocacy groups, industry will come to discover the value of these "neglected diseases" .According to the World Health Organization (WHO), just 10% of the global health research budget is spent on diseases that account for 90% of the global disease burden.

VIII.8. Consortium approach to R&D

Pharma preparations are evolving from pills, pills, and more pills for the (Western patient tested) non-specified masses of patients in a limited number of "broadband" therapy areas like cardiovascular, central nervous system, metabolism, respiratory tract, and some anti-infectives, towards personalized medicine in an emerging model of predict-and-prevent health attitude; of integrated therapies that involve diagnostics, drug-device combinations, and new delivery systems. And this in a sophisticated, cost/effectiveness-oriented, highly competitive environment, pushing individuals for real health-behavior modification, away from a paternalistic social welfare spirit towards more personal responsibility/co-payments, more selfdiagnosis/selfmedication.

Chapter Nine

*Targeted
Therapies*

IX.1. Drivers and Passengers on the Road to Cancer

Scientists at the Wellcome Trust Sanger Institute, where one-third of the human genome was sequenced, have now pioneered decoding the sequence of cancer genomes. They have carried out the broadest survey yet of the human genome in cancer by sequencing more than 250 million letters of DNA code, covering more than 500 genes and 200 cancers.

The survey[280] shows that the number of mutated genes that drive development of cancer is greater than previously thought[281]. Significantly, as well as driver mutations for cancer, each cell type carries many more passenger mutations that have hitchhiked along for the ride. The study shows that a challenge for cancer biologists will be to distinguish the drivers from the larger number of passengers.

"The human genome is a vast place and this, our first deep systematic exploration in cancer, has thrown up many surprises. We have found a much larger number of mutated driver genes produced by a wider range of forces than we expected" said Professor Mike Stratton, co-leader of the Cancer Genome Project at the Sanger Institute.

All cancers are believed to be due to mutations - abnormalities in genes. The availability of the human genome sequence has opened the door to analyzing hundreds to thousands of genes, which will ultimately allow scientists to acquire a complete catalogue of the mutations in individual cancers.

The team studied more than 500 genes of a type called kinases, some of which have been previously implicated in causing cancers. One example is the *BRAF* gene: the 2002 pilot phase of the team's work showed that *BRAF* was mutated in more than 60% of cases of malignant melanoma. That observation has driven discovery of new drugs to treat melanoma, some of which are in clinical trials. The current study[282] was much

---

[280] *Haber DA and Settleman J (2007) Drivers and passengers. Nature 446: 145-6 DOI: 10.1038/446145a*

[281] *Greenman C et al. (2007) Patterns of somatic mutation in human cancer genomes. Nature 446:153-8. DOI: 10.1038/nature05610*

[282] *participating centers were: Cancer Genome Project, Wellcome Trust Sanger Institute; EMBL-European Bioinformatics Institute; Molecular Pathology Unit, Neurosurgical Service and Center for Cancer Research, Massachusetts General Hospital and Harvard Medical School; Royal Brompton Hospital, London; Ludwig Institute for Cancer Research, Brussels; Laboratory of Pathology/Experimental Patho-Oncology, Erasmus MC University Medical Center Rotterdam, Daniel den Hoed Cancer Center, Josephine Nefkens Institute, Rotterdam;*

broader and included breast, lung, colorectal and stomach cancers, which are the most common cancer types.

Table 9.1.   Cancer is a disease of the genes[283]

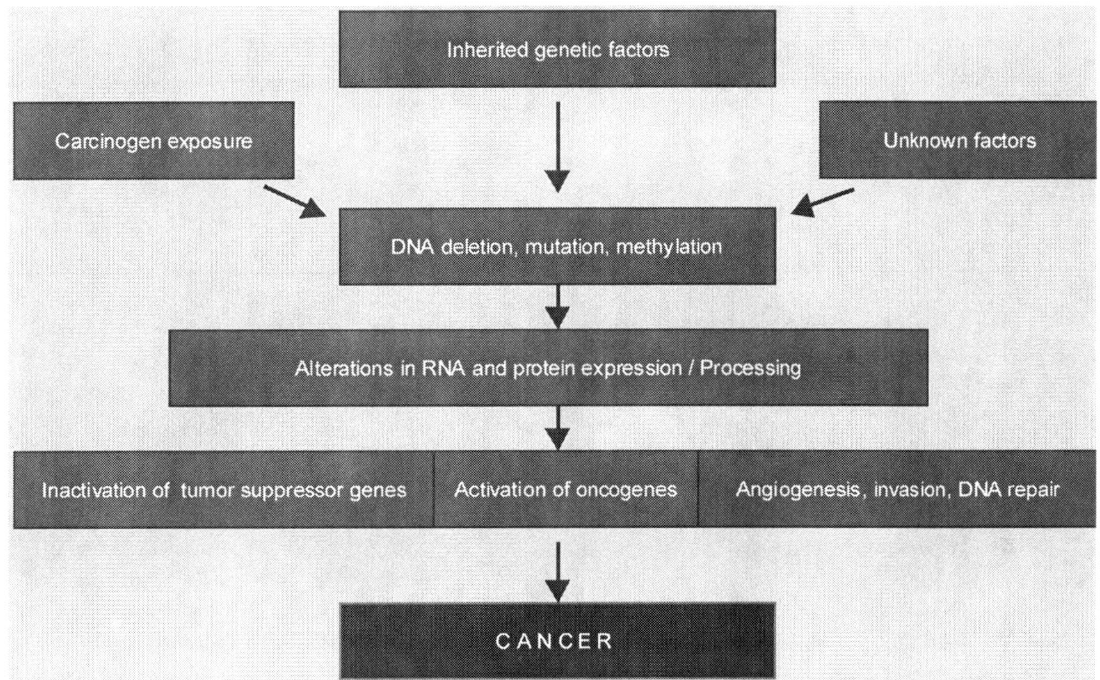

The study nicely shows the power of unbiased large-scale mutational analyses in human cancer, and points to a large number of unappreciated protein kinases that appear to be important in tumorigenesis. The new research showed that mutations in cancers can be divided into drivers or passengers. ***Driver mutations are the ones that cause cancer cells to grow, whereas passengers are co-travelers that make no contribution to cancer development.*** The team identified possible driver mutations in 120 genes, most of which had not been seen before.

"It turns out that ***most mutations in cancers are passengers***. However, buried amongst them are ***much larger numbers of driver mutations than was previously anticipated***. This suggests that many more genes contribute to cancer development than was thought. Our understanding of the roles of kinase proteins sheds light on some of the mutations. Kinases can act as a series of relays, switching on and off in our cells, to control cell behavior, such as cell division. For example, we found that a group of kinases involved in the Fibroblast Growth Factor Receptor signaling pathway was hit much more than we expected, particularly in colorectal cancers. The team also found that the mutations carry important coded messages within them. The type of mutation found varies markedly between individual cancers, reflecting the processes that

University of Pennsylvania Cancer Centre; Department of Gynecological Oncology, Westmead Hospital and Westmead Institute for Cancer Research, University of Sydney at the Westmead Millennium Institute; Institute of Cancer Research, Sutton; Department of Haematology, Addenbrooke's NHS Trust and University of Cambridge; Cancer Research UK Genetic Epidemiology Unit, University of Cambridge; Queensland Institute of Medical Research, Royal Brisbane Hospital; Van Andel Research Institute, Grand Rapids; Department of Pathology, The University of Hong Kong.

[283] *Adapted from: Carbone DP, Vanderbilt-Ingram Cancer Center, Comprehensive Technologies for Understanding Lung Cancer Biology and Identifying Novel Targets for Intervention. Presentation given in a series of Continuing Education articles, published online by Medscape. Accessed 19 March 2007*

generated the mutations, some of which were active decades before the cancer showed itself[284].

Table 9.2.   DNA sequence from, top, normal and, bottom, cancer tissue from a case of non-small-cell lung cancer (NSCLC)[285].

C CA T CA C G TA TG C T T C C T G G G G AC A A

C CA T C A C G TA TG C   CA T C A C G TA TG C T
T T C C T G G G G AC A A

A region of the *ERBB2* gene is duplicated in some cells, resulting in two sequences from the tumor sample - one normal and one cancer-causing form.

Patterns of mutations are an archaeological record written into the DNA of each cancer telling us about the factors that caused the cancer in the first place, which were often active many decades previously. Some of the patterns can be deciphered, such as the signatures of damage from ultraviolet radiation (sunlight) or cancer-causing chemicals in tobacco, but others are currently cryptic and will require decoding in the future.

This study vindicates all of the effort that went into the Human Genome Project. Understanding the mutations that cause cancer is crucial in order to develop accurately targeted treatments[286].

The new research shows cancers in a different light and highlights many different insights into how cancers develop. ***The challenge will be distinguishing the drivers from the passengers.***

In some cases this is straightforward. For many others, it appears, scientists will have to analyze much larger numbers of each cancer type. New, faster DNA sequencing technologies will play an important part in achieving the scale of study needed.

The time is right to apply the powerful tools of genomics to obtain a comprehensive view of what goes wrong at the DNA level in cancer. The important and interesting data on

---

[284] *Explanation given by Dr Andy Futreal, co-leader of the Cancer Genome Project.*
[285] *source: Wellcome Trust Sanger Institute*
[286] *Comment given in a statement to the press by Dr Mark Walport, Director of the Wellcome Trust.*

protein kinases in this further encourages the conclusion that a full assault on the cancer genome will yield many opportunities to revolutionize diagnosis and treatment[287].

IX.2. Intelligent innovation

A frontrunner in targeted medicine is Swiss pharmaceutical giant Hoffmann LaRoche. The company underwent far-reaching changes on account of a new generation of drugs, repositioning itself away from a mainly CNS (although without giving up the CNS franchise) and focusing on becoming an *innovative, dedicated targeted medicine player.* With wider access to eligible populations with longer treatment timeframes, combined with high returns on these type of products, many of the pipeline products should attain niche-blockbuster status.

Some analysts see this strategy as "temporary", since the firms size suggests, that it will opt back into the mass medicines market as soon as its research arm provides the kinds of products that the segment requires. This may well be the case. However, the company's strategy is not to rely solely on high-potential blockbuster drugs prescribed by general practitioners (the so-called mass medicines market blockbusters, preferred by Anglo-Saxon companies), due to the inherent risk of dependence on one or two blockbuster drugs for corporate success and...survival. The alternative model to the mass medicines market approach, intelligently chosen by Roche management several years ago, gives the company a clear advantage in risk management and competitiveness over those companies that focus solely on blockbuster drugs in the mass medicine markets

Roche now operates on a strong footing in the narrower competitive field of *severe diseases -market segments that require less commercial support.* These segments include hospital-based products and therapies prescribed by specialists and also allow for a much faster return on investment than products launched in the mass market sector. *Expect Roche to be among the top 3 global companies ten years from now.* The oncology and virology pillars of Roche's pharmaceutical business accounted for over 50% of revenues in 2006. Given the high returns on targeted therapies, their proportional contribution to operating profits is even far greater.

 At their turn, with the progress made in diagnosis and screening, which has broadened the eligible and treated population, Novartis, AstraZeneca, and Aventis (now Sanofi-Aventis) understood the potential of targeted therapies and decided to jump on the bandwagon of targeted cancer therapies.

IX.3. Being first to recognize advances in medical sciences

Roche has not only recognized before the others that advances in medical science would revolutionize the pharmaceutical industry business model, the company also acted before the others and clearly *included diagnostics in their business structure.*

Pharmaceutical companies have brought great pharmaceuticals to mankind. Their lifeblood-focus on therapeutics, understandable as it is, overlooked the role diagnostics play.

---

[287] *According to Dr. Francis S. Collins, director of the National Human Genome Research Institute at the National Institutes of Health.*

Table 9.3. Roche's business model: Focus on innovation, new technologies, diagnostics

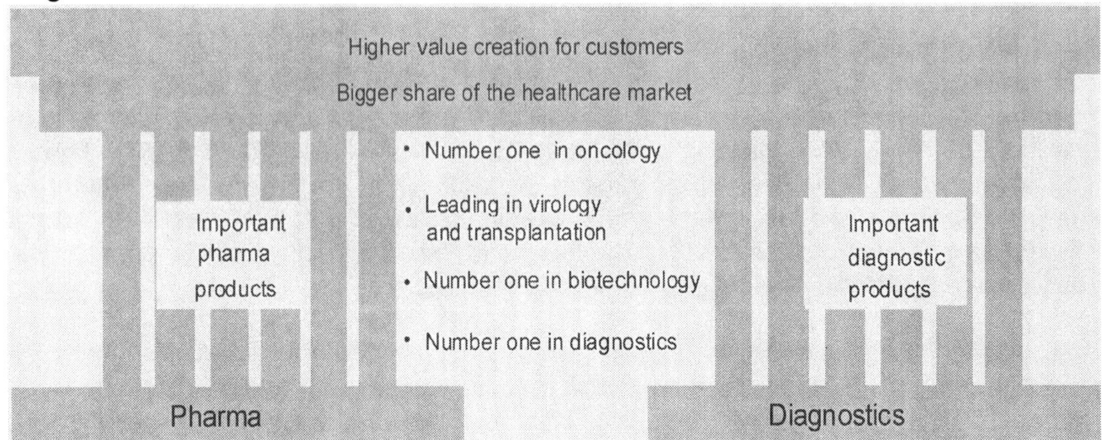

IX.4. An essential precursor for targeted medicines

Diagnostics are important in finding patients and getting them on the right therapy. The shift towards personalized therapy can only take place if new diagnostics become available. Allowing the diagnosis of disease at an extremely early stage in a way that cannot be determined by any other process, once a far-away scientific idea, is a near-term reality. Experts in the field of diagnostics are looking to molecular imaging and in vivo in-patient diagnostics to replace the existing gold standards. Roche had the right view on the future of scientific medicine, included diagnostics in its business structure, and took the necessary steps to restructure its operations. Much to the envy of much of its peers.

According to Amersham Health, which partnered with GE Medical Systems in November 2001, and which specializes in diagnostic pharmaceuticals for neurology, oncology, cardiology, and radiotherapy products for treatment of cancer, PET (Positron Emission Tomography; the key to unlocking a new generation of targeted molecular pharmaceuticals). PET scanners work by reading fluorodeoxyglucose, a radio-active fluorine isotope of glucose- that images fluorine and gives an exquisite picture of high levels of glucose uptake in rapidly growing tumors. The number of PET scans conducted is expected to rise from 400.000 in 2002 to 1.200.000 by 2010. *The growing use of PET in –mostly oncology and neurology- drug research* (many fluorine-based products remain to be discovered, the company states) will lead to products that can be useful in diagnosing a broad variety of diseases.

Roche Diagnostics, which concentrates on Diabetes Care, Molecular Diagnostics, and Immunochemicals, considered the alliance with Affymetrix as a next big step forward : the use of *tests based on DNA chip technology*.

Roche Diagnostics remained the global leader in 2006 in an increasingly competitive market, with a market share of 19%. Divisional sales increased 5% for the year in local currencies (6% in Swiss francs; 5% in US dollars), fuelled by new product launches. This was slightly above the market growth rate. The Centralized Diagnostics, Near Patient Testing and Applied Science units were the main contributors to growth.

IX.5. Centralized Diagnostics – cobas 6000 analyzer series

In 2006 Roche Centralized Diagnostics posted above-market sales growth of 5% and remained the industry leader with a market share of about 13%. The rollout of the medium-throughput cobas 6000 analyzer series and the European launch of the cobas c 111 analyzer for customers with small testing volumes marked important steps in a business strategy centered on making clinical chemistry and immunochemistry testing simpler and more efficient. An application for US marketing approval for the cobas c 111 analyzer was submitted to the FDA in late 2006. The cobas 6000 analyzer series is a fully automated, integrated system capable of handling more than 95% of the routine tests performed daily a medium-volume laboratory. Thanks to its flexible, modular design, it can be configured exactly to customers' individual needs, and new modules can be added at any time as those needs grow. Immunoassay sales continued to grow significantly faster than the market, advancing 13% in 2006 thanks to products like the Elecsys proBNP and Elecsys Troponin T assays for cardiac disorders. Sales of the NT-proBNP marker grew 28%, helped by additional US approval of the Elecsys proBNP assay for use in assessing the risk of cardiac events in patients with stable coronary artery disease.

IX.6. Molecular Diagnostics – focus on automation

Roche Molecular Diagnostics maintained its leading market share at about 38% as sales advanced 3% for the year. Virology – the largest segment by sales – grew 5% in 2006, in line with the virology market. Stepped-up sales efforts for the combined Cobas AmpliPrep/Cobas TaqMan platform and its menu of viral load tests for HIV and hepatitis B and C virus (HBV, HCV) drove product sales. Offering fully automated sample preparation and analysis, Cobas AmpliPrep/Cobas TaqMan enhances laboratory productivity and test result integrity. Monitoring viral load (the amount of virus in a patient's blood) is an important way of assessing disease progression and treatment response. In June Roche began rolling out the new fully automated cobas s201 modular blood screening system and cobas TaqScreen MPX multiplex test across Europe. The cobas TaqScreen MPX test, which simultaneously detects HIV, HCV and HBV in donated blood, received CE Mark (Conformité européenne) certification in March 2006. Large US laboratories signed on to offer the AmpliChip CYP450 Test, a microarray-based test that detects genetic variations which can affect the way patients respond to treatment with many widely prescribed drugs.

IX.7. Near Patient Testing –coagulation monitoring and hospital-based blood glucose monitoring

This business area reinforced its market leadership in 2006. Overall sales rose 7% for the year, helped by the continued trend towards decentralized testing. Roche Near Patient Testing's newest coagulation monitoring systems, CoaguChek XS for patient self-monitoring and CoaguChek XS Plus for healthcare professionals provide patients taking oral anticoagulants and their health professionals accurate, on-the-spot results from a single drop of blood. Their successful launch has strengthened Roche's global leadership in coagulation monitoring. Roche Near Patient Testing is also the clear leader in hospital-based blood glucose monitoring. The Accu-Chek Inform meter and Accu-Chek Advantage and Accu-Chek Sensor test strips are the core products driving Roche's growing market share in this segment.

IX.8. Applied Science – Genome Sequencer 20 opens up new market segment for Roche

Roche Applied Science's sales grew 12% in 2006, nearly twice the market growth rate. Growth was driven primarily by the LightCycler 480 instrument and Genome Sequencer 20 system. LightCycler 480 is a highly versatile high-throughput gene expression and mutation analysis platform based on the polymerase chain reaction (PCR) technology *pioneered by Cetus* (later acquired by Roche). The innovative Genome Sequencer 20 system, first launched in late 2005, marks Roche's successful entry into the attractive DNA sequencing research market. It can sequence long DNA fragments and entire genomes 60 times faster than conventional commercially available instruments. Roche Applied Science is also a supplier of industrial reagents and substrates, which account for a major part of its sales revenues. These products were important contributors to growth in 2006.

IX.9. Advances in diagnosing infections and tumors[288]

Major advances have been made in diagnosing infections and tumors, and these already have great practical significance. DNA analysis techniques have gained broad acceptance in the field of pathogen diagnostics and have probably had the most concrete benefits for medicine in this area to date. There are also diverse possibilities for using DNA analysis in differential diagnosis of cancers. As cancer always involves changes at the level of DNA, chromosome DNA analysis techniques are suitable for early identification, progress monitoring and controlling the success of therapy.

In the narrower field of genetic diagnostics, the main focus of hopes and fears is diagnosis of multifactorial diseases affected by a number of genes and environmental influences. Most of the widespread diseases –cancer, cardiovascular, metabolic (including diabetes) and neurodegenerative diseases (Alzheimer)- have multifactorial etiology. It is expected that research into the human genome will open up a new dimension of understanding of the role of genes in the incidence of the diseases listed. In clinical practice, however, diagnosis using human DNA is still very largely limited to single-gene inherited diseases. Examples of real use of possible tests for multifactorial diseases are still limited at present to a small number, the most frequently cited being proof of mutations in the bCRA genes causing breast cancer and the association of ApoE with a risk of Alzheimer's. Hopes are focused particularly on diagnosing genetic disposition to cancer.

IX.10. Screening tools

An example of a potential screening tool for cancer is the MitoChip from Affymetrix.
Mitochondria, the energy-producing organelles that power our cells, are unique because they are equipped with their own genetic instructions distinct from the DNA stored in the cell nucleus. Mitochondria are oblong, thread-like structures dispersed throughout the cell's cytoplasm. Hundreds to thousands of mitochondria exist in each human cell, occupying up to a quarter of their cytoplasm. Sometimes informally described as "cellular power plants," mitochondria convert organic materials into ATP, the cell's energy currency and without which life would cease.

As early as the 1920s, scientists uncovered clues that mitochondria might play a role in causing cancer. But like the other DNA in the cell nucleus, scientists lacked the needed

---

[288] TAB, Buero fuer Technikfolgen-Abschaetzung beim Deutschen Bundestag. Summary of TAB working report No.66, April 2000

research tools throughout most of the 20th century to systematically study the chemical composition of the mitochondrial genome, or complete set of genes, and its association to human disease. In the early 1980s, scientists in England performed the then-Herculean feat of sequencing the complete human mitochondrial genome. The genome consisted of 16,568 base pair, or units, of DNA and encoded 37 contiguous genes.

By 1996, new technology brought new opportunity. Scientists with the company Affymetrix in Santa Clara, Calif. developed the first mitochondrial sequencing microarray. Roughly the size of a quarter, the silicon chip had lithographically annealed to it up to 135,000 short, arrayed bits of DNA sequence that, collectively, spanned most of a single strand of mitochondrial DNA.

The chip exploited the fact that DNA exists naturally as a double-stranded molecule. By gathering mitochondrial DNA and breaking it into short, single-stranded bits, it was shown that each bit would pair, or hybridize, with its complementary sequence arrayed on the chip. By crude analogy, each bit is like a unique magnet that sticks to its mirror image. But if the extracted DNA contains mutations or other variations from the standard consensus sequenceannealed to the chip, the bits with those changes would appear abnormal to the specially designed computer software programs that read the chip. The software programs will read not only the identity of the expected bases of DNA in a process called "base calling" but those of the variations The chip enabled laboratories to resequence the mitochondrial genome much faster than the traditional manual and automated strategies. Just as importantly, like an iPod to music lover, the chip served as the broad technological platform for laboratories to customize arrays more attuned to their research interests. In 2004, Dr. Anirban Maitra and his colleagues at Johns Hopkins did exactly that with the MitoChip v1.0. It marked the first mitochondrial resequencing microarray designed as a potential screening tool for cancer.

In 2006, a second generation "lab on a silicon chip" called the MitoChip v2.0 was developed. This updated version the first time rapidly and reliably sequences all mitochondrial DNA. It is predicted to become a key tool in accelerating research on mitochondrial DNA, a growing area of scientific interest. This interest stems from data that suggests natural sequence variations and/or mutations in each person's mitochondrial DNA could be biologically informative in fields as diverse as cancer diagnostics, gerontology, and criminal forensics.

IX.11. A tremendous need for innovative therapeutics
Research & Drug development budgets have risen over the last decade. Numerous tools have been put in place within drug discovery. Risks for the research driven industry have increased as they search for first in class drug candidates. However, these risks are mitigated by the high potential of drugs that offer genuine benefits over and above existing drugs.

While certain pharmaceutical deliberately refused to or could not embrace the blockbuster model because they did not have the financial backing to focus uniquely on true research for novel therapies, and follow the second-in-class or even the third-in-class route, counting on me-too developments for survival, others favor a combination of follow-up compounds with genuine first-in-class products. In-licensing has also become increasingly popular, and as a result, few companies are willing or able to. Still others

Table 9.4. Where the high reward comes from...

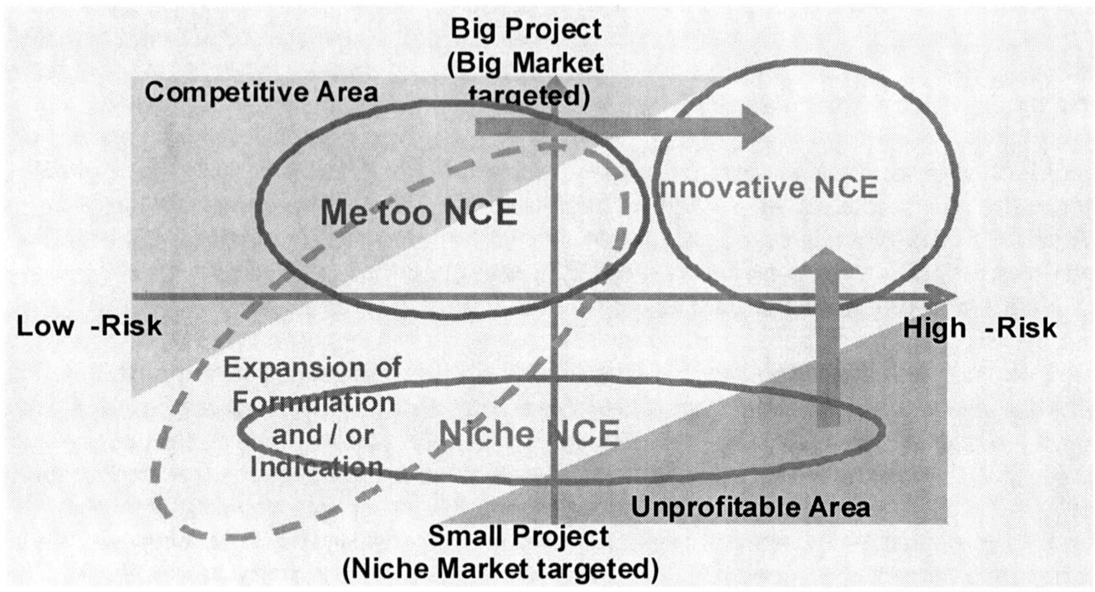

count on in-licensing (where the competition for picking the cherries on the cream is enormous). Few are able to focus uniquely on true research for novel therapeutics. Those that do and succeed in satisfying (part of) the tremendous expectations society today puts in medical progress, stand to benefit greatly.

All agree that developing new molecular targets would be their primary activity if just they could keep up with a sustained financial investment. The stakes are high. Whereas in the 1990s, primary care markets promised high returns (see the cholesterol-lowering and anti-depression drugs launched in those days), the future undoubtedly will be one of targeted medicines. With oncology topping the list of Phase I and Phase II new drug candidates (developments in cancer therapeutics outnumber the combined development of all other drug candidates in all other therapeutic fields!), industry clearly understood that it has to change and now prepares for another type of business; society can expect surprising and highly welcome progress in the relentless battle of fighting disease.

Table 9.5. Candidate drugs in early stage clinical development

Category	Projects in Phase I & II	Category	Projects in Phase I & II	Category	Projects in Phase I & II
Oncology	1.430	Hepatitis	83	Inflammatory Bowel Disease	55
Pain	131	Rheumatoid arthritis	82	Psoriasis	52
HIV	111	Asthma	78	Depression	51
Diabetes	109	Dementia	76	Multiple sclerosis	40
Immunomodulators	89	Inflammation	70	Thrombosis	39

IX.12. The race for new therapies

The race for new therapies in oncology, a therapeutic field where the need for better medicines is tremendous, is driven by discoveries made in immunology, cellular and molecular biology.

Table 9.6. Future therapy

Diseases amenable to gene therapy*	Available drug therapy (examples)
AIDS	RT inhibitors,...
Alzheimer's Disease	Nootropics, Neurotonics,...
Cancer	Cytokines Immunomodulators,...
Cystic Fibrosis	Mucolytic Agents,...
Growth Hormone Deficiency	Recombinant Growth Hormone,..
Hepatitis B	Vaccines
Hypercholesteremia	HGMCoA Reductas Inhibitors,...
Influenza Pandemic	Vaccines
Muscular Dystrophy	N/A
Parkinson's Disease	DOPA, Benzaseride,...

(*the list is non-exhaustive...)

The race for a new wave of innovative therapies is on. In particular in the field of oncology.

Progress in oncogenics

Cancer is a metaphorical example of the perfect invasion by a founder species. Like the first pregnant finch that landed on a deserted island in the Galapagos Archipelago, the first cancer cell in the human body has to undergo many mutations through many generations to establish itself as an invader of different organs in the body. But once it is there, like any newly stabilized species in different ecological niches, cancer is tough to get rid of.

Of course, the difference with cancer is that it destroys itself when it kills you off, whereas many new species stabilize. Obviously, no metaphor is the be-all of reality. But I'm hoping scientific debates will help people look at cancer in a new light.[289]

Table 9.7. Cancer pathologies (selected): incidence, mortality, prevalence, 5-year survival rate[290]

	Incidence	Mortality	Prevalence	5-year survival rate
Breast	215.990	40.954	2.420.213	90,3
Prostate	230.110	29.002	2.024.489	99
Lung & Bronchus	173.770	158.342	358.028	15,7
Colon/rectum	146.940	53.772	1.076.335	65,9
Urinary bladder	60.240	13.281	511.791	81,3
Melanoma of the skin	55.100	7.952	690.021	91,7
Non-Hodgkins lymphoma	54.370	20.723	381.129	64,8
Kidney & Renal pelvis	35.700	12.647	239.266	65,5
Leukemia	33.441	21.395	208.620	50,8
Pancreas	31.860	31.771	28.447	4,9
Oral cavity and pharynx	28.260	7.826	235.856	62
Ovary	25.580	14.067	172.765	44,7
Thyroid	23.600	1.409	296.414	97,2
Stomach	22.710	11.859	60.300	22,2
Liver & Intrahep	18.920	18.718	19.429	9,9
Brain & Other nervous	18.400	12.829	98.086	33,9
Myeloma	15.270	10.655	53.720	33,6
Cervic uteri	10.520	3.850	250.726	74,3
Testis	8.981	357	112.845	96,3
Lymphoma	7.879	1276	151.207	87,1

Note: Incidence & mortality estimated cancer cases in 2004; Prevalence: complete prevalence count, invasive cancers only, Jan 1, 2004; 5-year relative survival rates for the period 1996-2003. These figures reflect the situation in most Western countries. For detailed statistics on cancer in the USA, visit the Website of the National Cancer Institute (USA) at: http://seer.cancer.gov/csr/1975_2004/results_merged/sect_01_overview.pdf

---

[289] *Professor emeritus Robert C. von Borstel at the American Association for the Advancement for Science (AAAS) Conference, February 2007, where he delivered a talk about how cancer cell mutation and selection are metaphorically similar to how a new species begins its evolution. The former University of Alberta biologist has been working with DNA - the molecule that carries our cells' genetic information - ever since 1947, six years before its structure was described by Watson and Crick. His Natural Sciences and Engineering Research Council (NSERC) work on how radiation can kill cells, how the DNA molecule itself can control mutation, and other research has earned him a fellowship in the AAAS.*

[290] *National Cancer Institute. Statistics published online by the U.S. National Cancer institute. Accessed August2, 2007. http://seer.cancer.gov/csr/1975_2004/results_merged/sect_01_overview.pdf*

Cancer is a group of highly heterogeneous pathologies that have become a leading cause of mortality in the US, Japan and Europe.

Taking all cancers together, the following trends have been observed .

- In industrialzed nations, prevalence is rising due to longer average patient life spans (the incidence/mortality ratio is moving up).
- The cancer problem becomes a global problem: in developing countries, incidence and mortality ratio is rapidly rising (also due to better diagnostic possibilities).
- Except for a few cancers in which remission is usually achieved, therapeutic needs are still enormous for most patients with malignancies. The increase in cancer treatments will have to be absorbed by the Health Insurance systems. But: the need for bigger volumes of products will drive down their current high prices.
- Five-year survival, the therapeutic needs indicator used in this group of pathologies, is currently estimated at 62%. This is a relative gauge, since vital prognoses vary broadly from one type of cancer to another.

IX.13. A veritable pandemic

Cancer has become a veritable pandemic : more than six million people die of it each year worldwide, while ten million new case are reported. Cancer is responsible for no less than 12 percent of deaths worldwide (25% in the USA !). The WHO forecasts annual incidence of new cases of 15 million to 2020 and 10 million deaths. The US National Institute of Health estimates the cost of cancer to the community in 2002 at $171.6bn. Still in the US, cancer already causes 50 million visits to practitioners each year, representing one million interventions and 750.000 radiotherapies.

Table 9.8.   SEER Cancer Incidence and US Death Rates, 2000-2004. By Cancer Site and Race[291]

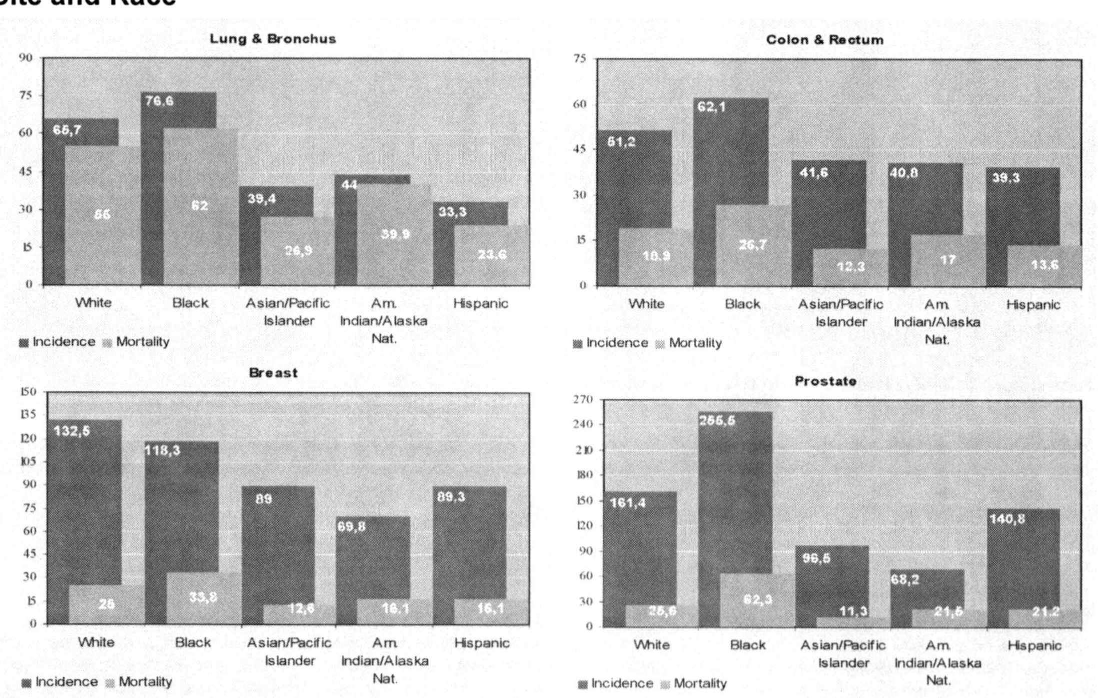

[291] *National Cancer Institute, USA. Data published online at http://seer.cancer.gov/csr/1975_2004/results_merged/sect_01_overview.pdf (page 85, accessed Aug 2, 2007). For detailed information, please visit this Website.*

IX.14. Natural selection as a driver of the evolution of cancer?

The dynamics of evolution are fully in play within the environment of a tumor, just as they are in forests and meadows, oceans and streams. This is the view of researchers in an emerging cross-disciplinary field that brings the thinking of ecologists and evolutionary biologists to bear on cancer biology.

Insights from their work[292] may have profound implications for understanding why current cancer therapies often fail and how radically new therapies might be devised.

A tumor cell population is constantly evolving through natural selection. The mutations that benefit the survival and reproduction of cells in a tumor are the things that drive it towards malignancy. Evolution is also driving therapeutic resistance. When you apply chemotherapy to a population of tumor cells, you're quite likely to have a resistant mutant somewhere in that population of billions or even trillions of cells. This is the central problem in oncology.

The reason scientists haven't been able to cure cancer is that they are selecting for resistant tumor cells. When we spray a field with pesticide, we select for resistant pests. It's the same idea.

There are three necessary and sufficient conditions for natural selection to occur and that all are met in a population of tumor cells.

The first requirement is that there be variation in the population. This variation is evident in tumors, which are a mosaic of many different genetic mutants.

The second condition is that the variation must be heritable. This, too, can be seen within a tumor-cell population. When mutant tumor cells divide to replicate, the daughter cells share the same mutations.

The final condition is that the variation has to affect fitness, the survival and reproduction of the cells. All of the characteristics that are considered hallmarks of cancer affect fitness, according to Maley. Among these are that cancer cells no longer need normal growth inhibition signals in their environment, they no longer require an external signal to divide as healthy cells do, and they are able to suppress a vital set of internal instructions that require cells to self-destruct when their genes are mutated beyond repair. This protective cell-suicide program carried by normal cells is known as apoptosis. Seeing a tumor in this light opens a window on new therapeutic strategies.

It's not just a metaphor to say tumor cell populations are evolving. Evolution is going on in the tumor. So let's think about how we might want to influence that evolution. Can we push it down paths that might be more beneficial to us?[293]

One idea might be to develop new drugs that would act as benign cell boosters. Such

---

[292] *A review by researchers at The Wistar Institute of current research in this new field, is published online and in the December 2006 issue of the journal Nature Reviews Cancer. The Wistar Institute is an international leader in biomedical research, with special expertise in cancer research and vaccine development. Founded in 1892 as the first independent nonprofit biomedical research institute in the USA, Wistar has long held the prestigious Cancer Center designation from the National Cancer Institute. Discoveries at Wistar have led to the creation of the rubella vaccine that eradicated the disease in the U.S., rabies vaccines used worldwide, and a new rotavirus vaccine approved in 2006. Wistar scientists have also identified many cancer genes and developed monoclonal antibodies and other important research tools. Today, Wistar is home to eminent melanoma researchers and pioneering scientists working on experimental vaccines against flu, HIV, and other diseases.*

[293] *Carlo C. Maley, Ph.D., an assistant professor in the Molecular and Cellular Oncogenesis Program at Wistar whose own research focuses on this area; senior author of the review.*

drugs would specifically target the more benign cells in a tumor to increase their relative fitness over their malignant neighbors. This would allow the benign cells to outcompete the malignant cells, leading to a less aggressive, less dangerous tumor.

Another idea the scientists are pursuing is to try to increase the fitness of chemosensitive cells so that they outcompete any resistant cells that are in the tumor. So they "sucker" the tumor into a vulnerable state and then hit it with chemotherapy.

In their review[294], the researchers also explored how the ecological ideas of competition, predation, parasitism, and mutualism unfold in tumors. Here again, they found that the concepts from another field helped to illuminate cancer biology. Mutant cells compete with each other for needed resources. The immune system often kills tumor cells like a predator hunting prey, and the tumor cells that develop defenses against the predation are the ones that survive and reproduce.

An example of parasitism in the tumor environment can be seen in angiogenesis, in which a subset of tumor cells send chemical signals to stimulate the host to generate new blood vessels to supply the tumor with nutrients. The neighboring cells that aren't investing resources in producing the signals take advantage of the nutrients nonetheless.

Mutualism describes a situation in which two organisms interact in a mutually beneficial way. Tumor cells send signals to stimulate the growth of the cells that form the scaffold in which the tumor cells grow, known as fibroblasts. The fibroblasts, in turn, send signals to the tumor cells to stimulate their growth. Recent studies suggest, too, that the fibroblasts in a tumor microenvironment begin to acquire mutations of their own.
They're co-evolving, and it becomes a dynamic, runaway process.

IX.15. The cost of patient time associated with cancer care.
Several studies have estimated the direct medical costs of cancer care, but few have attempted to include a patient's time associated with cancer care, such as time spent traveling to and from care, waiting for appointments, and receiving services and treatments, all of which represent time not spent working or pursuing day-to-day activities.

In the new study, researchers[295] quantified the patient time costs associated with cancer care. They used information from the Surveillance, Epidemiology, and End Results–Medicare database on more than 760,000 patients with 11 different types of cancer and from 1.1 million Medicare enrollees without cancer. Using data from 1995 to 2001, they estimated each patient's time spent at physician and emergency room visits, chemotherapy treatments, radiation therapy, hospitalizations, outpatient surgeries, and imaging procedures. They then estimated how long each patient spent traveling to, waiting for, and receiving care. The costs were calculated using a dollar value of $15.23 per hour, the median U.S. wage rate in 2002.

During the first 12 months after diagnosis, the average length of time for hospitalization was highest for patients with gastric and ovarian cancers (21.1 and 20.8 days) and shortest for patients with melanoma (2.2 days), prostate cancer (3.8 days) and breast cancer (4.0 days). Compared to similar people without cancer, cancer patients' net time

---

[294] *The lead author on the Nature Reviews Cancer article is Lauren M.F. Merlo, Ph.D., a postdoctoral fellow in the Maley lab. The co-authors are John W. Pepper, Ph.D., at the University of Arizona, Tucson, and Brian J. Reid, M.D., Ph.D., at the Fred Hutchinson Center. The work was initiated by the Santa Fe Institute and supported by the National Institutes of Health, the Commonwealth Universal Research Enhancement Program of the Pennsylvania Department of Health, and the Pew Charitable Trusts*

[295] *Center. The work was initiated by the Santa Fe Institute and supported by the National Institutes of Health, the Commonwealth Universal Research Enhancement Program of the Pennsylvania Department of Health, and the Pew Charitable Trusts*

associated with medical care varied, ranging from 17.8 hours for melanoma to 351.3 hours for gastric cancer and 368.1 hours for ovarian cancer. When the researchers applied the dollar costs to time spent on medical care in the first 12 months after diagnosis, they found that net patient time costs were lowest for melanoma ($271) and prostate cancer ($842) and highest for gastric ($5,348) and ovarian ($5,605) cancer.

In the last year of life, hospitalization time was longest for patients with gastric (35.4 days), lung (32.4 days), and ovarian (31.9 days) cancer. Estimates of patients' net time spent on medical care were lowest for melanoma (99.1 hours) and highest for ovarian (485.3 hours), lung (488.3 hours), and gastric (512.2 hours) cancer. They calculated that the net patient time costs during the last year of life ranged from $1,509 for melanoma to $7,799 for gastric, $7,435 for lung, and $7,388 for ovarian cancer. Hospitalizations were the largest component of patient time costs in both the initial year after diagnosis and in the last year of life.

For 2005, the estimated cost for the initial phase of care alone was approximately $2.3 billion, they write. These estimates could be combined with estimates of direct and indirect costs to better understand the overall burden of cancer in the United States.

What we see here is a measure of the patient's burden of commitment—measured in dollars—associated with receiving today's cancer therapy. We hope that policy makers recognize the substantial economic burden of cancer in the United States and that this cost derives from many sources. [296]. The editorial writers [297] also note that these calculations do not address the emotional cost cancer patients and their families endure.

IX.16. Europe's aging population causing major increase in cancer burden

Between 2004 and 2006, the number of new cases of cancer diagnosed each year in Europe has increased by 300,000 according to estimates published in 2007 [298]. It is estimated that in 2006 there were 3.2 million new cases of cancer (up from 2.9 million in 2004) and 1.7 million deaths from the disease in the whole of Europe.

With an estimated 3.2 million new cases (53% occurring in men, 47% in women) and 1.7 million deaths (56% in men, 44% in women) each year, cancer remains an important public health problem in Europe, and the aging of the European population will cause these numbers to continue to increase, even if age-specific rates of cancer remain constant. Evidence-based public health measures, such as screening, exist to reduce deaths from breast, cervical and colorectal cancer, while the incidence of lung cancer, and several other forms of cancer, could be diminished by tobacco control [299].

---

[296] *According to Dr. Larry G. Kessler, of the U.S. Food and Drug Administration and Dr. Scott D. Ramsey of the Fred Hutchinson Cancer Research Center in Seattle, in an accompanying editorial. Editorial: Kessler LG, Ramsey SD. The forest and the trees: the human costs of cancer. J Natl Cancer Inst 2007; 99:2–3*

[297] *Note: The Journal of the National Cancer Institute is published by Oxford University Press and is not affiliated with the National Cancer Institute.*

[298] *Estimates of the cancer incidence and mortality in Europe in 2006. Annals of Oncology. doi:10.1093/annonc/mdl498 [2] In 2006 the European Union comprised: Austria, Belgium, Cyprus, Czech Republic, Denmark, Estonia, Finland, France, Germany, Greece, Hungary, Ireland, Italy, Latvia, Lithuania, Luxembourg, Malta, Poland, Portugal, Slovakia, Slovenia, Spain, Sweden, The Netherlands and United Kingdom. Europe comprised the 25 EU countries plus Albania, Belarus, Bosnia Herzegovina, Bulgaria, Croatia, Iceland, Macedonia, Moldova, Norway, Romania, Russian Federation, Serbia and Montenegro, Switzerland, Ukraine.*

[299] *Professor Peter Boyle, Director of the International Agency for Research on Cancer (IARC) in Lyon, France, who prepared the report with IARC colleagues, and warned that despite better prevention and treatments, Europe faced a major increase in the cancer burden because of the aging population. He said urgent action was needed now to tackle cancer, particularly in Central and Eastern Europe, through measures such as tobacco control and more widespread screening for breast and colorectal cancer, as well as efforts to improve people's diet and exercise and reduce levels of obesity.*

IX.17. Incidence of cancer in Europe

The 25 EU countries accounted for nearly 2.3 million of the new cases and over one million cancer deaths. Lung, colorectal, breast and stomach cancers are the top four killers. Lung cancer remains the biggest killer, with an estimated 334,800 deaths in 2006 (19.7% of the total number of deaths from cancer), followed by colorectal cancer (207,400 deaths), breast cancer (131,900 deaths) and stomach cancer (118,200 deaths).

Lung cancer retains its status as the leading cause of cancer death in Europe in 2006. The overwhelming majority of lung cancer is caused by tobacco smoking, and tobacco control is clearly a number one priority in Europe, not only aimed at men, particularly the male populations of Central and Eastern Europe, but increasingly targeted towards women, especially in Northern Europe.

IX.18. Breast cancer has overtaken lung cancer as the commonest cancer diagnosed overall

Since the previous estimates for 2004, breast cancer has overtaken lung cancer as the commonest cancer to be diagnosed overall, with 429,900 new cases in 2006 (13.5% of all cancer cases). It was followed by colorectal cancer (412,900 cases, 12.9%) and lung cancer (386,300 cases, 12.1%). According to scientists, the rise in the number of breast cancer cases could be attributed partially to the introduction of organized mammography screening programs, which meant that more cancers were detected, and at an earlier stage. These programs have the short-term consequence of increasing the incidence, which has risen by 16% since our latest 2004 estimates. Despite the benefit of screening programs, they said that deaths from breast cancer were continuing to rise (130,000 in 2004, 131,900 in 2006) because of the aging population. In women, after breast cancer, colorectal cancer (195,400, 13.1% of the total) and cancer of the uterus (149,300, 10%) were the most commonly diagnosed cancers. In men, the widespread use of the Prostate Specific Antigen (PSA) tests was also having an effect on the numbers of prostate cancers being detected; prostate cancer was the most frequent cancer diagnosed (345,900, 20.3% of the total), followed by lung cancer (292,200, 17.2%) and colorectal cancer (217,400, 12.8%). Despite the widespread use of PSA testing in many European countries, the number of deaths from prostate cancer has increased by around 16% since 1995 due, in large measure, to the rapid increase in the numbers of men reaching older ages.

IX.19. The need for organized screening programs

The second most common cause of cancer death in both men and women was colorectal cancer. There remains hope that dietary modifications, increased physical activity and avoidance of obesity could lead to reductions in the incidence and mortality from colorectal cancer. However, progress has been very slow; the number of deaths has increased by 1.8% since our previous 2004 estimates. Screening for colorectal cancer has been shown to be effective, and clearly there is a need for organized colorectal cancer screening programs throughout Europe.

Deaths from stomach cancer are continuing to decline in men and women throughout Europe, although it still accounts for 5.6% of all new cases of cancer (5.9% in 2004) and 7.4% of all cancer deaths (8.1% in 2004). "However, higher incidence and mortality rates occur in the Central and Eastern European countries, possibly reflecting a lower level of affluence, a diet lower in fresh fruits and vegetables and higher rates of Helicobacter pylori infection", the report indicates.

Significant differences in the chances of surviving other cancers existed between the Eastern and Baltic European countries and other European countries. In the 25 member countries of the European Union an estimated 23,600 women died from cancer of the uterus in 2006 and 46,600 died in the whole of Europe. The number of years of life lost could be reduced in women living in Central and Eastern European countries if efficient national cervical cancer screening programs were in place. The increased burden of cancer incidence in Europe between 2004 and 2006, which is estimated to have risen by 300,000 to 3.2 million, demonstrates the impact of the aging of the European population and underlines the need for active and effective tobacco control measures and screening programs in Europe.

IX.20. New types of cancer treatments

As a widespread and pervasive health issue in Europe and the US, cancer has created a need for new technology that is going to define the exact, molecular basis for each type, and therefore bring in new types of treatment. Today, for example, cancer genetics is an exploding area in clinical research.

Currently, treatment for cancer depends on the type of cancer; the size, location, and stage of disease; and the person's general health. Drugs and biologics play an important role in the treatment of cancer. Attempts to decipher the human genome have launched an exciting new era in biomedical research with tremendous potential for cancer treatment. As a consequence, treatment will less depend on cancer type (organ location, histology) and be more driven by molecular features.

Table 9.9. New tumor-fighting targets[300]

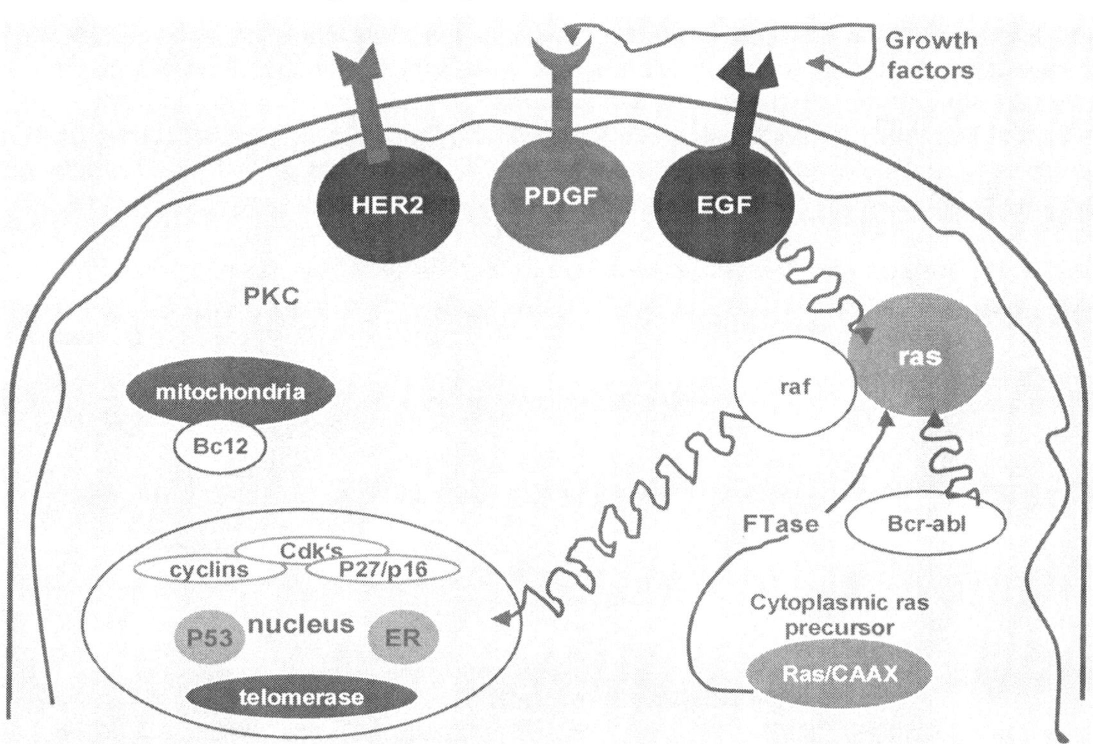

Key: PDGF: pletelet-derived epidermal growth factor; EGF: epidermal growth factor receptor; PKC: protein kinase C; Cdk's: cyclin dependent kinases; Er: estrogen receptor; Ftase. Farnesyltransferase.

---

[300] *Adapted from Science (vol.270, page 1970)*

Today, antibody based oncology drugs are linked to more than 140 different targets, divided into 35 classifications of molecular function.

Table 9.10. Antibody based oncology drugs, targets, molecular function:

→Carboxypeptidase activity →Catalytic activity →Cell adhesion molecule activity →Chaperone activity →Chemokine activity →Cofactor binding →Complement activity →Cytokine activity →DNA topoisomerase activity →Extracellular matrix structural constituent	→G-protein coupled receptor activity →Growth factor activity →Hormone activity → Hydrolase activity →Intracellular ligand-gated ion channel activity →Metallopeptidase activity →MHC class I receptor activity →Molecular function unknown →Oxidoreductase activity	→Peptide hormone →Protease inhibitor activity - Protein binding →Protein serine/threonine kinase activity →Receptor activity →Receptor binding →Receptor signaling complex scaffold activity →Receptor signaling protein tyrosine phosphatase activity	→RNA-directed DNA polymerase activity - Serine-type peptidase activity →T cell receptor activity →Translation regulator activity →Transmembrane receptor activity →Transmembrane receptor protein tyrosine kinase activity →Transporter activity →Unclassified

Early 2007, there are 354 antibody drugs in active development or more than 620 projects for cancer, of which 33 projects are in Phase III clinical trials, 154 projects in Phase II, 121 projects in Phase I, 312 projects in preclinical. In all targeting around 50 different cancer indications.

New drugs are now being designed to target specific molecular features characteristic of cancer cells, including genetic mutations, epigenetic factors causing changes in gene expression, structural changes in the proteins that are products of mutated genes, and derangements in signal transduction pathways. In essence, any specific difference in the molecular composition of tumor cells can become the basis for "targeted" therapy. In the future, the treatment for each patient's cancer will be individualized based on the unique repertoire of molecular targets expressed by their particular tumor[301]. Patient profile will also play an increasing role through identification of more precise prognostic factors.

Table 9.11. Impact of information technology on the practice of health care[302]

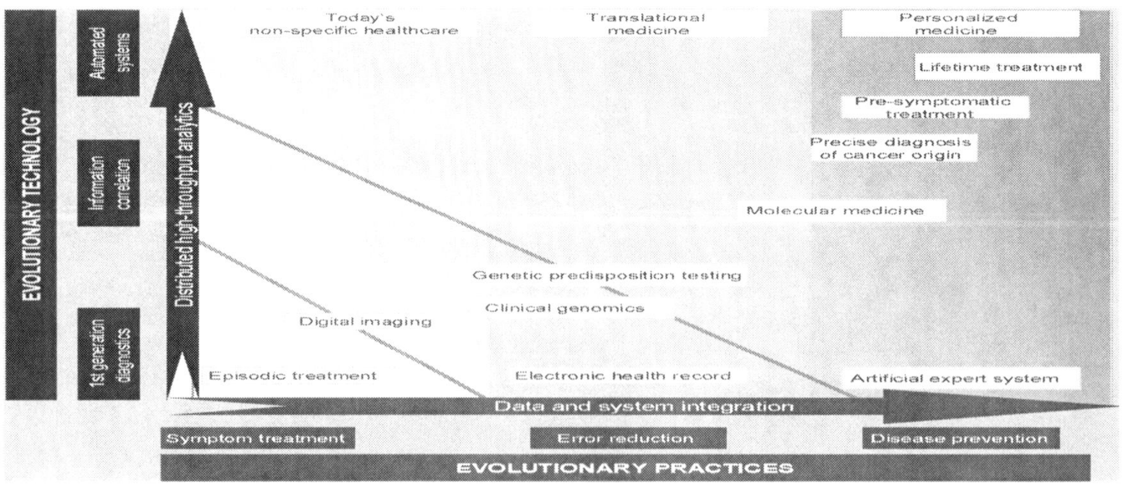

301 *Secretary Tommy G. Thompson, U.S. Department of Health and Human services (HHS). Report on medical innovation and seniors, July 11, 2002*
302 *Adapted from "E-Health. Informatik trends in der Medizin und im Gesundheitswesen". Powerpoint Presentation, chart 3, by Bernhald Tilg, UMIT, Innsbruck University Hospital, 2007*

IX.21. Scientific research efforts stimulate hope for new targeted medicines

Each day, scientists worldwide prove that they are coming closer to understanding the complex mechanism by which cancer develops. Each day, worldwide, their study results stimulate hope for finding new treatment possibilities. The following examples should illustrate this:

1. Hope for a more effective and less toxic cancer drug.
2. Opening a window on new therapeutic strategies.
3. A new way to control cell growth, potential targets for cancer treatments.

IX.21.1. Hope for a more effective and less toxic cancer drug

Detailed evaluation conducted at WEHI into a possible new cancer drug suggests that it may prove to be more effective and less toxic than current chemotherapeutic drugs[303].

Scientists from WEHI's cancer research divisions have been assessing the potential of a new compound, ABT-737, that was developed by US-based healthcare company, Abbott.

Under normal circumstances, human cells have a limited lifespan. They die when they are damaged, worn out or no longer needed by the body. When they die, these cells are replaced by new ones. The body depends on a normal and healthy process called programmed cell death - or apoptosis - to ensure that unwanted cells die on cue. If this process fails, then the damaged cells live on and multiply indefinitely and uncontrollably. This uncontrolled multiplication of rogue cells can lead to cancer.

Conventional chemotherapeutic drugs target and attempt to kill rapidly dividing cancer cells. This is sometimes successful in halting the disease, but these drugs inevitably damage many normal tissues. Hence, even when the chemotherapy works, the side effects for the patient can be very serious.

ABT-737 is a drug with a different strategy for attacking cancer. Rather than attempting to poison the rogue cells, the new drug attempts to reactivate the healthy and normal cell death program that failed to kill the unwanted cells on cue.[304] Normally, the cell death machinery is switched on when damaged cells need to be removed. The failure of the machinery to be turned on when it should can lead to cancer. ABT-737 is a 'switch flicker' that kicks the cell death machinery into action. Much more remains to be done to assess the drug's safety and effectiveness in patients, but early results from the laboratory are promising. Our hope is that the new drug will prove to be more effective while having fewer side effects..

IX.21.2. Opening a window on new therapeutic strategies

The dynamics of evolution are fully in play within the environment of a tumor, just as they are in forests and meadows, oceans and streams. This is the view of researchers in an emerging cross-disciplinary field that brings the thinking of ecologists and evolutionary biologists to bear on cancer biology. Insights from their work may have profound implications for understanding why current cancer therapies often fail and how radically new therapies might be devised[305].

---

[303] Brad Allen. Hope for a more effective and less toxic cancer drug. The Walter and Eliza Hall Institute of Medical Research (WEHI), Australia, November 20, 2006. The findings of the scientific assessment team are published in the 13 November 2006 issue of Cancer Cell

[304] Explanation given by WEHI's Dr. David Huang in a statement to the press

[305] Dr. Carlo C. Maley, assistant professor in the Molecular and Cellular Oncogenesis Program at Wistar. The review was made by researchers at The Wistar Institute of current research in this new field, appeared in the December 2006 issue of the journal Nature Reviews Cancer.

A tumor cell population is constantly evolving through natural selection.[306] The mutations that benefit the survival and reproduction of cells in a tumor are the things that drive it towards malignancy. Evolution is also driving therapeutic resistance. When you apply chemotherapy to a population of tumor cells, you're quite likely to have a resistant mutant somewhere in that population of billions or even trillions of cells. This is the central problem in oncology. The reason we haven't been able to cure cancer is that we're selecting for resistant tumor cells. When we spray a field with pesticide, we select for resistant pests. It's the same idea.

The scientist notes that there are three necessary and sufficient conditions for natural selection to occur and that all are met in a population of tumor cells. The first requirement is that there be variation in the population. This variation is evident in tumors, which are a mosaic of many different genetic mutants.
The second condition is that the variation must be heritable. This, too, can be seen within a tumor-cell population. When mutant tumor cells divide to replicate, the daughter cells share the same mutations.

The final condition is that the variation has to affect fitness, the survival and reproduction of the cells. All of the characteristics that are considered hallmarks of cancer affect fitness, according to Maley. Among these are that cancer cells no longer heed normal growth inhibition signals in their environment, they no longer require an external signal to divide as healthy cells do, and they are able to suppress a vital set of internal instructions that require cells to self-destruct when their genes are mutated beyond repair. This protective cell-suicide program carried by normal cells is known as apoptosis.

Seeing a tumor in this light opens a window on new therapeutic strategies. It's not just a metaphor to say tumor cell populations are evolving. Evolution is going on in the tumor. So let's think about how we might want to influence that evolution. Can we push it down paths that might be more beneficial to us?

One idea might be to develop new drugs that would act as benign cell boosters. Such drugs would specifically target the more benign cells in a tumor to increase their relative fitness over their malignant neighbors. This would allow the benign cells to outcompete the malignant cells, leading to a less aggressive, less dangerous tumor.

Another idea is to try to increase the fitness of chemosensitive cells so that they outcompete any resistant cells that are in the tumor, followed by a chemotherapy. So you sucker the tumor into a vulnerable state and then you hit it with your therapy.

In their review, the researchers also explored how the ecological ideas of competition, predation, parasitism, and mutualism unfold in tumors. Here again, they found that the concepts from another field helped to illuminate cancer biology. Mutant cells compete with each other for needed resources. The immune system often kills tumor cells like a predator hunting prey, and the tumor cells that develop defenses against the predation are the ones that survive and reproduce.

---

[306] *The Wistar Institute is an international leader in biomedical research, with special expertise in cancer research and vaccine development. Founded in 1892 as the first independent nonprofit biomedical research institute in the US, Wistar has long held the prestigious Cancer Center designation from the National Cancer Institute. Discoveries at Wistar have led to the creation of the rubella vaccine that eradicated the disease in the U.S., rabies vaccines used worldwide, and a new rotavirus vaccine approved in 2006. Wistar scientists have also identified many cancer genes and developed monoclonal antibodies and other important research tools. Today, Wistar is home to eminent melanoma researchers and pioneering scientists working on experimental vaccines against flu, HIV, and other diseases. The Institute works actively to transfer its inventions to the commercial sector to ensure that research advances move from the laboratory to the clinic as quickly as possible.*

318

An example of parasitism in the tumor environment can be seen in angiogenesis, in which a subset of tumor cells send chemical signals to stimulate the host to generate new blood vessels to supply the tumor with nutrients. The neighboring cells that aren't investing resources in producing the signals take advantage of the nutrients nonetheless. Mutualism describes a situation in which two organisms interact in a mutually beneficial way. Tumor cells send signals to stimulate the growth of the cells that form the scaffold in which the tumor cells grow, known as fibroblasts. The fibroblasts, in turn, send signals to the tumor cells to stimulate their growth. Recent studies suggest, too, that the fibroblasts in a tumor microenvironment begin to acquire mutations of their own[307].

IX.21.3. A new way to control cell growth, potential targets for cancer treatments

When cells go about the business of dividing, they can get sidelined. Maybe there aren't enough nutrients. Maybe there aren't the right signals to resume multiplying. Either way, cells go quiet.

What can restart cell division -- the process that drives the development of embryos, the renewal of hair, skin and blood, and the creation of cancer -- is a single transcription factor called GABP[308]

The work[309] introduces a new pathway that can be manipulated to control cell growth. Since cell growth is a fundamental biological process, the research may shed light on everything from miscarriages to muscular dystrophy. The main application, however, is cancer. Since a key characteristic of cancer cells is unchecked growth, the research identifies potential targets for new treatments.

This discovery not only adds to our basic understanding of cell division, it could lead to better cancer drugs. And they're needed. Cancer touches everyone[310].

During the cell cycle, the four-phase process of cell division, there is a period when the biochemical brakes are put on and cells become inactive. Then the process is kick-started and cells move into the so-called S phase, when DNA is duplicated. This is a critical juncture. If genes are missing or broken, these alterations are passed on to the new cell -- and could result in disability or in diseases such as cancer.

So biologists are keenly interested in identifying the accelerators that rev-up cell division. Ets transcription factors, a family of gene-regulating proteins that are major players in embryonic and cancer development, seemed obvious culprits. Rosmarin, a hematologist-oncologist, studies one member of the Ets family called GABP. This transcription factor helps make a variety of cells, including white blood cells. If those cells develop abnormally, leukemia results.

The investigators wanted to know the exact function of GABP in the cell cycle and created mice that carried a mutation -- tiny DNA sequences were inserted into their

---

[307] *The lead author on the Nature Reviews Cancer article is Lauren M.F. Merlo, Ph.D., a postdoctoral fellow in the Maley lab. The co-authors are John W. Pepper, Ph.D., at the University of Arizona, Tucson, and Brian J. Reid, M.D., Ph.D., at the Fred Hutchinson Cancer Research Center. The work was initiated by the Santa Fe Institute and supported by the National Institutes of Health, the Commonwealth Universal Research Enhancement Program of the Pennsylvania Department of Health, and the Pew Charitable Trusts.*

[308] *According to new research from The Warren Alpert Medical School of Brown University and Rhode Island Hospital.*

[309] *The work was published online in Nature Cell Biology, February 2007.Zhong-Fa Yang, an instructor in medicine at Brown and a postdoctoral research fellow at Rhode Island Hospital, was the lead author of the journal article. The National Heart, Lung and Blood Institute, the National Center for Research Resources and the Herbert W. Saint '49 Fund at Brown University funded the work.*

[310] *Dr. Alan Rosmarin, associate professor in the Department of Medicine and the Department of Molecular Biology, Cell Biology and Biochemistry at Brown and director of clinical oncology research for Lifespan, Rhode Island's largest health care system.*

GABP-making gene. These DNA bits would serve as a time bomb of sorts, deleting a critical piece of the gene when given a chemical signal. From these mice, the scientists grew fibroblasts -- common connective tissue cells -- in a Petri dish with nutrient-rich serum and watched them grow. When they detonated their time bomb, GABP was disrupted, and the fibroblasts' ability to divide was dramatically reduced. At the same time, other genes known to restart cell division were unchanged.

The team confirmed GABP's critical role in cell growth another way. Simply forcing dormant cells to make GABP, they found, was enough to rouse cells from their slumber and get them to grow again. The researchers stated that they found a new pathway to control cell growth and that, because a way is known to disrupt GABP and stop division, there is the possibility that a drug can be made to do the same thing in cancer cells.

IX.22.   Understanding tumor growth in cancer: an interdisciplinary approach

Cancer remains one of the most challenging diseases faced by mankind. Recent interdisciplinary research suggests that its pathophysiology is strongly influenced by the mind. What we are learning about this link may inform the development of biological and behavioral interventions to prevent and treat cancer in the future.

A recent review in the journal Nature[311] summarized molecular, cellular, and clinical studies that have elucidated many of the mechanisms underlying the links between biology and behavior in cancer. Evidence regarding links between psychosocial and behavioral factors and tumor growth include:

• Stress, depression, and lack of social support play a role in the growth and development of cancer. For example, the breakup of a marriage has been associated with a twofold increase in the risk of breast cancer, and long-term chronic depression appears to increase general cancer risks.

• Psychosocial factors have an impact on cellular and molecular processes that, in turn, contribute to the incidence and progression of cancer.

• Treatment of animals with drugs that block sympathetic nervous system (SNS) activity, a key component of the physiological response to stress, has been shown to inhibit the effects of behavioral stress on cancer.

Early results of this research indicate a complex matrix of psychological, social, and biological factors in cancer, ranging from social isolation to viral infection, which in turn affect known physiological processes that lead to specific types of cancers in animal subjects. Further research in this area may yield targeted interventions – for the mind, the body, or both – that use this knowledge to reduce the burden of cancer.

IX.23.   Which pathways to target?

The traditional way of approaching cancer biology is to identify genes in cancer cells altered by mutations or in their functions. A new method, developed by researchers at the U.S. National Cancer Institute, identifies additional genes that are not necessarily mutated or altered but that are nonetheless required for the cancer cell's survival. The researchers used the method, technically called a loss-of-function RNA interference genetic screen, to identify three genes not previously linked to cancer. These genes turn on a cellular process, or pathway, that is continuously activated in a type of lymphoma cell. Lymphoma is a cancer of the lymph nodes. The genes could become targets of therapies for a type of lymphoma called activated B cell-like diffuse large B-cell lymphoma (DLBCL). The genetic screen revealed a new mechanism in this lymphoma.

---

[311] Antoni, M.H. et al.,The influence of bio-behavioral factors on tumor biology: pathways and mechanisms. Nature (2006); 6, 240-248

More broadly, there is an opportunity to apply this genetic screen to all types of cancer in order *to create a new classification of the disease based not on cancer type, but on which pathways inside a cancer cell are critically required for its proliferation or survival.* The researchers state that this type of functional classification is critical because what is needed to know most about a cancer cell is which pathways should be targeted for any particular cancer.

The researchers plan to expand the screens to include cell cultures representing all types of human lymphomas and eventually all types of human cancers. The Achilles heel genetic screen is complementary to NCI-led efforts to sequence human cancer genomes because identifying critical pathways in cancer cells will help focus the search for relevant genetic mutations[312].

Table 9.12. Targeted signaling pathways that are being pursued in oncology drug development:

→Alpha6 Beta4 Integrin Signaling Pathway →Androgen Receptor Signaling Pathway →B Cell Receptor Signaling Pathway →EGFR1 Signaling Pathway	→IL-2 Signaling Pathway →IL-4 Signaling Pathway →IL-6 Signaling Pathway →Kit Receptor Signaling Pathway →Notch Signaling Pathway	→T Cell Receptor Signaling Pathway →TGF-beta Receptor Signaling Pathway →TNF-alpha Signaling Pathway →Wnt Signaling Pathway

IX.24. Altering lymphocytes, a new type of gene therapy

A team of researchers at the National Cancer Institute (NCI) has demonstrated sustained regression of advanced melanoma in a study[313] of patients by genetically engineering patients' own white blood cells to recognize and attack cancer cells.

These results represent the first time gene therapy has been used successfully to treat not only to melanoma, but also for a broad range of common cancers, such as breast and lung cancer[314].

Autologous lymphocytes -- a person's own white blood cells – have previously been used to treat metastatic melanoma. In a process called adoptive cell transfer, lymphocytes are first removed from patients with advanced melanoma. Next, the most aggressive tumor-killing cells are isolated, multiplied in the lab, and then reintroduced to patients who have been depleted of all remaining lymphocytes. While reasonably successful, this method can only be used for melanoma patients and only for those who already have a population of specialized lymphocytes that recognize tumors as abnormal cells.

Thus, the scientists[315] sought an effective way to convert normal lymphocytes in the lab into cancer-fighting cells. To do this, they drew a small sample of blood that contained normal lymphocytes from individual patients and infected the cells with a retrovirus in the laboratory. The retrovirus acts like a carrier pigeon to deliver genes that encode specific

---

[312] *NCI Media Communication, March 29, 2006// Ngo VN et al, A loss-of-function RNA interference screen for molecular targets in cancer. "Nature", online March 29, 2006.*

[313] *"Science", online edition, August 31, 2006.*

[314] *Dr. Elias A. Zerhouni, Director NIH in a statement to the press, "New method of gene therapy alters immune cells for treatment of advanced melanoma; technique may also apply to other common cancers.", August 2006.*

[315] *NCI researchers, led by Dr. Steven A. Rosenberg,*

proteins, called T cell receptors (TCRs), into cells. When the genes are turned on, TCRs are made and these receptor proteins decorate the outer surface of the lymphocytes. The TCRs act as homing devices in that they recognize and bind to certain molecules found on the surface of tumor cells. The TCRs then activate the lymphocytes to destroy the cancer cells.

In this study, newly engineered lymphocytes were infused into 17 patients with advanced metastatic melanoma. There were three groups of patients in this study. The first group consisted of three patients who showed no delay in the progression of their disease. As the study evolved, the researchers improved the treatment of lymphocytes in the lab so that the cells could be administered in their most active growth phase. In the remaining two groups, patients received the improved treatments. Two patients experienced cancer regression, had sustained high levels of genetically altered lymphocytes, and remained disease-free over one year. One month after receiving gene therapy, all patients in the last two groups still had 9 percent to 56 percent of their TCR-expressing lymphocytes. There were no toxic side effects attributed to the genetically modified cells in any patient.

Approaches to increase the function of the engineered TCRs – including the development of TCRs that can bind to tumor cells more tightly – and to further optimize delivery methods using retroviruses are under investigation. In addition, the researchers believe it may be beneficial to further modify lymphocytes by inserting molecules that assist in directing lymphocytes to cancerous tissues. Clinical trials are being conducted to enhance treatment effectiveness using total body radiation therapy to deplete a patient's supply of non-altered lymphocytes before replacing them with purely engineered cells. The researchers also have isolated TCRs that recognize common cancers other than melanoma.

IX.25. Blocking a group of genes of the HOX family: another novel therapeutic strategy

Cancer sometimes involves processes that should only occur in the embryo being mistakenly reactivated in adult cells. Researchers[316] have exploited this to develop a novel therapeutic strategy. This has led to the development of a new drug, HXR9 that blocks the activity of a group of genes known as the HOX family.

According to the project leader, Dr Richard Morgan, who started the work on HXR9, HOX genes are important in determining the identity of cells and tissues in the early embryo, but they are also expressed by cancer cells[317]. HXR9 blocks HOX activity thereby killing cancer cells in a highly specific manner'. HXR9 shows particular promise in treating malignant melanoma together with lung, prostate and kidney cancer.

Further work by the PGMS group has shown that embryonic genes expressed in cancer cells could also be important diagnostic markers. A number of embryonic proteins are secreted by prostate tumors and can be detected in blood serum, thereby providing a potential method for detecting prostate cancer and ultimately even assessing the extent of the disease without an invasive procedure.

---

[316] *at the Postgraduate Medical School at the University of Surrey.*

[317] *The results appeared in the June 2007 issue of Cancer Research. Work on HXR9 was started by Dr Morgan whilst he was a senior lecturer at St. George's, University of London, and the University of Surrey has agreed a licensing deal to allow Dr Morgan to continue with the work at the PGMS.*

322

HOX genes, which guide an organism's master development plan, may be turned on and off in the embryo by a special type of RNA, according to U.S. researchers.

If the team is right, the discovery could lead to new ways of understanding and treating a variety of diseases, including a type of childhood leukemia clinicians think begins with gene rearrangements in the womb. A third of the genome's RNA conveys the orders of DNA in the cell nucleus to the cell cytoplasm where the orders are carried out.

Researchers[318] have discovered that non-coding RNA (ncRNA), which comprises two-thirds of the available RNA in the genome, may control how the HOX gene functions in embryos.

The team discovered that HOX gene activity in fruit flies was regulated by the Trithorax region of their genome, and that ncRNA in that region regulated HOX expression by blocking the activity of gene coding elements, a process called transcription interference.

Since the area of the human genome that regulates the gene responsible for acute lymphocytic leukemia (ALL) corresponds to the Trithorax area in the fruit fly genome, the researchers think similar ncRNA activity may guide the gene rearrangements responsible for the disease.

IX.26 Studying HOX expression patterns...finding ncRNAs influence on gene expression patterns at distant locations.

One of today's hottest medical discoveries, is RNA interference, or RNAi. Scientists believe that RNAi evolved millions of years ago as nature's way of preventing harmful viruses from replicating inside cells. RNA, the 'chemical cousin' of DNA, functions like scratch paper inside cells, providing temporary sets of chemical instructions for making specific proteins. In RNAi, a tiny molecule attaches itself to a strand of RNA and tricks special enzymes in the cell into recognizing the RNA as a foreign invader. Those enzymes chop up the RNA, shredding the chemical instructions for making the protein. Today scientists are taking advantage of the RNAi phenomenon by designing tiny molecules that attach to an RNA of the scientist's choice, thereby impairing the cell's ability to make a specific protein. Researchers are using the technique to study the functions of genes and proteins and to develop new therapies[319].

Large, seemingly useless pieces of RNA - a molecule originally considered only a lowly messenger for DNA - play an important role in letting cells know where they are in the body and what they are supposed to become, researchers[320] have discovered. The finding implies that ancient RNA molecules can orchestrate gene activity across vast portions of the human genome - a cell's genetic blueprint. It also suggests they may be important in cancer development and stem cell maintenance. Overall, the work adds another brick to the growing wall of evidence suggesting that RNA is more than a mere genomic servant.

RNA is best known for ferrying protein-coding instructions from DNA, once thought to be the master molecule of the genome, to the cell's assembly factories. But cracks in this

---

[318] *at Thomas Jefferson University in Philadelphia, led by Alexander Mazo and Svetlana Petruk,*

[319] *Thomas Steitz, Sterling Professor of Molecular Biophysics and Biochemistry at Yale University at Duke University's 75th Anniversary Science Symposium, September 25-.26, 2006)*

[320] *at Stanford University School of Medicine, led by Dr John Rinn, postdoctoral scholar and Dr Howard Chang, assistant professor of dermatology.*

theory began to appear when it became evident that many RNA molecules aren't capable of making protein. While more recent research has shown that small bits of RNA can silence individual genes by interfering with their expression, longer pieces, called non-coding RNAs, have been more perplexing. These ncRNAs have long been molecules of mystery. They look just like they should code for proteins, but they don't[321].

Although ncRNAs have been shown to affect the expression of neighboring genes, the relative abundance of the molecules - accounting for about half of the DNA transcribed in the cell - suggests that they may have a wider sphere of influence than previously thought. The scientists have discovered that ncRNAs can influence gene expression patterns at distant locations in the cell. They were surprised to find that at least one of these molecules can suppress genes on a completely different chromosome. This opens up the whole genome to potential regulation by ncRNAs[322].

The researchers were investigating how human skin cells, or fibroblasts, know where they are in the body. They had previously shown in different types of cells that groups of genes known as HOX act as a sort of global positioning system by maintaining unique patterns of expression over many generations of cell division. But until the use of a new type of gene chip called a tiling array in the new study to home in on nearby regions of DNA, they didn't know how the HOX expression patterns themselves were determined.

The tiling array allowed the scientists to map the boundaries of the regions around four clustered sets, or loci, of HOX genes, known as HOXA through HOXD, to near-nucleotide resolution. That's somewhat like zooming in on a single home from a satellite map on Google Earth. Not only did they locate many previously unknown ncRNA genes nestled among the HOX genes, they also identified areas that serve as shared landing pads for proteins that either activate or suppress the neighboring regions.

The fact that the ncRNAs have remained virtually unchanged over millions of years suggests they may be playing non-traditional but vital roles in gene expression. Looking more deeply, the researchers found that depleting one ncRNA dubbed HOTAIR, in the HOXC region on chromosome 12 of a skin cell, significantly increased the expression of HOXD genes on chromosome 2. The finding marks the first time that ncRNA has been shown to affect gene expression on a chromosome other than its own.

The Stanford[323] researchers believe that HOTAIR functions by affecting chromosomal packing in the nucleus. Inactive chromosomal regions are tightly wound around proteins called histones and cannot be copied into RNA. Loss of HOTAIR in skin cells specifically frees the HOXD control region for binding by activating proteins.

Their next objective is to find out how these RNAs work structurally, and what upstream regulatory molecules might be controlling their expression. They have one clue: HOTAIR binds to and activates a group of enzymes called the Polycomb Repressive Complex 2 that modifies histones and helps them wind up the DNA.

---

[321] *Statement by Dr. John Rinn in a press release from Stanford University, June 28, 2007*

[322] *The research was published in the June 29, 2007 issue of the journal Cell. The research was supported by the National Institutes of Health, the Beckman Center Interdisciplinary Translational Research Program, the Damon Runyon Cancer Research Foundation and the Emerald Foundation.*

[323] *Stanford University Medical Center integrates research, medical education and patient care at its three institutions - Stanford University School of Medicine, Stanford Hospital & Clinics and Lucile Packard Children's Hospital at Stanford*

The researchers' interest is more than just theoretical. Polycomb proteins are improperly regulated in some types of cancers. HOX gene expression patterns are likely important to keep stem cells from improperly differentiating into skin, muscle, or other tissues. Understanding how ncRNAs affect these processes will have important implications for cancer therapies and stem cell research, they believe.

The medical research community is really interested in how ncRNA finds its putative target in the genome. There remains a whole level of biologicalcomplexity to be explored, including how HOTAIR knows where to go, how it talks to other factors and how it controls histones.

The work will also provide insight into the evolution of gene regulation. Because RNA is thought to have preceded DNA in the evolutionary timeline, it makes sense that it still plays a role in controlling DNA's function.

IX.27. The sudden run on RNA interference technology

With many multinational pharmaceutical companies scrabbling to fill their R&D pipelines, the latest spotlight seems to be focused on RNA interference technology, as both Switzerland-headquartered Roche and UK based AstraZeneca almost simultaneously announced major deals in this sector. Roche revealed that it has entered into a major alliance with US firm Alnylam Pharmaceuticals to obtain a nonexclusive license to its technology platform for developing RNAi therapeutics. The deal, which could be worth as much as $1.0 billion to Alnylam, is the biggest drug discovery collaboration in the sector to date. Alnylam's stock rocketed 50% to $23.22 on the news on July 9, 2007. Moreover, says Alnylam, it "highlights the increasing recognition of the revolutionary potential of RNAi to Big Pharma and the need to play what is clearly the hottest scientific space in biotechnology now." This technology, which won a Nobel Prize in 2006, is seen to have potential in the discovery of treatments for diseases such as cancer, blindness and AIDS[324].

SiRNA indeed is a 2006 Nobel prize winning technology that "silences" genes linked to the onset of disease. The value can be explained by a few examples of the research made by selected companies:

IX.27.1. AtuRNAi

Silence Therapeutics, formerly known as SR Pharma, is a leader in RNAi therapeutics. The Company has developed novel, chemically modified proprietary siRNA molecules ("AtuRNAi"), which have a number of advantages over conventional siRNA molecules including enhanced stability against nuclease degradation. In addition, the company has developed a proprietary delivery system ("AtuPLEX"), which increases bioavailability, circulation times and functional intracellular uptake of siRNA molecules.

SR Pharma announced it expects to begin the clinical development of its own AtuRNAi therapeutic molecules for systemic cancer indications in 2007. SR Pharma also has AtuRNAi compounds partnered with Quark Biotech and Pfizer, the lead program of which is due to commence Phase I trials for Age-related Macular Degeneration in the near future.

---

[324] *RNAi therapeutics in the Spotlight, as Roche invests up to $1B in Alnylam deal. Pharma Marketletter, July 16, 2007, p.3*

IX.27.2. ddRNAi technology

Benitec, an Australian biotechnology company develops therapeutics to treat serious diseases using its proprietary ddrnai technology. Its program is focused on infectious diseases and cancer. The company announced the commencement of its first HIV human clinical trial in collaboration with City of Hope.[325] The pilot study is designed to determine the safety and feasibility of RNA-based anti-HIV therapy with lentivirus-transduced hemapoetic progenitor cells (HPC) in patients undergoing hematopoetic stem cell transplantation (HCT) for intermediate and high grade AIDS lymphoma.

The lentivirus encodes 3 forms of anti-HIV RNA: RNAi in the form of a short hairpin RNA (shRNA) targeted to an exon in HIV-1 tat/rev (shl), a decoy for the HIV TAT-reactive element (TAR), and a ribozyme that targets the host cell CCR5 chemokine receptor (CCR5RZ). The vector, used to transduce autologous CD34-selected HPC, is called rHIV7-shl-TAR-CCR5RZ and was manufactured by the Center for Biomedicine and Genetics at City of Hope.

IX.27.3. An expressed interfering RNA (eiRNA) approach

The biotech company Nucleonics focuses on the development of novel expressed interfering RNA (eiRNA) based therapeutics. It has received clearance from the US Food and Drug Administration for its Investigational New Drug (IND) application to begin a Phase 1 human safety study of an investigational eiRNA therapy for the treatment of chronic hepatitis B (HBV) infection. This will be the first systemic delivery of an RNAi therapeutic to patients.

The company's approach, which simultaneously targets four different regions of the HBV genome, offers the opportunity to destroy all of the different RNA molecules produced by HBV within an infected cell through RNA interference. Therefore, if successful, it could offer a significantly more potent antiviral therapy than has ever before been achieved[326].

Nucleonics employs an expressed interfering RNA (eiRNA) approach whereby scientists insert plasmid DNA coding for relevant short hairpin RNAs (shRNA) into targeted cells, inducing the cells to produce and deliver specific shRNA sequences. Nucleonics' researchers have shown the ability of shRNA produced in this way to silence genes, including Hepatitis B, Hepatitis C, and HIV, in relevant cell lines for extended periods of time. Moreover, they have silenced multiple genes in adult mice without triggering an interferon response. Nucleonics' product pipeline includes eiRNA therapeutics directed against chronic Hepatitis B, Hepatitis C, pan-influenza (including H5N1 avian flu), as well as prostate and ovarian cancer.

IX.27.4. siRNA therapeutic candidate CALAA-01

Calando pharmaceutical's cyclodextrin-containing polymers form the foundation for its two-part siRNA delivery system. The first component is a linear, cyclodextrin-containing polycation that, when mixed with small interfering RNA (siRNA), binds to the anionic "backbone" of the siRNA. The polymer and siRNA self-assemble into nanoparticles of approximately 50-80 nm diameter that fully protect the siRNA from nuclease degradation in serum. The siRNA delivery system has been designed to allow for intravenous injection. Upon delivery to the target cell, the targeting ligand binds to membrane

---

[325] *City of Hope is a leading research and treatment center for cancer, diabetes and other life-threatening diseases. Founded in 1913, City of Hope is a pioneer in the fields of bone marrow transplantation and genetics and shares its scientific knowledge with medical centers locally and globally, helping patients battling serious diseases.*

[326] *Robert Towarnicki, CEO of Nucleonics Inc., in a statement to the press. Press release, May 2, 2007.*

receptors on the cell surface and the RNA-containing nanoparticle is taken into the cell by endocytosis. There, chemistry built into the polymer functions to unpackage the siRNA from the delivery vehicle. In addition to targeting tumors, the targeting of liver cells has also been accomplished *in vivo*.

The Pasadena based company presented positive results of preclinical efficacy testing in mice using multiple, systemic dosing with its lead siRNA therapeutic candidate CALAA-01, a nanoparticle containing non-chemically modified siRNA and a transferrin protein targeting agent formulated with the company's RONDEL (RNA/Oligonucleotide Nanoparticle Delivery) technology, in patients with unresectable or metastatic solid tumors.[327]

The formulation investigated contains Calando's proprietary delivery technology and utilizes RNA interference in cancer cells with an siRNA duplex targeting the M2 subunit of ribonucleotide reductase, a well-established cancer target. This duplex, developed at Calando, demonstrates potent anti-proliferative activity across multiple types of cancer types in vitro and in vivo.

Calando, together with City of Hope (COH) and the UCLA Cancer Center (UCLA), is planning a broad-based, open-label, dose-escalation Phase I clinical trial for its CALAA-01. The formulation to be investigated contains Calando's proprietary delivery technology and utilizes RNA interference in cancer cells with an siRNA duplex targeting the M2 subunit of ribonucleotide reductase, a well-established cancer target. This duplex, developed at Calando, demonstrates potent anti-proliferative activity across multiple types of cancer both in vitro and in vivo. The nanoparticle formulation has human transferrin protein (Tf) targeting agents decorating the surface. The transferrin receptor has long been known to be up-regulated in a broad spectrum of cancer cell types, and Tf binding to this receptor triggers the uptake of nanoparticles by cancer cells and release of the siRNA payload.

IX.27.5. Ensuring that the novel therapeutic modality will achieve its full potential

Sirna Therapeutics, a clinical-stage biotechnology company, was the first to take siRNA drugs into clinical trials. It had two drugs in clinical development –one for AMD and one for hepatitis C– when Merck & Co bought it, 2007, in a deal worth US$1.1bn (Euro 850m). Sirna Therapeutics has been at the forefront of efforts to create RNAi-based therapeutics, medicines which could significantly alter the treatment of disease. RNAi-based therapeutics selectively catalyze the destruction of the RNA transcribed from an individual gene. This enables an entirely novel approach to discovering drugs with the potential to produce highly specific, potent, and long-lasting effects.

The acquisition of Sirna complements the cutting-edge research on RNA expression that Merck has been doing since the 2001 acquisition of Rosetta Inpharmatics, Inc.

Sirna Therapeutics completed its Phase I clinical trial for Sirna-027, a chemically optimized, short interfering RNA (siRNA), in AMD in 2005 and with its strategic partner, Allergan, Inc., is moving Sirna-027 forward into Phase II clinical trials. The company has selected a clinical candidate for hepatitis C virus, Sirna-034. It has established an exclusive multi-year strategic alliance with GlaxoSmithKline for the development of siRNA compounds for the treatment of respiratory diseases. Sirna has a leading

---

[327] *At the American Association for Cancer Research (AACR) annual meeting on April 17, 2007 in Los Angeles*

intellectual property portfolio in RNAi covering over 250 mammalian gene and viral targets and over 200 issued or pending patents covering other major aspects of RNAi technology, including the microRNA technology.

Merck & Co sees in RNAi a powerful enabler of drug discovery in cells, in animals, and in humans. We can potentially use this technology to target the activity of genes which control the activity of cancer cells, and so produce their destruction without damaging normal cells[328].

IX.28. A promising therapeutic field to the pharmaceutical industry

Cancer is poorly served by traditional chemotherapies, is a major opportunity for pharma/biotech firms. This opportunity is magnified because the investment needed to develop a cancer drug is lower than other diseases: the field of oncology is a high priority for regulatory authorities who are willing to give it fast track status on the basis of smaller, and therefore cheaper clinical trials, because a few extra months of survival could be enough to win FDA approval. When compared to other therapeutic areas, this is still the case, although in oncology trials are becoming larger even for conditional approvals due to the need for large randomized trials.

Table 9.13.   New therapies: effective and predicted approval dates[329]

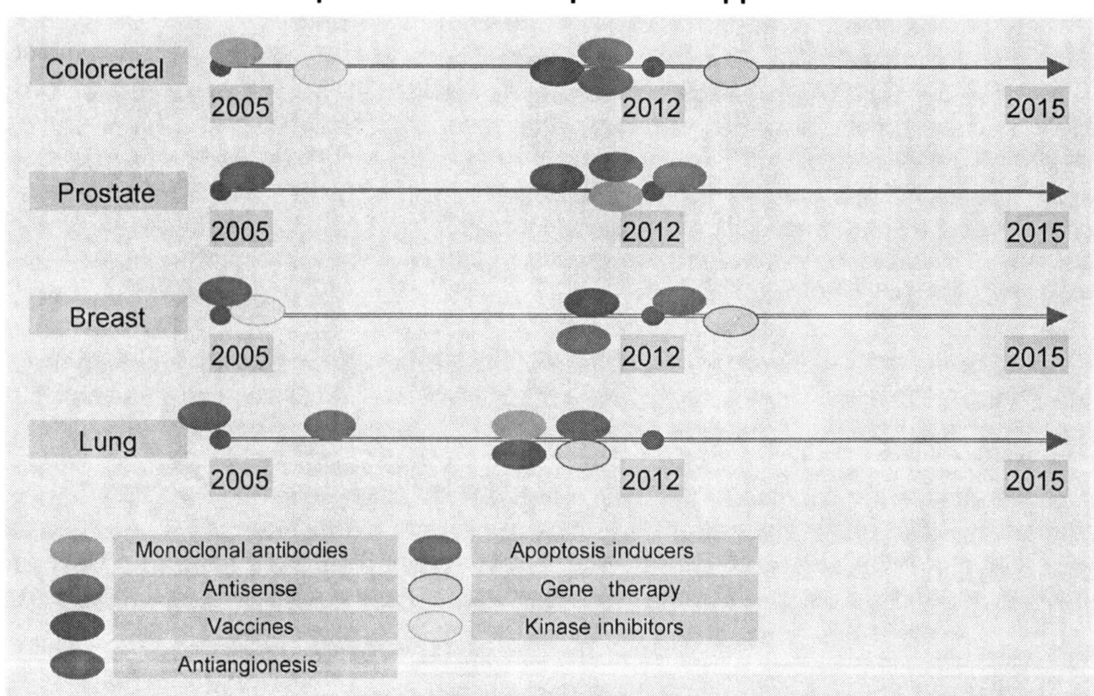

The field of oncology is also attractive because this therapeutic avenue has a price component that is boosted as the disease acquires chronic features, a factor of great influence in any company's R&D investment decision. Additionally, the cost of medical promotion is low when compared to primary care therapeutics that need an army of sales reps.

---

[328] Stephen H. Friend, M.D., Ph.D., executive vice president and franchise head, Oncology and Neuroscience, Merck Research Laboratories in a statement to the press, Merck & Co corporate press release, October 30, 2006.
[329] Source: adapted from McVie G.et al, Cancer 2025Chemotherapy, Expert Rev., Anticancer Ther. 4(3), Suppl.(2004):S44

The figures speak for themselves : 2000, Taxol and Leupron were the only two oncology products generating more than $1bn in sales. 2004, Rituxan reached the $2bn mark. Since their launch, several new oncology products have been put on the fast track to become "oncology-busters". Among them figure Taxotere, Gemzar, Gleevec, Zometa and Eloxatin.

IX.29. Gleevec continues to surprise

For example, Gleevec, also known as Glivec and Imatinib, was designed to treat chronic myelogenous leukemia (CML). In CML, ABL, an enzyme, goes into overdrive because of a chromosomal mix-up that occurs during blood cell development. The genes ABL and BCR become fused and produce a hybrid BCR-ABL enzyme that is always active. The overactive BCR-ABL, in turn, drives the excessive proliferation of white blood cells.

Gleevec revolutionized the treatment of chronic myelogenous leukemia: It has changed the disease from being a hard-to-treat disease into a condition that can usually be successfully managed. It is one of the first agents using this new approach that targets abnormal proteins fundamental to the cancer.[330]Unlike most current cancer therapies that kill both normal and cancer cells leading to unwanted side-effects, Gleevec and other drugs in this class are designed to zero in non specific cancer-causing molecules, eliminating cancer cells while avoiding serious damage to other, non-cancerous cells. Early studies of this drug have shown that in patients with chronic myelocytic leukemia, white blood-cell counts are restored to normal levels[331].

Although Gleevec (a compound that was discovered in 1980…and for which Phase I studies started only almost 20 years later, in 1998) is available in the U.S., other countries have restricted its use and/or reimbursement. For example, the preliminary review of Gleevec in the UK by officials at the government-sponsored National Institute for Clinical Excellence recommended that the drug only be used in patients who had already gone into the "accelerated phase" of their disease.[332]In the U.S., Gleevec is indicated for treatment of patients with CML in blast crisis, the accelerated phase or in chronic phase after failure of interferon-alpha therapy.

The drug, held up as an example of a 'targeted' cancer therapy, has shown to work by other mechanisms as well. The latest study[333], found that Gleevec activates a process inside cells called 'autophagy' - literally 'self-eating'.

Gleevec switched on autophagy in every type of cell the team studied, hinting that the drug would be effective against cancers other than CML.

Drugs are known to have possible downsides. Gleevec is no exception to the rule. Studies have linked the drug to heart failure in a small number of patients, and drug resistance continues to be a problem. Fortunately, the Kimmel Cancer Center at Thomas Jefferson University in Philadelphia have found that by reactivating a protein that is normally shut off in leukemia and in Gleevec-resistant cancer cells, leukemia development is halted.

---

[330] *Wall Street Journal May 16, 2003*

[331] *. Thompson, U. S. Department of Health and Human Services, Report on Medical Innovation and Seniors, (July 2002).*

[332] *Hawkes N. Patients are denied life-saving cancer drug. London Times. May 28, 2002*

[333] *led by Dr Hermann Shatzl's team at the University of Munich in Germany, as mentioned in: Research sheds light on Gleevec mechanism. Cancer Research UK , press release 06.03.07*

A drug that could turn on the gene that makes the protein C/EBP-alpha, a "transcription factor" (molecular switches that turn on genes when their function is needed) required for cells to differentiate, then, might control or even eliminate the cancer.

The fact that highly innovative drugs like Gleevec have been very successful in a limited number of patients confirms that there is no "one cure for all" in cancer treatments. But as yet researchers have not fully understood the molecular mechanism underlying these drugs' activity, they continue their search for information that might expand their usefulness to a broader patient population and address problems of resistance that can develop, thereby contributing to the discovery of ever more new aspects in e.g. the origins and working mechanisms of tumor cells. Improved diagnostic and predictive analysis will facilitate predictive targeting of treatments to patient populations most likely to respond.

One such recent finding is that targeted cancer therapy drugs like Gleevec (imatinib) and Tarceva (erlotinib), which destroy tumors by interfering with specific proteins or protein pathways, may disrupt the balance between critical cellular signals in a way that leads to cell death[334].

Known mechanisms of resistance to Gleevec include mutations in the protein sequence of the BCR-ABL tyrosine kinase, multidrug resistance gene overexpression, and the activation of alternate signaling pathways involving the SRC family kinases. FDA granted accelerated approval to Sprycel (dasatinib) for adults with chronic myeloid leukemia who are intolerant to Gleevec (imatinib) or become resistant to it.

The scenario of evolution of drug resistance has been most elegantly studied for two successful agents, methotrexate (the cytotoxic folate analogue) and imatinib (the highly effective inhibitor of the BCR–ABL tyrosine kinase). In some cases, targeted drugs and conventional antitumor agents can both be affected by a common resistance mechanism involving drug efflux (the MDR transporter)[335], or mutations in cell-death pathways. Because of resistance to single agents, combination therapy is essential for tumor eradication and cure. Even the most successful of the targeted molecules, imatinib, does not fully eradicate the malignant clone. Fortunately, clinical trials have demonstrated potent synergy between targeted molecules, particularly monoclonal antibodies such as rituximab (Rituxan)[336], bevacizumab[337] and trastuzumab (Herceptin)[338], and traditional chemotherapy[339].

IX.30. Disrupting the balance of cellular signals

Scientists[340] presented evidence for their theory, which runs counter to an alternative hypothesis called "oncogene addiction." Better understanding these drugs' mechanism

---

[334] *Sue McGreevey, University of Massachusetts General Hospital. Press release, November 17, 2006.*

[335] *Roninson, I. B. et al. Isolation of human mdr DNA sequences amplified in multidrug-resistant KB carcinomacells. Proc. Natl Acad. Sci. USA* **83**, *4538–4542 (1986).*

[336] *Coiffier, B. et al. CHOP chemotherapy plus rituximab compared with CHOP alone in elderly patients with diffuse large-B-cell lymphoma. N. Engl. J. Med. 346, 235–242 (2002).*

[337] *Hurwitz, H. et al. Bevacizumab plus irinotecan, fluorouracil, and leucovorin for metastatic colorectal cancer. N. Engl. J. Med.* **350**, *2335–2342 (2004).*

[338] *Slamon, D. J. et al. Use of chemotherapy plus a monoclonal antibody against HER2 for metastatic breast cancer that overexpresses HER2. N. Engl. J. Med.* **344**, *783–792 (2001).*

[339] *Bruce A. Chabner and Thomas G. Roberts Jr.,Chemotherapy and the war on cancer. Nature Reviews, Cancer, Volume 5, January 5, 2005. 65*

[340] *Massachusetts General Hospital, established in 1811, is the original and largest teaching hospital of Harvard Medical School. The MGH conducts the largest hospital-based research program in the United States, with an annual research budget of nearly $500 million and major research centers in AIDS, cardiovascular research, cancer, computational and integrative biology, cutaneous biology, human*

330

of operation could help surmount current limitations on their usefulness and lead to the discovery of additional protein targets[341].

"It looks like these drugs reduce the activity of their target proteins in such a way that cell-death signals remain high while survival signals drop. This model gives us clues that could lead to more successful treatment strategies and answer questions about the limited effectiveness these drugs have had[342].

It has become apparent that certain forms of cancer depend on mutations in specific genes, called oncogenes, for their development and survival. These include the EGFR gene in non-small-cell lung cancer and a gene called BCR-ABL in leukemia. Both of those genes code for proteins called kinases, which regulate the processing of key cellular signals.

The cancer-associated mutations overactivate the kinases in ways that lead to the uncontrolled growth of a tumor.

The "oncogene addiction" theory proposes that the internal circuitry of tumor cells becomes so reliant on the oncogenic protein or the pathway it controls that the cells die if kinase activity is suppressed. Since kinases controls two types of cellular signals - some leading to cellular survival, others to cell death - the MGH team proposed an alternative explanation: that survival signals drop quickly after kinase activity is suppress, releasing their control over persistent cell-death signals. To test this hypothesis, the scientists conducted several experiments using oncogene-expressing cell lines. In lines expressing tumor-associated versions of BCR-ABL, EGFR, or another kinase called Src, the survival-associated signals dropped quickly after kinase activity was suppressed, while cell-death signals were maintained.

Because the oncogenes had been artificially introduced into those cell lines, the researchers then tested their model in human lung cancer cells with the EGFR mutation. Again, kinase suppression, this time by application of Iressa, produced a rapid reduction in survival signals and eventual cell death as cell-death signals rose. A subsequent experiment with the Src cell line showed that cells pushed into a malignant form by expression of the mutant kinase could survive after Src activity was suppressed if a survival signal was supplied from another source, implying that the cells are not totally dependent on the oncogene's activity.

While all of these drugs have different targets, they appear to act in a similar way, causing a reduction in survival-promoting proteins while apoptotic [cell-death promoting] signals persist and drive the cells towards death. We suggest that the term 'oncogenic shock' may be a more accurate way to describe a process in which the very thing that kept the tumor alive - overexpression of a kinase - is turned against itself when the balance is disrupted to allow the cell-death signals to predominate.

---

genetics, medicalimaging, neurodegenerative disorders, regenerative medicine, transplantation biology and photomedicine. MGH and Brigham and Women's Hospital are founding members of Partners HealthCare System, a Boston-based integrated health care delivery system.

[341] The report's lead author is Dr. Sreenath Sharma, PhD, of the MGH Cancer Center, The study was supported by grants from the National Institutes of Health, the V Foundation and a Saltonstall Scholar Award.

[342] The researchers presented their opinion In the November 2006 issue of Cancer Cell; the researchers were led by Dr. Jeffrey Settleman, director of the Center for Molecular Therapeutics at the Massachusetts General Hospital Cancer Center

The new model also could explain why targeted drugs have not worked well in combination with standard chemotherapy drugs, which shut down the cell cycle and may actually halt the cell-death process, he adds. And if survival and apoptotic signals do drop and recover at different rates, giving these medications in a cyclic fashion, rather than continuously as currently prescribed, might better take advantage of the temporal windows of vulnerability and could possibly avoid drug resistance. Drugs that target the survival and cell death signals themselves may present another new strategy.

These findings explain why activated kinases are such good targets and support the importance of searching for more. More than 500 kinases have been identified, but we only have a half-dozen targeted kinase inhibitors. Finding new treatment targets and identifying the patients whose tumors have those kinases may bring us closer to the goal of truly personalized cancer treatment, the researcher concluded.

IX.31. Rapamycin: new findings
The latest generation of cancer chemotherapeutic drugs specifically targets mutant enzymes or "oncoproteins" that have run amok and now promote uncontrolled cell growth.

As promising as these drugs are, cancer cells with their backs against the wall have the tendency to fight back. A major goal of cancer research is to frustrate these acts of cellular desperation.

Investigators at the Salk Institute for Biological Studies uncovered one means cancer cells use to stay alive and in doing so suggest a strategy to overcome their recalcitrance. They showed that resistance to the chemotherapeutic drug rapamycin is mediated by the survival factor NF-kB.

Rapamycin, like Gleevec, is a so-called signal transduction inhibitor or STI, a small molecule that stifles inappropriate growth signals sent by mutant proteins in cancer cells. STIs may look like overnight successes, but they are actually the result of decades of hard work.

The Hunter lab previously showed that mouse cells lacking tumor suppressors known as TSC genes are more susceptible to the lethal effects of chemotherapeutic agents than are normal cells. Why cells from these TSC null mice were so poorly equipped to survive was not entirely clear. Researchers moved those mouse studies to the next level by tinkering with TSC activity in human cancer cells. They were able to extend their mouse model based on TSC null cells to different human cancer cell lines, where they knocked down TSC expression and showed that the same pattern held true.

Specifically, they found that human cells lacking TSC genes were vulnerable to chemotherapeutic attack because they couldn't activate a major line of defense mediated by the Nuclear Factor kappa B, known as NF-kB, which triggers both inflammatory and survival responses by inducing transcription of specific genes._ Not only did this explain why TSC null cells are vulnerable to insult, but it also provided biochemical evidence that there is crosstalk between two survival mechanisms. The findings showed for the first time that the TSC complex can regulate the NF-kB signaling cascade.

The experiments also explained a paradox: TSC null cells treated with rapamycin actually survived cellular insult better than untreated cells—a highly inauspicious outcome if the goal is to *kill* cancer cells. The scientists found that rapamycin did that by increasing NF-kB activity in the TSC null cells when they were exposed to chemotherapeutic drugs.

Rapamycin, an immunosuppressant used to block organ rejection after transplants, also inactivates proteins stimulating cell division and in clinical trials has been combined with other drugs to halt cancer cell growth.

Table 9.14.   Chronology of the development of protein kinase inhibitors

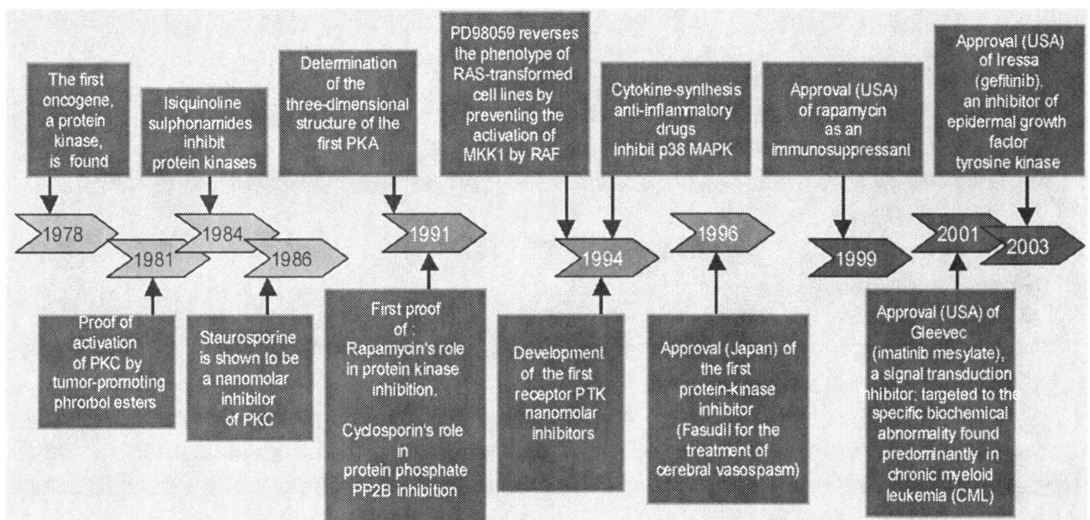

MAPK:Mitogen-activated protein kinase: CML: Chronic myelegenous lymphoma; PKC:Protein kinase C; PTK: Protein tyrosene kinase; PKA: Protein kinase A;MKK1: Mitogen-activated protein kinase 1.

But to cancer cells, rapamycin is both friend and foe. According to the researchers, rapamycin is not as successful as initially expected in treating cancer. Instead of killing cells, one ends up triggering a survival response in them. This study, however, suggests that taking NF-kB out of the game would make rapamycin less "friendly."

A major problem of chemotherapy is that sooner or later cancer cells develop resistance, which requires higher and higher doses of chemotherapeutics. Rapamycin-mediated killing of cancer cells could be increased by inhibiting the function of NF-kB proteins. These studies provide the basis for arriving at this very important conclusion, which has enormous bearing on cancer treatment. They suggest the potential use of NF-kB signaling inhibitors as adjuvants to maximize the effect of rapamycin-based therapeutics. The findings will have a significant impact on human health.

Researchers believe also that rapamycin may one day be beneficial as a potential therapeutic strategy to limit cell death caused by ischemia or reperfusion injury, and possibly long-term prevention of ventricular remodeling – the changes in size, shape and function that may occur to the left ventricle of the heart. Rapamycin blocks protein synthesis by inhibiting the mammalian target of rapamycin (mTOR), an essential component in the pathway of the cell cycle progression. The drug has been found to be important in transplant medicine and especially in kidney or heart transplantation. Because of the antibiotic's antigrowth properties, rapamycin effectively reduces coronary restonosis, the abnormal narrowing of a blood vessel. In coronary angioplasty, stents coated with rapamycin are implanted to reduce the risk of restonosis.

IX.32. Molecularly targeted therapies

Conventional anticancer drugs have tended to be non-selective, attacking both cancerous and healthy cells. Consequently, cancer chemotherapy is often accompanied by a variety of devastating short- or long-term side effects. Moreover, individual patient responses to conventional agents are highly variable, even in cases where specific cancers appear to be histologically identical. Molecularly targeted therapies based on recent progress in genomics and proteomics, however, hold out the promise of being far more selective, thereby drastically reducing the incidence of side effects in patients undergoing cancer treatment.[343] (Although some of these therapies have either specific toxicities :

> ~ Hypertension with bevacizumab (Avastin);
> ~ Skin rashes with cetuximab (Erbitux);
> > or can enhance cytotoxic toxicities :
> ~ Hematological toxicity with oblimersen (Genasense) when combined with
> > dacarbazine.

Efficacy can often be optimized when they are combined with cytotoxic chemotherapeutic agents.

IX.33.   Regulatory risk-benefit evaluation vs. industry's use of surrogate endpoints for clinical efficacy...

Progress in understanding the various aspects of the molecular biology of tumor cells and overall tumor pathology have led to the discovery of numerous candidate drugs aimed at playing a role in the various existing chemotherapeutic treatments. Regulatory bodies continue to see the risk-benefit evaluation as single therapy or in combination therapy based on the clinical benefit, which may be a variable criterion depending on the stage of the disease and the status of the patient, whereas industry sees the need of using novel surrogate endpoints for clinical efficacy based on pharmacological or pharmacodynamic features as well as population stratification or profiling[344].

Table 9.15.   Cancer market share of targeted therapy 2005-2015

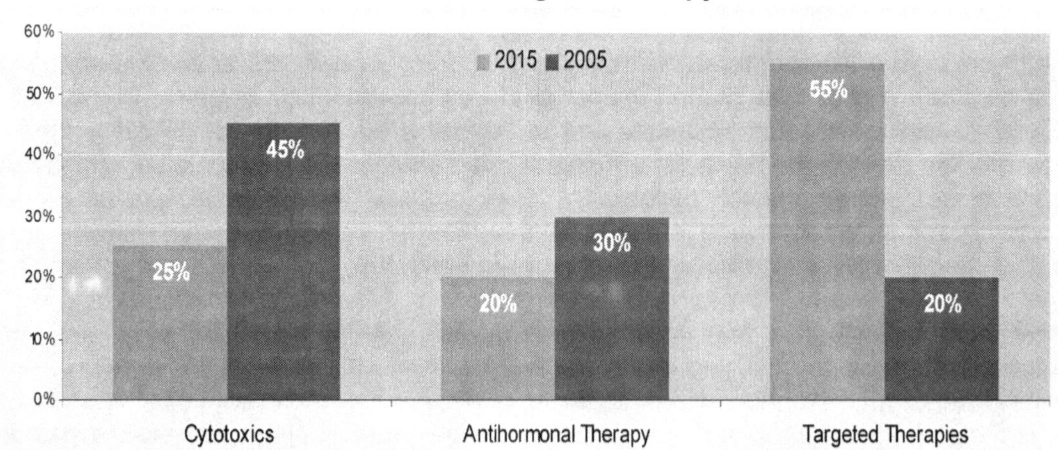

[343]Livingston DM, Shivdasani R. Toward mechanism-based cancer care. Journal of the American Medical Association . 285(5), 588-593, 2001.
[344] J.M. Alexandre, Hôpital Européen Georges Pompidou, France / S. André, Wyeth Research, at DIA ´s 16th Annual Meeting, Prague, 3/2004

The markets for cancer therapies have seen little growth in the period from 1980 to 2000, mainly because of the lack of significant progress in therapeutic innovation. In this market, where one cannot count with exponentially growing volumes, innovation, which justifies and guarantees high drug prices, came slowly with the launch of the taxan products (Taxol from BMS and Taxotere from Aventis, now Sanofi-Aventis), followed next by the appearance of the new molecules EPO, G-CSF for the treatment of anemia and neutropenia induced by chemotherapies. In less than a decade, total sales of this new class of hematopoietic factors of growth, equaled the total sales of chemotherapeutic treatments.

With 45% of the US$ 20 billion cancer therapy market in 2005, cytotoxics were by far the most used of the standard treatment regimens already in place. Patent expiries and the resultant generic copies competing for market share, will have a serious impact on the overall market significance of cytotoxics. By contrast, targeted therapies are expected to reach 55% of the top 20 cancer drug sales in 2015.

IX.34. Efficacy criteria

To be successful, the major points of differentiation for new cancer treatments remain to procure better tolerance, improving quality of life and, above all, improved survival time. Remarkable progress has been made in some pathology (breast cancer, prostate cancer), where five-year survival rates clearly demonstrate the benefits made by medical science. Furthermore, colorectal cancer is very likely the indication in which the most important progress has been made, both in adjuvant and in metastatic settings, during the last years.

Table 9.16.   Currently used chemotherapeutics

Mostly used molecules	Brandname	Main indication
Alkylant agents		
Busulfan	Generic	Chronic myeloid leukemia
Carboplatin	Paraplatin (Bristol-Meyers Squibb)	Solid-tumor cancers
Chlorambucil	Generic	Lymphoma, multiple myeloma, leukemia
Cisplatin	Generic	Solid tumor cancers
Cyclophosphamide	Generic	Lymphoma, multiple myeloma, leukemia
Dacarbazine	Generic	Melanoma
Ifosfamide	Generic	Cancer of the testicles
Oxaliplatin	Eloxatin (Sanofi-Aventis)	Colorectal cancer
Antimetabolites (alkaloids)		
Docetaxel	Taxotere (Aventis)	Hematological and solid-tumor cancer
Etoposide	Generic	Hematological and solid-tumor cancer
Irinotecan	Camptosar (Pfizer)	Colorectal cancer
Paclitaxel	Generic	Hematological and solid-tumor cancer
Topotecan	Hycamtin (GlaxoSmithKline)	Cancer of the ovaries and small cell lung
Vinblastine	Generic	Lymphoma, cancer of the testicles
Vincristine	Generic	Hematological and solid tumor cancer
Vinorelbine	Generic	Lung cancer
Antimetabilites (purin analogues)		
Cladribine	Leustatin (Ortho)	lymphoma, leukemia
Fludarabine	Fludara (Schering)	Lymphoma, leukemia
Mercaptopurine	Generic	Leukemia
Antimetabolites (pyrimidin analogues)		
Cytarabine (ARA-C)	Generic	Lymphoma, leukemia
Fluorouracil (5-FU)	Generic	Colorectal cancer and solid-tumor cancer
Gemcitabine	Gemzar (Eli Lilly)	Lung cancer, cancer of the pancreas
Antibiotics (anthracyclines)		
Bleomycin	Generic	Solid-tumor cancer
Daunorubicin	Generic	Leukemia
Doxorubicin	Generic	Hematological and solid-tumor cancer
Epirubicin	Ellence (Pfizer)	Leukemia
Mitomycin	Generic	Solid-tumor cancer

(non exhaustive list)

Table 9.17. **Currently used hormonal therapies**

Mostly used molecules	Brandname	Main indication
Androgen receptor antagonists		
Flutamide	Generic	Breast cancer
Tamoxifen	Generic	Prostate cancer
Aromatase inhibitors		
Aminoglutethimide	Generic	Breast cancer
Anastrazole	Arimidex (AstraZeneca)	Breast cancer
LH-RH analogue		
Goserelin	Zoladex (AstraZeneca)	Breast cancer
Leuprolide	Generic	Prostate cancer

(non exhaustive list)

On the premise that cancer derives from cell proliferation mechanisms, initial research focused on controlling the cell reproduction cycle. The first molecules to be developed, and still in use today, were cytotoxic molecules that follow two main principles .
-   the selective poisoning of cells in a cell growth cycle, eventually leading to the death of these cells (apoptosis)
-   the blocking of the cycle of dividing cells, also leading to apoptosis

Table 9.18. **Benefits of chemotherapy**[345]

High CR + High cure 5%	High CR + High cure 40%	Low CR + Low cure 55%
H D	A M L	N S C L C
Childhood ALL	Breast	Colon
Testis	Ovary	Stomach
Chrorio	SCLC	Prostate
Childhood BL	Sarcoma	Pancreas
	Myeloma	Glioma

(ALL: Acute lymphoblastic leukemia; AML: Acute myelogenous leukemia; BL: Burkitt's lymphoma; CR: Complete response; HD: Hodgkin's Disease; NSCLC: Non-small cell lung carcinoma; SCLC: Small cell lung carcinoma). Note: Prostate and colon cancer are moving towards high cure 40%.

For several decades, numerous classes of cytotoxics (chemotherapy) have been   the treatment of choice in solid-tumor type of cancers or in cancers of hematological origin. Hormonal therapy was the preferred option in breast and prostate cancer. The latest generation of cytotoxics offered cancer patients improved tolerance and, in case of for example, of CRC with oxaliplatin (Eloxatin)and irinotecam (Camto/Camptosar) the median survival moved from 12 to 20months+ in metastatic patients. Optimal therapeutical progress came from combinations (either concommittant or sequenced) of cytotoxics, administration protocols, and treatments for side effects.

The treatments of the 1990s combined several molecules to a new kind of therapeutic standards, and the practice continues. Academic research, setting the pace in oncology therapy, strongly contributes to refine the tolerance and efficacy profiles of combinations. The assessment of new molecules, by comparing them to existing ones, and done by influential scientists, helps the industry (e.g. to set up trials). It proves that the interplay between academic institutions (dominated by a more open-minded, co-operative spirit

---

[345] *Source: McVie G. et al; Cancer 2025: Chemotherapy, Expert Rev. Anticancer Ther. 4(3),Suppl. (2004):S44*

from the side of Anglo-American academia) and industry can be highly beneficial for the development of much needed cancer therapeutics.

Table 9.19.   Chronology of the various cytotoxic classes

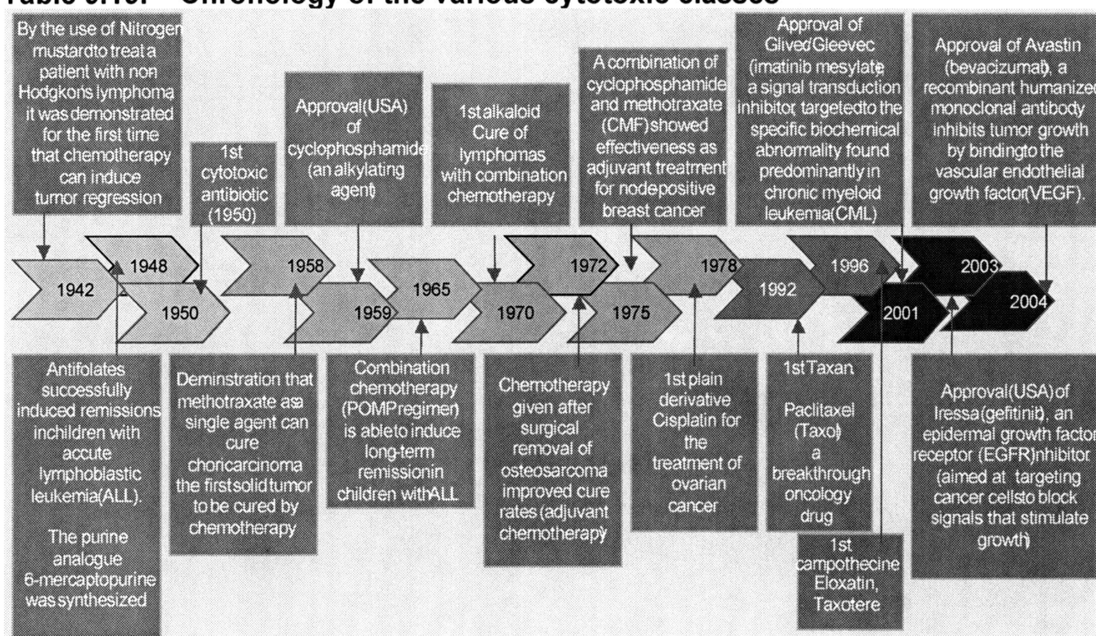

The survival rate for all cancers combined and for certain site-specific cancers has improved significantly since the 1970s. This is due, in part, to both earlier detection and advances in treatment. Survival rates markedly increased for cancers of the prostate, kidney, lymphoma, colon, ovary, breast, and liver. New diagnostic tools and treatments promise higher survival rates in both the short and long-term future.

Table 9.20.   Cancer survival rate 2004-2008

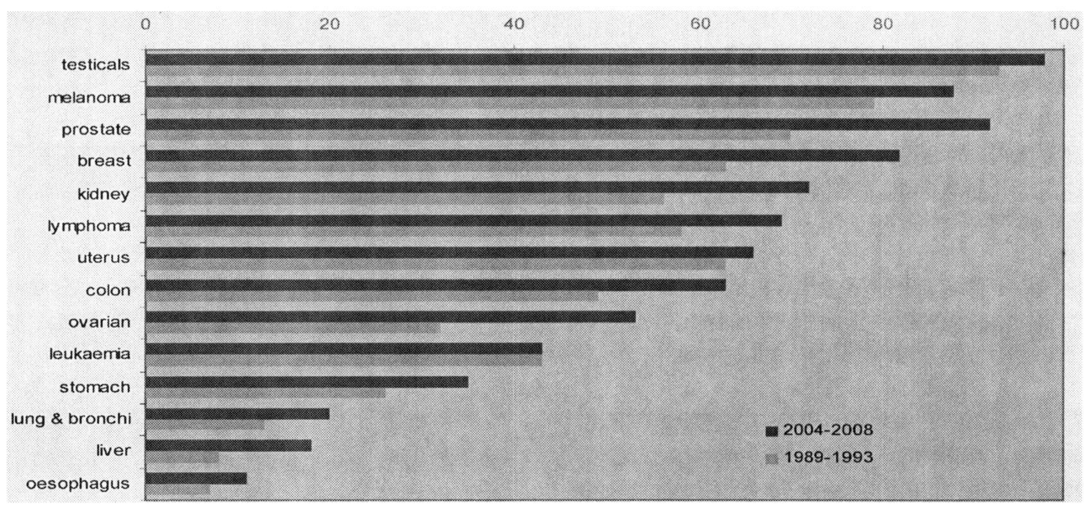

The entire oncology market including the supportive care classes across the seven major markets, was said to be worth $35 billion in 2005. Due to their long history in oncology and applicability for use across a range of tumor types, the cytotoxics accounted still for the majority of the 20 most used therapeutics. Experts expect that by

2015, the targeted therapies will constitute the majority of therapeutics as their inclusion into standard treatment continues and as line extension and horizontal expansion strategies are implemented.

Table 9.21.   Prediction of the cancer market 2010, 2015

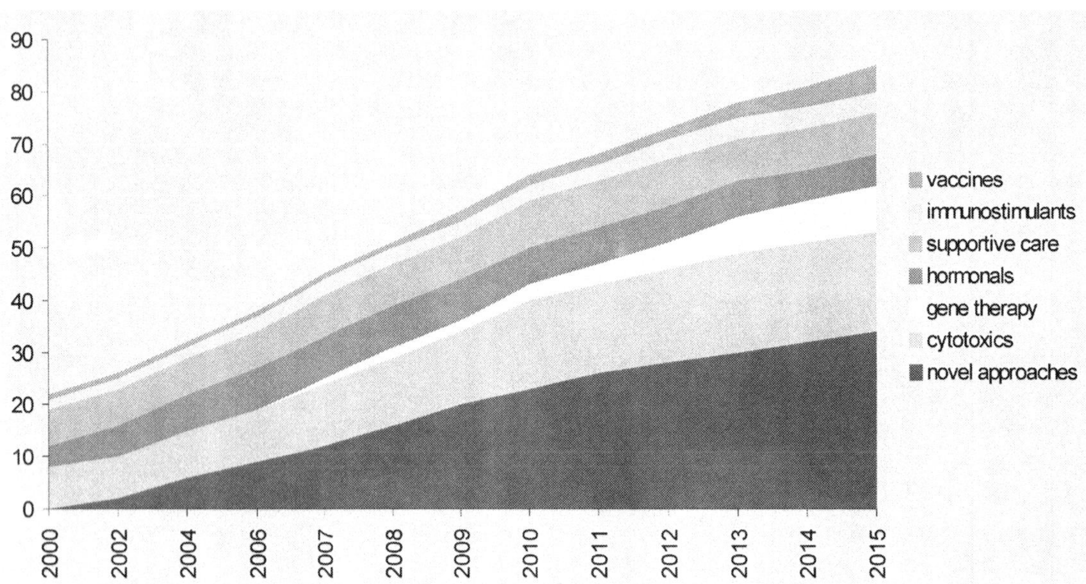

Annual growth will be driven by new technology -particularly biologically targeted therapies; earlier intervention; patient numbers (aging population - other diseases controlled). Worldwide sales in 2010 and in 2015 are predicted tot reach US$ 64bn, and US$ 85bn respectively[346]. We should keep in mind, that the figure represents sales in value, not volume.

IX.35.   Finding the mechanism by which cells resist chemotherapy.
A team of researchers from the Universidad Autonoma de Barcelona (UAB) Mutagenesis Group has identified one of the mechanisms used by cancer cells to resist chemotherapy [347].

They describe how proteins of the Fanconi/BRCA pathway recognize the presence of genetic mutations in order to repair them. The researchers also found that alteration of this mechanism makes tumor cells much more sensitive to certain drugs.

One of the main mechanisms responsible for repairing mutations in humans is the cancer-suppressing Fanconi anemia/BRCA pathway. This mechanism makes it possible for the cells to identify genetic mutations in order to correct them.

If this mechanism does not function correctly, it leads to Fanconi anemia, a rare genetic disorder characterized by progressive bone-marrow failure, various congenital malformations and a very high risk of cancer.

---

[346] *Adapted from: Future Drugs Expert Rev. Anticancer Ther. 4(3),Suppl.(2004):S29*

[347] *UAB Press release, March 10, 2007. Founded in 1968, the UAB has been rapidly recognized worldwide for its quality and innovation in research. It has become a breeding ground for quality researchers and a center for the dissemination of knowledge and technologies.*

338

Furthermore, the proteins of this pathway are largely responsible for the resistance of tumors to many antitumor agents such as cisplatin and other chemotherapeutic agents that kill tumor cells by producing DNA interstrand crosslinks. That is, they identify cellular alterations induced by chemotherapy and correct them, "accidentally" helping the tumor.

Many tumors have molecular anomalies in this pathway. These defects mean the tumors can be treated efficiently using certain antitumor agents. There are at least 13 genes involved in the pathway. Three of these (BRCA2, BRIP1 and PALB2) are responsible for the high proportion of hereditary breast cancers (between 5 and 10% of all breast cancers).

Understanding how this DNA repair pathway works is of great interest to biomedicine, not only for Fanconi anemia patients, but also for the general cancer population, since it determines the efficacy of chemotherapy in treating many tumors. However, the involvement of 13 genes in the same pathway makes the study more complex.

The researchers have found that the mutations block the DNA replication process, a process that is necessary, especially in tumor tissues, for the cells to be able to divide and proliferate. By blocking the replication process, the mutations activate a type of enzyme, the ATR kinase, which phosphorylates (introduces phosphate groups into) histone H2AX, a protein present in the chromatin that surrounds the damaged DNA. The phosphorylated histone H2AX indicates the location of the genetic damage to the Fanconi proteins and places them in exactly the right place to repair the DNA.

The researchers have shown that one of the 13 Fanconi proteins, the FANCD2, binds directly to the phosphorylated histone H2AX. The BRCA1 protein also plays a part in this process and, alongside the BRCA2, it is involved in most hereditary breast cancers. So these proteins cooperate in repairing the genetic damage, preserving the stability of the chromosomes and preventing the onset of tumors.

This research will have many implications on biomedicine. The increase in knowledge on this pathway will make it possible to design strategies for the chemosensitization of tumor cells- The team has also observed that many breast cancer cell lines are between two to three times more sensitive to chemotherapy if they have partially inhibited the Fanconi FANCD2 gene expression. The sensitivity or resistance to diverse range of chemotherapeutics is found to be linked to a common pathway.

Using a functional genomic screen, scientists have defined elements that impact the responsiveness of cancer cells to drugs commonly used as anticancer therapeutics. The research [348] identifies individual genes that are associated with resistance to chemotherapeutic drugs and sets the stage for future studies that may significantly enhance the ability to predict whether or not a particular tumor will respond to treatment.

Resistance to chemotherapeutic drugs is the primary cause of treatment failure in patients with metastatic cancer. The investigators[349] used RNA interference to directly examine the contribution of over 800 candidate proteins to the sensitivity or resistance of cancer cells to several drugs that are commonly used to treat cancer.

---

[348] *Haldar et al.: "A Conditional Mouse Model of Synovial Sarcoma: Insights into a Myogenic Origin." Publishing in Cancer Cell 11, 498-512, June 2007*
[349] *Cancer Research UK London, University College London, Baylor College of Medicine in Houston, Cambridge Research Institute. Their work was funded by Cancer Research UK.*

Using this technique, the researchers found that resistance to the chemotherapeutic agent paclitaxel, a member of the taxane family, as expected, involves genes that impair drug-induced mitotic arrest following knockdown. Silencing of these genes in many cases also induces polyploidy and multinucleation in the absence of drug treatment. The researchers conclude that specific disruption of the mitotic checkpoint promotes paclitaxel resistance and that chromosomal numerical heterogeneity may be a useful predictor of paclitaxel resistance in some cancers.

Ceramide metabolism was identified as a critical regulator of sensitivity to a wide range of chemotherapeutic drugs. Although ceramide has been associated with apoptosis for some time, the mechanisms have not been well understood. In this study, decreased expression of a ceramide transport protein, COL4A3BP, sensitized cancer cells to multiple cytotoxic agents. Further, expression of COL4A3BP was increased in drug-resistant tumor cells and in a small cohort of ovarian cancers following paclitaxel treatment.

The researchers suggest that paclitaxel-induced prolonged mitotic arrest may result in ceramide accumulation and initiation of apoptosis, while inhibition of this arrest, characterized by polyploidy, may suppress ceramide generation and promote cell survival.

These data suggest that the taxane class of drugs may lack efficacy in tumors with high levels of chromosomal instability characterized by chromosomal numerical heterogeneity and provide a rational basis for identification of patients likely to benefit from these drugs[350].

IX.36. From cytotoxic to cytostatic

While traditional chemotherapy agents kill cancer cells by being cytotoxic, the new agents work mostly by interrupting their growth *i.e.* are cytostatic. A few examples:

- Genentech and Partner Roche's Avastin (bevacizumab), a monoclonal antibody that targets vascular endothelial growth factor, a protein known to play a pivotal role in stimulating the growth of new blood vessels in tumors.
- Erbitux (cetuximab). ImClone Systems, US marketing partner Bristol-Myers Squibb, and European partner Merck KGaA, combined forces to market this IgG1 monoclonal antibody that targets the epidermal growth factor receptor and thereby inhibits the abnormal growth of cancer cells. ImClone and B-MS had their original application rejected by the FDA in 2001, prompting a collapse in the former's share price and followed by the well-documented insider trading scandal. However, the product was rehabilitated when Germany's Merck KGaA, which has licensed rights to Erbitux outside North America and has co-exclusive rights for Japan, released promising new data at ASCO in 2003. It received FDA approval for use in treatment of colorectal cancer in February 2004.
- Millennium and Johnson&Johnson's Velcade (bortezomib). Millennium Pharmaceuticals initiated a Phase II trial at the end of 2002 to evaluate the agent as a potential therapy for refractory metastatic colorectal cancer. The agent inhibits proteasomes, leading to disruption of the cell cycle and apoptosis of cancer cells, and won US approval in 2003 as a third-line treatment for patients with multiple myeloma.

---

[350] *Dr. Julian Downward from the Cancer Research UK London Research Institute and colleagues*

* AstraZeneca's Iressa (gefitinib), an EGFR blocker which was originally approved for the treatment of patients with advanced non-small cell lung cancer who have previously received treatment with chemotherapy.

Table 9.22.   Economic importance of cancer antibodies

Class of compound	Indication	Target Mechanism of action	Brand name (company)	Sales 2005 % vs 2004 US$mn
Rec chimeric mab	1st line treatment of diffuse, large B-cell CD20+ NHL plus chemother.	CD20	Rituxan; MabThera; rituximab (Roche/ Genentech Biogen IDEC, Chugai/Zenyaku	2.600 (+59) 3.176 (+22)
Rec mab	Metastatic cancer	HER2	Herceptin; (Roche/ Genentech)	2.146
see above	Metastatic colorectal cancer	VEGF	Avastin; bevacizumab	1.276 (+141)
Rec chimeric mab	Refractory metastatic colorectal cancer	EGF-R	Erbitux; ImClone, BMS Merck KGaA	675 (NN)

(figures for 2005)

IX.37. Fragmentation of the oncology therapeutic treatment segment

Advances in tumor understanding have initiated major R&D changes, with a resulting shift towards targeted therapies, which make up 45% of the new medicines pipeline. As specific tumor pathways are targeted, fragmentation of the oncology market will occur as patient differentiation becomes based on malignant growth drivers rather than primary tumor site. Future cancer treatment is predicted to employ a multidisciplinary approach, incorporating the forthcoming influx of targeted therapies into current standard regimens. The cytotoxics and anti-hormonals markets will become increasingly genericized, although cost of treatment will continue to increase due to the incorporation of targeted therapies. Breast, prostate, NSCLC, and colorectal cancer will remain the focus of R&D due to their high incidence, unmet needs and commercial potential. However, companies investing in niche markets such as pancreatic cancer expect to garner substantial benefits if an effective agent is developed, due to the high level of unmet needs.

IX.38. A huge unmet need for oncology drugs with prolonged survival rates

There is a huge unmet need for oncology drugs providing an advantage against standard treatments in terms of survival and that would therefore have access to large markets. Drug developments acting through new molecular mechanisms and that can be used for the major cancer indications, have nichebuster potential. Since 2004, unexpectedly positive results of approval-supporting trials have been released on Tarceva (Osi Pharmaceuticals/Genentech/Roche), Avastin (Genentech/Roche), Erbitux (ImClone/ and BMS/Merck KGaA). So far, there had been serious doubts about the efficacy of these drugs. As a result, the market capitalization of the relevant has more than doubled. In addition, companies were able to increase the profitability of development projects, as there are special regulations for the development of oncology drugs. This strengthened the medical research community's confidence in the future of targeted cancer therapies.

Unlike in the case of clinical trials on many other drugs, oncology drugs are usually tested on patients during their development, which means that an indication of their efficacy can already be derived during the first clinical phase. In addition, an analysis of the global product pipelines shows that the main emphasis in the development of already known and effective cytotoxic substances is on a new active-ingredient formulation, which aims to improve the drugs' effectiveness and tolerability. As most types of cancer are lifethreatening and insufficiently treatable, the FDA has introduced a number of regulatory special cases in order to speed up clinical development and the approval process. The FDA is awarding "fast-track" status to many drugs in clinical development (these must be drugs "intended to treat serious or life threatening conditions" that "demonstrate the potential to address an unmet medical need"), as a result of which companies are given the option among other things to file for approval on a rolling basis and not all at once, as is the norm. Once the approval documentation has been filed, the FDA decides whether it will be processed in six months as a "priority review" or in ten months as a "standard review." This is independent of an "accelerated approval," which allows approval on the basis of a surrogate end point, for example, if it is regarded as a suitable parameter for clinical benefit.

Many companies use these various options in order to be able to market their products as quickly as possible. Analysts at NatexisBleichroeder see an increasing focus on innovative drugs as a strength: although their development is more risky than that of drugs for the treatment of known targets, they often show a much better efficacy.

Opportunities are arising among other things as a result of demographic changes, as the incidence of cancer increases with age. In addition, the use of biomarkers – for patient stratification, for example, increases the success probability in clinical development, as illustrated by Herceptin.

The most recent successes with innovative drugs such as Avastin, Erbitux, Glivec, Iressa and Tarceva show impressively the use of molecular biological findings on the way cancer arises and on cancer growth to develop effective drugs, an area where biotech companies have established themselves as innovation drivers. In the future, the basic sciences will likely continue to form the scientific base for the development of new drugs, for example for tumor vaccines, which, it is hoped, would help strengthen the body's own immune system in its fight against cancer cells.

IX.39. Mathematics driven oncology research
Researchers of the Division of Mathematics at Dundee, have developed a mathematical model - similar in concept to weather forecasting but considerably more complex - which predicts how tumors grow and invade tissue. The results produced by the model have given startling insights into how cancerous tumors develop in the body. The findings show that the aggressiveness of cancer tumors may be determined by the tissue

environment in which they grow[351]. The combination of maths and laboratory research to develop such models has been hailed as a "new era in cancer research"

The scientists[352] envision a future when computer simulations like this will be used to predict a tumor's clinical progression and formulate treatment plans, in a fashion not dissimilar to how we forecast the weather now.

With their model for cancer invasion, the scientists would like to generate the kind of predictive power that we all know for the weather forecast.

According to the investigators [353], what their research shows is that the micro-environment in which the tumor grows acts like a Darwinian selective force upon how the tumor evolves.

Much of current biomedical research being carried out on cancer is done in isolation of the real environment in which the tumor naturally grows, but these results show that this environment could be the crucial determining factor in the tumor's development. The model also shows a clear relationship between the shape of a cancer tumor and how aggressive it is. Aggressive tumors tend to assume a spidery shape in the model, while more benign growths are generally more spherical in shape. This is important in terms of the surgical removal of tumors. A model like this could help predict how tumors will grow in different tissue environments, i.e. different areas of the body, and what the best strategy may be to treat them.

One interesting aspect of this is that if one makes the environment the tumor is growing in more harsh or barren, then the more likely it is that any surviving cancer cells will be the most aggressive and hardiest ones. This clearly has a potential to impact on how certain cancers are treated, since most of the current treatment strategies are focussed on making the tissue environment as harsh as possible for the tumor in the hope of destroying it. But as the investigator's research predicts this could allow the most aggressive cancer cells to dominate any residual tumor left after treatment and since these more aggressive cells tend to be the most invasive, this could result in an increased chance of metastasis. In the future this research could help tailor treatment in a patient specific manner, with the mathematical model being an additional weapon in the armory against cancer.

Mathematicians are bringing entirely new vistas to the field of cancer research, cancer is no longer an ugly beast to defeat, but rather it is a complex process that can be described rationally and conquered perhaps slowly, but surely. Understanding the individual components of cancer progression is still a necessary task, but it is no longer sufficient: It is how the components interact with each other that determines outcome. The study shows this clearly, the genetic changes in cancer cells interact dynamically with their immediate surroundings, with outcomes that remind us of the natural selection process in evolution.

The long-term goal of the scientists is that, with the tools of mathematical modeling and computer simulation, cancer treatment will no longer be a trial and error guessing game.

---

[351] *Tumor 'weather Forecast' Gives New Insights In Cancer. Dundee University press release, December 23, 2006.*

[352] *Professor Vito Quaranta, a leading American cancer biologist who is collaborating on the project., in a statement to the press, issued by Dundee University, December 23, 2006*

[353] *Dr Anderson's research is published in the scientific journal Cell, December 1st 2006, and is one of the few purely mathematical modeling papers to appear in the history of this prestigious biological journal.*

With mathematics-driven oncology research[354], they will be able to determine which drugs will work at which stage.

Professor Lisette de Pillis[355] is leading a team that is developing and testing new mathematically optimal approaches to controlling multiple simultaneous cancer treatment strategies, which include chemotherapy, immunotherapy and vaccine therapy. The scientists developed a series of specific mathematical models to address cancer cell division rates and other components. They seek to capture in their mathematical models the complex dynamics of the interactions among cancer cells, our immune system, and medical treatments.

The mathematical tools they are developing will help determine best treatment practices through simulated experiments at no risk to patients. According to the scientists, they can also allow them to customize treatments for individuals. The simulations, geometric visualization and treatment optimization tools we have created allow for virtual (computational) experiments to be run in a variety of cases. Harnessing the power of the body's own immune system is a promising approach to combating a growing cancer. However, precisely how cancer immunotherapies work, and how they should be administered optimally, either alone or in conjunction with chemotherapies, remains an open question of great interest and import to the medical community.

In this cross-disciplinary project, they are developing computational and mathematical tools capable of modeling the complex cascade of biological tumor-immune interactions, and of determining effective combination treatment strategies. Our tools have the potential to provide clinical guidance in the development of new treatment protocols through preliminary evaluations of simulated scenarios[356].

IX.40. Angiogenesis: a promising new approach to cancer treatment?

The growth of blood vessels, a process known as angiogenesis, is a normal process in the body that is essential for organ growth and repair. In many diseases, including most forms of cancer, this carefully regulated process becomes imbalanced, and normal blood vessel growth is redirected toward supplying nutrients and oxygen to feed diseased tissue, destroy normal tissues, and in the case of cancer, allow tumor cells to escape and travel to distant sites in the body. Researchers have tried to stop disease-related angiogenesis by identifying the molecules that stimulate blood vessel and developing new drugs to block their action. However, blocking angiogenesis requires a delicate balance between tumor and normal cells as most angiogenesis-related molecules are also critical for normal blood vessel growth in the body -- for example, during menstruation, pregnancy, or tissue repair. Thus, drugs that target critical angiogenesis molecules can cause a wide range of unintended side effects in healthy tissue.

---

[354] *Mathematical biology is increasingly influential in determining the behavior of cells in diseases such as cancer. The University of Dundee has played a pioneering role in the development of mathematical biology. Professor James D. Murray, widely acknowledged as the father of mathematical biology, did his PhD in Dundee and Professor Mark Chaplain developed the mathematical modeling of cancer tumors in the city.*

[355] *Professor de Pillis, a cancer researcher and mathematics professor at Harvey Mudd College in California, and her team presented their National Science Foundation-sponsored research project "Curing Cancer with Mathematics" at the 13th annual meeting of the Coalition for National Science Funding in Washington, D.C., June 26, 2007. The team is working to develop and test models of cancer growth and to implement mathematically optimal approaches to controlling multiple simultaneous cancer treatment strategies, which include chemotherapy, immunotherapy and vaccine therapy.*

[356] *Harvey Mudd College is a national (USA) leader in undergraduate education in engineering, science and mathematics, with a strong emphasis on humanities and the social sciences. The college's mission is to produce citizens sensitive to the impact of their work on society. Harvey Mudd College is a member of The Claremont Colleges consortium, which also includes Claremont Graduate University, Claremont McKenna College, Keck Graduate Institute of Applied Life Science, Pitzer College, Pomona College and Scripps College. "New research project: Curing cancer with mathematics", Academy Communications press release, Boxford, June 22, 2007*

A team of researchers[357] has uncovered a set of genes that are turned on, or expressed, at high levels only in the blood vessels that feed tumors in mice and humans. These genes, and the proteins they encode, are important new potential targets for novel drugs that could selectively cut off a tumor's blood supply without affecting the blood vessels of healthy tissues, overcoming one of the major concerns of current anticancer therapies targeted at blood vessel growth[358].

These results offer new insights into what is an important aspect of tumor development. How blood vessels grow, intertwined with normal tissue in a tumor's microenvironment, is not just an area of scientific interest; it's a research field that is continuing to develop potentially potent and specific anticancer agents that cut off the tumor from a vital support system.[359]

Angiogenesis is one of the most promising approaches to cancer treatment, the best example of which is Avastin, a therapeutic antibody against vascular endothelial growth factor (vEGF).

Angiogenesis inhibition can be addressed at different points of the cascade of reactions implicated in this mechanism. Three classes of molecules have been developed:
- inhibitors of the angiogenesis activation cascade
- inhibitors of endothelial growth
- inhibitors of matrix metalloproteinase (MMP)

Chemical entities that block signals to the vEGF receptor (vEGF-R) do so by blocking tyrosine kinase activity in the intracellular part of the vEGF receptor. By preventing the protein phosphorylation that results from vEGF/vEGF-R interaction, the signal to propagate endothelial cells is not sent.

Three vEGF receptors have been identified on the surface of vascular endothelial cells. ***It is not yet known which one triggers most the cell growth.***

If a drug is able to target only one receptor it may not be enough to block the transmission if the message is send via one or the other two vEGFRs. Monoclonal antibodies that block vEGF/vEGF-R interaction are directed against vEGF or its receptor. By doing so, this immune therapy seeks to block effector/receptor interaction and, by extension, cellular signalization. Avastin belongs to this category.

Some examples of angiogenis products include Angiostatin, Borrelidin, Endostatin, Fumagillin, Imiquimod, Thalidomide, Withaferin A, Bevacizumab:

Angiostatin, (human, recombinant) an inhibitor of tumor growth, metastasis, and angiogenesis, inhibits growth and induces apoptosis in tumor and endothelial cells. It is produced by the action of proteases on plasminogen, which is the inactive precursor of plasmin[360], and may function *via* its interaction with integrins[361].

---

[357] *At the National Cancer Institute (NCI), part of the National Institutes of Health (NIH). Researchers are from the Tumor Angiogenesis Section, Mouse Cancer Genetics Program, CCR, NCI, Frederick.*
[358] *Seaman S, Stevens J, Yang MY, Logsdon D, Graff-Cherry C, and St. Croix B. June 11, 2007. "Cancer Cell", Vol. 11, Issue 6.*
[359] *NCI Director Dr. John E. Niederhuber. National Institutes of Health press release, June 11, 2007.*
[360] *Morikawa W. et al. J. Biol. Chem. 2000 275 38912*
[361] *Wickstrom SA et al. Cancer Res. 2002 62 5580*

Borrelidin is a novel nitrile-containing macrolide antibiotic produced by Streptomyces species which inhibits threonyl-tRNA synthase (ThrRS)[362]. It displays pronounced antiangiogenic activity (IC_{50}=0.8 nM) and also induces the collapse of formed capillary tubes in a dose-dependent fashion..

Endostatin (human, recombinant), an inhibitor of tumor growth and angiogenesis, inhibits migration and induces apoptosis in tumor and endothelial cells[363]. It is produced by the action of proteases on the basement membrane component collagen XVIII, and may function *via* its interaction with integrins[364]

Fumagillin inhibits angiogenesis and endothelial cell proliferation. Exhibits antitumor activity in estrogen-induced pituitary adenoma. Inhibits methionine aminopeptidase type II (MetAP-2) by covalently modifying a conserved active-site histidine[365].

Imiquimod is a topical immune response modifier that inhibits angiogenesis. The mechanisms involved in its antiangiogenic activity include upregulation of antiangiogenic cytokines such as IL-18[366] and downregulation of MMP-9[367].

Thalidomide inhibits TNFα production by stimulated human monocytes and induces monocyte apoptosis[368]. It costimulates primary human T lymphocytes and inhibits angiogenesis. Thalidomide is finding applications in various immunologic diseases such as Crohn's disease and in angiogenesis-dependent diseases.

Withaferin A is a novel angiogenesis inhibitor found in the medicinal plant Withania somnifera. It is a potent inhibitor of endothelial cell (HUVEC) sprouting (IC_{50}=12 nM). It alters cytoskeletal architecture by covalently binding annexin II and stimulating its basal F-actin cross-linking activity which inhibits the migratory and invasive capability of endothelial cells[369]. Withaferin A is a novel neuronal regenerative agent. It induced significant regeneration of both axons and dendrites in an A β-induced axonal atrophy model[370]. Potently inhibits NFκB activation by preventing the TNF-induced activation of IκB kinase β (IKKβ)[371].

➔ Apoptosis in oncology: Cancer cells frequently and possibly invariably possess apoptotic defects. In this context, apoptotic pathways provide exciting molecular targets for new therapeutic agents to specifically promote apoptosis of cancer cells. There are today over 190 apoptopic drugs in active development or more than 466 clinical projects for the treatment of more than 40 different cancer indications.

IX.41. Bevacizumab (Avastin): first in class; first line indications

Avastin, a recombinant humanized monoclonal IgG1 antibody, was a first in class, yes, but after a study with the drug was negative and courageous management, who believed in the drug, proved its value. Avastin has received Fast Track status from the FDA for use in combination with Xeloda (capecitabine) for the treatment of metastatic

---

[362] Raun B. et al. J. Biol. Chem. 2005 280 571

[363] Yu Y.et al. Proc. Natl. Acad. Sci. USA 2004 101 8005

[364] Tarui T. et al. J. Biol. Chem. 2001 276 39562

[365] Lowther TW et al. Proc. Natl. Acad. Sci. USA 1998 95 12153

[366] SMajewshi S.et al. Int. J. Dermatol. 2005 44 14

[367] Hesling C. et al. Brit. J. Dermatol 2004 150 761

[368] Gockel HR et al. J. Immunol. 2004 172 5103

[369] Falsey RR et al. Nat. Chem. Biol. 2006 2 33

[370] Kuboyama T. et al. Br. J. Pharmacol. 2005 144 961

[371] Kaileh H. et al. J. Biol. Chem. 2006 Epub ahead of print, Dec. 6

breast cancer. The drug is the first product approved (as a first-line treatment for patients with metastatic colorectal cancer –cancer that has spread to other parts of the body) that works by preventing the formation of new blood vessels, a process known as angiogenesis. Avastin was shown to extend patient's lives by about five months when given intravenously as a combination treatment along with a standard chemotherapy drug for colon cancer (the "Saltz regimen" also known as IFL). IFL treatment includes irinotecan, 5-fluorouracil (5FU) and leucovorin. Avastin is a genetically engineered version of a mouse antibody that contains both human and mouse components. Special technology also allows it to be produced in large quantities in the laboratory.

Angiogenesis inhibitors such as Avastin have been studied, first in the laboratory and then in patients, for three decades with the hope they might prevent the growth of cancer. This is the first such product that has been proven to delay tumor growth and more importantly, significantly extend the lives of patients.

The approval of Avastin is the result of many years of research and development exploring a promising new approach to fighting cancer, and it is one of a number of recent new treatments for colorectal cancer that taken together, have significantly improved the armamentarium for fighting this disease. They are proof of the promise offered by biomedical innovation. The dedication of everyone involved in these efforts is making a real difference in the lives of cancer patients.

Avastin (bevacizumab), the first entrant in the anti-angiogenesis class, is by far the most advanced molecule and the one that has been most widely evaluated. Genentech focused the drug's development in 1st line indications.

The drug is now approved for use in combination with 5-fluorouracil chemotherapy treatment of first- or second-line metastatic colorectal cancer, as well as in combination with carboplatin and paclitaxel for first-line treatment of unresectable, locally advanced, recurrent or metastatic non-squamous non-small cell lung cancer. Due to its broad-acting anti-cancer mechanism, Genentech (Roche) has been conducting studies to broaden the number of indications that Avastin is approved for. Upcoming label expansion trial results include adjuvant colorectal cancer treatment and HER2-negative breast cancer.

Avastin has also been recently studied for efficacy at lower doses in the treatment of lung cancer. Data from the AVAIL study conducted by Roche show efficacy of Avastin at the 7.5mg/kg dose level, as compared to the current approved dose of 15 mg/kg. This may lead to physicians using the lower dose levels. TheAVAIL study evaluated the use of cisplatin and Gemcitabine in combination with Avastin atthe 7.5 mg/kg and 15 mg/kg dose levels versus cisplatin and Gemcitabine alone and included 1043 patients in Europe.

Today, there are more than 100 companies developing more than 150 vascular targeting and angiogenesis inhibitory drugs in more than 300 clinical trials targeting around 40 different cancer indications.

IX.42. VEGF: new medicines development
Inhibition of VEGF has become a popular target in treating cancer conditions and restricts formation of new blood vessels that are necessary for tumor growth and survival. The VEGF target represents a more novel mechanism than previous chemotherapies that targeted cell division. Numerous companies are developing VEGF-

based anticancer treatments. Aside from Avastin, the anti-VEGF products that are commercialized are primarily focused on wet AMD, and do not compete in the oncology space. To follow is a selection of anti-VEGF drug candidates that are currently in development for cancer indications.

Angiocept (CT-322) , from Adnexus Therapeutics, specifically targets VEGFR-2 and is currently being evaluated in Phase 1 studies in the U.S., with initiation of worldwide Phase 2 trials expected in late 2007.

Axitinib, from Pfizer, is an orally-active VEGFR Tyrosine Kinase Inhibitor that targets VEGFR-1, VEGFR-2, and VEGFR-3. Pfizer is evaluating the drug in Phase 3 studies for the treatment of thyroid and pancreatic cancers, as well as Phase 2 trials for various tumor types.

BIBF-1120, form Boehringer Ingelheim, is an orally active drug designed to target VEGFR, PDGFR, and fibroblast growth factor receptor (FGFR).

CDP-791, from UCB, targets VEGFR-2 by blocking the ligands VEGF-A, VEGF-C and VEGF-D, which bind to VEGFR-2. The drug is currently in Phase 2 trials for adjuvant treatment of NSCLC.

E7080, from Eisai, is an orally active agent in development by Eisai that targets VEGFR-2. E7080 is currently in Phase 1.

IMC-18F1, from ImClone Systems, targets VEGFR-1 and is in Phase 1 development for treatment of breast and colorectal cancers.

IMC-1121b, from ImClone Systems,   targets VEGFR- 2/KDR and is in Phase 1 development for treatment of solid and liquid tumors.

Motesanib diphosphate (AMG 706), from Amgen, is an orally-active agent that targets VEGFR-1, VEGFR-2, and VEGFR-3. It is also under evaluation for effects on platelet-derived growth factor (PDGFR) and c-kit signaling (stem cell factor receptor). The drug candidate is in Phase 2 trials as monotherapy for thyroid cancer and adjuvant therapy for NSCLC, breast cancer and colorectal cancer.

OSI-930, from Osi Pharmaceuticals, is an orally active agent that is in Phase 1 development. The drug targets c-kit and VEGFR-2. The company is currently conducting a Phase 1 dose escalation study.

Pazopanib, from GlaxosSmithKline, is an investigational, oral angiogenesis inhibitor targeting vascular endothelial growth factor receptor (VEGFR), platelet-derived growth factor receptor (PDGFR), and c-kit, important proteins in the angiogenic process[372][373] Pazopanib is currently in Phase III development for the treatment of advanced or metastatic RCC, after completing patient enrollment several months ahead of schedule, and is also being studied in a number of other trials across various tumor types, more precisely in Phase 2 for breast cancer in combination with Tykerb and Phase 1 for adjuvant NSCLC and colorectal cancer

---

[372] *Sonpavde, G. and Hutson, T. Pazopanib: A Novel Multitargeted Tyrosine Kinase Inhibitor. Curr Oncol Rep. 2007; Mar;9(2):115-9*

[373] *Jain RK. Antiangiogenic therapy for cancer: current and emerging concepts. Oncology. 2005 Apr;19(4 Suppl 3):7-16*

PTC-299, from PTC Therapeutics, is an orally active small molecule that inhibits VEGF production, as opposed to targeting VEGF receptors like Avastin. The molecule is in Phase 1 studies.

Recentin, from AstraZeneca, is in Phase 2/3 and is a VEGFR-TKI inhibitor. Recentin is under evaluation for treatment of NSCLC and colorectal cancer.

Vatalanib, a co-development from Novartis and Schering, also known as PTK/ZK, is a multi-VEGFR inhibitor and targets VEGFR-1, VEGFR-2 and VEGFR-3. PTK/ZK is currently being evaluated in numerous indications in Phase 1 and Phase 2 studies. Vatalanib is a new oral antiangiogenic molecule that inhibits all known vascular endothelial growth factor (VEGF) receptors. The agent is under investigation for the treatment of solid tumors.

Disposition and biotransformation of vatalanib were studied in an open-label, single-center study in patients with advanced cancer, by Loren Jost et al., who reported eight patients were given a single oral ^{14}C-radiolabeled dose of 1000 mg vatalanib administered at steady state obtained following 14 consecutive daily oral doses of 1000 mg non-radiolabeled vatalanib. Vatalanib was well tolerated. The majority of AEs corresponded to common toxicity criteria grade 1 or 2. A rapid onset of absorption was indicated. . Terminal elimination half-lives in plasma were 4.6 ± 1.1 h for vatalanib. The agent cleared mainly through oxidative metabolism. Vatalanib and its metabolites were excreted rapidly and mainly via the biliary-fecal route[374].

Large randomized phase III trials have demonstrated the efficacy of sunitinib and sorafenib in the treatment of patients affected by gastrointestinal tumors and renal cancer refractory to standard therapies, respectively. Positive results also have been reported with the combination of ZD6474 and chemotherapy in previously treated non-small cell lung cancer patients. For other agents, such as vatalanib, contrasting outcomes in metastatic colorectal cancer patients have been reported. However, several key questions remain to be addressed, regarding the choice of an adequate dose or schedule, the presence of "off-target" effects, the safety of long-term administration, and the research of new clinical end points or methodological approaches for the optimal clinical development of these agents[375].

VEGF-Trap, from Regeneron ans Sanofi-Aventis, is an alternative to antibody, in sequestering VEGF. The drug is in Phase 2 clinical trials.

Veglin, from VasGene Therapeutics, is a VEGF antisense oligonucleotide that binds VEGF-A, VEGF-B and VEGF-C mRNA. The mechanism of Veglin leads to mRNA degradation and thus inhibits signaling by VEGF-A, VEGF-B and VEGF-C. It is currently being evaluated in Phase 1 studies both as stand alone therapy and in combination with other anticancer drugs.

Zactima, from AstraZeneca, is a VEGF/EGF TKI inhibitor with RET kinase activity and currently in Phase 3 trials for NSCLC. An NDA for Zactima is expected in 2H08.

Angiogenesis inhibitors show promise, but evaluation for optimal efficacy has been a problem, given that the mechanisms of action of these agents differ from conventional

---

[374] *Lorenz J. et al., Metabolism and disposition of vatalanib (PTK787/ZK 222584) in cancer patients. First published on August 1, 2006; DOI: 10.1124/dmd.106.009944. For more information, contact: hans-peter.gschwind@novartis.com*

[375] *Morabito A. et al, Tyrosine kinase inhibitors of vascular endothelial growth factor receptors in clinical trials: current status and future directions. Oncologist. 2006 Jul-Aug;11(7):753-64.*

cytotoxic agents and surrogate markers for inhibition of angiogenesis are not available.[376]

IX.43. Sorafenib (Nexavar) targets two important processes that enable cancer growth

Nexavar is an oral multi-kinase inhibitor that targets both the tumor cell and tumor vasculature. In preclinical models, Nexavar targeted members of two classes of kinases known to be involved in both cell proliferation (growth) and angiogenesis (blood supply) – two important processes that enable cancer growth. These kinases included RAF kinase, VEGFR-1, VEGFR-2, VEGFR-3, PDGFR-B, KIT, and FLT-3. Preclinical models have also demonstrated that Raf/MEK/ERK has a role in HCC, therefore blocking signaling through Raf-1 may offer therapeutic benefits in HCC.

Nexavar is currently approved in nearly 50 countries, including the U.S. and the European Union, for the treatment of patients with advanced kidney cancer. In addition, Nexavar has been evaluated as a single agent in a Phase III clinical trial for the treatment of advanced hepatocellular carcinoma (HCC), or primary liver cancer, a study that was stopped early due to positive outcome[377].

In the trial, the sorafenib group lived on average 10.7 months compared with 7.9 months for the placebo group –a more than 40 percent improvement in survival time. Sorafenib has been described as the first convincing and rigorously tested treatment to show a survival benefit for these patients[378].

The next steps include testing sorafenib earlier in the disease and eventually in combination with other targeted therapies. Bevacizumab (Avastin), erlotinib (Tarceva), and (sutent) have all been tested for liver cancer in phase II studies[379]. In addition to company-sponsored trials, there are a number of Nexavar studies being sponsored by government agencies, cooperative groups, or individual investigators, as a single agent or combination treatment in a wide variety of cancers, including adjuvant RCC, advanced liver cancer, metastatic melanoma, non-small cell lung cancer and breast cancer.

The focus of efforts to improve on the efficacy of sorafenib[380] is on use with IFN, bevacizumab, or temsirolimus. Preliminary evidence with this approach is promising and will be the subject of the next generation of randomized trials in renal cell carcinoma.

The treatment of metastatic renal cell carcinoma is deeply evolving. After the cytokines, interleukin 2 and interferon era, which raised great expectations followed by deep disillusions (related to an efficacy limited to a small group of patients and an important toxicity for a majority of them), the time of targeted therapies has come. These new therapies mainly work on the VHL-HIF way, which controls several key molecules involved in the tumor angiogenesis phenomena, especially the VEGF[381]. Several anti-VEGF agents, including ligand-binding agents such as bevacizumab and the small

[376] Zakarija A., Soff G., Update on angiogenesis inhibitors.Curr Opin Oncol. 2005 Nov;17(6):578-83.

[377] Nexavar Trial Stopped Early Due to Positive Outcome. Press release. Bayer Pharmaceuticals Corporation and Onyx Pharmaceuticals Inc. February 12, 2007.

[378] Dr. Philip J. Johnson of the Institute for Cancer studies at Birmingham University, who discussed the results in a special session at ASCO, Society of Clinical Oncology annual meeting, June 1-5, 2007.

[379] NCI Cancer Bulletin, ASCO 2007 Annual Meeting Coverage,p.5

[380] Flaherty KT, Sorafenib in renal cell carcinoma. Clin Cancer Res. 2007 Jan 15;13(2 Pt 2):747s-752s.

[381] Escudier B., Metastatic renal cell carcinoma and its treatments. Ann Urol (Paris), 2006 Nov;40 Suppl 3:S82-5

molecule inhibitors of VEGF and related receptors such as sunitinib and sorafenib, have demonstrated clinical activity in patients with metastatic RCC [Other agents that inhibit alternative targets such as the mammalian target of rapamycin (mTOR) have also demonstrated activity][382].

Recent progress in our understanding of the genetic alterations that occur in the pathogenesis of melanoma provides exciting opportunities for therapy. The most important signaling pathways in melanoma lie downstream of NRAS: the RAS-BRAF-MAPK pathway. A great deal of attention has been focused on the high mutation rate in the BRAF oncogene, which approaches 60%, because BRAF itself is an appealing drug substrate and because of the central contribution of BRAF function to melanoma development that the mutation rate signifies. Agents that specifically target BRAF, such as sorafenib, as well as new molecules that function both upstream and downstream of BRAF, are being actively investigated[383].

Researchers have shed light on a new mechanism of action of the anti-cancer drug sorafenib, which could stimulate the development of novel regimens in which it is combined with other molecularly targeted agents for patients with blood cancers and solid tumors. In the new study[384], they, identified a mechanism by which sorafenib inhibits protein translation, and which may be involved in reducing expression of pro-survival factors, such as Mcl-1, and other proteins.

Previous findings showed that in human leukemia cells, sorafenib lethality was less a consequence of Raf inhibition, but rather reflected interference with the synthesis of Mcl-1. Research found that sorafenib interfered with Mcl-1 translation, a process in which proteins are synthesized from their constituent amino acids. However, the mechanism by which protein translation was inhibited by sorafenib remained largely unknown.

In the new work[385], the team found that in human leukemia cells, sorafenib induces a process known as endoplasmic reticulum (ER) stress, which results from accumulation of misfolded proteins in the ER. The ER is a subcellular structure which plays a key role in cellular protein disposition. When stressed in this way, the cell responds to the protein load by reducing protein synthesis, increasing levels of protein chaperones, and by accelerating protein degradation. However, according to the lead researcher, when ER stress exceeds a certain threshold, the ER stress response is converted from an adaptive to a pro-death response.

The team observed that exposure of cells to sorafenib resulted in the pronounced phosphorylation of a protein known as eIF2Ãi, a process that serves as a critical brake on protein translation in cells subjected to ER stress. Interestingly, they also found that sorafenib, by virtue of its ability to inhibit Raf, also prevented an increase in expression of a chaperone protein known as Grp78, which is classically induced in the ER stress response, and which helps to resolve stresses associated with increased protein loads.

[382] Garcia JA, Rini BI, Recent progress in the management of advanced renal cell carcinoma. CA Cancer J Clin. 2007 Mar-Apr;57(2):112-25.

[383] Haluska FG, Ibrahim N., Therapeutic targets in melanoma: map kinase pathway. Curr Oncol rep- 2006 sep;8(5):400-5

[384] led by Dr. Steven Grant, Virginia Commonwealth University Massey Cancer Center's associate director for translational research and co-leader of the cancer center's cancer cell biology program. Grant, who is a professor of medicine and the Shirley Carter and Sture Gordon Olsson Professor of oncology, collaborated with a team from the Departments of Medicine, Biochemistry, and Pharmacology.

[385] This work was supported by awards from the National Cancer Institute, the Leukemia and Lymphoma Society of America, the Department of Defense and the V Foundation. The findings were published online in the journal Molecular and Cellular Biology, June 2007. Dr. Mohamed Rahmani, from the VCU Department of Internal Medicine, was the lead author of the paper.

The net effect of these actions was to induce a shutdown of protein synthesis accompanied by a dramatic increase in cell death.

IX.44. Sunitinib: counting on new evidence to support the routine use of systemic treatment

Sunitinib maleate (Pfizer Inc.'s Sutent/SU11248) is a highly selective, multi-targeted tyrosine kinase inhibitor that starves tumors of blood and nutrients needed for growth and simultaneously kills cancer cells that make up tumors. more than doubled survival and significantly reduced tumor growth and spread in a Phase III study in patients with Gleevec-resistant gastrointestinal stromal tumors (GIST). Encouraging Phase II results also were observed in other tumor types, including metastatic renal cell carcinoma (mRCC), metastatic breast cancer and neuroendocrine tumors. These data support Sutent's activity against hard-to-treat tumors, particularly GIST and mRCC, both potentially life-threatening and with few, if any, other options[386].

Advances in the understanding of the biology of renal cell cancer have identified tumor angiogenesis as a target for drug therapy. New therapies have therefore emerged aimed at vascular endothelial growth factor and other growth factors mediating angiogenesis. These include bevacizumab, an antibody against vascular endothelial growth factor, and the oral drugs sunitinib and sorafenib. Until recently there was little evidence to support the routine use of systemic treatment. Chemotherapy has response rates (defined as a reduction of more than 50% in tumor size) of under 8% and is therefore of little value. This is because of the multidrug resistance protein found in proximal tubule cells, from which clear cell and papillary renal cell carcinoma are thought to originate.[387]

➔➔Today, there are more than 150 companies developing more than 260 protein kinase inhibitory drugs in more than 450 clinical trials targeting around 45 different cancer indications. The number of protein kinase targets has increased to over 50 different molecular targets in just a few years. Examples of targets are:

9.22. Protein kinase inhibitor drugs: clinical trials & cancer indications

➔Aurora kinase A&B; ➔CD22 antigen; ➔Cell division cycle 2, G1 to S and G2 to M; ➔CHK1 & CHK2 checkpoint homologue (S pombe); ➔Cyclin-dependent kinase 1A, 2, 4, 6, 7,9;	➔Epidermal growth factor receptor; ➔FK506 binding protein 12-rapamycin associated protein 1; ➔fms-related tyrosine kinase 1,3 & 4; ➔met proto-oncogene;	➔Mitogen-activated protein kinase 14; ➔Phosphoinositide-3-kinase, catalytic, alpha & gamma polypeptide; ➔Protein kinase C, alpha 6 beta 1;	➔TEK tyrosine kinase, endothelial; ➔Topoisomerase (DNA) II alpha 170kDa; ➔v-raf-1 Murine leukemia viral oncogene homologue 1; ➔v-src Sarcoma viral oncogene homologue

IX.45. mTOR inhibitors Temsorilus, AP23573

Cytokine-based immunotherapy was the only viable option in metastatic, nonresectable renal cell carcinoma (RCC) for many years. Systemic immunotherapy has become increasingly established as a standard therapy during the last 15 years. In this context, interleukin-2 (IL-2) and interferon-alpha (IFN-alpha) turned out to be the most effective single agents in RCC. Subsequently, the approved subcutaneous application of these compounds was the preferred administration route.

---

[386] Dr. John LaMattina, Pfizer president of worldwide research and development, in a statement to the press, "New Data Show Pfizer's Sutent/SU11248 Extends Overall Survival in Gleevec-resistant GIST", May 17, 2005
[387] Pavlakis N., Drug treatment of renal cancer. Aust Prescr 2006;29:151–3

Response rates with cytokine combination therapy were almost similar to those of more aggressive concepts using additional chemotherapeutic agents[388].

The year 2006 marked a turning point in the daily management of patients with metastatic renal cell carcinoma. The growing understanding of molecular mechanisms involved in the pathogenesis of the disease, especially clear-cell carcinoma, has led to the development of multiple targeted therapies with significant clinical benefits[389].

The discovery of a relationship for the VHL tumor suppressor gene, hypoxia inducible factor-1 alpha, and vascular endothelial growth factor in the growth of clear-cell renal cell carcinoma (RCC) has identified a pathway for novel targeted therapy. Several agents, including the small-molecule targeted inhibitors sunitinib, temsirolimus, sorafenib, and the monoclonal antibody bevacizumab, have demonstrated antitumor activity in randomized trials. Superior activity was found with sunitinib and temsirolimus versus cytokines in first-line therapy. [390]

IX.45.1 Temsorilus (Torisel): superior therapy vs. Cytokines in first-line therapy of RCC

Temsorilus is an intravenous analog of rapamycin, presents immunosuppressive properties and also antiproliferative activity. Its principal target is the mTOR serine/threonin kinase which controls the initiation of the transcription of many ARNm implicated in carcinogenesis. Blocking mTOR interferes with cell growth, division, metabolism, and angiogenesis.

Breast cancers, glioblastoma and renal cell carcinoma were particularly studied with response rates from 10 to 20 %. In hematology, mantle-cell lymphoma is of particular interest because of constitutional activation of cyclin D1 (response rate of 40 %)[391].

Torisel is the first mTOR inhibitor to be filed for approval for the treatment of a cancer. It is an investigational drug that specifically inhibits the mTOR kinase, a protein that regulates cell proliferation, cell growth and cell survival. The registration dossier contains interim data from a three-arm, Phase 3 clinical trial of 626 patients who had received no prior systemic therapy. The results showed that Torisel significantly increased overall survival by 49 percent in patients with advanced RCC compared with interferon-alpha, a treatment for advanced RCC. The U.S. Food and Drug Administration (FDA) has accepted the file and granted priority review status to Wyeth Pharmaceutical's New Drug Application (NDA) for this investigational drug[392].

Renal cell carcinoma continues to be an extremely difficult-to-treat form of cancer and there continues to be a need for new therapies. Immunotherapy confers a small but significant overall survival advantage in metastatic renal cell carcinoma (RCC) but a need exists to develop more effective systemic therapies[393]. Scientists consider sunitinib, sorafenib and temsirolimus as promising new drugs.

---

[388] *Autenrieth M. et al., Systemic therapy of metastatic renal cell carcinoma. Urologe A. 2006 May;45(5):594-9.*

[389] *Pouessel D, Culine S., Targeted therapies in metastatic renal cell carcinoma: the light at the end of the tunnel. Expert rev Anticancer Ther., 2006 Dec;6(12):1761-7.*

[390] *Motzer RJ, Bukowski RM, Targeted therapy for metastatic renal cell carcinoma. J Clin Oncol., 2006 Dec 10;24(35):5601-8.*

[391] *Mounier N. et al, Update on clinical activity of CCI779 (temsirolimus), mTOR inhibitor. Bull Cancer, 2006 Nov 1;93(11):1139-43.*

[392] *Wyeth Pharmaceutical, press release of December 19, 2006. Note: A priority designation can be given to an NDA for a drug that, if approved, would be a significant improvement compared with existing treatments.*

[393] *Larkin JM, Eisen T., Kinase inhibitors in the treatment of renal cell carcinoma. Crit Rev Oncol Hematol 2006 Dec;60(3):216-26. Epub 2006 July 24.*

IX.45.2. AP23573

AP23573, from Ariad, is a novel small-molecule inhibitor of the protein mTOR, a "master switch" in cancer cells. It is currently in phase 1 and 2 clinical trials in patients with solid tumors and hematologic cancers. AP23573 has been designated both as a fast-track product and an orphan drug by the US FDA and as an orphan drug by the EMEA for the treatment of soft-tissue and bone sarcomas. Despite advances in cancer therapy, prolonged cancer remission remains difficult to achieve in many types of solid tumors. Additional treatment options are needed for patients whose cancer is progressing and unresponsive to currently available therapies. Novel molecularly targeted agents -such as AP23573- combined with traditional cytotoxic agents –such as paclitaxel and capecitabine- are being studied in new regimens to identify more effective treatments for patients with advanced cancers. ARIAD and Merck are codeveloping AP23573 for oncology indications globally. Additionally, Ariad is working with Medinol Ltd to develop stents and other medical devices that deliver AP23573 to prevent reblockage at sites of vascular injury following stent-assisted angioplasty.

IX.46. Selective MMP inhibitor BMS 275291: early enthusiasm

BMS275291 is a matrix metalloproteinase inhibitor (MMPI). It targets one of the steps involved in angiogenesis.

Several classes of agents now exist that target the different steps involved in angiogenesis. Among them are drugs inhibiting matrix breakdown such as marimastat, prinomastat, BMS275291 and BAY12-9566. Despite early enthusiasm for many of these agents, clinical trials have not yet demonstrated significant increases in overall survival and toxicity remains an issue[394].

IX.47. Integrin inhibitors Vitaxin and Cilengitide

IX.47.1. Vitaxin, also known as MEDI-522, is a humanized monoclonal IgG1 antibody that specifically binds a conformational epitope formed by both the integrin alpha V and beta 3 subunits. It blocks the interaction of alpha V beta 3 with various ligands such as osteopontin and vitronectin[395].

Integrins are cell surface adhesion molecules coupling the extracellular environment to the cytoskeleton as well as receptors for transmitting signals important for cell migration, invasion, proliferation, and survival[396].

IX.47.2. Cilengitide, a cyclic Arg-Gly Glu (RGD) pentapeptide, targets two specific integrins, which are thought to be particularly important for angiogenesis. The targets are integrin alpha5, beta3 and integrin alpha5, beta5 –so called because each protein is formed from two distinct subunits.

Integrins are a family of protein receptors found in membranes, with many cells having many different types on their surface. They are used to regulate cell signaling and control certain cell functions such as growth, differentiation and motility. Integrins are crucial to the process of tumor angiogenesis.

---

[394] *Sridhar SS. Shepherd FA, Targeting angiogenesis: a review of angiogenesis inhibitors in the treatment of lung cancer. Lung cancer, 2003 Dec;42 Suppl 1:S81-91*

[395] *Wilder RL, Integrin alpha V beta 3 as a target for treatment of rheumatoid arthritis and related rheumatic diseases. Ann Rheum Dis. 2002 Nov;61 Suppl 2:ii96-9.*

[396] *Tucker GC, Integrins: molecular targets in cancer therapy. Curr Oncol Rep. 2006 Mar;8(2):96-103*

The molecule has been designated orphan drug status in both the USA and Europe for the treatment of malignant glioma and glioma, respectively.

Anti-angiogenic treatment is believed to have at least cystostatic effects in highly vascularized tumors like pancreatic cancer. Researchers[397] compared the treatment effects of the angiogenesis inhibitor cilengitide and gemcitabine with gemcitabine alone in patients with advanced unresectable pancreatic cancer and concluded that there were no clinically important differences observed regarding efficacy, safety and QoL between the groups. The observations lay in the range of other clinical studies in this setting. The combination regimen was well tolerated with no adverse effects on the safety, tolerability and pharmacokinetics of either agent[398].

Surprisingly, despite the broad theoretical impact of such molecules on integrin function, and thus on pathology, the clear identification of discrete clinical niches for their use remains to be defined. The parallel development of integrin antagonists as imaging tools for patient selection may accelerate the discovery of new avenues for their use.[399]

IX.48. Signal transduction inhibitors

Using confocal microscopy, the scientists have succeeded in live imaging of this process. In an article published in Science, the researchers describe how the Wnt signal induces the formation of large protein complexes close to the cell membrane composed of phosphate labeled LRP6 and adaptor protein axin. Another component of the Wnt signaling pathway, the Disheveled protein (Dvl), is vital for the formation of these aggregates. The investigators have shown that these are real protein complexes, rather than chance assemblies in a transport vesicle of the cell.

They have developed a model to explain the formation of the aggregates termed "LRP6 signalosomes". Thus, Wnt binding leads to a clustering of receptor proteins LRP6 and Frizzled, which form polymers with the aid of Dvl. This leads to a high density of the LPR6 receptor, which, in turn, facilitates LRP6 phosphorylation by kinase CK1gamma and promotes the binding of axin.

The findings now provide a completely new concept of intracellular communication. Like in many other biochemical processes - such as protein degradation by proteasomes or apoptosis - here, too, large protein polymers, like molecular machines, are required to bring together the participating reaction partners in a confined space.

Datamonitor believes the angiogenesis and signal transduction inhibitor classes will provide the greatest revenue potential for the pharmaceutical industry, congruent with their combined clinical and commercial appeal. The crowded and competitive signal transduction class will result in a high degree of fragmentation and cannibalization, limiting potential revenue generation. In one of Datamonitor´s reports (August 2004) on targeted therapy, Reuters Business Insight notes that the cytostatic nature of many molecular-targeted treatments means that significant tumor regressions are only likely to be seen in the context of combinatorial treatment approaches alongside 'classic' cytotoxic treatments and alternate molecular-targeted treatments.

---

[397] *Researchers of the Department of General Surgery, University of Heidelberg, Germany, led by Dr. H.Friess*
[398] *Friess H. et al., A randomized multi-center phase II trial of the angiogenesis inhibitor Cilengitide (EMD 121974) and gemcitabine compared with gemcitabine alone in advanced unresectable pancreatic cancer. BMC Cancer, 2006 Dec 11;6:285.*
[399] *Tucker GC, Integrins: molecular targets in cancer therapy. Curr Oncol Rep. 2006 Mar;8(2):96-103*

The financial impact of adopting novel molecular-targeted treatments may be prohibitive to market uptake. Mitigating this, improved diagnostic and predictive analysis will facilitate predictive targeting of treatments to patient populations most likely to respond. A convergence of scientific advances has enabled the identification of molecular targets specific to cancer cells, resulting in therapies with enhanced selectivity, efficacy and reduced toxicity. Developers can rise to the clinical, regulatory and economic challenges to commercialization by improving strategic decision-making in progressing drugs through the developmental pipeline.

IX.49. Cetuximab (Erbitux): especially effective when used in combination with other therapy forms

Oncology has "nichebuster" potential...
Merck KGaA entered this area through a number of research agreements and the acquisition of Lexigen Pharmaceuticals in the USA. Of particular interest to Merck KGaA were Lexigen's range of –innovative in their mode of action- developmental immunocytokines (or fusion proteins consisting of a recombinant antibody attached to a cytokine), which Lexigen had in an advanced stage of research.

Merck KGaA's Erbitux, licensed from ImClone Systems, is active against epidermal growth factor receptors (EGF-R); and especially effective when used in combination with other therapy forms, such as chemotherapy and radiotherapy.

Combination therapy is being described as the most successful in diseases such as AIDS and cancer. Cancer itself is a very complex set of diseases. Cancerous tumors are usually the result of several different genetic mutations. Erbitux, in the same way as other, well known cancer drugs like Herceptin, only addresses one of them. Erbitux is not a mass market miracle drug. Early tests have shown that the drug works in only 23 percent of patients who possess a dysfunctional element (EGF-R is present in about 70% of the patients), known to be errant in a broad range of cancers. However, for those patients whose tumors shrink, and disappear, the drug definitely is a miracle drug.

The best strategy to fight cancer, experts state, is to understand the tumor as it sits in the patient, get a good handle on what are all the good targets and hit them, all at once. That is where one can get a cancer cure. To obtain it, an arsenal of drugs, each of them targeting a different malfunction, is considered to be the best available of therapies. With the acquisition of Switzerland based Serono, the company "reshuffles" its product portfolio and intends to become a specialty medicines player.

IX.50. Erlotinib (Tarceva) reversibly and selectively inhibits tyrosine kynase activity

Progress in identifying the biochemical and molecular causes of cancer has led to discovery of abnormalities that characterize cancer cells and represent targets for development of drug therapies. TK receptors represent one such target when these are present in elevated quantities and/or aberrant forms. Abnormalities in these cell surface receptors have been correlated with development and progression of cancer, poor response to chemotherapeutic agents, and low survival rates[401].

---

[401] *Hamid O., Emerging Treatments in Oncology : Focus on Tyrosine Kinase (erbB) Receptor Inhibitors, J. Am. Parm. Assoc.2004;44:52-58*

Several tyrosine kinase inhibitors have been developed and evaluated over the past decade, of which the majority are reversible competitors with ATP for binding to the intracellular catalytic domain of the tyrosine kinase. One such EGFR tyrosine kinase inhibitors that has shown excellent antitumor activity is erlotinib hydrochloride, an oral quinazoline derivate that reversibly and selectively inhibits tyrosine kinase activity.

Roche/Osis/Genentech's Tarceva[402] is the second entrant in the EGFR-TKI class behind Iressa. The drug underwent a clinical development similar to that of Iressa. However, Roche's new drug benefits from a much more powerful study design than its direct competitor did (the IDEAL2 trial) when filing for a 3rd line indication in NSCLC:
-a population of 700 patients in the BR21 trial versus 216 patients in the IDEAL2 study;
-a Best supportive Care comparison arm;
-an overall survival rate as the primary endpoint, in line with FDA recommendations for registration.

Tarceva in combination with gemcitabine is indicated for the first-line treatment of patients with locally advanced, unresectable or metastatic pancreatic cancer.

IX.51. Gefitinib (Iressa): a once a day pill designed to shrink tumors without the side effects of chemotherapy

Iressa is indicated for the treatment of NSCLC. The medicine's active ingredient is a potent and selective inhibitor of epidermal growth factor tyrosine kinase. The compound treats NSCLC that has progressed despite treatment with platinum-based and docetaxel chemotherapy –two drugs that are standard of care for the disease. It is a once-a-day pill designed to shrink tumors without the harsh side effects of chemotherapy. Iressa has undergone testing for activity in combination with conventional cytotoxic treatments such as fluorouracil and leuvocovorin in patients with metastatic colorectal cancer. Although Iressa has considerable promise in colorectal cancer, the product showed exceptionally strong responses in NSCLC, so much so that Iressa was moved directly to Phase III clinical trials, bypassing Phase II[403]. The tolerability and drug-drug interaction profile of this product was equally beneficial, with no drug interactions reported in patients on chemotherapeutic treatment regimens such as paclitaxel and carboplatin.

After discussions with the U.S. Food and Drug Administration (FDA), the company announced that it is making a labeling change to gefitinib tablets. Based on the lack of survival benefit in the Phase III Trial 709 (ISEL) comparing gefinitib to placebo in advanced recurrent NSCLC and the availability of other drugs that do prolong life, the revised label indicates that Iressa is only to be used in patients who are benefiting or have benefited from Iressa. After September 15, 2005, no new patients are allowed access to Iressa unless they are being enrolled into a clinical trial that was approved by an Institutional Review Board (IRB) prior to June 17, 2005, or they are part of a clinical study that is being conducted under an investigational new drug application (IND). Clinical trials approved by an IRB after June 17, 2005, must be conducted under a valid IND, subject to prior review and approval of the proposed protocol by AstraZeneca to ensure drug supply[404].

---

[402] *The compound was discovered by Pfizer and later transmitted to OSI Pharmaceuticals who co-developed the product with Genentech/Roche.*
[403] *Natexis Bleichroeder pharmaceutical market analysis, March 30, 2004*
[404] *http://www.iressa-us.com/prof.asp accessed August 1, 2007*

IX.52. Panitumumab (ABX-EGF) : the 1st fully human monoclonal antibody to target the EGFr

This compound[405] is the first fully human monoclonal antibody that targets the epidermal growth factor receptor (EGFr), a protein that plays an important role in cancer cell signaling. Panitumumab, an IgG2 monoclonal antibody, binds with high affinity to the EGFr. Panitumumab was generated with Abgenix's XenoMouse technology, which creates a fully human monoclonal antibody that contains no murine (mouse) protein. The body's immune system can recognize the mouse protein found in chimeric antibodies as foreign and launches an immune response in the form of infusion reactions, allergic reactions or anaphylaxis. The goal of developing fully human monoclonal antibodies, which by definition contain no mouse protein, is to offer effective, high affinity therapies that minimize the potential for this type of immune response. Panitumumab is being evaluated in clinical trials as both a monotherapy and in combination with other agents for the treatment of various types of cancer, including colorectal, lung and kidney[400]. Panitumumab (Vectibix from Amgen) received an accelerated approval from the FDA[401] (after showing effectiveness in slowing tumor growth and, in some cases, reducing the size of the tumor) for the treatment of patients with colorectal cancer that has metastasized following standard chemotherapy. FDA approved Vectibix on the basis of the results of a randomized, controlled clinical trial of 463 patients with metastatic cancer of the colon and the rectum after undergoing treatment with chemotherapy drugs, fluoropyrimidine, oxaliplatin and irinotecan.

Although EGFr normally helps regulate the growth of many different cells in the body, EGFr also can stimulate cancer cells to grow. In fact, many cancer cells actually require signals mediated by EGFr for their survival. Residing on the surface of these tumor cells, EGFr is activated when naturally occurring proteins in the body, such as epidermal growth factor (EGF) or transforming growth factor alpha (TGF-alpha), bind to it. This binding changes the shape of EGFr, which, in turn, triggers internal cellular signals that stimulate tumor cell growth. Panitumumab binds to EGFr, preventing the natural ligands such as EGF and TGF-alpha from binding to the receptor and interfering with the signals that would otherwise stimulate growth of the cancer cell and allow it to survive.

The discovery of the mutations that confer sensitivity to EGFR inhibitors represents an important advance in the strategy of cancer drug development. Most agents in current use, including cytotoxics and the newer targeted drugs, have relatively low response rates in patients with any given histological class of solid tumors.

The use of molecular or genomic tests, or proteomics, to identify reliable markers for drug sensitivity will undoubtedly become one of the key objectives early in the clinical development of new drugs.[402]

Because it contains no mouse immunoglobulin (IG) sequences, Panitumumab is expected to show a safety profile with a low incidence of infusion reactions, antigenicity and allergic response.

---

[405] *Co-developed by Amgen and Abgenix*

[400] *Amgen Abgenix company press release :Panitumumab Significantly Improves Progression-Free Survival in Phase 3 Randomized Metastatic Colorectal Cancer Study, Nov 7, 2005*

[401] *FDA Press release, September 27, 2006*

[402] *Roberts, T. G. Jr. & Chabner, B. A. Beyond fast track for drug approvals. N. Engl. J. Med. 351, 501–505 (2004). Quoted in: Bruce A. Chabner and Thomas G. Roberts Jr.,Chemotherapy and the war on cancer. Nature Reviews, Cancer, Volume 5, January 5, 2005. 65*

IX.53. Lapatinib (Tykerb, GW-572016): the ability to cross the blood-brain barrier

Tykerb is an epidermal growth factor receptor (EGFR) and ErbB-2 (Her2/neu) dual tyrosine kinase inhibitor. The FDA approval (March 2007) of Lapatanib is described as incredibly significant as it provides a treatment for an often fatal area cancer where more options are needed. Tykerb targets two significant genes: the ErbB2 receptor and the similar ErbB1 receptor, which is associated with cancer growth. For women who are HER2 positive, the risk of cancer spreading to the brain is much greater. Tykerb, which is given in an oral pill form, addresses this with its ability to cross the blood-brain barrier[403].

IX.54. Dasatinib (Sprycel) Effective Against Difficult-to-Treat ALL

Two new targeted drugs- nilotinib and dasatinib - can be safely given to patients with chronic myelogenous leukemia and Ph-positive acute lymphocytic leukemia who cannot take imatinib (Gleevec) or who have become resistant to the older drug, according to results from early clinical trials. The two new drugs also showed encouraging anticancer activity in these diseases. E.g.: The results[404] of a 36-patient phase II clinical trial indicate that the multitargeted drug dasatinib effectiveness did not appear to be hampered by most of the mutations in a key protein that have been associated with imatinib resistance. Patients in the trial had a specific chromosomal translocation, often referred to as the Philadelphia chromosome, that is associated with a rapid course of disease after ALL diagnosis and poor survival. The translocation creates a fused protein known as BCR-ABL, the same protein typically seen in chronic myelogenous leukemia, for which imatinib has proven to be an effective treatment.

These data are highly significant given the refractory nature of patients enrolled in this trial to current treatment modalities, including imatinib[405]

IX.55. Farnesyltransferase inhibitor Tipifarnib (Zarnestra), Ionafarnib (Sarasar): making chemotherapy much more effective

Tipifarnib is an oral nonpeptidomimetic farnesyl transferase inhibitor developed to inhibit a variety of farnesylated targets potentially relevant to the therapy of various malignancies. The agent has, thus far, been tested in a wide array of both solid tumors and myeloid malignancies[406].

Farnesyltransferase is required to set in motion a series of molecular events involved in the development of cancer, e.g. the activation of proteins, such as one called Ras, that contribute to the growth of cancer cells. Scientists at the Montefiore Medical center believe that by switching off the farnesylation-dependent pathways, chemotherapy can be made much more effective.

In advanced transitional cell carcinoma, scientists focus their clinical research on identifying systemic treatment options and new agents in order to improve survival of patients. Cisplatin-based chemotherapy is standard treatment of patients with metastatic urothelial cancer; however, despite regimens as the cisplatin-gemcitabine combination, the overall response rates vary between 40% and 65%, with complete response in 15%-

---

[403] Susan G. Komen for the Cure. Press release. FDA Announces Approval of New Treatment for Advanced Breast Cancer, March 13, 2007.

[404] http://www.cancer.gov/cancertopics/druginfo/dasatinib. Accessed June 3, 2007

[405] According to the study leader Dr. Olivier Ottmann from Johann Wolfgang Goethe University in Germany and colleagues in a May 11, 2007 early online release in Blood; reported by the NCI in its Cancer Research Highlights, May 29, 2007 • Volume 4 / Number 18

[406] Mesa RA, Tipifarnib: farnesyl transferase inhibition at a crossroads. Expert Rev Anticancer Ther. 2006 Mar;6(3):313-9.

25% with survivals up to 16 months. This survival is frequently achieved with severe and life-threatening side effects. None the less, almost all responding patients relapse within the first year[407].

Tipifarnib has displayed the most interesting activity in the myeloid malignancies of myelodysplastic syndrome, myelofibrosis with myeloid metaplasia and elderly/high-risk acute myeloid leukemia. Overall clinical response rates of approximately 20-30% have been reported in myelodysplastic syndrome and acute myeloid leukemia patients who have few alternative therapeutic options[408].

Myelodysplastic syndrome (MDS) is a disorder of hematopoietic stem cells characterized by ineffective hematopoiesis. The result is pancytopenia leading to transfusion-dependent anemia, an increased risk of infection or bleeding, and a potential to progress to acute myeloid leukemia (AML). MDS is most prevalent among older individuals, many of whom also suffer from other medical conditions[409].

Patients with high-risk myelodysplastic syndromes (MDS) should be treated with low-intensity rather than high-intensity therapy. Low-intensity therapy (LIT) would result in a lower probability of treatment-induced death than would high-intensity therapy (HIT). Patients with high-risk MDS are generally treated with HIT, which is also associated with an average risk of treatment-induced death of 25%, the researchers explained. They hypothesized that LIT, involving such agents as decitabine or tipifarnib with or without low-dose cytosine arabinoside (ara-C), might reduce the rate of treatment-induced death[410].

As said before: the prognosis for any patient with progressive or recurrent invasive transitional cell carcinoma remains poor. In this context, the focus of clinical research in these invasive cancers concentrates on identifying; therefore, the need for development of new and tolerable agents is urgent[411]. FTIs have demonstrated activity in a variety of hematologic diseases, including acute myeloid leukemia, myelodysplastic syndrome, chronic myeloid leukemia, and multiple myeloma.

Several new agents are being tested, including lonafarnib, another farnesyltransferase.

IX.56. Proteasome inhibitor Bortezomib: showing impressive activity in patients with relapsed multiple myeloma

As therapy for hematologic malignancy evolves, new regimens and novel agents that target specific cellular processes allow a more optimistic prognosis for many patients. Tipifarnib and Bortezomib are two new, targeted treatments for hematologic malignancies. Bortezomib, a proteasome inhibitor, has shown impressive efficacy in patients with relapsed multiple myeloma and as initial treatment, including before autologous stem cell transplantation. It has been studied as monotherapy and in combination with standard treatments such as dexamethasone, and with newer agents such as the immunomodulators thalidomide and lenalidomide; response is encouraging, even in patients who have relapsed after previously receiving components of a regimen

---

[407] *Perabo FG, Muller SC.New agents for the treatment of advanced transitional cell carcinoma.Best Pract res Clin Haematol,2006;19(2):293-300*

[408] *Mesa RA, Tipifarnib: farnesyl transferase inhibition at a crossroads. Expert Rev Anticancer Ther. 2006 Mar;6(3):313-9.*

[409] *Larson RA, **Myelodysplasia: when to treat and how. Best Pract Res Clin Haematol.** 2006;19(2):293-300.*

[410] *Reported by researchers from the University of Texas MD Anderson Cancer Center at the 8th International Symposium on Myelodysplastic Syndromes (MDS), held May 12th to the 15th, 2005. Presentation title: Comparison of "High" and "Low" Intensity Approaches for High-Risk MDS. Abstract O-67*

[411] *Perabo FG, Muller SC, New agents for treatment of advanced transitional cell carcinoma. Ann Oncol. 2006 Oct 3; [Epub ahead of print]*

as single agents. Bortezomib is generally well tolerated, including in combination with novel and conventional agents[412].

International research[413] explores combination treatment of a number of mouse tumor types with bortezomib (Velcade) and an agonist antibody against the Apo2L/TRAIL "death receptor." As part of the collaboration, the Australian and Japanese investigators have supplied NCI with a quantity of the purified antibody and with lymphocyte - tumor hybrid cells called "hybridomas" that produce it.

The Apo2L/TRAIL receptor is one of several members of the tumor necrosis factor superfamily of receptors that are able to induce apoptosis (programmed cell death) when activated. This receptor has received considerable attention lately because of the finding that many cancer cell types are sensitive to Apo2L/TRAIL-induced apoptosis, whereas most normal cells are not.

Bortezomib works by blocking the action of structures inside cells called "proteasomes," which are large enzyme complexes that degrade abnormal or misfolded proteins and proteins that are normally targeted for destruction. Some proteins are normally targeted for orderly destruction to maintain cellular growth control. Research has shown that proteasome inhibition can induce apoptosis and that cancer cells are more sensitive to this effect of bortezomib than normal cells.

The antibody against the Apo2L/TRAIL receptor alone has shown some efficacy against tumor cells, but preliminary experiments have indicated that bortezomib enhances its activity. This combination treatment will hopefully allow a reduction of tumor burden in the absence of the major immunosuppression that can occur during traditional chemotherapy or radiation therapy. The investigators' aims are to promote local tumor destruction and, hopefully, to improve the natural immune response against the tumor cells.

As mentioned before, misfolded and disused proteins are eliminated by a cellular shredder called the proteasome. The cell labels the proteins it wants to dispose with Ubiquitin (Ub) in order to avoid the unwanted degradation of still needed proteins. Malfunctions in the ubiquitin-proteasome system can be fatal for the organism. In particular cancer and immunological disorders but also developmental defects are the consequences. The knowledge about the ubiquitination reaction as well as the effect of ubiquitination on diverse cellular functions has already led to the development of highly promising drugs like Bortezomib and Herceptin. The molecular mechanism of Ub conjugation and its biological meaning has been discovered by Avram Hershko, Aaron Ciechanover and Irwin Rose who have been awarded the noble price for their work in 2004. Scientists now describe a novel Ub conjugation reaction that might allow a more efficient manipulation of key proteins in the treatment of cancer and other diseases and state that our current view on protein ubiquitination as it is found in the textbooks has to be revised[414].

---

[412] Armand JP et al. The emerging role of targeted therapy for hematologic malignancies: update on bortezomib and tipifarnib. Oncologist 2007 Mar;12(3):281-90.

[413] The National Cancer Institute (NCI) is collaborating with investigators at the Peter McCollum Cancer Centre in Melbourne, Australia and Juntendo University in Tokyo, Japan http://www.cancer.gov/nci-international-portfolio (accessed Aug 6, 2007)

[414] Prof. Ivan Dikic and his team from the institute of biochemistry II at the University of Frankfurt. Press release Johann Wolfgang Goethe-University Frankfurt, June 26, 2007.

According to the scientists, virtually every process within an eukaryotic cell is directly or indirectly controlled by ubiquitin. This includes fundamental cellular programs such as protein degradation, DNA replication, signal transduction or protein trafficking. Up to now it was believed that the cooperation of three enzymes is needed to attach Ub to another protein: E1, E2 and E3. E1 activates Ub and transfers it to E2, which in turn cooperates with an so-called Ub ligase (E3) that couples Ub to a specific target protein. However, they discovered that certain proteins can be ubiquitinated independently of E3 ligases which caused a stir among experts.

A prerequisite for the E3-independent reaction is the presence of a Ub-binding domain (UBD) in the protein that is able to recognize Ub. This type of protein is of utmost importance for the cell because it can distinguish between ubiquitinated and non-ubiquitinated proteins. UBD-proteins are the key to the effects of Ub - both in normal and malignant cells.

The researchers demonstrated that proteins equipped with an UBD can ubiquitinate themselves by directly recruiting Ub-loaded E2 enzymes. The discovery provides the basis for novel therapeutic approaches that are more specific than drugs like Bortezomib.

IX.57. Signal transductor inhibitors: bexaroten, perifosine
IX.57.1. Bexarotene (Targretine)
Bexarotene is an oral retinoid therapy that is effective for the treatment of early and advanced-stage cutaneous T-cell lymphoma (CTCL) in patients who have failed on other therapies. However, bexarotene treatment is associated with unavoidable side-effects, in particular hypertriglyceridaemia and hypothyroidism, which are manageable with adequate concomitant medications and are reversible on cessation of treatment[415].

IX.57.2. Perifosine
Perifosine [octadecyl-(1,1-dimethyl-piperidinio-4-yl)-phosphate] is a novel heterocyclic alkylphosphocholine signal transduction inhibitor, showing activity against multiple cell types. Many studies have been undertaken to establish the therapeutic benefit of this compound. The outcome has given mixed results. Examples of positive findings:

The compound has displayed significant antiproliferative activity in vitro and in vivo in several human tumor model systems and has recently entered phase II clinical trials. Other alkylphospholipids have been previously used as antileishmanial agents, and miltefosine (Impavido) is now established as the first oral drug for the treatment of visceral and cutaneous leishmaniasis. Perifosine showed the higher activity against all tested strains. This study demonstrates for, the first time, an in vitro activity of perifosine against different species of Leishmania in the promastigote stage[416].

Researchers [417] found that synthetic alkyl-lysophospholipids (ALPs) edelfosine and perifosine induced apoptosis in multiple myeloma (MM) cell lines and patient MM cells, whereas normal B and T lymphocytes were spared. ALPs induced recruitment of Fas/CD95 death receptor, Fas-associated death domain-containing protein, and

---

[415] *Assaf C et al., Minimizing adverse side-effects of oral bexarotene in cutaneous T-cell lymphoma: an expert opinion.Br J Dermatol., 2006 Aug;155(2):261-6.*
[416] *Cabrera-serra MG, In vitro activity of perifosine: a novel alkylphospholipid against the promastigote stage of Leishmania species. Parasitol Res. 2007 Apr;100(5):1155-7. Epub 2007 Jan 6.*
[417] *from the University Hospital of Salamanca in Spain*

procaspase-8 into lipid rafts, leading to the formation of the death-inducing signaling complex (DISC) and apoptosis[418].

Combined modality treatment has improved outcome in various solid tumors. Besides classic anticancer drugs, a new generation of biological response modifiers has emerged that increases the efficacy of radiation. Perifosine enhances radiation-induced cytotoxicity, as evidenced by reduced clonogenic survival and increased apoptosis induction in vitro and by complete tumor regression in vivo[419].

Alkylphospholipid agents target cell membranes and inhibit Akt activation. Research showed that baseline phosphorylation of Akt in multiple myeloma (MM) cells is completely inhibited by perifosine in a time- and dose-dependent fashion, without inhibiting phosphoinositide-dependent protein kinase 1 phosphorylation. Perifosine induces significant cytotoxicity in both MM cell lines and patient MM cells resistant to conventional therapeutic agents. Importantly, Perifosine induces apoptosis even of MM cells adherent to bone marrow stromal cells. Perifosine triggers c-Jun N-terminal kinase (JNK) activation, followed by caspase-8/9 and poly (ADP)-ribose polymerase cleavage. Interestingly, phosphorylation of extracellular signal-related kinase (ERK) is increased by perifosine. Furthermore, perifosine augments dexamethasone, doxorubicin, melphalan, and bortezomib-induced MM cell cytotoxicity. Finally, perifosine demonstrates significant antitumor activity in a human plasmacytoma mouse model, associated with down-regulation of Akt phosphorylation in tumor cells[420].

IX.58. Apoptosis stimulators: oblimersen, aprinocarsen, exisulind
IX.58.1. Oblimersen (Genasense)
Oblimersen is an experimental agent that inhibits the production of a protein known as Bcl-2 in cancer cells. This protein can stop a cell from destroying itself, and is often over-expressed in cancer. As an antisense drug, oblimersen provides a complementary genetic strand to the messenger RNA that produces Bcl-2, inactivating it and preventing the protein from being produced.

The first "antisense" drug to be tested in chronic lymphocytic leukemia (CLL) shows benefit in a phase III clinical trial for a specific subset of patients - those who are still sensitive to a chemotherapy drug often used to treat this cancer.
Researchers[421] found that oblimersen produced a four-fold increase in "CP/nPR," a clinical response defined by no definitive evidence of disease, in patients who were sensitive to the chemotherapy drug fludarabine, compared to patients who no longer responded to fludarabine.

The addition of oblimersen to fludarabine plus cyclophosphamide significantly increases the CR/nPR rate in patients with relapsed or refractory CLL (particularly fludarabine-

---

[418] Gajate C., Mollinedo F., Edelfosine and perifosine induce selective apoptosis in multiple myeloma by recruitment of death receptors and downstream signaling molecules into lipid rafts. Blood, 2007 Jan 15;109(2):711-9. Epub 2006

[419] Vink SR et al., Radiosensitization of squamous cell carcinoma by the alkylphospholipid perifosine in cell culture and xenografts.Clin Cancer Res. 2006 Mar 1;12(5):1615-22.

[420] Hideshima T et al., Perifosine, an oral bioactive novel alkylphospholipid, inhibits Akt and induces in vitro and in vivo cytotoxicity in human multiple myeloma cells. Blood, 2006 May 15;107(10):4053-62. Epub 2006 Jan 17.

[421] Research at The University of Texas M. D. Anderson Cancer Center and reported in the early on-line edition of the March 20, 2007 Journal of Clinical Oncology

sensitive patients), as well as response duration among patients who achieve CR/nPR[422].

More than 15,000 new cases of the disease will be diagnosed 2007 in the USA, according to the American Cancer Society, and about 4,500 people will die from the cancer.

IX.58.2. Aprinocarsen (Affinitak)

Affinitak is the antisense oligonucleotide that targets protein kinase C (PKC). Research found that PKC-alpha is involved in critical tumor cell processes, and has proliferation, malignant, and metastatic potential. It is involved in growth factor signaling and drug resistance. PKC-alpha is abnormally expressed in many human tumors and is a negative prognostic factor[423].

IX.58.3. Exisulind

Exisulind is a selective, apoptotic antineoplastic drug, a GMP phosphodiesterase inhibitor. The agent is expected to provide an alternative method to treating or preventing prostate cancer. A phase II evaluation of docetaxel plus exisulind in patients with androgen independent prostate carcinoma study[424] showed rather disappointing results. The researchers concluded that overall, their trial indicated that the toxicity profile and efficacy of this regimen is unlikely to be substantially better than single agent docetaxel.

IX.59. Cell cycle regulators: Seliciclib (CYC-202; R-roscovitine), Indisulam (E-7070), UCN –01

IX.59.1. Seliciclib (CYC-202; R-roscovitine)

Seliciclib is the first selective, orally bioavailable inhibitor of cyclin-dependent kinases 1, 2, 7 and 9 to enter clinical trial. Preclinical studies showed antitumor activity in a broad range of human tumor xenografts[425].

IX.59.2. Indisulam (E-7070)

Indisulam (N-(3-chloro-7-indolyl)-1,4-benzenedisulfonamide, E7070) [426] is a novel sulfonamide anticancer agent in clinical development for the treatment of solid tumors. Its mechanism of action is multifactorial, as indisulam arrests cell cycle in the G1 phase, strongly inhibits carbonic anhydrase, a critical enzyme involved in many physiological processes and whose association with cancer became obvious in the last period, and markedly alters gene expression levels of at least 60 transcripts. Four Phase I clinical trials gave promising results, showing the compound to possess nonlinear pharmacokinetics[427].

---

[422] O'Brien S. et al, Randomized phase III trial of fludarabine plus cyclophosphamide with or without oblimersen sodium (Bcl-2 antisense) in patients with relapsed or refractory chronic lymphocytic leukemia. J Clin Oncol. 2007 Mar 20;25(9):1114-20. Epub 2007 Feb 12.

[423] Dr. Branimir I. Sikic, „AFFINITAK (LY900003, ISIS 3521), an Antisense Inhibitor of PKC-Alpha, in the Therapy of Non-Small Cell Lung Cancer", slide presentation explaining clinical trial results at Stanford on Affinitak (as a single agent and in combination with carboplatin and paclitaxel therapy and with gemcitabine/cisplatin therapy) in the therapy of NSCLC

[424] Sinibaldi et al., Am J Clin Oncol, 2006 Aug;29(4):395-8

[425] Benson C et al., A phase I trial of the selective oral cyclin-dependent kinase inhibitor seliciclib (CYC202; R-Roscovitine), administered twice daily for 7 days every 21 days. Br J Cancer, 2007 Jan 15;96(1):29-37. Epub 2006 Dec 19

[426] from the Japanese pharmaceutical company Eisai

[427] Supuran CT, Indisulam: an anticancer sulfonamide in clinical development. Expert Opin Investig Drugs, 2003 Feb;12(2):283-7.

Indisulam is currently being evaluated in phase II clinical studies. Sixteen patients received study treatment and were eligible. Thrombocytopenia was the major dose limiting toxicity followed by neutropenia. Both drugs contributed to the myelosuppressive effect of the combination. Indisulam 500 mg m(-2) in combination with carboplatin 6 mg min ml(-1) was identified not to cause dose limiting toxicity, but a delay of re-treatment by 1 week was required regularly to allow recovery from myelosuppression. The recommended dose and schedule for an envisaged phase II study in patients with non-small cell lung cancer is indisulam 500 mg m(-2) in combination with carboplatin 6 mg min ml(-1) repeated four-weekly. Patients who do not experience severe thrombocytopenia at cycle 1 will be permitted to receive an escalated dose of indisulam of 600 mg m(-2) from cycle 2 onwards[428].

The findings of another study, aimed to measure the objective tumor response rate following treatment with indisulam [E7070; N-(3-chloro-7-indolyl)-1,4-benzenedisulfonamide] as second-line therapy in patients with advanced non-small cell lung cancer, showed that despite evidence of tumor-specific indisulam-induced apoptosis, neither of the treatment schedules has single-agent activity as second-line treatment of non-small cell lung cancer[429].

IX.59.3. UCN-01

UCN-01 is an anticancer drug that belongs to the family of drugs called staurosporine analogs which have demonstrated an ability to inhibit multiple kinases involved in cell-cycle progression and apoptosis, including Chk-1 and PDK1. In pre-clinical studies, UCN-01 has demonstrated synergistic effect with DNA-damaging agents including chemotherapy and radiation therapy. In-vitro, UCN-01 has been shown to be synergistic with agents affecting the PI3-K pathway including perifosine and mTOR inhibitors. In clinical trials, as reported by investigators at the National Cancer Institute, durable single-agent responses have been seen in patients with anaplastic large-cell lymphoma.

Researchers [430] found that the coadministration of the farnesyltransferase inhibitor L744832 promotes UCN-01-induced apoptosis in human multiple myeloma cells through a process that may involve perturbations in various survival signaling pathways, including extracellular signal-regulated kinase, Akt, and STAT3, and through a process capable of circumventing conventional modes of myeloma cell resistance, including growth factor- and stromal cell-related mechanisms. They also raise the possibility that combined treatment with farnesyltransferase inhibitors and UCN-01 could represent a novel therapeutic strategy in multiple myeloma.

Clinical trials are ongoing with novel drugs that interfere with cell cycle regulation and signaling molecules in CLL, including Alvocidib, UCN-01, bryostatin 1, depsipeptide, and oblimersen. It remains to be seen whether these chemotherapeutic approaches offer real benefit for patients by prolonging survival with an improved quality of life[431].

---

[428] *Dittrich C et al., A phase I and pharmacokinetic study of indisulam in combination with carboplatin. Br J Cancer, 2007 Feb 26;96(4):559-66. Epub 2007 Feb 6.*

[429] Talbot DC, A randomized phase II pharmacokinetic and pharmacodynamic study of indisulam as second-line therapy in patients with advanced non-small cell lung cancer.Clin Cancer Res., 2007 Mar 15;13(6):1816-22.

[430] *Researchers Pey and colleagues, at the Department of Medicine, Virginia Commonwealth University/Medical college of Virginia*

[431] *Wendtner CM, Advances in chemotherapy for chronic lymphocytic leukemia. Semin Hematol. 2004 Jul;41(3):224-33.*

IX.60. Monoclonal antibodies targeting unprecedented receptors: MLN2704, ibritumomab (Zevalin)

IX.60.1.　MLN2704 A large number of promising agents for the management of high risk metastatic prostate cancer are in varying stages of development. Phase III results have been reported for the endothelin-A receptor antagonist atrasentan. Several immunotherapies are currently in phase II/III trials, namely the GM-CSF transduced tumor cell vaccine GVAX, the prostatic acid phosphatase loaded dendritic cell vaccine Provenge and the prostate specific antigen expressing poxvirus vaccine PROSTVAC-VF. Another immunotherapy, the prostate specific membrane antigen immunoconjugate MLN2704 is in phase I/II study. The first clinical inhibitors of survivin are in early phase I studies. Several of these agents, including atrasentan, have shown statistically significant but modest effects in the advanced disease setting in which they have been studied. [432].

MLN2704 is composed of a deimmunized monoclonal antibody (MLN591) directed at prostate specific membrane antigen (PSMA) conjugated to the chemotherapeutic agent DM1. MLN2704 binds to PSMA, which is rapidly internalized into the cell delivering a lethal dose of chemotherapy directly to prostate cancer cells. Unlike the enzyme PSA which circulates in the blood, PSMA is a protein expressed on the cell surface of virtually all prostate cancer cells and its abundance on the cell surface increases as the disease progresses and becomes refractory to hormonal therapy. PSMA has little expression in normal tissues.

To develop MLN2704, Millennium Pharmaceuticals conjugates the chemotherapeutic agent DM1 -- licensed from ImmunoGen, Inc.-- to the MLN591 antibody, which targets PSMA. DM1 was developed by ImmunoGen specifically for antibody-directed delivery to tumor cells.

In December 2003, MLN-2704 was granted "Fast Track" status by FDA and has been selected for the Continuous Marketing Application Pilot 2 program. This latter innovative program is designed to facilitate scientific exchange and speed approval time.

IX.60.2.　90Y-Ibritumomab tiuxetan (Zevalin)

Zevalin was the first [433] radioimmunotherapy to receive FDA approval. Zevalin is indicated for the treatment of relapsed or refractory low grade, follicular, or transformed B-cell non-Hodgkin's lymphoma (NHL). This indication includes patients with Rituxan (rituximab)-refractory follicular NHL. Zevalin has been approved as part of a therapeutic regimen involving Rituxan.

Antibody-based therapeutic approaches have had a significant impact in the treatment of non-Hodgkin's lymphoma (NHL). Rituximab's development as an anti-CD20 antibody heralded a new era in treatment approaches for NHL. While rituximab was first shown to be effective in the treatment of relapsed follicular lymphoma, it is now standard monotherapy for front-line treatment of follicular lymphoma, and is also used in conjunction with chemotherapy for other indolent, intermediate and aggressive B-cell lymphomas. The development of rituximab has led to intense interest in this type of therapeutic approach and to development and approval of the radioimmunoconjugates

---

[432] *Brand TC, Tolcher AW, Management of high risk metastatic prostate cancer: the case for novel therapies. J Urol., 2006 Dec;176(6 Pt 2):S76-80; discussion S81-2.*
[433] *In February 2002. Bayer-Schering's Zevalin is licensed from Biogen Idec.*

of rituximab, (90)Y-ibritumomab tiuxetan and (131)I-tositumomab, which have added to the repertoire of treatments for relapsed follicular lymphoma and increased interest in developing other conjugated antibodies[434].

Zevalin consists of a monoclonal antibody linked to the radioactive isotope yttrium-90. After infusion into a patient, the monoclonal antibody targets the CD20 antigen, which is found on the surface of mature B cells and B-cell tumors. In this manner, cytotoxic radiation is delivered directly to malignant cells.

The addition of yttrium 90 ibritumomab tiuxetan (Zevalin) to a standard condition regimen for patients undergoing autologous stem cell transplantation improves overall survival and progression-free survival among older patients with diffuse large B-cell lymphoma. The 2-year overall survival was 93% among patients with diffuse large B-cell lymphoma who received Zevalin along with a conditioning regimen comprising carmustine, cytarabine, etoposide, and melphalan (known as BEAM)[435].

IX.61. innovative early stage projects (examples): efaproxiral, lomeguatrib, MDX 447, Cortistatin A, AP25375, Azixa, MDX-010,

IX.61.1. Efaproxiral

Efaproxiral (efaproxyn) is an effective addition to whole-brain radiation therapy (WBRT) in patients with brain metastases. Metastasis of cancer to the brain is common, affecting as many as 170,000 individuals annually in the US alone. WBRT produces local control/response rates of 50–75%, and other treatments have also shown some success; however, post-diagnosis survival is poor, at a median of approximately 4.5 months. Because tumor resistance to radiation is increased by hypoxia, treatments to reduce hypoxia could be valuable. Efaproxiral changes the conformational structure of hemoglobin, thereby assisting oxygen release and improving tumor oxygenation. In the REACH/RT-009 study, Suh et al. tested whether addition of efaproxiral to WBRT and supplemental oxygen resulted in longer survival than combined WBRT and oxygen[436].

IX.61.2. Lomeguatrib (Patrin)

PaTrin-2 is an inhibitor of the DNA repair enzyme alkylguanine alkyltransferase (known as AGT or ATase) that is a major cause of resistance of tumors to cytotoxic drugs that cause DNA damage by O^6 methylation of DNA. Chemotherapeutic agents of this type include cancer treatments such as the alkylating agents carmustine and dacarzabine (melanomas are treated with dacarzabine). The newer agent 'Temodar' temozolomide (Schering-Plough) is the most recently introduced treatment of this type and this drug can be administered orally.

The efficacy of O^6 methylating agents is reduced by AGT which in some types of tumor is expressed at higher levels than others. It is believed that the higher levels of AGT are responsible for sub-optimal efficacy of alkylating agents in some tumor types. This potent inactivator of Atase is undergoing trials[437] in melanoma in order to establish if PaTrin (as an adjuvant) is effective in increasing the sensitivity of cancers to treatment

[434] *Fanale MA, Younes A, Monoclonal Antibodies in the Treatment of Non-Hodgkin's Lymphoma. Drugs, 2007;67(3):333-50.*

[435] *Presentation given by Amrita Krishnan, MD, associate professor and physician, the department of hematology and hematopoietic cell transplantation, City of Hope Cancer Center, Duarte, California, who led the research team, at the American Society of Hematology 48th Annual Meeting and Exposition (ASH), December 2006. Presentation title: A Comparison of BEAM and Yttrium 90 Ibritumomab Tiuxetan (Zevalin) in Addition to BEAM (Z-BEAM) in Older Patients Undergoing Autologous Stem Cell Transplant for B-Cell Lymphomas: Impact of Radioimmunotherapy on Transplant Outcomes. Abstract 3043*

[436] *Nature Clinical Practice Oncology (2006) 3, 116doi:10.1038/ncponc0409*

[437] *Under the auspices of KuDOs (acquired by AstraZeneca in 2005).*

by Temozolomide patients. Pharmacodynamic analyses of MGMT inactivation in peripheral blood cells and brain, colorectal and prostate tumors are being undertaken in order to establish the dose required for complete MMGMT inactivation[438][439].

IX.61.3. MDX 447

MDX447 is a bispecific antibody constructed by cross-linking F(ab') fragments of monoclonal antibody (MoAB) H22 to FcgammaRI and MoAB H425 to the epidermal growth factor receptor (EGFR). It combines a cancer targeting component provided by Merck KGaA (EMD82633) with an immune system triggering component developed by Medarex. Specific for both the high-affinity immunoglobulin G (IgG) receptor CD64 and epidermal growth factor receptor (EGFR), MDX447 may enhance cellular immune responses against EGFR positive cells, resulting in increased tumor cell lysis

IX.61.4. Cortistatin A

The Shair group, Cambridge, is working on what they call the unique natural product cortistatin A. which they describe as an extremely potent inhibitor of the migration and proliferation of HUVECs (Human Umbilical Vascular Endothelial Cells); IC50 = 180 pM (VEGF-stimulated HUVECs proliferation). It was also shown, they write[447], that cortistatin A is 3000 times more potent at inhibiting the proliferation of HUVECs versus other cancer and normal cell lines, suggesting that cortistatin A may be a highly selective angiogenesis inhibitor. Cortistatin A has an unprecedented structure, with a particularly challenging oxabicyclooctene ring system. The group is working on a distinct metal-catalyzed cascade reaction to construct cortistatin A.

IX.61.5. Azixa

EpiCept Corporation announced the results of scientific studies that
1) demonstrate that Azixa is effective in treating multiple types of human tumors in animal models,
2) reveal a mechanism by which the compound exerts its effects, and
3) show that it is not affected by cellular proteins known to be involved in cancer drug resistance.

Azixa is one of two compounds currently in clinical trials discovered through EpiCept's Anti-cancer Screening Apoptosis Program (ASAP). It is part of the EP90745 series of apoptosis inducers, which was licensed by EpiCept to Myriad Genetics, Inc. as part of an exclusive, worldwide development and commercialization agreement. It has a second mode of action due to vascular disruption activity (VDA). The compound is currently being evaluated in two Phase II human clinical trials, one in patients with primary brain cancer and the other in brain metastases due to melanoma.

EpiCept's proprietary apoptosis screening technology can efficiently identify new cancer drug candidates and molecular targets that selectively induce apoptosis in cancer cells through the use of chemical genetics and its proprietary live cell high-throughput caspase-3 screening technology. Chemical genetics is a research approach investigating the effect of small molecule drug candidates on the cellular activity of a

---

[438] *Ranson M., Lomeguatrib, a Potent Inhibitor of O⁶-Alkylguanine-DNA-Alkyltransferase: Phase I Safety, Pharmacodynamic, and Pharmacokinetic Trial and Evaluation in Combination with Temozolomide in Patients with Advanced Solid Tumors. Cancer Research Vol. 12, 1577-1584, March 2006.*

[439] *PaTrin-2 was developed jointly by teams led by Dr G. Margison of the Paterson Institute, Manchester, England in a program supported by Cancer Research UK and Professor T.B.H. McMurry and Dr R.S. McElhinney of Trinity College Dublin, Ireland. The Health Research Board, Ireland partially funded the research at Trinity College Dublin.*

[447] *http://www.chem.harvard.edu/research/faculty/matthew_shair/matthew_shair.html. Accessed July 11, 2007*

protein, enabling researchers to determine the protein's function. Using this approach with its proprietary caspase-3 screening technology, the company's researchers can focus their investigation on the cellular activity of small molecule drug candidates and their relationship to apoptosis.

This combination of chemical genetics and caspase-3 screening technology allows them to discover and rapidly test the effect of small molecules on pathways and molecular targets crucial to apoptosis, and gain insights into their potential as new anticancer agents. This screening technology is particularly versatile and can be adapted for almost any cell type that can be cultured, as well as measure caspase activation inside multiple cell types (e.g., cancer cells, immune cells, or cell lines from different organ systems or genetically engineered cells). This allows researchers to find potential drug candidates that are selective for specific cancer types, which may help identify candidates that provide increased therapeutic benefit and reduced toxicity.

IX.61.6. Ipilimumab

Ipilimumab (also known as MDX-010) is a fully human antibody against human CTLA-4, a molecule on T cells that suppresses the immune response. Medarex and Bristol-Myers Squibb are investigating the potential of ipilimumab to enable the immune systems of certain types of cancer patients to help suppress tumor growth. Ipilimumab is currently in three separate registrational studies for metastatic melanoma as a second-line monotherapy treatment, as a first-line treatment in combination with dacarbazine, and as a second-line treatment in combination with a melanoma-peptide vaccine. Ipilimumab is also involved in multiple Phase II clinical trials to investigate the product's potential activity in other tumor types, as well as in combination studies with other chemotherapy, immunotherapy and vaccines[448].

IX.61.7. MT201

MT201 (Adecatumumab) from is a human monoclonal antibody targeting EpCAM expressing tumors. Adecatumumab is being developed by Micromet in collaboration with Merck Serono in a phase 1b clinical trial evaluating MT201 in combination with docetaxel for the treatment of patients with metastatic breast cancer.

If successful, this drug could be a very large opportunity for pharmaceutical companies because it addresses a large proportion of breast cancer patients. Up to 80% of breast cancer patients are EpCAM positive.

IX.61.8. MAGE-A3 ASCI

GSK's ASCIs represent a novel class of medicines designed to train the immune system to recognize and eliminate cancer cells in a highly specific manner. These novel cancer immunotherapeutics combine tumor antigens, delivered as purified recombinant proteins, and the company's proprietary Adjuvant Systems which are specific combinations of immunostimulating compounds selected to increase the anti-tumor immune response. ASCIs may be used to reduce the risk of tumor recurrence following surgery, or to impact tumor growth in an early metastatic setting.

The highly specific mode of action of these ASCIs may not only avoid harming the normal tissue but also aid in selecting patients eligible for the treatment, depending on the expression of the tumor antigens.

---

[448] *Medarex press release, Jan 26, 2007*

MAGE-A3, an investigational compound[449], is a tumor-specific antigen that is expressed in a large variety of cancers, including Non-Small Cell Lung Cancer, Head and Neck Cancer, Bladder Cancer, with no expression in normal cells. Expression of the MAGE-A3 gene has been observed in testicular cells but without antigen presentation capabilities.

IX.61.9. Epratuzumab

Epratuzumab is a novel humanized antihuman CD22 IgG1 antibody , which has undergone preclinical and phase I/II clinical evaluation in patients with indolent or aggressive lymphoma. Data suggest that this agent is well tolerated, and can induce tumor regressions. Trials are currently evaluating its safety and activity in combination with rituximab (chimeric anti-CD20) and standard chemotherapy are ongoing. Initial results suggest that these regimens have acceptable toxicity, and that epratuzumab warrants further evaluation as an adjunct to standard lymphoma treatment regimens.

IX.61.10. Certolizumab Pegol (Cimzia)

Anti-TNF-alpha-blocker infliximab (Remicade) or adalimumab (Humira) are being used to treat Chrohn's Disease. Certolizumab expands the therapeutical war chest against CD.

New data from a post hoc analysis of the PRECiSE clinical trial program demonstrated that the anti-TNF certolizumab pegol maintained response and remission in patients with moderate to severe Crohn's disease, regardless of whether or not they had been previously treated with infliximab. These results build on the PRECiSE 2 studies.

The PRECiSE Program, composed of four studies (PRECiSE 1, 2, 3, and 4), represents an innovative, large, comprehensive development program for certolizumab pegol in Crohn's disease, including over 1,300 patients, with a planned follow-up phase of up to five years.

PRECiSE 1 is a unique trial in patients with active Crohn's disease - the first reported Phase III double-blind, placebo-controlled study of an anti-TNF extending to 26 weeks, in which eligible patients were randomized at study baseline without pre-selection of responders.

In the PRECiSE 2 study, patients responding at Week 6 to open-label induction therapy with certolizumab pegol were randomized to either placebo (n=210) or certolizumab pegol (n=215) and followed for a total of 26 weeks. In this trial, 62.8% of certolizumab pegol patients, compared to 36.2% of placebo patients, maintained clinical response at Week 26 (p<0.001). Clinical response was defined as a ?100 point decrease in CDAI. Similarly, 47.9% of certolizumab pegol patients were in clinical remission at week 26 compared to 28.6% on placebo (p<0.001).4 Remission was defined as CDAI < 150 points.

PRECiSE 3 and 4 are both long-term (up to five years) open-label trials assessing the longer-term efficacy, safety and tolerability of certolizumab pegol (Cimzia) in patients from PRECiSE 1 and PRECiSE 2, and are currently ongoing.

Belgian pharmaceutical company UCB continues to study the clinical profile of

---

[449] *MAGE-A3 antigen has been in-licensed by GSK from the Ludwig Institute for Cancer Research, the largest international academic institute dedicated to understanding and controlling cancer.*

certolizumab pegol in Crohn's disease. A new clinical trial involving 600 patients called WELCOME will further examine the effects of certolizumab pegol in patients failing or intolerant to infliximab. In addition, the MUSIC study will investigate the impact of certolizumab pegol on endoscopic and mucosal healing, and the CONCISE trial will examine the corticosteroid-sparing effect of certolizumab pegol in Crohn's disease.

Certolizumab pegol is the first and only PEGylated Fab' fragment of a humanized anti-TNF-alpha antibody (TNF-alpha; Tumor Necrosis Factor), evaluated as once-monthly dosing administered subcutaneously. The engineered Fab' fragment retains the biologic potency of the original antibody without the cytotoxicity mediated by the Fc portion present in the original monoclonal antibodies. Certolizumab pegol has a high affinity for human TNF-alpha, selectively neutralizing the pathophysiological effects of TNF-alpha. Over the past decade, TNF-alpha has emerged as a major target of basic research and clinical investigation. This cytokine plays a key role in mediating pathological inflammation, and excess TNF-alpha production has been directly implicated in a wide variety of diseases.

IX.61.11. PI-88

PI-88, from Progen, retards the growth of primary tumors by inhibiting angiogenesis in two ways:
1) Heparan sulfate mimicry causes inhibition of heparanase, which prevents the release of angiogenic growth factors from the extracellular matrix, and
2) interaction with angiogenic growth factors VEGF, FGF-1 and FGF-2, reduces their functional activity. PI-88 is in Phase 2 trials for lung cancer, liver cancer, multiple myeloma and melanoma. Multi-target action may help differentiate from single mechanism drugs.

IX.61.12. Catumaxomab (Removab): trifunctional antibody therapy

Trifunctional antibodies are a new class of intact bispecific antibodies that show unmet effector qualities by activation of not only T cells but also simultaneous activation of Fegamma receptor type I/III+cells (macrophages, NK-cells and DC). These trifunctional antibodies (trAb) lead to efficient specific killing of targeted tumor cells without any pre- or co-stimulation[440]. Thus, the antibody induces tumor specific cell mediated cytotoxicity in vitro and in vivo. Its intraperitioneal administration has demonstrated safety and led to efficient tumor cell killing and reduction of malignant ascites in ovarian cancer patients.[441]

IX.61.13. WX-G250 (Rencarex)

Tumor antigens such as CA-IX (Carbonic Anhydrase IX) are becoming increasingly important in the diagnosis and treatment of cancer. CA-IX is the most specific antigen known for clear cell renal cell carcinomas, for example and its potential importance in other tumor types is beginning to be recognized. Munich/Germany based Wilex AG, in collaboration with the Ludwig Institute for Cancer Research, is developing a kappa light chain chimeric antibody derived from the variable region of the mouse monoclonal antibody G250 (development name WX-G250 or Rencarex) which specifically binds to the CA-IX antigen. The drug candidate Rencarex is currently in a clinical Phase III trial to prevent the development of metastatic disease in kidney cancer patients following partial or complete removal of the affected kidney.

---

[440] Heiss MM et al, Immunotherapy of malignant ascites with trifunctional antibodies. Int J Cancer, 2005 Nov 10;117(3):435-43
[441] Sebastian M et al, Treatment of malignant pleural effusion with trifunctional antibody catumaxomab (anti-EpCAM x anti-CD3): results of a Phase I/II study. J Clin Oncology,2006 Asco annual meeting proceedings, Vol 24,No18S (June 20 supplement),2006:2548

IX.61.14. Panzem

Panzem (2-methoxyestradiol or 2ME2), from EntreMed, targets endothelial cell growth to induce anti-angiogenis activity. EntreMed is developing Panzem in capsule and liquid formulation forms, which are currently in Phase 2 trials. ENMD-1198 is a modified version of 2- methoxyestradiol designed to increase anti-tumor and anti angiogenic activity, as well as decrease its rate of metabolism. A pre-clinical study of oral daily treatment with ENMD-1198 showed the disruption of microtubules within tumor cells and a substantial decrease in tumor cell proliferation and angiogenesis.

IX.61.15.   Angiostatic Cocktail

Angiogen has developed the "Angiostatic Cocktail" that has been shown to produce Angiostatin4.5. Angiostatin4.5 inhibits angiogenesis by inducing selective apoptosis of endothelial cells. The Angiostatic Cocktail combines two drugs that are already approved by the FDA for other indications – a plasminogen activator (urokinase) and a pharmaceutical-free sulfhydryl donor (such as Mesna).

IX.61.16. ATN224

ATN-224 from Attenuon, is an orally active small molecule that inhibits multiple kinase signaling pathways that are important for tumor growth, angiogenesis, and metastasis. ATN-224 has been observed to inhibit VEGF and FGF-2 signaling in endothelial cells, IGF-1, EGF, and integrin signaling in tumor cells, PDGF signaling in pericytes, and NF-κB. ATN-224 inhibits copper-zinc superoxide dismutase (SOD1) by removing copper from the enzyme. The inhibition of SOD1 causes a decrease in intra-cellular $H2O2$ levels, which increases the activity of phosphatases that turn off kinase signaling. ATN-224 has been evaluated in two Phase 1 trials; one involving patients with solid tumors and the other involving patients with advanced hematologic malignancies.

IX.61.17.   Combrestastatin A4 Prodrug

Combretastatin A4 Prodrug, from OXiGENE, works by affecting the microtubules that form the cytoskeleton of endothelial cells lining the tumor vasculature. When this tubulin structure is disrupted, endothelial cells change shape from flat to round and stop blood flow, thus starving the tumor of nutrients and inducing tumor cell death. Combretastatin acts primarily on tumors because tubulin alone is responsible for maintaining the structure in newly formed endothelial cells (mature endothelial cells are protected by Actin).

IX.61.18. Volociximab

Volociximab, a co-development from PDL BioPharma and Biogen Idec is a chimeric monoclonal antibody that inhibits the functional activity of α5β1 integrin, a protein found on activated endothelial cells that are involved in the angiogenesis. Volociximab is in Phase 2 clinical trials in patients with advanced solid tumors. PDL BioPharma and Biogen Idec are co-developing volociximab.

IX.61.19. LY2275796

LY2275796, from Isis Pharmaceuticals, is an antisense anti-cancer drug that is licensed to Eli Lilly and is currently in Phase 1 studies. This drug targets eukaryotic initiation factor-4E (eIF- 4E), a protein involved in tumor progression, angiogenesis and metastases. It is believed that eIF-4E may act as a critical regulator of cancer progression.

IX.62.   Differentiation therapy of cancer: developing   HDAC inhibitors as anti-cancer drugs.

Research activity around histone deacetylase inhibitors (HDAC Inhibitors) has been driven by their ability to modulate transcriptional activity. Cancers commonly involve genes that ultimately affect gene transcription through their interaction with the DNA scaffold – histones and chromatin structures.

The causal involvement of nuclear receptors in the etiology of cancer is documented by mutations in the retinoic acid receptor (RAR),found in acute promyelocytic leukemia (APL),hepatocellular carcinomas and lung cancer.

Such alterations may lead to the deregulated recruitment of enzymes having histone deacetylase (HDAC) activity which might result in inappropriate repression of gene expression. HDAC inhibition causes acetylated nuclear histones to accumulate in both tumor and normal tissues, providing a surrogate marker for the biological activity of HDAC inhibitors in vivo.

The effects of HDAC inhibitors are highly selective, leading to transcription activation of certain genes such as the cyclin-dependent kinase inhibitor p21WAF1/CIP1 but repression of others. HDAC inhibition not only results in acetylation of histones but also transcription factors such as p53, GATA-1 and estrogen receptor-alpha[442].

Inhibition of HDACs could thus relief this block of gene transcriptional activity and result in the induction of a cellular differentiation program of tumor cells, subsequently preventing the cells from further growth or even induce cell death.

One of the most important reasons this therapeutic class is attracting investment is that HDAC inhibitors are able to improve the efficacy of existing cytostatics. The second reason is that HDAC inhibitors are able to target the transcription of specific disease-causing genes. Research indicates that these molecules will be able to modulate various cellular functions, including cell differentiation, cell cycle progression, apoptosis, cytoskeletal modifications and angiogenesis. They sensitize cancer cells to overcome drug resistance when used in combination with other anti-cancer agents. However, drug development is at an early stage. Only a small number of companies are involved in HDAC inhibitors. One of these is G2M Cancer Drugs AG, which identified a small molecular drug, G2M-777, to have HDAC inhibitory activity. Based on its differentiation-inducing and HDAC inhibitory properties, promising results with tumor cells derived from different tissue origins indicate that certain solid tumors may be treated with this compound.

A second company with a big interest in this new field of experimental products is Aton Pharma (acquired by Merck & Co in a move to offset the loss of revenue from existing drugs that are losing patent protection[443]). The company was created by three leading cancer investigators from Memorial Sloan-Kettering Cancer Center (MSKCC), along with a scientist of Columbia University, who, together, and over the course of two decades, have performed research with the objective to develop safer, more effective cancer treatments. Their pioneering work focused on the emerging class of anti-tumor agents-

[442] Vigushin DM, Coombes RC, Histone deacecytlase inhibitors in cancer treatment. Anticancer Drugs, 2002 Jan;13(1):1-13

[443] Merck plans to buy developer of cancer drugs.The New York Times. February 24, 2004

histone deacetylase (HDAC) inhibitors[444]. Aton's leading product, suberanilohydroxamic acid (SAHA)[445], a cytodifferentiating agent and histone deacetylase inhibitor, is a potential cancer chemopreventive in midstage, or Phase 2, clinical trials. The drug is being studied for treatment of cutaneous T-cell lymphoma, a slow-growing form of cancer in which some of the body's white blood cells become malignant.

A team of Japanese scientists[446], in order to find novel nonhydroxamate histone deacetylase (HDAC) inhibitors, synthesized a series of thiol-based compounds modeled after suberoylanilide hydroxamic acid (SAHA), and evaluated their inhibitory effect on HDACs. Compound 6, in which the hydroxamic acid of SAHA was replaced by a thiol, was found to be as potent as SAHA, and optimization of this series led to the identification of HDAC inhibitors more potent than SAHA[447].

In cancer cell growth inhibition assay, S-isobutyryl derivative 51 showed strong activity, and its potency was comparable to that of SAHA. The cancer cell growth inhibitory activity was verified to be the result of histone hyperacetylation and subsequent induction of p21(WAF1/CIP1) by Western blot analysis. Kinetical enzyme assay and molecular modeling suggest the thiol formed by enzymatic hydrolysis within the cell interacts with the zinc ion in the active site of HDACs[448].

Researchers[449] provided evidence that hydroxamic derivative LBH589, an HDAC inhibitor from pharmaceutical giant Novartis, induces a wide range of effects on endothelial cells that lead to inhibition of tumor angiogenesis. These results support the role of HDAC inhibitors as a therapeutic strategy to target both the tumor and endothelial compartment and warrant the clinical development of these agents in combination with angiogenesis inhibitors[450].

Research[451] indicate that the antitumor effect of FK228, an HDAC inhibitor from Gloucester, in clinical development, in esophageal cancer cells is shown at least in part through Prdx1 activation by modulating acetylation of histones in the promoter, resulting in tumor growth inhibition with apoptosis induction[452].

Another HDAC inhibitor drug candidate is Titan Pharmaceuticals' Pivanex AN-9 (pivaloyloxymethyl butyrate). In Lung Cancer[453], Reid et al reported on the results of a multicenter phase II trial confirming the therapeutic activity and safety profile of Pivanex as a single agent in refractory non-small cell lung cancer. A further Phase IIb study of Pivanex in combination with docetaxel in non-small cell lung cancer was terminated

---

[444] Slowik H., "Aton Pharma". Start-Up (a Windhover publication), December 2002. (Art# 2002900224)

[445] Drugs.2004 Jul;7(7):674-82

[446] *lead by Dr. T. Suzuki, Graduate School of Pharmaceutical Sciences, Nagoya City University, 3-1 Tanabe-dori, Mizuho-ku, Nagoya, Aichi 467-8603, Japan. suzuki@phar.nagoya-cu.ac.jp*

[447] *Suzuki et al. Thiol-based SAHA analogues as potent histone deacetylase inhibitors. Bioorg Med Chem Lett. 2004 Jun 21;14(12):3313-7.*

[448] *Suzuki T. et al, Novel inhibitors of human histone deacetylases: design, synthesis, enzyme inhibition, and cancer cell growth inhibition of SAHA-based non-hydroxamates. J Med Chem. 2005 Feb 24;48(4):1019-32.*

[449] *from the Sidney Kimmel Comprehensive Cancer Center at Johns Hopkins.*

[450] *Qian DZ. Et al, Targeting tumor angiogenesis with histone deacetylase inhibitors: the hydroxamic acid derivative LBH589.Clin Cancer Res. 2006 Jan 15;12(2):634-42*

[451] *By Dr. Hoshino, Department of Frontier Surgery (M9), Graduate School of Medicine, Chiba University, Japan.*

[452] *Hoshino I et al., Histone deacetylase inhibitor FK228 activates tumor suppressor Prdx1 with apoptosis induction in esophageal cancer cells. Clin Cancer Res. 2005 Nov 1;11(21):7945-52.*

[453] *Reid et al., Phase II trial of the histone deacetylase inhibitor pivaloyloxymethyl butyrate (Pivanex, AN-9) in advanced non-small cell lung cancer. Lung Cancer. 2004 Sep;45(3):381-6*

early due to significant safety issues. Titan decided to investigate the drug candidate's efficacy and safety in the treatment of refractory chronic lymphocytic leukemia.

Although the market potential for drugs developed to treat chronic lymphocytic leukemia is extremely small compared to that of non-small cell lung cancer, targeting this indication does have the advantage that Titan may be able to apply for orphan drug classification. This status is awarded by the FDA to select approaches that offer potential therapeutic value in the treatment of rare diseases and conditions and enables companies to receive pre-filing regulatory guidance as well as reduced filing fees, and provides for market exclusivity in the US for a period of seven years if the orphan drug approval is granted by the FDA. Success in this niche indication could reopen the door for Pivanex in more common cancers[454].

In an announcement[455] of the review of interim efficacy and safety data from the ongoing open-label trial evaluating belinostat (PXD101) monotherapy for the treatment of CTCL, the observed preliminary objective response rates have met the predefined criteria for expansion of the CTCL arm of the trial. The predefined criteria in the protocol required at least 2 objective responses be observed in the first 13 patients treated. The trial will continue to further evaluate the single agent activity of intravenous belinostat (PXD101) for the treatment of this cancer.

CuraGen describes belinostat (PXD101) as a novel, investigational small molecule HDAC inhibitor; as a promising drug candidate that has been shown in preclinical studies to have the potential to treat a wide range of solid and hematologic malignancies either as a monotherapy or in combination with other active agents. The company is currently conducting 5 proof-of-concept clinical trials with intravenous belinostat (PXD101), including Phase II monotherapy studies in multiple myeloma and T-cell lymphomas, and Phase Ib/II trials evaluating belinostat (PXD101) in combination with 5-FU for colorectal cancer, belinostat (PXD101) in combination with paclitaxel and carboplatin for ovarian cancer, and belinostat (PXD101) in combination with Velcade (bortezomib) for Injection for multiple myeloma. The company is also evaluating an oral formulation of belinostat (PXD101) in a Phase I safety and dose-escalation trial.

NCI-sponsored[456] clinical trials are being conducted in parallel to the ongoing studies sponsored by CuraGen. Data from all of these studies will help identify the most promising cancer indication(s) and define the subsequent steps in the clinical development of belinostat (PXD101)[457].

MethylGene's scientists and collaborators observed that isotypic selective inhibition of certain HDAC enzymes (histone deacetylases are a family of 11 functionally enzymes) may have the potential to impact diseases including cancer, diabetes, inflammation, cardiovascular and neurodegenerative diseases[458]. The researchers are developing MGCD0103, a rationally designed, orally available, isotypic-selective, small molecule inhibitor of histone deacetylases (HDACs).. Aberrant DNA methylation is also believed to

---

[454] Trials & Tribulations: Advances in the Development of Histone Deacetylase (HDAC) Inhibitors for the Treatment of Cancer. Lead Discovery DailyUpdates-Oncology, September 10, 2004

[455] In an announcement by CuraGen Corporation and TopoTarget A/S, December 2006.

[456] A clinical trials agreement (CTA) was signed with the National Cancer Institute (NCI) in August 2004, under which the NCI is conducting a number of additional studies evaluating the safety and potential efficacy of belinostat (PXD101).

[457] http://www.curagen.com/get_prod.php?prod_id=2 (accessed March 22, 2007)

[458] MethylGene company press releases, 14th April 2005, 14th November 2005, 8th December 2005, 5th June 2006, 15th September 2006

play a role in cancer. Vidaza (azacitidine, marketed by Pharmion), a demethylating agent, an approved first-line monotherapy treatment for MDS. Vidazza was the first drug approved in MDS. In completed phase III clinical trials, the complete response rate was 6%, the overall response rate was 16%, when partial response was included. HDAC inhibitors, such as MGCD0103, and demethylating agents both act by turning on tumor suppressor genes that have been inappropriately turned off. Tumor suppressor genes are the body's natural defense against cancer. Preclinical studies demonstrate that MGCD0103 and demethylating agents synergistically kill cancer cells.

The combination of azacitidine with MGCD0103 has encouraging safety, PK and clinical activity profiles and appears to be well-tolerated in patients with advanced AML/MDS[459].

Sept 2006, a trial[460] started with MGCD0103 in patients with relapsed or refractory B-cell lymphomas. Specific patient populations include patients with diffuse large B-cell lymphoma (DLBCL) and follicular lymphoma, two tumor types that are classified as non-Hodgkin's lymphomas (NHL). Key objectives of the study will be to determine the effectiveness of MGCD0103 as a treatment option for these patients. Secondary objectives include determining the safety profile, as well as assessing biomarkers and predictive markers for MGCD0103. The trial is expected to last up to 24 months. Many cases of DLBCL and follicular lymphoma exhibit disrupted HDAC-dependent epigenetic regulation of cancer-related genes. Small molecule HDAC inhibitors are thought to have anticancer activity by regulating aberrant gene expression and restoring gene expression to normal levels[461].

HDAC inhibition represents a novel mechanism for treatment of non-Hodgkin's lymphoma since key targets such as BCL-2 and BCL-6 have been shown to be regulated by HDAC

For the big pharmaceutical companies, the key issue is the prospect of alliances between the small number of companies specializing in HDAC inhibitors, this new class of mechanistic anticancer therapeutics, and the much larger group of companies that are involved in the development of cytostatics.

Table 9.24.   HDAC inhibitors

HDAC inhibitors in clinical development	Ongoing chemical development of HDAC inhibitors			
CI-994 (Pfizer)	Short-chain fatty acids	Butirate & phenylbutirate	Epoxyketone-containing cyclic tetrapeptides	Chlamydocin
FK228 (Gloucester)		Valproate		Cyl-1 & Cyl-2
LBH589 (Novartis)	Hydroxamic acids	ABHA		Diheteropeptin
MGCD0103 (MethylGene)		Oxamflatin		HC-toxin
NVP-LAQ824 (Novartis)		Propenamides		Trapoxins
Pivanex (Titan)		Pyroxamide		WF-3161
PXD101 (CuraGen, TopoTarget)		SAHA	non-epoxyketone-containing cyclic tetrapeptides	Apicidin
SAHA (Merck & Co)		Scriptaid		FR901228
		Trichostatins		CHAPs
	Other structures	Depudecin	Benzamides	CI-994
(non-exhaustive lists)		Organosulfur compounds		MS-275

<hr>

[459] *"Phase I/II HDAC poster (Trial 005): "Phase I/II study of the oral isotype-selective histone deacetylase (HDAC) inhibitor MGCD0103 in combination with azacitidine in patients with high-risk Myelodysplastic Syndromes (MDS) or Acute Myelogenous Leukemia (AML)"and also "Phase I HDAC poster (Trial 004): "A Phase I Study of MGCD0103 Given as a Twice Weekly Oral Dose in Patients with Leukemia (AML, ALL or CML) or Myelodysplastic Syndromes".* The clinical data was presented at two poster sessions at the 48th Annual Meeting of the American Society of Hematology (ASH) in Orlando on Sunday, December 10th, 2006.

[460] *Sponsored by MethylGene and Pharmion.*

[461] *Dr. Michael Crump, Associate Professor of Medicine, University of Toronto Lymphoma Site Leader at Princess Margaret Hospital and principal investigator for this trial in a statement to the press (Sept 15, 2006. MethylGene company press release)*

IX.63. Colorectal Cancer

Colorectal cancer includes the development of malignant tumors within the colon and rectal segment of the gastrointestinal tract. The development of CRC is a multi-step process in which, over the course of five to 15 years, normal colonic mucosa progress to benign adenomatous polyps and ultimately carcinoma. Typical symptoms of CRC include a change in bowel habits, blood in the feces or development of anemia due to bleeding, leading to fatigue. The most common sites of disease beyond the bowel wall are the abdominal lymph nodes, lungs, liver, and bone Blockages due to a colorectal tumor may cause pain and discomfort. Screening programs are available for colorectal cancer. The extent of cellular differentiation of a tumor provides information on the aggressiveness of the cancer. CRC tumors are graded on a scale of 1 to 4, with the higher-grade tumor representing a less differentiated tumor. For example, tumors that are graded as G-1 are well-differentiated cancers and resemble normal colon tissue. These tumors are considered less aggressive than tumors with a higher grade and are associated with a better prognosis. The extent of tumor spread for CRC is classified by the stage of disease at diagnosis. Over the years, a number of staging systems have been used to the categorize CRC. Currently, the American Joint Committee on Cancer staging system is used within the US.[462]

A Phase II trial comparing the effectiveness of celecoxib, taken alone or in combination with eflornithine, in preventing colorectal cancer in patients with familial adenomatous polyposis (FAP) has been initiated in the United States and the United Kingdom. FAP is an inherited disorder that is characterized by the development of numerous polyps in the colon and rectum. People diagnosed with FAP are at increased risk of colon cancer. Although most FAP patients undergo colectomy (surgical removal of all or part of the colon), researchers are interested in developing drugs that may offer an additional measure of protection to individuals with this condition. In the trial, 120 patients between the ages of 18 and 65 who have been diagnosed with FAP have been randomly assigned to receive celecoxib alone or celecoxib plus eflornithine. Eflornithine, also known as alpha-difluoromethylornithine, is an inhibitor of the enzyme ornithine decarboxylase (ODC). Inhibitors of ODC have been shown to suppress tumor formation in experimental models of bladder, breast, colon, and skin carcinogenesis[463].

Table 9.25. Current treatments in colorectal cancer

Stage I	Stage II	Stage III	Stage IV
Surgery	Surgery	Surgery	Surgery if possible
	Adjuvant chemotherapy in trials and off-label	1) Folfox 2) 5-FULV / Capecitabine	Folfox / Folfiri chemotherapy 1st line/ 2nd line.

Researchers have provided new information about a protein responsible for colorectal cancer and the target of a potential drug against this cancer. Called clusterin, this protein has been linked to the development of tumor cells and resistance to cancer therapy, but how it works is not well understood. Pending questions include how this protein is expressed in normal and cancer cells, how it helps cancer cells escape ionizing radiation and chemotherapy, and which patients will benefit from treatment with a drug targeting clusterin.

---

[462] *Colon and rectum. In: Am. Joint Committee on Cancer, eds. AJCC Cancer Staging Manual. 6th ed. New York, NY: Springer; 2002:113-124.*

[463] *http://www.cancer.gov/nci-international-portfolio (accessed Aug 6, 2007)*

The researchers discovered that clusterin is not expressed in normal cells, while in 25 percent of colorectal tumors, the cancer cells contained clusterin. They also showed that the protein is actually made by the cancer cells themselves. These new cancer, especially for patients with tumors producing clusterin. The findings should help improve current therapies against colorectal cancer.

IX.64. Improving median survival rates

Launched in the USA in 1962, the antimetabolite fluorouracil (5-FU) has a long and distinguished history in the treatment of CRC. Initially, 5-FU was used as a single agent; later modulation with agents such as leucovorin (folinic acid) enhanced the cytotoxicity of 5-FU. Enormous efforts have been made to improve fluorouacrils efficacy through the use of biochemical modulators and alternative dosing schedules, albeit mainly as a consequence of the lack of activity of other cytotoxic agents in colorectal cancer. However, fluorouacril's clinical benefit, whether used as an adjuvant to surgery or in the palliation of symptoms, remained comparatively modest. Combination 5-FU and leucovorin (5-FU/LV) has become the cornerstone of chemotherapeutic regimens for the management of metastatic CRC. In metastatic disease, 5-FU/LV increases overall survival when compared with best supportive care, although the median survival for individuals receiving 5-FU/LV therapy was only about 11 months.[464]

One target for colorectal cancer, in addition to other solid and hematologic malignancies, is tumor angiogenesis. Animal models have shown that for a tumor to expand beyond 1 to 2 mm it must develop a vascular network that supports cell growth and metastases.[465] During the last decade a number of new agents, notably irinotecam and oxaliplatin, have entered the CRC therapy market. These treatments are now used both alone and in combination with fluorouracil in the treatment of CRC.

Roche's Xeloda (capecitabine) shows comparable efficacy rates to fluorouacril, but it is orally administered.

IX.65. Potential for bispecific antibodies containing two different binding sites?

CRC presents a reasonable five-year overall survival rate of 61% (SEER Cancer Statistics 1998), but less than 10% of patients diagnosed with metastatic disease can expect to survive for five years. This highlights the need for an effective screening program for CRC, which receives less attention than diseases such as prostate and breast cancer, despite an incidence rate in 2003 that ranks it as the second most common cancer in the seven major markets. There is also a need for increased education regarding CRC prevention.

Drug treatments for CRC exhibit similar limitations to those for other cancers in terms of excessive side effects and insufficient survival benefit. In the absence of a curative treatment (in metastaic setting), there is also a need for agents that can improve the quality of life for patients with unresectable tumors. Furthermore, existing treatments tend to follow protracted delivery schedules and are relatively expensive, highlighting the need for a treatment that is not only efficacious, bit also inexpensive and convenient to deliver. Even in the absence of improved efficacy, treatments that are less toxic and more convenient are likely to experience rapid uptake by physicians.

---

[464] *Advanced Colorectal Cancer Meta-Analysis Project. Modulation of fluorouracil by leucovorin in patients with advanced colorectal cancer: evidence in terms of response rate. J Clin Oncol. 1992;10:896-903.*
[465] *Hicklin D, Ellis L. Role of the vascular endothelial growth factor pathway in tumor growth and angiogenesis.J Clin Oncol. 2005;23:1011-1027.*

IX.66. Regional differences in treatment preferences

In 1999, eloxatin was approved in the EU. In 2000, Camptosar (irinotecan) became the first new colorectal cancer therapeutic to be approved in the US (for 1st line treatment of patients with metastatic disease) after more than four decades of domination by injectable 5-FU. It was the start signal for a series of innovative competitors (and several more are in development).

Treatment preferences differed for many years between Europe and the USA. In the USA, the IFL regimen (5-FU delivered via a bolus injection, although criticized for the acute toxicity of this regime; in Europe, the mode of administration is infusion) had over 50% new Rx in 2002, but was taken over by the FOLFOX4 regimen in 2003, with 40% of new Rx in 1st line treatment and 47% new Rx in 2nd line, because studies demonstrated superior efficacy with this regimen over the 5FU/LV or over the IFL (5FU/LV/Camptosar) regimen. Currently, the treatment preferences in the USA and in Europe are as follows:

Table 9.26. Treatment preferences in metastatic setting

	Frequency of use	Treatment
1st line	Most used	Folfox4 (Eloxatin) or Folfiri (Camptosar + 5-FU infusional + LV) or Avastin + Irinotecan based regimen (Approval of Avastin restricted its use with Irinotecan) or Avastin + 5-FU based regimen primarily Folfox 60% or
	Least used	Xeloda / 5-FU LV
2nd line	Most used	Folfox4 if Folfiri in 1st line or Folfiri if Folfox4 in 1st line or
	Least used	Xeloda
3rd line		Camptosar (Irinotecan) + Erbitux

IX.67. Breast Cancer

Breast cancer incidence continues to rise in many countries, and it is the most common type of malignancy in women. It accounts for one in three cancers diagnosed. However, the last decade has seen a fall in mortality rates from beast cancer, reflecting both more effective treatments and increased screening and public awareness of the disease. Despite improvements in treatment, breast cancer is still particularly prevalent, with an estimated 15 million women seeking medical treatment each year, resulting in 1.3 million biopsies - the vast majority of which prove to be benign.

Survival statistics are good, reflecting the predominantly early detection of the disease, and the success of treatment at this stage.

A vast amount of new clinical research data has emerged and several new clinical trials have been initiated and others generated new results. Protein kinase inhibitors and epothilones have generated substantial amount of new research data in this field.

Access to treatment differs from country to country. With approximately 600 cancer specialists, serving a population of 59m, the average UK cancer consultant sees 550 patients a year –and in some cases sees as many as 1.000/year- compared with an average of 250 in France and Germany.

Once cancer is diagnosed, care in the UK can vary significantly. The National health Service (NHS) is divided into about 115 local health authorities, each of which receives a fixed sum from central government and decides independently how to spend it. A woman in the UK with breast cancer has a 67% chance of surviving for 5 years after treatment, compared with a 72% chance for a woman in Germany and 84% in the USA. In the UK, owing to differences in decisions made by local health authorities, a patient may be denied specific types of treatment on ground of costs by a system that will not pay for the drugs needed –drugs that, in some cases, have been developed in the UK itself.

The National Institute for Clinical Excellence (NICE) recommendations are positive for women with breast cancer in the UK. However, this will impact patients suffering from other conditions, who will be affected by the transfer of available resources to the treatment of breast cancer. The situation reflects the financial burden on the system in the UK –and illustrates the cost in human suffering as a result from prioritizing specific healthcare areas over others. Is this a system which has lost track with human reality and which, as a consequence, urgently needs a profound overhaul ?

IX.68. Risk factors, key predictors

The risk of developing breast cancer increases with age, with most breast cancers occurring in women over the age of 50. White women are at greater risk of developing the disease than black, Asian and Hispanic women, although the prognosis for white women is better. Exposure to estrogen over time is thought to increase the risk of breast cancer, and therefore women who started menstruation early, experienced menopause late or have never had children may be at increased risk. A number of genetic factors have been identified that confer increased risk of developing the disease, and presence of one or more of these factors can be anticipated from a family history of the disease.

The majority of breast cancer cases are caused by genetic changes that occur during a woman's lifetime and not by genetic mutations inherited from her parents. However, researchers estimate that inherited mutations play a role in anywhere from 5 to 27 percent of all breast cancer cases.

In the mid 1990s, researchers found that mutations in the tumor suppressor genes BRCA1 and BRCA2, localized on chromosome 17 and 13, respectively, are a major cause of the hereditary form of the disease. Women inheriting these mutations have a 40 to 85 percent lifetime risk of developing breast cancer, as well as an increased risk of ovarian cancer. Till 2006, most of the studies on BRCA1 and BRCA2 mutations have focused on families known to be at high risk for breast cancer and on women who develop breast cancer at a relatively young age.

Research identified key predictors of whether a woman with breast cancer is likely to carry a BRCA1 or BRCA2 mutation. Such information is important because it can help to improve means of assessing which women may benefit the most from genetic testing, increased breast cancer screening and other measures aimed at early detection, treatment or prevention. The most significant predictors for BRCA1 mutations were: Jewish ancestry, a family history of ovarian cancer and a family history of breast cancer occurring before age 45.

380

For BRCA2 mutations, researchers uncovered fewer predictors, and they had more modest effects. Among the breast cancer patients studied, the only significant predictors of a BRCA2 mutation were early age of onset (before age 45) in the patient herself or early onset of breast cancer in mother, sisters, grandmothers or aunts.

IX.69. A landmark therapeutic

The proto-oncogene HER2/neu localized on chromosome 17q21 is implicated in several types of cancers, approximately 20-30% of all breast cancers overexpress the HER2 gene and amplification or overexpression of HER/neu correlates with poor prognosis.

Mutation in the p53 tumor suppressor gene is the most common genetic abnormality seen in cancer and up to 50% of all primary breast carcinomas have this mutation. Inflammatory breast cancer, the most deadly form of the disease, has been directly linked to the RhoC GTPase gene; this oncogene may also be involved in other aggressive forms of breast cancer[466].

In 1998, the FDA licensed a new biologic approach for the treatment of metastasic breast cancer, or cancer that has spread beyond the breast and lymph nodes under the arm. The new intravenous product, Herceptin (Trastuzumab), is a humanized anti-HER2/Neu antibody, which specifically targets the protein product of the gene known as HER-2, HER2/neu, or c-erbB2. It was approved for use alone for certain patients who have tried chemotherapy with little success or as a first-line treatment for metastatic disease when used in combination with Taxol (paclitaxel), and Taxotere (docetaxel) in the EU.

Herceptin was the second monoclonal antibody approved to treat cancer. The first, Rituxan (Rituximab) was developed by Genentech and its Partner IDEC Pharmaceuticals Corp., for patients with one type of non-Hodgkin's lymphoma, a cancer of the immune system. Herceptin is a monoclonal antibody bioengineered from part of a mouse antibody which is altered to closely resemble a human antibody. It binds to a protein called HER2 which is found on the surface of some normal cells and plays a role in regulating cell growth. In laboratory experiments, Herceptin inhibited tumor cell growth by this binding action.

In the case of metastatic breast cancer cells, approximately 30% of tumors produce excess amounts of HER2. Only patients who have tumors with this characteristic have been studied and shown to benefit from the new, targeted approach using Herceptin. In one of four major national studies that examined early use of trastuzumab, they discovered that the drug could change the natural history of the disease. Trastuzumab cut cancer recurrence-the return of the disease by 52 percent, compared to standard therapy

Herceptin, approved for use in the European Union for advanced (metastatic) HER2-positive breast cancer in 2000 received approval for early HER2-positive breast cancer in November 2006.

In the advanced setting, Herceptin is approved for use as a first-line therapy in combination with paclitaxel where anthracyclines are unsuitable, as first-line therapy in

---

[466] Sorbera L.A. et al, Pipendoxifene. *Drugs of the Future 2002, 27(10):942-947*

combination with docetaxel, and as a single agent in third-line therapy. In the early setting, Herceptin [467]is approved for use following standard (adjuvant) chemotherapy. To October 2006, over 310,000 patients with HER2-positive breast cancer have been treated with Herceptin worldwide.

Using the readily available original breast cancer tissue sample, the Oncor Inform HER-2/neu test identifies the presence or absence of increased copies of the HER-2/neu gene. This indicates whether or not breast cancer is likely to return.

In clinical trials, 31 percent of patients with originally localized breast cancer that had a positive HER-2/neu test died within five years of surgery, while 97 percent of patients with negative test results survived at least five years. Conventional detection procedures must wait until the disease recurrence is present, allowing the cancer to advance before treatment begins.

Three different clinical trials showed that adding trastuzumab (Herceptin) to standard adjuvant chemotherapy significantly reduced the risk of recurrence in women with the early-stage breast cancer, HER-2/neu positive, which has an over expression of protein in the gene. Approximately 50,000 women in the United States are diagnosed with HER-2/neu positive breast cancer each year, representing about 20% of invasive breast cancers[468].

Trastuzumab revolutionized the treatment of HER2-positive breast cancers and represents an effective therapy for some women with one of the most aggressive forms of breast cancer.[469] The main side effect of trastuzumab is that it can damage heart muscle cells. Heart abnormalities have been detected in 2 to 7 percent of women taking the drug, and about one in ten women cannot take the drug because preexisting heart problems put them at greater risk for heart damage. To date, there appear to be fewer cardiac effects associated with lapatinib therapy.

Both drugs are prescribed to women whose cancerous breast cells have HER2 genes that are overactive. Approximately one in four women with breast cancer have this overactive gene, which is associated with increased cancer recurrence and worse outcomes. Lapatinib was approved in March 2007 for use in women who have not responded to trastuzumab therapy.

According to the scientists, their experiments in isolated human heart muscle cells indicate that lapatinib activates a pathway that protects cardiac muscle cells from the death-promoting effects of mediators of inflammation, which are activated in cancer patients, particularly those who have received chemotherapies that damage heart tissue. In contrast, trastuzumab does not activate this protective pathway.

---

[467] *Herceptin is marketed in the United States by Genentech, in Japan by Chugai and internationally by Roche.*

[468] *Testimony delivered Apr. 6, 2006, before the U.S. House Subcommittee on Labor-HHS-Education Appropriations, to discuss the NCI budget request for Fiscal Year 2007, by Dr John E. Niederhuber, Deputy Director, National Cancer Institute, and Richard Turman, Deputy Assistant Secretary, Office of Budget, Department of Health and Human Services. http://www.cancer.gov/aboutnci/FY07-budget-request. (Accessed Aug 6, 2007)*

[469] *Statement by Duke University Medical Center oncologist Dr. Neil Spector, first author of a study that appeared early online in the journal Proceedings of the National Academy of Science. The study was supported in part by the Duke Comprehensive Cancer Center. Other members of the team included scientists from Duke, the Weizman Institute, Cell Signaling Technologies, and Targeted Molecular Diagnostics. Press release from Duke University, June 13 2007.*

Using this system, they claim that they could theoretically screen drugs that are in the development phase to see what their effects may be on heart muscle cells, stating that they may be able to select the drug candidates that have the fewest cardiovascular side effects and theoretically would be safer for patients. (A way to find out about some of these potential problems long before the drugs even make it to market).

Additionally, there is the potential for the development of similar drugs that can be used as protective agents in situations where the heart is stressed for periods of time, such as during heart attacks, coronary artery bypass surgery or angioplasty. Such a drug could even be used to preserve cardiac function in hearts being harvested for transplant.

Both chemotherapies and hormonals are used in the treatment of breast cancer. Tamoxifen was the prototype hormonal agent of the class of selective estrogen receptor modulators, approved 1977 and still widely used today in the treatment of the disease. Other typical hormonal therapies include goserelin (an estrogen suppressor) and anastrazole (an aromatase inhibitor).

Table 9.27.   Breast cancer market opportunities

Chemotherapy is also utilized as an adjuvant treatment to surgery, despite the unpleasant side effects accompanying cytotoxic medicines. Agents such as cyclophosphamide, 5-fluoroucaril and methotrexate are most commonly used as cytotoxics for breast cancer.

An international clinical trial led by the National Cancer Institute of Canada Clinical Trials Group (NCIC CTG), showed that the estrogen-suppressing drug letrozole (Femara) reduced the risk of breast cancer recurrence and the incidence of new breast cancer in the opposite breast by 42 percent compared to placebo in women whose tumors were hormone receptor-positive.

Previous studies had shown that 5 years of treatment with the drug tamoxifen after surgery, radiation therapy, and chemotherapy for early-stage breast cancer could reduce the risk of recurrence by almost half in women whose tumors were estrogen receptorpositive. Some of these studies also showed, however, that no additional benefit is obtained by continuing tamoxifen treatment beyond 5 years. Because more than 50 percent of the women who experience a recurrence of their cancer do so more than 5 years after diagnosis, additional treatment options are necessary. This trial showed that a 5-year course of letrozole, when given after 5 years of tamoxifen therapy, significantly reduced the risk of local recurrence and metastasis (distant recurrence)[470].

Novel and targeted cancer therapies are starting to make an impact on the breast cancer market, with the monoclonal antibody Herceptin rapidly moving toward gold-standard therapy for a subset of patients whose tumors over-express a certain protein.

Table 9.28.   Showing promise in the treatment of cancer[471]

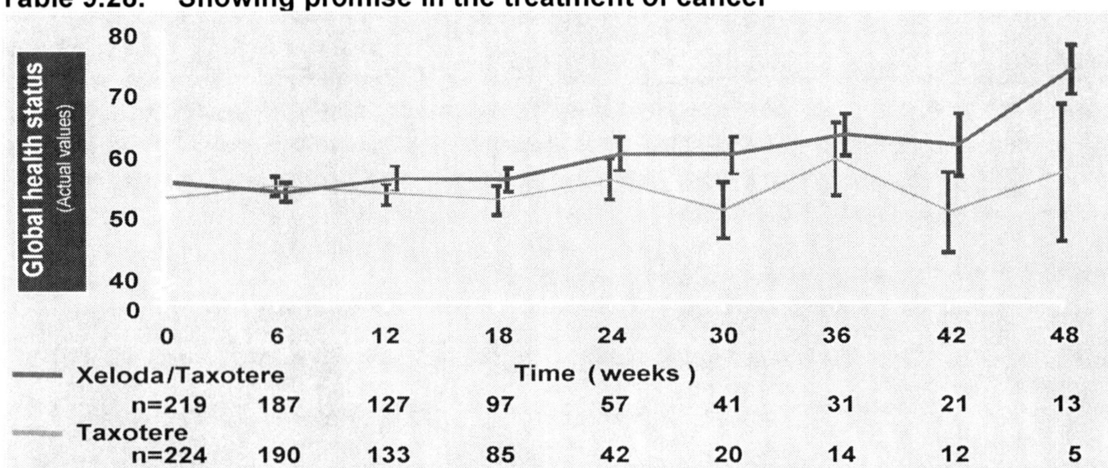

Xeloda is a useful chemotherapy for metastatic breast cancer, both as monotherapy and in combination with other cytotoxic medicines[472].

Table 9.29.   Superior survival with Xeloda/Taxotere

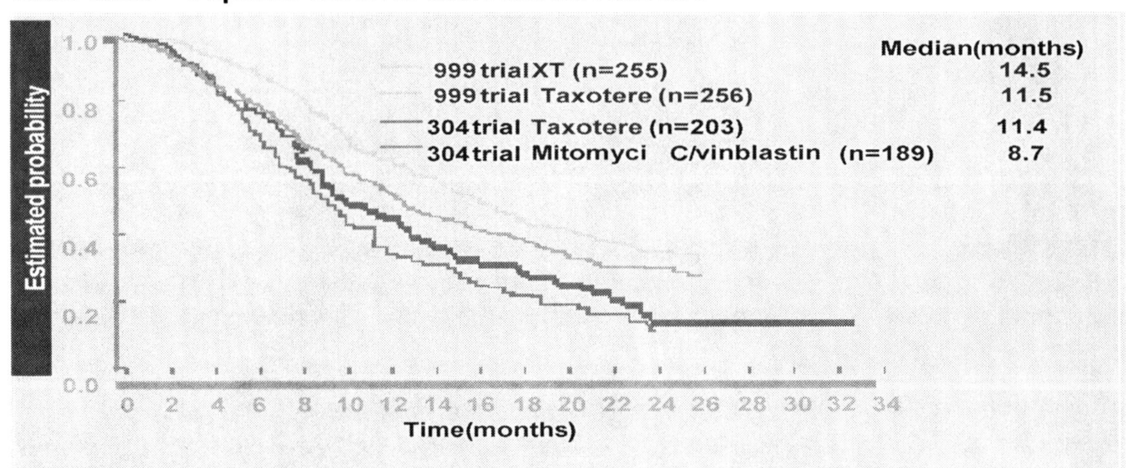

[470] with support from the Canadian Cancer Society (CCS), and in partnership with NCI and Novartis Pharmaceuticals, In addition to patients recruited through the NCIC CTG and NCI, European participants were enrolled in the letrozole study by the European Organization for Research and Treatment of Cancer (EORTC) and the International Breast Cancer Study Group (IBCSG). Participants in the trial were enrolled through hospitals, cancer centers, and institutes throughout Canada, the United States, England, Belgium, Ireland, Italy, Poland, Portugal, and Switzerland. The trial's participants will continue to be followed for 10 to 15 years. http://www.cancer.gov/nci-international-portfolio/page7(accessed Aug 6, 2007)

[471] Source : O'Shaugnessy, San Antonio Breast Cancer Symposium 2001

[472] According to researchers of EORTC Investigational Drug Branch for Breast Cancer, Chemotherapy Unit, Institut Jules Bordet, Brussels, Belgium. Hospital Universitario San Carlos, Servicio de Oncologia Medica, Madrid, Spain. Cancer Research UK, Department of Medical Oncology, University of Glasgow and Beatson Oncology Centre, Glasgow, United Kingdom. The research was published in an article "Moving forward with Capecitabine", December 7, 2002 issue of Oncology

The proven activity of capecitabine has provided the rationale to explore its use earlier in the course of the disease and in combination with other agents, particularly those known to further up regulate thymidine phosphorylase (TP) concentrations in tumor issue[473].

IX.70. Regulating the effectiveness of Taxol chemotherapy

Taxotere (docetaxel) is a semi-synthetic with slight chemical differences versus Taxol.. Taxol (also known as paclitaxel) was originally derived from the Pacific yew tree, These taxanes stabilize a cell's "microtubules," the road-like protein structures that send chemical signals to all parts of the cell, and which must be flexible if a cell is to divide. Taxanes lock these structures into place, not allowing them to change when the cell begins to divide - which is necessary for tumor growth. Research has also indicated that the drugs induce programmed cell death (apoptosis) in cancer cells by inactivating an "apoptosis stopping protein" called BCL2, thus stopping it from inhibiting cell death.

Cancer researchers[474] have taken a step towards understanding how and why a widely used chemotherapy drug works in patients with breast cancer. In laboratory studies, the researchers isolated a protein, caveolin-1, showing that in breast cancer cells this protein can enhance cell death in response to the use of Taxol, one of two taxane chemotherapy drugs used to treat advanced breast and ovarian cancer. But in order to work, they found the protein needs to be "switched on," or phosphorylated. Their finding suggests it may eventually be possible to test individual breast cancer patients for the status of such molecular markers as caveolin-1 in their tumors to determine the efficacy-to-toxicity ratio for Taxol. Because breast tumors are not all the same, it is important to know the cancer's molecular makeup in order to increase the efficiency, and lower the toxicity, of chemotherapy drugs, and this work takes us some steps forward in this goal. It also offers insights into why some breast cancer cells can become resistant to therapeutic drugs[475].

Additionally, the study identifies caveolin-1 as a new molecular target for increasing the efficacy of taxanes[476]. Caveolin-1 is a protein that is found in most cells under normal conditions and it is involved in an array of cellular events that ranges from vesicle trafficking to cell migration. It is, therefore, as a key regulator of multiple events within the cell. In cancer, the expression level of caveolin-1 can vary depending on cell type. However, the precise role of caveolin-1 in cancer has been controversial: whether it acts as a suppressor or facilitator of tumor formation depends on the cell type. In human breast cancer, caveolin-1 has been known to act as a tumor suppressor since caveolin-1 expression is down-regulated during the primary stages of breast cancer. More recent studies indicate that caveolin-1 expression is increased in more aggressive types of breast cancer.

The scientists sought to determine factors that regulate expression and function of caveolin-1 in the breast. In this study, they show that in their breast cancer cell model, phosphorylated caveolin-1 increased cell death by activating other key regulators vital to both breast cancer progression and cell death, including BCL2, the same protein that

---

[473] *Biganzoli L. et al, Oncologist 2002 Dec;7 suppl 6:29-35*

[474] *at Georgetown University Medical Center.*

[475] *Dr Ayesha Shajahan, of Lombardi Comprehensive Cancer Center at Georgetown, the study's first author in a statement to press by Georgetown University Medical Center, February 25, 2007.The results of the research were reported in the Journal of Biological Chemistry. The study was supported by grants from the National Institute of Health, and the Department of Defense to Clarke and a postdoctoral fellowship award from Susan G. Komen Breast Cancer Foundation to Shajahan.*

[476] *Dr. Robert Clarke, Professor of Oncology and Physiology & Biophysics (see reference 287, above)*

Taxol works on; p21, which controls cell cycle progression; and the tumor suppressor p53. If caveolin-1 isn't phosphorylated, breast cancer cells appear to be resistant to Taxol treatment, the researchers conclude.

Despite success in treating some cancers with chemotherapy, many tumors are naturally resistant to anticancer drugs or become resistant to chemotherapy after many rounds of treatment. Researchers at NCI and elsewhere have discovered one way that cancer cells become resistant to anticancer drugs: they expel drug molecules using pumps embedded in the cellular membrane. One of these pumps, called P-glycoprotein (P-gp), is the protein product of the MDR1 gene and contributes to drug resistance in about 50 percent of human cancers. P-gp prevents the accumulation of powerful anticancer drugs, such as etoposide and Taxol, in tumor cells. The same pump is also involved in determining how many different drugs, including anticancer drugs, are taken up or expelled from the cell[477].

While the monoclonal antibody drug Herceptin has been very successful in treating ErbB2-positive breast cancer, patients can relapse, other drug targets are needed. A newly found secreted factor may prove to be such a target. Research showed that Akt1 causes cancer cells to secrete a factor "CXCL16" that promote breast cancer cell migration. There goal is to find a way of blocking CXCL16 production.

IX.71. Genes set scene for metastasis

Studies of human tumor cells implanted in mice have shown that the abnormal activation of four genes drives the spread of breast cancer to the lungs. New studies reveal that the aberrant genes work together to promote the growth of primary breast tumors. Cooperation among the four genes also enables cancerous cells to escape into the bloodstream and penetrate through blood vessels into lung tissues.

Although shutting off these genes individually can slow cancer growth and metastasis, the researchers found that turning off all four together had a far more dramatic effect on halting cancer growth and metastasis. Metastasis - the leading cause of mortality in cancer patients - entails numerous biological functions that collectively enable cancerous cells from a primary site to disseminate and overtake distant organs. It is the deadliest transformation that a cancer can undergo, and therefore researchers have been looking for specific genes that propel metastasis.

In the newly published experiments, the researchers also found that they could reduce the growth and spread of human breast tumors in mice by simultaneously targeting two of the proteins produced by these genes, using drugs already on the market. The researchers[478] are exploring clinical testing of combination therapy with the drugs - cetuximab (trade name Erbitux) and celecoxib (Celebrex) - to treat breast cancer metastasis.

In an earlier study, the same scientists had identified 18 genes whose abnormal activity is associated with breast cancer's ability to spread to the lungs. In the new study the

---

[477] *Kimchi-Sarfaty C. et al, A "silent" polymorphism in the MDR1 gene changes substrate specificity. "Science Express", December 21, 2006.*

[478] *The research team, led by Howard Hughes Medical Institute investigator Joan Massagué, Chair of the Cancer Biology and Genetics Program at the Memorial Sloan-Kettering Cancer Center, published its findings in articles in the journal Nature and in the online early edition of the Proceedings of the National Academy of Sciences. Press release from the Howard Hughes Medical Institute , April 15, 2007. And also the Memorial Sloan-Kettering press release of April 15, 2007*

investigators [479] focused on four of these genes. The gene set comprises EREG (epiregulin or epidermal growth factor receptor ligand), cyclooxygenase COX2, MMP1 and MMP2 (matrix metalloproteinases 1 and 2 that are expressed in human breast cancer cells), were already known to help regulate growth and remodeling of blood vessels.

They found that depriving aggressive metastatic tumor cells of the genes for these four proteins decreased both their ability to grow large aggressive tumors in the mouse mammary gland and also the ability to release cells from these tumors into the circulation. The remarkable thing was that while silencing these genes individually was effective, silencing the quartet nearly completely eliminated tumor growth and spread. Microscopic analysis of blood vessel structure in the tumors revealed that knocking down all four genes greatly reduced growth of the tangle of blood vessels typically seen in tumors. Their findings provide a beautiful explanation for how the genes that they identified in breast cancer patients as being associated with lung metastasis manipulate blood vessels to give them an advantage both in the primary tumors and in the lung.

The two drugs Cetuximab and Celecoxib[480] act directly on proteins produced by the genes the biologists had been studying. The researchers also tested whether cetuximab and celecoxib would work effectively in concert to reduce metastasis in mice. The researchers found that the combination of these two inhibitory drugs was effective, even though the drugs individually were not very effective. This really nailed the case that if these genes can be inactivated in concert, it will affect metastasis.

While clinical trials of the drug combination are being discussed, there are already treatments to diminish the chance of metastasis in breast cancer, so such trials would have to be designed very carefully to understand how and whether the new drug combination would be of additional benefit. The researchers explored how the entire group of 18 genes, called the 'lung metastasis gene-expression signature' (LMS) influenced both breast tumor growth and spread to the lungs[481].

After analyzing 738 human breast cancer tumors, the researchers concluded that those in which the LMS genes were abnormally active were, indeed, more likely to develop lung metastases. They also found that the activity of these LMS genes gave cancer cells a growth advantage by allowing tumors to develop a rich network of blood vessels to deliver oxygen and nutrients. Although large tumors are more likely to metastasize, the findings indicated that the activity of the LMS genes was also critical to the metastasis process. As the tumors grow and become enriched with LMS-positive cells, because the genes give them an advantage, they reach a point where the tumor becomes richly vascularized. Then, they can massively execute the advantage the LMS genes provide them to metastasize to the lung.

IX.72. No two patients are identical. Each cancer has a different blueprint.
Johns Hopkins Kimmel Cancer Center scientists have completed the first draft of the genetic code for breast and colon cancers. Their report   identifies close to 200 mutated

---

[479] *Massagué and his colleagues at Sloan-Kettering, along with researchers from Hospital Clinic de Barcelona and the Institute for Research in Biomedecine in Spain (the study is published in the journal „Nature")*
[480] *Cetuximab is an antibody that blocks the action of epiregulin and is used to treat advanced colorectal cancer. Celecoxib is an inhibitor of COX2 that is used as an anti-inflammatory, and is being tested in clinical trials against many types of cancer.*
[481] *In an article published in the Proceedings of the National Academy of Sciences by Massagué and his colleagues. Co-authors on the paper were from the University of Chicago, The Netherlands Cancer Institute, Veridex L.L.C., The Cleveland Clinic and the Erasmus Medical Center in The Netherlands.*

genes, now linked to these cancers, most of which were not previously recognized as associated with tumor initiation, growth, spread or control. Just as sequencing the human genome laid the groundwork for subsequent research in genetics, these data lay the foundation for decades of research on colon and breast cancers. In total, the team combed through 465 million nucleotides – several encyclopedias' worth of letters – to find approximately 1,500 DNA nucleotides that differed from the normal code in important ways. Virtually all these mistakes were mere single-nucleotide "typos."

Some 200 genes were significantly mutated; the mutated genes in breast and colon cancers were almost completely distinct, suggesting very different pathways for the development of each of these cancer types.

This gives us some understanding of why breast and colon cancers, and most likely other cancers as well, are very different diseases and develop through different processes. Now researchers will study how these mutations occur in breast and colon cancers, perhaps searching for environmental agents or cellular processes that drive these changes.

The team also found that the average number of mutant genes in each cancer is about 100, and at least 20 are likely to be crucial for tumor formation. Each cancer has a different blueprint, they conclude.

No two patients are identical.

IX.73.   Reverting cancer cells back to a pre-malignant state.
Scientists[482] have shown that the activity of a gene that commandeers other cancer-causing genes, returning them to normal, can predict the prognosis of an individual with breast cancer.

The gene, Dachshund, normally regulates eye development and development of other tissues, in essence playing a role in determining the fate of some types of cells. The researchers looked at cancer cells from more than 2,000 breast cancer patients and found that this commandeering or "organizing" ability is increasingly lost in cancer cells and associated with the progression of disease. The more the gene is expressed in breast cancer, the researchers saw, the better the patient did.

This is a new type of gene in cancer that commandeers the cancerous genes and returns them to normal. The standard cancer treatment strategy has been to block the proliferation of cancer cells or cause them to die. This is quite different. They have shown that the Dachshund gene reverts the cancerous phenotype and turns the cell back to a pre-malignant state. Cells don't die, but rather, they revert.

In the work, the researchers showed that Dachshund could block breast cancer growth in mice and also could halt breast cancer from invading other tissues in cell culture. They also found that the gene inhibits the expression of the cyclin D1 gene, a cancer-causing gene that is overexpressed in about half of all breast cancers.

---

[482] *at the Kimmel Cancer Center at Thomas Jefferson University in Philadelphia*

388

IX.74. Genetic signature predicts recurrence of cancers

Researchers working with breast cancer stem cells at the Stanford University School of Medicine have found 186 genes that together can predict the risk of recurrence in breast cancer patients. Additionally, the same genes predict the recurrence of prostate cancer, lung cancer and medulloblastoma, the most common form of childhood brain cancer. These data suggest that there are some fundamental properties of the malignancy that are shared between many types of tumors.

The Stanford scientists also found they could predict aggressive cancers with even greater accuracy when they combined the 186-gene signature with another previously identified group of genes in the cells surrounding the tumor. The findings are a significant step toward using insights from cancer stem cell research to develop better tools for diagnosing and treating the disease.

Their research finding demonstrates, for the first time, the clinical value of isolating pure human cancer stem cells. It is now shown that the human breast cancer stem cell reveals information not evident when the whole tumor is analyzed[483].

Dr Clarke[484] has already begun working with surgeons to help translate these findings into tools that will help doctors zero in on the best treatment for individual cancer patients. For instance, he hopes that within five years a surgeon would be able to leave more breast tissue intact in a patient whose genetic profile showed that her tumor was unlikely to return. Likewise, a woman at high risk of relapse would get more aggressive treatment.

Cancer stem cells are believed to be the cells that continually replenish cancer, like the spring at the source of a creek. Although other cancer cells can damage the body through sheer bulk, they can't form new cancers.

This newest breast cancer study isn't the first to identify a group of genes that can predict the risk of cancer recurrence[485]. However, according to the Stanford scientists, these genes are particularly effective at flagging risky cancers, especially when combined with a second group of genes[486], active in the normal cells surrounding the tumor, called the stroma.

In essence, combining the two gene profiles is like examining both the soil (the stroma) and the seed (the cancer stem cells) to find out if the combination is likely to produce a tumor that can recur. The finding shows that malignancy is the result of an interaction between the cancer cells in the tumor and the environment, as many researchers had previously surmised.

Over time, researchers will likely be able to refine the list of genes needed to most accurately determine a patient's risk of recurrence. A smaller list of genes could make a genetic profiling test more cost-effective for widespread use.

---

[483] *Dr. Irving Weissman, director of the Stanford Institute for Stem Cell Biology and Medicine, in a statement to the press, January 16, 2007*

[484] *Dr. Michael Clarke, the Karel H. and Avice N. Beekhuis Professor in Cancer Biology, and the lead author of the study, published in the Jan. 18, 2007 issue of the New England Journal of Medicine. Stanford University Medical Center press release, January 16, 2007.*

[485] *Dr. Clarke found the first cancer stem cell in a solid tumor in 2003 while at the University of Michigan. Until he found the deadly cells in breast cancer, researchers had thought they might be restricted to blood cancers. Since then, other researchers have found the cells in brain and prostate cancer. January 2007, Dr. Mark Prince, an assistant professor at the University of Michigan, reported finding cancer stem cells in head and neck cancers.*

[486] *originally identified by a team led by Dr. Patrick Brown, professor of biochemistry.*

IX.75. Variation in the Caspase 8 gene

A large-scale analysis of data on breast cancer risk has concluded that a common variation in the gene caspase-8 (CASP8) is associated with a somewhat lower risk of the disease. Variants are small changes that occur in a gene sequence. The results are from the second study published by the Breast Cancer Association Consortium (BCAC).

The consortium, which was launched in 2005, has helped speed the discovery of genes and variants involved in breast cancer through the pooling of data from many studies and should prove invaluable to future research studies that explore the genetic aspects of cancer[487].

Mutations in genes with large effects on breast cancer risk have been identified, yet most single population-based studies lack the statistical power to detect more common variations in genes that contribute only small amounts to breast cancer risk. To confirm a previously identified association with a variant of the CASP8 gene, for example, the Consortium used data on 33,000 women from 14 studies.

This study[488] provides proof of principle that consortia like the BCAC are valuable for understanding the contributions of genetic factors in complex diseases. A better understanding of the biology of breast cancer is likely to come from the identification of these variants and future studies that investigate the mechanisms underlying the associations.

To date, the BCAC has evaluated about 20 single nucleotide polymorphisms (SNPs), the most common type of gene variant in which a single unit of DNA may vary from one person to the next. Each SNP investigated has been linked to breast cancer risk by at least two studies that included more than 10,000 individuals. The BCAC uses essentially all the available data on a SNP, including unpublished results, to either confirm or refute an association.

Two previous studies suggested that a SNP in CASP8, called D302H, is associated with a reduction in breast cancer risk. The researchers caution that population-based data alone cannot prove that this particular variant is responsible for the association. Other studies are underway to determine whether D302H or another variant nearby might be responsible.

Most of the 14 studies that contributed data to the Consortium's analysis included women of predominantly white European ancestry. The variant is estimated to be present in 13 percent of white women of European descent. (The variant was not present in two Asian populations studied.)

Future studies will also investigate the biology of how variation in this gene may protect against the disease. The protein produced by the CASP8 gene participates in programmed cell death, or apoptosis, a defense mechanism that allows cells to commit suicide rather than develop into a tumor. DNA damage can trigger apoptosis, and one

---

[487] *NIH Director Dr. Elias A. Zerhouni in a statement to the press, February 11, 2007.*
[488] *Cox A, Dunning AM, Garcia-Closas M, et. al. A common coding variant in CASP8 is associated with breast cancer risk. "Nature Genetics",*
Vol. 39, No. 2. February 11, 2007

hypothesis is that the CASP8 SNP may enhance the body's ability to clear cancerous cells from the body and thereby lower the risk of breast cancer.

While mutations in some genes such as BRCA1 have previously been linked to cancer risk, the genetic variation found in the CASP8 gene is the first common variant to be definitively associated with breast cancer risk. In addition to this variant, the study found some support for an association with breast cancer risk for a variant in the gene TGFB1(Transforming Growth Factor Beta 1).

Effective strategies for finding variants in genes that contribute modestly to breast cancer risk are needed. Mutations in genes such as "BRCA1" or "BRCA2" account for less than 25 percent of the excess familial risk of breast cancer. Much of the remaining variation in genetic risk is likely to be explained by the cumulative effect of multiple variants in genes that individually confer relatively small amounts of risk.

The SNPs in CASP8, and perhaps in TGFB1, are the first examples of such variants. Given that only a small number of genes have been investigated in this depth, the results suggest that large numbers of breast cancer susceptibility variants will eventually be found[489].

IX.76 A novel cellular mechanism may guide cell movement and possibly cancer invasion

Scientists have long thought that microtubules, part of the microscopic scaffolding that the cell uses to move things around in order to hold its shape and divide, originated from a tiny structure near the nucleus, called the centrosome.

But: researchers[490] revealed a surprising new origin for these cellular "highways". They report that the Golgi apparatus -- a stack of pancake-shaped compartments that sorts and ships proteins out to their cellular destinations -- is the source of a particular subset of these microscopic fibers. The findings point to a novel cellular mechanism that may guide cell movement and possibly cancer cell invasion.

Microtubules are the largest of the three main types of filaments that make up the cytoskeleton -- a web of microscopic fibers inside the cell.

They form when two globular proteins, alpha- and beta-tubulin, polymerize into long chains, which then assemble into long, hollow tubes. In order to gain a foothold, nascent microtubule "seeds" must be anchored at a structure near the cell's nucleus called the centrosome or microtubule-organizing center (MTOC). From the MTOC, the growing microtubules launch out in all directions to the cell's periphery. Their rapid assembly and disassembly helps transport proteins throughout the cell and generate polarized (directional) signal distribution that causes cells to move. While microtubules in some specialized cells can originate from non-centrosomal structures, the centrosome has

[489] Dr. Douglas F. Easton, Cancer Research United Kingdom Genetic Epidemiology Unit, one of the leaders of the Breast Cancer Association Consortium (BCAC). (The first results published by the BCAC appeared in the October 4, 2006 issue of the Journal of the National Cancer Institute).
[490] Dr. Irina Kaverina, and colleagues at Vanderbilt University Medical Center. Contributing authors included scientists from Vanderbilt, from the Austrian Academy of Sciences, New York State Department of Health, University of Melbourne (Australia), Erasmus Medical Center (Netherlands), University of Porto (Portugal), and Scripps Research Institute. The research was supported by the department of Cell and Developmental Biology at Vanderbilt University Medical Center, and grants from the Austrian Science Fund, Fundao para a Cincia e a Tecnologia of Portugal, Luso-American Foundation for Development Work, and the National Institutes of Health.

been considered the main origination point for microtubule "nucleation" in most cells. Until now.

The investigators found that microtubules originating at the Golgi are directed toward the cell "front," or the leading edge, of motile cells. Since such an orientation is needed for directional migration, they hypothesized that this subset of microtubules may influence cell motility by facilitating the transport of proteins needed for movement to the cell front.

In addition to identifying this novel site of microtubule nucleation, they also examined the molecular mechanisms governing the process and found that proteins normally associated with the plus ends of microtubules, called CLASPs, localize to a specific compartment of the Golgi (the Trans Golgi Network) and stabilize the microtubule "seeds" at the Golgi. Golgi-originating microtubules could also be an important factor influencing how cancer cells invade distant tissues.

Because microtubules play a central role in cell division, cancer drugs like colchicine, vincristine and paclitaxel (Taxol) can block cell division by altering microtubule dynamics.

IX.77. New treatments in breast cancer: predictions

Despite the launch of several new agents over the past few years and continued interest in breast cancer therapeutics, there still remains a lack of curative treatments. This is particularly true for the advanced stages of the disease where treatments focus on delaying the spread of cancer, extending survival and enhancing quality of life for the patient. In particular, there is a need to improve the efficacy of agents in treating those cancers that have developed resistance to current cytotoxics (e.g. through multi-drug resistance) or hormonals such as Tamoxifen. The innovatives class may yield solutions to some unmet needs over the course of the next decade.

In view of this lack of curative treatments for late stage breast cancer long-term survival for breast cancer patients still depends on early diagnosis, which has been improved through the use of large scale screening programs and public awareness. However, there still exists some debate about the reliability of screening programs and there are global differences in the availability of such programs, which could be remedied to some extend by the availability of cheaper diagnostics.

Another major issue with the current treatments is the toxicity of many of the drugs employed. Hormonal treatments are linked to a reduction in estrogen levels, leading to many symptoms commonly associated with menopause, and even inducing menopause early in some women. Other side effects include an increased risk of developing cataracts and potentially increased risk of uterine cancer. The side effects of cytotoxic treatments are also well documented and tend to significantly reduce the quality of life of the patient. It is possible that some of these problems may be resolved by improved drug delivery and dosing schedules, but innovative therapies can capitalize on the need for drugs with a lower side effect profile, as well as the possibility of offering improved survival rates in combination with existing therapies.

There are at least 144 candidates in the breast cancer drug candidates pipeline of which, 26% are molecular targeted therapeutics (MTTs). Due to the cytostatic nature of several MTTs, significant tumor regressions will likely be achieved when combined with cytotoxics. As cytotoxics comprise one-third of the potential drug candidates, they will remain the backbone of treatment.

Datamonitor believes Genentech/Roche's Avastin (bevacizumab) will earn blockbuster status within the breast cancer market. Unlike Herceptin, Datamonitor predicts that Avastin, which is not limited to HER-2 positive tumors and will have a major impact on first-line metastatic disease, will dominate the breast cancer market with 2016 sales approaching $1.5bn.

Tykerb (lapatinib) is an epidermal growth factor receptor (EGFR) and ErbB-2 (Her2/neu) dual tyrosine kinase inhibitor developed by GlaxoSmithKline as a treatment for solid tumors such as breast and lung cancer. Tested on women with the aggressive HER2 breast cancer, which is present in about 20-25 percent of breast cancers, Tykerb appears to arrest the development of breast cancer in some patients with metastatic, treatment-resistant. Findings were supported by a large phase III trial which halted enrollment in April of 2006 due to the unanimous recommendation of an Independent Data Monitoring Committee[491]. Tykerb's sales could reach over $500m by 2016 in this setting alone.

The US FDA has approved Tykerb to be used in combination with capecetabine (Xeloda), another cancer drug, for patients with advanced, metastatic breast cancer that is HER2 positive (tumors that exhibit HER2 protein). The combination treatment is indicated for women who have received prior therapy with other cancer drugs, including an anthracycline, a taxane, and trastuzumab (Herceptin).

This approval is a step forward in making new treatments available for patients who have progression of their breast cancer after treatment with some of the most effective breast cancer therapies available. New targeted therapies such as Tykerb are helping expand options for patients[492].

Microtubule-targeting agents such as the taxanes are highly active against breast cancer and have become a cornerstone in the treatment of patients with early and advanced breast cancer.

A number of novel microtubule-targeting agents are currently under investigation. These agents can potentially evade the mechanisms underlying the development of the multidrug resistance (MDR) phenotype commonly associated with recurrent breast cancer. Epothilones are among the most advanced of the new agents in clinical development. Structurally unrelated to taxanes, Epothilones may be poor substrates for MDR, and the expression of MDR proteins is not altered in epothilone-resistant in vitro models. Cross resistance between epothilones and taxanes is not observed in vitro or in vivo[493].

The natural epothilones and their analogs are a novel class of microtubule-stabilizing agents that bind tubulin and result in apoptotic cell death. Several compounds of this new family of agents are in clinical development. Among them are ixabepilone (BMS247550, a semi-syntethic analog of epothilone B), patupilone (EPO 906, epothilone B), KOS-862 (epothilone D), XRP6258 (a new taxoid from Sanofi-Aventis. In Phase II studies, it showed activity against tumor progression in patients with metastatic bfreast cancer having failed previous taxane containing chemotherapy. A phase III study has

[491] Susan G. Komen for the Cure Perspective on the Benefits of Tykerb, press release, March 13, 2007

[492] Dr. Steven Galson, Director of FDA's Center for Drug Evaluation and Research in a statement to the press, March 2007.

[493] Lee JJ, Swain SM, Development of novel chemotherapeutic agents to evade the mechanisms of multidrug resistance (MDR). Semin Oncol., 2005 Dec:32(6suppl 7):S22-6

also been initiated in the treatment of hormone-resistant prostate cancer after failure of docetaxel treatment), and Larotaxel (a new taxoid derivative designed to overcome reistance in the existing agents if this class, docetaxel and paclitaxel. A phase III program has been started in the treatment of pancreatic cancer after failure of gemcitabine chemotherapy. A program evaluating the compound in combination with other anticancer agents in the treatment of metastatic breast cancer is also ongoing).

The National Cancer Institute is committed to exploit its rapidly increasing knowledge of genetics, molecular biology, and immunology to develop even more effective and less toxic treatments for breast cancer. The NCI will expand its ability to target and disrupt the effects of molecular changes that cause breast cells to become cancerous. In addition, the Institute will use this knowledge to personalize breast cancer therapy. For example:

Gene expression analysis has led to the identification of five subtypes of breast cancer that have distinct biological features, clinical outcomes, and responses to chemotherapy. This knowledge should allow the development of treatment strategies based on an individual's tumor characteristics.

A patient's response to chemotherapy is influenced not only by the tumor's genetic characteristics but also by inherited variation in genes that affect a person's ability to absorb, metabolize, and eliminate drugs. This knowledge should allow prediction of tumor response to and the likelihood of severe adverse effects from individual chemotherapy drugs or classes of drugs. It should also aid in the design of more effective and less toxic chemotherapeutic agents[494].

- Improved screening techniques will allow earlier diagnosis of breast cancer;
- Research in molecular biology is in its infancy: the multiple tissue-specific interactions of the ER (estrogen receptor) and its response to antiestrogen therapy will be studied extensively;
- The development of pure antiestrogen agents which completely block estrogenic activity will improve the efficacy of breast cancer treatment;
- Cytostatic therapy, combination therapy (of antiestrogens and aromatase inhibitors) will increasingly be used;
- Many of the new antiestrogens will be used for both osteoporosis and breast cancer prevention.

IX.78. Access to cancer therapy in various geographies

To some degree, discussion of advances in treatment and diagnosis are secondary to the cost of cancer treatments, and the differences in funding between different countries. Away from the focus on developed world cancer burden, developing countries account for 60% of all cancer patients in the world and yet only have access to 5% of available resources. The World Health Organization (WHO) has estimated that by the year 2020 the incidence of cancer will have risen to 15 million with nearly two thirds of the cases attributed to developing countries. If survival rates remains the same then there will be nearly 10 million people dying from cancer at that time -of which three quarters will be in the less developed areas. In other words there is likely to be a two-fold increase in cancer related deaths in developing countries in the next twenty years. The major killers

---

[494] *http://www.cancer.gov/aboutnci/cancer-advances-in-focus/breast (accessed Aug 6, 2007)*

394

will be lung, breast and gastro-intestinal cancers. Together they will account for a third of all new cases.

Even within the developed world there are significant differences in survival rates for cancers.

Table 9.30. 5-year survival rate in breast cancer, by region (in %)

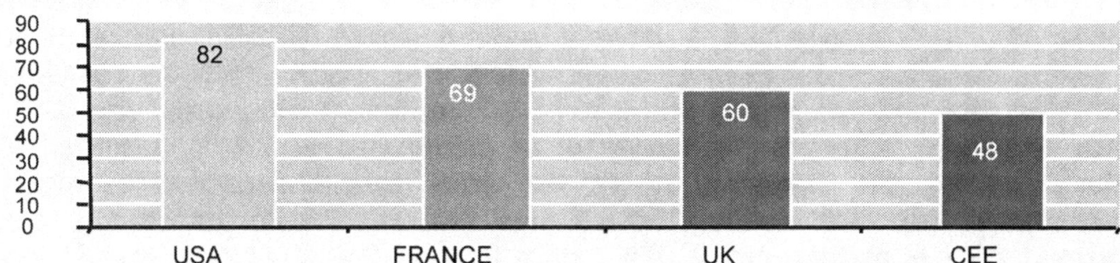

The survival rates in testicular cancer show a similar pattern, in which the US leads both Europe and Eastern European countries in effective treatment of cancer. A recent report has also noted that the reduction in death rates from testicular cancer in the last thirty years has only been just over 20% for Central Eastern Europe (CEE) compared to nearly 70% in the US and Europe.

Table 9.31. Reduction in death rate from testicular cancer, by region (in %)

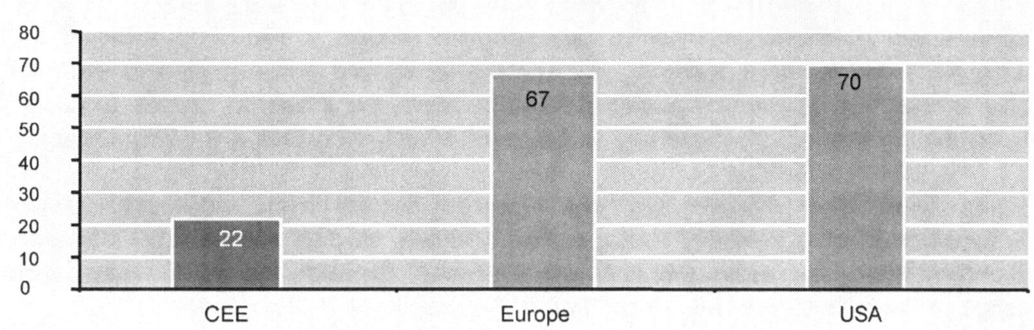

IX.79. Tailoring the therapy to the cancer

A single receptor molecule can perform different functions in different cancer types, thereby complicating approaches to therapy. A study compared the functionality of the HER2/neu receptor in the cancer cells of breast and ovarian cancer tissue. Supported by the Austrian Science Fund FWF[495], the team of scientists involved have shown that the cellular process regulated by this receptor vary greatly between different cancer types. As HER2/neu is the target of successful breast cancer therapy, this result is of major significance for the treatment of ovarian cancer.

Breast and ovarian cancers can both be hereditary, can both be traced back to the same genetic defect and consequently can both possess a large number of HER2/neu receptors. Why therefore do both cancer types not react in the same way when this receptor is blocked? An approach that has proved to be the biggest success of the past

---

[495] *This was the key finding of a study published in the British Journal of Cancer (BJC), as announced by the FWF/Austrian Science Fund. Press release, January 29, 2007*

20 years in the treatment of breast cancer has proved unsuccessful in therapies for ovarian cancer. Scientists[496] have achieved a major breakthrough in finding an answer to this puzzling question.

One Receptor. Two Effects. The team compared tissue samples from 148 ovarian cancers with results from breast cancer tissue samples and the available patient data. This comparison uncovered interesting differences between the two tissue types. While around 25% of ovarian cancer samples also exhibited a high occurrence of the HER2/neu receptor (a known fact), a different signal molecule (CXCR4) was unaffected in the ovarian cancer tissue. However, breast cancer cells, which exhibit elevated levels of HER2/neu, also produce greater amounts of CXCR4 than healthy cells. The CXCR4 molecule has been linked to the formation of metastases and it is assumed that HER2/neu induces the formation of CXCR4 while simultaneously protecting the molecule against degradation caused by enzymes, thus enabling the cancer to become more aggressive (i.e. metastasizing). The results from the Medical University of Vienna now show that the signaling effect produced by HER/2neu is not involved in ovarian cancer.

Molecular Diagnostics Optimize Therapy: On the significance of these results, and according to Professor Krainer, "for almost ten years scientists have been able to identify hereditary breast cancer using molecular diagnostics and rely on monoclonal antibodies for therapy. The first antibody to be approved for use as a medicine blocks precisely the HER2/neu receptor, thus impeding the cancer's growth. This is a perfect example of a tailor-made approach to therapy". His work now reveals just how important it is to carry this differentiation further forward in the development of cancer therapies. After all, in the case of ovarian cancer cells, although the same monoclonal antibody fits this receptor, it has little effect. His laboratory is using findings such as these to create a basis for optimizing the treatment of cancer and to discover where therapies are going wrong.

This study clearly demonstrates just how important results from fundamental research can be for state-of-the-art cancer therapy. Furthermore, studies such as this also enable health professionals to choose the optimum treatment for each individual patient from a vast range of therapies. After all, there is no one-size-fits-all treatment for cancer

IX.80. Ovarian cancer

Ovarian cancer is commonly referred to as "the silent killer," as it usually is not discovered until its advanced stages. It is the most deadliest of all the gynecological diseases. If diagnosed and treated while the cancer is confined to the ovary, the 5-year survival rate is more than 90 percent, according to the American Cancer Society.

Unfortunately, only 19 percent of all cases are found in the early stages.

One woman in 58 will develop ovarian cancer during her lifetime, the American Cancer Society said (2006). Two factors account for the dismal mortality outcomes. One is the absence of reliable early detection markers, and the other is inadequacy of present therapy for advanced disease. To improve patient survival, it is critical to identify new biomarkers for early detection. According to Datamonitor, carboplatin plus paclitaxel remains firmly established as the treatment of choice not only in first line but also in

---

[496] *Dr. Dietmar Pils, a member of the laboratory headed by Prof. Michael Krainer, an oncologist at the Department of Internal Medicine I, Medical University of Vienna,*

second line given the significant proportion of patients retaining platinum sensitivity. It will be difficult for cytotoxics in development to penetrate the ovarian cancer market unless product differentiation can be ascertained.

Vascular endothelial growth factor (VEGF) inhibitors show some promise for the treatment of ovarian cancer. In particular, Genentech/Roche's Avastin is considered to have the most potential, so much so that it is believed to be used significantly off-label despite the halting of a Phase II trial due to safety concerns.

This is not the case for the VEGF Trap, a novel antiangiogenic agent acting as a trap for circulating Vascular Endothelial Growth Factor. Five Phase III trials were initiated in 2007, investigating the synergy of VEGF Trap with chemotherapy regimens in the treatment of patients with solid tumors.

According to estimates for Canada [497], in 2006, more than 22,000 women were diagnosed with breast cancer and 5,300 died of it; approximately 2,300 new cases of ovarian cancer were diagnosed and about 1,600 women died from the disease. These data clearly show that the survival rate of ovarian cancer is only about 25%, compared to 75% for breast cancers.

Although these figures are impressive, the pharmaceutical industry has yet to "discover" the importance of ovarian cancer. Their new drug candidate research efforts continue to treat ovarian cancer as a secondary indication (as a first indiciation targeted with the VEGF Trap, developed by Regeneron in cooperation with Sanofi-Aventis).

IX.81. Ovarian cancer's deadly secrets
Women and their doctors at present have no effective ways to detect ovarian cancer at an early, treatable stage. There are few if any early physical symptoms of ovarian cancer and no tests to detect cellular changes that might indicate precancerous lesions, as Pap smears do for cervical cancer. By the time a woman with ovarian cancer experiences symptoms, tumors are typically large and often metastatic.

Standard therapy for ovarian cancer now usually involves surgery to remove the tumor, followed by chemotherapy, which is initially effective but not curative. The disease frequently comes back in a drug-resistant form. New drugs are badly needed that can target the distinct molecular defects in the different types of ovarian cancer, which may be more accurately seen as not one disease, but several related ones.

A National Cancer Institute (NCI)-sponsored trial reported that women who received chemotherapy directly in their abdomens as part of treatment for advanced ovarian cancer lived more than a year longer than women who received the same chemotherapy intravenously. The findings confirm and expand recent research showing that intraperitoneal (IP) chemotherapy, which delivers drugs directly to the abdominal cavity through a catheter, can significantly increase survival for some women with the disease. About 400 women in the study were given chemotherapy after successful removal of their tumors. Half received intravenous (IV) cisplatin and paclitaxel, and the others received IV paclitaxel, plus IP cisplatin and paclitaxel. The women who received IP chemotherapy lived on average 16 months longer than women who had IV chemotherapy alone, an unusually large survival benefit for a clinical trial. One year after

---

[497] *released by the Canadian Breast Cancer Foundation and the Canadian Cancer Society*

treatment, both groups of patients reported a similar quality of life. As the results were made public, NCI issued a rare clinical announcement to raise awareness about IP chemotherapy for ovarian cancer among physicians and patients. The NCI announcement - the first since 1999 - was warranted because IP chemotherapy is widely regarded as an old technology and previous trials have generated little interest among physicians. Ovarian cancer causes the most deaths of any gynecological cancer in the United States and frequently goes undetected until tumors spread beyond the ovaries.

A new study helped clarify the relationship between low malignant potential (LMP), low-grade, and high-grade serous ovarian tumors. The study suggests that LMP tumors are not early precursors of aggressive ovarian cancer, as had been suspected, but may be part of an entirely different class of tumors. Moreover, the study showed that low-grade serous tumors are more similar to LMP tumors than to high-grade tumors. These findings suggest the need to reexamine treatment options for women with low-grade disease, who currently receive the same therapy as patients with high-grade disease.[498].

A study [499] shed light on ovarian endometrioid adenocarcinoma, the second most common form of a baffling, deadly disease for which early detection methods and effective treatments have been elusive so far, and puts forth a promising new mouse model that already is being used for preclinical drug testing

The new mouse model developed in the U-M lab (University of Michigan Health System) is based on molecular defects shown to be present in human ovarian tumor cells. This, and other existing mouse models, if designed to mimic the four major types of ovarian cancer, should provide key tools for learning how gene mutations and cell changes lead to disease, and for finding treatments during ovarian cancer's early stages, when treatments are most likely to be effective[500].

There is a need for models to do preclinical testing of new drugs that target the specific molecular defects in a patient's tumor cells. Using the genetically engineered mice developed, one preclinical study is already under way, testing an existing drug called Rapamycin. The lab's mouse model can also be used to test new drug candidates that inhibit the cell-messaging systems defective in ovarian endometrioid adenocarcinoma.
The researchers[501] report that defects in two cellular signaling pathways occur together in a subset of ovarian endometrioid adenocarcinomas and appear to cooperate in the development of human tumor cells. They examine the effects of altered genes in the Wnt/Ã¢-catenin and PI3K/Pten signaling pathways, each implicated in several types of cancer[502].

The researchers analyzed gene mutations and signaling pathway defects in human ovarian tumor cells, then created a strain of genetically engineered mice with the same defects to see if ovarian tumors would develop. In all of the mice altered to possess both

---

[498] *Testimony delivered Apr. 6, 2006, before the U.S. House Subcommittee on Labor-HHS-Education Appropriations, to discuss the NCI budget request for Fiscal Year 2007, by Dr John E. Niederhuber, Deputy Director, National Cancer Institute, and Richard Turman, Deputy Assistant Secretary, Office of Budget, Department of Health and Human Services. http://www.cancer.gov/aboutnci/FY07-budget-request. (Accessed Aug 6, 2007)*
[499] *Cancer Cell, Vol 11, 321-333, 10 April 2007*
[500] *According to senior author Kathleen R. Cho, who treats patients as a member of the U-M Comprehensive Cancer Center, in a statement to the press, April 14, 2007*
[501] *Cho, lead authors Rong Wu and Neali Hendrix-Lucas, and other members of Cho's team, as well as scientists at the Van Andel Research Institute in Grand Rapids and Kumamoto University in Japan.*
[502] *Genes that are mutated in ovarian cancer, as in other cancers, result in the production of proteins that alter the normal function of signaling pathways in cells. Defects in these pathways can prevent the normal action of tumor suppressor genes and allow cancer to develop and grow.*

pathway defects, ovarian tumors - similar in morphology and biological behavior to human ovarian endometrioid adenocarcinomas - rapidly developed and often metastasized.

To create the new mouse model for ovarian endometrioid adenocarcinoma, the scientists used two existing types of transgenic mice to breed mice that expressed the two signaling defects they wanted to study.

Next, the researchers plan to tackle the "very big challenge" of dissecting the molecular basis of serous ovarian cancer, the most common form of ovarian cancer and the one responsible for most ovarian cancer deaths.

That work, combined with ongoing projects on ovarian endometrioid adenocarcinoma, could help researchers develop screening tests that can detect the most common types of ovarian cancer early, and design more effective treatments for women with advanced disease.

A collaboration between the US National Cancer Institute and the Karolinska Institute in Sweden to examine ovarian cancer risk using markers of hormone exposure during pregnancy.

Factors associated with reproduction likely play an important role in the development of ovarian cancer. During pregnancy, the placenta produces large amounts of the female sex hormones estrogen and progesterone. A number of factors, including gestational age of the developing fetus, affect the levels of these hormones during pregnancy. Although research has shown that increasing parity (number of births) reduces a woman's risk of ovarian cancer, no study to date has examined the associations between indictors of hormonal exposures during pregnancy and subsequent ovarian cancer risk.

NCI is collaborating with the Karolinska Institute to conduct a large cohort study aimed at exploring these associations. Investigators are using information gathered on more than 1.2 million women who delivered their first infant between 1973 and 2000 and who were included in the nationwide Swedish Medical Birth Register.
Researchers will:
1) study ovarian cancer risk using markers of hormone exposures during pregnancy, including birth weight, gestational age, single or multiple birth, pregnancy-induced hypertensive diseases, and placental weight;
2) examine whether the protective effects of increasing parity and a high age at first birth are influenced by markers of hormone exposures during pregnancy; and
3) assess the importance of other factors on ovarian cancer risk.

The investigators plan to gather information about maternal characteristics, pregnancy complications, placental weight, and birth characteristics for all births associated with the women in the cohort. In addition, information about gynecologic surgeries, vital status, and ovarian cancer incidence for these women will be collected from population-based registries.

Use of the Swedish research registries in this study allows a large, cost-effective study to be conducted where information has been prospectively collected about markers of hormone exposure during pregnancy[503].

Not-for-profit institutional research contributes to new findings:

IX.82. RNA splicing factor implicated in ovarian tumor cell growth

Researchers in the UIC College of Pharmacy found that interfering with the production of a splicing factor can inhibit the growth and invasiveness of tumor cells in test-tube experiments.   In a previous study, they observed that human ovarian tumors overexpressed polypyrimidine tract-binding protein, or PTB, and another splicing factor compared to normal matching ovarian tissues. In the new study, they showed that knocking down PTB expression with small, interfering RNA substantially impairs ovarian tumor cell growth, colony formation and invasiveness[504].

Polypyrimidine tract-binding protein is a key regulator of splicing. Defects in pre-messenger RNA splicing have been shown to cause a variety of human diseases. Evidence suggests that altered splicing is associated with and possibly involved in tumor progression or metastasis.

This is the first time anyone has shown that interference with the production of a splicing factor can inhibit tumor cell growth and tumor invasiveness. Alternatively-spliced genes and splicing factors are likely to play a key role as therapeutic targets and diagnostic markers in the next decade.

IX. 83. Characterizing the aggressiveness of cancer cells

Levels of a small non-coding RNA molecule called let-7 appear to define different stages of cancer better than some of the "classical" markers for tumor progression. By suppressing genes that are active in the developing embryo, silenced just before birth, and re-activated years later in many advanced cancers, the let-7 family of "microRNAs" - tiny snippets of RNA that can put the brakes on expression of selected genes - appears to prevent human cancer cells from reasserting their prenatal capacity to divide rapidly, travel and spread.

Since they were first discovered in 1993, there had been growing interest in microRNAs and their role in gene regulation. Hundreds of these tiny molecules, about 20 nucleotides in length, have been discovered, scattered throughout the human genome. They act in most cases by attaching themselves to specific sites on messenger RNA, where they block ribosome access and thus prevent production of that protein.

There may be no human cancer that is not regulated by microRNAs, and among microRNAs, let-7 appears to be a key player in preventing a cancer from becoming more aggressive.

The expression levels of let-7 can discriminate more effectively between more and less advanced stages of cancers than any other microRNA. Loss of members of the let-7 family may be a major determinant of cancer progression. Understanding how microRNAs such as let-7 keep cancers in check could also point toward a whole new

---

[503] *http://www.cancer.gov/nci-international-portfolio/page5 (accessed Aug 6, 2007)*

[504] *William Beck, professor and head of biopharmaceutical sciences University of Illinois at Chicago. Press release, April 13, 2007. The research has been published online in the journal Oncogene*

400

class of anti-cancer therapies. The data suggests that human tumors can be divided into two major subtypes, the let-7hi and let-7lo-expressing tumor cells. This separation may not be restricted to ovarian cancer, or to the NCI60 panel of tumor cells, but could apply to a multitude of tumor types.

There is growing evidence that large-scale gene-expression patterns can be regulated by microRNAs. Many of them are beginning to be expressed shortly before birth, where they turn off genes that were necessary for the rapidly developing embryo. Probably a number of embryonic genes, after being turned off for decades, are reexpressed in cancer cells, enabling those cells to regain their embryonic capacity to move around and invade other tissues.

The loss of let-7 could be seen as one crucial step in this process of tumor progression. One of its functions is to maintain differentiated states by preventing the expression of embryonic genes such as HMGA2

IX.84. Podocalyxin, a new culprit in cancer metastase

Thanks to a discovery by a team of researchers at the University of British Columbia, new non-toxic and targeted therapies for metastatic breast and ovarian cancers may now be possible[505]. In a collaboration between UBC stem cell and cancer scientists, it was found that a protein called podocalyxin - which the researchers had previously shown to be a predictor of metastatic breast cancer - changes the shape and adhesive quality of tumor cells, affecting their ability to grow and metastasize.

The discovery demonstrated that the protein not only predicted the spread of breast cancer cells, it likely helped to cause it.

The researchers found that podocalyxin significantly expands the non-adhesive face of cells, allowing individual cells to brush aside adhesion molecules situated between tumor cells. The "freed" cells then move away from the original site to form new tumors at other sites. Also, the protein causes tumor cells to sprout microvilli, or hair-like projections, that may help propel cancer cells to other sites.

In addition, when the protein expands the non-adhesive face of cells it drags along with it a second protein called NHERF-1 - a protein shown by others to be implicated in cell growth and invasion. The researchers now believe the mechanism applies to difficult-to-treat invasive breast and ovarian cancers.

They are now tapping into what causes the characteristic cell shape changes seen in cancerous tumors and possibly how these cells grow and metastasize. It gives them a whole new target for therapy. They believe that, if they can block the protein, they may be able to stop the spread of cells[506].

Next steps involve advancing the research in animal models, designing antibodies to block the function of the protein and working with the UBC-based Center for Drug Research and Development to identify new therapies to combat metastasizing cancer.

---

[505] Hilary Thomson, University of British Columbia, Potential 'Smart' Therapies For Breast, Ovarian Cancer from UBC Discovery. March 21, 2007. The findings were published online by the Public Library of Science.

[506] Assoc. Prof. of Medical Genetics and stem cell expert Kelly McNagny, University of British Columbia, co-senior principal investigator, in a statement to the press, March 21, 2007 (Other researchers involved in the project include: BC Cancer/ Vancouver Hospital investigators David Huntsman and Blake Gilks; as well as Wayne Vogl and Jane Cipollone of UBC's Dept. of Cellular and Physiological Sciences).

The researchers say information from this discovery may speed development of new therapies to within 10 years.

IX.85. Cervical cancer

Cervical cancer is a major global health problem, with nearly 500,000 new cases occurring each year worldwide. It is the second most common cancer - and the third leading cause of cancer deaths - in women worldwide. Each year an estimated 270,000 women die from the disease, and it is the leading cancer killer of women in the developing world.[507]

HPV infection is very common; every sexually active woman is at risk of contracting a type of HPV, which may cause cervical cancer. While there are many different types of HPV that may cause cancer, HPV 16 and 18 account for more than 70 per cent of all cervical cancers globally[508].

Table 9.32. HPV vaccines sales forecast 2010

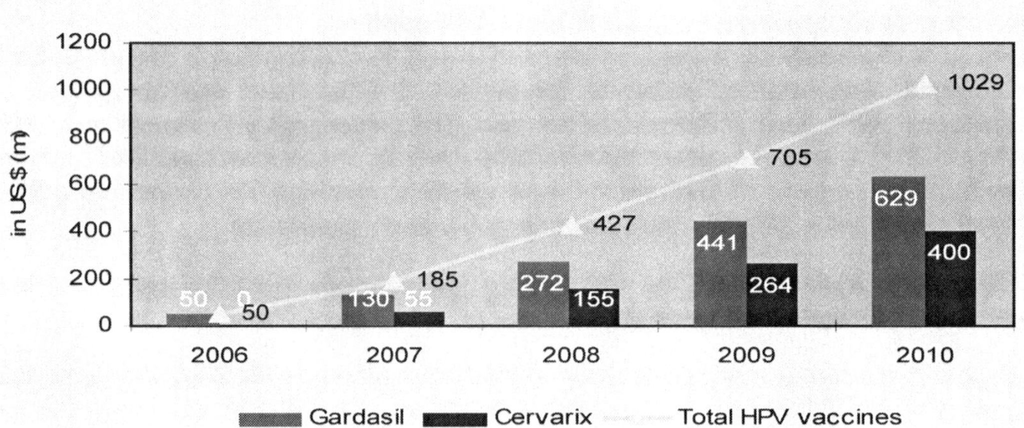

Following FDA's approval of Merck's Gardasil in June 2006, which followed a 12000-patient trial, healthcare consultants Wood Mackenzie published their sales development estimates for Gardasil (a vaccine for human papilloma virus) and for GlaxoSmithKline's Cervarix, said to be Gardasil's closest rival. GSK has estimated that the global market for HPV vaccines will be worth US $ 2billion to US $ 4 billion by 2010. Wood Mackenzie[509] is more conservative and forecasts some US $ 1 billion for both the vaccines in 2010. Gardasil has first to market advantage in all major markets. Cervarix's novel adjuvant may provide cross-protection against other "high-risk" PPV strains. The compound is approved for females aged 9-26 years old; Cervarix for females aged 10-55. It is expected that both companies will have to work hard to raise patient and physician awareness of the vaccines, with Merck laying the foundations and GSK having to overcome Gardasil's first-to-market advantage.

Cervarix was developed to prevent infection from the two most prevalent cancer-causing types of the human papillomavirus (HPV), specifically HPV 16 and 18. In clinical trials, the compound demonstrated excellent protection from persistent infection against both

[507] Ferlay J, Bray P, Pizani P, Parkin DM. Cancer incidence, mortality and prevalence worldwide. Available at: GLOBOCAN 2002. Accessed September 20, 2005

[508] De Villier EM, Heterogenicity of the Human Papilloma virus infections. J Clin Virology 189; 63: 4898-903

[509] Wood Mackenzie, Pharmaceutical Executive Europe, September 2006

HPV 16 and HPV 18 and associated precancerous lesions[510]. Preliminary evidence of broader protection against other cancer-causing strains of HPV, in addition to HPV 16 and 18, was shown[511]. Cervarix is formulated with the innovative proprietary adjuvant AS04 selected to ensure that the compound confers high and sustained antibody protection levels[512].

In the next five years, GSK expects to launch more major new vaccines: a vaccine against rotavirus, a vaccine to prevent pneumococcal disease, an improved flu vaccine for the elderly, and a meningitis combination vaccine for infants.

Approximately 1,100 women from Canada, Brazil, and the United States participated in the HPV vaccine trial. These women were randomly assigned to receive doses of the HPV-16/18 vaccine or a placebo initially and then at 1 month and 6 months after their initial injection. The researchers found that the GSK vaccine was 91.6 percent effective in protecting against incident infections with HPV-16 or HPV-18 in women treated according to the trial protocol. The vaccine was 100 percent effective in preventing persistent viral infections during the study period, and it was 93.5 percent effective in preventing cervical cell abnormalities associated with HVP-16 or HPV-18 infection. Additional follow-up of participants in this trial has demonstrated that effective protection against infection remains high for up to 4 years[513]. Longer-term efficacy of the vaccine is still not known.

In another clinical trial, the vaccine developed by Merck was found to be nearly completely effective in preventing incident and persistent infections with HPV-16 and HPV-18 and against cervical cell abnormalities associated with HPV-16 or HPV-18[514]. This vaccine - called Gardasil - also protected against two other HPV strains, HPV-6 and HPV-11, which cause 90 percent of genital warts. More than 12,000 women from 13 countries participated in the trial. Gardasil was administered in three doses over 6 months and provided 100 percent protection against HPV infection for the 18-month duration of the study. The long-term efficacy of the vaccine is still not known, and follow-up studies will be required to answer this question.

The US National Cancer Institute (NCI) is now conducting a Phase III clinical trial in Costa Rica - where cervical cancer rates are high - to further test the HPV 16/18 vaccine developed by GSK. Over 7,400 women have been enrolled in the trial. They will be followed for at least 4 years to allow investigators to gather information about the vaccine's long-term safety and efficacy. NCI investigators also plan to evaluate other potential effects of the vaccine, including:
1) its effectiveness against additional HPV strains;
2) its ability to speed the healing of established cervical infections; and

---

[510] Harper DM et al., Efficacy of a bivalent L1 virus-like particle vaccine in prevention of infection with human papillomavirus types 16 and 18 in young women: a randomized controlled trial. Lancet 2004; 364: 1757-1765.

[511] Dubin G et al. Cross-protection against persistent HPV infection, abnormal cytology and CIN associated with HPV-16 and 18 related HPV types by a HPV 16/18 L1 virus-like particle vaccine.

[512] Giannini SL et al. Superior Immune Response Induced by Vaccination with HPV 16/18 L1 VLP Formulated with AS04 Compared to Aluminum Salt Only Formulation. Poster B68, presented at the 4th Annu.Int.Conf. on Frontiers in Cancer Prevention Research (AACR), October 30 - November 2, 2005, Baltimore, MD, USA

[513] Harper D, et al. on behalf of the HPV Vaccine Study group. Sustained efficacy up to 4.5 years of a bivalent L1 virus-like particle vaccine against human papillomavirus types 16 and 18: follow-up from a randomised control trial. Lancet, April 15, 2006;367(9518):1213-1290.

[514] Finn S, et al. Prophylactic quadrivalent human papillomavirus (HPV) (types 6, 11, 16, 18) L1 virus-like particle (VLP) vaccine (Gardasil) reduces cervical intraepithelial neoplasia (CIN) 2/3 risk. Oral abstract LB-8a. Infectious Diseases Society of America meeting. San Francisco, CA. October 7, 2005.

3) 3) evaluate the immune mechanisms of long-term protection. The NCI trial in Costa Rica will also provide important information that will be useful to evaluate the cost-effectiveness of HPV vaccination and combined vaccination and screening prevention efforts[515].

Scientists believe that HPV vaccination will significantly reduce not only the health burden caused by cervical cancer, but also the anxiety suffered by women around Pap smears and particularly around abnormal results. It has the potential to revolutionize women's health. A cervical cancer vaccine is described as being a major innovation and an important development for women everywhere[516].

A study at the Karolinska Institute in Stockholm, Sweden of the influence of human papilloma virus (HPV) viral load on the progression of localized cervical carcinoma to invasive cervical cancer. Although HPV infection is an established cause of cervical cancer, it is not known whether viral load influences the progression from localized cervical carcinoma in situ (CIS) to invasive cancer and/or interacts with host genetic factors. Since clinical intervention precludes direct observation of this progression, unconventional approaches are required. The investigators are attempting to: 1) quantify the absolute and relative risks for CIS and invasive cancer as a function of time since the detection of HPV and high viral load of HPV strain 16 (HPV-16); 2) assess whether a persistent HPV-16 high viral load is a determinant of CIS and invasive cancer development; 3) assess whether a specific histocompatibility antigen genotype is associated with risks for CIS and invasive cancer and if the association is mediated via a higher viral load and/or persistence of HPV infection; and 4) assess whether infection with the bacterium Chlamydia trachomatis is associated with risks for CIS and invasive cancer. The researchers will take advantage of Sweden's extensive documentation in computerized registries of its population-based Pap smear screening program, its ascertainment of all incident cases of cervical CIS and invasive cancer, and archived Pap smears and tissue specimens[517].

IX.86. Lung cancer
Lung cancer is of major concern. It is a leading killer for men and women. It is estimated that 174,470 new cases of lung cancer were diagnosed in the United States in 2006[518].

The lungs contain an upper and lower lobe, with the larger right lung also containing a middle lobe. Oxygen exchange is performed by millions of tiny air sacs, called alveoli, which are connected to the windpipe by bronchioles, which lead into the main bronchial tubes. There are a number of different types of tumors that can form in the lung, of which the most commonly diagnosed are small cell (SCLC) and non-small cell (NSCLC) lung cancers. Collectively these make up close to 95% of diagnosed cases of lung cancer, with NSCLC accounting for 75% of all cases.

---

[515] *NCI international portfolio: Addressing the global challenge of cancer http://www.cancer.gov/nci-international-portfolio/page4 (accessed Aug 6, 2007)*

[516] *Dr Anne Szarewski, from Cancer Research UK in a comment to the press at the occasion of GSK's submittal for review a marketing application to the EMEA, March 11, 2006.*

[517] NCI international portfolio: Addressing the global challenge of cancer. Visit http://www.cancer.gov/nci-international-portfolio/page4 (accessed Aug 6, 2007)

[518] *National Institutes of Health press release, September 2006*

Table 9.33.   Color-enhanced image of a lung cancer cell dividing[519]

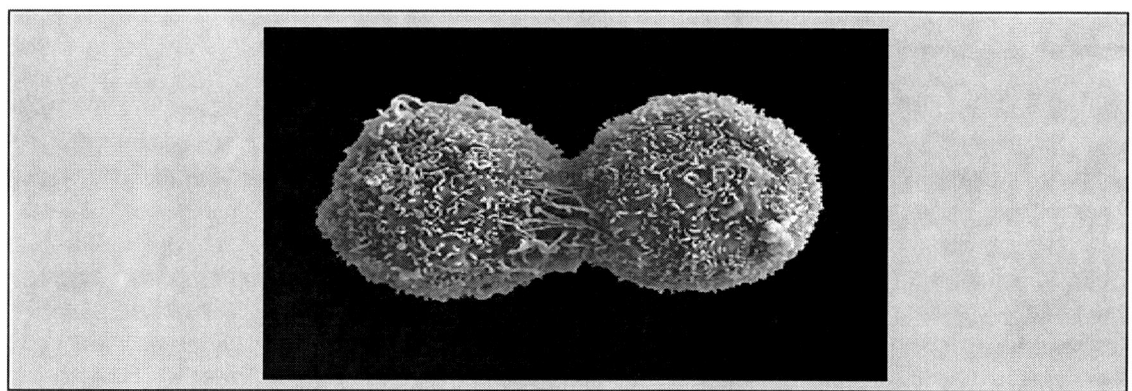

NSCLC can be subdivided into three further types of cancer:

Squamous cell carcinoma is formed from round cells that replace damaged cells along the epithelium of the main, lobar or segmental bronchi. It is relatively slow growing and carries a good prognosis;

Adenocarcinoma usually arises from mucus-producing cells of the lung, and is the most common type of lung cancer, accounting for about 50% of all cases. Adenocarcinomas tend to be detected at a late stage and quite often carry a poor prognosis;

Large cell carcinoma is the most rare form of NSCLC and develops near the surface of the lung

SCLC is more aggressive than NSCLC, with a greater tendency to metastasize. The disease is closely associated with smoking, and is classified on the basis of two stages (limited and extensive) due to its rapid progression. As with NSCLC, there are different subtypes of SCLC, however all exhibit a similar prognosis.

A third, rare form of lung cancer is malignant mesothelioma, in which the tumor is found in the sac lining the chest (pleura). The incidence of malignant mesothelioma is low, with about 14 cases per million in the USA, and this form of lung cancer is primarily due to exposure to asbestos. Prognosis is poor, with median survival rates of only six to eight months[520].

Over 65% of newly diagnosed lung cancer patients are male, although an increasing number of female patients are been diagnosed with lung cancer, and more women now die of the disease than from breast cancer.

Lung cancer has a more limited range of risk factors than many other cancers, the most important of which is a history of smoking.

Treatment regimens for lung cancer currently are based on the use of the cytotoxic platinum analogues, cisplatin and carboplatin. Both compounds exhibit similar efficacy, although carboplatin has demonstrated more favorable toxicity. However, due to the availability of generic cisplatin compounds, cisplatin is the cheaper treatment and

---

[519] *source: Anne Weston, Wellcome Trust Sanger Institute*
[520] *SEER Cancer statistics 1975-2000)*

favored in most countries. In the treatment of NSCLC, the gold-standard treatment involves cisplatin/carboplatin prescribed in combination with another cytotoxic agent, such as a taxane (paclitaxel or docetaxel), vinorelbine or gemcitabine.

Preclinical and clinical studies support that Xyotax metabolism by lung cancer cells may be influenced by estrogen, which could lead to enhanced release of paclitaxel and efficacy in women with lung cancer compared to standard therapies. Xyotax (paclitaxel poliglumex)[521] is a biologically-enhanced chemotherapeutic that links paclitaxel, the active ingredient in Taxol, to a biodegradable polyglutamate polymer, which results in a new chemical entity. When bound to the polymer, the chemotherapy is rendered inactive, potentially sparing normal tissue's exposure to high levels of unbound, active chemotherapy and its associated toxicities. Blood vessels in tumor tissue, unlike blood vessels in normal tissue, are porous to molecules like polyglutamate. Based on preclinical studies, it appears that Xyotax is preferentially distributed to tumors due to their leaky blood vessels and trapped in the tumor bed allowing significantly more of the dose of chemotherapy to localize in the tumor than with standard paclitaxel. Once inside the tumor cell, enzymes metabolize the protein polymer, releasing the paclitaxel chemotherapy.

By the time lung cancer is diagnosed, 85 percent of patients are inoperable, often due to serious coexisting health conditions or poor respiratory function. For these patients, minimally invasive Interventional Radiology procedures can provide effective treatment of the tumor with minimal discomfort and a maintained quality of life.

During thermal ablation with RFA or Microwaves, the interventional radiologist uses imaging to guide a small needle through the skin into the tumor; energy is then transmitted to kill the tumor with heat or cold. According to a recent study, fifty-seven percent of lung cancer patients who were treated with thermal ablation survived an additional three years, two years beyond the average life expectancy[522].

Radiofrequency ablation (RFA) offers a non-surgical, localized treatment that kills the tumor cells with heat, while sparing nearby healthy lung tissue. As a result, ablation is much easier on the patient than systemic therapy, and does not affect overall health. Most people can resume their usual activities in a few days after this outpatient treatment.

Microwave ablation utilizes electromagnetic microwaves to agitate the water molecules in the tumor and surrounding tissue, ultimately reversing the cells' polarity. This change in polarity causes the cells to rotate back and forth, causing friction and heat which kills the cell (coagulation necrosis)[523].

IX.87. Extension of patient survival

It has now been almost 30 years since Dr J. Folkman first proposed that inhibition of angiogenesis could play a key role in treating cancer; however, it is only recently that anti-angiogenesis agents have entered the clinical setting. The search for novel

---

[521] *Cell therapeutics Inc., announcing that Xyotax qualifies for fast track designation for the treatment of PS2 (poor performance status) women with first-line advanced non-small cell lung cancer (NSCLC). April 12, 2007.[Designation was initially granted on June 11, 2003 for Xyotax for NSCLC PS2 patients (male and female). Confirmation from the FDA was obtained on January 24, 2006 that the existing fast track designation encompasses the treatment of female patients with advanced NSCLC with PS2]*

[522] *Journal of Vascular and Interventional Radiology, July 2006*

[523] *Dr. John Rundback, Director, Interventional Institute, Holy Name Hospital, Teaneck, New Jersey; Associate Professor, Columbia and HolyName Hospital ; Chair, Society of Interventional Radiology Foundation Clinical Trials Network. Lung Cancer Treatment Advances, Anniversary Of Peter Jennings Death. 28 Aug 2006*

therapies is particularly important in lung cancer, where the majority of patients succumb to their disease despite aggressive treatments. There is strong interest in drugs that block endothelial cell signaling via vascular endothelial growth factor (VEGF) and its receptor (VEGFR) including rhuMAb VEGF, SU5416, SU6668, ZD6474, CP-547,632 and ZD4190. Drugs that are similar to endogenous inhibitors of angiogenesis including endostatin, angiostatin and interferons. There has also been renewed interest in thalidomide. Drugs such as squalamine, celecoxib, ZD6126, TNP-470 and those targeting the integrins are also being evaluated in lung cancer. Despite early enthusiasm for many of these agents, Phase III trials have not yet demonstrated significant increases in overall survival and toxicity remains an issue. Scientists hope that as their understanding of the complex process of angiogenesis increases, so will their ability to design more effective targeted therapies[524].

In the absence of radically more effective anti-cigarette smoking awareness/prevention campaigns, the major unmet need with lung cancer is extension of patient survival. Despite significant improvements in pharmacotherapy for lung cancer in the last 20 years, the current five-year survival figure is less than 15%, dramatically less than the rates for other major cancers. These figures mask the difference in survival expected between patients suffering from SCLC and NSCLC, with five-year survival figures of approximately 10% and 20% respectively. However, in either case, the long-term outlook for the patient is poor and there is a drastic need for drugs that can offer significant extension in survival and, ideally, extended disease-free survival. There is also a need for improved second-line treatments for patients who fail to respond to first-line treatments, or suffer a relapse after receiving first-line therapy.

In oncology research, data from pivotal studies of Zactima (ZD6474) in second- and third-line non-small cell lung cancer are anticipated in mid-2008, and the drug has been given orphan status by US and European regulatory authorities. AstraZeneca is also pushing AZD2171 into Phase III. This compound's combination of potency, selectivity and pharmacokinetics gives it the potential to be the best-in-class VEGFR-TKI with activity in all forms of lung cancer.

A comprehensive understanding of the expression and alteration patterns in tumor cells will allow the identification of biomarkers, the prediction of biological behavior, the prediction of response to intervention and the identification of novel therapeutic targets[525].

Early 2007, researchers identified an 80-gene biomarker to help in the early detection of lung cancer. Prolonging bronchoscopy by only about 5 minutes, the new biomarker is designed to help evaluate smokers suspected of having cancer. The study[526] found that combining cytopathology with the gene-expression biomarker improves the diagnostic sensitivity of the overall bronchoscopy procedure from 53% to 95%. But the scientists said they hope the biomarker will one day prove to be an effective stand-alone screening tool.

During an interview with Medscape[527], the lead study author[528] predicted that the test,

[524] Sridhar SS, Shepherd FA, Targeting angiogenesis: a review of angiogenesis inhibitors in the treatment of lung cancer. Lung Cancer, 2003 Dec;42 Suppl 1:S81-91
[525] Dr. David P. Carbone, Vanderbilt-Ingram Cancer Center. Comprehensive Technologies for Understanding Lung Cancer Biology and Identifying Novel Targets for Intervention. Medscape Continuing Education.
[526] Nat Med. Published online before print March 4, 2007.
[527] Allison Gandey, Interview author, Lung Cancer Biomarker may increase sensitivity of bronchoscopy. Medscape news release, March 7, 2007.

coupled with bronchoscopy, is about 1 or 2 years from the clinic and, as an independent screening method, perhaps about 10 years away.

IX.88 Cancer stem cells looking and behaving like normal stem cells...

Current cancer therapies often succeed at initially eliminating the bulk of the disease, including all rapidly proliferating cells, but are eventually thwarted because they cannot eliminate a small reservoir of multiple-drug-resistant tumor cells, called cancer stem cells, which ultimately become the source of disease recurrence and eventual metastasis. Research suggests that for chemotherapy to be truly effective in treating lung cancers, for example, it must be able to target a small subset of cancer stem cells, which they have shown share the same protective mechanisms as normal lung stem cells. Scientists used cell surface markers and dyes to identify cancer stem cells as well as normal adult stem cells and their progeny in samples obtained from normal lung and lung cancer tissue samples. They identified a very small, rare set of resting cancer stem cells in the lung cancer samples that looked and behaved much like normal adult lung tissue stem cells. Both the cancer and normal stem cells were protected equally by multiple drug resistance transporters, even if the bulk of the tumor responded to chemotherapy.

The very fact that cancers can and do relapse after apparently successful therapy indicates the survival of a drug-resistant, tumor-initiating population of cells in many types of refractory cancers. Because of the similarities between the way that normal stem cells and cancer stem cells protect themselves, cancer therapies have to be designed specifically to target cancer stem cells while sparing normal stem cells, the researchers stated[529].

IX.89. Molecular profiling of lung cancer

Scientists seek to better understand the mechanisms underlying this disease, and microRNAs may provide a way to examine the regulation of cancer-related genes. miRNAs are small segments of genetic material called ribonucleic acid, or RNA, and are thought to control gene expression. Their actions could change the expression of cancer-related genes within a cell and lead to malignancies.

Researchers[530] have discovered that a tiny piece of genetic code apparently goes where no bit of it has gone before, and it gets there under its own internal code.

MiRNAs latch on to and gum up larger strands of RNA that carry instructions for making the proteins that do all the cell's work. They are like molecular rheostats that fine-tune how much protein is being made from each gene. That's why normally miRNAs always have appeared to stick close to the cell's protein-making machinery.

But during a survey of more than 200 of the 500 known miRNAs found in human cells, the researchers discovered one lone microRNA "miles away" in cellular terms from all the others. The findings demonstrate for the first time that despite their tiny size, miRNAs

---

[528] Dr. Avrum Spira from Boston University Medical Center, in Massachusetts,

[529] Dr. Vera Donnenberg, professor of surgery and pharmaceutical sciences, University of Pittsburgh Schools of Medicine and Pharmacy, The University of Pittsburgh researchers, who led the study. The researchers presented this ground-breaking research at the Tissue Engineering and Regenerative Medicine International Society (TERMIS) North American Chapter meeting at the Westin Harbor Castle conference center in Toronto. June, 2007

[530] at McKusick-Nathans Institute of Genetic Medicine at Johns Hopkins

contain elements consisting of short stretches of nucleotide building blocks that can control their behavior in a cell.

A team of researchers has found that the expression pattern of certain miRNAs, may predict tumor aggressiveness in some patients with lung cancer. These findings indicate that miRNAs may represent a new class of diagnostic and prognostic tools for lung cancer. The researchers identified two miRNAs -- "has-mir-155" and "has-let-7a-2" -- that could be used as prognostic indicators in patients with adenocarcinoma, a malignancy of the mucous glands in the lungs. High levels of "has-mir-155" or low levels of "has-let-7a-2" were associated with poor prognosis. Specifically, overexpression of "has-mir-155", or the signaling of the miRNA to change the amount of a protein produced, was the most significant indicator of this prognosis, independent of tumor stage. Although these miRNAs have been identified in other cancers, this is the first evidence linking "has-mir-155" to lung cancer.

A tumor with an overexpression of "has-mir-155" or reduced expression of "has-let-7a-2" would indicate the need for aggressive chemotherapy or radiation treatments. Other tumors that do not show high "has-mir-155" or low "has-let-7a-2" levels are less aggressive, and those patients might not require more therapy.

This study is significant because it provides another tool for studying prognosis that is independent of tumor stage. Following surgery, 50 to 60 percent of patients with stage I lung cancer will develop metastatic disease within five years. This may indicate that there are micrometastases that have not been detected by imaging, scanning or pathology. In the future, scientists can use miRNAs and other biological predictors to select patients who may need more aggressive treatment versus those who may not. Additional studies confirming these results are the next step before incorporating miRNA analysis into routine clinical practice. miRNAs are going to be important biomarkers of not only diagnosis and prognosis, but therapy, as well. The next step is to identify the genes that the miRNAs are affecting; they could be used as potential targets for developing novel therapies[531].

In the study[532], a total of 104 pairs of primary tumor tissues and corresponding noncancerous lung tissues were examined. The tumor and corresponding normal tissues were obtained from the same patient to eliminate genetic differences between tumor and normal tissues.

Researchers focused primarily on the more common adenocarcinomas in their analysis; adenocarcinomas and squamous cell carcinomas. In the study, adenocarcinomas of the lung comprised 63 percent of the tumors studied and squamous cell carcinomas comprised 37 percent. Patterns of miRNA expression in each tumor and normal tissue pair were studied by microarray analysis. A microarray allows the measurement of hundreds of miRNAs simultaneously in a single sample. Five miRNAs displayed different expression levels in tumor tissues versus their corresponding noncancerous controls and, thus, were selected for further study. Upon statistical analysis, data showed that patients with high "has-mir-155" or low "has-let-7a-2" had poorer survival than patients

---

[531] *Dr. Curtis Harris, chief of the Laboratory of Human Carcinogenesis at the National Cancer Institute (NCI), and co-leader of the study, in a statement to the press, issued by the National Cancer Institute, March 2006*

[532] *The study is a collaboration among researchers at The Ohio State University Comprehensive Cancer Center, Columbus, Ohio; the Jikei University School of Medicine, Tokyo, Japan; National Cancer Center Research Institute, Tokyo, Japan; and the Center for Cancer Research at the National Cancer Institute (NCI), which is part of the National Institutes of Health. The study results appear in the March 13, 2006, issue of "Cancer Cell".*

showing low "has-mir-155" or high "has-let-7a-2" expression. The difference in the prognosis of these two groups was highly statistically significant.

After examining tissue from lung cancer patients, and following each patient to see how long they lived, researchers found that miRNA expression patterns were independent of tumor stage. When the scientists combined all clinical and molecular factors, they found that a high level of "has-mir-155" or a low level of "has-let-7a-2" was the most significant prognostic factor for an unfavorable patient outcome.

IX.90. New antibody for EGFR causes lung cancer regression

Mutant forms of the protein EGFR are important for the development of lung cancer in a substantial proportion of individuals with this disease. However, not all individuals express the same mutant EGFR, for example, some have a mutation that affects the intracellular part of EGFR and some have a mutation that affects the extracellular part of EGFR (known as the EGFRvIII mutant). The potential of therapeutics to benefit individuals with lung cancer caused by the distinct EGFR mutants can be examined using two mouse models of lung cancer, one driven by expression of EGFR with a mutation in the intracellular part of the protein and one driven by expression of the EGFRvIII mutant.

Scientists from the Dana-Farber Cancer Institute in Boston show that a mouse antibody that binds EGFR (mAb806) causes the regression of lung tumors in both models of lung cancer. By contrast, a second antibody that binds EGFR (cetuximab), and that is already in clinical investigation for the treatment of a specific subset of patients with lung cancer, only induced tumor regression in mice with lung tumors driven by expression of EGFR with a mutation in the intracellular part of the protein. Importantly, the humanized form of mAb806 (ch806) caused regression of lung tumors in both models of lung cancer. The data presented in this study therefore indicate that ch806 might provide a new therapeutic for the treatment of patients with lung cancer driven by mutant forms of EGFR, in particular the EGFRvIII mutant[543].

IX.91. Bcl-2 Family Protein Inhibitors (ABT-263)

Researchers have been interested in the Bcl-2 family of proteins since their role in preventing apoptosis -- the natural process by which damaged or unwanted cells die and are cleared from the body -- was proven more than a decade ago. Discovered by Abbott scientists, ABT-263 restores programmed cell death, a natural mechanism for the elimination of cancerous cells, by inhibiting the function of Bcl-2 proteins.

Bcl-2 proteins play a central role in regulating apoptosis, as well as tumor formation, tumor growth and resistance to treatment. Pioneering work in structural biology at Abbott established how the Bcl-2 family of proteins interact with one another, leading researchers to develop a novel compound that causes cancer cells to self-destruct.

ABT-263 recently entered Phase I clinical trials for lymphomas and solid tumors, including small cell lung cancer. Preclinical data has shown that Abbott's Bcl-2 family protein inhibitors bind to Bcl-2 proteins, restoring cell death to cancerous cells. Additionally, the compounds were found to enhance the effects of chemotherapy and radiation used to treat other types of cancer, such as non-small cell lung cancer.

---

[543] *Wong K-K, Dana-Farber Cancer Institute, Harvard Medical School. Therapeutic anti-EGFR antibody 806 generates responses in murine de novo EGFR mutant-dependent lung carcinomas. Online in advance of publication in the February 2007 print issue of the Journal of Clinical Investigation. J Clin Invest. 2007 February 1; 117(2): 346–352. Press release by Karen Honey, Journal of Clinical Investigation, January 30, 2007*

IX.92. Prognostic tool to alter clinical decision making

Researchers report evidence of a promising new example of personalized medicine[544]. Scientists are moving away from treating cancer patients as a population and are beginning to focus instead on single patients with individual characteristics. Tests such as the lung metagene model may soon help clinicians decide which patients require chemotherapy and which may be at lower risk and can enjoy a better quality of life by avoiding that route.

Even if the test can be used to increase patient survival by only 5%, 10.000 lives a year would be saved. The test should do better than that.

The researchers looked at 89 patients with early-stage non–small-cell lung cancer. They evaluated gene-expression profiles that predicted the risk of recurrence in 2 independent groups of 25 patients from the American College of Surgeons Oncology group study and 84 patients from the Cancer and Leukemia Group B study.

The investigators report that the lung metagene model predicted recurrence for individual patients significantly better than did clinical prognostic factors and was consistent across all early stages of non–small-cell lung cancer. Applied to the cohorts from the 2 trials, they report that the model had an overall predictive accuracy of 72% and 79%. The predictor also identified a subgroup of patients with stage IA disease who were at high risk for recurrence and who might be best treated by adjuvant chemotherapy.

Using the unique genomic signatures from each tumor, the new test predicted with up to 90% accuracy which early-stage lung cancer patients would suffer a recurrence of their cancer and which patients would not. Scientists now have a tool that can be used to move these high-risk patients from the 'no-chemotherapy' group into the aggressive-treatment group[533]. The researchers hope genomic tests like theirs will be used not only to predict patient outcomes but also to select individual drugs that will best match the tumor's molecular makeup.

IX.93 Kidney cancer

There is currently no treatment available for patients with advanced kidney cancer. Scientists at pharmaceutical companies, the National Cancer Institute, laboratories, and other universities are investigating whether drugs that inhibit HIF-1 may be useful for cancer therapy.

Studying a rare inherited syndrome, researchers have found that cancer cells can reprogram themselves to turn down their own energy-making machinery and use less oxygen, and that these changes might help cancer cells survive and spread.

The scientists report that the loss of a single gene in kidney cancer cells causes them to stop making mitochondria, the tiny powerhouses of the cell that consume oxygen to generate energy.

---

[544] *N Engl J Med. 2006; 355:570-580.*

[533] *Dr. Anil Potti, Duke University in Durham,lead author of the study, as emphasized to Medscape's Allison Gandey during an interview, published online, August 18, 2006.*

Instead, the cancer cells use the less efficient process of fermentation, which generates less energy but does not require oxygen. As a result, the cancer cells must take in large amounts of glucose. The appetite of cancer cells for glucose is so great that it can be used to identify small groups of tumor cells that have spread throughout the body.

Although changes in mitochondria have been described in many cancers, the Hopkins study shows for the first time how a cancer-causing mutation can block their production. There must be a strong advantage to cancer cells to stop using a highly efficient process in favor of one that generates much less energy.[534]

But turning down the "thermostat" in a sense, may give the cancer cell a survival edge. The scientists[535] found that if they reversed the switch and forced kidney cancer cells to start making mitochondria again, the cells produced increased amounts of free radicals, which can cause cells to stop dividing or even die.

They uncovered the mitochondrial mechanism in a study of Von Hippel-Lindau (VHL) syndrome, caused by a single gene mutation and characterized by the tendency to develop tumors in many parts of the body, including the kidney, brain and adrenal glands. They then measured mitochondria content and oxygen use in kidney cancer cells that contain no VHL protein and in the same cells with VHL "engineered" back in. Restoring VHL caused the cells to make two to three times more mitochondria and use two to three times more oxygen.

VHL normally blocks the action of HIF-1, a protein that the research group discovered in 1992. Cells normally make HIF-1 only under low oxygen conditions, when fermentation is necessary to make energy. However, in the absence of VHL, HIF-1 is active even when oxygen is plentiful and switches on genes that help a cell take up more glucose.

This work shows that excess HIF-1 counteracts a protein called MYC, which normally stimulates cells to make mitochondria. Because MYC is turned on in many other cancers, these results suggest that shutting down the mitochondria must be a very important event in kidney cancer, the scientist note.

The study of kidney cancers associated with rare genetic disorders is limited by the infrequent occurrence of the individual diseases and the small numbers of patients available for study in any one nation. Therefore, international collaboration is essential.

One such collaboration between the National Cancer Institute NCI scientists and investigators at the University of Manitoba in Winnipeg, Canada and the University of Birmingham in the United Kingdom led to the successful cloning of the Birt Hogg Dube (BHD) gene, which is associated with a rare hereditary syndrome characterized, in part, by a high predisposition to malignant kidney tumors that are often bilateral and multifocal.

---

[534] *Dr. Gregg Semenza, professor of pediatrics and director of the vascular biology program in the Institute for Cell Engineering at Johns Hopkins.*

[535] *Reporting in the May 8 issue of Cancer Cell. Announced by a press release from Johns Hopkins, May 11, 2007, entitled " Cancer Cells Can Reprogram Themselves To Turn Down Their Own Energy-Making Machinery In Order To Survive". The research was funded by the National Institutes of Health.*

412

Ongoing efforts include the identification of additional patients with familial kidney cancers of undetermined etiology worldwide to increase our ability to identify major genes that contribute to the development of kidney tumors[536].

IX.94. Prostate cancer

Prostate cancer represents the most common noncutaneous malignancy in men, with an estimated 234,460 new cases and 27,350 deaths in 2006[537]. With the widespread use of prostate-specific antigen screening, as many as one in six men in the USA will be diagnosed with prostate cancer. Significant healthcare resources are currently devoted to the treatment of this disease, specifically aimed at improving the side effects of successful treatment. Surgery or radiation therapy provides the best chance of cure from this disease. However, as many as 50% of patients treated with curative intent will develop a recurrence 10-15 years following treatment. Hormonal ablation via medical or surgical castration provides disease control, but is associated with significant hot flushes, loss of libido and impotence[538].

Although the genetic picture of prostate cancer has been far from crystal clear, scientists from Harvard Medical School have used a new genetic mapping technique[539] known as "admixture mapping" to identify a strong genetic link to prostate cancer. The findings confirm a previously reported genetic association to prostate cancer and help to focus future research efforts to a defined "genetic neighborhood" — one that likely contains a major, and as yet unidentified, prostate cancer risk gene.

Admixture mapping can be a particularly efficient and powerful method for dissecting the genetic underpinnings of complex diseases, which, like prostate cancer, often vary in frequency among different populations. The technique was first conceived more than 50 years ago, but has only recently been possible due to scientific advances in detecting and analyzing human genetic variation on a comprehensive, genome-wide scale. It involves systematically searching the genome for small genetic differences in individuals who have a particular disease and originally descended from two or more distinct geographic regions — such as African Americans, whose ancestry lies in both West Africa and Europe, or Hispanic Americans, whose ancestry lies in the Americas and Europe. Such genetic variants are present across all human populations, but their precise distribution can provide clues about the genetic factors that underlie common and genetically complex diseases.

The prostate cancer risk locus highlighted in this study contains a total of 9 genes and represents about one-thousandth of the human genome. While it carves out a substantial patch of genetic ground to be covered in the search for the gene (or genes) responsible, the work is notable for unearthing one of the first common genetic risk factors for cancer.

Indeed, specific inherited risk factors for other forms of cancer have been identified — in breast cancer and in colon cancer, for example — yet, unlike the prostate cancer risk

---

[536] NCI international portfolio: Addressing the global challenge of cancer http://www.cancer.gov/nci-international-portfolio/page (accessed Aug 6, 2007)

[537] Jemal A, Siegel R, Ward E, et al. Cancer statistics, 2006. CA Cancer J Clin. 2006;56:106-130.

[538] Webster WS, Leibivich BC, Exisulind in the treatment of prostate cancer. Expert Rev Anticancer Ther. 2005 Dec;5(6):957-62.

[539] Broad Institute of MIT and Harvard. Freedman M et al. (2006) Admixture mapping identifies 8q24 as a prostate cancer risk locus in African American men. Proceedings of the National Academy of Sciences August 21 advance online edition; doi:10.1073/pnas.0 605832103.

locus on chromosome 8, they tend to be rare in the overall population and account for a relatively small proportion of their respective cancers.

IX.95. Screening
Improvements in screening and diagnosis have led to significant reductions in mortality since the introduction of the prostate-specific antigen (PSA) test in 1994.

The lack of a definitive randomized trial means that we do not know whether PSA screening can prolong survival. Still, indirect evidence suggests that in populations in which prostate cancer screening is available, earlier diagnosis of disease and stage migration have resulted in reduced morbidity and mortality. Although still imperfect, PSA is a powerful biomarker that can help identify men at risk for prostate cancer and help guide treatment decision making.

PSA remains the best biomarker for prostate cancer. However, several abstracts found that the PCA3 molecular assay may be beneficial for both screening modalities. Expression of the *PCA3* gene is upregulated in prostate cancer and can be detected in urinary sediment after DRE. A group of investigators from California evaluated the PCA3 test in 147 men with either treated cancer (post prostatectomy), benign prostatic hypertrophy, untreated prostate cancer, and normal prostates; the last 3 were based on transurethral ultrasonography (TRUS) biopsies.[30] There was near-complete separation of the groups on the basis of only PCA3 results. With biopsy as the reference method, the sensitivity was 62% and the specificity 82%. These promising results indicate that PCA3 may be used in conjunction with PSA serology for prostate cancer detection. Investigators from The Netherlands examined the results of PCA3 testing in 299 patients, 114 of whom were diagnosed with prostate cancer and 48 who were treated with surgery. Although there was no association between PCA3 and pathologic stage or Gleason grade, there was a correlation with prostate cancer volume[540].

IX.96. Advances in predicting the probability of developing prostate cancer
Cancer occurs more frequently in the prostates of men than in any organ other than the skin. While DNA damage caused by exposure to the sun is likely the cause of many skin cancers, the cause of prostate cancer remains largely unknown. Research[541] points to absence of critical mechanisms protecting prostate cells from DNA damage as a key contributor to the development of prostate cancer.

The absence of at least two key checkpoint elements in the DNA damage response pathways may predispose human prostatic epithelial cells to accrual of DNA lesions and provide a mechanistic basis for the high incidence of cancer in the prostate.

Why prostate cells lack these mechanisms is unknown, but discovery of ways to restore these checkpoints controls might protect against prostate cancer.

Age is the biggest risk factor for prostate cancer, with incidence rising dramatically above the age of 50. In the USA, less than 10% of diagnosed patients are younger than the age of 60. Geographically, European and North American males are at higher risk. Cases have been reported on the early onset of prostate cancer as a result of using

---

[540] *Dr. Mitchell H. Sokoloff. Screening for Prostate Cancer: 2006 AUA Highlights. Continued Medical Education.*

[541] *Research conducted by the group of Marikki Laiho, M.D., Ph.D., a Professor at the University of Helsinki, Finland, in collaboration with Donna Peehl, Ph.D., an Associate Professor (Research) at Stanford University, US. Results of the study are published in the online Early Edition of the Proceedings of the National Academy of Sciences, April 2007.*

steroids, and the hormone DHEA (dehydroepiandrosterone) is thought to increase the risk in a similar manner.

A study[542] has identified seven genetic risk factors-DNA sequences carried by some people but not others-that predict risk for prostate cancer. According to the study's findings, these risk factors are clustered in a single region of the human genome on chromosome 8 and powerfully predict a man's probability of developing prostate cancer.

The identification of combinations of genetic variants that predict more than a fivefold range of risk for prostate cancer is an important step in helping scientists understand the higher risk for prostate cancer in African Americans compared with other U.S. populations and, more importantly, why some men develop prostate cancer and others do not[543].

While the HMS/USC team identified seven genetic variants on chromosome 8, two other studies[544] highlight the importance of this region in prostate cancer and each provides independent support for the findings presented by the HMS/USC team. Together, the three studies provide robust evidence of the role genetic variants play in prostate cancer.

According to the HMS/USC study, the seven genetic variants each independently predict risk for prostate cancer, with the predictive strength varying depending on the variant. Because almost all the risk factors were of highest frequency in African Americans, they may contribute to the known higher rate of prostate cancer among African Americans compared with other U.S. populations.

The study also produced novel biological findings. It revealed that each genetic contributor to prostate cancer risk is located outside of the coding regions of genes, in regions previously designated as junk DNA.

Two papers in 2006 highlighted this same chromosomal region as important in prostate cancer. The company deCODE genetics in Iceland first identified two specific genetic variants that contributed to risk for prostate cancer. The HMS/USC group then published a paper that identified a small region (about 4 million nucleotides, or 1/1,000th of the genome) as likely to contain important genetic risk factors. As part of this study, the HMS/USC group carried out a whole-genome screen for prostate cancer genes in about 2,500 African Americans. The paper suggested that more variants were likely in the region.

To find the additional risk factors in the current study, the team systematically tested genetic changes in this region of the genome in roughly 7,500 African American, Japanese American, Native Hawaiian, Latino, and European American men with and without prostate cancer. The majority of the men studied were drawn from the

---

[542] *led by researchers at the Keck School of Medicine of the University of Southern California (USC) and Harvard Medical School*

[543] *Christopher Haiman, assistant professor of preventive medicine at the Keck School of Medicine of USC, lead author of the study, published in the online edition of Nature Genetics, April 2007, in a statement to the press "Study Identifies Multiple Genetic Risk Factors For Prostate Cancer", April 6, 2007*

[544] *One of the studies is from deCODE Genetics in Iceland while the other is led by Dr. Gilles Thomas and Dr. Stephen Chanock at the National Cancer Institute.*

Multiethnic Cohort Study (MEC)[545], an epidemiological study of more than 215,000 people from Los Angeles and Hawaii created in 1993[546].

Despite the knowledge that prostate cancer occurs with high frequency in men of African descent in the Americas, little information is available regarding the epidemiology of prostate cancer in native African men, even though prostate cancer seems to be prevalent in that population as well. The objective of a study at the Hospital General de Grand Yoff in Senegal, initiated by the US National Cancer Institute, is to examine the role of genes that regulate the physiological disposition of testosterone in the development of prostate cancer and to evaluate whether these genes explain, in part, ethnic differences in prostate cancer rates. An understanding of the complex interplay of genetic variability at multiple loci and of environmental agents may improve our understanding of ethnic differences in prostate cancer development and risk prediction[547].

Researchers report[548] that a variation in a portion of DNA strongly predicts prostate cancer risk and that this common variation may be responsible for up to 20 percent of prostate cancer cases in white men in the United States. The research was conducted by investigators from the National Cancer Institute (NCI), part of the National Institutes of Health, and their partners in the Cancer Genetic Markers of Susceptibility (CGEMS) initiative. CGEMS researchers are scanning the entire human genome to identify common, inherited gene mutations that increase the risks for breast and prostate cancers. Building on this finding may lead to the identification of men at highest risk for prostate cancer, diagnose the disease earlier, and hopefully prevent it all together.

This gene variation was discovered on chromosome 8. Humans normally have 46 chromosomes in each cell, divided into 23 pairs. Two copies of chromosome 8, one inherited from each parent, form one of the pairs. Chromosome 8 spans about 146 million base pairs (the chemicals that comprise DNA), represents about 5 percent of the total DNA in cells, and contains an estimated 700 to 1,100 genes.

The region the CGEMS study identified on chromosome 8 is marked by a number of single nucleotide polymorphisms (SNPs), including rs6983267. SNPs are the most common type of gene variant in which a single unit of DNA may vary from one person to the next. The rs6983267 SNP is located in a segment of DNA that has few known or predicted genes for prostate cancer.

The researchers also confirm that a previous finding of a different variant, marked by SNP rs1447295, is also associated with prostate cancer. The rs1447295 SNP is located nearby on the same arm of chromosome 8. The old and the new susceptibility loci, or gene locations, appear to act independently; a change in one region did not affect the degree of risk conferred by the other. The rs1447295 location could be responsible for about seven percent of prostate cancer cases in white men of north European descent. Thus, taken together with rs6983267, these two genetic changes could account for as much as one quarter of prostate cancer cases in white men. The increased risk conferred by these loci was observed for all age groups studied.

---

[545] *The genetic study of the MEC grows out of a multiyear, ongoing collaboration between USC, the University of Hawaii, and collaborators at HMS and the Broad Institute of Harvard and MIT*

[546] *by USC's Henderson and Laurence Kolonel of the University of Hawaii.*

[547] *NCI international portfolio: Addressing the global challenge of cancer http://www.cancer.gov/nci-international-portfolio/page4 (accessed Aug 6, 2007)*

[548] *Yeager M, Orr N, Hayes RB, et al. Genome-wide association study of prostate cancer identifies a second locus at 8q24. "Nature Genetics". Online April 1, 2007, in advance of the May 1, 2007 print issue.*

An initial genome-wide association study was conducted in 2,329 men from across the United States who are participating in the NCI's Prostate, Lung, Colorectal, and Ovarian Cancer Screening Trial (PLCO) that began in 1993. The PLCO analysis compared 1,172 men with prostate cancer to 1,157 who did not have cancer.

CGEMS results were further confirmed by a number of other studies, including the American Cancer Society Cancer Prevention Study II, the Health Professionals Follow-up Study, the CeRePP French Prostate Case-Control Study, and the Alpha-Tocopherol, Beta-Carotene Cancer Prevention Study. Combined, these studies enrolled 6,266 men.

Similar data on breast cancer, the second-leading cause of cancer-related deaths in women in the U.S., are now being generated and are expected to be released soon. The CGEMS database will soon contain close to 2.5 billion genotypes, allowing researchers to identify genetic risk factors for breast and prostate cancers using 540,000 SNPs across the genome. By comprehensively surveying for common genetic variations and following-up promising findings in confirmatory studies, researchers hope to identify and verify associations that increase or decrease the risk of these cancers[549].

IX.97. Therapy options

Prostate cancer has a high cure rate in the early stages (stage I/II), where active treatment will usually consist of surgery or radiotherapy. Many cases of early-stage prostate cancer may be managed with observation only, if the disease is causing no side effects and treatment would have a greater impact on quality of life than living with the disease would. Drug therapy may be incorporated in the advanced stages of the disease, where there tends to be a good, albeit contemporary (approximately two years) response to hormonal treatments.

The goal of treatment is destruction of the tumor or blockade of tumor growth signals. In cases managed with surgery, treatment may involve removal of the entire prostate (radical prostatectomy), part of the prostate to relieve blockage of the urethra (transurethral prostatectomy) or in some cases castration (orchiectomy) to reduce testosterone levels. Cryosurgery is under examination as a relatively inexpensive, noninvasive and efficacious treatment, involving the use of liquid nitrogen to freeze the localized tumor. Radiotherapy is generally utilized in stage I, II and III prostate cancer as an alternative to surgery; both external beam radiation and internal brachytherapy are employed. Radiotherapy is selected over surgery based on patient and physician preference, geographical patterns of care and patient candidacy for radiotherapy versus surgery on the grounds of other co-morbidities.

Drug therapy typically consists of the use of hormonals to slow the progression of the disease by reducing the level of androgens or blocking their effect. In general, hormonal therapies cause regression or stabilization of the tumor for approximately two years, after which the tumor may resume active growth. Cytotoxic chemotherapy is employed for advanced tumors where conventional therapies have failed or the tumor has become hormone-refractory. Chemotherapy had mixed results in the treatment of prostate cancer;

---

[549] *Analyses and data from the CGEMS study will be available through NCI's caBIG(tm) (Cancer Biomedical Informatics Grid(tm)), at <http://calntegrator.nci.nih.gov/cgems/>. The summary results from the scan in PLCO are freely available to other researchers at this website.*
More information on NCI's Cancer Genetic Markers of Susceptibility (CGEMS) initiative and NCI's Cohort Consortium, including collaborating studies and institutions, is available at: <http://cgems.cancer.gov>, and <http://epi.grants.cancer.gov/Consortia/cohort.html>.

in early trials, many of the chemotherapy options led to stabilization of disease and improvement of symptoms but did not increase survival[550] [551]. The PSA response rates were often between 20% and 30%, with median survival not exceeding 12 months following the development of androgen-independent prostate cancer (AIPC). Over the past years, most chemotherapy agents were at best palliative[552].

Hormonal drug therapy is routinely used in the treatment of prostate cancer. Testosterone levels can be reduced by using luteinizing hormone-releasing hormones (LHRHs) or LHRH agonists, or by blocking the effect of androgens with anti-androgens. These two drug classes are often combined to form a total androgen blockade (TAB), which is effectively chemical castration. TAB has failed to conclusively establish survival benefit in clinical trials and remains used in a minority of patients in advanced stage. Typical LHRH agonists are leuprolide and goserelin, with bicalutamide and flutamide used as anti-androgens. The use of bicalutamide (Casodex from AstraZeneca) has significantly decreased in localized stage between 2003 and 2005 since the EPC trial demonstrated increased mortality with bicalutamide 150mg. This has not only undermined physicians' confidence in the drug in localized stage but it has also affected the uptake of the drug in locally advanced and advanced stage disease.

Since bone metastases occur in 85% of patients with prostate cancer[553] a treatment strategy that targets both the bone and the disease is the optimal approach[554].

For doublets with docetaxel or second-line chemotherapy, multiple studies have shown interesting and promising results with calcitriol, thalidomide, bevacizumab, satraplatin, vaccines, ixabepilone, and atrasentan.

IX.98. Provenge, GVAX

Improvements in screening and diagnosis have led to increased early-stage diagnosis, but there remains a need for more expansive screening programs and increasing patient awareness, particularly in older men who are more at risk. Innovative treatments such as gene therapy raise the possibility of reduced side effects, while novel cytotoxics that are more specific should also produce a reduction in associated toxicity, however such treatments are still developmental.

Research is extremely active in this area and new agents with novel modes of action are being evaluated in clinical trials, including vaccines such as Provenge (Dendreon) and GVAX for hormone refractory PC. Although physicians are hopeful of molecular-targeted agents for HRPC, primary research suggests that its use will remain relatively low in 2010.

Provenge (sipuleucel-T or APC-8015) is an investigational product that may represent the first in a new class of active cellular immunotherapies that are uniquely designed to stimulate a patient's own immune system. The compound is in late-stage clinical

[550] *Tannock IF, Osoba D, Stockler MR, et al. Chemotherapy with mitoxantrone plus prednisone or prednisone alone for symptomatic hormone resistant prostate cancer: a Canadian randomized trial with palliative end points. J Clin Oncol. 1996;14:1756-1764.*

[551] *Kantoff PW, Halabi S, Conaway M, et al. Hydrocortisone with or without mitoxantrone in men with hormone refractory prostate cancer: result of the Cancer and Leukemia group B 9182 study. J Clin Oncol. 1999;17:2506-2513.*

[552] *Winston WT, Novel Agents and Targets in Managing Patients With Metastatic Prostate Cancer . Cancer Control. 2006;13(3):194-198.*

[553] *Koeneman KS et al., Osteomimetic properties of prostate cancer cells: a hypothesis supporting predilection of prostate metastasis and growth in the bone environment. Prostate. 1999;39:246-261.*

[554] *Baweja M, Tan W. Bone metastases from prostate cancer: known and novel targets for palliation. Support Palliat Cancer Care. 2005;1:47-58.*

development for the treatment of patients with advanced prostate cancer. In clinical studies, patients typically received three infusions over a one-month period as a complete course of therapy. An FDA advisory committee[555] found substantial evidence of activity and safety of Provenge for the treatment of patients with asymptomatic, metastatic, androgen-independent (also known as hormone refractory) prostate cancer.

IX.99. PARP Inhibitors (ABT-888)

DNA damaging agents remain some of the most successful treatments for cancer. The enzyme Poly(ADP-ribose)polymerase (abbreviated PARP) can help repair DNA damage caused by these agents used to treat cancer and render them ineffective. As PARP activity is often increased in cancer cells, it provides these cells with a survival mechanism.

ABT-888 is an oral PARP-inhibitor developed by Abbott researchers to prevent DNA repair in cancer cells and increase the effectiveness of common cancer therapies such as radiation and alkylating agents.

ABT-888 is to undergo trials for melanoma. Preclinical data indicates ABT-888 has improved the effectiveness of radiation and many types of chemotherapy in animal models of cancer.

IX.100.   Revealing genetic activity of tumors

Peering into the body and visualizing its molecular secrets, once the stuff of science fiction, is one step closer to reality[556].

A research team reported that by looking at images from radiology scans - such as the CT scans a cancer patient routinely gets - radiologists can discern most of the genetic activity of a tumor. Such information could lead to diagnosing and treating patients individually, based on the unique characteristics of their disease.

Potentially in the future one can use imaging to directly reveal multiple features of diseases that will make it much easier to carry out personalized medicine, where you are making diagnoses and treatment decisions based exactly on what is happening in a person[557].

In some ways, the work brings to mind a device that science fiction fans may recall from the TV series, "Star Trek". In almost every episode of 'Star Trek,' there is a device called a tricorder, which they used noninvasively to scan living or nonliving matter to determine its molecular makeup. Something like that would be very, very useful.

In real life, this approach would avoid the pain and risk of infection and bleeding from a biopsy and would not destroy tissue, so the same site could be tested again and again. This work will help doctors obtain the molecular details of a specific tumor or disease

---

[555] Recommendation to the US FDA by the US Food and Drug Administration's (FDA) Office of Cellular, Tissue and Gene Therapies Advisory Committee, March 27, 2007.Dendreon corporate press release.

[556] A study from researchers at the Stanford University School of Medicine and the University of California, San Diego School of Medicine, published May 21, 2007 in the advance online edition of Nature Biotechnology. Additional contributions came from researchers at UCSD, the University of Southern California, the University of Hong Kong and UC San Francisco. The work was funded by the National Institutes of Health, Radiological Society of North America, Damon Runyon Cancer Research Foundation, the Israel Science Foundation and UCSD.

[557] Statement to the press by co-senior author Dr. Howard Chang, assistant professor of dermatology at Stanford, who led the genomics arm of the study, released by Stanford University, May 17, 2007

without having to remove body tissue for a biopsy. Ideally, scientists would have personalized medicine achieved in a noninvasive manner.

Radiology - while making great technological advances towards capturing more and more information - seemed to be largely oblivious to a fundamental shift in medicine towards genomic, personalized medicine that was beginning to take place[558].

A problem with using biopsied material for microarrays is that the tissue is destroyed in the process. Thus, there is no opportunity to re-test the same tissue after, say, a course of chemotherapy. Imaging through MRI or CT, however, leaves all organs intact and functioning.

The scientist compared their process-translating genetic activity patterns into medical imaging terminology - to the breakthrough that occurred when archeologists uncovered the Rosetta Stone in 1799. On the stone was the same text written in three versions: hieroglyphics, Egyptian and Greek languages. Every time certain letters showed up in Greek, a certain set of symbols would show up in hieroglyphics. That correspondence allowed previously undecipherable hieroglyphic writing to be understood.

The first step for the researchers was the equivalent of finding words for the hieroglyphics: to define the language of radiology. The initially defined mutually agreeable terminology for more than 100 features that appeared on scans. As their work progressed, they found they only needed 28 of them to capture maximal information. They then matched those imaging features with a vast stockpile of microarray data generated from human liver cancer samples. They also could compared their data with how the cancer patient fared.

What they found is that two very different aspects of cancer - how it looks by imaging and how it behaves on a molecular level - have a strong connection. Out of the 5,000 or more genes that have different activity in cancerous tissue, the researchers could reconstruct 80 percent of gene expression based on looking at standard CT scans the patients had undergone.

The fact that they saw strong connections between the imaging features and the molecular gene activity data suggests that this could be a promising and fruitful research direction, the scientists stated. Much like being able to identify the aromas from wine once the lexicon of wine-tasting is realized, radiologists - already experts in recognizing the visual differences between normal and pathological tissues - simply need to know what to look for and what it means.

IX.101. Detecting key mutations in tumors

Cancer is exceedingly complex, and includes more than 200 different diseases.

It is universally recognized that cancer is a disease of the genome, of mutations within genes responsible for cell growth and survival, and a great deal of effort has gone into finding those mutations, to the point where several hundred to a thousand are now

---

[558] Dr. Michael Kuo, assistant professor of interventional radiology at University of California San Diego, and director of the Center for Translational Medical Systems at UCSD, the study's other senior author, in a statement to the press, released by Stanford University, May 17, 2007. At the time the project started at Stanford in 2001, the medical school was ground-zero for studies of DNA microarrays - the lab tools that can screen thousands of genes at a time, developed by biochemistry professor Dr. Patrick Brown. Microarrays have proven to be extremely useful for identifying groups of genes that are more active or less active in a disease such as cancer, compared with normal tissue.

known. The challenge has been how to determine which of them are involved in each of the hundreds of kinds of cancer that occur in humans — and to develop accurate, affordable methods of detecting key mutations in tumor samples[559].

A study led by researchers at Dana-Farber Cancer Institute[560] and Broad Institute of the Massachusetts Institute of Technology and Harvard University may help relieve a bottleneck between scientists' expanding knowledge of the genetic mutations associated with cancer and the still nascent ability of doctors to use that knowledge to benefit patients. The findings provided the first demonstration of a practical method of screening tumors for cancer-related gene abnormalities that might be treated with "targeted" drugs. The study[561] results constitute an important step toward the era of "personalized medicine," in which cancer therapy will be guided by the particular set of genetic mutations within each patient's tumor, the authors suggest.

The authors[562] took advantage of a scientific serendipity to devise a simple test to detect important cancer mutations. Mutations in oncogenes (genes linked to cancer) do not occur randomly; rather, they seem to arise most frequently in certain regions of the oncogenes. As a result, researchers didn't necessarily have to scan the entire length of each gene, but could focus instead on the sections most likely to harbor mutations.

They performed these screenings with a technology known as high-throughput genotyping, a fast, relatively inexpensive way of profiling gene mutations within cells. It involves extracting DNA from a tumor sample, copying this material thousands of times, depositing segments of it in tiny "wells" on a small plate, and mixing in reagents that reveal whether each segment carries a specific mutation. Automated equipment then reads the plates to determine which mutations are present in each sample.

In the study, the researchers scanned 1,000 human tumor samples[563] for 238 known mutations in 17 specific oncogenes (those 17 were chosen because they are mostly "classic, well-known" contributors to cancer). They found at least one mutation in 298 of the samples, or 30 percent of the entire group, which was in keeping with the rates reported in scientific literature for the types of cancer examined.

According to the researchers, mutations were identified in the percentages they expected, which indicates this technique is on-target for the mutations they were interested in. Overall, the technique worked very well: they were able to obtain mutation profiles that were accurate, sensitive, and cost-effective.

---

[559] *Dr. Levi Garraway, Dana-Farber and the Broad Institute, lead author of a study on a practical technology of screening tumors for cancer-related gene abnormalities that might be treated with "targeted" drugs; published online on the Nature Genetics web site, February 12, 2007*

[560] *Dana-Farber Cancer Institute is a principal teaching affiliate of the Harvard Medical School and is among the leading cancer research and care centers in the United States. It is a founding member of the Dana-Farber/Harvard Cancer Center (DF/HCC), designated a comprehensive cancer center by the National Cancer Institute.*

[561] *Major funding for the study was provided by grants from Genentech, Inc., the American Cancer Society, the National Cancer Institute, the Prostate Cancer Foundation, the Burroughs-Wellcome Fund, the Robert Wood Johnson Foundation, and the Novartis Institute for Biomedical Research.*

[562] *Co-authors include Dr. Matthew Meyerson of Dana-Farber and the Broad Institute, and researchers from Dana-Farber; the University of Texas Southwestern Medical Center; Nagoya City University Medical School, Japan; the David Geffen School of Medicine at the University of California at Los Angeles; Brigham and Women's Hospital; University Hospital of Zurich, Switzerland; Northwestern University Feinberg School of Medicine; the National Cancer Institute; National Naval Medical Center; the Medical University of Vienna, Austria; the Center of Molecular Medicine at the Austrian Academy of Sciences; the Wistar Institute in Philadelphia; the Novartis Institutes for Biomedical Research; Harvard Medical School; the Max-Planck Institute for Neurological Research in Cologne, Germany; and the University of Cologne.*

[563] *The co-lead authors of the paper are Dr. Roman Thomas, formerly of Dana-Farber and the Broad Institute, and Alissa Baker of Dana-Farber, who contacted cancer researchers around the world to obtain the 1,000 tumor samples used in the study.*

The scans produced some surprises as well. Mutations were found in several types of tumors where they had not been previously recognized. Researchers also discovered an unexpectedly large number of instances where the same set of mutations co-occurred within tumor cells, suggesting that oncogenes often work in partnership.

It is a step toward the day when cancer patients will routinely have their tumors scanned for specific mutations, and treatment will be based on the cancer's particular genetic profile.

In the first decade of the 21st century biologists have access to an unprecedented and ever expanding quantity of information about the 30,000 genes that make up the human genome and the genomes of model organisms. With the sequence information at hand, the major challenge is to understand how it directs the formation of a complex multicellular organism and how the genetic makeup of an individual may predispose them to certain diseases[564].

The first line of research concerns the normal and healthy process of programmed cell death or "apoptosis". Under ideal circumstances, about one million of our old and damaged cells die during every second of our lives, to be replaced by the same number of new and healthy cells. The failure of apoptosis allows cells to multiply uncontrollably, which may lead to the formation of cancerous tumors. But since most cells still retain most of the "machinery" for apoptosis, a drug that could "switch on" the process could provide a new approach to cancer therapy.

The second line of research concerns stem cells: rare cells with the remarkable ability to generate an entire tissue. The almost unlimited regenerative capacity of stem cells has a built-in danger. If a stem cell acquires the ability to proliferate excessively, it can go on to form a tumor. If tumor growth is maintained by stem cells, it will be essential to develop new forms of therapy that target these rare cancer stem cells, in addition to the current standard therapy of targeting the bulk of tumor cells[565].

IX.102. Assessing all of the important genomic changes involved in cancer

NCI and NHGRI launched The Cancer Genome Atlas (TCGA) in December 2005 as a collaborative three-year pilot project to test the feasibility of using large-scale genome analysis technologies to determine all of the important genomic changes involved in cancer.

The overall goal of TCGA is to delve more deeply into the genetic origins that lead to this complexity, in order to enable the discovery and development of a new generation of therapies, diagnostics, and preventive strategies for all cancers. Results from the TCGA Pilot Project may provide the results we need to detect cancer early, in its most treatable stage, and provide new targets for the development of specific therapies[566].

TCGA will use cutting-edge technologies and knowledge from the Human Genome Project and other genomic studies to assess the range of genomic changes that cause the uncontrolled cell growth that characterizes cancer. TCGA will analyze hundreds of

---

[564] *WEHI, Molecular Medicine Division, August 7, 2006. http://www.wehi.edu.au/facweb/indexresearch.php?id=65, accessed March 13, 2007*

[565] *http://www.wehi.edu.au/WEHI_Press/index_single_press.php?id=104, October 16, 2006. Accessed March 12, 2007.*

[566] *. Acting NCI Director Dr. John E. Niederhuber in a statement to the press, September 13, 2006*

422

tumor specimens with multiple technologies, including the comparison of genome sequences from the cancers with the normal DNA sequence derived from the same patients, in order to identify changes that are specifically associated with cancer[567].

The Cancer Genome Atlas[568] Pilot Project is an important opportunity as we survey the potential future of cancer research. We must put aside our specific disease interests and channel our energy into the larger issues that can empower the cancer research enterprise to find new ways to improve outcomes for all cancer patients[569].

Success factors for the TCGA Pilot Project include completion of genomic analysis of the three initial cancer types; identification of specific alterations in genes associated with cancer; and differentiation of cancer subtypes based on genomic changes. In addition, the pilot project will establish a publicly available integrated database that scientists can use to generate new knowledge through research. TCGA data will be made available through public databases supported by NCI's cancer Biomedical Informatics Grid (caBI) and the National Library of Medicine's National Center for Biotechnology Information (NCBI). TCGA data will be provided in a manner that meets the highest standards for protection and respect of the research participants.

The Cancer Genome Atlas is a revolutionary project with the potential to provide cancer researchers the information needed to generate new hypotheses that can be tested in the laboratory and the clinic. Achieving the long-term goal of TCGA to identify all of the significant genomic changes in all cancers will benefit patients by enabling the discovery and development of the molecular biomarkers needed to develop targeted interventions to prevent and cure cancer[570].

Data from the TCGA Pilot Project will provide researchers and clinicians with an early glimpse of what promises to become an unprecedented, comprehensive "atlas" of molecular information describing the genomic changes in all types of cancer. TCGA will ultimately enable researchers throughout the world to analyze and employ the data to develop a new generation of targeted diagnostics, therapeutics, and preventives for all cancers, and pave the way for more personalized cancer medicine.

The cancers to be studied in the TCGA Pilot Project are lung [571], brain [572] (glioblastoma[573]), and ovarian [574]. The three cancers selected[575] for study by TCGA were identified in a process that began in the fall of 2005 with a Request for Information (RFI) from NCI and notification to NCI Cancer Centers. The goal of the RFI was to identify candidate biospecimen collections that employed the highest level of ethical, technical, biologic, pathologic, and bioinformatics standards in the development of their biorepository.

---

[567] National Human Genome Research Institute Director Dr. Francis S. Collins in a statement to the press, September 13, 2006

[568] For more details about The Cancer Genome Atlas, visit: http://cancergenome.nih.gov.

[569] Doug Ulman, cancer survivor, chief mission officer of the Lance Armstrong Foundation, and chair of the NCI Director's Consumer Liaison Group in a statement to the press, September 13, 2006

[570] NCI Deputy Director Dr. Anna D. Barker in a statement to the press, September 13, 2006

[571] It estimated that 174,470 new cases of lung cancer were diagnosed in the United States in 2006.

[572] Brain tumors account for nearly 90 percent of all primary central nervous system tumors. It is estimated that 18,820 new cases of brain cancer were diagnosed in the United States in 2006, and 12,820 patients died from the disease.

[573] Glioblastomas, also called glioblastoma multiforme or grade IV astrocytoma, are often fatal, malignant brain tumors that grow and spread very aggressively, and are the most frequently occurring type of brain cancer.

[574] Often detected late, ovarian cancer causes more deaths than any other cancer of the female reproductive system. 20,180 new cases of ovarian cancer and 15,310 deaths from the disease were expected in the United States in 2006.

[575] These cancers, which collectively account for more than 210,000 cancer cases each year in the United States, were selected because of the availability of biospecimen collections that met TCGA's strict scientific, technical, and ethical requirements.

The collections that qualified were evaluated in a three-stage process. The primary criteria addressed minimal requirements for the quality and quantity of the samples and the associated clinical information. Secondary criteria were then applied to effectively rank those tumor collections that met the primary criteria. Site visits to the biorepositories were then conducted by NCI and NHGRI staff. Critical factors for TCGA, such as timing, logistics, and the need for re-consent of patients, were factored into selection of the biorepositories to be considered. Finally, the biorepositories that emerged from this tiered process were further reviewed by an expert panel that included representatives from the surgery, research, pathology, bioethics, and patient advocate communities[576].

When fully operational, TCGA will consist of four integrated components: a Biospecimen Core Resource (BCR); Cancer Genome Characterization Centers; Genome Sequencing Centers; and a Principal Bioinformatics Resource.

September 2006, the two institutes also announced the biorepositories that will provide biospecimens of the first three cancer types to be studied as part of the TCGA Pilot Project. The source of the lung cancer biospecimens will be the Lung Cancer Tissue Bank of the Cancer and Leukemia Group B (CALGB) clinical trials group, which is housed at the Brigham and Women's Hospital in Boston. The source of the brain tumor (glioblastoma) biospecimens will be MD Anderson Cancer Center in Houston. The ovarian cancer biospecimens will be provided by the Gynecologic Oncology Group tissue bank at the Children's Hospital of the Ohio State University in Columbus.

The International Genomics Consortium, in collaboration with the Translational Genomics Research Institute, was selected to establish and manage TCGA's BCR. The BCR will collect, store, process, and distribute biomolecules from cancerous and normal samples to the Cancer Genome Characterization Centers and Genome Sequencing Centers for genomic analysis.

TCGA's Cancer Genome Characterization Centers will analyze samples from the BCR to identify key genomic alterations, such as copy number changes and/or chromosomal rearrangements, some of which are known to contribute to cancer development and/or progression. Selected genes will be sequenced by the Genome Sequencing Centers using high-throughput methods to identify small genomic changes, such as single base mutations and small insertion/deletions. TCGA Cancer Genome Characterization Centers and the Genome Sequencing Centers will be selected in the coming months.

IX.103. The discovery of biomarkers: undervalued and under-funded
"Biomarkers" are defined as endogenous molecules (such as proteins or metabolites) or injected agents (such as imaging agents) whose presence or state correlates with important physiological processes, disease outcomes and treatment response (including toxicity and efficacy).Biomarkers are molecular, biological, or physical characteristics that indicate a specific underlying physiologic state. By other words: Biomarkers are substances sometimes found in the blood, other body fluids, or tissues that measure biological processes, and in addition to genes and proteins, can be complex carbohydrate (sugar) structures that are attached to protein and lipid (fat) molecules.

---

[576] The primary, secondary, and additional criteria are described in detail at http://cancergenome.nih.gov/media/qanda.asp

The genes contained within the chromosomes of cells serve as the hereditary blueprints for the construction of proteins. Each gene specifies a distinct protein with a unique sequence of amino acids. This amino acid ultimately dictates the three-dimensional structure of a protein, which determines its biochemical interactions and specialized function. It is estimated that of the 32.000 human genes, approximately 1200 are responsible for 1600 disorders. The challenge now is to connect the proteins encoded by these genes and the final metabolites to establish how physiological processes are mediated and regulated, in order to identify biomarkers for the in vitro or in vivo diagnosis and treatment of disease in a new era of personalized medicine.

We now understand that cancer arises in a single cell as a result of genetic changes that alter a number of cellular processes — growth control, immortality, apoptosis, somatic evolution, angiogenesis, metastasis — and many cancers appear to have activated a wound healing genetic expression program. These changes are driven by abnormal methylation or a high rate of mutation. The proteins that function in each of these cellular circuits provide not only potential drug targets, but also signals that may allow scientists to non-invasively visualize and monitor physiology.

Moreover, new advances continue at an astonishing rate. In just the last few years we have seen: the sequencing of the human genome, providing a catalogue of all human genes; the development of RNAi technology, permitting the sophisticated loss of function analysis of human cells, and the identification of cancer stem cells, defining a potential new paradigm for cancer etiology.

Such recent advances, however, have been translated into effective diagnostics in only a few cases to date – for example, imaging agents that detect DNA replication, apoptosis, or proteolysis. In some respects, the discovery of new biomarkers appears to have been undervalued and under-funded relative to drug discovery. For example, the NCI Early Detection Research Network (EDRN), charged with discovering and validating new biomarkers, has not yet brought new agents to patient care.

It is time to unleash the diagnostic and informational content of our knowledge of altered molecular circuits into improved diagnostic agents for cancer patients. Effective biomarkers will improve patient outcomes. Individuals at risk for cancer or with cancer would benefit enormously by better methods for...
(i)   determining cancer risk,
(ii)   detecting and localizing cancer at its earliest stage,
(iii)   profiling for therapeutic decision making, and
(iv)   monitoring response to therapy in real time.

It is already evident that molecular diagnostics can improve diagnosis and treatment. Genetic translocations or transcript array profiles allow stratifying many organ-specific cancers (breast, leukemia, lymphoma, sarcoma) into different subtypes that have distinctive therapeutic outcomes. For example, *Myc* gene amplification status predicts the outcome for childhood neuroblastoma. The quantity of Bcr-Abl transcript predicts disease recurrence in chronic myelogenous leukemia long before clinical symptoms recur [577].

The rapid development of new techniques and the generation of results concerning the use of genomic biomarkers for in vitro and in vivo cancer detection, accompanied by the recent application of new in vivo gene expression markers, demonstrates that different genomic approaches can be effectively used to deter specific cancer biomarkers. The genomic approach can be further applied for the estimation of prognosis in cancer disease, for therapy planning, and for monitoring therapy in vitro and in vivo. However, these advances represent only a small step in the development of clinically applicable biomarkers [578] [579].

IX.104. Biomarkers: greatly accelerating basic and translational research

Biomarker research already has identified biological indicators that have had immense impact in prevention and treatment of disease. For example, blood pressure and cholesterol biomarkers have enabled diagnostics and therapies which have contributed to a 50 percent decrease in cardiovascular mortality in the U.S. over the last 30 years. Biomarkers measured in a variety of patient samples, including blood, tissue, urine and cerebrospinal fluid, are used in a diverse array of clinical settings. They are used to identify risk for disease, to make a diagnosis, to assess the severity of a disease and which organs are involved, and to guide treatment. Biomarkers will greatly accelerate basic and translational research.

Biomarkers serve a wide range of purposes in drug development, clinical trials, and therapeutic assessment strategies and can be used to measure the natural history of a disease, indicate and compare the effectiveness of new pharmaceutical treatments, or define the state of a given disease or condition.

Although many successful biomarkers have been developed to date, advances in genetics and proteomics promise to usher in a new era of abundant, informative biomarkers that could transform the application of molecular biology to human disease.[580] The application of biomarkers to cancer is leading the way because of the unique association of genomic changes in cancer cells with the disease process. Consequently, DNA-based biomarkers are already becoming incorporated into routine patient management and are providing lessons on the value added by appropriate diagnostic tests. Moreover, cancer management illustrates the complexity of the disease process, which can potentially be distinguished through appropriate biomarkers applied to different individuals, different types of disease, the progression of disease states and the multi-step nature of cancer treatment.

Scenarios for the use of biomarker-based diagnostics for cancer include the following: risk assessment, noninvasive screening for early-stage disease, detection and localization, disease stratification and prognosis, response to therapy and, for those in remission, screening for disease recurrence. Cost and potential morbidity increase as we progress along this continuum.

---

[577] Hartwell LH, Lander ES, Co-chairs, A Strategic Plan for Improving Biomarkers for cancer. Report of Working Group on Biomedical Technology. Report to National Cancer Advisory Board NCAB Working Group on Biomedical Technology. Presented: February 16, 2005.

[578] Pennisi E., The human genome. Science 291(5507),1177-1180 (2001).

[579] Oeler P, "omics"-based imaging in cancer detection and therapy. Future medicine (2006) 3 (1),19-32

[580] Hartwell L. et al. Biomarkers for cancer screening, diagnosis, and treatment: a systems approach. Abstract. Nature Biotechnology, August 2006.

426

The goals in applying diagnostic tests are

(i)   to identify persons harboring potentially life-threatening cancers at the earliest stage possible,

(ii)   (ii) to avoid false-positive tests and diagnosing of cancers that would otherwise not threaten a person's well-being to avoid psychological stress and unnecessary treatments, and

(iii)   (iii) to minimize the overall cost of the program. It is unlikely, however, that any single test will perfectly meet all of these goals.

IX.105. Glycobiology of cancer: The search for glycan-based biomarkers

The U.S. National Cancer Institute (NCI) is funding a new $15.5 million, five-year initiative to discover, develop, and clinically validate cancer biomarkers by targeting the carbohydrate (glycan) part of a molecule[581].

Scientists have long recognized that certain sugar structures, which are attached to protein and lipid molecules, may be important as markers for cancer development. While this area has compelling scientific interest, its biological and chemical complexities have often discouraged investigation. Today, with the advent of advanced technologies to conduct protein and carbohydrate chemistry, research into this intriguing area has experienced renewed interest."

Numerous studies comparing normal and tumor cells have shown that changes in the glycan structures of cells correlate with cancer development. Glycans are extremely abundant, but recent advances in technology have only now allowed a systematic study of these structures. Many protein biomarkers also have glycan components and analysis of these two molecular structures together may improve the value of tests such as those for prostate-specific antigen (PSA), CA-125, and carcinoembryonic antigen, which are sometimes used in prostate, ovarian, and colon cancer detection, respectively.

Looking at different types of biomarkers and new ways to identify them is critically important to both the basic understanding of cancer and the ability to identify early cancer and risk for cancer[582].

IX.106. Moving medicine from a curative model of today to a preemptive era

The US Foundation for the National Institutes of Health (FNIH), the US National Institutes of Health (NIH), the US Food and Drug Administration (FDA), and the Pharmaceutical Research and Manufacturers of America (PhRMA) launched a major public-private biomedical research partnership, *The Biomarkers Consortium*, to search for and validate new biological markers -biomarkers- to accelerate dramatically the delivery of successful new technologies, medicines, and therapies for prevention, early detection, diagnosis, and treatment of disease.

---

[581] *The NCI's Tumor Glycome Laboratories are the principle component of the new trans-NIH Alliance of Glycobiologists for Detection of Cancer and Cancer Risk. The other components of the alliance are the Consortium for Functional Glycomics funded by the National Institute of General Medical Sciences and several Glycomics and Glycotechnology Resource Centers supported by the National Center for Research Resources. The NCI's Early Detection Research Network (EDRN) is also an alliance member, providing support for design and statistical analysis, patient accrual, and collection of clinical specimens to facilitate validation studies using EDRN's existing components.*

[582] *Dr. Sudhir Srivastava, chief of the Biomarkers Research Group in NCI's Division of Cancer Prevention. More information Can be found at: The National Institute of General Medical Sciences Consortium for Functional Glycomics at http://functionalglycomics.org, and The National Center for Research Resources' Glycomics and Glycotechnology Research Centers at http://www.ncrr.nih.gov/Glycomics. (accessed Auguast 22, 2007).*

The most useful biomarkers known today took several decades to develop. Moreover, harmonization to arrive at common approaches to biomarker assessment and evaluation has not been accomplished. FNIH, CMS, FDA, NIH, and the pharmaceutical, biotechnology, diagnostics, and medical device industries recognize the critical need for

Table 9.34. The seven Tumor Glycome Laboratories projects:

Project	The discovery and clinical validation of cancer biomarkers using glycan array
Objective	Determine the diagnostic or prognostic anti-glycan auto-antibody signatures in patients. For breast cancer, determine how many years prior to diagnosis that progression to cancer can be predicted.
Institution	Cellexicon
Cancer type under study	breast, lung, melanoma, ovary.
Project	Immunogenic sugar moieties of prostate cancers
Objective	Identify anti-glycan autoantibody signatures in prostate cancer patients.
Institution	Stanford University
Cancer type under study	prostate
Project	Early cancer detection and prognosis through glycomics
Objective	Identify biomarkers from glycans released from serum glycoproteins and develop high-throughput platforms to measure biomarkers suitable for the clinic.
Institution	Indiana University
Cancer type under study	colon, lung, ovary, prostate
Project	Glycan markers for the early detection of breast cancer
Objective	Identify breast cancer biomarkers based on aberrant glycan modifications on defined amino acid residues of serum glycoproteins
Institution	Northeastern University
Cancer type under study	breast
Project	Tumor glycomics laboratory for discovery of pancreatic cancer markers
Objective	Identify glycoprotein and glycolipid biomarkers for pancreatic cancer in pancreatic ductal fluid that can also be found in serum. Develop assays for promising biomarkers.
Institution	University of Georgia
Cancer type under study	pancreas
Project	Autoantibodies against glycopeptide epitopes as serum biomarkers of cancer
Objective	Determine auto-antibody signatures to mucin glycopeptides in pancreatic and breast cancer patients
Institution	University of Nebraska
Cancer type under study	breast, pancreas
Project N	eu5Gc and Anti-Neu5Gc antibodies for detection of cancer and cancer risk
Objective	Expand on research showing that cancer patients express cell surface glycans containing the sialic acid N-glycolylneuraminic Acid (Neu5Gc) and produce autoantibodies to these structures
Institution	University of California San Diego
Cancer type under study	lung, ovary, pancreas

a coordinated cross-sector partnership effort to more rapidly identify and qualify biomarkers to support basic and translational research, guide clinical practice and, ultimately, support the development of safe and effective medicines and treatments for a wide range of diseases. The Biomarkers Consortium will harmonize approaches to identifying viable biomarkers, verifying their individual value, and formalizing their use in research and regulatory approval.

Rapid realization of the aims of The Biomarkers Consortium is beyond the capacity of any single sector of the US nation's health enterprise, much less the single-institution, or single-investigator, science research approach. This initiative is large-scale and complex. It requires the expertise of all stakeholders—government, industry, patient groups, academia, and other private groups[583].

---

[583] *Dr. Charles A. Sanders, chairman of the FNIH board of directors and of The Biomarkers Consortium Executive Committee in a statement to the press, October 5, 2006. The Foundation for the National Institutes of Health was established by the United States Congress to support the mission of the National Institutes of Health – improving health through scientific discovery. The foundation identifies and develops opportunities for innovative public-private partnerships involving industry, academia, and the philanthropic community. A non-profit, 501(c)(3) corporation, the Foundation raises private-sector funds for a broad portfolio of unique programs that complement and enhance NIH priorities and activities.*

The core challenge is to move medicine from a curative model of today to a preemptive era when we can identify and track a disease process as early as possible. The identification of biomarkers is an essential element for the new era of predictive, preemptive, personalized medicine. The consortium enables government, industry, and philanthropy to come together to explore and develop common tools for a common purpose for everyone's benefit[584].

The Foundation[585] for NIH will manage The Biomarkers Consortium under direction of the consortium executive committee, with guidance from leading scientists within and outside the member organizations, plus input from a public representative and funding partners. The principal funding for activities of The Biomarkers Consortium will be by contributions to the Foundation for NIH, manager of the consortium, from private-sector partners representing industry, non-profit and advocacy organizations, and the philanthropic community. The foundation is actively seeking additional partners to facilitate funding for the consortium's operations ("funding members") and for individual research projects—each of which will emerge as a distinct scientific initiative under the consortium's administrative umbrella. The Biomarkers Consortium Executive Committee[586], comprising public and private representatives from the founding partner organizations and other relevant stakeholders, is the primary decision-making body of The Biomarkers Consortium.

For America's pharmaceutical companies to partner with the other health research sectors is a pioneering effort that will advance biomarker research and development as never before. This research is critically important to the development of new cures, treatments, and diagnostics addressing the most pressing diseases facing patients today[587].

The creation of The Biomarkers Consortium is an important landmark. It represents a new step in sharing both the burdens and the fruits of fundamental scientific work and can help identify areas of opportunity, clarify responsibilities, and make important new findings openly available[588].

Among the first projects and project concepts will be several focusing on lymphoma, lung cancer, depression and diabetes.

These markers can be used in clinical studies to assess whether a drug is safe and effective as well as in studies to learn more about health and disease. In addition, the FDA can use biomarkers to determine whether drugs can safely and effectively treat disease. One of the key priority public health challenges singled out in the FDA's *Critical*

---

[584] *Dr. Elias A. Zerhouni, NIH Director, in a statement to the press at the occasion of the announcement of the creation of the Biomarkers Consortium, October 5, 2006. The National Institutes of Health (NIH) -- "The Nation's Medical Research Agency" -- includes 27 Institutes and Centers and is a component of the U. S. Department of Health and Human Services. It is the primary Federal agency for conducting and supporting basic, clinical, and translational medical research, and it investigates the causes, treatments, and cures for both common and rare diseases.*

[585] *Other partners in The Biomarkers Consortium include NIH, FDA, and the Center for Medicaid and Medicare Services (CMS).*

[586] *The consortium's funding members include: the Alzheimer's Association; AstraZeneca; the Biotechnology Industry Organization; Bristol-Myers Squibb; GlaxoSmithKline; The Leukemia & Lymphoma Society; Johnson & Johnson; Eli Lilly & Company; Pfizer Inc; the Pharmaceutical Research and Manufacturers of America; and F. Hoffmann-La Roche*

[587] *Billy Tauzin, president and CEO of PhRMA which represents the USA's leading pharmaceutical research and biotechnology companies, in a statement to the press, October 5, 2006. The Pharmaceutical Research and Manufacturers of America (PhRMA) represents the country's leading pharmaceutical research and biotechnology companies. PhRMA companies are leading the way in the search for new cures. Members alone invested an estimated $39.4 billion in 2005 in discovering and developing new medicines.*

[588] *Health and Human Services Secretary Mike Leavitt in a statement to the press, October 5, 2006 regarding announcement of Biomarker Consortium.*

Path to New Medical Products report is the need for biomarker qualification and standards to support innovation in health care.

One of the most pressing goals under the FDA's Critical Path Initiative is to improve developmental science and technology to deliver better diagnostics and more personalized treatments to patients more quickly and ultimately improve outcomes. The identification and use of biomarkers in drug development can have a catalytic impact on this mission[589].

Information and results from consortium projects will be used broadly by researchers to promote the appropriate use of biomarkers to improve the public health worldwide.

Promising project concepts that to date have been approved in principle for further development are:
* Lung Cancer and Non-Hodgkin's Lymphoma Projects;
* Whole Genome Association in Major Depressive Disorder: Identifying Genomic Biomarkers for Treatment Response;
* Diabetes and Pre-Diabetes Biomarker Project;
* Evaluate the Utility of Adiponectin as a Biomarker Predictive of Glycemic Efficacy by Pooling Existing Clinical Trial Data from Previously Conducted Studies.

The projects in particular:

IX.106.1. Lung cancer and non-Hodgkin's lymphoma projects

Two of the consortium's initial projects, which will be conducted by the National Cancer Institute (NCI) [590], will assess the use of Fluorodeoxyglucose-Positron Emission Tomography (FDG-PET) as a potential biomarker for clinical trials conducted in non-Hodgkin's lymphoma and non-small cell lung cancer. The use of FDG-PET, a promising imaging technology, will be evaluated as a potential predictive biomarker of tumor response and patient outcome. These FDG-PET studies could have enormous impact on patient management by validating a tool that can measure responses to treatments and enable more rapid drug development.

The rationale for using FDG-PET relies on the fact that tumor cells require substantial energy to survive and spread. Cells metabolize glucose for energy, and, in tumor cells, this metabolism is generally enhanced. FDG-PET is an effective way to measure glucose metabolism. This imaging procedure involves first injecting a patient with FDG that is attached to a radioactive tracer (F18), followed by imaging of the patient in a PET machine. Tumor cells consume significantly larger amounts of the FDG than normal cells, which enables this imaging technique to detect tumors as small as 1 cm. In both of the planned FDG-PET clinical trials, the levels of "tumor uptake" of FDG are measured/imaged before treatment and again at intervals thereafter. A decrease in uptake of FDG would indicate a decline in the number of tumor cells.

---

[589] *Dr. Andrew C. von Eschenbach, FDA Acting Commissioner of Food and Drugs, in a statement to the press, October 5, 2006. The U. S. Food and Drug Administration (FDA) is responsible for protecting the public health by assuring the safety, efficacy, and security of human and veterinary drugs, biological products, medical devices, our nation's food supply, cosmetics, and products that emit radiation. The FDA is also responsible for advancing the public health by helping to speed innovations that make medicines and foods more effective, safer, and more affordable; and helping the public get the accurate, science-based information they need to use medicines and foods to improve their health.*

[590] *The National Cancer Institute (NCI) is a component of NIH. The NCI coordinates the National Cancer Program, which conducts and supports research, training, health information dissemination, and other programs with respect to the cause, diagnosis, prevention, and treatment of cancer, rehabilitation from cancer, and the continuing care of cancer patients and the families of cancer patients*

The FDG-PET lung and lymphoma clinical trials, which emerged from the Oncology Biomarker Qualification Initiative (OBQI), an interagency collaboration, were announced by NCI, FDA, and CMS in February 2006. The OBQI was designed through the NCI-FDA Interagency Oncology Task Force (IOTF) to evaluate specific biomarkers in cancer clinical trials for potential incorporation by the FDA as outcome measures for parameters such as treatment response. It is anticipated that these first two trials to evaluate FDG-PET in non-Hodgkin's lymphoma and lung cancer will inform both the regulatory review process for these cancers and also provide CMS with evidence-based measures to make informed reimbursement decisions.

Although we are early in the development of imaging tools as biomarkers, this research holds the potential, over time, to be used not only in the diagnosis of cancer, but in monitoring and predicting response to therapy[591].

IX.106.2. Whole genome association in major depressive disorder: identifying genomic biomarkers for treatment response

This project would be an extension of the National Institute of Mental Health[592] Sequenced Treatment Alternatives to Relieve Depression (STAR*D) Study, a nationwide public health clinical trial to determine the effectiveness of different treatments for people with major depressive disorder. This study would expand the search for genes associated with effective outcomes in the treatment of the disorder.

Depression is a serious illness affecting millions of Americans, some who have depression as well as their families, friends, and co-workers. The cost of depression in human suffering, lost work, and medical expenditures rises every year. Anti-depressants help many patients, but no one medicine helps everyone, and it may take weeks before knowing whether the medicine is working. This study will seek genetic markers of treatment response, thereby providing a potential for a simple test to predict who will respond to which medicine-shortening response time, diminishing suffering, and helping patients function better at home and at work.

IX.106.3. Diabetes and pre-diabetes biomarker project

This project would build upon an existing National Institute of Diabetes and Digestive and Kidney Diseases[593] pilot study, and would seek to discover new biomarkers related to type II diabetes and pre-diabetes that could be used in developing a new assay and speed up and improve the translation of this assay from the bench to the bedside. This new methodology of diagnosis could lead to a more reliable, cheaper, and faster diabetes test.

Diabetes mellitus is a devastating disease affecting not only blood sugar levels, but also causing kidney disease, blindness, heart disease, and peripheral vascular disease. It usually begins with elevated blood sugars that may be present for months or years before a diagnosis is made. To determine if a patient has diabetes, the glucose

---

[591] NCI Director Dr. John E. Niederhuber in a statement to the press, October 5, 2006

[592] The National Institute of Mental Health (NIMH) is one of 27 components of the National Institutes of Health (NIH), the Federal government's principal biomedical and behavioral research agency. NIH is part of the U.S. Department of Health and Human Services. The NIMH is the lead Federal agency for research on mental and behavioral disorders. Its mission is to reduce the burden of mental illness and behavioral disorders through research on mind, brain, and behavior.

[593] The NIDDK, a component of the NIH, conducts and supports research in diabetes and other endocrine and metabolic diseases; digestive diseases, nutrition, and obesity; and kidney, urologic and hematologic diseases. Spanning the full spectrum of medicine and afflicting people of all ages and ethnic groups, these diseases encompass some of the most common, severe, and disabling conditions affecting Americans.

tolerance test is done, but is inconvenient and not terribly reliable. This project will seek to identify biomarkers of diabetes from before the disease really begins to the time of full-blown diabetes. This can allow earlier treatment, better monitoring, and ultimately reduce morbidity and mortality from diabetes as well as providing the potential for huge savings in health care costs.

IX.106.4. Evaluate the utility of adiponectin as a biomarker predictive of glycemic efficacy by pooling existing clinical trial data from previously conducted studies

The primary objective of this study would be to determine whether a soluble protein, adiponectin, has utility as a predictive biomarker of glycemic control in normal non-diabetic subjects and patients with type II diabetes following treatment with a novel and promising new class of compounds, Peroxisome Proliferator-Activated Receptor (PPAR) agonists.

Insights into the mechanism driving cancer cell growth have led to some success in the development of targeted therapeutics. In general, these agents exert their anti-cancer effects by specifically blocking signaling that: promotes tumor cell proliferation, obstructs cell death, hampers cellular differentiation, or facilitates angiogenis. However, the molecular pathways that underlie these cellular processes are multifaceted and often redundant. As well, the genetic or epigenetic aberrations that drive the development of cancers are heterogeneous, making it a challenge to devise a successful cancer therapeutic by targeting a single gene. Obviously, a more profound understanding of the complexity of these signaling pathways and new approaches with which to unravel the mechanisms underlying tumoregenic genetic changes are needed to improve targeted cancer therapies[594].

---

[594] *Cancer Therapeutics: The Road Ahead October 8-10, 2007, Capri, Italy. The Consiglio Nazionale delle Ricerche (CNR), Cell Death and Differentiation, and Nature organized the conference "Cancer Therapeutics: The Road Ahead; a translational meeting, that summarized identification of novel targets for the development of new anti-cancer agents.*

Chapter Ten

Adding a New Dimension

in

Healthcare

The Biomarkers Consortium is an important landmark on the journey to personalized medicine. It represents a new step in sharing both the burdens and the fruits of fundamental scientific work. It can help identify areas of opportunity, clarify responsibilities, and make important new findings openly available. Most of all, it can help propel us toward an era of Personalized Health Care and improved development of medical treatments[595].

Collaboration is one key. There is a crucial need to learn to work across sectors and across corporate barriers to make best use of fundamental new information. Likewise, there is the need for common standards to make information manageable[596].

X.1. The new era of personalized healthcare

The past few years have witnessed a revolution in the understanding of health and disease, brought on by the sequencing of the human genome and the creation of a map of human genetic variation. This revolution has been given a name: personalized medicine. The roots of personalized medicine are based in providing the right diagnosis and treatment to the right patient at the right time at the right cost.

People vary from one another in many ways - what they eat, the types and amount of stress they experience, exposure to environmental factors, and their DNA. Many of these variations play a role in health and disease. For example, the natural variations found in our genes could influence our risk of developing a certain disease, as well as how our bodies respond to that disease. The combination of these variations across several genes can affect each individual's risk of developing a disease or reacting to something in the environment, and can be one of the reasons why a drug works for one patient and not another[597].

---

[595] *Health and Human Services Secretary Mike Leavitt in a statement to the press, October 5, 2006. Announcement of Biomarker Consortium.*

[596] *Health and Human Services Secretary Mike Leavitt in a statement to the press, October 5, 2006. Announcement of Biomarker Consortium.*

[597] Personalized Medicine: Promises and Prospects. Conference brochure. Harvard Center for Genetics and Genomics, November 3-4, 2005)

Table 10.1.   The right drug, the right dose, the right patient

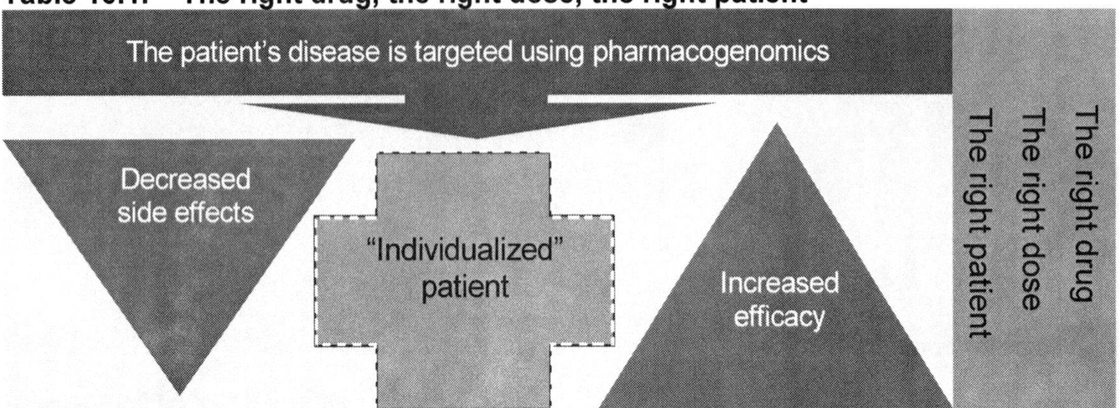

X.2. Understanding disease in ways never before possible

For many years, Americans have understood that advances in basic biomedical science have the potential to bring about a new era in medical care. The sequencing of the human genome, combined with dramatic new capabilities at the molecular level of biology, hold out the promise that we will be able to understand and manage disease in ways never before possible[598].

A goal of personalized medicine is the a priori identification and matching of patients with therapies that maximize clinical outcome and/or minimize deleterious side effects[599].

For the patient, the goal is Personalized Health Care -- the recognition that every one of us is a unique biological being, and the capability to deliver highly individualized care tailored to genetic and molecular features that are central to health outcomes.

For the health care system, we are aiming toward much more effective development of diagnostics, drugs and other therapies. Growing understanding of biological mechanisms can mean not only targeting therapies with much greater precision to each individual, but also safer, faster and more cost-effective development of these products.

But before these changes can occur for patients and for the health care system, new ways must also be developed to carry out the business of information discovery and sharing. As scientists seek to translate basic discoveries into bedside treatment, they need new tools for managing complex information on a new scale.

X.3. Achieving optimal medical outcomes

Personalized medicine refers to using methods of molecular analysis to better manage disease or predisposition toward disease. It aims to achieve optimal medical outcomes by helping physicians and patients choose among the disease management approaches likely to work best in the context of a patient's genetic and environmental profile. It avoids the unbelievable enormous waste generated by today's medicines for the masses; medicines that work for part of the patients only.

---

[598] *Health and Human Services Secretary Mike Leavitt in a statement to the press, October 5, 2006. Announcement of Biomarker Consortium.*
[599] *Fossella JA, Genetic structure in human populations: implications for the personalized medicine value chain. Future Medicine (2006) 3(1),1-7*

The failures of one-size-fits-all medicine are a driver for the new paradigm of personalized medicine. While the transition to personalized medical healthcare will be difficult, it has to be done to benefit the patient population[600].

We recognize that many diseases we aim to treat have a strong genetic component. Additionally, we know some patients taking medicines show efficacy or experience side effects, while others do not. Our understanding of all this has improved greatly but these areas remain as significant challenges in drug development[601].

One of several reasons of this high failure rate is indeed the complexity of the diseases to be tackled, with both the genetic and environmental influences, and only limited models available[602].

The decision process within drug discovery is difficult, and the experimental systems are very divergent. Conclusions between experimental systems (e.g. molecular, cellular, animal, and human) are not obvious, and hence drugs still fail to a significant extend in clinical trials, or, even worse, in the market place[603].

We all recognize the challenges and risks that the drug discovery and development poise. Industry-wide, many new drugs have been discontinued in late-stage trials, after hundreds of millions of dollars have been invested. Many times this results because a subset of patients responded differently from the rest. And as recent events have shown, even FDA/EMEA approved, marketed drugs can run into problems when a small number of patients have an unexpected response. Drugs developed based on genetic information will be directed more specifically towards the patients who are most likely to benefit from them[604].

Risk-prediction information would allow patients the best access to the appropriate medicines by grouping them according to these predictors[605].

X.4. The right drug at the right time

Personalized medicine hopes to use the variations in drug efficacy - both in the patient and in the molecular underpinnings of the disease itself - to develop new diagnostic tests and treatments and to identify the sub-groups of patients for whom they will work best. It can also help determine which groups of patients are more prone to developing some diseases and, ideally, help with the selection of lifestyle changes and/or treatments that can delay onset of a disease or reduce its impact.

Genetic testing in practice is inherently valuable in predicting risk, and could be used to personalize healthcare. "We have dumb medicine today," It has been pointed out that

---

[600] *Dr. Janet Woodcock, Deputy Commissioner of Operations and Chief Operating Officer, U.S. Food and Drug Administration (FDA), at Harvard center for genetics and Genomics, Nobember3, 2005. She suggested that Gouvernement is in a position to either facilitate or impede the practice of personalized medicine, and that her organization, FDA, would "lead in that area." For her pioneering role in advancing personalized medicine through programs such as the Critical Path Initiative and the Guidance on Pharmacogenomic Data Submissions, Dr Woodcock has been honored by the Personalized Medicine Coalition (PMC) with the first Leadership in Personalized Medicine Award*

[601] *John LaMattina, Sr VP Global Research and Development, Pfizer Inc.,in a statement to the press at the occasion of the company's announcement of an investment in Perlegen Sciences, 2006.*

[602] *Strohman R., Maneuvering in the complex path from genotype to phenotype. Science 296, 701-703. 2002*

[603] *Knowles J, Gromo G. Target selection in drug discovery. Nature drug Discov. 2. 63-69, 2003*

[604] *Brad Margus, co-founder and CEO of Perlegen in a statement to the press, January 2006.*

[605] *Dr. Reed Tuckson, Senior Vice President, Consumer Health and Medical Care Advancement, UnitedHealth Group, the conference "Personalized Medicine: Promises and Prospects". Harvard Center for Genetics and Genomics, November 3-4, 2005*

only 11 out of 100 people taking statins would ever experience a cardiac event. Here, physicians are relying on a surrogate, LDL [cholesterol], that's not necessarily connected

Table 10.2.   The way forward

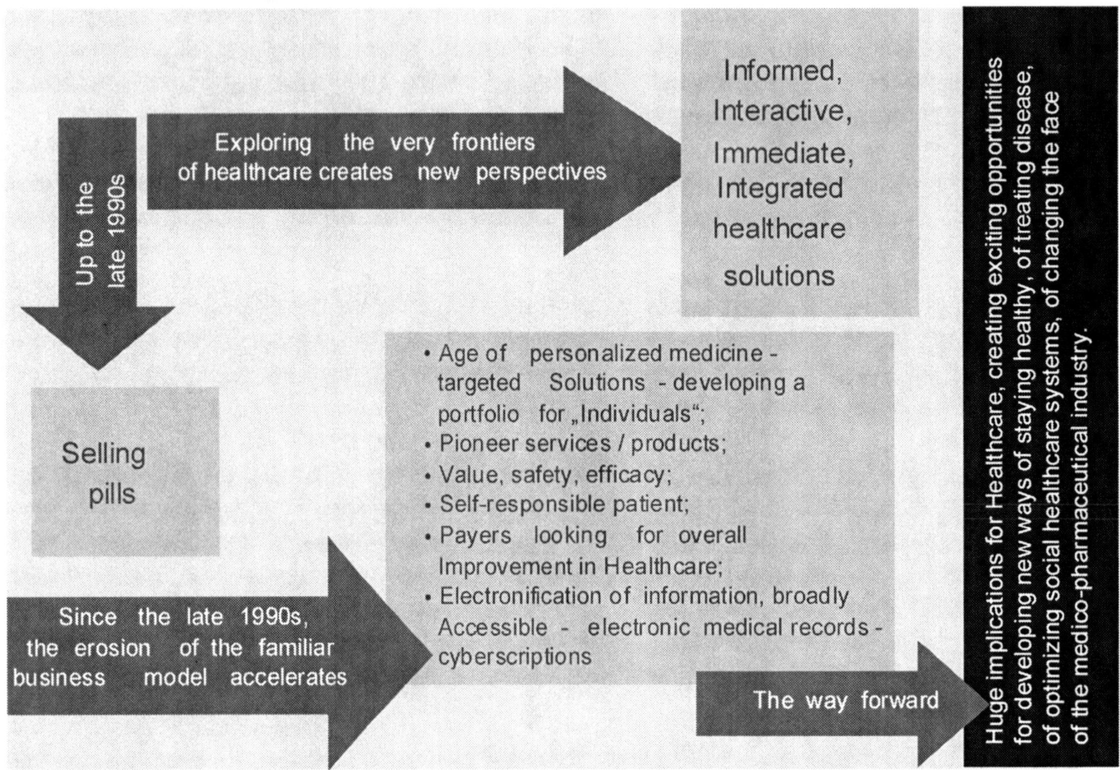

to heart attack or stroke. Doctors have learned to treat and lower a lab value, but its effectiveness on patient care is still in question. "Treatment" could happen in a different way.

Here is an example: In one family, where several members of a family who suffered from heart attack, genetic studies revealed a penetrant gene mutation known to predispose carriers to heart attack in a grandfather, father and two-year old son. The preventative health care strategy for the father was a change in his lifestyle to better manage the risk. With the knowledge of the two-year old's risk, and the fact that atherosclerotic plaque begins accumulating in adolescence, steps could be taken early and continued throughout the next five or six decades of the child's life to improve the likelihood of a positive outcome[606].

Diagnostic tests comprise only three percent of overall medical costs, yet impact 70 percent of medical decision-making at the level of the provider. In addition, the combination of suitable diagnostics during clinical trials could prevent lengthy and expensive trials by predicting and creating a stratified patient pool genetically inclined to respond favorably to a drug.

---

[606] *Dr. Eric Topol, Provost, Cleveland Clinic Lerner College of Medicine, the Harvard Medical School - Partners Healthcare Center for Genetics and Genomics, December 2005, p3*

Table 10.3. Reshaping the "traditional" healthcare model

Pre-genomics „one size fits all medicines "

Gold standard model of the late 20th century: diagnose and treat

- Disease symptom
- Diagnosis (in vitro diagnostics, imaging)
- Therapy selection (medicines for the masses)
- Treatment
- Outcome

Targeted medicine

New millennium model: Early intervention, predict & prevent

- Disease predisposition profiling (genetics, functional genomics)
- Monitoring & prevention
- Diagnosis (in vitro diagnostics, imaging)
- Therapy selection (personalized medicines)
- Treatment
- Treatment monitoring (in vitro diagnostics, imaging)

The cost of creating a drug and the low rate of treatment success of drugs on the market are both factors that contribute to the healthcare financial crisis. In due time, personalized medicine will become the new "traditional" medicine, providing patients "the right drug at the right time."[607]

X.5. Far-reaching mental and structural transformation is inevitably going to happen.

This much is not new to many executives. However, the ossified corporate organizational structure still has a lethargic influence on traditional BigPharma's workforce. On those who show an entrepreneurial spirit.

When business thrives on apparent success (success of today is based on the decisions of yesterday), it means that when inflated expectations about drug candidate pipelines - together with sustained sales of older products- create the impression of doing things right, there is little or no room left for "the real thing". In such cases, proposing initiatives and opting for change becomes an even more risky undertaking than it already is.

It is foreseeable that independent, state funded organizations, such as the US National Institutes of Health or the National Cancer institute(NCI), could become centers of research excellence and provide pharmaceutical companies with their findings. As a result, pharmaceutical companies will become "virtual companies".

On the way to becoming "virtual", these companies are expected to develop so-called Integrated Therapeutic Firms that combine the targeted pharmaceutical, biotechnological and, where appropriate, also the diagnostic, medico-technological and nutritional "know-how" under one common motive of a shared "corporate vision/identity" and a joint "brand image" to the integrative cover of specific therapeutic areas.

---

[607] Mara Aspinall, President, Genzyme Genetics Inc., at the conference "Personalized Medicine: Promises and Prospects". Harvard Center for Genetics and Genomics, November 3-4, 2005

The NCI, among others, is a good example of what can be achieved by governmental and/or not-for-profit organizations.

The National Cancer Institute's (NCI's) record of achievement in conducting, fostering, and funding cancer research is evident in the major advances that have been generated through its intramural and extramural research programs. These advances have helped define our understanding of human physiology, genetics, and cell biology; how normal cellular and physiologic activities can be subverted during the malignant process; and our ever improving ability to use this knowledge to prevent, diagnose, and treat cancer. It is particularly notable that NCI has supported the research efforts of at least 20 Nobel Prize winners. For approximately half of these individuals, NCI supported the awarded research. For many of the remaining Nobel laureates, NCI supported important ongoing research. Some of the advances arising from these efforts include the discovery of the segmental nature of eukaryotic genes, RNA splicing, oncogenes, and reverse transcriptase, as well as clinical applications such as magnetic resonance imaging, allogeneic bone marrow transplantation, and hormonal treatments for cancer.

NCI supports large-scale studies to identify exposures, lifestyles, and genes that affect cancer risk. For example, scientists linked recent, rapidly rising rates of esophageal cancers to two factors: increasing rates of obesity and gastrointestinal acid reflux, which increases the risk of a pre-malignant condition called Barrett's esophagus.

NCI also supports research to understand the many factors involved in a normal cell becoming cancerous and spreading to other parts of the body. Basic laboratory and animal research provides clues about the inner workings of cells that would otherwise be beyond our reach.

The investment in identifying molecular profiles, or signatures, of cancer has been producing impressive results. As early as 2000, NCI researchers were using gene expression profiling to distinguish between different subtypes of lymphoma. Two years later, investigators were reporting preliminary success using proteomic technologies to detect and diagnose some cancers at early stages. Since then, researchers have been discovering and validating molecular profiling techniques that many experts believe are setting the stage for improved cancer detection and diagnosis.

An NCI research team confirmed the effectiveness of using a new cell-based immunotherapy approach combined with chemotherapy for treating advanced stage metastatic melanoma. The goal is to fight cancer tumors by stimulating and reintroducing a patient's own T cells (rare disease-fighting immune cells). The promise of this therapy is that a patient's own immune system may be used to effectively treat existing tumors[608].

Through the years, NCI has been a leader in exploiting new communications technologies to share cancer information. For example, the Institute's Physician Data Query (PDQ) database became the first publicly available, disease-specific, electronic information resource when it went online in 1982. NCI is now pioneering another frontier

---

[608] *Testimony delivered Apr. 6, 2006, before the U.S. House Subcommittee on Labor-HHS-Education Appropriations, to discuss the NCI budget request for Fiscal Year 2007, by Dr John E. Niederhuber, Deputy Director, National Cancer Institute, and Richard Turman, Deputy Assistant Secretary, Office of Budget, Department of Health and Human Services. Visit http://www.cancer.gov/aboutnci/FY07-budget-request (accessed August 6, 2007)*

in information sharing through its cancer Biomedical Informatics Grid (caBIG) project[609]. NCI's caBIG is creating a unifying technology platform or "world-wide web" for cancer research. caBIG is well on the way to its goal to create a network of interconnected data, applications, individuals, and institutions that will redefine how cancer research is conducted and care is provided. This initiative has also whetted considerable commercial interest.[610].

Another notable advance came with the announcement of results from the NCI-sponsored Digital Mammographic Imaging Screening Trial (DMIST). The study found that digital mammography is more accurate than film mammography for women with dense breasts, as well as for several other groups of women, including women under 50 and pre- and perimenopausal women. Overall, DMIST offers a model case study of how NCI can be an agent of change, pursuing new approaches to research, partnering with the private and public sectors, and fueling the development of technologies to achieve an important advance. It is particularly noteworthy that NCI and the American College of Radiology Imaging Network (ACRIN) secured the involvement in DMIST of four companies that developed and manufactured digital mammography machines for our use in clinical trials: Fischer Medical, Fuji Medical, General Electric Medical Systems, and Hologic.

The technology revolution is speeding up and enabling the discovery process. Nanotechnology has emerged as a key strategy for imaging molecular features of cancer and will ultimately lead to personalized medicine. NCI's investment in nanotechnology is a powerful example of leveraging resources from the private sector through its Centers of Cancer Nanotechnology Excellence.

NCI is taking an important role in PROMIS, a National Institutes of Health Roadmap Initiative. PROMIS aims to develop ways to measure patient-reported symptoms, such as pain and fatigue, and aspects of health-related quality of life across a wide variety of chronic diseases and conditions, including cancer. Two years into development, researchers are designing and building an innovative technology that promises to revolutionize how patients report clinically important symptoms and outcomes. This technology will provide clinical research communities with a consistent and validated approach to measure these clinically relevant, but subjective and difficult-to-measure outcomes.

Of equal significance, is NCI's and the National Human Genome Research Institute's (NHGRI) The Cancer Genome Atlas (TCGA) Pilot Project. As described before in this book, TCGA is a comprehensive effort to accelerate understanding of the molecular basis of cancer and which evolved from the Human Genome Project (HGP). The TCGA Pilot Project will develop and test the science and technology needed to systematically identify the genetic changes in a small number of cancers.

Through CanCORS, NCI is supporting the largest ever observational study of cancer care delivered in diverse, population-based health care settings. This prospective cohort study has enrolled 10,000 patients with newly diagnosed lung or colorectal cancer. Vital

---

[609] *http://www.cancer.gov/aboutnci/ncia (accessed August 6, 2007)*

[610] *Testimony delivered Apr. 6, 2006, before the U.S. House Subcommittee on Labor-HHS-Education Appropriations, to discuss the NCI budget request for Fiscal Year 2007, by Dr John E. Niederhuber, Deputy Director, National Cancer Institute, and Richard Turman, Deputy Assistant Secretary, Office of Budget, Department of Health and Human Services. Visit http://www.cancer.gov/aboutnci/FY07-budget-request (accessed August 6, 2007)*

information will be collected on how clinical practices affect outcomes, and what influence certain characteristics—of patients, providers, and community health care delivery systems—have on the services that patients eventually receive. CanCORS is providing a unique opportunity to examine community practices regarding palliative (non-curative symptom control) and end-of-life care. This study takes into account the perspective of the patient, caregiver, and providers among a diverse group of patients followed over time. NCI is supporting developmental research to test the feasibility of collecting such measures routinely within clinical practice[611].

Building on NCI-funded research, large-scale clinical trials yielded results that will have profound effects in preventing and treating many cancers. NCI has been a major leader in the molecular metamorphosis of biomedical medicine that has benefited all fields of medical research. Without NCI's pioneering role in funding research - including basic science, clinical trials, and translational investigations - into the molecular and genetic processes that underlie all disease and the training of new cancer researchers, it is unlikely that the advances we are seeing today in many health areas - from AIDS to macular degeneration - would have occurred at the pace they have.

X.6. Integrated care providers

Personalized medicine requires more/better monitoring. Focus will be put on early detection of risks through regular monitoring. What means : avoiding disease or allowing treatment of disease in its initial phase.

In the new pharmaceutical environment, companies become system providers. The salvation of the large companies (at least of a few of them) will be their effective marketing creativity and clout (today, the industry spends 25% of its revenue on marketing and sales), leaving the discovery itself to more nimble entrepreneurs and the development to international institutions.

We may end up with very few "pharmaceutical centers of excellence". MegaPharmaBiotech outsources its research for more efficiency. Funding/control is made by large, independent organizations. The Sick Funds/Managed care organizations, pharmaceutical companies and hospitals become integrated care providers.

Table 10.4. The integrated virtual MegaPharma-Biotech company

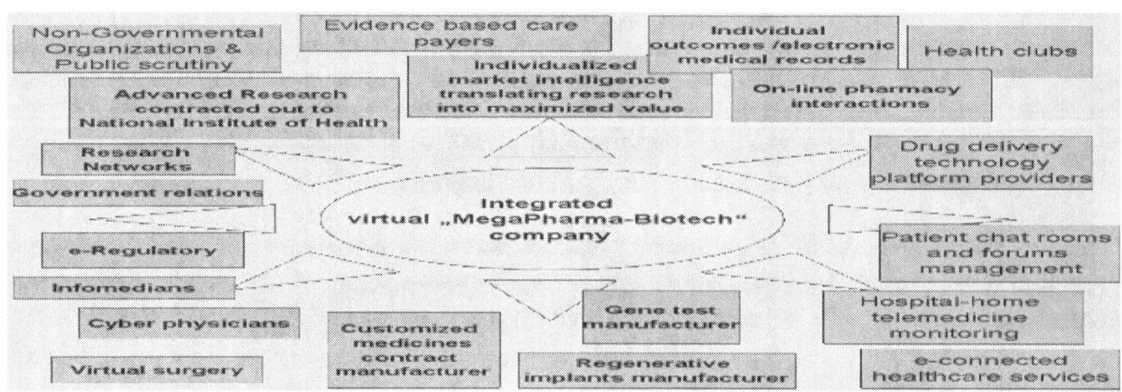

[611] The Nation's Investment in Cancer Research: A Plan and Budget Proposal for Fiscal Year 2008. Prepared by Dr. John E. Niederhuber, Director, National Cancer Institute, as mandated by The National Cancer Act of 1971 (Public Law 92-218), October 2006. http://plan.cancer.gov/ (accessed Aug 6, 2007)

440

Research-oriented companies are in a better position to adopt the new challenge at an early stage and thus create competitive advantages for themselves, even for the future (what CEOs, trapped by quarterly financial performance pressure, rarely think of !). These companies often possess the required know-how to develop new strategies for all points of the new value addition chain. However, what they lack are the organizational, operative structures as well as adequate incentive systems. As the new system will in any event lead to considerable additional costs in the early stages –on account of the measures that need to be taken for education, prophylaxis and regular healthcare measures for the individual –these companies would require partners for "pre-financing" the new system

The various company health insurance funds would be the first to be most inclined to assume such a role. Company health funds often operate on shoestring budgets and lower expense levels than the big health insurance funds for the public at large. Therefore, they frequently offer financial advantages to insured persons. These advantages could then be guided towards those insured under those funds for developing and implementing the new system and thus creating a limited and transparent frame as a prerequisite for the new system. In this sense, focused experiences could then serve as the blueprint for the entire health system.

The pharmaceutical companies who follow this path would be able to establish themselves as "masters" of the entire value chain and bring about "health" and also an unexpected competitive edge.

X.7. Tomorrow's "winning" healthcare model

No one yet knows which healthcare model will survive and thrive. But one thing is for sure : The future mentality will be based on medicines for "My" body instead of drugs for "Every" body. It will speed up the drive of individuals towards more lifestyle/health-management, including to products and services offered by new healthcare categories.

Leo Mauren[612], former Roche executive, explains that it would be necessary to reposition all the persons involved in the health system so that, on the one hand, "health services for all" can still be financed and, on the other, the individual/patient/consumer occupies the notable position that has been their due for a considerable time.

The starting point for the thought process is the fact that the service providers currently active in the health systems consider "health" to be a "limited" and "isolated" part of the value addition chain and are therefore trying to optimize their own interests. For example, the pharmaceutical industry positions itself factually only in a two-dimensional way :
1) research oriented companies – a high degree of innovation, and relatively high cost of medicines;
2) generics vendors – a low degree of innovation (if any), but a relatively low cost of medicines.

The companies are prisoners of their own uni-dimensional ROI philosophy and of the "disease"-based system that helps them to stay healthy thanks to the reimbursement of medicines (either by state-run compulsory "social" security plans or by private health insurance) . For pharmaceutical companies, value creation is restricted to a company and/or product.

---

[612] *Correspondence with the author, 2005-2007*

A less isolated or restricted view of the "value creating chain of health" would place the patients at the beginning and center of all activities. First, this "point of view" would be free from all the hindrances of individual interests that partly block the view today. Thus, the two dimensions of the pharmaceutical companies positioning matrix would be supplemented by a third dimension that one could term as the **"efficiency of health"**.

Table 10.5.1. A two-dimensional view of the pharmaceutical industry

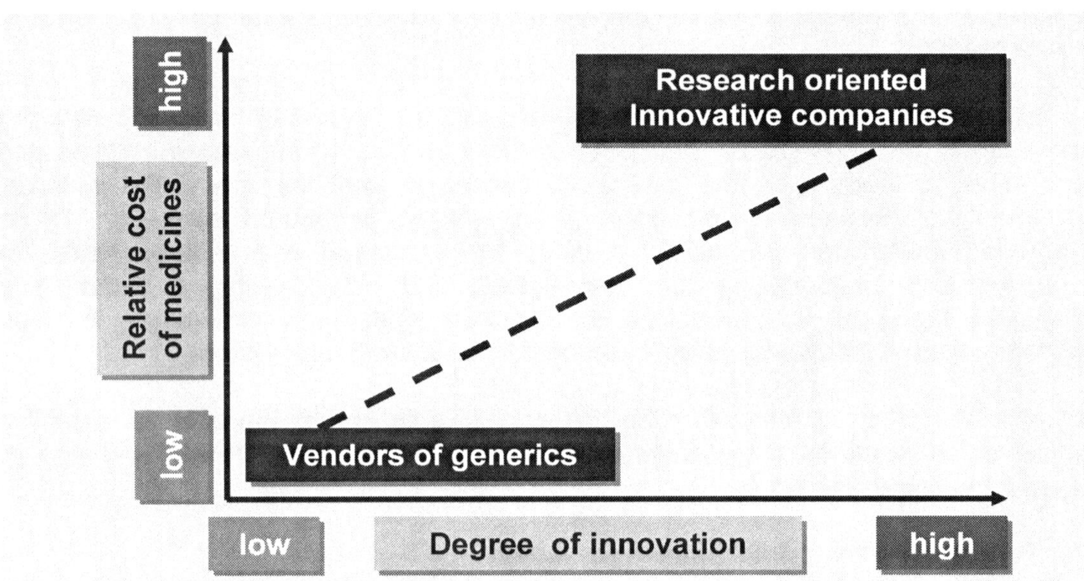

Table 10.5.2. Adding a dimension to the matrix

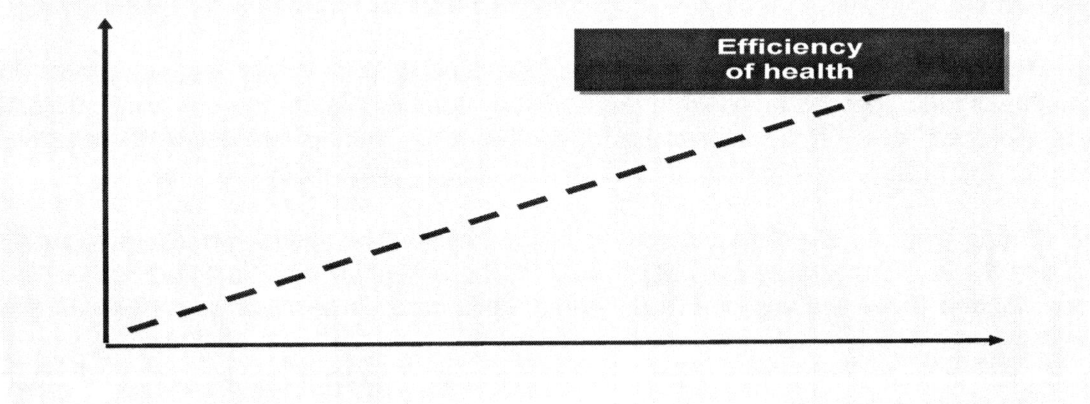

Traditionally, this dimension, has no place in the matrix because it this places the individual at the center of all considerations. If one personalizes the individual and assigns all diseases, illnesses and dispositions to him/her, the question that then arises is: How can chronic illnesses such as diabetes, coronary disease, cancer and others be "handled" (i.e. not necessary "treated") optimally at an "individual" level? If one defines "handling" in this sense as a new, integrated value addition chain, then it will guide the service providers covering the entire value chain:

    -prophylaxis about the;
    -early detection for;

-therapy;
-monitoring the diseases

X.8. Service providers covering the entire health prediction-prevention-therapy value chain

The individual or all individuals would be integrated into a system where almost, right from birth, it demands a high degree of self-responsibility, consciousness and knowledge about one's own health. For this, corresponding training and education in the early years of life would be unavoidable. It would be necessary to have correspondingly "positive" behavioral patterns that have a positive effect on the health and would make a person liable for "penalty" if they were not conformed to. Such a system would make a major contribution to avoiding certain diseases, or at least detecting them at an early stage and treating them efficiently (read : cost-effectively).

If today the existence of disease is a prerequisite for many service providers (like the pharmaceutical industry, pharmacies, hospitals) to optimize their interests, the same would not, however, be the case in the new system. Here, a service provider would have to cover the entire value chain of one or more potential diseases and thus optimize its financial interests only be reducing costs over the entire value addition chain. The incentive system would thus have to be designed in almost the same way as we find it in the health insurance funds today.

X.9. A health system that is really worth its name.

The results that would ensue for the individual can be easily gauged : Either way it would be an improvement on system we have today, *which takes illness or disease as the necessary prerequisite for its own existence.*

Service providers in the entire value addition chain would of course be needed even in the future. The changed self-consciousness and the changed incentive pattern of the new system would also bring about a change in the role of each individual. The new system will "enforce" a change in behavior among service providers and individuals and could, in turn, bring forth an efficient health system that is really worthy of its name.

X.10. Raising questions and questioning existing truths

However, it would be necessary to raise seriously a few questions and also question strategies:

1) Can the patient, as a passive receiver of care be turned into an active(co-)decision maker?

2) Can the astronomic amount of money spent on research –by the remaining handful of global R&D powerhouses targeting the same therapeutic categories - be better used ?

3) Can a massive system of doctors and hospitals be changed from a purely isolated therapy provider into a system that takes a holistic view ?

If one takes the third dimension to be the decisive step into the future in order to think afresh --to be the ways and means of how the health of an individual can be ensured in an economical and efficient way-- then both the first and second dimensions of this interpretation can be viewed as the initial stages of the new thought process :

The first dimension : From the 1950s to 1975 (unrestricted health for all is a much hailed postwar social achievement; "no limits" creates a free self-service health-supermarket for everyone involved in the healthcare provider system);

The second dimension : From 1975 to 2010 (cost containment measures have no more than homeopathic effect; a "system in crisis" = collapse looms, profound renewal imposes itself);

The third dimension ? Pharmaceutical companies become service providers.

BigPharma, most of which companies are rather slow to innovate, faces the erosion of the large indication areas by the entrepreneurial outfits. With every major progress the larger companies find themselves progressively more dependent on outside expertise. In their search for higher efficiency and profit, they may resist by creating a network of satellite supplier-companies, by more cross-border alliances, by evolving out of their existing franchises and defend their territory through non-drug means, such as services for the patient conditional on the use of the company's drug. That's for the short-term.

Imagine a future where technological innovation occurs almost exclusively outside established companies but is nevertheless acquired by them because BigPharma is systematically ineffective at incubating new ideas and small companies lack the marketing cloud to effectively bring new ideas to market. The fully integrated giant of today is doomed to disappear in its present configuration.

Personalized medicine also offers the possibility of improved health outcomes and has the potential to make healthcare more cost-effective.

Personalized medicine is poised to transform healthcare over the next several decades. New diagnostic and prognostic tools will increase our ability to predict the likely outcomes of drug therapy, while the expanded use of biomarkers - biological molecules that indicate a particular disease state - could result in more focused and targeted drug development.

Table 10.6. Towards medicines for "my body"

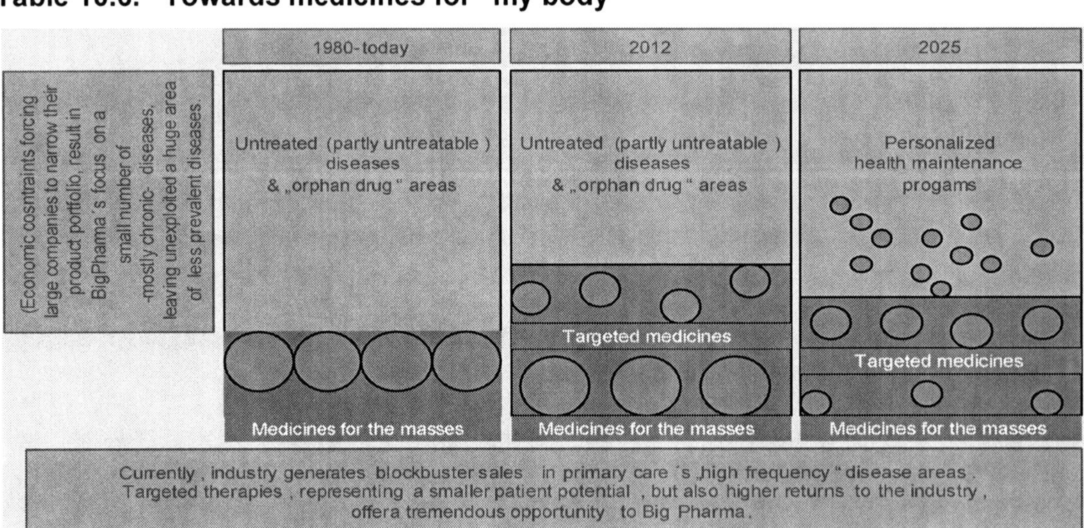

Personalized medicine is a target-based approach poised to change the way medicine is practiced, from basic biomedical research to maximum individualized patient care.

Society is to witness breakthroughs in understanding causes of major diseases within the next decade.

Advances in technology create new measurement possibilities of predisposition and progression. The treatment of diseases with underlying biological causes that remain largely mysterious, such as cancer, diabetes, depression and arthritis, will be facilitated by the identification of the genetic material that predisposes to these difficult-to-treat illnesses. This knowledge will allow physicians to tailor medical treatment to patients more exactly than it has ever been possible. Although the large majority of drugs that are brought into clinical trials never reach the market, the medical potential of new decade gene-based drugs is enormous.

The future holds great promise. Politicians, the medico-pharmaceutical industry, patient advocacy groups, we all have to mentally prepare for a world of huge opportunities in treating disease, in preventing disease, in... predicting disease.

Chapter One: The Operating Environment. Past, Present and Future

Advances in technology and better understanding of disease management is ensuring that the aging population will consume 5% or more per year. This can be forecast to grow from7-8% to an average of 12.14% of GDP. Most countries' current systems only fund half this amount. It could therefore be argued that we have barely started to address the impact of healthcare costs. In systems where healthcare is effectively free at the point of delivery, it is no surprise that the consumer of that service feels no pressure to control costs. Governments in industrialized countries have tried a variety of discriminatory processes to introduce some element of control –largely without success- but with the consequence of effectively rationing healthcare, usually unfairly.

The dangers of focusing a disproportionate amount of resources on high technology, curative care to the near total exclusion of primary health care; the increasing administrative burden on the system; drifting intergenerational relations; the unstoppable "social" healthcare systems financial deficits, amplified by Government's appetite for all sort of taxes on medicines and medical services; the accelerating rhythm in genetic discoveries promising new dimensions in treatment of health disorders...the perception and possibilities of maintaining the welfare state, leading to reform after reform of the Western nation's Health Care systems in an aging world and a globally linked economy is one of today's core social and economic issues, and will remain at the center of the western post-industrialized countries' political agenda for years to come.

The challenge is to eliminate wastage and abuse, to penalize socially driven cost escalation (tobacco, alcohol, drugs, etc.) and to encourage life cycle financial planning, to ensure better preparations are made by all individuals to manage their own care as they age, and...to think beyond cost containment measures.

The golden future of healthcare is ahead of us, not behind. Genomics open the horizons for discovery of treatments for unconquered diseases.   The future of gene-based drug therapy will allow patients to be classified into genetically defined subsets based on disease susceptibility and drug responsiveness, with the ultimate goal of optimizing drug therapy by improving efficacy and decreasing toxicity.

Preparing for a future of unseen opportunity in healthcare requires a profound shift in thinking from all the stakeholders: A new era of personalized medicine, ushered in by advances in biotechnology and genetic engineering (among other technologies), will radically transform the nature of the healthcare systems world-wide..

Chapter Two: Medical Community at a Crossroads

Ethics and human suffering run headlong into direct conflict with a healthcare system in need of general overhaul. The driving force behind the development of health economics is the increase in healthcare costs worldwide, no matter which health care system is concerned.

Not too long ago, the cost of medicines did not matter that much, and only the efficacy, safety and quality of the pharmaceuticals involved were the parameters any decision was based on. But particularly for pharmaceuticals, costs have become a key issue (although prescription drugs take only a 10% share of the total healthcare spending, they receive a 90% share in the attention of health authorities). And, consequently, pressure on prices for the pharmaceutical industry's products has increased dramatically. From their side, doctors face unprecedented cuts in their already modest (when compared to other professions and to their responsibility) revenues; the pharmaceutical industry is being blamed for its "excessive " promotion and prices.

Turning scientific progress in therapies to prevent or cure disease is what the medical community expects from the pharmaceutical industry. In view of the many unresolved health problems, Society expects the industry to continue playing an important role in the discovery process of new medicines. At an accelerated pace. Is the industry ready? What about the vision of those who have the experience of making drugs?
Is the pharmaceutical industry ready to tune in to a changed new healthcare world? What changes will the medical community itself undergo?

Calculations are easily made of how one can create and develop blockbuster drugs, but the reality is that barely 1% of all drugs reach a turnover of US 1bn. The rapid decrease of the patent period makes it less likely in the future that billion dollar drugs will dominate top-line growth. It is more likely that ever more money will need to be invested in marketing in order to grow each of the products to its maximum sales. The salvation of the large companies will be their effective product development, leaving the discovery itself to external entrepreneurs. In the search for higher efficiency and profit, the larger companies find themselves progressively more dependent on outside expertise, alliances, cooperations, mergers.

The drug industry has not much time left to decide how it will metamorphose itself. Instead of recognizing and responding to the fundamental changes facing their sector of activity, many companies will stick to their tired ideas. Expect and prepare for exciting times in a health care industry with new dimensions.

Chapter Three: Medicines for the Masses

The pharmaceutical industry creates brands, invests in brands, and sees the investment disappear in smoke in a relatively short period of time, due to its own success of constantly creating new, better medicines. Society became used to a constant flow of new, better medicines.

The healthier Western society became, the more it regarded maximum access to medicine as a right and duty, the more medicine it craved and the more it required from that super-medicine. Everyone wanted the best medicine…provided someone else pays the bill. High expectations from society and investors alike, pushed industry leaders to focus their efforts on few, high selling primary care drugs (by preference for the chronically ill).

The "one size fits all" primary care blockbuster medicines model is primary an American one if we consider the healthcare system, the US market and the number of patents that will expire in the very near future. The consequences of „blockbuster" patent loss for pharmaceutical companies are a key consideration. This system presupposes a launch of 2 to 3 new chemical entities (NCE) per year to maintain double-digit growth of the „blockbuster-based" company. Mission impossible. From the medicines supply side, society will experience two major categories of drugs: generic compounds for the masses (generics will assume a more central role as patients bear a greater percentage of their healthcare costs and payers seek to restrict the growth of healthcare expenditures), and more targeted therapies for „niche" patient groups.

For science to advance, researchers need to have a vision of what the future may hold. Many scientists have built their careers on being able to foresee the critical trends in medical science, in fields ranging from stem cells to genetics to health-care policy. Europe's regulatory overreach, resulting from its view that society can be rationally planned and directed, thereby making the union of 27 selfish nations (with 27 national health authorities defending their specific interests) the tax and spend champions of the world, forced the medical community (researchers, physicians, pharmaceutical companies, biotech entrepreneurs) to leave for more supportive regions. It resulted in a reversal of the European dominance of drug discovery that had held sway since the Industrial Revolution. It allowed America's medical-industrial sector to rise to dominance.

Chapter Four: A Healthcare Industry Reinventing Itself

In order to meet the demands of tomorrow's healthcare, as well as those of the financial community, the monolithic pharmaceutical industry, unable to respond to the demand of a multitude of specialty has to fundamentally rethink its value proposition.

The medical product development process is no longer able to keep pace with the tremendous basic scientific innovation. Only a concerted effort to apply the new biomedical science to medical product development will succeed in modernizing the critical path. The new science is not being used to guide the technology discovery process in the same way that it is accelerating the technology discovery process. For medical technology, performance is measured in terms of product safety and effectiveness. Not enough applied scientific work has been done in creating new tools to get fundamentally better answers about how the safety and effectiveness of new products can be demonstrated, in faster time frames, with more certainty, and at lower costs. As a result, the vast majority of investigational products that enter the clinical trials fail.

In the 1960s, scientists discovered three different classes of clinical drug, each of which recognized DNA in a different way. Subsequent drugs have used only these three ways to recognize the DNA.

New research, announced early 2006, has isolated a fourth, which is completely different and opens up entirely new possibilities for drug design. This discovery will revolutionize the way that drug developers think about how to design molecules to interact with DNA. It will send chemical drug research off on a new tangent. By targeting specific structures in the DNA scientists may finally start to achieve control over the way our genetic information is processed and apply that to fight disease.

The India-China region will come more on its own in the twenty-first century and their enthusiastically developing pharmaceutical industry, aware of the huge challenges and opportunities it faces, will respond with increased investment in R&D and new technologies in order to participate in tomorrows smaller but in number rapidly proliferating, high return-on-investment generating, and more individually tailored custom business. The industry has already moved far from its origins as the producer of copycat drugs, has extended its scientific network by e.g. cross-border alliances, and is evolving quickly towards becoming a highly specialized sector, resulting in a number of companies that will be among the global leading experts in well-defined areas of a more sophisticated drug discovery and development process

By 2012, and largely due to the billion dollar cost of basic drug research, only five of today's major Western pharmaceutical companies, joined by a few newly arrived industry giants from India and China, are expected to be responsible for 80% of global innovative drug research. Society will have to decide if this is in the world community's interest? *Will the world leave the choice of therapeutic research areas to a handful of company CEO's?*

To the advantage of patients worldwide, progress in the battle against disease will come from those basic research driven drug developers that, by capitalizing on their knowledge and expertise in targeted medicines, and encouraged by patient advocacy groups and academics recognize the signs on the wall and will come to discover the real value of ‚neglected diseases‘; of less prevalent diseases. The focus of their drug discovery will move away from a small number of mostly chronic diseases that have been traditionally targeted by the medical research community and towards other, numerous areas of less prevalent diseases- This will, in turn, create a targeted medicines sector and, in time, lead to the development of personalized health maintenance programs.

Biotechnology is often called the "Third Industrial Revolution. This is probably an understatement. The first two revolutions affected merely products and processes, material aspects of the quality of modern life.

By many measures, medicines have played a major role in the improved health of the world during the 20th century. This becomes obvious when one simply looks at the 1920´s and the role medicines played in eradicating diseases during that period. In the 1960s again, medicines played a very substantial role in medical progress. They helped reduce mortality and improve quality of life. The biotechnology revolution addresses the individual himself as well as his individual aspects and perceptions. Biotechnology has already begun to change our society and culture persistently, with so far unseen speed and dynamics.

A scientific revolution is taking place. We are reaching new heights in our understanding of life and disease. In the past, the goal was to develop new drugs –proteins or small molecules—that helped correct the chemical imbalances within a body suffering from disease. Now, with the availability of genetic information, that paradigm has changed completely and forever. The drug providers of the future do not simply produce medicines anymore but advance and apply the use of scientific knowledge itself. This fact transforms not only the research driven new drug developer's world but the practice of medicine as well.

The medical doctor will no longer be the main repository of medical expertise for a sophisticated patient. The accelerating rate of change in medicine makes the "half-life" of knowledge too short. We are not far from the time when the most up-to-date information will be obtained from a database run by a computer expert or dedicated server and where the last missing link to semi-automated medical care is long-distance accurate diagnosis, something that has already been proven to be feasible.

Scientific research and new drug development is to provide huge benefits over the next 20 years:
- Therapeutic advances will prevent 4.400.000 deaths from heart attacks and strokes;
- Lung cancer deaths will be cut by 409.000 as a result of new drug discoveries;
- New medicines will reduce leukemia deaths by 83.000;
- Prescription drugs will decrease the severity of rheumatoid arthritis cases by as much as 50%;
- For many diseases, such as Alzheimer's, and arthritis, new drug research and development will likely generate nearly all of the future medical progress;
- Breakthroughs in biotechnology will advance the fight against cancer and viral diseases

Contract Research Organizations, alliances between all stakeholders are needed to turn high expectations in new medicines into reality.

Modern biology right now is in metamorphosis, much like physics at the beginning of the last century. A transition takes place from a rather descriptive subject into a discipline that is coming to understand its molecular foundations. Preconditions to this were technological developments in the 1970s and 1980s that made it possible to register the molecular base for relevant categories of molecules (DNA, RNA, proteins, metabolites) qualitatively and quantitatively. Most fundamental modern technological developments (such as recombinant DNA, sequencing, computer technologies) were developed in the US and are being commercialized over there faster and more consequently than anywhere else. Biotechnology has been responsible for a disproportionate number of the new drugs introduced to the market in recent years, accounting for nearly 60 percent of new molecular entities (NMEs) during 2006. Biotechnology's greatest potential for curing chronic and currently "incurable" diseases lies in gene therapy.

Chapter Six: Differentiating Strategies in Drug Delivery Systems

This chapter highlights the protein-centric nature of drug development: almost all drugs on the market, as well as almost all drug candidates in clinical trials, target proteins, with the next largest category, nucleic acids, comprising only 2% of drug targets.

Researchers are a step closer to being able to deliver life-saving drugs through tiny molecules that would travel through the bloodstream and destroy only disease-ridden cells

Tomorrow's therapy model will be based on drugs that target higher severity conditions. Certain of them will be based on large molecules. Although their degree of therapeutic clinical advantage will replace the current focus on size, these new targeted therapeutics will be big in their respective niches. A variety of new generations of drug delivery systems will contribute to optimal therapy outcome.

 Although drug delivery is still young, today 10% of medicines sold world wide incorporate one form of patented drug delivery technologies. Proprietary Drug Delivery technologies are adding real value to drug products by making them more convenient, efficacious and safe. In the future, an increasing number of medicines will incorporate drug delivery technologies. This boom will be driven by NCEs incorporating melt tablets, water insoluble, injectable depot, PEGylation, inhalation, and targeted delivery technologies, as well as the life cycle management medicines from large pharmaceutical companies and super generics from the drug delivery companies/specialty pharma.

Stomach retentive technologies of the future will allow scientists to develop best modified release drug products in terms of the timing and the speed of drug release. In the oral delivery field we should expect to see more products using melt tablets as well as water insoluble drug technologies. Because of its importance, investment and efforts in the development of effective delivery technologies for the oral delivery of macromolecules will continue. Injectable depot formulations will keep adding real value to injected macromolecular drugs and provide both convenience and improved efficacy and safety to patience. Novel delivery systems to transport drugs from circulating blood into the central nervous system (current drug delivery is restricted by the blood barrier and the blood-cerebrospinal fluid barrier) will emerge. Researchers work on micropsheres, liposomes, polymeric micelles, vector-mediated delivery, niosomes, nanoparticles, solid lipid nanoparticles, noninvasive gene therapy, and nasal administration to try to circumvent the blood - brain barrier. Targeted drug delivery has the chance to revolutionize drug treatment by providing safer and more efficacious delivery for cancer and immune diseases. Delivery of large molecular drugs from the skin by poration technologies may be a reality within the next 5-10 years.

Chapter Seven: An Ongoing Search for Better Medicines

The medicines currently marketed reach only about 10% to 20% of the body's possible drug targets, or 400 to 500 protein receptors and enzymes out of an estimated 5.000. In the next twenty years, all 5.000 targets will be accessible.

Genomics is expected to lead to the identification of thousands of new targets for development. Whether these will be amenable to small molecule intervention or require protein and nucleic acid based therapies remains to be seen. To overcome the difficulty of biological and chemical diversity, the message is "creativity is everything."

The application of pharmacogenomics and pharmacogenetics in drug development holds great promise to shed scientific light on the often risky and costly process of drug development, and to provide greater confidence about the risks and benefits of drugs in specific populations. Pharmacogenomics' promise lies in its potential ability to individualize therapy by predicting which individuals have a greater chance of benefit or risk, thus helping to maximize drugs' effectiveness and safety.

Using genomic testing to guide drug therapy will constitute a significant shift from the current practice of population-based treatment towards 'fine-tuning' individual therapy.

It is hoped that pharmacogenomic testing will help identify diseases that have a high probability of responding to a particular medication or regimen and that it may also be used to help track down the cause of certain rare, serious drug side effects.

The formation of the genome database and the identification of the significance of genes are leading the discovery process away from the old ‚symptom-indication‘-based R&D approach to one where a protein and a gene may have significance in disease states that have nothing to do with the ‚specialized‘ expertise of a pharmaceutical company. The growing ability to more accurately define the target population through DNA screening will help to make medicines more efficient and largely eliminate the side-effects seen when a drug is used in a less suitable patient. It also means that older, previously rejected or strongly restrained drugs can be reconsidered for the right patient group –identified on the basis of their genome.

Researchers think there are over two million mutations directly causing disease in the human genome. Some 100,000 mutations have been discovered. But there is no systematic global system for collecting and sharing complete and accurate information on these mutations with clinicians around the world. And 95 percent of human mutations are yet to be discovered. Fast, reliable access for researchers to information on mutations, and the damage they cause, would transform genetic medicine by: 1) enabling doctors to rapidly diagnose and inform patients with rare diseases; 2) allowing new diagnostics; 3) helping researchers develop new treatments for hundreds of genetic diseases including cystic fibrosis and thalassemia; 4) assisting in uncovering the causes of common diseases such as breast cancer and asthma.

This chapter gives a sampling of scientific developments and fascinating trends to watch in the coming years. E.g.: Engineering RNA for medical purposes; A role for dueling RNAs; The promise of RNAi in gene therapy; The tumor-targeting capability of nanoparticles; Magnetic nanocrystals; Nanocrystalline coating; Tailor-made molecular medicines; Molecular "cages" to deliver drugs; Microarrays: cancer to get personal; Natural electric fields in regenerative medicine; Stem cell future shaped by epigenetics; Tackling diseases influenced by different types of T cells; Vaccines for people with compromised immune systems; Genomics for treating people with depression; Changing the way heart disease is detected and treated in women; Detecting heart disease long before the disease presents itself; Using bi-functionalized dendrimers to discover disease-causing proteins.

All the information point in the same direction: the end of the small-molecule, oral-formulation drug era and the emergence of therapies based on New Biology, including gene replacement and protein-based therapeutics in combination with various drug delivery systems. Challenging times with big rewards are awaiting the medical world beyond 2015.

Chapter Eight: Coming Up with Medicines not Imagined Even a Decade Ago

Medicines that will make their own way through the body and attack precisely the diseased cells on reaching their destination – such has been the dream of physicians and pharmacists since time immemorial. Each day, researchers come a little closer to reaching this goal.

Disease, as diagnosed by symptoms is said to be replaced by genomics, which identifies people who will definitely respond to a specific medicine. The future is said to offer prevention, protection and treatment of disease on a personalized base. High-speed computers and DNA filters called microarrays, which made it possible to quickly identify varieties in genes, allow for more forays into personalized medicine. A new "one-size-fits-one" approach instead of "one-size-fits-all"-approach will develop with more drugs like Herceptin and Iressa to treat disease.

The understanding of the genetic basis of gene activity will help medical research to provide individuals with information about their personal predisposition to disease.

The promise of stem cells and regenerative medicine discoveries and their potential to transform the treatment of disease, continues to drive biological research and technology development in the life sciences. Analysts predict that a market based on stem cell therapies could grow anywhere from $10-30 billion by 2012

Regenerative medicine may well be the future of health care: Its goal is to refurbish diseased or damaged tissue using the body's own healthy cells. It's a term that is often used synonymously with tissue engineering, though those involved in regenerative medicine place more emphasis on the use of stem cells to produce tissues.

Advanced therapies herald new forms of medical treatment. Tissue engineering, through its focus on the regeneration of human tissues, is expected to have a major impact on future medical practice. One of the longer term perspectives is to be able to regenerate full organs and hence to tackle the issue of donor shortages.

By 2012, it is likely that predictive genetic tests will be available for as many as a dozen common conditions, enabling individuals to take preventive steps to reduce their risks of developing such disorders. Doctors will also begin tailoring prescribing practices to each patient's unique genetic profile, choosing medications that are most likely to produce a positive response.

"Molecular" techniques have become firmly entrenched in the diagnostic armamentarium of pathologists and clinicians, and are a bridge to the future of medical diagnostics and therapeutics.

Molecular testing for the complex genetic diseases can be expected by 2015. The testing researchers are doing in 2007 is for single gene defects that are not very common diseases. For other diseases, like diabetes or atherosclerosis, which also have a genetic component, they just don't know all the genes yet. ..but are confident that within the near future, they will at least know a majority of them. And probably through some kind of microarray technology, they'll be able to do some kind of predictive testing for coronary artery disease, hypertension, diabetes, etc..

By 2020, the impact of personalized medicine is likely to be far more important than any of us can envision today. New gene-based drugs will be developed for diabetes, heart disease, Alzheimer's disease, schizophrenia, and many other conditions.

Nanotechnological advances will allow the effective introduction of "personalized therapy", such as administration of a small dose of a drug to obtain maximum effect; the reduction of secondary effects by means of drugs that act only on the therapeutic target; or the development of new more effective administration routes.

Chapter Nine: Targeted Therapies

This chapter focuses on cancer as an example of targeted therapies.

Oncogenics can provide one of the key routes to development of successful targeted therapies. It does this by: accelerating the stream of novel targets for small molecule drugs; reducing the timeline between discovery of new gene targets and new drugs from 10-15 years to 5 years; and enabling high throughput screening, combinatorial chemistry and microarrays that should speed up drug development.

Molecularly targeted therapies based on recent progress in genomics and proteomics hold out the promise of being far more selective, thereby drastically reducing the incidence of side effects in patients undergoing treatment.

Improved diagnostics are central to the ongoing development and success of targeted therapies. Trastuzumab is one recent example of an innovative drug for which the development of an appropriate diagnostic test has proven essential to strong uptake within the breast cancer therapy segment and allowed the drug to be appropriately targeted and demonstrate cost-effectiveness for certain patients.

There has been a clear move away from cytotoxics in favor of cytostatic drugs, which has been integral to the development of targeted cancer therapies. While traditional chemotherapy agents kill cancer cells by being cytotoxic, many new agents work mostly by interrupting their growth i.e. are cytostatic.

For breast cancer, new treatments will be driven by: improved screening techniques; research in into the multiple tissue-specific interactions of the ER (estrogen receptor) and its response to antiestrogen therapy will be studied extensively; the development of pure antiestrogen agents which completely block estrogenic activity; increasing use of cytostatic therapy, combination therapy (of antiestrogens and aromatase inhibitors).

The challenges present an opportunity for targeted cancer drugs that may reduce the side effect profile and outperform the efficacy of existing drugs.

Drug treatments for colorectal cancer exhibit similar limitations to those for other cancers in terms of excessive side effects and insufficient survival benefit. In the absence of curative treatment, there is also a need for agents that improve the quality of life for patients with unresectable tumors. Existing treatments tend to follow protracted delivery schedules and are expensive.

The major unmet need with lung cancer is extension of patient survival. The current five-year survival rate is less than 15%, dramatically less than the rate for other major cancers. There is also a need for improved second-line treatments for patients who fail to respond to first-line treatments, or suffer a relapse after receiving first-line therapy.

While cure rates for early-stage disease are high, there remains a critical unmet need for a curative treatment for advanced metastatic prostate cancer or hormone-refractory prostate cancer. There is also a need for new treatments with fewer side effects than current cytotoxics and hormonals.

Currently, treatment for cancer therapies depends on the type of cancer; the size, location, the stage of disease and the person's general health. Advances in research mean the treatment will less depend on cancer type (organ, location,, histology) and be more driven by molecular features.

Innovatives are the most promising area for all cancers and may create the answer to reduced side effects associated with cancer treatment and have the vast potential of dramatically increasing disease free survival and overall survival rates. The promise behind innovatives and reduced side effects arises from the potential of specifically targeting cancer cells, avoiding killing normal healthy cells, a common problem associated with cytotoxics and anti-metabolites.

For decades, scientists have looked for new therapies that target cancer. Researchers now know that cancer cells have several unique characteristics -- they require new blood vessels to deliver oxygen and nutrients (angiogenesis), they grow uncontrollably (proliferation), they travel throughout the body (metastasis), and they escape programmed cell death, a natural process by which the body rids itself of damaged or unwanted cells (apoptosis). The medical research community is dedicated to finding the next generation of small molecule and biologic therapies that will have an impact on this horrible disease.

This chapter describes the role of specific drugs in cancer therapy, e.g.: sorafenib, sinutinib, rapamycin, bevacizumab, temsirolimus, vatalanib, cilengitide, cetuximab, gefitinib, erlotinib, panitumumab, lapatinib, tipifarnib, lonafarnib, bortezomib, bexaroten, perifosine, oblimersen, aprinocarsen, exisulind, seliciclib, dasatinib, indisulam,UCN-01, ibritumomab, and innovative early stage drug development projects.

Chapter Ten: Adding a New Dimension in Healthcare

From the Greeks to the First World War, medicine's tasks were simple : to grapple with lethal diseases and gross disabilities, to ensure live births and manage pain. It performed these as best it could but with meager success.

A new (second) dimension in healthcare emerged in the 1930`s, with the sulfonamides as the first medicines to cure disease. Progress in biomedicine provided doctors with the tools to ease suffering rather than watch patients die.
 A tremendous shift in ideas about medicine occurred. It is to the credit of conventional medicine that, when science revealed progress, it has abandoned products of a former generation of medicines for better ones.

The postwar healthcare thinking resulted in generous socialized Statutory Healthcare Systems. Universal healthcare was a fantastic social ideology based on the willingness of bringing relief to the suffering, whenever illness strikes. At the same time, the manipulation of power by parties involved in health provision increased. The focus started to shift away from universal healthcare for the suffering patient to the care of an ailing healthcare system, which itself suffered from budget overruns.
Currently, the economics of care, the cost-benefit analysis, leading to a rational drug therapy philosophy, dominates the debate.

Genomics introduces many more objectives for the future therapy providers, so makes the links in the "value chain" far more complex. The new linked areas are:

456

- gene sequencing (how do the proteins made by the identified gene sequences cause the effects of the disease and how can this be altered by individualized drug therapy?);
- population genotyping (will genotyping -the prediction of future morbidity- ever become routine for individuals and populations with no history of disease? Will the health outcomes of these patients be improved? Does knowledge and fear of a future disease improve or diminish your present quality of life? Do individuals have a right to genetic information about family members, if this data affects their own health and future?);
- health economic modeling (Models which demonstrate that the probable increased cost of the new innovative therapies and diagnostic tools leads to a decreased overall cost of healthcare spend per patient by reducing inpatient stays, reducing the onset and severity of disease, and providing therapy that should provide effective treatment as it has been designed to target the identified genetic causative factor of the disease);
- drug discovery and development (Further basic research into how the gene mutation alters the proteins produced and the effect proteins have on causing the disease state, resulting in innovative drug therapies, adjunct therapies, nutrigenomics);
- diagnostics and diagnosis (development and validation of genetic test kits, both for the identification of future morbidity, and of the specific identification of the exact genetic cause of a disease state); information management (management of an individual's genetic data on a scale and of a complexity which today is almost unimaginable);
- disease management (development of total health/wellness and disease management packages –with he potential of healthcare services, including education of the individual, the measurement of health outcomes, as will the personalized drug therapies supplied).

With further research into the information provided by the human genome project, scientists will be able to compare individual variations against a general blueprint for the human race. The result will be tailored treatments and the ability to pinpoint the mechanisms that can result in problems later in a person's life. Scientists believe that one day in the not so far future, they will be able of predicting whether a person will develop diabetes, mental disorders, cancer, heart disease and a range of other conditions.

We are entering an exciting new era of predictive health where an individuals personal genetic code will provide guidance on healthcare decisions. Maps of insertions and deletions will be used together with SNP maps to create one big unified map of variation that can identify specific patterns of genetic variation to help us predict the future health of an individual. The next phase is to figure out which changes correspond to changes in human health and develop personalized health treatments. This could include specific drugs tailored to each individual, given their specific genetic code.

Ultimately, each person's genome could be re-sequenced in a doctor's office and his or her genetic code analyzed to make predictions about their future health. The traditional government approval processes will prove inadequate to service the demand for an ever increasing number of specialized drugs.

On the way to predictive medicine, a less isolated or restricted view of the "value chain of health" will have to take a place at the center of all activities. Visionary stakeholders prepare for a healthcare system of the third dimension.

On the way to becoming 'virtual', basic research driven drug developers are expected to develop so-called Integrated Therapeutic Providers that combine the genomic, pharmaceutical, biotechnological and diagnostic know-how' to take advantage of the significant unmet medical need and the rapid pace of scientific advances in the treatment of health problems.

In the new healthcare environment, the leftovers of today's drug research companies become system providers. Their salvation (at least of a few of them) will be their effective marketing creativity and clout, leaving the discovery itself to more nimble entrepreneurs and the development to government funded international institutions (NIH, NCI, and others).